Trichomonads Parasitic in Humans

B.M. Honigberg
Editor

Trichomonads Parasitic in Humans

With 79 Illustrations

Springer-Verlag
New York Berlin Heidelberg
London Paris Tokyo Hong Kong

B.M. Honigberg, PhD
Professor, Center for Parasitology
Department of Zoology
University of Massachusetts
Amherst, MA 01003, USA

Cover illustrations (left to right): Dorsal view of *Trichomonas vaginalis;* dorso-lateral view of *Trichomonas vaginalis;* right view of *Pentatrichomonas hominis.* These figures appear on pp. 10 and 28 of the text.

Library of Congress Cataloging-in-Publication Data
Trichomonads parasitic in humans / B.M. Honigberg, editor.
    p.   cm.
  Includes bibliographies and index.
  ISBN-13:978-1-4612-7922-8
  1. Trichomoniasis. 2. *Trichomonas vaginalis.*
 I. Honigberg, B. M.
  [DNLM: 1. Trichomonas Infections. 2. *Trichomonas vaginalis.* QX
70 T823]
QR201.T55T68   1989
616.9'36—dc19
DNLM/DLC   88-39320

© 1990 by Springer-Verlag New York Inc.
Softcover reprint of the hardcover 1st edition 1990

All rights reserved. This work may not be translated or copied in whole or in part without the written permission of the publisher (Springer-Verlag, 175 Fifth Avenue, New York, NY 10010, USA), except for brief excerpts in connection with reviews or scholarly analysis. Use in connection with any form of information storage and retrieval, electronic adaptation, computer software, or by similar or dissimilar methodology now known or hereafter developed is forbidden.
The use of general descriptive names, trade names, trademarks, etc. in this publication, even if the former are not especially identified, is not to be taken as a sign that such names, as understood by the Trade Marks and Merchandise Marks Act, may accordingly be used freely by anyone.
While the advice and information in this book are believed to be true and accurate at the date of going to press, neither the authors nor the editors nor the publisher can accept any legal responsibility for any errors or omissions that may be made. The publisher makes no warranty, express or implied, with respect to the material contained herein.

9 8 7 6 5 4 3 2 1

ISBN-13:978-1-4612-7922-8   e-ISBN-13:978-1-4612-3224-7
DOI: 10.1007/978-1-4612-3224-7

Acknowledgments

Several persons helped in the preparation of this book for publication. Special thanks are due to my wife, Rhoda Springer Honigberg; my assistant, Mr. Jonathan A. Turner; Mr. Lawrence M. Feldman, Coordinator, Biological Sciences Library, University of Massachusetts at Amherst; and Ms. Theresa Kornak, Supervising Production Editor, Springer-Verlag, New York.

Contents

1. **Introduction**.. 1
   G. Piekarski

2. **Taxonomy and Nomenclature** 3
   B. M. Honigberg

3. **Structure** 5
   B. M. Honigberg and G. Brugerolle

4. **Immunologic Aspects of Human Trichomoniasis** 36
   J. P. Ackers

5. **Biochemistry of *Trichomonas vaginalis*** 53
   Miklós Müller

6. **Nucleic Acid Metabolism in *Trichomonas vaginalis*** 84
   Ching C. Wang

7. **Cultivation of Trichomonads Parasitic in Humans** 91
   David Linstead

8. **Employment of Experimental Animals in Studies
   of *Trichomonas vaginalis* Infection** 112
   Jaroslav Kulda

9. **Host Cell–Trichomonad Interactions and Virulence Assays
   Using In Vitro Systems**.. 155
   B. M. Honigberg

10. Microflora Associated with *Trichomonas vaginalis* and Vaccination
    Against Vaginal Trichomoniasis 213
    Carol A. Spiegel

11. Clinical Manifestations of Urogenital Trichomoniasis in Women 225
    Michael F. Rein

12. Epidemiology and Clinical Manifestations
    of Urogenital Trichomoniasis in Men 235
    John N. Krieger

13. Urogenital Trichomoniasis in Children 246
    Alicja Kurnatowska and Alina Komorowska

14. Cytopathology and Histopathology of Female Genital Tract
    in *Trichomonas vaginalis* Infection 274
    Prabodh K. Gupta and John K. Frost

15. Pathology of Urogenital Trichomoniasis in Men 291
    William A. Gardner, Jr., and Donald E. Culberson

16. Laboratory Diagnostic Methods and Cryopreservation of Trichomonads 297
    Alexander McMillan

17. Epidemiology of Urogenital Trichomoniasis 311
    Joseph G. Lossick

18. Therapy of Urogenital Trichomoniasis 324
    Joseph G. Lossick

19. Trichomonads Found Outside the Urogenital Tract of Humans 342
    B. M. Honigberg

20. Symptomatology, Pathology, Epidemiology, and Diagnosis
    of *Dientamoeba fragilis* 394
    Gunter Ockert

    Index 411

Contributors

John P. Ackers, M.A., Ph.D., Senior Lecturer, Department of Parasitology, London School of Hygiene and Tropical Medicine, London, United Kingdom

Guy Brugerolle, Sc.D., Researcher, Laboratory of Zoology and Protistology—CNRS, Blaise Pascal University of Clermont-Ferrand, Aubiere, France

Donald E. Culberson, Ph.D., Instructor, University of South Alabama, College of Medicine, Department of Pathology, Mobile, Alabama 36617, USA

John K. Frost, M.D., Professor of Pathology; Director, Division of Cytopathology, with Joint Appointment in Gynecology and Obstetrics, The Johns Hopkins University School of Medicine, Baltimore, Maryland 21205, USA

William A. Gardner, Jr., M.D., Professor and Chairman, Department of Pathology, University of South Alabama, College of Medicine, Mobile, Alabama 36617, USA

Prabodh K. Gupta, M.B., M.D., Professor of Pathology and Laboratory Medicine, University of Pennsylvania School of Medicine; Director—Cytopathology and Cytometry, Hospital of the University of Pennsylvania, Philadelphia, Pennsylvania 19104, USA

B. M. Honigberg, Ph.D., Professor, Center for Parasitology, Department of Zoology, University of Massachusetts, Amherst, Massachusetts 01003, USA

Alina Komorowska, M.D., Professor (Emeritus), Institute of Gynecology and Obstetrics, Medical Academy, Łódź, Poland

John N. Krieger, M.D., Associate Professor, Department of Urology, University of Washington, School of Medicine, Seattle, Washington 98195, USA

Jaroslav Kulda, Ph.D., Senior Research Scientist, Department of Parasitology, Faculty of Science, The Charles University, Prague, Czechoslovakia

Alicja Kurnatowska, M.D., Professor, Department of Medical Biology and Parasitology, Institute of Biology and Morphology, Medical Academy, Łódź, Poland

David Linstead, Ph.D., Department of Molecular Sciences, Wellcome Research Laboratories, Beckenham, Kent, United Kingdom

Joseph G. Lossick, D.O., M.S., Chief, Clinical Investigation Section, Epidemiology Research Branch, Sexually Transmitted Disease Division, Centers for Disease Control, Atlanta, Georgia 30333, USA

Alexander McMillan, M.D., F.R.C.P., Consultant Physician, Department of Genito-Urinary Medicine, Royal Infirmary of Edinburgh, Edinburgh, United Kingdom

Miklós Müller, M.D., Associate Professor, The Rockefeller University, New York, New York 10021, USA

Günter Ockert, Sc.D., University Lecturer for Parasitology; Head, Reference Laboratory for Malaria and Other Protozoan Diseases of the German Democratic Republic, Halle Institute of Hygiene, Halle, German Democratic Republic

Gerhard Piekarski, Ph.D., Professor and Director (Emeritus) of Medical Parasitology, Institute for Medical Parasitology, University of Bonn, Bonn, Federal Republic of Germany

Michael F. Rein, M.D., Associate Professor of Medicine, Division of Infectious Diseases, University of Virginia School of Medicine, Charlottesville, Virginia 22908, USA

Carol A. Spiegel, Ph.D., A.B.M.M., Director, Clinical Microbiology, University of Wisconsin Hospital and Clinics, Madison, Wisconsin 53792; Assistant Professor, Department of Medical Microbiology, University of Wisconsin, Madison, Wisconsin 53706, USA

C. C. Wang, Ph.D., Professor of Chemistry and Pharmaceutical Chemistry, University of California, Department of Pharmaceutical Chemistry, San Francisco, California 94143, USA

1

Introduction

G. Piekarski

Forty years have passed since Trussell[1] published an important book in the English language on *Trichomonas vaginalis* Donné and urogenital trichomoniasis. During the intervening four decades, much information has been accumulated on trichomonads parasitic in humans and on the diseases they cause. In light of this, many parasitologists and clinicians believe that the time has come for a complete review, in book form, of various aspects of these parasites and of the trichomonad parasitemias. This need has been further reinforced by the finding that, despite the use of effective antitrichomonal drugs during the past years, the prevalence of human urogenital trichomoniasis, the world's most common sexually transmitted disease, has been increasing significantly. As might have been expected, therefore, discussion of various aspects of *T. vaginalis* and of the disease it causes occupies much space in this book.

Few clinicians or parasitologists doubt any longer that *T. vaginalis* includes strains inherently virulent for humans. However, the expression of virulence potential of the strains may be affected by a variety of factors in the host organism. Admittedly, some of these factors are poorly understood, and their study ought to constitute an important aim of future research. With the apparent exception of *Dientamoeba fragilis* Jepps & Dobell, trichomonads other than *T. vaginalis* have not been proved to be potentially virulent for the human host.

Trussell[1] and some of his predecessors and contemporaries emphasized the fact that each trichomonad species parasitizing humans is structurally and physiologically distinct from all the others and that, in most cases, each is restricted to a unique site in the host. Yet this appears not to have been brought home to at least some clinicians. Of special importance in this connection are *T. vaginalis* and *Pentatrichomonas hominis* (Davaine), the intestinal trichomonads of many mammalian species, including humans. Some practitioners still tell their patients about the possibility of an intestinal origin of trichomonal infection of the urogenital tract.

Trichomonas vaginalis infection of newborns, infants, and young children constitutes an interesting, although not extensively pursued, area of investigation. For hitherto incompletely understood reasons, the period of pregnancy appears to favor the increase of symptomatic trichomoniasis; there is, therefore, little doubt that infants can be infected during birth. Pediatric urogenital trichomoniasis is discussed in Chapter 13 by two of the few experts in this field.

Aside from urogenital trichomoniasis of newborns, infants, and young children, which, as a rule, is not transmitted by sexual intercourse, there are very few adequately documented cases of other methods of *T. vaginalis* transmission. This does not mean, however, that urogenital trichomoniasis in men is well understood; Chapters 12 and 15 are devoted to this topic. It is clear from the presently available data that much remains to be learned about *T. vaginalis* infection in men before the epidemiology of the disease can be fully understood.

As was pointed out in the summaries of the proceedings of the recent International Symposium on Trichomonads and Trichomoniasis,[2] important advances have been made in our knowledge of the biochemistry of trichomonads. This information is summarized in Chapters 5 and 6. Although less well understood than their biochemical aspects, the immunologic attributes of trichomonads, and of *T. vaginalis* in particular, have been the subject of many studies. By using modern methods, deeper insight is being gained into the immunology of trichomonads. Chapter 4 is devoted to this subject, and antigenic characteristics of the parasites and various aspects of immunologic host responses are mentioned also in several other chapters.

Although 5-nitroimidazoles are very effective antitrichomonal drugs, and although the potential carcinogenicity of these compounds appears exaggerated, nitroimidazole-resistant strains have been encountered, albeit relatively rarely. In light of this, there appears to be a need for development of new effective drugs. Another approach to control of trichomoniases is vaccination. Some advances seem to have been made in this direction, but the mode of action of the allegedly effective vaccine, Solco-Trichovac,® "manufactured from inactivated abnormal strains of lactobacilli,"[3] is poorly understood[2]; claims of the efficacy of this vaccine are still not generally accepted.

The authors of the various chapters included in this book attempt to throw new light on the many aspects of human trichomoniases. Their primary aim is to acquaint the clinician and the parasitologist with the most recent advances in our understanding of these infections, with special emphasis on those affecting the urogenital tract. Furthermore, the information presented here should provide sound bases for future research leading ultimately to control and eradication of infections caused by trichomonads.

References

1. Trussell RE: *Trichomonas vaginalis* and trichomoniasis. Springfield, Ill: Charles C Thomas, 1947.
2. Honigberg BM, Ackers JP: Summaries of the Proceedings of the International Symposium on Trichomonads & Trichomoniasis, Prague, July 1985. Published as Introductions to the various sections. In Kulda J, Čerkasov J (eds): Proceedings of the International Symposium on Trichomonads and Trichomoniasis, Prague, July 1985 (Post-Symp Publ Pt 1, 2.) *Acta Univ Carolinae (Prague) Biol* 30(3,4): 180, 219, 247, 319 (1986) 1987; (5,6): 397, 457, 469 (1986).
3. Ruttgers H (ed): Trichomoniasis. Scientific papers of the Symposium on Trichomoniasis, Basel. *Gynäkol Rundsch* 22(suppl 2):1-91, 1983.

Editor's Note: While this book was in press, a large monograph was published on urogenital trichomoniasis, written in Slovak, by M. Valent and M. Klobušický [*Urogenitálna Trichomoniaza* (Edícia pre Postgraduálne Štúdium Lekárov; Dérerova Zbierka, Zväzok 103), 251 pp. Bratislava: Osveta, 1988]. About 80% of the text deals with epidemiology, clinical and pathologic aspects, diagnosis, and therapy of *Trichomonas vaginalis* infection. Unfortunately, the content of the monograph is accessible only to those who can read Slovak (the very brief English, Russian, and German summaries do not do justice to the work). However, the numerous references to various aspects of urogenital trichomoniasis (some of them rarely, if ever, quoted in Western European and North American literature) render the book valuable to those interested in this infection.

2

Taxonomy and Nomenclature

B. M. Honigberg

Introduction

This brief chapter provides guidelines to the taxonomy and nomenclature of the trichomonad species parasitizing humans. It is not intended to be an exhaustive treatment of the systematic and nomenclatural problems of these species. Such a treatment can be found in several publications, most of which are cited in this chapter.

Taxonomic Status and Nomenclature

Taxonomic Status

All the species included in this book are members of the zoomastigophorean order Trichomonadida Kirby, 1947,[1] emend. Honigberg, 1974.[2] Four of the species found in humans, *Trichomonas vaginalis* Donné, 1836,[3] *Trichomonas tenax* (O. F. Müller, 1773),[4] *Pentatrichomonas hominis* (Davaine, 1860),[5] and *Trichomitus fecalis* (Cleveland, 1928),[6] belong to the subfamily Trichomonadinae Chalmers & Pekkola, 1918,[7] emend. Honigberg, 1963[8] of the family Trichomonadidae Chalmers & Pekkola, 1918,[7] emend. Kirby, 1947.[1] The fifth species, *Dientamoeba fragilis* Jepps & Dobell, 1918,[9] belongs to the subfamily Dientamoebinae Grassé, 1953,[10] emend. Honigberg, 1974[2] of the family Monocercomonadidae Kirby, 1944,[11] emend. Honigberg, 1974.[2]

Nomenclature

Trichomonas vaginalis Donné, 1836

There has not been nor is there any controversy with regard to the name *Trichomonas vaginalis* Donné, 1836,[3] emend. Ehrenberg, 1838[12] as the proper designation for the urogenital trichomonad of women and men.

Trichomonas tenax (O. F. Müller, 1773)

Since Dobell's[13] extensive discussion of the nomenclature of the oral flagellate, the name *Trichomonas tenax* (O. F. Müller, 1773)[4] Dobell, 1939[13] generally has been accepted for this species. Some of the relatively minor nomenclatural problems related to invasion of the respiratory tract by trichomonads, one of which might have been *T. tenax*, have been considered in a previous work[14]; they are touched on also in Chapter 19, dealing with trichomonads found outside the urogenital tract of humans.

Pentatrichomonas hominis (Davaine, 1860)

There have been many nomenclatural problems with *Pentatrichomonas hominis* (Davaine, 1860)[5] Wenrich, 1931.[15] These problems have been addressed previously by several workers[8,15-19]; they are also mentioned briefly in Chapter 19.

Trichomitus fecalis (Cleveland, 1928)

Nothing needs to be said about the nomenclature of this flagellate found in only one patient.

It is not even certain that humans are the primary hosts of *Trichomitus fecalis* (Cleveland, 1928)[6] Honigberg, 1963.[8]

Dientamoeba fragilis Jepps & Dobell, 1918

Although the taxonomic status of this organism was officially changed from that of an intestinal ameba to that of a trichomonad flagellate,[2] there has never been any question about the name of this species, which is *Dientamoeba fragilis* Jepps & Dobell, 1918.[9]

Clearly, the fields of medicine and parasitology can be served best if the correct nomenclature of the trichomonad species found in humans is universally employed.

References

1. Kirby H: Flagellate and host relationships of trichomonad flagellates. *J. Parasitol* 33:214-228, 1947.
2. Honigberg BM, in Camp RR, Mattern CFT, Honigberg BM: Study of *Dientamoeba fragilis* Jepps & Dobell. II. Taxonomic position and revision of the genus. *J Protozool* 21:79-82, 1974.
3. Donné A: Animalcules observés dans les matières purulentes et le produit des sécrétions des organes génitaux de l'homme et de la femme. *C R Acad Sci Paris* 3:385-386, 1836.
4. Müller OF: *Vermium terrestrium et fluviatilium, seu animalium infusorum, helmithicorum et testaceorum, non marinorum, succincta historia,* vol 1, pt 1 and 2. Havniae et Lipsiae: Heineck et Taber, 1773.
5. Davaine CJ: *Traité des entozoaires et des maladies vermineuses de l'homme et des animaux domestiques,* 1st ed. Paris: JB Baillière et Fils, 1860.
6. Cleveland LR: *Tritrichomonas fecalis* nov. sp. of man; its ability to grow and multiply indefinitely in faeces diluted with tap water and in frogs and tadpoles. *Am J Hyg* 8:232-255, 1928.
7. Chalmers AG, Pekkola W: *Chilomastix mesnili* (Wenyon, 1910). *Ann Trop Med Parasitol* 11:213-264, 1918.
8. Honigberg BM: Evolutionary and systematic relationships in the flagellate order Trichomonadida Kirby. *J Protozool* 10:20-63, 1963.
9. Jepps MW, Dobell C: *Dientamoeba fragilis* n. g., n. sp., a new intestinal amoeba from man. *Parasitology* 10:352-367, 1918.
10. Grassé P-P: Famille des Dientamoebidae Grassé, nov. In Chatton E, Ordre des Amoebiens Nus ou Amoebaea. In Grassé P-P (ed): *Traité de Zoologie,* vol 1, Fasc 2, pp. 50-54. Paris: Masson et Cie, 1953.
11. Kirby H: Some observations on cytology and morphogenesis in flagellate protozoa. *J. Morphol* 75:361-421, 1944.
12. Ehrenberg CG: *Die Infusionsthierchen als vollkommene Organismen.* Leipzig: Leopold Voss, 1838.
13. Dobell CC: The common flagellate of the human mouth, *Trichomonas tenax* (O.F.M.): Its discovery and its nomenclature. *Parasitology* 31:138-146, 1939.
14. Honigberg BM: Trichomonads of importance in human medicine. In Kreier JP (ed): *Parasitic Protozoa,* vol 2 pp. 275-454. New York: Academic Press, 1978.
15. Wenrich DH: Morphological studies on the trichomonad flagellates of man. *Arch Soc Biol Montevideo Suppl Actas Cong Int Biol Montevideo* Fasc 5:1185-1204, 1931.
16. Wenrich DH: Morphology of the intestinal trichomonad flagellates in man and of similar forms in monkeys, cats, dogs and rats. *J Morphol* 74:189-211, 1944.
17. Wenrich DH: The species of *Trichomonas* in man. *J Parasitol* 33:177-188, 1947.
18. Kirby H: The structure of the common intestinal trichomonad of man. *J Parasitol* 31:163-175, 1945.
19. Flick EW: Experimental analysis of some factors influencing variation in the flagellar number of *Trichomonas hominis* from man and other primates and their relationship to nomenclature. *Exp Parasitol* 3:105-121, 1954.

3

Structure

B. M. Honigberg and G. Brugerolle

Introduction

The five trichomonad species found in humans—*Trichomonas vaginalis, Trichomonas tenax, Pentatrichomonas hominis, Trichomitus fecalis,* and *Dientamoeba fragilis*—have the general structural characteristics of the order Trichomonadida. The first four species possess also the attributes of the subfamily Trichomonadinae of the family Trichomonadidae,[1] while *Dientamoeba* has those of the atypical subfamily Dientamoebinae of the family Monocercomonadidae.[2]

Unquestionably, the most widely studied is *Trichomonas vaginalis,* the urogenital trichomonad, but except for *Trichomitus fecalis,* the structure, including the fine structure, of the remaining species is also rather well understood. However, some of the modern methods of cytologic research have not been applied to all trichomonad species found in humans. The relationships between structure and function of the cell organelles, many of which were discussed by Kulda et al,[3] should be investigated in greater depth.

Trichomonas vaginalis Donné, 1836

Light Microscopy

The account of *Trichomonas vaginalis* structure based on light microscopy (Fig. 3.1) is taken from the report of Honigberg and King.[4]

The urogenital trichomonads vary in size and shape. In fixed and stained preparations (Fig. 3.1e–g) they measure, on average, 9.7 ± 0.07 [4.5 to 19 (32 in one strain)] $\times$ 7.0 ± 0.04 [2.5 to 12.5 (14.5 in one strain)] μm ($n = 1000$ in 10 strains; 100 per strain); living organisms are about one-third larger (Fig. 3.1a–c). The shape is variable in fresh and in fixed and stained preparations. Physicochemical conditions (e.g., pH, temperature, oxygen tension, and ionic strength) affect the shape of trichomonads; however, shape tends to be more uniform among cells grown in nonliving culture media than among those observed in vaginal secretions and urine. Details regarding variations in shape can be found in the references listed by Honigberg.[5] In general, in axenic cultures grown in liquid media, for example, trypticase-yeast-maltose (TYM),[6] with or without low concentrations of agar and without solid food particles, the organisms are usually ellipsoidal, ovoidal, or spheroidal (Fig. 3.1a–c,e–g). In the presence of higher concentrations of agar, of certain food particles, and of various substrate layers (cells or tissues) in vivo and in vitro, trichomonads tend to be ameboid (Fig. 3.1h). In some instances, several ameboid organisms are applied to one another by their pseudopods, forming groups of about two to five (Fig. 3.1j). The two organisms shown in Figure 3.1j appear to phagocytose one particle.

There are four anterior flagella that in some organisms are arranged in two groups, each

---

The investigations of B.M. Honigberg that yielded many of the results summarized in this chapter were supported by Research Grants AI 00742 and AI 16176, from the National Institute of Allergy and Infectious Diseases, U.S. Public Health Service. The studies of G. Brugerolle were supported by Centre National de la Recherche Scientifique and the University of Clermont-Ferrand, France.

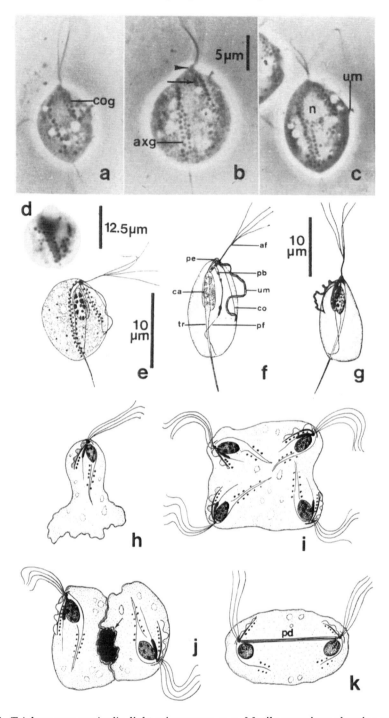

Fig. 3.1. *Trichomonas vaginalis,* light microscopy. **a–c.** Motile organisms showing nuclei (*n*), anterior flagella, waves of the undulating membranes (*um*), as well as paraxostylar (*axg*) and paracostal (*cog*) granules (hydrogenosomes) marking the positions of the axostyles and costae. Note the parallel rows of the paraxostylar granules, which constitute a diagnostic characteristic of *T. vaginalis.* The flagella appear to emerge from a small anterior cone (*arrowhead* in **b**) that marks the region of the periflagellar canal. The small arrow in **b** points to the

containing two organelles. The longest flagella in nondividing organisms are 12.5 ± 0.2 (6.5 to 18.5) μm long ($n = 100$, in two strains; 50 per strain). The recurrent flagellum, originating dorsal to the anterior locomotor organelles, is incorporated into the free margin of the undulating membrane along the accessory filament (Fig. 3.1f); this flagellum, which constitutes the inner component of the margin, has no free part continuing beyond the posterior end of the membrane. The undulating membrane extends, on average, for about the anterior one-half to two-thirds of the cell body length (Fig. 3.1e–g). The shortest membranes are found in the longest mature organisms (Fig. 3.1g). The membrane is supported by a slender costa.

There is an anteriorly located, typically rather elongate nucleus (Fig. 3.1e–g), with its left ventral surface applied to the spatulate axostylar capitulum (Fig. 3.1e,f). The anterior part of the capitulum is continuous with the pelta, a crescent-shaped membranous structure that surrounds the area of emergence of the anterior flagella from the cell (Fig. 3.1f,g). Posterior to the nucleus, the capitulum continues as a slender, hyaline, somewhat attenuating axostylar trunk that courses near the anteroposterior axis of the organism, its terminal segment projecting for less than one-third to over one-half the cell length beyond the posterior surface of the flagellate (Fig. 3.1a,b,e–g). The projecting segment attenuates usually to a sharp point (Fig. 3.1e,g).

The typically V-shaped parabasal apparatus, situated dorsally and to the right of the nucleus, consists of a parabasal body and two parabasal filaments, each associated with one arm of the body. Of the filaments, the one related to the dorsal arm of the parabasal body is typically stouter and longer (Fig. 3.1g).

Paraxostylar (*axg*) and paracostal (*cog*) granules (hydrogenosomes) are seen in living organisms (Fig. 3.1a–c) and in hematoxylin-stained preparations (Fig. 3.1d,e). Among the most striking constant features of *T. vaginalis* is the arrangement of the paraxostylar granules. In all instances in which the granules are seen clearly they are aligned in three rows that parallel the axostyle from its anterior end downward to the area of its emergence from the posterior surface of the flagellate. The two lateral rows of granules seem to be in one optical plane and the central row in another. Any cell seen in various living or fixed and stained preparations, including Papanicolaou smears, that corresponds in size and/or shape to *T. vaginalis* and that may show virtually no other organelles except for paraxostylar granules in the typical arrangement can be diagnosed with considerable confidence as belonging to this species.

---

region of the parabasal apparatus. [Dark phase contrast; electronic flash (300 W/sec). The scale in **b** applies also to **a**, and **c**].

d. Organism showing the typical arrangement of paraxostylar granules. (Weak Flemmings's fixative; iron hematoxylin stain.)

e–g. Line camera lucida drawings of fixed and stained organisms. **e.** Left view. All structures, except for the pelta, are evident. The anteriorly located kinetosomal complex is quite large and the parabasal body applied to the dorsal surface of the nucleus appears rod- or sausage-shaped. Note also the paraxostylar and paracostal dark-staining granules (hydrogenosomes) and their arrangement in association with the axostyle and the costa. (Hollande's fixative; iron hematoxylin stain.) **f,g.** Left and right views. All the mastigont organelles, including the anterior flagella (*af*), pelta (*pe*), V-shaped parabasal body (*pb*) with unequal arms and parabasal filaments (*pf*), undulating membrane (*um*), as well as the capitulum (*ca*) and trunk (*tr*) of the axostyle are evident in both figures. The free margin of the undulating membrane consists of the attached part of the recurrent flagellum (inner) and the accessory filament (outer) (see **b**). (Hollande's fixative; protargol stain.)

h,k. Diagrams (drawn to a scale similar to that of **e** and **f**). **h.** Ameboid form. **i.** Organism that underwent nuclear and mastigont division not followed by cytokinesis ("polymastigote form"). **j.** "Agglomeration" of two individual organisms. Evidently they are trying to phagocytize a large particle (*arrow*). **k.** Organism in late division, showing the extracellular spindle [paradesmose (*pd*)]. **a–g:** Reproduced with permission of the American Society of Parasitologists, from Honigberg BM, King VM: Structure of *Trichomonas vaginalis* Donné. *J Parasitol* 50: 345-364, 1964. **h–k:** Original diagrams, G. Brugerolle.

Under certain, usually unfavorable, conditions the trichomonads tend to become very large, being mononucleate or, more often, polynucleate. Judging from the size and contents of the nuclei, the mononucleate giants are polyploid. In many instances, nuclear divisions accompany those of the mastigont systems (Fig. 3.1i); in some they do not. It is thought that the "polymonad" organisms resulting from divisions of the mastigont systems (with or without the nuclei) not followed by cytokinesis do not represent developmental stages in the life cycle of trichomonads.

Division in *T. vaginalis*, as in all trichomonads, is of the cryptopleuromitotic type, with a conspicuous extranuclear spindle (the paradesmose) connecting the division poles (Fig. 3.1k).

Under unfavorable environmental conditions the organisms round up, forming pseudocysts in which typically the locomotor organelles are internalized; however, no cyst wall surrounds such forms. As pointed out by researchers describing the pseudocysts of *Trichomitus batrachorum*, most students of trichomonads have felt that pseudocysts represent degenerated stages. (For references, see Mattern et al.[7] and Jírovec and Petrů[8].) Subsequently, however, pseudocysts of *Tritrichomonas muris* were found capable of giving rise to normal motile trophozoites.[9] No such phenomenon was ever described from *T. vaginalis*; also, as far as can be ascertained, true cysts, similar to those reported from a few trichomonads of amphibians and invertebrates,[10] have not been reported from the urogenital trichomonad of humans. However, some investigators observed in this species nonmotile, presumably resistant stages occurring in natural and experimental infections. Recently, such forms were mentioned in several abstracts.[11-13] In some instances, the allegedly resistant forms may have been pseudocysts; however, the identity of the organisms devoid of all organelles, as discussed by a number of workers,[5,14-19] is in question.

Electron Microscopy

The fine structure of *T. vaginalis* has been the subject of several studies in which scanning electron microscopy (SEM), freeze-fracture electron microscopy (FEM), and transmission electron microscopy (TEM) of ultrathin sections were employed (Fig. 3.2).

Scanning Electron Microscopy

The observations made with the aid of SEM[20] (Fig. 3.2a–d) represent an extension of those reported from TEM studies, for example, Nielsen et al,[21] Brugerolle,[22] and original micrographs included in this chapter. SEM studies revealed the relationships among the various organelles seen on the organism's surface, for example, the recurrent and anterior flagella emerging from the periflagellar canal (Fig. 3.2b), as well as details of the margin of the undulating membrane, such as the spatial relationships between the recurrent flagellum and accessory filament (Fig. 3.2c,d). Some of the scanning electron micrographs show the site of emergence of the flagella from the periflagellar canal, whose walls are reinforced by the pelta (Fig. 3.2d). The anterior flagella emerge very close to one another, while the recurrent locomotor organelle exits from the canal dorsal to the four anterior ones and its emerging part appears to be separated from them by a lamellar partition (Fig. 3.2d). In some organisms one can see pseudopods (Fig. 3.2a). In all views, the cell surface appears velvety (Fig. 3.2a–d); at least some of the small pits in the surface probably are openings of pinocytotic canals.

Freeze-Fracture Electron Microscopy

Freeze-fracture electron microscopy[23] aided in the analysis of structures made up of membranes. The nature of the external faces (EFs) and protoplasmic faces (PFs) of the membranes revealed by the freeze-fracture method are not discussed here, although these faces are indicated in the relevant electron micrographs (Fig. 3.2e–g,k,m,n) and noted in the corresponding legends.

Among the most striking findings were those pertaining to the surface membranes of the anterior flagella, which were seen to contain rosettelike formations of intramembranous particles (IMPs), each rosette consisting of about 9 to 12 particles (arrows in Fig. 3.2e). Such rosettes were present also in *Tritrichomonas foetus* (Fig. 3.2f). Furthermore, Bardele,[24] on the basis of his examination of an additional trichomonad and of two hypermastigotes, both groups belonging to the superorder Parabasilidea, and of several nonparabasiliid flagellates, concluded that the flagellar rosettes were limited to the members of the aforementioned superorder. Although no rosettes were found on the membrane of the recurrent flagellum (*RF*), the area of attachment of this flagellum to the cytoplasmic membrane fold of the undulating membrane [proximal to the part of the membrane referred to as the "accessory filament" (AcF)* by light microscopists] was marked by three parallel rows of intramembranous particles (arrow in Fig. 3.2g). Details of the undulating membrane and the spatial relationships among the dorsal fold, the marginal lamella-containing accessory filament, and the recurrent flagellum are also evident in freeze-fracture electron micrographs (Fig. 3.2g,h). Of interest is a comparison between micrographs of a transversely fractured undulating membrane (Fig. 3.2h) and of such membranes sectioned transversely (Fig. 2i,j). The latter micrographs reveal the structural details of the recurrent flagellum (Fig. 3.2i) and of the marginal lamella (Fig. 3.2i,j), the latter organelle showing periodicity (Fig. 3.2j). The freeze-fracture technique reveals the structure of the nuclear envelope with its pores (*P*) (Fig. 3.2k); endoplasmic reticulum (*ER*) canals surround the nucleus (Fig. 3.2k). This technique is also useful for demonstrating the cisternae and vesicles of the Golgi complex (*Go*) [parabasal body (PB)] (Fig. 3.2k,l) and of the hydrogenosomes (*H*) with their inner (*In*) and outer (*Ou*) limiting membranes [Fig. 3.2m (From *Trichomonas vaginalis*), n (from *Tritrichomonas foetus*)].

Transmission Electron Microscopy

An original schematic diagram (Fig. 3.3) reconstructed from ultrathin sections examined by TEM is included to show the structure of *T. vaginalis*. In analyzing the various structures of this organism in the electron micrographs of such sections (Figs. 3.4a-c and 3.5a-c), the reader should consult the schematic diagram.

A low magnification micrograph at a nearly frontal section of the organism (Fig. 3.4a) reveals many of the organelles, being especially useful for understanding their spatial relationships. However, this micrograph does not include the undulating membrane or adequate representation of the rootlet organelles, for example, the costa or parabasal filaments. Various structures are shown in their relative positions; most of these are illustrated and discussed in greater detail in subsequent parts of the section dealing with *T. vaginalis*. The following structures are seen in Figure 3.4a: At the anterior end is the periflagellar canal (*PC*) supported by the pelta (*Pe*) composed of microtubules. Anterior flagella (*AF*) are seen in the lumen of the canal. On the observer's right, near the anterior end of the organism, the pelta is connected with the capitulum of the axostyle (*CaAx*), also consisting of a sheet of microtubules, along the peltar-axostylar junction (*J*), seen here in cross section. The nucleus (*N*) enclosed in a double-membraned envelope containing nuclear pores is apposed to the inner surface of the capitulum. It is surrounded by endoplasmic reticulum (*ER*). Posterior to the nucleus the capitulum narrows into the tubular axostylar trunk (*TrAx*), which courses posteriad near the anteroposterior axis of the flagellate. The trunk projects from the posterior surface of the organism being surrounded by the cell membrane. A transverse section of the trunk (Fig. 3.4b) reveals that this supporting organelle is formed by a sheet of microtubules turned on itself. On either side of the axostyle there is a row of electron-dense rounded bodies (hydrogenosomes), each of which is delimited by a membrane and contains dense granular matrix. Pinocytotic canals (*PiC*) and pinocytotic vesicles (*PiVe*) are seen at or near the cell surface, the vesicles being present also deeper in the cytoplasm. In addition, food vacuoles (*V*) of various sizes are distributed throughout the cytoplasm. A thin cell coat (*Ct*) covers the cell membrane.

Many fine-structural details of *T. vaginalis* are seen in Figs. 3.2i,j; 3.4b-d; and 3.5a-d. The various structures are discussed in several sections.

Kinetosomal Complex

Distribution of the kinetosomes, five in *T. vaginalis*, conforms to the pattern characteristic of trichomonads in general. All these basal bodies are embedded in microfibrillar material of electron density higher than that of the surrounding cytoplasm. Kinetosome R of the recurrent flagellum is situated at an approximately right angle to those, #1 to #4, of the anterior flagella (Fig. 3.3; see also Figs. 12 and 14 of *Trichomo-*

---

*The term "accessory filament" is employed throughout the subsequent sections of this chapter without further explanation.

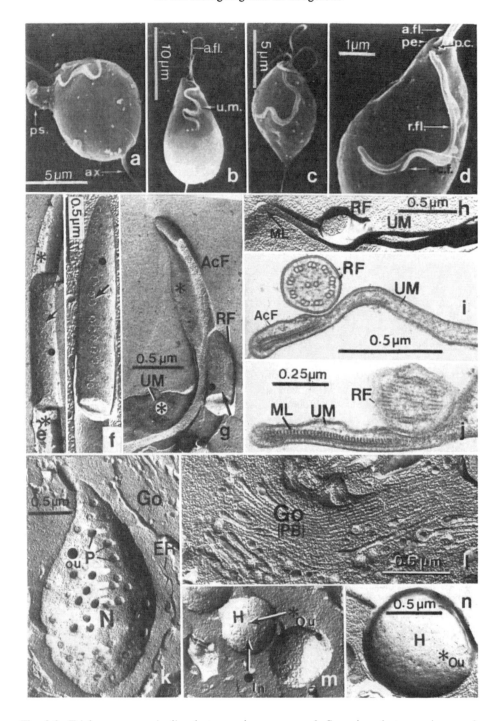

Fig. 3.2. *Trichomonas vaginalis*, electron microscopy. **a–d.** Scanning electron micrographs. Dorsolateral (**a**) or dorsal (**b–d**) views. In all figures, note the anterior flagella (*a. fl.* in b) emerging from the periflagellar canal (*p.c.* in **d**). The flagella are closely apposed to one another for some distance. The recurrent flagellum (*r. fl.* in **d**) emerges from the periflagellar canal posterior and dorsal to the anterior ones, the area of its emergence being separated by a narrow cytoplasmic partition from the anterior part of the canal. The wall of the canal is reinforced by the pelta (*pe*), the outline of which is evident in **d**. The margin of the undulating

nas gallinae in reference 25). Kinetosomes #1 and #3 are associated with hook-shaped lamellae (filaments), F_1 and F_3 (Fig. 3.5a,b) that course clockwise and possess a periodic structure. However, the periodicity of these two lamellae is not shown clearly in any of the figures included in this chapter (for the periodic striations of F_1, see Fig. 12 in Ref. 26 dealing with *Pentatrichomonas hominis*). Kinetosome #2 is connected to the peltar-axostylar complex by several sigmoid lamellae or fibers (F_2) that course from the ventral surface of the kinetosome and terminate by spreading over the dorsal surface of the complex in the area that includes the peltar-axostylar junction (J); however, as might be anticipated, in sections cut at many levels the junction appears to be located to the left of the F_2 filament (Fig. 3.5a,b).

Undulating Membrane, Costa, and Parabasal Apparatus

The undulating membrane (*UM*) is formed by a finely periodic (Fig. 3.2j) marginal lamella membrane (*u.m.* in **b**), typically shorter than the body, consists of the attached part of the recurrent flagellum (*r. fl.*) (inner component) and the outer accessory filament (*ac. f.*) (outer component); the relationships of the two components are seen in **d**. (For details of the preparation method in Figs. **a–d**. see reference no. 20.)

e–h and **k–n**. Freeze-fracture electron micrographs. **e** and **f**, Anterior flagella (**e** from *Trichomonas vaginalis;* **f** from *Tritrichomonas foetus*). In the external [EFs (*)] and protoplasmic [PFs (●)] faces, the intramembranous particles (IMPs) are arranged into rosettes (*arrows*) each consisting of about 9 to 12 units. The rosettes vary in their numbers and distribution on the flagella. For the scale to **e** see **h**. **g, h**. Undulating membrane (*UM*) and the apposed recurrent flagellum are fractured in an oblique **g** and transverse **h** plane. In **g** note the cytoplasmic membrane fold whose surface represents the EF (*). The part of the fold distal to the area of attachment of the recurrent flagellum (*RF*) corresponds to the accessory filament (*AcF*) of the light microscopist (here and in all subsequent electron micrographs showing the undulating membranes of the several species). Neither the PFs (●) nor the EFs (*) of the recurrent flagellum (*RF*) contain any rosettes; however, the area of attachment of the flagellum to the surface of the membrane is marked typically by three rows of IMPs (*arrow* in **g**). In a transverse section **h** there is evident the periodic marginal lamella (*ML*) with its loop enclosed in the part of the fold distal to the recurrent flagellum. **k**. Nucleus (*N*) is shown. Note the pores (*P*) in the nuclear envelope. The PF of the outer nuclear membrane (●*Ou*) is evident in this micrograph. The Golgi complex (*Go*) (parabasal body) is seen in the upper right corner of the figure. The nucleus is surrounded by the endoplasmic reticulum (*ER*). **l**. Golgi complex (*Go*) (parabasal body, *PB*). In one of the arms of the parabasal body note the details of the stack of tightly packed cisternae, apparently fractured in their longitudinal planes, and of the round vesicles. **m** and **n**. Hydrogenosomes (*H*). A double membrane surrounds these rounded organelles. On the surface of the hydrogenosomes shown in **m** there is evident the EF of the outer membrane (* *Ou*) (seen also on the organelle from *Tritrichomonas foetus* in **n**) and the PF of the inner membrane (● *In*). (For the scale to **m**, see **k**; for details of the preparation method, see reference no. 23.)

i and **j**. Transmission electron micrographs. Transverse **i** or nearly transverse **j** ultrathin sections through the undulating membrane (*UM*) show all the structures constituting this organelle. The microtubules of the recurrent flagellum (*RF*) are clearly evident in **i**, while the material possibly attaching the flagellum to the membrane is seen in **j**. The appearance of the loop of the marginal lamella is shown in **i**. The distinctly striated marginal lamella (*ML*) (**j**) is thought to be an extension of a rootlet filament of kinetosome #1. (*AcF*, accessory filament.)

(For methods of preparation, see various reports published by G. Brugerolle, for example, reference 22 and the published papers cited therein, and reference 31.)

a–d: Reproduced with permission of the Society of Protozoologists, from Warton A, Honigberg BM: Structure of trichomonads as revealed by scanning electron microscopy. *J Protozool* 26: 56-62, 1979. **e–h, k–n**: reproduced with permission of the Society of Protozoologists, from Honigberg BM et al: A freeze-fracture electron microscope study of *Trichomonas vaginalis* Donné and *Tritrichomonas foetus* (Riedmüller). *J Protozool* 31:116-131, 1984. **i,j**: original electron micrographs prepared by G. Brugerolle.

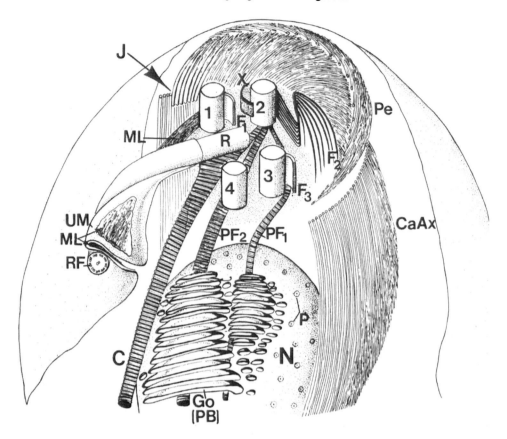

Fig. 3.3. Schematic diagram of *Trichomonas vaginalis* karyomastigont (i.e., system including the nucleus), representing a reconstruction based on examination of ultrathin sections by electron microscopy. The following structures are shown:

Kinetosomal complex, including the parallel kinetosomes #1 to #4 of the anterior flagella and kinetosome R of the recurrent flagellum (*RF*). The rootlet filaments and/or lamellae originating from these kinetosomes are the clockwise coursing filament F_1 and likely the filament that continues as the marginal lamella (*ML*) of the undulating membrane (*UM*), both originating from kinetosome #1 (the periodicity of these filaments is not shown in the diagram); the sigmoid F_2 filaments or lamellae that course from the ventral surface of kinetosome #2 toward the peltar-axostylar complex (*Pe-CaAx*) spreading over its dorsal surface including the area of the peltar-axostylar junction (*J*); the X filament, originating from kinetosome #2 and coursing between this kinetosome and F_1; the clockwise-turning periodic F_3 originating from kinetosome #3. The filaments or lamellae connect the kinetosomes with one another or with other organelles, and anchor the complex to various areas of the cytoplasm.

Costa (*C*), with the Type-B periodic structure, originating from kinetosomes R and #2 as the broad costal base and supporting the undulating membrane (*UM*) along the dorsal surface of the cell—there are no physical connections between the costa and the membrane.

Undulating membrane consisting of a dorsal fold of the cytoplasmic membrane which encloses the marginal lamella (*ML*) and of the attached recurrent flagellum (*RF*).

Parabasal apparatus composed of parabasal filaments 1 and 2 (PF_1, PF_2), and Type-A periodic structure, which emerge from the kinetosomal complex somewhat to the left of F_3 and near kinetosome #4, respectively, and of a parabasal body (*PB*), two Golgi complexes (*Go*). Each complex, which is applied to one of the filaments at some distance posterior to the kinetosomal complex, consists of stacks of cisternae surrounded by vesicles. The relationship of the long axes of the cisternae to the PFs shown in the figure has been chosen arbitrarily and, in actuality, as seen in some of the sections, need not necessarily conform to the relationship suggested in the diagram (cf. the diagram of *T. gallinae* in Fig. 1 in Reference 25).

Peltar-axostylar complex consists of the pelta (*Pe*) supporting the wall of the periflagellar

(*ML*) (Fig. 3.2i,j), connected to kinetosome #1 (not shown). According to some workers[26] the marginal lamella is the continuation of F_1, while according to others[22] the lamella originates independent of F_1 in kinetosome #1 (see Fig. 3.3). In any event, the lamella probably raises the cell membrane, thus forming a thin, finlike dorsal fold to which comes to adhere the recurrent flagellum (*RF*) (Fig. 3.2i,j). The connection of the flagellum to the membrane, not seen clearly in many, if not indeed most, ultrathin sections, appears in some preparations to involve electron-dense material between the outer membrane of the flagellum and the fold (Fig. 3.2j); however, the means of the aforementioned connection have not been fully elucidated by studies of ultrathin sections. (The physical connection between the recurrent flagellum and the membrane is easier to visualize by FEM (Fig. 3.2g).

The costa (*C*), coursing near the cell surface directly under the undulating membrane, is not connected physically with the latter organelle. This major rootlet filament has the Type-B periodic pattern.[27] Its periodicity is estimated at about 42 nm, on the average, in ultrathin longitudinal sections of conventionally stained material (Figs. 3.5a–c; 3.6a–d; see also references 22, 25, 26), or about 60 nm in preparations of whole costae subjected to negative staining (Fig. 3.6f; see also reference 22). The ultrastructural details of the Type-B costa were discussed in a report dealing with *P. hominis*.[26] Since as far as can be ascertained this organelle is virtually identical in the intestinal species and in *T. gallinae*[25] and *T. vaginalis*, the remarks pertaining to the former two apply equally to the urogenital species. Longitudinal sections of the costa reveal the presence of very dense, about 10-nm thick,[25] major transverse bands that form the 42-nm period (Fig. 3.6a–d). In addition to the major complete bands, the costa contains incomplete secondary transverse bands. There are also fine longitudinal filaments arranged in a "herringbone" pattern (Fig. 3.6b,d). There is an alternation of direction in the adjacent horizontal rows at an angle to the longitudinal axis (Fig. 3.6b,d). A discontinuity in the alternating pattern, seen in Fig. 3.6b,d, is more clearly evident in a highly enlarged segment (Fig. 3.6d). In the latter segment also note that the major transverse striations consist of electron-dense filaments cut in cross-sections. A plausible explanation of the pattern is presented in a diagram (Fig. 3.6e)—the network may consist of a connected set of laterally flattened hexagons, with a single fibril projecting through each hexagonal area (for additional discussion of the ultrastructure of the Type-B costa, see reference 26). In somewhat oblique sections (Fig. 3.6c), the major transverse bands appear to consist of dense parallel fibers that course at about a 45° angle to the anteroposterior axis. In such sections, less dense filaments parallel to one another and to the anteroposterior axis of the costa are evident in the spaces between the major bands (Fig. 3.6c). The anterior end of the costa together with the parabasal filament 2 (*PF₂*) constitute the broad costal base (*CBa*) that originates from kinetosome R and at least one of the other kinetosomes, namely #2 (Fig. 3.5a–c). The relationships between the two major rootlet filaments have been discussed by several workers.[22,25] As far as can be ascertained, the costa is noncontractile in Trichomonadinae, including *T. vaginalis*.

The two parabasal filaments, PF_1 and PF_2, have periodicity virtually identical to that of the costa (Figs. 3.5b, 3.6g–i); however, their band pattern is quite different. Here, each of the electron-dense areas consists of four lines; these areas are separated from one another by electron-lucent bands (Fig. 3.6g,h). Parabasal filament 1 (*PF₁*), connected to kinetosome #2, courses to the right of the lamella F_3 associated with kinetosome #3; in many sections it appears slenderer than PF_2. The latter filament which,

---

canal from which emerge the flagella and of the axostyle. Only the anterior part of the axostyle, the wide, somewhat spoon-shaped capitulum (*CaAx*), is shown. The left ventral surface of the nucleus, with an envelope that contains numerous pores *(P)*, is apposed to the dorsal concave surface of the capitulum. The pelta and axostyle are composed of sheets of 20-nm microtubules, the two parts of this skeletal organelle being connected to each other along the peltar-axostylar junction (*J*). The pelta constitutes the inner and the capitulum the outer layer of the junction.

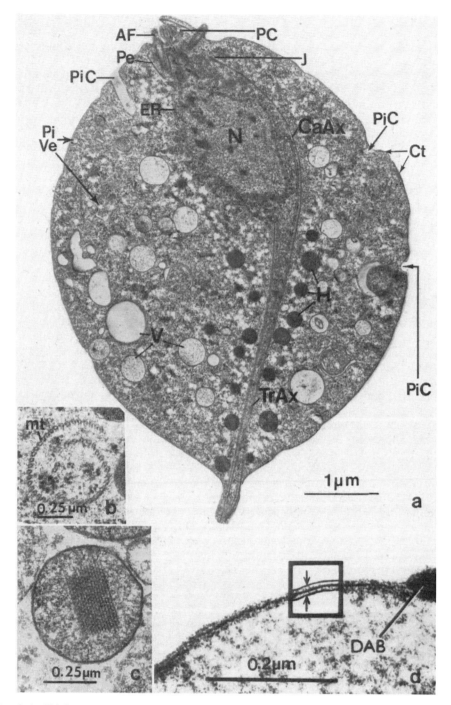

Fig. 3.4. *Trichomonas vaginalis*, transmission electron micrographs. **a.** Nearly frontal section of an entire organism. From the periflagellar canal (*PC*), with its wall supported by the pelta (*Pe*), seen on both sides of the canal, emerge the flagella; one of the anterior flagella (*AF*) is seen in an oblique section; the one to the right of it is shown in a nearly transverse section. The nucleus (*N*), containing several chromatin condensations, is surrounded by a nuclear envelope with pores; it is lodged in the depression of the axostylar capitulum (*CaAx*). Endoplasmic reticulum (*ER*) is seen around and close to the nucleus. The axostylar capitulum narrows posterior to the nucleus into the axostylar trunk (*TrAx*) that projects beyond the

as stated previously, has a common base with the costa, originates from kinetosomes #2 and R, then appears to pass under kinetosome #4 (Figs. 3.3, 3.5b). The two parabasal filaments support the assemblies of Golgi cisternae and vesicles (*Go*) (parabasal bodies) (Fig. 3.5c). This relationship is evident in Fig. 3.8m of *T. tenax,* the arrangement being identical in the latter species and in *T. vaginalis.* The filaments together with the Golgi complexes constitute the parabasal apparatus.

Peltar-Axostylar Complex

This complex consists of two microtubular sheets (in each, the microtubules have 20-nm diameters) that overlap anterior to the nucleus, forming the peltar-axostylar junction. In the junction, the pelta constitutes the dorsal (inner) layer (Fig. 3.5a-c; see also Fig. 3.8h-k of *T. tenax*). The microtubules of the sheets, arranged side by side, are interconnected by microfibrillar bridges. The pelta appears quite long in some sections (e.g., Fig. 3.8h of *T. tenax*). It is evident from the schematic diagram (Fig. 3.3) and the section in Fig. 3.4a that the anterior part of the axostyle, whose anterior-most segment connects with the pelta, is broad, forming the somewhat concave capitulum into which fits the nucleus (Fig. 3.4a). As stated before, posterior to the nucleus the microtubular sheet turns upon itself forming a tube, the axostylar trunk (Fig. 3.4b), which appears to narrow slightly but progressively until its terminal segment, covered by the cell membrane, protrudes from the posterior cell surface (Fig. 3.4a).

Most of the structures discussed in the preceding section and in this section are cytoskeletal elements found in all typical trichomonads studied to date by electron microscopy. The peltar-axostylar complex, consisting of microtubules, appears to be a supportive entity, the pelta, for example, reinforcing the wall of the periflagellar canal. The costa, the largest rootlet organelle, is thought to support the overlaying undulating membrane. As suggested by Kulda et al,[3] the parabasal filaments and the smaller rootlet filaments or lamellae may perform an important function in preserving the integrity of the trichomonad cell by connecting the kinetosomes to one another and to other parts of the cell, some of which are more readily identifiable than others. For example, the F_2 lamellae anchor the array of kinetosomes (kinetosomal complex) to the peltar-axostylar complex, and a lamella originating from kinetosome #1 continues as the marginal lamella

---

posterior surface of the cell; the projecting part is surrounded by the cell membrane. Hydrogenosomes (*H*) are lined up mainly along the axostylar trunk, and numerous vacuoles (*V*) are distributed throughout the cytoplasm. Depressions probably representing openings into the pinocytotic canals (*PiC*) are seen on the cell surface and these canals extend into the cytoplasm. Near the invagination leading into a pinocytotic canal (on the right), note the cell coat (*Ct*). Pinocytotic vesicles (*PiVe*) are seen in the cytoplasm. (*J*, peltar-axostylar junction.) **b.** In a transverse section, the axostylar trunk is seen as a sheet of interconnected microtubules turned upon itself (cf. Figs. 1, 2 and 6 in Reference 25). **c.** Hydrogenosome limited by a membrane, which appears to be double in some regions. The organelle contains dense granular matrix within which there is a large proteinaceous crystalline structure.

(For details of the preparation methods and material used in **a-c**, see various reports by Brugerolle, noted in the legend for Fig. 3.2i,j.)

d. Part of the double limiting membrane of a hydrogenosome as seen in an ultrathin section of a preparation obtained by differential centrifugation of *Tritrichomonas foetus* cell homogenate. The two closely apposed unit membranes (*arrows*) that limit the organelle are evident in the square area in the center of the micrograph. Part of an intramembranous vesicle formed between the two unit membranes is seen at the right margin of the figure. The contents of the vesicle give a positive 3,3'-diaminobenzidine (*DAB*) reaction; however, the granular matrix of the hydrogenosome is DAB-negative.

a-c: Original electron micrographs prepared by G. Brugerolle. **d:** Reproduced with permission of Pergamon Press, from Čerkasov J et al: Carbohydrate metabolism of *Tritrichomonas foetus* with particular respect to enzyme reactions occurring in hydrogenosomes. In Vitale LJ, Simeon V (eds): Industrial and Clinical Enzymology, Proc FEBS Spec Meet on Enzymes, Cavtat, Dubrovnik, Yugoslavia, 1977. *Trends in Enzymology* 61:257-275, 1980.

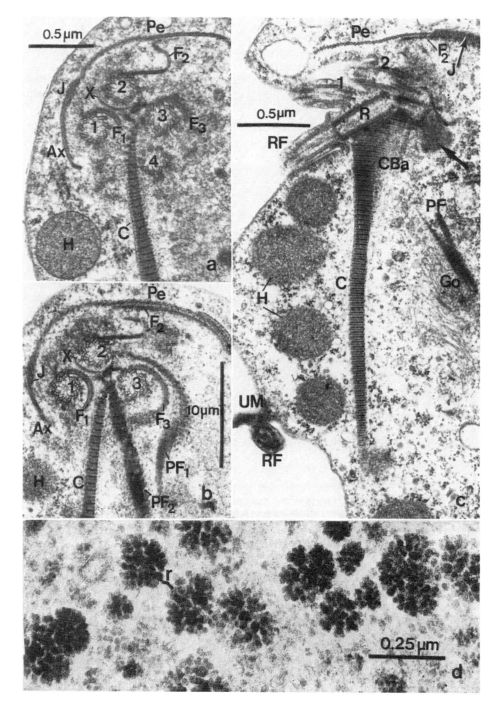

Fig. 3.5. *Trichomonas vaginalis*, transmission electron micrographs. **a–c.** Sections through the anterior part of organisms showing the peltar-axostylar complexes as well as the kinetosomal complexes, various rootlet filaments, and cytoplasmic inclusions in dorsal (**a,b**) and lateral and somewhat dorsal (**c**) views. **a** and **b.** Cross-sections of kinetosomes #1 to #4. Their typical microtubular structures and connecting fibrils are seen in **a**, which reveals also the clockwise-coursing filaments (lamellae) F_1 and F_3 of kinetosomes #1 and #3. The sigmoid F_2 filaments (lamellae) extend between the ventral surface of kinetosome #2 and the dorsal surface of the pelta (*Pe*) on which, in the views included in **a** and **b**, they are seen to be attached to the right of the peltar-axostylar junction (*J*). Filament X, which evidently also originates

into the dorsal finlike fold of the cell membrane which, together with the recurrent flagellum, constitutes the undulating membrane.

The biochemical and biophysical properties of the trichomonad cytoskeleton were investigated recently by Cappuccinelli et al.[28] Tubulin from the axostylar microtubules of *T. vaginalis* was studied by the use of microtubule inhibitors and quite extensively with the aid of fluoresceinisothiocyanate (FITC)-conjugated monoclonal antibodies (MAbs) specific for various antigenic sites of the tubulin molecule. Axostylar tubulin reacted with MAbs developed against sheep brain tubulin and pig brain tubulin. The results suggested also that different tubulins were present in a single trichomonad cell. Actin from the urogenital trichomonad, purified by anion exchange chromatography and polymerization-depolarization cycles, was shown to differ in some biochemical and biophysical characteristics from skeletal muscle actin. The protozoan actin migrated slightly behind the muscle actin in polyacrylamide gel electrophoresis and the former was more acidic. When cleaved with proteases, trichomonad actin showed peptide sequences different from those of the muscle actin. This protein from *T. vaginalis* had a very low affinity for phalloidin, inhibited DNase (deoxyribonuclease) I activity, and activated Mg^{2+}-ATPase less than actin from muscle.

Structure of the Cytoplasm and Cell Membrane and Inclusions Other Than the Mastigont Organelles

In ameboid cells, the rather clear peripheral zone of the cytoplasm, the "ectoplasm," in which only a few large inclusions are seen, con-

---

from kinetosome #2, is interposed between this kinetosome and the F_1 filament with which it may be connected. The costa (*C*), with its Type-B structural pattern, is seen at the level of these sections (**a,b**) to originate from the kinetosomal complex dorsal and somewhat to the right of kinetosome #2 and to the left of kinetosome #3; a connection of its anterior part with parabasal filament 2 (PF$_2$) can be observed in **b**. A hydrogenosome (*H*) with granular matrix and a well-defined membrane is seen to the left of the costa (*C*) in **a**; note also the junction (*J*) between the pelta (*Pe*) and the axostyle (*Ax*). Kinetosome #4, seen in **a**, is not shown in **b**. However, parabasal filaments 1 (PF$_1$) and 2 (PF$_2$), with the Type-A structural pattern, are evident in the latter figure. PF$_1$, usually less robust than the one seen here, originates from the kinetosomal complex near the right side of kinetosome #2 and then courses clockwise around kinetosome #3 to the right of F_3; it continues posteriad into the cytoplasm. The typically stouter PF$_2$ originates near kinetosome #2 dorsal to PF$_1$. It courses to the left of kinetosome #3, close and evidently ventral to kinetosome #4 (see diagram in Fig. 3.3), then continues posteriad (to the right of the costa) into the cytoplasm.

    c. In a lateral and somewhat dorsal view (longitudinal section at an angle to the "parasagittal" plane), the following structures are seen (going from the anterior toward the posterior end): the pelta (*Pe*), in cross section, is underlain by the F_2 filament; further toward the right margin of the figure, the pelta is overlain by the microtubular sheet of the axostylar capitulum. Thus in this section the junction (*J*) may be thought of as consisting of three layers. Below the peltar-axostylar complex, note kinetosomes #1 and #2 (very near to the origin of their flagella) in oblique sections. Below these kinetosomes and tilted at an acute angle to them is kinetosome R (in a longitudinal section) from which originates the recurrent flagellum (*RF*). The broad periodic costal base (*CBa*) is apposed to the ventral surface of kinetosome R from which it originates (in part). Ventral to the costal base is an atractophore (*black arrow*), the presence of which suggests that the organism is in early division. The costal base continues into the costa with its Type-B structural pattern. Dorsal to the costa is a row of hydrogenosomes (*H*). At the level parallel to the posterior visible end of the costa, note the undulating membrane (*UM*) with the recurrent flagellum (*RF*) in a nearly transverse section. Ventral to the costa is a segment of a parabasal filament (*PF*) with the parabasal body material [Golgi complex (*Go*) vesicles and cisternae] applied to it.

    (For details of the preparation procedures of the material in **a-c**, see various reports by Brugerolle noted in the legend for Fig. 3.2i,j.)

    d. High magnification electron micrograph of glycogen granules arranged in rosettes (*r*), typical of this carbohydrate storage product in trichomonads.

    a-c: Original electron micrographs prepared by G. Brugerolle. **d.** Courtesy of Dr. M. Müller, The Rockefeller University, New York.

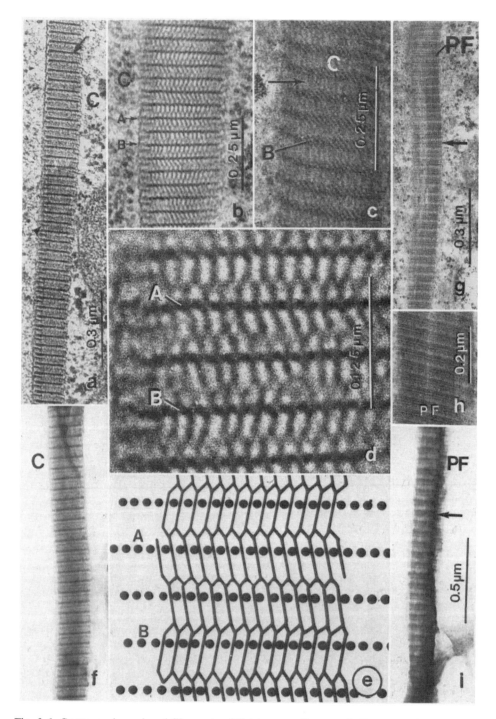

Fig. 3.6. Costae and parabasal filaments of Trichomonadinae. **a–f.** Costae. (**a** from *T. gallinae;* **b–e** from *P. hominis;* **f** from *T. vaginalis*). Most species of the subfamily Trichomonadinae have Type-B costae with essentially the same periodic structure. In longitudinal sections of fixed material **a–e**, the periodicity is about 42 nm; it appears to be about 60 nm in material processed by the negative staining method (uranyl acetate) of a whole costa (**i**). **a.** In each repeating unit, the incomplete secondary transverse band (*arrowhead*) is located closer to the thick major transverse band (*arrow*) anterior to it than to the major band posterior to it. **b.** In

tains numerous microfilaments, tangled or arranged in bundles (* in Fig. 3.7a). To date, the nature of these filaments is not fully understood (see Honigberg[5] for discussion and relevant references).

The cell membrane is invested with a very thin *cell coat* (*Ct*). This coat can be seen more clearly on some parts of the surface (Fig. 3.4a; and see references 3 and 5 for additional information). Endocytosis involves both *phagocytosis* and *pinocytosis*. Phagocytosis occurs preferentially in the posterior part of the cell[29] resulting in the formation of food vacuoles of various sizes (*V* in Fig. 3.4a of *T. vaginalis* and in Fig. 3.11f of *D. fragilis*). For obvious reasons, phagocytosis can be studied at a greater advantage in organisms obtained directly from the host or grown in xenic cultures (Fig. 3.11f of *D. fragilis*) or in axenic cultures in media containing many solid particles. Pinocytotic canals (*PiC*) are formed anywhere on the flagellate's surface and coated pinocytotic vesicles (*PiVe*) are seen in the cytoplasm (Fig. 3.4a). (For additional information and references, see Kulda et al[3] and Honigberg.[5])

Rough endoplasmic reticulum is clearly evident around the nucleus in ultrathin sections (Fig. 3.4a) and in freeze-fracture preparations (Fig. 3.2k). *Glycogen granules* are arranged typically in large rosettes (*r*) (Figs. 3.5d, 3.7a). (For biochemical aspects, see Chapter 5 of this book.)

The electron-dense, microbody-like inclusions, about 0.5 to 1.0 μm in diameter, are known as *hydrogenosomes* (paraxostylar and paracostal granules of the light microscopist) (Fig. 3.4c; see also the freeze-fracture electron micrographs in Fig. 3.2m,n). They are distributed in the cytoplasm of all stages (Fig. 3.7b,c), but primarily along the axostyles and costae of nondividing organisms (Fig. 3.4a). These organelles, which play an important role in the metabolism of trichomonads (see Chapter 5), have a unique ultrastructure. The matrix of these typically spherical bodies is finely granular and often contains a dense, amorphous, or

---

addition to the complete major and incomplete transverse bands, the costae have longitudinal filaments arranged in a "herringbone" pattern. A discontinuity in this pattern is evident in about the middle (*A-B*) of the segment shown in **b**. **c.** In an oblique section, densely staining parallel filaments, coursing at about a 45° angle to the anteroposterior axis of the costa, are evident in the major transverse bands (*B*). Between these bands note less dense filaments (*arrow*) parallel to one another and to the anteroposterior axis. **d.** Details of the network of longitudinal filaments and the discontinuity (*A-B*) in their pattern, as seen in **b**, are shown at a much higher magnification in **d**. At this magnification, the major transverse bands appear to be composed of discrete dark-staining dots, presumably cross-sections of filaments coursing at a right angle to the anteroposterior axis of the costa. **e.** Line diagram of the segment of the costa shown in **d** represents a possible explanation of the Type-B structural pattern. **f**, Segment of a costa processed by the negative staining method. For scale, see **i**.

g-i. Parabasal filaments from *T. gallinae* in longitudinal sections of fixed material (**g,h**) and a segment of a negatively stained whole filament (**i**) from *T. vaginalis*. All species of the order Trichomonadida studied to date by electron microscopy have parabasal filaments with the Type-A pattern of the transverse bands. The periodicity of the parabasal filaments is the same as that of the Type-B costae, that is, about 42 nm. However, in the Type-A pattern each band consists of four thin transverse parallel lines.

(For details of the preparation methods of the material in **a-d,g,h**, see reference nos. 25 and 26; for **f,i** see reports by Brugerolle noted in the legend for Fig. 3.2i,j.)

a,g,h: Reproduced with permission of the Society of Protozoologists, from Mattern CFT et al: The mastigont of *Trichomonas gallinae* (Rivolta) as revealed by electron microscopy. *J Protozool* 14:320-339, 1967; **b-e** reproduced with permission of the Society of Protozoologists, from Honigberg BM et al: Structure of *Pentatrichomonas hominis* (Davaine) as revealed by electron microscopy. *J Protozool* 15:419-430, 1968. [Figs. 3.6a-e,g,h were included in an arrangement similar to the present one in Honigberg BM: Trichomonads of veterinary importance. In Kreier JP (ed): *Parasitic Protozoa*, vol 2, pp 163-273. New York: Academic Press, 1978. These figures were reproduced with permission of Academic Press, New York.] Figures **f** and **i** are electron micrographs, prepared by G. Brugerolle, of a whole costa (**f**) and parabasal filament (**i**) of *T. vaginalis* negatively stained with uranyl acetate.

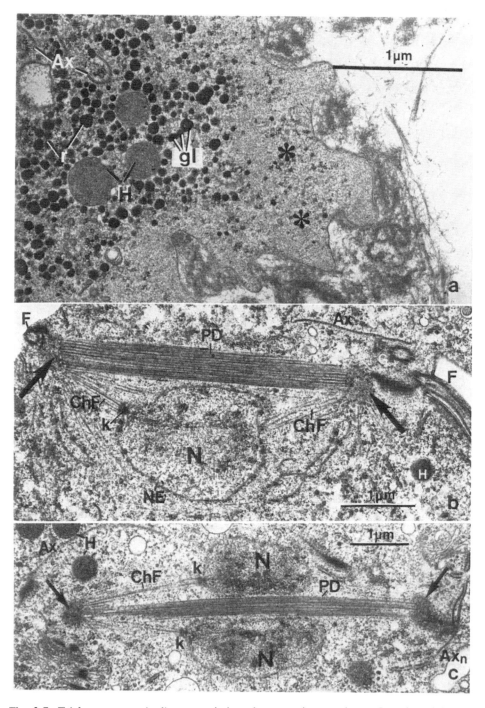

Fig. 3.7. *Trichomonas vaginalis,* transmission electron micrographs. **a.** Pseudopod formed by the peripheral cytoplasm, "ectoplasm" of the organism contains microfilaments, evident in the areas marked by asterisks (*). Numerous glycogen granules (*gl*), typically in rosette configurations (*r*) (see also Fig. 3.5d) are evident in the region of the cytoplasm adjacent to the ectoplasm. Hydrogenosomes (*H*) are also present in this zone. Axostylar microtubules (*Ax*) are seen above and to the left of the hydrogenosomes.

b,c. Organisms dividing by crytopleuromitosis. Early in division there develops in each daughter mastigont a "bell-clapper"-like structure, the atractophore (*large arrows*). The at times crystalline core (Fig. 3.4c). They are bounded by two very thin unit membranes closely apposed to each other (Fig. 3.4c,d), "except in certain osmotically sensitive areas, where they [are] separate [d], forming flat vesicles"[30] marked by 3, 3'-diaminobenzidine (DAB) deposits in preparations treated with this compound (Fig. 3.4d). The details, that is, the two closely apposed limiting membranes (*arrows*) and part of the intermembranous pocket with DAB-positive contents (near the right margin of the figure), shown in Fig. 3.4d were seen in an ultrathin section of a preparation obtained by centrifugation of *T. foetus* homogenate.[30] The hydrogenosome matrix fails to react with DAB, which indicates that catalase, the marker enzyme for peroxisomes, is absent. On the other hand, the contents of the intermembranous pockets are DAB-positive; however, the reaction is nonenzymatic and its significance remains unclear.[30] The fact that the organelle shown in Figs. 3.2n and 3.4d came from *T. foetus* is of little importance, because the hydrogenosomes in all trichomonads studied to date appear to be identical (for pertinent references, see refs. 3 and 5 and Chapter 5). There is evidence that hydrogenosomes divide by binary fission.[3,23,30]

Nucleus and Cell Division

The nondividing nucleus, surrounded by an envelope with typical pores (Figs. 3.2k, 3.4a), contains dispersed chromatin granules (Fig. 3.4a).

As in many other protozoa, the nuclear envelope (*NE*) does not disappear during mitosis (Fig. 3.7b,c). In addition, an extranuclear spindle (paradesmose, *PD*) is formed, a feature characteristic of trichomonads and hypermastigotes (Fig. 3.7b,c). The onset of division is marked by the appearance of a supplemental kinetosome that gives rise to a ventrally located flagellum, and by the development of two rod-shaped organelles, the atractophores, each with a distal enlargement, the "bell-clapper" rods ("battachios" of the French), which become the poles of the division figure (Figs. 3.5c, 3.7b,c,). Microtubules originating from the atractophores form one or more bundles of microtubules, the paradesmose. The chromosomal microtubules (fibers) (*ChF*) also originate from atractophores; they are directed toward the nucleus, attaching to the chromosomal kinetochores (centromeres) that are inserted on the nuclear envelope (Fig. 3.7b,c). The details of the relationships among the chromosomal fibers, chromosomes, and kinetochores can be seen in Figure 13 of reference 31. By elongation of the extranuclear spindle, the daughter mastigont systems separate and migrate to the opposite poles (Fig. 3.7b,c), as do the daughter chromosomes (Fig. 3.7c). Ultimately two daughter nuclei are formed in telophase (Fig. 3.7c). In each daughter mastigont, the normal number of flagella is reconstituted

---

extranuclear spindle, paradesmose (*PD*), made up of microtubules that originate from the atractophores, extends between these two structures, which become positioned at the division poles. As the division progresses, the paradesmose elongates (**b,c**) and comes to lie against the nuclear envelope (*NE*), often in a depression of this envelope (c; see also Fig. 14 in reference 31). From the base of each atractophore originate also small bundles of microtubules, the chromosomal fibers (*ChF*), which attach themselves to the kinetochores (*k*) (centromeres) of the chromosomes, the kinetochores being inserted on the nuclear envelope. Some of the chromosomal fibers appear to terminate free in the cytoplasm. In addition to the missing kinetosomes and rootlet filaments, new axostyles develop in each new mastigont system—one of these cytoskeletal organelles (Ax_n) is evident in the right mastigont in **c**. A segment of the old (parental) degenerating axostyle (*Ax*) is still present in the organism shown in **b**. The chromatin material in the nucleus (which remains surrounded by the envelope during the entire process of mitosis) becomes condensed into chromosomes that are rather poorly defined and take light stain (they are seen more clearly in **c**).

(For details of the preparation method of the material and a more detailed, albeit still incomplete, description of cryptopleuromitosis, see Brugerolle[31].)

All figures are original electron micrographs prepared by G. Brugerolle. Parts **b** and **c** were published previously and are reproduced with permission of Gustav Fischer Verlag, Stuttgart, from Brugerolle G: Etude de la cryptopleuromitose et de la morphogénèse de division chez *Trichomonas vaginalis* et chez plusieurs genres de Trichomonadines primitives. *Protistologica* 11:457-468, 1969.

by the development of the missing kinetosomes and rootlet filaments lost to one or the other of the daughter mastigonts in early division. The parental peltar-axostylar complex is lost during division (see a remaining segment of the parental peltar-axostylar complex in Fig. 3.7b) and a new complex develops in each daughter cell (Ax_n in Fig. 3.7c).

Trichomonas tenax (O. F. Müller, 1773)

As indicated in Chapter 2, since the extensive discussion of the nomenclature of the oral trichomonad of humans was published by Dobell,[32] the generally accepted name for this species is *Trichomonas tenax* (O. F. Müller, 1773). The possible relationships of this presumably commensal organism with those (strains of *T. tenax* and *T. vaginalis*) reported from lesions of the respiratory passages and lungs of humans are discussed in Chapter 19.

Light Microscopy

Wenrich,[33-35] on the basis of iron hematoxylin-stained preparations, described many structural details of *T. tenax* as revealed by light microscopy. Similar preparations were employed by Hinshaw[36] in his studies of the division process in this species. In the late 1950s Honigberg and Lee[37] examined organisms from five xenic and monoxenic cultures of *T. tenax* using phase-contrast optics with living organisms and bright field for fixed preparations, and either iron hematoxylin-stained or protargol-stained flagellates. By employing the phase-contrast and protargol staining techniques, in addition to the classic hematoxylin method, they were able to extend and amplify the information available at the time of their study. The following account is based on their observations.[37]

In protargol-stained preparations, the body of the flagellate measures 7.1 ± 0.06 (4.0 to 13.0) × 4.7 ± 0.05 (2.0 to 9.0) μm ($n = 300$, in three strains grown in xenic cultures; 100 per strain). The organism is variable in shape, but typical flagellates are ellipsoidal or ovoidal (Fig. 3.8a–c). Normal nondividing organisms have four anterior flagella of unequal length, often arranged in two groups, each containing a pair of subequal locomotor organelles (Fig. 3.8a–c). The longest anterior flagellum measured in two strains ($n = 100$; 50 per strain) averaged 10.9 ± 0.15 (7.0 to 15.0) μm. The recurrent flagellum which, together with the accessory filament, originates from the kinetosomal complex dorsal to the anterior flagella, is incorporated into the free margin of the undulating membrane (Fig. 3.8c). As in other Trichomonadinae, the flagellum constitutes the inner and the accessory filament the outer component of the margin. The membrane, with relatively few waves, together with its supporting slender costa, extends from less than one-half to over three-fourths of the body length (Fig. 3.8b,c). There is no free posterior flagellum. The left ventral side of a typical trichomonad nucleus containing usually small chromatin granules and a spheroidal nucleolus (Fig. 3.8a,c) is applied to the spatulate capitulum of the axostyle (Fig. 3.8a–c). The anterior part of the capitulum connects to a crescent-shaped pelta that surrounds the area in which the flagella emerge from the cell (Fig. 3.8b,c). Posterior to the nucleus, the capitulum narrows progressively, giving rise to a slender axostylar trunk that projects from the posterior cell surface. The projecting segment narrows more or less gradually to a sharp point (Fig. 3.8a–c). In living organisms and in those stained with hematoxylin there are rows of dense granules (hydrogenosomes) along the costa (paracostal granules), but not along the axostyle (Fig. 3.8a).

The parabasal apparatus consists of a rod-shaped parabasal body applied to the proximal segment of a parabasal filament that frequently extends far into the cytoplasm, in some instances terminating near the posterior end of the cell (Fig. 3.8b,c).

A relatively extensive account of division of the oral trichomonad was also published by Honigberg and Lee; see Fig. 12-25 in their account.[37] It suggests that this process conforms to that described from other trichomonad species whose division has been studied.

Electron Microscopy

Although several recent reports of the fine structure of *T. tenax* as revealed by SEM and TEM[38-42] have been published, no thorough description of this species based on electron microscope studies is available. For the most part, the accounts have been sketchy and not illustrated by micrographs meeting current standards of electron microscopy. Of these reports, that by Filice et al,[39] although short on micrographs, contains a satisfactory description of the organism; some of the few micrographs that were included in the paper are among the most satisfactory published to date (e.g., Figs. 3 and 4 in reference 39). In light of the fact that *T. tenax* was the only trichomonad parasitic in humans that had not been adequately described on the basis of electron microscope observations, a study of this organism by TEM was undertaken by one of our students, T. P. Poirier, in collaboration with S. C. Holt and B. M. Honigberg. A report of this investigation is now ready for press.[43] With the single exception of Fig. 3.8g (by G.B.), all the TEM illustrations of *T. tenax* included in this chapter are from the latter report. In studying the electron micrographs the reader should compare them with those of *T. vaginalis,* shown in the preceding section of this chapter, with those of *P. hominis* dealt with in the immediately following section, and with figures included in the extensive account of *Trichomonas gallinae,*[25] the organism most closely resembling *T. tenax.* In general, the schematic diagram of *T. vaginalis* shown in Fig. 3.3 is applicable to *T. tenax.* Except for *P. hominis,* all the aforementioned species are very similar, especially when examined by TEM, a method rather unsatisfactory for differentiation of congeneric species. Among the micrographs of *T. tenax* the transverse sections through the undulating membranes (Fig. 3.8f,g) should be compared with those of *T. vaginalis* (Fig. 3.2i,j) and *P. hominis* (Fig. 3.9f), for often slight structural differences among the marginal lamellae can be noted among the membranes of the subfamily Trichomonadinae. However, to be taxonomically useful the differences in the fine structure of the undulating membranes must be the subject of extensive comparative studies in many genera and species of this subfamily.

In transverse and more or less oblique sections passing from the anterior parts of the ventral surfaces to the more posterior areas of the dorsal surfaces of organisms (Fig. 3.8d,e,i–l) one can find a variety of structures and visualize their spatial interrelationships. The nearly transverse section in Fig. 3.8h, which shows the entire or almost entire length of the pelta, appears to be tilted toward the left (note the periflagellar canal around the right part of the flagellum originating from kinetosome #3). Sections cut through more superficial planes of the cells reveal mixtures of sections of flagella and kinetosomes (Fig. 3.8e,h,i). They are useful for analyzing the spatial relationships among the kinetosomes, flagella, and periflagellar canals (e.g., Fig. 3.8i). Furthermore, in many such sections (Fig. 3.8e,h), and even in those in which the flagella are no longer present (Fig. 3.8d), fine fibers may be seen radiating from some of the kinetosomes. These are the transitional fibers, known to emerge from the anterior ends of kinetosomes (for a brief discussion, see Honigberg et al).[26] Thus, the presence of the transitional fibers in a section indicates that it was cut in a plane passing close to the cell's surface.

Sections cut further away from the organism's anterior surface reveal the clockwise curved rootlet filaments (lamellae) F_1 and F_3 of kinetosomes #1 and #3, respectively (Fig. 3.8j–l). A suggestion of weak periodic cross-striations is evident in F_3 shown in Fig. 3.8l. Filament X connects with kinetosome #2 and is seen coursing between this basal body and F_1 in Figure 3.8k, while the large sigmoid F_2 lamellae originating on the ventral side of kinetosome #2 and terminating on the dorsal surface of the peltar-axostylar complex can be observed in two micrographs (Fig. 3.8j,k).

The Type-B costa is present in all the ultrathin sections of the organism shown in Fig. 3.8 save one (Fig. 3.8h); its structure can be resolved best in Fig. 3.8e. The connections of the costa with kinetosomes R (Fig. 3.8d,e,i,l) and #2 (Fig. 3.8k) is also evident. Parabasal filaments, with Type-A periodic structure are evident in some of the sections. Parabasal filament 1 (PF_1) can be seen in Fig. 3.8k, while parabasal filament 2 (PF_2) is visible in Fig. 3.8d,k. The structure of the Golgi complex (parabasal body) applied to a parabasal filament is shown clearly in the nearly frontal section through the anterior part of the organism (Fig. 3.8m), which includes also the nucleus and some of the components of the mastigont system. Of interest with regard to the Golgi complex is the longitudinally cut stack of cisternae, closely apposed to one another, and the surrounding vesicles, some of which are filled with electron-dense material (for a discussion of the Golgi complex, see Kulda et al[3]).

As far as can be ascertained, there is to date

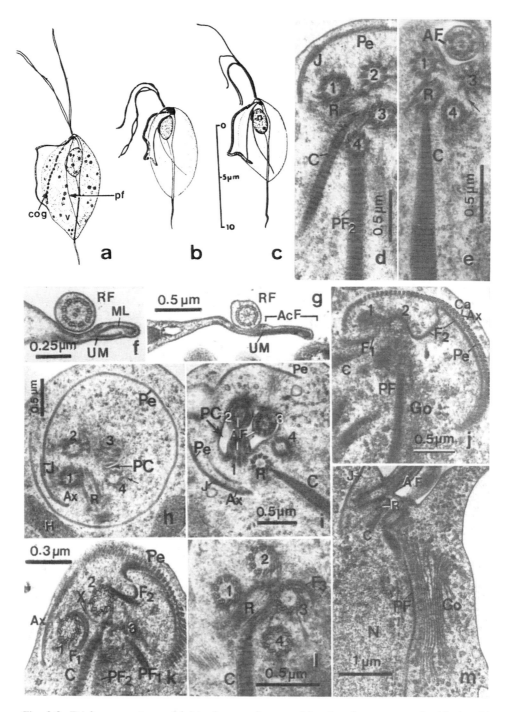

Fig. 3.8. *Trichomonas tenax*. Light micrographs. **a–c.** Line drawings, prepared with the aid of camera lucida, of organisms fixed in Bouin's fluid (*B*) and stained with iron hematoxylin (**a**) or with protargol (**b,c**). The scale to the left of **c** is applicable to all three figures. (For the abbreviated designations of most of the organelles, see Fig. 3.1f and 3.9a.) **a.** Right view showing most of the structures. The paracostal granules (*cog*) (hydrogenosomes), demonstrable in iron hematoxylin–stained organisms, mark the position of the costa. Some of the other rounded cytoplasmic inclusions are various ingested particles, including bacteria enclosed in food vacuoles or storage products. The nuclear structure is seen clearly in iron hematoxylin-stained flagellates. The parabasal body is usually not visible in such organisms, but the para-

basal filament (*pf*) is often evident. The free margin of the undulating membrane consists of two filaments, whose identities and relative positions are like those described for *Trichomonas vaginalis*. The four anterior flagella are arranged in two groups, each containing two subequal motor organelles. **b,c.** Right views. All the mastigont organelles, including the rod-shaped parabasal body of variable length, but typically shorter than the nucleus, as well as the often stout (**c**) parabasal filament, are readily demonstrated by protargol staining. No hydrogenosomes or details of cytoplasmic structure or inclusions are usually discernible in such preparations. In occasional organisms (e.g., **c**), the nuclei show their structure, in this instance a uniform distribution of small accumulations of chromatin and a single spherical nucleolus surrounded by a clear halo, a pattern typical of trichomonads. Note the differences in structure of the parabasal bodies and in the relative lengths of the undulating membranes of the organisms shown in **b** and **c**.

d-m. Transmission electron micrographs. **d,e,h-l.** Kinetosomal regions with the various rootlet filaments (lamellae) (F_1, F_2, F_3) are shown in approximately dorsal views in sections cut at different levels and at a variety of angles to transverse planes.

Judging from the appearance of and relationships among the structures seen in the sections shown in **d, e, h,** and **i**, these sections include relatively superficial regions of the organisms. Furthermore, the sections in **d, e,** and **i** appear to pass through planes tilted toward the posterior surfaces of the cells; the section in **h** passes through a plane tilted toward the left surface. In this figure, note the periflagellar canal (*PC*) to the right of the base of the anterior flagellum and kinetosome #3 as well as the transitional fibers (*small arrow*) emerging from the anterior end of kinetosome #4. **e** and **h** show one, and **i** three anterior flagella (*AF*) in periflagellar canals (*PC*) and kinetosomes associated with the transitional fibers (**e**, *arrow*) include sections passing through very superficial areas near the anterior ends of the cells. A nearly transverse section through almost the entire length of the pelta (*Pe*) and part of the axostylar capitulum (*Ax*) are seen in **h**; note also the peltar-axostylar junction (*J*). This junction is evident also in **i** and at a more posterior level in **d**. The considerable length of the peltar-axostylar junction is shown in **j**—the layer dorsal to the actual junction consists of the distal segments of the F_2 lamellae spreading on its dorsal layer, that is, the pelta (*Pe*). The Type-B costa (*C*) is visible in all the aforementioned figures, except for **h**, which includes a section made at a level above the origin of this major rootlet filament, whose ultrastructure can be resolved quite well in **e**. In somewhat deeper sections, even in those in which some of the kinetosomes show the transitional fibers (e.g., kinetosome #2 in **d**), indicating that the sections are quite oblique, note the parabasal filaments with their Type-A periodic pattern)—PF_2 in **d**, and both PF_1 and PF_2 in **k**. In **j** the Golgi complex (*Go*) is apposed to the right side of the parabasal filament. Figures **k** and **l** reveal the clockwise filaments of lamellae F_1 and F_3 of kinetosomes #1 and #3, respectively. Filament X originating from kinetosome #2 and coursing between this basal body and F_1 is evident in **k**.

f,g. Transverse sections of undulating membranes (*UM*). In both figures, note the fold of the cytoplasmic membrane enclosing the marginal lamella (*ML*) visible for some distance proximal to the area of adherence of the recurrent flagellum (*RF*) on the side of the fold opposite this area. Distal to the recurrent flagellum the marginal lamella forms a loop (more typical in **g**). The physical connection between the RF and the fold is difficult to resolve, but a suggestion of a more electron-dense, presumably cementing substance between the two structures can be seen in **f** (*AcF*, accessory filament).

m. Nearly frontal section. Note the peltar-axostylar complex of microtubules and their junction (*J*) (on the observer's left), with the pelta (anterior to the junction) supporting the wall of the periflagellar canal that contains two of the anterior flagella (*AF*). The parabasal filament (*PF*), to the right of the nucleus (*N*), as well as the Golgi complex (*Go*) (parabasal body), consisting of tightly stacked elongate cisternae and vesicles around them, are also evident in this figure.

[For transmission electron microscopy (**d-m**), the organisms were fixed in 2% glutaraldehyde in 0.028 M veronal acetate, 0.2 M cacodylate, or 0.2 M phosphate buffer, pH 7.2, at 4 to 6 °C. They were postfixed in buffered osmium tetroxide. The fixed material was embedded in Epon, sectioned with a diamond knife, and stained with uranyl acetate and lead citrate. Sections were examined in a JEOL 100S electron microscope operating at 80 kV.]

a-c. Reproduced with permission of the American Journal of Hygiene, from Honigberg BM, Lee JJ: Structure and division of *Trichomonas tenax* (O. F. Müller). *Am J Hyg* 69:177-201, 1959; **d-f** and **h-m** reproduced with permission of the authors, from Poirier TP et al: Fine structure of *Trichomonas tenax* (O. F. Müller)[43]; **g** is an original micrograph prepared by G. Brugerolle.

Pentatrichomonas hominis (Davaine, 1860)

Although there are compelling reasons to use the name *Pentatrichomonas hominis* (Davaine, 1830) Wenrich, 1931 for the species inhabiting the colon of humans, even today some workers refer to this organism by other names. The ones used most often are *Trichomonas intestinalis* Leuckart, 1879; *Trichomonas hominis* (Davaine, 1860) Grassi, 1888; and *Pentatrichomonas ardindelteili* (Derrieu & Raynaud, 1914) Mesnil, 1914. The reasons for the use of *Pentatrichomonas hominis* have been discussed on several occasions[1,5,33,44]; a brief consideration of the nomenclatural problems with regard to the intestinal trichomonad of humans and a variety of other mammals is included in Chapter 2. Natural host distribution was discussed in several reports,[45-47] while infections of experimental hosts were recorded in the references listed by Levine.[48]

Light Microscopy

The descriptive account of *P. hominis* as observed with the light microscope given below is based on various reports, but especially those of Wenrich,[33-35,45] Kirby,[49] and Wenrich and Saxe.[47] It includes also some unpublished observations (of B. M. H.). The composite diagram in Fig. 3.9a is based on the results of studies of protargol-stained organisms,[49] and unpublished observations.

The body of the flagellate is ellipsoidal, ovoid, or pyriform. Living organisms have erratic rapid swimming movements. The trichomonad is highly plastic, capable of changing its shape; however, pseudopod formation is less common than in *T. vaginalis*. Parts of the cytoplasm can detach from the posterior end of the body by constriction. This process, known as autotomy, is common among trichomonads. Pseudocysts, but not true cysts, can be formed under unfavorable environmental conditions.

In fixed preparations stained with iron hematoxylin, *P. hominis* averages in length 7.7 (6.0 to 14.0) μm (n = 225, from four populations), and in width 5.3 (4.0 to 6.5) μm (n = 50, from two populations). Four of the five anterior flagella, of unequal to subequal length, are arranged in one group that in its appearance and type of movement resembles the homologous organelles found in other trichomonads. The longest flagellum of this group averages in length 8.7 (7.5 to 12.0; even up to 18.0) μm (n = 25 in a single population). The independent (fifth anterior) flagellum (*i.fl.*) originates somewhat ventral to the group of typical flagella. The details of the kinetosomal complex cannot be resolved by light microscopy. The independent flagellum, ranging in length from 6.0 to 13.0 μm is directed posteriad; its movement resembles that of the recurrent flagellum found in other trichomonads. In both its position and type of movement, this flagellum is thought of as a precocious recurrent flagellum, although the steps in division that would be involved in its incorporation into the margin of the undulating membrane, as described for *Tetratrichomonas prowazeki*[50] and for *Trichomonas tenax*,[37] have not been studied to date. The five anterior flagella are said to have the "4 + 1" arrangement limited to the genus *Pentatrichomonas*. This arrangement of anterior flagella constitutes a more important characteristic of the genus than their number, from which it gets its name.

The undulating membrane and costa originate from the kinetosomal complex dorsal to the anterior flagella. Both these organelles, which exhibit a slight counterclockwise torsion, are as long as or somewhat longer than the body. The membrane is rather well developed, its free double margin being composed of the distally located (external) accessory filament and the proximally located (internal) attached segment of the recurrent flagellum. The recurrent flagellum continues beyond the end of the membrane as a free posterior flagellum (*p.fl.*) ranging in length from 4.5 to 8.0 μm, with an average of 5.7 μm (n = 50, in two populations). In hematoxylin-stained organisms, the costa seems to be wider than the flagella; it tapers toward both ends. In protargol-stained preparations this organelle looks like a rod, about as wide as the flagella, which tapers at both ends. In living organisms examined with the aid of phase contrast and in those stained with hematoxylin one can see dark-staining granules (hydrogenosomes) arranged in a row dorsal to the costa; they correspond to the paracostal granules of the light microscopist.

The spatulate capitulum of the axostyle is

somewhat wider than the axostylar trunk. The capitulum is situated very near the left ventral surface of the nucleus and connects anteriorly to a well-developed, crescent-shaped pelta that surrounds the anterior end of the organism. The trunk of the axostyle, a hyaline rod of medium diameter, courses near the anteroposterior axis of the flagellate. Its terminal segment projects from the posterior cell surface and, as seen in fixed and stained preparations, tapers more or less gradually to a point. No paraxostylar granules have been reported from *P. hominis*.

The parabasal body, evident in protargol-impregnated preparations, but not in those stained with hematoxylin, is a small uniformly argentophilic disc.[49] In preparations bleached before staining, the central part of the disc appears dark, the peripheral part remaining free of stain.[51,52] The parabasal filament, the second component of the parabasal apparatus, is a relatively short fibril (also demonstrated best in protargol-treated organisms) to which is applied the dorsal side of the argentophilic disc. The entire parabasal apparatus is located dorsal and to the right of the broadly ellipsoidal or spheroidal nucleus, characterized by a uniform distribution of small chromatin granules and the presence of a nucleolus surrounded by a clear halo. The nuclear structure can be studied most satisfactorily in iron hematoxylin–stained organisms, like those shown in the figures included in Wenrich's reports.[33-35,45]

Electron Microscopy

As far as can be ascertained, the only extensive accounts published of the fine structure of *P. hominis* are those by Wartoń and Honigberg[20] (SEM) and Honigberg et al[26] (TEM). A schematic diagram of this species is included in Brugerolle's thesis.[22]

Scanning Electron Microscopy

Scanning electron micrographs provide useful information about the structure of the cell surface and about the spatial relationships of various organelles located on the surface. In *P. hominis* these micrographs show the composite structure of the free margin of the undulating membrane, which consists of the peripheral accessory filament and the attached part of the recurrent flagellum situated under this filament (Fig. 3.9b,c). Of interest also is the area in which the recurrent flagellum detaches from the accessory filament and becomes a free posterior flagellum (*arrowhead* in Fig. 3.9b). The relationships among the several flagella in the area of their emergence from the periflagellar canal, whose wall is supported by the pelta, are also evident in some micrographs (Fig. 3.9d). The four anterior flagella, which correspond to the homologous organelles of other trichomonads, emerge very close to one another. The recurrent flagellum leaves the cell through what appears to be a separate opening in the wall of the periflagellar canal located slightly dorsal and posterior to the emergence area of the anterior locomotor organelles (Fig. 3.9c). It becomes attached to the accessory filament a very short distance posterior to the area of its emergence (Fig. 3.9c,d). At 180° from the recurrent flagellum, and ventral to it, one can see the cover of the area from which emerges the independent flagellum (Fig. 3.9d). The position of this latter organelle and the projecting terminal segment of the axostyle are evident in Fig. 3.9b. The surface of *P. hominis*, like that seen in scanning electron micrographs of *T. vaginalis*, appears velvety (Fig. 3.9b–d).

Transmission Electron Microscopy

In most respects, the fine structure of the intestinal trichomonad of humans resembles that of other Trichomonadinae. In the figures (Fig. 3.9e,g,h,j) included in this chapter, the kinetosomes and various mastigont organelles are shown in nearly dorsal views of the kinetosomal region. The section in Figure 3.9i is more longitudinal, with some of the mastigont structures in side views. Certain aspects of the Golgi complexes (parabasal bodies) are seen in several micrographs (Fig. 3.9e,g,j). As in other trichomonads, the compact Golgi complex of *P. hominis* consists of a stack of elongate, flattened, and nodular cisternae made of smooth endoplasmic reticulum. Around the cisternae are spheroidal vesicles, some of them filled with electron-dense material (Fig. 3.9g,j). The relationship between the parabasal body (*Go*) and parabasal filament 1 is evident in Fig. 3.9j. Nuclei enclosed in their envelopes can also be seen (Fig. 3.9e,i). With the exception of parabasal filament 2, identifiable with confidence only in Fig. 3.9j), all the rootlet filaments, including the Type-B costa and the quite broad parabasal

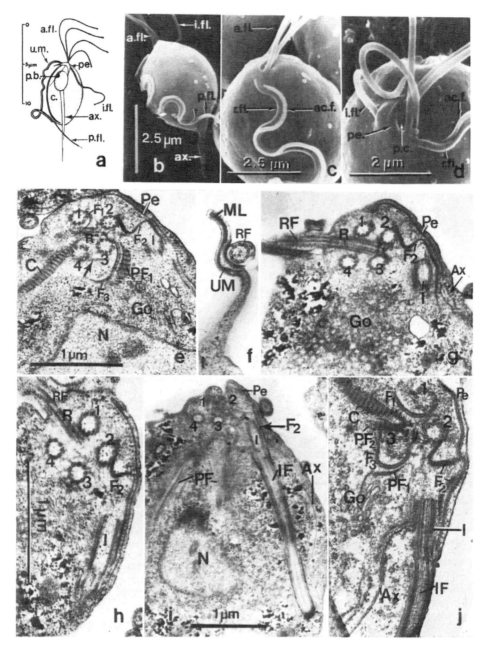

Fig. 3.9. *Pentatrichomonas hominis.* **a.** Organism in right view. A composite diagram based on light microscopic observations of protargol-stained preparations (Hollande's fixative and protargol). The anterior flagella (*a.fl.*), including the independent flagellum (*i.fl.*) and the parabasal apparatus, consisting of a small discoid or drop-shaped parabasal body (*p.b.*) applied to a fine parabasal filament, are evident. The diagram also reveals the rather large crescent-shaped pelta (*pe.*) at the anterior end of the body and the relatively stout trunk of the axostyle (*ax.*). A medium-width costa (*c*), which exhibits slight clockwise torsion, supports a well-developed undulating membrane (*um.*) reaching the posterior end of the body, with the free posterior flagellum (*p.fl.*) extending beyond the end of the membrane. The free margin of the membrane consists of the accessory filament (outer component) and the attached segment of the recurrent flagellum (inner component).

b–d. Scanning electron micrographs. **b.** Right view. The typical four anterior flagella (*a.fl.*), filament 1 with Type-A periodic structure, are evident in Figure 3.9e. Sections through the peltar-axostylar complex, which consists of sheets of microtubules, are also shown (Fig. 3.9e,g–j).

The structure differentiating the genus *Pentatrichomonas* and the species *P. hominis* from other trichomonads is kinetosome I (Fig. 3.9e,g–j) of the independent flagellum (*IF*) and this flagellum (Fig. 3.9h–j). The kinetosome is located slightly dorsal to (just behind) the area at which the sigmoid F_2 lamellae, arising from kinetosome #2, spread on the dorsal surface of the peltar-axostylar complex (Fig. 3.9g,h,j). The independent flagellum proceeds from its kinetosome posteriad and some-

---

the independent flagellum (*i.fl.*), the undulating membrane, and the terminal projecting segment of the axostyle (*ax.*) are seen. The posterior end of the accessory filament, which coincides with the end of the undulating membrane, is indicated by the arrowhead. The recurrent flagellum continues beyond the end of the undulating membrane as a free posterior flagellum (*p.fl.*). **c.** Dorsal view. The typical four anterior flagella (*a.fl.*) emerge from the periflagellar canal. The recurrent flagellum (*r.fl.*) leaves the cell through an opening situated dorsal and somewhat posterior to the main periflagellar canal. The accessory filament (*ac.f.*) appears to emerge from underneath and to the right of the recurrent flagellum; it parallels the attached segment of the latter flagellum for the entire length of the membrane constituting its outer margin. **d.** Top (anterior) view showing the emergence area of all six flagella. The periflagellar canal (*p.c.*) is filled with the proximal segments of the four typical anterior flagella. The recurrent flagellum emerges posterior and dorsal to the anterior flagella; the accessory filament (*ac.f.*) is seen to the right of the recurrent flagellum. The outline of the pelta (*pe.*) is evident on the ventral side of the organism (to the observer's left), its upper (dorsal) margin forming the rim of the periflagellar canal. The independent flagellum (*i.fl.*) emerges ventral to the pelta at about a 180° angle to the recurrent flagellum.

(For methods of preparation in **b–d**, see reference 20.)

e–j. Transmission electron micrographs. **e,g,h.** Nearly dorsal views in sections, cut at different levels and tilted to a greater or lesser degree in reference to the transverse planes, that include the kinetosomal complexes. The section in **i** appears to exhibit the largest deviation from the transverse plane. The four kinetosomes (#1 to #4) of the typical anterior flagella are seen in **e,g–i**, but kinetosome #4 is missing at the level of the section shown in **j**. Parts of the sigmoid F_2 filaments (lamellae) are visible in all figures; they are seen in many sections (e.g., **i**, *large arrow*) to be applied to the dorsal layer of the peltar-axostylar junction. Kinetosome R is present in nearly longitudinal sections in **e,g,h**: the recurrent flagellum (*RF*) is seen to originate from this kinetosome. The independent flagellum (*IF*) originates from kinetosome I (**e,g–j**) situated at an about 180° angle in relation to kinetosome R (i.e., near the ventral cell surface) (**e,g,h**). In some organisms, the proximal segment of the IF appears to lie in a cytoplasmic groove or canal **i,j**. The clockwise-coursing filaments (or lamellae) F_1 and F_2 are associated with kinetosomes of the corresponding numbers (**e,j**). Type-B costa (*C*) (**e,j**). Type-A parabasal filaments (*PF*) (**e,j**) Golgi complex [*Go* (parabasal body)], in oblique (**e,j**) and transverse (**g**) sections, are also evident. An unidentified structure can be observed near kinetosome #4 (**e**, *arrow*).

f. Cross-section of an undulating membrane (*UM*). The recurrent flagellum (*RF*) is applied to the dorsal body fold with which it appears to be physically connected. The part of the fold distal to the RF contains the loop of the marginal lamella (*ML*), seen to extend for some distance proximal to the area of attachment of the RF.

(For methods of preparation in **e–j**, see Honigberg BM et al[26]; also various reports published by G. Brugerolle, as given after the legends to Fig. 3.2i,j).

a: Reproduced with permission of Academic Press, from Honigberg BM: Trichomonads of importance in human medicine. In Kreier JP (ed):*Parasitic Protozoa*, vol 2, pp 275–454, 1978; **b–d** reproduced with permission of the Society of Protozoologists, from Wartoń A, Honigberg BM: Structure of trichomonads as revealed by scanning electron microscopy. *J Protozool* 26:56–62, 1979; **e–i** reproduced with permission of the Society of Protozoologists, from Honigberg BM et al: Structure of *Pentatrichomonas hominis* (Davaine) as revealed by electron microscopy. *J Protozool* 15:419–430, 1968; **j** is an original electron micrograph prepared by G. Brugerolle.

what to the right. In some sections the proximal segment of this flagellum appears to lie in a groove or canal (Fig. 3.9i,j).

The ultrastructural details of the Type-B costa of *P. hominis* are shown in the section of this chapter dealing with *T. vaginalis* (Fig. 3.6b–d). As mentioned in that section, our attempt to explain the architecture of this type of costa (Fig. 3.6e) is based largely on very detailed observations of the organelle in the intestinal trichomonad of humans.

The costa supports a well-developed undulating membrane consisting of a dorsal body fold, which contains the marginal lamella with its terminal narrow loop, and the adherent segment of the recurrent flagellum (Fig. 3.9f).

Trichomitus fecalis (Cleveland, 1928)

The history and nomenclature of this flagellate described by Cleveland[53] were discussed elsewhere.[1,5]

Trichomitus fecalis, isolated by Cleveland[53] from the feces of one man, could be grown for prolonged periods in simple media (hay infusion; saline or tap water supplemented with sera from various animals; Loeffler's saline-citrate medium) in the presence of bacteria, which were said to be indispensable for cultivation of the organism.[54] Cultivation in Loeffler's serum-citrate medium was successful at 20°C and 36°C, growth being faster at the higher temperature. The trichomonad was said to be infective for frogs and tadpoles.[53]

The following account is based mainly on Cleveland's[53] light-microscope observations of living organisms and of preparations fixed in Schaudinn's fluid and stained with iron hematoxylin. Cleveland's description is supplemented by data published by Tanabe[54] and Wenrich.[33] It also includes personal observations (of B. M. H.) made from organisms that, after being destained, were restained with protargol. Nearly all measurements given below are from Cleveland's report.[53] A composite diagram of *T. fecalis* (Fig. 3.10) is included to illustrate the descriptive account.

The body of the organism averages 9.5 (5.0 to 13.5) × 5.7 (4.0 to 6.0) μm (n = 300). Three anterior flagella, ranging in length from 18.0 to 27.0 μm (even up to 40.0 μm), originate from the kinetosomal complex. The well-developed

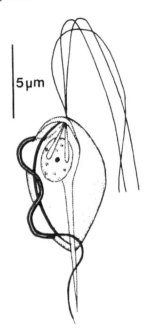

Fig. 3.10. *Trichomitus fecalis*. A composite diagram of an organism in right view. All the karyomastigont organelles are evident. (For descriptions of the organelles, see Fig. 3.9a and its legend.) The anterior flagella appear to be exceptionally long; the parabasal body is V-shaped. The peltar-axostylar complex is very well developed in the region of the pelta and the axostylar capitulum. [Based on the diagram included in Kirby H: *Protozoa of Man*. Berkeley: University of California Press, 1945, and on the modified version of this diagram in Honigberg BM: Trichomonads of importance in human medicine, from Kreier JP (ed): *Parasitic Protozoa*, vol 2, pp 275–454, 1978, reproduced with permission of Academic Press, and on original observations made by B. M. Honigberg using Cleveland's original preparations of *T. fecalis*, which were destained, then restained with protargol.]

undulating membrane is supported by a costa of moderate width. The undulating membrane and the costa, at least as long as the body, exhibit a slight counterclockwise torsion. The free external margin of the undulating membrane is composed on the outside of the accessory filament and on the inside of the attached part of the recurrent flagellum which continues beyond the end of the membrane as a free posterior flagellum.

The relatively broad and complex axostylar capitulum is apposed to the left ventral surface of a large typically trichomonad nucleus. The anteriormost segment of the capitulum is connected to a broad pelta. Posterior to the nucleus, the capitulum narrows. The narrower segment is the axostylar trunk of moderate diameter. The trunk courses near the anteroposterior axis of the organism and projects from the posterior body surface for a distance equal to from one-third to one-half the cell's length. The projecting terminal segment of the axostyle tapers more or less gradually to an often needle-like point.

The V-shaped parabasal apparatus is applied to the right dorsal surface of the nucleus.

As pointed out by one of us,[5] on morphologic grounds *T. fecalis* belongs in the *Trichomitus batrachorum* complex found in amphibians and squamate reptiles (lizards and snakes) (cf. the figures of nondividing organisms in reference 56). This affinity is also suggested by the infectivity of Cleveland's species for tadpoles and frogs. A careful perusal of the literature indicates that *T. fecalis* was isolated from the feces of only one human host, by Cleveland in 1928. In light of this and because of its structural and certain physiologic characteristics, this species may best be considered a parasite of uncertain host origin. Indeed, it may well be an incidental parasite of humans.

Dientamoeba fragilis Jepps & Dobell, 1918

The trichomonad affinities of *D. fragilis* were considered in some detail by one of us (B. M. H. in part II of reference 2). The taxonomic implications of these affinities were also discussed there. That discussion need not be repeated in the present chapter—the interested reader is referred to the earlier report.[2]

Light Microscopy

The following description is based on the accounts of Dobell[57] and Wenrich,[58]* both of whom studied *D. fragilis* in iron hematoxylin-stained preparations by light microscopy. It includes also data obtained in such preparations with the aid of light microscopy by Camp et al[2] (Fig. 3.11a-e). The measurements are those published by Dobell[57] and Wenrich[58] of organisms found in fecal and culture smears.

Rounded cells average about 9.0 (7.0 to 12.0) μm in diameter, some being as small as 3.5 or as large as 22.0 μm. Usually no more than one-fifth to one-third of a population consists of mononucleate organisms. Most of the parasites have two nuclei, being in arrested telophase. (For accounts of division, see Dobell[57] and Wenrich.[58]) Organisms with three or four nuclei are rather rare. The nuclei, unlike those seen in the more typical trichomonads, are rounded, averaging about 2 μm in diameter; however, nuclei as large as 5 μm in diameter have also been observed. In binucleate organisms, the nuclei appear to contain four[2,58] (according to Dobell,[57] six) chromatin bodies (CB), presumably chromosomes[58] (Fig. 3.11a,c). Cytoplasmic vacuoles and various inclusions are also evident (Fig. 3.11a-e). In all binucleate cells, the extranuclear spindle (*s*) (paradesmose), characteristic of trichomonads, is seen extending between the nuclei (Fig. 3.11a-e).

Electron Microscopy

The following account is based on the TEM study of Camp et al.[2] We shall employ a low magnification micrograph (Fig. 3.11f) to illustrate most of the structures found in *D. fragilis* in the stage of arrested telophase. No kinetosomes or flagella are seen in this figure, and presumably none are to be found in *D. fragilis*. Among the structures visible in the figure note the extranuclear spindle (paradesmose, *PD*) consisting of 20-nm microtubules. The spindle extends between the two polar complexes (*PC*). In addition to the spindle microtubules (*Mt*), three kinds of structures, all of them paired, are associated with a polar complex (Fig. 3.11f,g,h), the nonperiodic atractophores (*At*), the parabasal filaments (*PF*), and the parabasal bodies (*PB*), which are the Golgi complexes. Camp et al[2] stated that in *Dientamoeba* atractophores are the "paired nonperiodic elements (At) . . . which in their structure and spatial relationship with the spindle microtubules and parabasal filaments appear to correspond to the atractophores of Hollande and Carruette-

---

*Wenrich's paper[58] includes a review of his and others' previous findings. It also includes additional original observations. The report contains an extensive list of references to the cytology of *D. fragilis*, including the division process, that were published up to 1944, and little was added between 1944 and 1974 (see the list of references in Camp et al[2]).

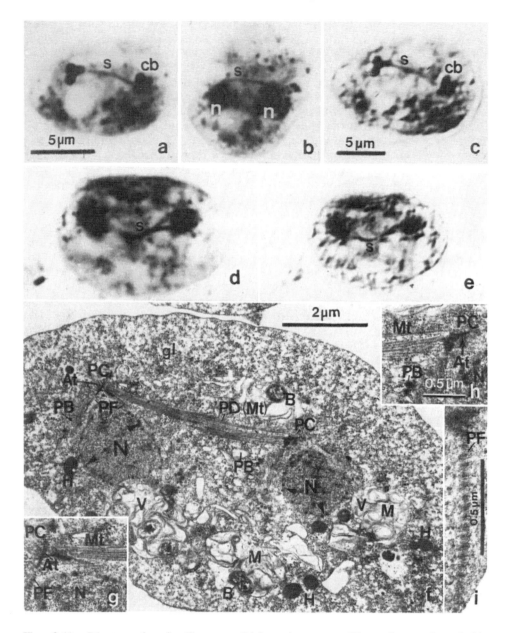

Fig. 3.11. *Dientamoeba fragilis*. **a–e**. Light microscopy. Photomicrographs of binucleate organisms. Four chromatin bodies (*CB*) (probably equivalents of chromosomes) are evident in the telophase nuclei of the organisms shown in **a** and **c**. The extranuclear spindle (*s*) (paradesmose) extends between the nuclei (*n*) in all figures. Branching of the spindle is seen near the right nucleus in **d**, Bouin's fluid and iron hematoxylin; **a,b,d**, bright field; **c,e**, Nomarski differential interference. Scale in **a** should be used also for **b** and **d**; that in **c** applies also to **e**.

f–i. Transmission electron microscopy. The view in **f**, a low magnification electron micrograph, is comparable to those seen in optical sections in **a–e**. The spindle (paradesmose) (*PD*), made up of microtubules (*Mt*), extends between the two nuclei (*N*), each of which is surrounded by an envelope with numerous pores (*arrowheads*). The microtubules originate from the nonperiodic atractophores (*At*), located in the polar complexes (*PC*). The paired atractophores (*At*) (**f–h**) continue as the parabasal filaments (*PF*), which extend on either side of each nucleus. The Golgi complexes [the parabasal bodies (*PB*)] are applied to the atracto-

Valentin." They went on to say that "each ramus of the composite periodic (parabasal filaments) and atractophore part of the polar complex appears to be spoon-shaped, with its convexity facing inward." The structure of the atractophores in dividing *Dientamoeba* as visualized by Camp et al[2] is shown partially in Figure 3.11f–h and in the figures illustrating their report (Figs. 7, 10 to 12 in reference 2). That atractophores found in various taxa of Parabasilidea are not identical (in their case the Hypermastigida) has been emphasized by Hollande and Carruette-Valentin[59] who divided into three categories these organelles, said to play a potential double role: forming of the spindle, and eventually providing for its contact with the nucleus. The atractophores described by Camp et al seem to agree best with those placed by the French workers[59] in the category of "atractophores with periodic structure associated with parabasal lamellae," except that they have no periodicity. The identity of the periodic structures extending from the atractophores as parabasal filaments can be established by examining them at a higher magnification (Fig. 3.11i) and by their association with the Golgi complexes (Fig. 3.11f–h). Actually, atractophores, the proximal segments of the parabasal filaments, the proximal parts of the parabasal bodies, and the microtubules arising from the atractophores constitute the polar complexes. Just below each polar complex there is a nucleus surrounded by an envelope with nuclear pores (*arrowheads* in Fig. 3.11f). Also evident in Fig. 3.11f are microbody-like inclusions, hydrogenosomes (*H*), glycogen granules (*gl*), and food vacuoles (*V*). The vacuoles contain bacteria (*B*), myelin configurations (*M*), or both. Additional details of *D. fragilis* fine structure can be found in the report of Camp et al.[2].

References

1. Honigberg BM: Evolutionary and systematic relationships in the flagellate order Trichomonadida Kirby. *J Protozool* 10:20-63, 1963.
2. Camp RR, Mattern CFT, Honigberg BM: Study of *Dientamoeba fragilis* Jepps & Dobell. I. Electronmicroscopic observations of the binucleate stages. II. Taxonomic position and revision of the genus. *J Protozool* 21:69-82, 1974.
3. Kulda J, Nohýnková E, Ludvík J: Basic structure and function of the trichomonad cell. In Kulda J, Čerkasov J (eds.): Proceedings of the International Symposium on Trichomonads & Trichomoniasis, Prague, July 1985 (Post-Symp Publ Pt 1). *Acta Univ Carolinae (Prague) Biol* 30(3,4):181-198, (1986) 1987.
4. Honigberg BM, King VM: Structure of *Trichomonas vaginalis* Donné. *J Parasitol* 50:345-364, 1964.
5. Honigberg BM: Trichomonads of importance in human medicine. In Kreier JP (ed): *Parasitic Protozoa*, vol 2, pp 275-454. New York: Academic Press, 1978.
6. Diamond LS: The establishment of various trichomonads of animals and man in axenic cultures. *J Parasitol* 43:488-490, 1957.
7. Mattern CFT, Honigberg BM, Daniel WA: Fine-structural changes associated with pseudocyst formation in *Trichomitus batrachorum*. *J Protozool* 20:222-229, 1973.
8. Jírovec O, Petrů M: *Trichomonas vaginalis* and trichomoniasis. In Dawes B (ed): *Advances in Parasitology*, vol 6, pp 117-188. New York: Academic Press, 1968.
9. Mattern CFT, Daniel WA: *Tritrichomonas muris* in the hamster: pseudocysts and the infection of newborn. *J Protozool* 27:435-439, 1980.
10. Brugerolle G: Sur l'existence de vrais kystes chez les Trichomonadines intestinales. Ultrastructure des kystes de *Trichomitus batrachorum* Perty 1852, *Trichomitus sanguisugae* Alexeieff 1911, et *Monocercomonas tipulae* Mackinnon 1910. *C R Acad Sci (Paris)* 277:2193-2196, 1973.

---

phores and parabasal filaments. The parabasal filaments (*PF*) and the overlying parabasal bodies (*PB*) correspond to the trichomonad parabasal apparatus. Rounded electron-dense cytoplasmic inclusions, most probably functional hydrogenosomes (*H*); glycogen granules (*gl*); and food vacuoles (*V*), some containing bacteria (*B*), myelin configurations (*M*), or both, are distributed in the cytoplasm. Polar complexes (*PC*), consisting of atractophores (*At*), spindle fibers (*Mt*), proximal segments of PFs, and proximal parts of PBs, are shown at a higher magnification in **g** and **h**. The Type-A-like periodic pattern of the transverse bands, characteristic of parabasal filaments, is evident in a higher magnification micrograph (**i**). Scale in **h** applies also to **g**. (For methods of preparation, see Camp RR et al.[2])

Reproduced with permission of the Society of Protozoologists, from Camp RR et al: Study of *Dientamoeba fragilis* Jepps & Dobell. I. Electronmicroscopic observations of the binucleate stages. II. Taxonomic considerations. *J Protozool* 21:69-82, 1974.

11. Kazanowska W, Kuczyńska K, Skrzypiec R: Pathology of *T. vaginalis* infection in experimental animals. *Wiad Parazytol* 29:63-66, 1983.
12. Hollander DH: Cultivation of *Trichomonas vaginalis* in a low carbohydrate medium. Abstracts Int Symp Trichomonads & Trichomoniasis, Prague, July 1985, p 17, 1985.
13. Fari A, Trevoux R, Verges J: Diagnosis and significance of non-motile forms of *Trichomonas vaginalis*. Abstracts Int Symp Trichomonads & Trichomoniasis, Prague, July 1985, p 28a, 1985.
14. Third Polish Symp on Trichomoniasis, Białystok, November 1965. *Wiad Parazytol* 12:137-508, 1966.
15. Fourth Polish Symp on Trichomoniasis, Zakopane, November 1968. *Wiad Parazytol* 15:213-496, 1969.
16. Sixth Polish Symp on Trichomoniasis, Karpacz, September 1975. *Wiad Parazytol* 23:477-693, 1977.
17. Seventh Polish Symp on Trichomoniasis, Łódź, June 1980. *Wiad Parazytol* 27:153-369, 1981.
18. Symp Int on Trichomoniasis, Białystok, July 1981. *Wiad Parazytol* 29:1-239, 1983.
19. Abstracts Int Symp Trichomonads & Trichomoniasis, Prague, July 1985, pp 1-113, 1985.
20. Wartoń A, Honigberg BM: Structure of trichomonads as revealed by scanning electron microscopy. *J Protozool* 26:56-62, 1979.
21. Nielsen MH, Ludvík J, Nielsen R: On the ultrastructure of *Trichomonas vaginalis* Donné. *J Microsc* 5:229-250, 1966.
22. Brugerolle G: *Contribution à l'étude cytologique des Protozoaires Flagellés parasites: Proteromonadida, Retortamonadida, Diplomonadida, Oxymonadida, Trichomonadida*, thesis. University of Clermont-Ferrand II for the degree of Docteur Es-Sciences Naturelles, Series E, Order 227, 1976.
23. Honigberg BM, Volkmann D, Entzeroth R, Scholtyseck E: A freeze-fracture electron microscope study of *Trichomonas vaginalis* Donné and *Tritrichomonas foetus* (Riedmüller). *J Protozool* 31:116-131, 1984.
24. Bardele CF: Functional and phylogenetic aspects of the ciliary membrane: a comparative freeze-fracture study. *Biosystems* 14:403-421, 1981.
25. Mattern CFT, Honigberg BM, Daniel WA: The mastigont system of *Trichomonas gallinae* (Rivolta) as revealed by electron microscopy. *J Protozool* 14:320-339, 1967.
26. Honigberg BM, Mattern CFT, Daniel WA: Structure of *Pentatrichomonas hominis* (Davaine) as revealed by electron microsopy. *J Protozool* 15:419-430, 1968.
27. Honigberg BM, Mattern CFT, Daniel WA: Fine structure of the mastigont system in *Tritrichomonas foetus* (Riedmüller). *J Protozool* 18:183-198, 1971.
28. Cappuccinelli P, Sellitto C, Zicconi D, Juliano C: Structural and molecular organization of *Trichomonas vaginalis* cytoskeleton. In Kulda J, Čerkasov J (eds): Summary of the Proceedings of the International Symposium on Trichomonads & Trichomoniasis, Prague, July 1985 (Post-Symp Publ Pt 1), *Acta Univ Carolinae (Prague) Biol* 30(3, 4):211-217, (1986) 1987.
29. Nielsen MH: Phagocytosis by *Trichomonas vaginalis* Donné. Substructural and electron-histochemical observations. In Savard P (ed): VII Cong Int Microsc Electr Grenoble 2:389-390, Ivre-Sur-Seine: Societé Française de Microscopie Electronique, 1970.
30. Čerkasov J, Čerkasovová A, Kulda J: Carbohydrate metabolism of *Tritrichomonas foetus* with particular respect to enzyme reactions occurring in hydrogenosomes. In Vitale LJ, Simeon V (eds): Industrial and Clinical Enzymology, Proc FEBS Spec Meet on Enzymes, Cavtat, Dubrovnik, Yugoslavia, 1979. *Trends in Enzymol* 61:257-275, 1980.
31. Brugerolle G: Etude de la cryptopleuromitose et de la morphogénèse de division chez *Trichomonas vaginalis* et chez plusieurs genres de Trichomonadines primitives. *Protistologica* 11:457-468, 1975.
32. Dobell CC: The common flagellate of the human mouth, *Trichomonas tenax* (O.F.M.): Its discovery and its nomenclature. *Parasitology* 31:138-146, 1939.
33. Wenrich DH: Morphological studies on the trichomonad flagellates of man. *Arch Soc Biol Montevideo Suppl Actas Cong Int Biol Montevideo Fasc* 5:1185-1204, 1931.
34. Wenrich DH: Comparative morphology of the trichomonad flagellates of man. *Am J Trop Med Hyg* 24:39-51, 1944.
35. Wenrich DH: The species of *Trichomonas* in man. *J Parasitol* 33:177-188, 1947.
36. Hinshaw HC: On the morphology and mitosis of *Trichomonas buccalis* (Goodey) Kofoid. *Univ Calif, Berkeley, Publ Zool* 29:159-174, 1926.
37. Honigberg BM, Lee JJ: Structure and division of *Trichomonas tenax* (O.F. Müller). *Am J Hyg* 69:177-201, 1959.
38. DeRysky S, Sapelli PL, Carnevale G, Carosi G, Dei Cas A, Filice G, Gatti S, Scaglia M: Ultrastructure des protozoaires buccaux. *Bull Group Int Rech Sci Stomatol Odontol* 20:229-258, 1977.
39. Filice G, Scaglia M, Carnevale G, Maccabruni A, Carosi G: Ultrastruttura di *Trichomonas tenax*. *Riv Parassitol* 40:305-315, 1979.
40. Ribaux CL: Étude du protozoaire buccal *Trichomonas tenax* en microscopie électronique à balayage et en transmission. *J Biol Buccale* 7:157-168, 1979.
41. Ribaux CL, Magloire H, Joffre A: Données complémentaires à l'étude ultrastructurale de *Trichomonas tenax*. Localisation intra-cellulaire

de la phosphatase acide. *J Biol Buccale* 8:213-228, 1980.
42. Brooks B, Schuster FL: Oral protozoa: Survey, isolation, and ultrastructure of *Trichomonas tenax* from clinical practice. *Trans Am Microsc Soc* 103:376-382, 1984.
43. Poirier TP, Honigberg BM, Holt SC: Fine structure of *Trichomonas tenax* (O. F. Müller), to be published.
44. Kirby H: Flagellate and host relationships of trichomonad flagellates. *J Parasitol* 33:214-228, 1947.
45. Wenrich DH: Morphology of the intestinal trichomonad flagellates in man and of similar forms in monkeys, cats, dogs and rats. *J Morphol* 74:189-211, 1944.
46. Reardon LV, Rininger BF: A survey of parasites in laboratory primates. *Lab Anim Care* 18:577-580, 1968.
47. Wenrich DH, Saxe LH: *Trichomonas microti*, n. sp. (Protozoa, Mastigophora). *J Parasitol* 36:261-269, 1950.
48. Levine ND: *Protozoan Parasites of Domestic Animals and of Man*, 2nd ed. Minneapolis: Burgess, 1973.
49. Kirby H: The structure of the common intestinal trichomonad of man. *J Parasitol* 31:163-175, 1945.
50. Honigberg BM: Structure and morphogenesis of *Trichomonas prowazeki* Alexeieff and *Trichomonas brumpti* Alexeieff. *Univ Calif, Berkeley, Publ Zool* 55:337-394, 1951.
51. Kirby H, Honigberg BM: Intestinal flagellates from a wallaroo, *Macropus robustus* Gould. *Univ Calif, Berkeley, Publ Zool* 55:35-66, 1950.
52. Jensen EA, Hammond DM: A morphological study of trichomonads and related flagellates from the bovine digestive tract. *J Protozool* 11:386-393, 1964.
53. Cleveland LR: *Tritrichomonas fecalis* nov. sp. of man; its ability to grow and multiply indefinitely in faeces diluted with tap water and in frogs and tadpoles. *Am J Hyg* 8:232-255, 1928.
54. Cleveland LR: The separation of a *Tritrichomonas* of man from bacteria; its failure to grow in media free of living bacteria; measurement of its growth and division rate in pure cultures of various bacteria. *Am J Hyg* 8:256-278, 1928.
55. Tanabe M: Morphological studies on *Trichomonas*. *J Parasitol* 12:120-130, 1926.
56. Honigberg BM: Structure, taxonomic status, and host list of *Tritrichomonas batrachorum* (Perty). *J Parasitol* 39:191-208, 1953.
57. Dobell CC: Researches on the intestinal protozoa of monkeys and man. X. The life history of *Dientamoeba fragilis*: Observations, experiments, and speculations. *Parasitology* 32:417-461, 1940.
58. Wenrich DH: Studies on *Dientamoeba fragilis* (Protozoa). IV. Further observations, with an outline of present-day knowledge of this species. *J Parasitol* 30:322-338, 1944.
59. Hollande A, Carruette-Valentin J: Les atractophores, l'induction du fuseau et la division cellulaire chez les Hypermastigines. Etude infrastructurale et révision systématique des Trichonymphines et des Spirotrichonymphines. *Protistologica* 7:5-100, 1971.

4

Immunologic Aspects of Human Trichomoniasis

J. P. Ackers

Introduction

This is a good time to be writing this chapter. The study of human trichomonads has gone through many phases since they were first described over a century ago, and much has been achieved. The three well-established species, *Trichomonas vaginalis, Trichomonas tenax,* and *Pentatrichomonas hominis,* are defined and their nomenclature finally settled. *Trichomonas vaginalis* is accepted as the genuine cause of a common and sometimes serious sexually transmissible disease; reliable diagnostic methods and effective treatments are widely available. All this was true 8 years ago and yet, writing then, we stated that "it cannot really be claimed that the study of the immunology of *T. vaginalis* is in a particularly fruitful phase, and this may well be the cause of the recent decline in published work on the subject."[1] An enormous amount of work had been done, mainly with the aim of developing serodiagnostic tests, but there was a feeling that progress had come to a halt—and at a very frustrating point where the tests worked but not quite well enough to be clinically useful. In the last 10 years, however, the whole picture has changed, with the explosion of new immunologic and molecular biologic techniques allowing new problems to be tackled and old ones taken up again. I hope that some of the excitement now pervading the field will come through what is, inevitably, still an account based mainly on older results; but there is no doubt that in another 10 years the whole subject will have changed beyond all recognition.

Trichomonas vaginalis

Of the three human trichomonads the clinical importance of *T. vaginalis* is so much greater than that of *T. tenax* and *P. hominis* that it justifies the enormous preponderance it enjoys in the published literature. Because the three organisms occupy quite different habitats, and because so little, relatively speaking, is known about the immunology of the oral and the large bowel parasites, a comparative treatment did not seem helpful. Most of this chapter, therefore, is concerned with *T. vaginalis,* followed by much briefer mention of the other two organisms.

Antigenic Structure

The first attempts at antigenic analysis of *T. vaginalis* were concerned with establishing the organism as a valid species, distinct from other human trichomonads. In very early work, hampered by the lack of methods for axenic culture, Tokura[2] showed that trichomonads were antigenic in experimental animals and that agglutination tests with the resulting sera could distinguish between *T. vaginalis* and *P. hominis.* After bacteria-free cultivation had been achieved, antigenic differences between *T. vaginalis* and *Tritrichomonas foetus* were conclusively shown by a number of workers; an excellent summary of this early and often inaccessible work has been given by Honig-

---

All research carried out in my laboratory was supported by the Medical Research Council of Great Britain.

berg,[3,4] and the details will not be repeated here. Another important report was that of Kott and Adler,[5] who demonstrated the existence of both unique and common antigens in *T. vaginalis, T. tenax,* and *P. hominis.*

Serotypes

Later work was more concerned with the demonstration of antigenic differences between different isolates of the same species—in almost all cases, *T. vaginalis*; because these studies were usually aimed at developing serologic tests for trichomoniasis, these antigenically defined populations were referred to as "serotypes." The existence of serotypes among isolates of *T. vaginalis* was first reported in 1956,[6] but has been extensively investigated by the Estonian group of Teras and collaborators.[7,8] These authors reported the presence of four serotypes among several hundred clinical isolates and demonstrated that the results were consistent whether determined by agglutination, indirect hemagglutination, or complement fixation.[9] Kott and Adler[5] found even more heterogeneity, detecting no fewer than eight different serotypes among only 23 isolates. The principal interest of the Estonian group was in developing diagnostic methods based on serum antibody detection, in which they were significantly more successful than were most other workers; they were and are of the opinion that much of the reason for this success is their use of a mixed antigen containing all four serotypes in all their assays. It is interesting, therefore, that this serotype specificity has not been apparent in some more recently described diagnostic tests employing indirect fluorescence,[10] an enzyme-linked immunosorbent assay (ELISA),[11] and direct agglutination (Table 4.1) (Ackers, unpublished observations), all of which failed to show higher titers or more positive results when serum from infected women was tested with homologous rather than heterologous antigen. Since, however, the existence of antigenic heterogeneity in *T. vaginalis* has been demonstrated directly (see below), it seems likely that the total polyclonal response, which is what these tests are measuring, is dominated by antibodies directed against common determinants.

Antigenic Heterogeneity

In recent years, several groups have readdressed the question of the antigenic diversity of *T. vaginalis*. Using quantitative fluorescent antibody methods, Su-Lin and Honigberg[12] examined five cloned strains of apparently differing pathogenicity (shown in this case by both the subcutaneous mouse assay[13] and by the clinical symptoms of the women from whom they had been isolated), and determined the relative amounts of common antigen present in each by measuring the reduction in fluorescence (compared with unabsorbed homologous rabbit antiserum) when the staining serum was absorbed with homologous and heterologous antigen. The results showed that each strain possessed unique antigens as well as antigens shared with the other strains examined. As a result of preliminary work,[14] it had been suspected that in *T. vaginalis,* as in *Trichomonas gallinae*[15] and *Histomonas meleagridis,*[16] a relationship existed between pathogenicity and antigenic complexity, but in this later and more thorough study[12] no definite correlation could be seen.

In the first of a number of papers that describe the application of modern immunologic methods to the antigenic analysis of *T. vagi-*

TABLE 4.1. Agglutination of nonkilled *T. vaginalis* by heated sera from infected women (inverse titers).

Sera	Organisms								
	434	991	8215	30	115	8217	415	7689	855
434	<1	1	<1	4	1	<1	<1	<1	<1
991	2	32	4	4	8	8	16	8	2
8215	4	8	8	8	8	2	2	8	1
30	4	8	8	8	16	2	8	4	1
115	4	8	4	32	16	4	16	8	2
8217	2	2	2	4	8	2	8	4	2
415	2	2	4	<1	4	4	8	4	2
7689	ND	2	2	8	1	ND	2	1	<1
8555	4	16	8	2	16	8	16	4	4

ND, Not done.

nalis, Alderete[17] examined the total trichloracetic acid precipitable proteins of five strains of the parasite by electrophoresis in sodium dodecyl sulfate containing polyacrylamide gels (SDS-PAGE). A complex pattern of protein bands was seen, covering a range of molecular weights (MWs) from 20 to 200 kilodalton (kDa), but apart from small variations in intensity there were no significant differences between the five strains; the pattern of bands from *T. foetus* was, however, considerably different. Radioimmunoprecipitation (RIP) was then used to discover which bands were recognized as immunogens by rabbits; it was shown that almost the same patterns were obtained with heterologous as with homologous sera, and that at least six proteins were precipitated by antiserum raised against *T. foetus*. No change in pattern was seen when organisms were reexamined after 1 year in laboratory culture; a similar result was reported by Delachambre[18] using the relatively insensitive technique of precipitation in gel, again in contrast with the results reported for *T. gallinae* and *H. meleagridis*.[19] In a second report Alderete[20] described the characterization of about 20 surface proteins accessible to antibody; the range of MWs was the same as before, but two bands with values of 65 and 92 kDa were particularly prominent. Subsequent work with two monoclonal antibodies has shown that, although both were directed against major surface proteins, they differed in that one reacted with all isolates tested, while the other was more selective, binding to some isolates and not to others. Nevertheless, within a particular culture not all organisms react even with the nonselective monoclonal antibody; the absence of this particular antigen, however, appears to be temporary rather than permanent, since some antigen-negative organisms when separated and recultured again produce a mixture of antigen-negative and antigen-positive cells.[21] In fact, Alderete and his colleagues[17,22,23] believe that all strains possess the full complement of antigens and that serotypic differences can be largely explained by the failure of particular vaccinated rabbits and infected mice and humans to respond to all of the antigens present. Antigenic heterogeneity and/or variability in immune response was also seen when Western blots of several *T. vaginalis* isolates were probed with a collection of sera from infected women.[24]

Monoclonal antibodies were also used by a third group, working in Seattle, to analyze the antigenic structure of *T. vaginalis*. In their first report, Torian and coworkers[25] described the preparation and characterization of eight antibodies that were able to divide nine parasite strains into two groups. One group of four strains was uniformly stained (body and flagella) by the antibodies, while the other was not; interestingly, all eight antibodies identified the same four strains, suggesting that the antigen, when present, is a powerful immunogen. A ninth antibody reacted not only with all eight strains of *T. vaginalis* but also with *T. gallinae, T. foetus,* and *Giardia lamblia*. On immunoblotting, this antibody reacted with a single polypeptide of approximately 62 kDa in all strains of *T. vaginalis* and with an antigen of 60 to 68 kDa in the other species. In a second report, Connelly and coworkers[26] described a detailed examination of the reaction of three of the antibodies described above with four strains of *T. vaginalis*—three of which were stained by the monoclonal antibodies and one (CDC) that was not. The three antibodies were bound to beads and used to affinity-purify detergent-solubilized organisms of one of the (reacting) parasite strains; purified antigen was examined by immunoblotting. In each case two main bands were recognized by polyvalent rabbit antiserum: a single 115-kDa antigen and one or more components in the range of 58 to 64 kDa. Further characterization of the 115-kDa antigen showed it to be sensitive to both pronase and periodate oxidation, and the authors believe it to be a surface glycoprotein that, in the living organism, has exposed carbohydrate groups. What seemed originally to be an inaccessible polypeptide chain is now believed also to be exposed (Torian, personal communication). Affinity purification of the nonreacting (CDC) strains yielded no antigen detectable by immunoblotting.

Finally, Krieger and others[27] examined the geographic variation among isolates of *T. vaginalis* from various areas of the United States by reacting them (using indirect immunofluorescence) with nine monoclonal antibodies; two of these antibodies reacted with all isolates, but four had a much narrower range of specificity. The patterns of reactivity were significantly different in isolates from the four geographic regions studied; interestingly and in contrast to the results of Torian and others,[25] all four of these "narrow range" antibodies reacted with *T. foetus*.

Undoubtedly the application of these powerful techniques will produce new and significant results that will soon render this account out of date. At present only two conclusions appear to be reasonably firmly established: first that *T. vaginalis* isolates possess a large number of common antigens and a much smaller number that are shared with other Trichomonadidae, and also that antibodies raised against *T. vaginalis* recognize not only common but also strain-specific antigens. Whether this represents genuine antigenic diversity or variability of immune responsiveness, and whether this phenomenon is important or irrelevant in the development of serodiagnostic tests is not yet clear.

Other Intraspecific Variability

Although it is not yet clear whether antigenic differences between strains of *T. vaginalis* are due to differences in genotype or to levels of expression, their existence is accepted. In fact, such differences are expected, if only because other apparently stable phenotypic differences between isolates have already been shown.

The best studied and most important of these differences is inherent virulence, which is discussed in detail elsewhere in this volume, but so far no unequivocal relationship between high or low virulence and antigenic structure has been discovered.[12] This is at first rather surprising, since recent studies by Warton and Honigberg[28,29] have shown that differences in exposed sugar residues in part correlate with differences in pathogenicity. Their final conclusion, based on the assessment of the amount of fluorescein-labeled soybean agglutinin bound to various strains of differing virulence, was that the "differences between the pathogenic and mild *T. vaginalis* strains reflected the *levels* of D-lactosyl residues on the cell surfaces—these residues were more abundant on strains having higher pathogenicity levels" (my italics). If differences in inherent pathogenicity are related to the amount (rather than the presence or absence) of one or more surface molecules, then the failure so far to identify an antigen specific for strains of high or low virulence is explicable. It also should be mentioned that, as described above, there is some evidence for the transient absence of one or more major surface proteins during growth of the organism. Alderete and coworkers[22] have reported that clones or subpopulations expressing this antigen (recognized by the monoclonal antibody C20A3) had approximately one-third of the cytotoxicity for cultured cell monolayers of antigen-negative organisms.

It has also proved possible to divide isolates of *T. vaginalis* into groups by the technique of isoenzyme electrophoresis.[30-32] In our own work[30] there was no obvious correlation between zymodeme and pathogenicity to the extent that this could be assessed by examining the patient's clinical notes. Antigenic differences between the various zymodemes were not sought, but could now be looked for; in an earlier study, Teras and others[33] found a correlation between the total activity of the enzyme hexokinase and the virulence, but not with the serotype, of a group of isolates.

Absorbed and Shed Antigens

It has been suspected by several workers that trichomonads obtained by in vitro culture possess medium components bound to their surface, and the absorption of many different components now has been directly demonstrated.[34] Interestingly, while most molecules can be removed by simple washing, others, including α_1-antitrypsin, α_2-macroglobulin, lactoferrin, and lipoproteins, appear to be firmly bound,[35-37] although some workers have claimed to be able to remove some of these molecules by prolonged washing in phosphate-buffered saline.[38] Although these bound molecules may have nutritional or protective rather than immunologic functions, there is no doubt that absorbed medium components are antigenic, because they have elicited monoclonal antibodies.[25] The extent to which *T. vaginalis* grown in vivo in its normal host possesses bound human proteins has not yet been discovered, but it seems unlikely that they are not present. To what extent, if any, such proteins assist the parasite's survival by conferring upon it a degree of "immunologic invisibility" is completely unknown.

As well as absorbing components, *T. vaginalis* sheds antigens into the medium in which it is grown[39]; and there is some evidence that this is an active and selective process rather than the result of parasite lysis. It is not known whether this release of antigens occurs in vivo, but Hampton and Honigberg, using SDS-PAGE followed by immunoblotting found large qualitative and quantitative differences between the protein sets present in cell-free supernatant fluids of trichomonad cultures and of peritoneal fluids from mice injected intraperitoneally.[38]

To what extent the serum and local antibodies produced by many but not all infected women are directed against these shed antigens is not yet known, but they clearly have the potential to be immunopathogenic and to divert the host's immune response away from the organism itself.

Antigenic Variation

One question that has not been considered seriously is whether antigenic variation, of the kind displayed by the African trypanosomes, occurs during the often chronic course of untreated trichomoniasis. While there is apparently no information available on this point, it appears at first sight unlikely that the relatively modest immunologic response that this parasite elicits and its physical separation from many of the effector mechanisms would provide sufficient selective pressure to favor such a complex survival mechanism, but the matter would certainly repay investigation. A related question is that of antigenic variation induced by prolonged in vitro culture. Several workers have failed to observe this phenomenon,[5,17] but Honigberg and his collaborators[4] have presented convincing evidence for its occurrence in *T. vaginalis* and other trichomonad species.

Effect of Immune Mediators on *T. vaginalis* In Vitro

Antibody

Although the work described in the preceding section, as well as the large number of at least reasonably successful serodiagnostic techniques that have been developed, shows clearly that *T. vaginalis* can elicit serum antibodies that will at least bind to it, there is much less evidence for consequences of this for the living organism. The effects of antibody plus complement are difficult to recognize because of the ability of the parasite to activate complement by the alternative pathway (see p. 41); the question of whether specific antibody can enhance lysis by way of the classic pathway is under investigation,[40] but there is some evidence that natural antibodies (see p. 44) may do this.

Although *T. vaginalis* growing within the mouse peritoneal cavity can scarcely be described as an in vitro system, it is sufficiently artificial to be included here. Using this system, the Estonian group of Teras and his coworkers studied the effect of serum from immunized rabbits in alleviating the damage produced by the infection. A series of reports, usefully summarized by Honigberg,[3] described significantly less extensive pathologic changes in the peritoneal cavities of animals pretreated with immune rather than control serum. Because the latter presumably contained normal levels of complement, these reports do suggest a direct effect by specific antibody, even though the effect was not serotype-specific, and there was no correlation with agglutinin titer. Apart from related studies by the same group with human sera[3] (see below) this effect does not seem to have been investigated further and the mechanism remains unknown. In general, however, it is difficult to point to any lethal effect of serum upon *T. vaginalis* that can be unambiguously attributed to induced humoral antibodies.

Very recently, however, it has been shown that monoclonal antibodies raised against *T. vaginalis* and *T. foetus* do possess effector functions. The same antibody (C20A3 of IgG2a subclass) that was shown by Alderete and coworkers[22,23] to recognize an antigen of *T. vaginalis* subject to phenotypic variation (see above) can apparently kill antigen-positive parasites in a complement-independent manner[41]; and Burgess[42] has described nine surface-reactive anti-*T. foetus* monoclonal antibodies, six of which promote complement-mediated lysis and one that acts as an opsonin for monocyte phagocytosis.

Given that *T. vaginalis* is a natural parasite of mucous surfaces, it is important to know if specific secretory IgA (sIgA) has any effect on the organism. In fact, there is little or no evidence on this point; complement-mediated cytotoxicity is not a normal property of this class of immunoglobulin, and therefore a directly lethal effect is unlikely. The two principal natural functions of sIgA, at least in the intestinal tract, appear to be the prevention of pathogen adhesion and the blocking of the uptake of antigenic molecules by intestinal epithelial cells. However, a recently described cytotoxic mechasnism involves the specific sensitization by sIgA of bacterial cells for killing by means of antibody-dependent cell-mediated cytotoxicity (ADCC) by gut-associated lymphoid tissue (but not thymus or popliteal lymph node) lymphocytes.[43] It will be of great interest to see if *T. vaginalis* (and, indeed, other enteric protozoan parasites) are susceptible to this form of killing; there is some evidence that *G. lamblia* is.

Cell-Mediated Killing

No evidence appears to have been found for the direct killing of *T. vaginalis* organisms by cytotoxic T cells, and since it is now well-known that such killing is major histocompatibility complex (MHC)-restricted, this is hardly surprising. Both macrophages and neutrophils have, however, been shown to kill this parasite, while the effect of eosinophils does not seem to have been investigated yet.

Landolfo and others[44] showed, by means of the release of [^{3}H]thymidine from prelabeled organisms, that a population of unstimulated cells from the peritoneal cavity of BALB/c mice was capable of killing *T. vaginalis* in vitro. Adherence experiments and the effect of the removal of the phagocytic cells strongly suggested that the effector cells were macrophages or macrophage-like; the tissue distribution of the activity was quite unlike that of natural killer (NK) cells. Nineteen strains of mice were tested for the activity of their resident peritoneal cells in this assay, and four consistently showed low levels of cytotoxicity. Interestingly, in an earlier report,[45] the same group had demonstrated strain differences in the susceptibility of mice to intraperitoneal infection with *T. vaginalis;* BALB/c mice were highly susceptible while three strains with A backgrounds were relatively resistant. By contrast, the one A-background strain tested for peritoneal cell cytotoxicity was in the low activity group, while BALB/c mice were the source of the most active cells. In a recent extension of this work[46] it has been shown that the cytotoxicity of macrophages from normal BALB/c mice could be greatly enhanced by incubation with T cells from mice that had been vaccinated against *T. vaginalis,* together with trichomonal antigen. Incubation with T cells from mice vaccinated with bovine serum albumin produced significantly less activation, and T cells from both groups of immunized mice had no cytotoxic activity by themselves.

The killing of *T. vaginalis* by polymorphonuclear neutrophils in vitro was described by Rein and others,[47] using a high number of neutrophils per parasite. The process was shown to depend on the presence of oxygen and complement but not, apparently, specific antibody, and the authors suggest that alternative pathway activation leads to the formation of C3b that binds to the surface of the parasite. This in turn leads to the binding of neutrophils through their C3b receptors. The most fascinating part of this report is the description of the process whereby a group of neutrophils surround and fragment a single parasite that is too large to be ingested whole, subsequently phagocytosing the pieces.

Complement

The activation of complement has been mentioned several times; the first explicit statement that *T. vaginalis* activates it by the alternative pathway was made in the report by Gillin and Sher[40] and soon after confirmed by Holbrook and others.[48] Both reports, which are very similar, describe the lysis of *T. vaginalis* by normal, antibody-free human and guinea pig serum and the prevention of lysis by heat inactivation, but not by selective inhibition of the classic pathway of activation. The fact that both C3- and C8-deficient sera do not lyse the parasite is interpreted as showing that the full terminal membrane attack complex is required. More recently, Demeš and Gombošová[49] have shown a great variability of fresh isolates in their susceptibility to lysis by serum complement; all strains, however, become uniformly sensitive after several in vitro passages. This interesting result is in some respects similar to that described for *Entamoeba histolytica*.[50]

Animal Models

The study of human trichomoniasis has long been hampered by the lack of a satisfactory animal model; progress in achieving this is described in another chapter in this volume. The use that has been made of experimental animals in the study of trichomonad immunology may be summarized by saying that the models that have provided useful information are very unphysiologic, while accurate models of intravaginal infection have yet to be developed to the point at which they are able to answer useful questions.

Intraperitoneal, intradermal, or subcutaneous infection of mice using a suitable strain of parasite produces a reproducible and destructive infection. Although hardly resembling the natural disease, it has been widely used as a target for attempts at protection by vaccination. In general, considerable protection can be produced by injection of living organisms[51-53] Protection appeared to be relatively long-lasting and not to be correlated with serum antibody titer. The challenge infection did not always have to be at the same site as the vaccination, but not all combinations gave good protection. In a later study Baba[54] showed that it was possible to vaccinate successfully with heat-killed organisms, but that large doses were required. A surprisingly high degree of protection against heterologous species was seen when *T. vaginalis*, *T. gallinae* and *T. foetus* were used for vaccination and challenge.

Although these results are interesting in themselves, they relate to a system that is too artificial to give any real idea whether active immunization of humans against *T. vaginalis* will ever be possible. Attempts to establish a model of intravaginal infection that plausibly resembles the human disease have run into so many difficulties that few attempts to modify the outcome by vaccination have yet been reported. Martinotti and her coworkers,[55] however, have been more successful than most in infecting mice intravaginally and have reported attempts to protect them against such an infection. Antigen was prepared from sonicated, freeze-dried organisms and administered by a variety of routes; a combination of the intraperitoneal and the intravaginal was found to be most effective. Repeated doses of vaccine were given followed by estradiol pretreatment and intravaginal challenge. Twelve of 16 control animals became infected, compared with 2 of 11 in the optimally protected group.

Human Defense Mechanisms

Human trichomoniasis, although it can be extremely unpleasant and may possibly have long-term harmful effects, is virtually never immediately fatal. There can be no doubt, however, from an examination of the lesions produced in mice by intraperitoneal or subcutaneous inoculation, and from the effects of the parasite on cells in culture, that *T. vaginalis* possesses considerable invasive and cytopathic potential. The almost invariable failure of *T. vaginalis* to spread out of the human genitourinary system may therefore be attributed to host defense mechanisms in the broadest sense. The nature of these defense mechanisms is far from completely understood, and several may represent physical, nutritional, or other nonimmunologic barriers. Some immunologic mediators, however, such as complement, neutrophils, and natural antibodies are probably very important and are considered below. All of the above are, however, "nonacquired," in the sense that they are presumably fully effective in the naive host suffering his or her first infection. What is far less clear is the extent, if any, of acquired immunity in human trichomoniasis.

A cursory glance at the records of any sexually transmitted disease (STD) clinic will show that a single attack of trichomoniasis does not confer any sort of solid immunity, because repeated infections (many at intervals too long to be easily explained as treatment failures) are common. This situation is, however, frequently found in protozoal infections, and it is probably more relevant to inquire whether a more gradual acquisition of resistance occurs, as, for example, in human malaria and, probably more relevant, giardiasis.[56] Provided that the intensity of exposure to infection remains constant, this kind of immunity results in a steady fall with age of detectable infections. If such a form of acquired resistance does occur in human trichomoniasis, one would predict a rapid rise in incidence following the commencement of sexual activity, followed after a few years by a steady decline as immunity develops. As far as I am aware, however, such a pattern has not been clearly apparent in the published statistics for the infection in women; the low incidence and diagnostic difficulties associated with the infection in men make it difficult do draw any conclusions in the latter case. Recent surveys of patients attending STD clinics in Zimbabwe,[57] as well as older statistics collected from several countries by Jírovec and Petrů,[58] show clearly the onset of infection in the middle to late teens, but whatever the local incidence, it seems to persist with no really significant variation until it begins to decline sharply (but certainly not to zero) in women over 50 years of age. Similarly, I have not discovered any reports suggesting a higher incidence or severity of *symptoms* in young women with no or only a few past infections. Finally, accounts of the clinical course of trichomoniasis before the discovery of effective chemotherapeutic agents describe graphically the misery of women enduring ineffective treatments for months or even years for an infection that often showed no sign of a spontaneous cure.[59] In all these respects trichomoniasis differs sharply from human giardiasis, in which the number of cases declines steadily after the age of peak incidence is passed,[60] in which first infections, whatever the age of the patient, are more likely to be symptomatic and to be accompanied more by heavy shedding of cysts than by infections in those who have had several previous attacks,[56] and in which most (but not all) patients cease shedding detectable numbers of cysts within 3 months.[56] It is difficult to avoid the conclusion that repeated infections with G. lamblia cause immunity to at least symptomatic infection, but that there is really no evidence for any degree of acquired immunity in human trichomoniasis. Both T. vaginalis and G. lamblia are well adapted, highly successful flagellate protozoan parasites of human mucosal surfaces. Their immunology, in fact, has been compared directly in a recent review,[38] but this does not contain a specific consideration of possible reasons why the outcome of repeated infection by the two parasites should differ so much. It may therefore be useful to consider the various human immunologic responses to T. vaginalis before trying to explain their apparent ineffectiveness.

Nonimmunologic Host Defense Mechanisms

Although not strictly the subject of this chapter, it is very likely that in human trichomoniasis, as in most infections, the responses of the host that prevent every infection from becoming overwhelming and fatal are made up of at least three components—nonimmunologic factors, immunologic but nonspecific defense mechanisms, and specific immunologic responses—and the last is not necessarily the most important.

In the case of the trichomoniasis, zinc has been suggested as a trichomonacidal component of prostatic fluid. The concentration found in normal men is lethal for most isolates of T. vaginalis, but the existence of both zinc-resistant parasites and of men with subnormal levels of prostatic Zn^{2+} has been postulated as a cause of the minority of cases of male trichomoniasis that are symptomatic.[61,62] In fact, the whole question of how long viable, infectious trichomonads persist in the infected but asymptomatic man (and why they do so in such low numbers) is one of the most important of the many that still remain to be answered about the epidemiology of human trichomoniasis.

It has recently been suggested that T. vaginalis, like many other pathogens, is limited in its growth by the ability to acquire iron and that the binding of lactoferrin to specific receptors on the parasite surface is a biochemical adaptation to achieve this.[37] The inability of T. vaginalis to bind transferrin, and so acquire iron from the bloodstream, may explain, in part, its inability to cause parasitemia. Certainly in mice, intraperitoneal T. foetus was reported to become extremely virulent with the supply of additional iron.[63]

Nonspecific Immunologic Defense Mechanisms

Complement

It has already been stated that T. vaginalis in vitro is capable of activating (and being killed by) complement via the alternative pathway (see above), and this alone would seem to provide a satisfactory explanation for the failure of the organism to disseminate by the bloodstream. In an attempt to see if complement alone could control or eliminate the infection from the cervix or vagina, Demeš and Gombošová[49] measured the amount of complement present in cervicovaginal secretions and also in menstrual blood. In the first, only occasional traces of complement were present in a few samples; nevertheless, 7 of 63 specimens possessed trichomonacidal activity owing to an unknown (but heat-stable) component. In the case of menstrual blood, complement was present, although in variable concentrations in different samples [0 to 20 CH_{50} (50% hemolytic unit of complement)]. These levels were correlated with trichomonacidal activity but there was no correlation with severity of clinical symptoms.

Rare cases of T. vaginalis spreading outside the urogenital system have been described—either widely disseminated infections in moribund patients or, more controversially, infections of other organ systems in otherwise normal people. These are discussed fully and critically in another chapter in this volume, but since it is already known that bacteria recovered from disseminated Neisseria gonorrhoeae infections[64] and invasive strains of E. histolytica[50] are relatively resistant to complement-medicated lysis, it would be very interesting to discover whether strains of *T. vaginalis*, if they could be recovered from these extragenital sites, show the same property.

Polymorphonuclear Neutrophils

The trichomonacidal activity of neutrophils, described above, is intimately linked to the activation of complement. In vitro, the lethal effect of these cells is observed only in the presence of serum containing noninactivated complement, and the process is believed to occur by the adherence of neutrophils through their C3b receptors to C3b bound to the target trichomonad.[47] The fact that C3a fragments, also liberated by the activation of complement, are chemotactic for neutrophils must surely mean that these cells (which form the largest proportion of the inflammatory cells present in the vaginal discharge of symptomatic, infected women) play an important part in the host defenses involved in human trichomoniasis.[65] Moreover, in two recent reports Mason and Forman[66,67] have shown that a so-far unidentified soluble substance produced by *T. vaginalis* is chemotactic for neutrophils. In their first report[66] these authors suggested that the active molecule was a heat-labile high molecular weight protein, but their more recent results[68] suggest a polypeptide with a molecular weight of about 900 daltons. The chemotactic effect of *T. vaginalis*-secreted products has since been confirmed by at least two other groups;[69,70] activity was alternatively reported as being greater[67,70] or not different[68,69] when more or less pathogenic isolates of *T. vaginalis* were tested. The effect was not related to the activation of complement.

Whether the attracted neutrophils kill a significant number of parasites in vivo, as they certainly can in vitro, is not known; but, since their presence can hardly benefit the parasite, it is reasonable to assume that the secreted attractant is an unavoidable byproduct of parasite metabolism.[66] The fact that the host's neutrophils have evolved the ability to migrate to the source of this molecule suggests that these cells may play a useful and significant role in the human response to this infection, particularly in view of the recent important description of foci of acute and chronic inflammation surrounding trichomonads within the prostate gland.[71]

Macrophages

Unstimulated murine macrophages have been shown to be efficient killers of *T. vaginalis* in vitro (see above). Their abilities varied considerably with the strain of mouse from which they were obtained, but not in a way that correlated with susceptibility to intraperitoneal infection. The same group of workers subsequently extended their studies to humans and showed that adherent cells obtained from peripheral blood of healthy volunteers of both sexes have appreciable levels of spontaneous cytotoxicity against *T. vaginalis*.[72] Various critiera were used to show that the active cells belong to the monocyte macrophage lineage. Although macrophages are of overwhelming importance in host defense against a number of protozoan parasites such as *Leishmania* spp., *Toxoplasma gondii*, and *Trypanosoma cruzi*, these are all intracellular pathogens, and it is not clear how important they will prove to be in combating an extracellular organism.

Natural Antibodies

The existence of agglutinating and (complement-mediated) lytic factors in the sera of humans unexposed to *T. vaginalis* has been recognized for many years—in fact, ever since it became necessary to define the lower limit of positivity of the diagnostic tests based on serum antibody that were being developed. In general, and depending on the technique being used, titers from 1:20 up to 1:160 were regarded as nonspecific (for a summary of this work, see Honigberg[3]). The activity was generally assumed to be due to antibody, although Weld and Kean[73] regarded it as due to carbon dioxide in the serum. An attempt to study the phenomenon in its own right was made by Reisenhofer,[74] who investigated the lytic and agglutinating activities of a wide variety of sera against *T. vaginalis*. Her conclusions in the case of human sera were, first, that unheated serum had a powerful lytic activity that did not depend on age, sex, or previous or current infection with the parasite and, second, that heating the sera for 30 minutes at 56°C reduced, but did not abolish, the activity. The agglutinating activity of human serum was very modest, but that of bovine serum much greater.[74] A major study of agglutinins in heat-inactivated sera by Samuels and Chun-Hoon[75] was chiefly concerned with *Tritrichomonas augusta*, but included the information that, when sera were active against a number of trichomonad species, including *T. vaginalis*, activity against one species could be absorbed out, with organisms of that species without affecting the titers against the others. This specificity convinced the authors that the agglutination observed was due to antibodies, and this opinion has not been challenged since. A large part of the previously reported lytic activity in unheated serum, however, was now attributed to alternative pathway activation of complement (see above).

As already stated, the agglutinin titers in human serum seem to be rather low, not exceeding 1:4, as reported by Tatsuki[76] and observed in our own unpublished work, although Nigesen regarded all titers below 1:160 as due to natural antibody (translated and summarized by Honigberg[3]). The origin of these natural antibodies in human and other sera is not known, but has been attributed to cross-reaction with components of the normal flora, infection with commensal trichomonads, or genetic predisposition.[75] In the case of humans, infections with *Trichomonas tenax* and *P. hominis* are relatively rare, particularly in youth, and the first explanation is perhaps the most probable.[1]

The role, if any, of natural antibodies in host defense mechanisms is not known, but specific natural antibody is known to be able to play a stimulatory role in the activation of complement by the alternative pathway, and thus may contribute to the effectiveness of that potent antitrichomonal system.[40] Finally, it should be noted that natural antibody detected in the serum of an apparently healthy subject using the ELISA technique did not react with parasite antigens when tested by radioimmunoprecipitation.[77]

Specific Immunologic Defense Mechanisms

Serum Antibody

Although the existence of specific serum antibody in at least some cases of human trichomoniasis is undoubted, there is very little evidence that it has any effect on the parasite in vivo. In fact, examination of an exhaustive survey[3] of the older results of attempts to develop serodiagnostic tests reveals that two of the three most-used techniques (complement fixation and indirect hemagglutination) give a higher percentage of positive results in patients with chronic long-lasting infections. Only agglutination tests gave similar results with acute and chronic cases. The interpretation of all such tests is complicated by the need to allow for preexisting natural antibody, but to the extent that new specific antibody is formed it is clearly ineffective in eliminating the infection. More recent serologic tests,[10,11,77-80] while confirming the existence of serum antibody and showing that IgM, IgG, IgA, but not, apparently, IgE are present, have shed no further light on their protective function, if any. The results of Chipperfield and Evans[81] showed that an increase in the number of IgM-, IgG-, and IgA-secreting plasma cells occurred in the endocervix following infection with *T. vaginalis, N. gonorrhoeae,* and *Candida albicans*. The greatest increase in IgM-bearing cells was found in patients with trichomoniasis. The fact that some sexual partners of infected men, who had escaped becoming infected, also showed increased numbers of endocervical plasma cells was interpreted as evidence that local production of antibody could have a protective effect.

The only directly beneficial effect of serum from infected patients appears to be its ability to protect mice against the destructive effects of intraperitoneal *T. vaginalis*.[3,82] The extent of the protective effect was assessed by numerically scoring the degree of damage to the internal organs of the animal. Summarizing the Estonian work, Honigberg[3] states that serum from noninfected men and women has no protective effect, while that from infected persons and from their sexual partners significantly reduces the extent and severity of the damage caused by the parasite. Additionally, there was no clear correlation between the protective effect and the agglutination or complement-fixing titer of the serum, the age or sex of the donor, or the clinical severity of the patient's symptoms. These results are of interest for two reasons: first, they provide a rare example of a protective effect of serum from infected patients that is not easily explained by nonantibody-mediated mechanisms and second, as with the plasma cell results of Chipperfield and Evans (see above), those who had been exposed to infection seemed to have as much of whatever protective property is being measured as those who actually became infected, although it is not absolutely clear whether the sexual partners whose serum was effective in the above assay were definitely free from infection. For both these reasons this reaction deserves to be reinvestigated, as does the whole question of the protective role of circulating antibody in human trichomoniasis.

Local Antibody

Logically a nondisseminating parasite can only be attacked by local antibody, so that in this context the word has two meanings: the local effects of serum antibody that has leaked into the genital tract, and the production and effect of locally synthesized antibody.

In the last 10 years the presence of antitrichomonal antibody in cervicovaginal secretions has been demonstrated by radioimmunoassay,[83] ELISA,[11,77] and immunofluoresence[84] methods. In general IgG and IgA antibodies were detected in a majority of, but by no means all, infected women; specific IgM and/or IgE was found either in a small percentage only or not at all. The last result is rather surprising in view of the findings of Chipperfield and Evans,[81] and raises the suspicion that a large part of the detected antibody could have been derived from the serum.

Although in our original work we suggested the existence of a possible correlation between the presence of local antibody and low numbers of parasites, the effect was far from definite, and in vitro samples containing some of the highest detected levels of antibody had no apparent harmful effects on cultured organisms (Ackers, unpublished data). Again, if local antibody is an effective defense mechanism its presence could explain the rarity of symptomatic trichomoniasis in men, but we were unable to find significant amounts of specific antibody in either infected men or male partners of infected women.[85] If in fact locally synthesized IgA antibody is important in host defense, it may possibly be by a recently described mechanism whereby specific IgA potentiates the effect of opsonization by IgG on the phagocytosis of target cells by human peripheral blood polymorphonuclear leukocytes.[86] On balance, however, there is no more evidence that either locally synthesized immunoglobulins or transuded serum antibody plays a major part in controlling *T. vaginalis* infections than there is for a role for circulating antibody.

Specific Cell-Mediated Immunity

The activities of neutrophils in trichomonal infections have already been described, but a small number of additional reports deal with classic cell-mediated immunity—the production of and effects on lymphokines by sensitized T cells in the presence of antigen, and delayed-type hypersensitivity.

Yano and his colleagues[87] examined the response of peripheral blood lymphocytes from patients with trichomonal vaginitis to parasite antigen. Proliferation of such cells, which can be detected by the incorporation of [^{3}H]-thymidine, took place when they were exposed to a soluble antigen produced by sonication of cultured organisms of a single strain of *T. vaginalis*. The response was apparently dependent on the presence of the helper/inducer subset (Leu1 and Leu3a-positive) of T cells. Cells from seven of ten infected women, but not those from healthy controls, gave a positive response in the assay, but only three of ten gave even a weak reaction (immediate, delayed, or both) on skin testing. Gel filtration of the soluble antigen showed that the most active fraction was eluted in the 100 kDa region, although some activity was spread over a wide range of molecular weights. Almost identical data and results were presented in another paper from the same group.[88] Recently Mason and Patterson[89] have confirmed and extended this work by showing that a patient's lymphocytes may be stimulated by excreted antigens as well as by those prepared by sonication and that antigen prepared from highly pathogenic and less pathogenic organisms (as assessed by the patient's symptoms) were equally effective. They also showed that neither excreted antigen nor the sonicated preparation were nonspecific mitogens—an important result since this is believed to be one of the many immunodepressive strategies adopted by parasitic protozoa.

In a wide variety of microbial infections, a key element in immunologic defense is provided by the ability of activated macrophages, stimulated by lymphokines (produced by specifically sensitized T cells in response to the presence of antigen) to kill pathogens with greatly enhanced efficiency. This process has been demonstrated with *T. vaginalis* and murine peripheral blood monocytes (see above) and has now been very briefly reported with human cells, but only from normal, healthy donors.[46] Further study of the role of specifically activated macrophages is clearly warranted, but as already stated, it may be that they are not of particular importance in responding to an extracellular parasite.

Long before the results described above had been obtained, attempts were being made to use skin testing as a diagnostic tool. Although the first reasonably successful experiments were carried out by Adler and Sadowsky;[90] in this area, as in so many others, the most extensive investigations were carried out in the 1960s by Teras's group and their results translated and summarized by Honigberg.[3] As far as its development as a practical diagnostic tool is concerned this test appears to be no better or worse than the majority of contemporary serum antibody methods, giving the correct answer in a majority of cases but giving false-positive and false-negative results too frequently to be clinically useful. However, as a very tentative pointer to the existence and even the importance of cell-mediated immune responses in human trichomoniasis it should be noted that, in contrast to most antibody based tests, far fewer positive results were obtained by one group of investigators from chronically infected women than from those with acute infections. If confirmed, these results could be read as suggesting a role for cell-mediated immunity in eliminating low-level infections, but at this stage such a conclusion would be highly speculative.

Conclusions

Any reading of the above must leave a strong impression that *specific* host immune responses play little or no role in combating human *T. vaginalis* infections. The fortunate failure of this really quite destructive pathogen to spread outside the genitourinary tract under normal circumstances can be explained by a number of factors of which complement is likely to be the most important, while within the genital tract infiltrating neutrophils seem capable of keeping parasite numbers within tolerable levels by employing a fascinating cooperative strategy against an organism too large to be ingested whole. Other immunologic (e.g., natural antibody) and nonimmunologic (e.g., zinc) factors will assist in the process and there really seems no need to invoke specific immunity at all. Yet the female reproductive tract, at least, *is* responsive to introduced antigen,[91] although its responsiveness is naturally modified by the necessity not to reject the immunologically alien fetal and spermatozoal antigens, and hints of a specific response to *T. vaginalis* recur throughout this chapter. Although, like the Cheshire cat, they tend to disappear when examined closely, attempts to pin them down will no doubt continue.

Immunosuppression by *T. vaginalis*

Unlike bacteria and viruses, parasitic protozoa tend to live in their hosts for a considerable length of time—quite long enough for the host to mount an effective immune response. The protozoa have therefore evolved an elaborate array of adaptations to permit them to survive in the immunized host.[92] As we have seen, *T. vaginalis,* by virtue of its extracellular existence in what is in any event a partially privileged site, seems to be little, if it all, affected by these immune responses, and, therefore, probably does not need to employ the complex tactics required of more exposed parasites. However, one commonly employed strategy that it may make use of is nonspecific immunosuppression. Martinotti and associates[93] studied the effects of intraperitoneal infection with *T. vaginalis* on the immunologic capacity of BALB/c mice, and were able to show depression of serum antibody response to sheep red-blood cells and the development of contact hypersensitivity to 2-phenyl-4-ethoxymethylene oxalone in such animals. They also showed that the same intraperitoneal infection accelerated the growth of a transplantable adenocarcinoma in mice.[94] As with all work carried out on intraperitoneally infected mice, it is difficult to know what significance these findings have for the very different human infection; so far evidence for significant immunodepression in man has not been reported.

Immunoprophylaxis

In view of all that has been said so far it is not surprising that the prospects for immunoprophylaxis of human trichomoniasis are generally regarded as poor. In fact, until recently, the only published account related to some very surprising procedures carried out by Aburel and coworkers.[95] One hundred women with refractory trichomoniasis unresponsive to chemotherapy were treated by the inoculation of increasing numbers of heat-killed *T. vaginalis* intravaginally; 40 were completely cured, and an additional 49 lost virtually all their symptoms. Three of the women who obtained no benefit were finally treated with direct local injections of hyperimmune rabbit anti–*T. vagi-*

nalis serum; all showed great clinical improvement, and one lost her infection completely. As far as I am aware, these rather drastic procedures have not been used since, even in the growing number of women infected with metronidazole-resistant trichomonads.

A completely different but equally controversial approach to immunoprophylaxis is represented by the proprietary product SolcoTrichovac®. This material consists of killed organisms of the so-called "aberrant coccoid forms" of lactobacilli and is administered to women as a course of three intramuscular injections at fortnightly intervals, followed by annual booster injections.[96] The rationale for this treatment is the theory that invasion by *T. vaginalis* is accompanied by the disappearance of the normal bacterial flora of the vagina and its replacement by aberrant organisms. Vaccination with these bacteria is said to induce antibodies that help to eliminate the abnormal flora and allow its replacement by normal lactobacilli and that also cross-react with *T. vaginalis* and provide an immune response that is effective both therapeutically and prophylactically.[96]

It is fair to say that these claims have been met with a great deal of skepticism by scientists and clinicians alike. Evidence has been produced that, for example, the claimed cross-reaction between the bacteria in the vaccine and *T. vaginalis* is unlikely to occur.[97] Nevertheless, some clinical trials did seem to show a genuine reduction in the incidence of reinfection in vaccinated women,[98] or even elimination of the infection,[99] although the latter trial was uncontrolled and included only 20 patients. SolcoTrichovac® remains highly controversial, but has been the subject of a number of symposia[100] and minisymposia.[101] It is to be hoped that a consensus on its effectiveness and place, if any, in the treatment of trichomoniasis will soon be established.

Pentatrichomonas hominis and *Trichomonas tenax*

It is disappointing, after describing such an upsurge of interest in the immunology of *T. vaginalis* to have to report that virtually no work has been done on the immunologic aspects of the other two human trichomonads in the last 10 years. Since the earlier work has been expertly reported and discussed in detail,[3,4,38] there is little point in repeating it here. The essential conclusions of a considerable amount of research, virtually all concerned with antigenic identity, are that all three human trichomonads possess both unique and shared antigens and that antigenic types (four in the case of *T. tenax* and "several" in the case of *P. hominis*) exist in these two species as in *T. vaginalis*. Two recent results, peripheral to other investigations, support the concept of a minority of antigens shared between different species. Alderete[77] found a low but definite positive result when rabbit anti-*T. tenax* serum was tested in an ELISA system with a *T. vaginalis* antigen, and we[102] found some degree of reactivity when two strains of *P. hominis* were tested in a *T. vaginalis* inhibition enzyme immunoassay.

Much of the early work with these two organisms was carried out to gather evidence for controversies long since settled—were there one or two human large-bowel trichomonads, and did one of them cause trichomonal vaginitis by autoinfection? What *is* disappointing is that the current controversies, those over the existence and importance of extraoral infections with *T. tenax* and extragenital infections with trichomonads generally, are as much in need of modern immunologic research to resolve them as were the disagreements of the past—and yet nothing is being done. Teras[103] claims that specific circulating antibodies to *P. hominis* and *T. tenax* do occur and signal the transformation of the parasites from harmless commensals to pathogenic invaders. This is certainly what appears to happen with *Entamoeba histolytica* and the results urgently need to be confirmed by other workers. Immunologic methods would also be very helpful in identifying the species of trichomonad present in those occasional specimens where they appear to be present far from their normal homes. As well as these practical considerations, both parasites are interesting in their own right. *Pentatritrichomonas hominis* is very unusual among the human large-bowel protozoa in not forming a resistant cyst, and *T. tenax*, believed to be transmitted principally by kissing, is suprisingly common among elderly persons with poor dental hygiene.[4] Neither organism deserves its current neglect.

References

1. Ackers JP, Lumsden WHR: Immunology of genito-urinary trichomoniasis. *Bull Mem Soc Med Hop Paris* 181:190-113, 1978.
2. Tokura N: Biologische und immunologische

Untersuchungen über die meschenparasitären Trichomonaden. *Igaku Kenkyu* 9:1-13, 1935.
3. Honigberg BM: Trichomonads. In Jackson GJ, Herman R, Singer L (eds): *Immunity to Parasitic Animals,* vol 2, pp. 469-550. New York: Appleton-Century-Crofts, 1970.
4. Honigberg BM: Trichomonads of importance in human medicine. In Kreier JP (ed): *Parasitic Protozoa,* vol 2, pp 275-454. New York: Academic Press, 1978.
5. Kott H, Adler S: A serological study of *Trichomonas* sp. parasitic in man. *Trans R Soc Trop Med Hyg* 55:333-344, 1961.
6. Schoenherr KE: Serological investigations of trichomonads. *Z Immunitaetsforsch Allerg Klin Immunol* 113:83-94, 1956.
7. Teras JK: On the varieties of *Trichomonas vaginalis*. Prog Protozool, Proc 2nd Int Conf Protozool. *Excerpta Med Found Int Congr Ser* 91:197-198, 1965.
8. Teras JK: Differences in the antigenic properties within strains of *Trichomonas vaginalis*. *Wiad Parazytol* 12:357-363, 1966.
9. Teras JK, Jaakmees H, Nigesen U, Rõigas E, Tompel H: On the agglutinogenic properties of *Trichomonas vaginalis*. *Wiad Parazytol* 12:370-377, 1966.
10. Mason PR: Serodiagnosis of *Trichomonas vaginalis* infection by the indirect fluorescent antibody test. *J Clin Pathol* 32:1211-1215, 1979.
11. Street DA, Taylor-Robinson D, Ackers JP, Hanna NF, McMillan A: Evaluation of an enzyme-linked immunosorbent assay for the detection of antibody to *Trichomonas vaginalis*. *Br J Vener Dis* 58:330-333, 1982.
12. Su-Lin K-E, Honigberg BM: Antigenic analysis of *Trichomonas vaginalis* strains by quantitative fluorescent antibody methods. *Z Parasitenkd* 69:162-181, 1983.
13. Honigberg BM: Comparative pathogenicity of *Trichomonas vaginalis* and *Trichomonas gallinae* to mice. I. Gross pathology, quantitative evaluation of virulence, and some factors affecting pathogenicity. *J Parasitol* 47:545-571, 1961.
14. Su-Lin K-E, Honigberg BM: Antigenic analysis of *Trichomonas vaginalis*. *J Protozool* 23:18A, 1976.
15. Stepkowski S, Honigberg BM: Antigenic analysis of virulent strains of *Trichomonas gallinae* by gel diffusion methods. *J Protozool* 19:306-315, 1972.
16. Dwyer DM, Honigberg BM: Immunologic analysis by quantitative fluorescent antibody methods of effects of prolonged cultivation on *Histomonas meleagridis* (Smith). *Z Parasitenkd* 39:39-52, 1972.
17. Alderete JF: Antigen analysis of several pathogenic strains of *Trichomonas vaginalis*. *Infect Immun* 39:1041-1047, 1983.
18. Delachambre D: Etude critique comparée des antigènes de deux clones d'une même souche d'un flagellé parasite (*Trichomonas vaginalis*). *Ann Parasitol Hum Comp* 55:1-11, 1980.
19. Dwyer DM: Concomitant antigenic and pathogenicity changes in *Histomonas meleagridis*. III Int Cong Parasitol, Munich. 6:1568, Vienna, Facta 1974.
20. Alderete JF: Identification of immunogenic and antibody-binding membrane proteins of pathogenic *Trichomonas vaginalis*. *Infect Immun* 40:284-291, 1983.
21. Alderete JF, Suprun-Brown L, Kasmala L, Smith J, Spence M: Heterogeneity of *Trichomonas vaginalis* and discrimination among trichomonal isolates and subpopulations with sera of patients and experimentally infected mice. *Infect Immun* 49:463-468, 1985.
22. Alderete JF, Kasmala L, Metcalfe E, Garza GE: Phenotypic variation and diversity among *Trichomonas vaginalis* isolates and correlation of phenotype with trichomonal virulence determinants. *Infect Immun* 53:285-293, 1986.
23. Alderete JF, Suprun-Brown L, Kasmala L: Monoclonal antibodies to a major surface glycoprotein immunogen differentiates isolates and subpopulations of *Trichomonas vaginalis*. *Infect Immun* 52:70-75, 1986.
24. Garber GE, Proctor EM, Bowie WR: Immunogenic proteins of *Trichomonas vaginalis* as demonstrated by the immunoblot technique. *Infect Immun* 51:250-253, 1986.
25. Torian BE, Connelly RJ, Stephens RS, Stibbs HH: Specific and common antigens of *Trichomonas vaginalis* detected by monoclonal antibodies. *Infect Immun* 43:270-275, 1984.
26. Connelly RJ, Torian BE, Stibbs HH: Identification of a surface antigen of *Trichomonas vaginalis*. *Infect Immun* 49:270-274, 1985.
27. Krieger JN, Holmes KK, Spence MR, Rein MF, McCormack WM, Tam MR: Geographic variation among isolates of *Trichomonas vaginalis*: Demonstration of antigenic heterogeneity by using monoclonal antibodies and the indirect immunofluorescence technique. *J Infect Dis* 152:979-984, 1985.
28. Warton A, Honigberg BM: Lectin analysis of surface saccharides in two *Trichomonas vaginalis* strains differing in pathogenicity. *J Protozool* 27:410-419, 1980.
29. Warton A, Honigberg BM: Analysis of surface saccharides in *Trichomonas vaginalis* strains with various pathogenicity levels by fluorescein-conjugated plant lectins. *Z Parasitenkd* 69:149-159, 1983.
30. Soliman MAI, Ackers JP, Catterall RD: Isoenzyme characterization of *Trichomonas vaginalis*. *Br J Vener Dis* 58:250-256, 1982.
31. Chyle M, Schneiderka P, Stepan JJ, Chyle P: Some enzyme and isoenzyme activities in

Trichomonas vaginalis influenced by low temperature preservation and virus infection. Abstracts Int Symp Trichomonads & Trichomoniasis, Prague, July 1985, p 20, 1985.
32. De Jonkheere JF: Isoenzymes and total protein patterns separated by agarose isoelectric focussing of *Trichomonas vaginalis*. Abstracts Int Symp Trichomonads & Trichomoniasis, Prague, July 1985, p 21, 1985.
33. Teras JK, Kazakova I, Tompel H, Mirme E: On the correlation between the biochemical activity, virulence and antigenic properties of the strains of *T. vaginalis*. Wiad Parazytol 19:389-391, 1973.
34. Peterson KM, Alderete JF: Host plasma proteins on the surface of pathogenic *Trichomonas vaginalis*. Infect Immun 37:755-762, 1982.
35. Peterson KM, Alderete JF: Acquisition of α_1-antitrypsin by a pathogenic strain of *Trichomonas vaginalis*. Infect Immun 40:640-646, 1983.
36. Peterson KM, Alderete JF: Selective acquisition of plasma proteins by *Trichomonas vaginalis* and human lipoproteins as a growth requirement for this species. Mol Biochem Parasitol 12:37-48, 1984.
37. Peterson KM, Alderete JF: Iron uptake and increased intracellular enzyme activity follow host lactoferrin binding by *Trichomonas vaginalis* receptors. J Exp Med 160:398-410, 1984.
38. Honigberg BM, Lindmark DC: Trichomonads and *Giardia*. In Soulsby EJL (ed): *Immune responses in parasitic infections: Immunology, Immunopathology, and Immunoprophylaxis*, vol 4, pp 99-139. Boca Raton, Fl: CRC Press, 1987.
39. Alderete JF, Garza GE: Soluble *Trichomonas vaginalis* antigens in cell-free culture supernatants. Mol Biochem Parasitol 13:147-158, 1984.
40. Gillin FD, Sher A: Activation of the alternative complement pathway by *Trichomonas vaginalis*. Infect Immun 34:268-273, 1981.
41. Alderete JF, Kasmala L: Monoclonal antibody to a major glycoprotein immunogen mediates differential complement-independent lysis of *Trichomonas vaginalis*. Infect Immun 53:697-699, 1986.
42. Burgess DE: *Tritrichomonas foetus*: Preparation of monoclonal antibodies with effector function. Exp Parasitol 62:266-274, 1986.
43. Tagliabue A, Nencioni L, Villa L, Keven DF, Lowell GH, Boraschi D: Antibody dependent cell-mediated antibacterial activity of intestinal lymphocytes with secretory IgA. Nature 306:184-186, 1983.
44. Landolfo S, Martinotti G, Martinotti P, Furni G: Natural cell-mediated cytotoxicity against *Trichomonas vaginalis* in the mouse. J Immunol 124:508-514, 1980.
45. Landolfo S, Martinotti MG, Martinotti P, Furni G: Genetic control of *Trichomonas vaginalis* infection. I. Resistance or susceptibility among different mouse strains. Boll Ist Sieroter Milan 58:48-51, 1979.
46. Martinotti MG, Jemma C, Giovarelli M, Musso T: Induction of macrophage activation by immune T lymphocytes for *Trichomonas vaginalis* killing. Abstracts Int Symp Trichomonads & Trichomoniasis, Prague, July 1985, p 49, 1985.
47. Rein MF, Sullivan JA, Mandell GL: Trichomonacidal activities of human polymorphonuclear neutrophils: Killing by disruption and fragmentation. J Infect Dis 142:575-585, 1980.
48. Holbrook TW, Boackle RJ, Vesely J, Parker BW: *Trichomonas vaginalis:* Alternative pathway activation of complement. Trans R Soc Trop Med Hyg 76:473-475, 1981.
49. Demeš P, Gombošová A: Factors of nonspecific immunity in human trichomoniasis. Abstracts Int Symp Trichomonads & Trichomoniasis, Prague, July 1985, p 47, 1985.
50. Reed SL, Sargeaunt PG, Braude AI: Resistance to lysis by human serum of pathogenic *Entamoeba histolytica*. Trans R Soc Trop Med Hyg 77:248-253, 1983.
51. Kelly DR, Schnitzer RJ: Experimental studies on trichomoniasis. II. Immunity to reinfection in *T. vaginalis* infection of mice. J Immunol 69:337-342, 1952.
52. Schnitzer RJ, Kelly DR: Short persistence of *Trichomonas vaginalis* in reinfected mice. Proc Soc Exp Biol Med 82:404-406, 1953.
53. Kelly DR, Schumacher A, Schnitzer RJ: Experimental studies on trichomoniasis. III. Influence of the site of the immunizing infection with *Trichomonas vaginalis* on the immunity of mice to homologous reinfection by different routes. J Immunol 70:40-43, 1954.
54. Baba H: Immunological studies on trichomonads. II. On the protection of mice from trichomonas infections by the immunization with heat-killed trichomonads. Nisshin Igaku 45:16-19, 1958 (in Japanese, English summary).
55. Martinotti MG, Cagliani I, Lattes C, Cappuccinelli P: Immune response and degree of protection in mice immunized with *Trichomonas vaginalis* antigen. G Batteriol Virol Immunol 70:3-12, 1977
56. Erlandsen SL, Meyer EA (eds): Giardia *and Giardiasis*. New York: Plenum Press, 1984.
57. Mason PM, Patterson B, Latif AS: Epidemiology and clinical diagnosis of *Trichomonas vaginalis* infection in Zimbabwe. Cent Afr J Med 29:53-56, 1983.
58. Jírovec O, Petrů M: *Trichomonas vaginalis* and trichomoniasis. Adv Parasitol 6:117-118, 1968.
59. Keighly EE: Trichomoniasis in a closed community: Efficacy of metronidazole. Br Med J 1:207-209, 1971.
60. Ravdin JI, Guerrant RL: A review of the para-

site cellular mechanisms involved in the pathogenesis of amebiasis. *Rev Infect Dis* 4:1185-1207, 1982.
61. Krieger JN, Rein MF: Canine prostatic secretions kill *Trichomonas vaginalis. Infect Immun* 37:77-81, 1982.
62. Krieger JN, Rein MF: Zinc sensitivity of *Trichomonas vaginalis:* In vitro studies and clinical implications. *J Infect Dis* 146:341-345, 1982.
63. Kulda J, Budilová M, Dohnalová J: The effect of ferric-ammonium citrate on the virulence of *Tritrichomonas foetus* to the laboratory mouse. *J Protozool* 24:15A-16A, 1977.
64. Schoolnik GK, Buchanan TM, Holmes KK: Gonococci causing disseminated infection are resistant to the bactericidal action of normal human serum. *J Clin Invest* 58:1163, 1976.
65. Ackers JP: The immunology of *T. vaginalis* infections. *Wiad Parazytol* 29:41-44, 1983.
66. Mason PR, Forman L: *In vitro* attraction of polymorphonuclear leucocytes by *Trichomonas vaginalis. J Parasitol* 66:888-892, 1980.
67. Mason PR, Forman L: Polymorphonuclear cell chemotaxis to secretions of pathogenic and nonpathogenic *Trichomonas vaginalis. J Parasitol* 68:457-462, 1982.
68. Chikunguwo S, Mason PR, Read JS: Investigation of *Trichomonas vaginalis* secretions and chemotactic factor. Abstracts Int Symp Trichomonads & Trichomoniasis, Prague, July 1985, p 57, 1985.
69. Rasmussen SE, Rhodes JM: Chemotactic effect of *Trichomonas vaginalis* on polymorphonuclear leucocytes *in vitro*. Abstracts Int Symp Trichomonads & Trichomoniasis, Prague, July 1985, p 50, 1985.
70. Brasseur P, Ballet JJ, Savel J: *Trichomonas vaginalis* derived chemotactic activity for human polymorphonuclear leucocytes. Abstracts Int Symp Trichomonads & Trichomoniasis, Prague, July 1985, p 51, 1985.
71. Gardner WA Jr, Culberson DE, Bennett BD: *Trichomonas vaginalis* in the prostate gland. *Arch Pathol Lab Med* 110:430-432, 1986.
72. Mantovani A, Polentarutti N, Peri G, Martinotti G, Landolfo F: Cytotoxicity of human peripheral blood monocytes against *Trichomonas vaginalis. Clin Exp Immunol* 46:391-396, 1981.
73. Weld JT, Kean BH: A factor in serum of human beings and animals that destroys *T. vaginalis. Proc Soc Exp Biol Med* 98:494-496, 1958.
74. Reisenhofer V: Über die Beeinflussung von *Trichomonas vaginalis* durch verschiedene Sera. *Arch Hyg Bakteriol* 146:628-635, 1963.
75. Samuels R, Chun-Hoon H: Serological investigation of trichomonads. I. Comparison of "natural" and immune antibodies. *J Protozool* 11:36-46, 1964.
76. Tatsuki T: Studies on *Trichomonas vaginalis.* II. Immunoserological reactions of *Trichomonas vaginalis* by sera and colostra from women infected therewith. *Nagasaki Igakkai Zasshi* 32:983-993, 1957. (in Japanese, English summary).
77. Alderete JF: Enzyme-linked immunoabsorbent assay for detecting antibody to *Trichomonas vaginalis:* Use of whole cells and aqueous extract as antigen. *Br J Vener Dis* 60:164-170, 1984.
78. Mathews HM, Healy GR: Evaluation of two serological tests for *Trichomonas vaginalis* infection. *J Clin Microbiol* 17:840-843, 1983.
79. Cogne M, Brasseur P, Ballet JJ: Detection and characterization of serum antitrichomonal antibodies in urogenital trichomoniasis. *J Clin Microbiol* 21:588-592, 1985.
80. Wos SM, Watt RM: Immunoglobulin isotypes of anti-*Trichomonas vaginalis* antibodies in patients with vaginal trichomoniasis. *J Clin Microbiol* 24:790-795, 1986.
81. Chipperfield EJ, Evans BA: The influence of local infection on immunoglobulin formation in the human endocervix. *Clin Exp Immunol* 11:219-223, 1972.
82. Teras JK: On the protective effect of the blood sera of patients with trichomoniasis of the genito-urinary tract on white mice infected intraperitoneally with cultures of *Trichomonas vaginalis. Izv Akad Nauk Est SSR (Biol Ser)* 10:19-26, 1961.
83. Ackers JP, Lumsden WHR, Catterall RD, Coyle R: Antitrichomonal antibodies in the vaginal secretions of women infected with *T. vaginalis. Br J Vener Dis* 51:319-323, 1975.
84. Su-Lin K-E: Antibody to *Trichomonas vaginalis* in human cervicovaginal secretions. *Infect Immun* 37:852-857, 1982.
85. Ackers JP, Catterall RD, Lumsden WHR, McMillan A: Absence of detectable local antibody in genitourinary tract secretions of male contacts of women infected with *Trichomonas vaginalis. Br J Vener Dis* 54:168-171, 1978.
86. Goldstine SN, Tsai A, Kemp CJ: Role of IgA antibody in phagocytosis by human polymorphonuclear leukocytes. In McGhee JR, Mestecky J (eds): The Secretory Immune System. *Ann NY Acad Sci* 409:824, 1983.
87. Yano A, Yui K, Aosai F, Kojima S, Kawana T, Ovary Z: Immune response to *Trichomonas vaginalis.* Immunochemical and immunobiological analyses of *T. vaginalis* antigens. *Int Arch Allergy Appl Immunol* 72:150-157, 1983.
88. Yano A, Aosai F, Yui K, Kojima S, Kawana T: Antigen-specific proliferation responses of peripheral blood lymphocytes to *Trichomonas vaginalis* antigen in patients with *Trichomonas vaginalis. J Clin Microbiol* 17:175-180, 1983.
89. Mason PR, Patterson BA: Proliferative re-

89. sponse of human lymphocytes to secretory and cellular antigens of *Trichomonas vaginalis*. *J Parasitol* 71:265-268, 1985.
90. Adler S, Sadowsky A: Intradermal reaction in trichomonad infections. *Lancet* 252:867-868, 1947.
91. Cineder B, De Weck A (eds): *Immunological Responses of the Female Reproductive Tract.* WHO Workshop, Geneva, Jan 1975. Copenhagen: Scriptor, 1976.
92. Wakelin D: *Immunity to Parasites.* London: Edward Arnold, 1984.
93. Cappuccinelli P, Giovarelli M, Landolfo G, Martinotti LV: Depressione della riposta immunitoria in topi infettati con *Trypanosoma congolense* e *Trichomonas vaginalis*. *G Mal Infett Parasit* 27:788-791, 1975.
94. Martinotti G, Naresio L, Landolfo S, Giovarelli M, Cappuccinelli P: Immunodepressione de *Trypanosoma congolense* e *Trichomonas vaginalis* e crescita di tumori nel topo. Atti XVII Cong Nat Soc Ital Microbiol, 1975, pp 471-477.
95. Aburel E, Zervos G, Titea V, Pană S: Immunological and therapeutic investigations in vaginal trichomoniasis. *Rom Med Rev* 7:13-19, 1963.
96. SolcoTrichovac Manufacturers Product Information. Solco Basle 4127, Birsfelden-Basle, Switzerland.
97. Gombošová A, Demeš P, Valent M: Immunotherapeutic effect of the lactobaccillus vaccine Solco Trichovac in trichomoniasis is not mediated by cross-reacting antibodies against *Trichomonas vaginalis*. Abstracts Int Symp Trichomonads & Trichomoniasis, Prague, July 1985, p 65, 1985.
98. Harriss JWR, Higy-Mandic L, McManus TJ: A double-blind comparative study in *Trichomonas vaginalis* infection: SolcoTrichovac vs. placebo. *Eur J Sex Trans Dis* 2:27-29, 1984.
99. Nevembi PM, Nyakeri LN: Clinical experience with SolcoTrichovac in the treatment of vaginal trichomoniasis. *East Afr Med J* 61:372-374, 1984.
100. Rüttgers H (ed): Trichomoniasis, Scientific Papers of the Symposium on Trichomoniasis, Basle, Oct 20, 1981. *Gynäkol Rundsch* 23 (suppl 2); 1-91, 1982.
101. Mini-symposium on SolcoTrichovac. Abstracts Int Symp Trichomonads & Trichomoniasis, Prague, July 1985, pp 63-68, 1985.
102. Yule A, Ackers JP: Development of enzyme immunoassays for the detection of *Trichomonas vaginalis* antigen. Abstracts Int Symp Trichomonads & Trichomoniasis, Prague, July 1985, p 22, 1985.
103. Teras J: Extraurogenital infections of man by Trichomonadidae. Abstracts Int Symp Trichomonads & Trichomoniasis, Prague, July 1985, p 83, 1985.
104. Alderete JF: *Trichomonas vaginalis* NYH286 phenotypic variation may be coordinated for a repertoire of trichomonad surface immunogens. *Infect Immun* 55:1957-1962, 1987.
105. Alderete JF: Alternating phenotypic expression of two classes of *Trichomonas vaginalis* surface markers. *Rev Infect Dis* 10 Suppl 2:S408-412, 1988.
106. Alderete JF, Demeš P, Gombošová A, Valent M, Yánoška A, Fabušová H, Kasmala L, Garza GE, Metcalf EC: Phenotypes and protein-epitope phenotypic variation among fresh isolates of *Trichomonas vaginalis*. *Infect Immun* 55:1037-1041, 1987.
107. Demeš P, Gombošová A, Valent M, Jánoška A, Fabušová H, Petrenko M: Differential susceptibility of fresh *Trichomonas vaginalis* isolates to complement in menstrual blood and cervical mucus. *Genitourin Med* 64:176-179, 1988.
108. Mathews HM, Moss DM, Callaway CS: Human serologic response to subcellular antigens of *Trichomonas vaginalis*. *J Parasitol* 73:601-610, 1987.
109. Moav N, Draghi E, David A, Gold D: Anti-*Trichomonas vaginalis* monoclonal antibodies including complement-dependent cytotoxicity. *Immunology* 63:63-69, 1988.
110. Torian BE, Connelly RJ, Barnes RC, Kenny GE: Antigenic heterogeneity in the 115,000 M_r major surface antigen of *Trichomonas vaginalis*. *J Protozool* 35:273-280, 1988.

5

Biochemistry of *Trichomonas vaginalis*

Miklós Müller

Introduction

Trichomonad flagellates are eukaryotic organisms (see Chap. 3) similar to other eukaryotic cells in many ways. They differ, however, from the majority of eukaryotic cells in several significant aspects, notably, in their nutritional requirements and energy metabolism. Trichomonads depend on a large number of preformed metabolites as nutrients, revealing the absence of major biosynthetic pathways. The character and subcellular organization of their energy metabolism have few parallels among eukaryotic cells and reveal striking affinities to primitive anaerobic bacteria, which are responsible for the efficacy of 5-nitroimidazole derivatives in the treatment of trichomoniasis. A unique feature is the presence of hydrogenosomes instead of mitochondria. It is tempting to regard these characteristics as adaptations to specialized environments. It would be premature, however, to go beyond pointing out that the nutrient-rich, largely anaerobic environment, and the properties of the organism living in it, are well matched to each other.

This chapter presents an overview of the biochemistry of *Trichomonas vaginalis,* the only species parasitizing humans about which detailed information is available. Other species, especially the extensively studied cattle parasite, *Tritrichomonas foetus,* are mentioned only to provide information on certain biochemical properties that were not explored in *T. vaginalis.* It has to be emphasized that there are striking biochemical differences among various trichomonad groups. Thus, results obtained with one species cannot be assumed to pertain to all, without actual experimental evidence.

However fascinating these differences might be, they are not explored here and no attempt is made to discuss the comparative biochemistry of trichomonad flagellates. The list of references is not comprehensive, but an attempt was made to document all statements, emphasizing more recent contributions. Abstracts were not used as a source of information.

The topic of this chapter has been reviewed repeatedly,[1-4] most recently in the monographic treatments of pathogenic trichomonad flagellates by the editor of this volume,[5,6] providing excellent sources of earlier information. It was gratifying to see, however, how many additional new findings and conceptual advances had to be evaluated for this chapter.

Biochemical studies on trichomonad flagellates began when bacteria-free (axenic) cultures of these organisms became available in the 1940s (see Chap. 7). Almost all data in this area have been obtained with the aid of mass cultures necessary to provide sufficient material for biochemical and physiologic studies. It is not known, however, to what extent cultured cells differ from organisms living in their hosts.

Research on trichomonads carried out in the author's laboratory is being supported by Research Grants AI 11942 and RR 07069, from the National Institute of Allergy and Infectious Diseases and the Division of Research Resources, National Institutes of Health, U.S. Public Health Service, respectively, as well as by Research Grants DMB 8415003 and PCM 8309294, from the National Science Foundation. In the past, support for these studies was provided by grants from the U.S. Public Health Service, National Science Foundation, and G.D. Searle & Co. as well as by contracts from the Centers for Disease Control, Atlanta, Georgia.

In experiments, the organisms are studied under conditions that differ from those at the natural sites of infection. This chapter, then, describes the biochemical properties of organisms cultivated and examined under artificial conditions. These often represent the extreme limits of what might occur under natural conditions involving strict anaerobiosis or equilibration with air and absence of all nutrients or medium abundant in nutrients. Most, but probably not all, of what is described corresponds to what could be observed in the host, were appropriate methods available. Differences most likely do exist and present an exciting challenge to future investigators.

The major conclusions of this chapter are based partly on well-established observations and partly on less detailed or reliable evidence. Therefore, the experimental basis of certain key conclusions is mentioned to help the reader assess the strength of the point made. An extensive search was made in *T. vaginalis* for a number of constituents, enzymes, and organelles with negative results. Such data are also presented in some detail, because this species is characterized as much by the absence of certain common enzymes and constituents of other eukaryotes as by the presence of others.

Studies on *T. vaginalis* have been and are being performed by different investigators on various strains and isolates grown in different media. Some are tested soon after isolation from the patient, while others only after extended cultivation in vitro. The experimental conditions also differ from study to study. Not surprisingly different reports, sometimes even from the same laboratory, give different values for the same constituent or process. It is reassuring, however, that reported values for several important constituents and processes are of the same order. The quantitative data presented in this chapter provide information only on the magnitude of the value in question. No greater significance should be attached to them than what is warranted in light of the above considerations.

Much information on *T. vaginalis* was obtained with the aid of histochemical (topochemical) methods, both at the light and electron microscopic levels. Although certain results agree with findings obtained with biochemical methods, a number of others, especially reports on the existence and subcellular localization of several enzymes, do not. For this reason, few histochemical data are included in this chapter.

Major Constituents

Data on the general biochemical composition of *T. vaginalis* are presented (Table 5.1) to provide a quantitative basis for the evaluation of specific contributions.

Trichomonas vaginalis contains numerous proteins. One- and two-dimensional gel electrophoreses have been used to map the major proteins of this organism. To increase the sensitivity of detection of the proteins, the trichomonads were radiolabeled in vivo either intrinsically with labeled amino acids[13] or extrinsically by iodination.[14] These methods reveal an expectedly complex picture. After two-dimensional separation of extracts of whole cells, more than 200 intrinsically labeled proteins and about 30 surface-iodinated components can be detected.[15] A number of proteins located on the cell surface are glycosylated, as shown by binding of lectins to intact organisms[16,17] and by characterization of proteins identified and separated with specific antibodies.[18,19] The more abundant proteins, as re-

TABLE 5.1. Major constituents of *T. vaginalis*[a].

Dry weight (μg)	100–130[b]	(7)	76[f] (8)	
Protein (μg)	60[b]	(7)	83[e] (9)	100[d] (10)
DNA (ng)	81[c]	(11)	300[b] (7)	530[f] (9)
RNA (μg)	20[b]	(7)		
Glycogen (μg)	17[b]	(12)	18[b] (7)	
Lipid (μg)	10[c]	(8)		

[a]In all instances weight is given per 10^6 cells, but the quantities are obtained by different ways: [b]calculated from percent composition with dry weight of 10^6 cells taken as 100 μg; [c]calculated from genetic complexity. [d]assumed, no details given; [e]calculated from N content; [f]direct determination; References in parentheses.

vealed by these methods, show little isolate-dependent variation.[15] The biochemical basis of differences in pathogenicity, antigenicity, and other properties of various isolates is probably correlated with proteins present in lower quantities.[15] Differences in lectin binding levels have been found in isolates of different levels of pathogenicity.[17]

Proteins separated by electrophoresis have been analyzed also by immunologic techniques to define components that elicit antibody response in infected humans or in animals during experimental infections.[13,14,20,21] The antigenic properties of *T. vaginalis* proteins and their role in host–parasite interactions are detailed in Chapter 4.

Low-temperature electron paramagnetic resonance (EPR) spectra of *T. vaginalis* (Fig. 5.1) indicate the presence of significant amounts of iron-sulfur proteins,[10,22] of which an about 12-kDa [2Fe-2S] ferredoxin has been purified and characterized.[23] Properties and the functional role of *T. vaginalis* enzymes and electron transport proteins are discussed later in the chapter.

Electrophoresis of cell extracts of *T. vaginalis* combined with detection of enzyme activities in the gels revealed isoenzyme heterogeneity of several enzymes.[24-29] Grouping of isolates exhibiting similar isoenzyme patterns (zymodemes) has great promise as an approach to the analysis of populations at the subspecific level, with potential applications in epidemiologic studies.

DNA of *T. vaginalis* has been isolated and characterized.[7,9,11,30] Its guanine plus cytosine content is low, between 29% and 33%. No indication was found for the existence of extranuclear DNA.[11,31] RNA of *T. vaginalis* has been studied less extensively. *Trichomonas vaginalis* nucleic acids, their biosynthesis, and function are discussed in more detail in Chapter 6.

The major nutrient reserve of *T. vaginalis* is glycogen, representing up to 20% of the dry weight of the organism.[7] Typical glycogen rosettes are a dominant feature of the ultrastructure of the cell (see Chap. 3, Fig. 3.5d). In cells incubated in the absence of carbohydrates, the glycogen content decreases; if glucose or other utilizable carbohydrate is present in the medium, it increases.[32] *Trichomonas vaginalis* glycogen has not been studied in any detail. Partially purified preparations have typical absorption spectra when stained with iodine and have an optical rotation ($[\alpha]D^{20}$) of 176 to 178,[7] lower than most glycogens but similar to *T. foetus* glycogen.[33] In cells exposed to maltose and certain nitrocompounds, spectral data indicated the presence of an additional polysaccharide that was not amylopectin.[7]

Although few studies have dealt with lipids of *T. vaginalis*,[34-36] its lipid composition is well characterized.[35] The major fatty acids are palmitic, stearic, oleic, linoleic (50% of the total), and α-linolenic. Cholesterol is the only sterol detected (more than 99% of the total). The major lipids are neutral lipids (30% of the total), phospholipids (65%), and glycolipids (5%). The following neutral lipids are present (in order of decreasing amounts): free cholesterol, cholesteryl and wax esters, triacylglycerols, and unesterified fatty acids. Phospholipids are (in descending order): phosphatidylethanolamine,

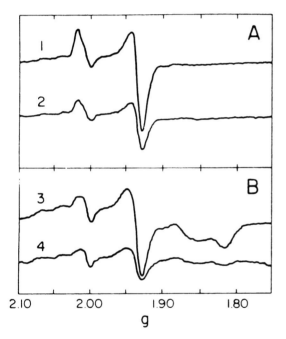

Fig. 5.1. Electron paramagnetic resonance (EPR) spectra of *T. vaginalis* cells incubated anaerobically with 50 mM glucose. Cells were grown in normal (20 μM iron, lines 2 and 4) or iron-enriched (200 μM iron, lines 1 and 3) medium. A, Spectra recorded at 25 K showing a signal primarily due to [2Fe-2S] ferredoxin. B, Spectra recorded at 10 K showing complex features probably due to several iron-sulfur centers. (Reproduced with permission of the author and the American Society for Microbiology, from Gorell TE: Effect of culture medium iron content on the biochemical composition and metabolism of *Trichomonas vaginalis*. J Bacteriol 161:1228-1230, 1985.)

phosphatidylcholine, phosphatidylglycerol, an unidentified acidic lipid, phosphatidylinositol, phosphatidylserine, N,N-dimethylphosphatidylethanolamine, sphingolipid, lysophospholipids, and phosphoinositides.

The occurrence and levels of some other constituents are discussed in other parts of this chapter, where appropriate in terms of their functional significance.

Nutritional Requirements and Uptake of Nutrients

Data on in vitro cultivation (see Chap. 7) clearly show that *T. vaginalis* requires as nutrients an unusually large number of organic compounds, some of which still remain to be identified. Lack of a complete list of the needed nutrients correlates with the fact that no defined growth medium has been developed, although major strides were made in recent years to achieve this goal for *T. vaginalis*[37] and for other trichomonads, for example, *T. foetus*.[38] Further work aimed at development of a completely defined medium and analysis of the metabolic reactions of *T. vaginalis* is needed to gain a better understanding of its minimal nutritional requirements.

Many data on nutritional requirements of *T. vaginalis* are based on studies in which putative nutrients, vitamins, or antagonists of nutrients were added to undefined media to determine their effect on growth. Although some of the information obtained by these means is valid, the results can be difficult to evaluate since they are rarely backed up by analytic data on the composition of the medium.

According to present views the following nutrients are indispensable:

1. A carbohydrate, primarily as energy source. Maltose, glucose, and galactose are utilized most readily,[39] but glycogen or starch added to the medium can also support growth. Some carbohydrates (e.g., sucrose, mannose) are not utilized.
2. Amino acids or a protein digest as a source of amino acids. No information is available about the essential or nonessential nature of various amino acids. Cysteine, a component of several media, in addition to being incorporated into proteins and participating in sulfur metabolism, may also play a role as a reducing compound, since it can be partly replaced by ascorbate.
3. Purines and pyrimidines as precursors for nucleotide and nucleic acid synthesis. Indeed, *T. vaginalis* lacks the capacity for de novo biosynthesis of purine and pyrimidine and even its capacity for interconversions of these compounds is unusually limited (see Chap. 6). Of the purines, both adenine and guanine or their corresponding nucleotides are required and, in contrast to the situation in *T. foetus,* they cannot be substituted by hypoxanthine or inosine. Similarly, of the pyrimidines, both cytidine or uridine and thymidine are required.
4. Saturated and unsaturated fatty acids and cholesterol, in view of the lack of corresponding biosynthetic capabilities in *T. vaginalis.*
5. Several vitamins of those present in the culture media used (see Chap. 7). Although little definitive information is available, present views on the metabolism of *T. vaginalis* permit some plausible guesses. It is likely that thiamine, nicotinic acid, pantothenic acid, and a flavin are necessary. Pyruvate:ferredoxin oxidoreductase of anaerobic bacteria contains thiamine,[40] and therefore should be present in the same enzyme of *T. vaginalis.* Thiamine-deficient cultures of *T. foetus* produce increased amounts of pyruvate,[41] indicating a block in its conversion by pyruvate:ferredoxin oxidoreductase. Pyridine nucleotides (NAD and NADP) are important cofactors of carbohydrate metabolism and possibly of other processes. Although their synthesis has not been studied in *T. vaginalis,* a requirement for nicotinic acid or nicotinamide is likely. Several oxidoreductases of *T. vaginalis* are flavoproteins, rendering the need for riboflavin or another flavin plausible. Pyridoxamine might be necessary for transamination reactions, although these reactions are not necessarily essential. Coenzyme A plays a major role in the urogenital trichomonad, implying a dependence on pantothenate. Vitamin B_{12} was found to be a growth requirement for this parasite,[42] but its physiologic role remains unknown. Biotin might be a cofactor in some reactions, but there is no evidence for this. In contrast, folate-dependent path-

ways[43] and quinones are absent; thus folate, p-aminobenzoate and coenzyme Q are probably unnecessary.

6. Several inorganic salts present in all nutrient media to satisfy obvious requirements. The dominant metabolic role of iron-sulfur proteins implies a marked iron requirement. Organisms grown in iron-enriched medium[10] or exposed to iron-saturated lactoferrin[44] have increased activities of certain iron-sulfur enzymes, indicating that the commonly used media are relatively iron-deficient.

Neither the rates nor the mechanisms of uptake have been studied for most nutrients. There is no doubt that a number of transport mechanisms exist in *T. vaginalis,* but these remain to be investigated. The rate of glucose uptake by cells suspended in non-nutrient medium containing 50 mM glucose is about 60 nmol min^{-1} (mg protein)$^{-1}$ under anaerobic conditions and increases by 50% under aerobic conditions.[32] The process might be fully saturated in the range from 3 to 30 mM, as indicated by equal stimulation of respiration by glucose in this range.[45] Of the amino acids, the concentration of arginine decreases most rapidly in a semisynthetic medium,[46] indicating a high rate of uptake and utilization.

The role of carrier proteins and surface receptors in the uptake of iron and lipids is discussed later in the chapter. *Trichomonas vaginalis* is an actively endocytic organism, and endocytic processes are likely to participate in the uptake of lipids. The organism is able to phagocytose larger particles, including bacteria.[47,48] This process is obviously of minor importance for organisms multiplying in liquid media, but its contribution to the nutrition of the trichomonad in the host remains to be investigated.

Release of End Products of Metabolism and Other Constituents

It was recognized some time ago that *T. vaginalis* is a fermentative organism that produces significant quantities of gaseous and acidic end products. These products have been quantitated in numerous metabolic studies. According to present views, the major end products are glycerol, lactate, acetate, CO_2, and H_2. In addition, malate and alanine have been reported as significant end products. Since in most reports only selected products are accounted for, their comparative evaluation is fraught with difficulties.

Some results from recent reports are presented in Tables 5.2 and 5.3. The overall rate of end product formation and the proportions of individual products depend on the conditions used, and for comparable conditions they differ from one report to another. Practically all variables in the conditions used affect end product formation. Acidic end products are formed anaerobically as well as aerobically; therefore, *T. vaginalis* is a fermentative organism under either condition. Hydrogen is a major product under anaerobic conditions, but the organism forms none of this gas in the presence of O_2.[45,52] There are only minor shifts elicited by aerobiosis in the production of glycerol, lactate, and acetate, although according to one report, a significant enhancement of lactate production was noted under aerobic conditions.[32] The presence of exogenous carbohydrate stimulates the overall rate of end product formation under either anaerobic or aerobic conditions. The absence or presence of CO_2 in the gas phase has, however, little effect.[32,49]

TABLE 5.2. Anaerobic metabolic end products of *T. vaginalis*.*

Utilized	Formed					
Glucose	Lactate	Glycerol	Acetate	H_2	CO_2	Reference
	nmol min^{-1} (mg protein)$^{-1}$					
nd†	97	24	22	19	19	49
	mM in medium after 75 min incubation					
14	9	9.7	11.5	nd	nd	50

*Data obtained from reports in which glycerol production was also given.
†Not determined.

TABLE 5.3. Glucose metabolism of *T. vaginalis* strain C-1.

Conditions	Utilized O$_2$	Formed products		
		Lactate	Acetate	H$_2$
	nmol min^{-1} (mg protein)$^{-1}$			
Anaerobic	—	47	12	14
Aerobic	24	42	9	—

Selected data from Müller M, Gorrell TE: Metabolism and metronidazole uptake in *Trichomonas vaginalis* isolates with different metronidazole susceptibilities. *Antimicrob Agents Chemother* 24:667-673, 1983.

A striking observation is that the rate of lactate formation and its relative contribution to total carbon elimination can vary significantly.[10,50,51] According to some reports, there occurs almost exclusively homolactic fermentation,[53] whereas in other investigations lactate is shown as only a minor end product.[10] Organisms that are resistant to metronidazole because of a deficiency of hydrogenosomal enzymes produce lactate exclusively.[54] Another important observation is that glycerol and acetate are produced in nearly equimolar amounts, as observed in the few studies in which glycerol was also determined.[49,50]

The large production of acidic end products is the basis of recent improved drug susceptibility assays for *T. vaginalis*, in which culture growth is monitored with the aid of acid–base indicators included in the assay medium[55] or through detection of lactate production.[56]

All major end products have been identified by reliable methods, even though methods used in earlier studies might have given less definitive results (Table 5.4). Malate as an end product needs separate consideration, however. This compound, identified by one-dimensional paper chromatography, has been described earlier as an important end product.[12,57] However, in recent studies using several isolates, malate was detected only in insignificant quantities (less than 1% of the total).[32,49,50] These discrepancies could be due to strain differences, but their explanation more likely depends on an identification error in the earlier studies.

In addition to the major end products listed above, *Trichomonas vaginalis* releases a number of compounds. Besides H$_2$ and CO$_2$, various other end products were detected in the "headspace" of cultures by gas chromatographic-mass spectrometric methods,[53,58] including a

TABLE 5.4. Methods used to identify and quantitate the major end products of *T. vaginalis*.

Product	Method	Selected References
Lactate	Colorimetric	45
	Enzymatic	32
	Paper or thin-layer chromatography	32,57
	Ion-exchange and ion-exclusion chromatography	32,49,51
	Gas chromatography	50
	^{13}C-NMR	50
Glycerol	Enzymatic	49,50
	^{13}C-NMR	50
Acetate	Enzymatic	32,50
	Ion-exchange and ion-exculsion chromatography	32,51
	Thin-layer chromatography	32
	^{13}C-NMR	50
H$_2$	Microrespirometry (gas not absorbed by alkali)	32,45
	Gas chromatography	49,51
	Mass spectroscopy	52,53
CO$_2$	Microrespirometry (gas absorbed by alkali)	32,45
	Gas chromatography	49,51
	Mass spectroscopy	52,53

number of carbon, nitrogen, and sulfur derivatives in trace amounts. The presence of some of these compounds is unexpected and needs either explanation or verification. Media used to grow trichomonad cells are complex; therefore, "spent" media cannot be subjected as easily to equally detailed analysis as the "headspace" of the culture.

Of the nitrogen-containing compounds among the products, alanine is the major amino acid,[32,50] although its quantities are low. Polyamines also appear in the medium during growth of cultures. Putrescine reaches extracellular concentrations that far exceed intracellular ones.[59] No quantitative information is available on the production of ammonia, although it is known to arise from arginine[46] and possibly from other nitrogen-containing compounds.

Release of proteins and other macromolecules by living *T. vaginalis* received little attention. The presence of a number of proteins with antigenic reactivity has been demonstrated in the medium of logarithmic phase cultures. A number of these originate from the cell surface as demonstrated by surface iodination of cells before the cultures are initiated.[60] Antigens released into the medium have also been found to induce specific proliferation of lymphocytes from patients with trichomoniasis but not from uninfected persons.[61]

Metabolism and Enzymes

Anaerobic Carbohydrate Metabolism

It is evident from the data presented elsewhere in this chapter that the main source of energy for *T. vaginalis* is carbohydrate. These data include demonstration of the dependency of the organism on exogenous carbohydrate for multiplication, its high glycogen content, and the proved or implied lack of enzymic mechanisms for the utilization of most amino acids and fatty acids as energy substrates.

Most of the available information concerns the exogenous glucose and endogenous glycogen utilization by the urogenital trichomonad. The entry of other carbohydrates into the energy metabolism has not been studied in any detail. Glucose or glycogen is converted to glycerol, lactate, acetate, malate, and CO_2. Under anaerobic conditions H_2 is also formed. Although endogenous glycogen reserves support a high rate of metabolism, the metabolism is markedly stimulated by the availability of utilizable carbohydrates in the medium.[32,45]

The pathways leading to the formation of the above end products by *T. vaginalis* are shown on a metabolic map (Fig. 5.2), constructed on the basis of evidence from enzymic and cell fractionation studies as well as of studies on the metabolism of living organisms. The processes discussed here take place in two compartments, the cytosol and the hydrogenosomes, separated by the envelopes of the latter organelles. Only certain functional aspects of this compartmentation are discussed here, while other properties and functions of hydrogenosomes are described later in the chapter.

As shown on the metabolic map (Fig. 5.2), glucose phosphorylated by hexokinase (Step 1) or glycogen phosphorylated by glycogen phosphorylase (Steps 2, 3) enters the glycolytic pathway (Steps 4 to 6) and is converted to equimolar amounts of dihidroxyacetone phosphate and glyceraldehyde 3-phosphate. The latter compound is then metabolized to phosphoenolpyruvate and finally pyruvate (Steps 8 to 12). It should be noted that reactions of this pathway correspond to those of classic glycolysis. Furthermore, it is likely that a small proportion of the hexose phosphates is processed by the pentose phosphate shunt (not shown in Fig. 5.2).

Various intermediates of the glycolysis give rise to the end products of *T. vaginalis* metabolism. Glycerol is produced by reduction of dihydroxyacetone phosphate to sn-glycerol 3-phosphate (Step 13), which is subsequently hydrolyzed to glycerol and inorganic phosphate (Step 14). Lactate is the product of pyruvate reduction (Step 15). Acetate, CO_2, and H_2 arise from pyruvate in the following sequential reactions. Pyruvate is oxidatively decarboxylated and the electrons are transferred primarily to protons with the formation of H_2 (Steps 19, 22, 23). In this reaction CO_2 is liberated and the acyl moiety is transferred to coenzyme A (CoA) to form acetyl-CoA. The latter product is converted to acetate and the energy of the thioester bond is conserved in two successive steps resulting in substrate level phosphorylation. In the first step the CoA moiety is transferred to succinate with the formation of succinyl-CoA and the liberation of acetate (Step 23). Succinyl-CoA serves as substrate for a reaction in which ADP is phosphorylated to ATP and succinate

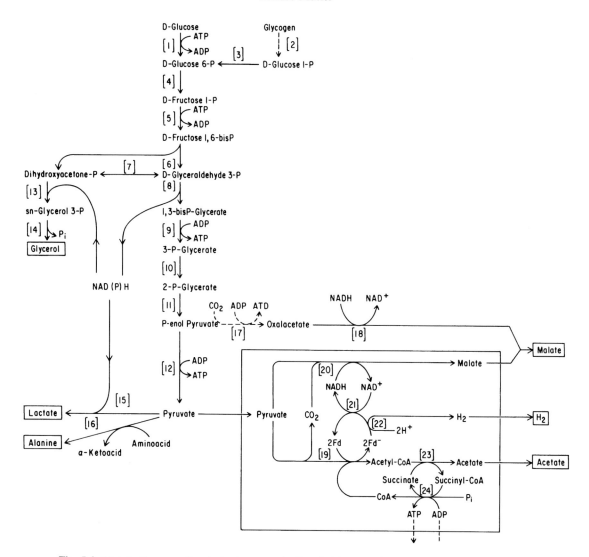

Fig. 5.2. Metabolic map of carbohydrate catabolism in *T. vaginalis*. [1]-Hexokinase, [2]-glycogen phosphorylase, [3]-phosphoglucomutase, [4]-phosphoglucose isomerase, [5]-phosphofructokinase, [6]-aldolase, [7]-triose phosphate isomerase, [8]-glyceraldehyde phosphate dehydrogenase, [9]-phosphoglycerate kinase, [10]-phosphoglyceromutase, [11]-enolase, [12]-pyruvate kinase, [13]-glycerol 3-phosphate dehydrogenase, [14]-glycerol 3-phosphatase, [15]-lactate dehydrogenase, [16]-alanine aminotransferase, [17]-phosphoenolpyruvate carboxykinase, [18]-malate dehydrogenase, [19]-pyruvate:ferredoxin oxidoreductase, [20]-malate dehydrogenase (decarboxylating), [21]-NAD:ferredoxin oxidoreductase, [22]-hydrogenase, [23]-acetate:succinate CoA-transferase, [24]-succinate thiokinase. Arrows indicate the assumed physiologic directions of the reactions. Dashed arrows indicate postulated reactions. The box designates the hydrogenosomal location of reactions [19] to [24], but does not reflect the relationship of the enzymes to the organelle's envelope. (Modified with permission of Elsevier Science Publishers, from Steinbüchel A, Müller M: Glycerol, a metabolic end product of *Trichomonas vaginalis* and *Tritrichomonas foetus*. *Mol Biochem Parasitol* 20:45-55, 1986 and Steinbüchel A, Müller M: Anaerobic pyruvate metabolism of *Tritrichomonas foetus* and *Trichomonas vaginalis* hydrogenosomes. *Mol Biochem Parasitol* 20:57-65, 1986.)

and free CoA are released (Step 24). Malate can arise by reductive carboxylation of pyruvate (Step 20). Reducing equivalents for this process originate from pyruvate oxidation (Step 19). An alternative possibility of malate formation is the carboxylation of phosphoenolpyruvate to oxalacetate followed by reduction to malate (Steps 17, 18).

All the enzymes of the glycolytic pathway leading to the formation of phosphoenolpyruvate and pyruvate have been detected in *T. vaginalis*: hexokinase,[62,63] phosphoglucomutase,[62-64] phosphoglucose isomerase,[63] phosphofructokinase,[62,63] aldolase,[62,63,65] triosephosphate isomerase,[62,63] glyceraldehyde 3-phosphate dehydrogenase,[63] phosphoglycerate kinase,[63] phosphoglyceromutase,[63] enolase,[63] and pyruvate kinase.[63] Despite the high glycogen content of the urogenital trichomonad, glycogen phosphorylase activity has not been detected.[62] However, the activities involved in the formation of glycerol from dihydroxyacetone phosphate, sn-glycerol 3-phosphate dehydrogenase,[49,64] and sn-glycerol 3-phosphate phosphatase,[49] as well as those involved in the formation of lactate from pyruvate via lactate dehydrogenase have been demonstrated.[62-64,66] According to cell fractionation studies, all these enzymes are localized in the nonsedimentable cytoplasmic compartment.[49,63,67] Few of them received more than cursory attention,[65,66] possibly because preliminary information indicated no unusual properties.

Three enzymes of the pentose phosphate shunt, namely, glucose 6-phosphate dehydrogenase, 6-phosphogluconate dehydrogenase, and transaldolase have been detected and localized in the cytosol.[63]

The enzymes possibly involved in malate production, cytosolic phosphoenolpyruvate carboxykinase (unpublished results) and malate dehydrogenase,[29,63] as well as hydrogenosomal malate dehydrogenase (decarboxylating),[28] have also been detected. Of these enzymes, malate dehydrogenase has been characterized in some detail.[68]

The enzymes responsible for the oxidation of pyruvate to acetate and H_2, accompanied by substrate level phosphorylation, have also been demonstrated in *T. vaginalis*. Localized in the hydrogenosomes, these are pyruvate:ferredoxin oxidoreductase,[69] hydrogenase,[69] acetate:succinate CoA transferase (unpublished results), and succinate thiokinase (unpublished results). These enzymes and the pathway catalyzed by them in *T. vaginalis* represent a significant deviation from typical eukaryotic pyruvate metabolism and therefore deserve detailed comments. The nature of the pathway is a unique characteristic distinguishing trichomonads and also certain other anaerobic protozoa, such as *Giardia lamblia*,[70] *Entamoeba* spp.,[71,72] and rumen ciliates[73] from most other eukaryotes. It also suggests a metabolic relationship of these protozoa to primitive anaerobic bacteria.[74] This relationship, suggested for *T. foetus* as early as 1955,[75] received direct experimental support in 1971.[76]

The first step of acetate formation, the oxidative decarboxylation of pyruvate with CoA as acyl-acceptor leading to the production of acetyl-CoA, is catalyzed in most prokaryotic and eukaryotic organisms by pyruvate dehydrogenase, a multienzyme complex that transfers the electrons to NAD. Pyruvate oxidation by this enzyme is an irreversible process. In contrast, oxidative pyruvate decarboxylation in *T. vaginalis* is catalyzed by a different enzyme, pyruvate:ferredoxin oxidoreductase.[69] Pyruvate:ferredoxin oxidoreductases have been purified from several anaerobic and halophilic bacteria[77] and detected in a number of other microorganisms. These enzymes are iron-sulfur proteins, are not multienzyme complexes, and form acetyl-CoA with ferredoxin as electron acceptor or synthesize pyruvate with reduced ferredoxin as electron donor. Ferredoxin can be replaced by flavins or the low redox potential dye, methyl viologen. These enzymes, however, do not transfer electrons to and from pyridine nucleotides. Most of these properties have been demonstrated for the activity responsible for pyruvate oxidation in *T. vaginalis*[69] and other trichomonads.[78,79]

The natural electron acceptor for pyruvate:ferredoxin oxidoreductases is ferredoxin, which when purified from *T. vaginalis*,[23] was found to be a 12-kDa protein containing a [2Fe-2S] cluster with axial symmetry, as detected by EPR spectroscopy. The optical absorbance spectrum differentiates *T. vaginalis* ferredoxin from other known ferrodoxins. The trichomonad protein can accept electrons from pyruvate:ferredoxin oxidoreductase and hydrogenase and also from ferredoxin-linked enzymes of other organisms. *Trichomonas vaginalis* ferredoxin represents the key electron transport component of the hydrogenosomal compartment, but plays no role in cytosolic reactions.

Ferredoxin reduced by pyruvate:ferredoxin oxidoreductase is reoxidized with protons as terminal electron acceptors through the action of a ferredoxin-linked hydrogenase[69] which produces molecular hydrogen. This activity has an electron transfer specificity similar to that of pyruvate:ferredoxin oxidoreductase, as tested in the direction of oxidation of H_2, that is, in the reversed direction from its in vivo function. Ferredoxins of different organisms, methyl viologen, and flavins, but not pyridine nucleotides, were found to be reduced with H_2 as electron donor. Hydrogenases of other organisms are iron-sulfur proteins[80] and it is likely that the trichomonad enzymes will be no exception.

Among the most striking characteristics of the aforementioned hydrogenosomal pathway are the participation of several iron-sulfur proteins in it and the low midpoint oxidation-reduction potential of its ferredoxin-linked reactions.

An NAD:ferredoxin oxidoreductase activity, localized in the hydrogenosomes,[81] probably links pyruvate oxidation and reductive carboxylation of pyruvate to malate by hydrogenosomal malate dehydrogenase (decarboxylating).

Acetyl-CoA is converted to acetate by acetate:succinate CoA transferase, which transfers the CoA moiety to succinate with the formation of succinyl-CoA (reported from *T. foetus*[81,82]). The latter serves as substrate for succinate thiokinase catalyzing the substrate level phosphorylation of ADP to ATP with the release of free CoA and succinate (unpublished results). Through this step the energy gained in the oxidation of pyruvate is conserved with an additional one mole of ATP formed for each mole of pyruvate that is not reduced to lactate but rather oxidized further.

As discussed above, the proportions of the end products vary considerably, depending on the strain and the experimental conditions used. Two general conclusions can be drawn, however. First, the production of lactate seems not to be correlated with that of other end products and, second, glycerol, acetate, and H_2 are produced in nearly equimolar amounts, suggesting an interdependence of the processes leading to their formation. The redox relationships of the above processes aid materially in explaining the ability of *T. vaginalis* to exhibit the variations noted in the proportions of its end products.

Production of lactate is in redox balance. NAD reduced by glyceraldehyde 3-phosphate is reoxidized by pyruvate, and thus lactate formation can be independent of the production of other end products. In contrast, acetate production is obligately coupled to the formation of other end products. This process presents the need for the removal of two pairs of electrons, since the further oxidation of pyruvate makes it unavailable as electron acceptor. In view of the nearly equimolar production of glycerol, acetate, and H_2 (see Tables 5.2, 5.3),[49,50] and considering the distribution of the corresponding enzymes between the cytosol and the hydrogenosomes, it may be assumed that reducing equivalents produced in the cytosol by the oxidation of glyceraldehyde 3-phosphate are quantitatively utilized in that compartment for the formation of glycerol. On the other hand, those originating from the oxidation of pyruvate are removed as H_2 in the hydrogenosomes. The foregoing considerations indicate that hydrogenosomal pyruvate metabolism in *T. vaginalis* is supported predominantly by cytosolic glycerol formation, without the participation of other mechanisms for the removal of reducing equivalents. In other words, they indicate that redox balance is maintained independently in the two major compartments of the cell, the cytosol and the hydrogenosomes and that no electron transport occurs across the hydrogenosomal envelope. However, this assumption needs further experimental testing.

The mechanisms underlying the differences in the ratio of lactate to the other end products remain to be elucidated. A factor in determining this ratio could be the rate of hydrogenosomal pyruvate metabolism. *T. vaginalis* grown in medium with high iron content and containing high levels of hydrogenosomal enzymes produces less lactate than acetate,[10] whereas organisms rendered deficient in hydrogenosomal functions produce almost exclusively lactate.[54]

Although the enzymes potentially involved in the carboxylation of C_3 compounds are present, reactions mediated by such enzymes apparently play a subordinate role in *Trichomonas vaginalis,* as indicated by the low level of malate production[32,49,50] and the absence of excretion of additional C_4 compounds. This is in contrast to *Tritrichomonas foetus,* in which succinate arising by carboxylation of C_3 compounds is a major end product.[75] The insignificant effect of pCO_2 on multiplication[83] and end product formation by *T. vaginalis*[32,49] supports the above conclusion. An early report on CO_2 fixation by *T. vaginalis*[84] detected the label of the fixed $^{14}CO_2$ in the carboxyl group of lactate,[84] an observation incompatible with CO_2 fixation by phosphoenolpyruvate carboxykinase or malate dehydrogenase (decarboxylating). Fixation via a CO_2 exchange reaction catalyzed by pyruvate:ferredoxin oxidoreductase[78] is more likely than the proposed carboxylation of ribulose 1,5-bisphosphate.[84] Regrettably no quantitative data are available on the magnitude of this process, which could well be negligible.

Aerobic Carbohydrate Metabolism

Trichomonas vaginalis is O_2-tolerant and survives under variable pO_2 conditions. Unfortunately no systematic survival studies with controlled O_2 levels are available.

Growth of in vitro cultures is greatest under anaerobic conditions.[83,85] Oxygen elicits a concentration-dependent inhibition if the culture medium is kept equilibrated with the gas phase. No growth is observed in cultures equilibrated with air.[83] Routinely employed media contain reducing compounds, thus being fairly anaerobic at inoculation. Cells also respire intensely in the cultures, removing residual O_2 and definitely rendering rich cultures anaerobic. These considerations account for the apparent contradictions between the everyday experience that cultures can be initiated and maintained in the absence of rigorous anaerobiosis and the finding that air-equilibrated medium does not support growth.

Respiration, that is, reduction of O_2 by living *T. vaginalis*, has been studied extensively.[32,39,51,86,87] The endogenous respiration rate of cells suspended in buffered saline in the absence of nutrients falls in the range of 10 to 50 nmol min^{-1} (mg protein)$^{-1}$ at 37°C.[39,51,87] Despite the anaerobic nature of *T. vaginalis*, the observed rate is of the order typically expected of an aerobic protozoon with the same protein content or size (estimated from Fig. 1 in reference no. 88). The oxygen affinity of the respiratory system of intact *T. vaginalis* is high, with a K_m for O_2 of about 3 to 5 μM.[89] This affinity is similar to that of cytochrome-mediated respiration and much higher than that observed for flavoprotein direct oxidases.

The rate of O_2 uptake increases by 50% to 150% in the presence of certain carbohydrates, for example, glucose, fructose, and maltose.[39,51] Malate also stimulates respiration. Compounds that do not include a number of other carbohydrates, and more significantly most intermediates of the tricarboxylic acid cycle and other organic acids as well as most amino acids tested.[87,90]

Respiration of living cells is unaffected by most metabolic inhibitors. Most important, practically none of the inhibitors of mitochondrial respiration, including cyanide, have any effect in concentrations corresponding to those used in studies on mitochondria. This indicates that the terminal oxidases involved in the reduction of O_2 are different from those of the mitochondria.[45,86,87,90]

Trichomonas vaginalis produces the same metabolic end products in the presence of O_2 as in its absence, except that H_2 production becomes undetectable at relatively low pO_2 values (less than 250 nM O_2).[52] The total rate of acidic end product formation and the proportion of lactate and acetate are not significantly different under aerobic and anaerobic conditions (see Table 5.3),[51] although some results indicate that aerobiosis might stimulate glycolysis and lactate formation.[32] Glycerol formation is also largely unaffected by aerobiosis.[49]

Enzymes or enzyme systems capable of transferring electrons to O_2 are present in the non-sedimentable part of the cytoplasm and in the hydrogenosomes. However, the relative contribution of these systems to the overall respiration of *T. vaginalis* remains to be elucidated.

The cytosol of *T. vaginalis* contains oxidases that use NADH and NADPH as substrates; these enzymes have been purified.[91,92] The NADH oxidase has a higher activity than the NADPH oxidase,[51] and both have high affinity for their substrates (K_m for O_2 is less than 10 μM), and have been characterized as flavoproteins, possibly containing Fe or another transition metal. The NADH oxidase has a tightly bound flavin and reduces O_2 to H_2O, an unusual property for a flavoprotein direct oxidase.[91] The activity of the NADPH oxidase is markedly stimulated by added FMN. This enzyme produces H_2O_2 in vitro, but its in vivo end product remains to be ascertained.[93] The physiologic role of the two oxidases is not firmly established, but they may be involved in lowering the intracellular pO_2, thus providing a mechanism for oxygen tolerance of *T. vaginalis*.[93]

The nature of the hydrogenosomal terminal oxidase is not known. It has a high affinity for O_2 ($K_m < 10$ μM),[89] probably produces H_2O_2, and is not sensitive to inhibitors of mitochondrial respiration (for data on *T. foetus*, see references 94 and 95). The mechanism of transfer of electrons from pyruvate:ferredoxin oxidoreductase to the hydrogenosomal terminal oxidase is not known.

The defensive mechanisms of *T. vaginalis* against the toxic products of O_2 reduction (e.g., $HOH^{\cdot}$, $O_2^{\cdot-}$, H_2O_2) are not known. Superoxide dismutase, a ubiquitous enzyme, described from *T. foetus* and *Monocercomonas* sp.,[96] has also been found in *T. vaginalis* (unpublished results). This enzyme has been examined in some detail in *T. foetus*.[96,97] Since cyanide does not inhibit this enzyme but H_2O_2 does, it is likely to contain iron.[97] Most of the superoxide dismutase activity is in the cytosol, but about 15% of the total is localized in the hydrogenosomes.[96] In contrast to *T. foetus*,[75,98] catalase cannot be detected in *T. vaginalis*.[75,86] There is no information on the presence in the latter species of other systems known to participate in the protection of other organisms against toxic O_2 derivatives, for example, the glutathione reductase–glutathione peroxidase system.

Amino Acid and Polyamine Metabolism

Although amino acids obviously play a major role in *T. vaginalis*, there is little information available on this subject. The essential or unessential nature of individual amino acids is not known, but in view of the probably limited capacity of *T. vaginalis* for amino acid biosynthesis and conversion it is likely that a number of them will be found essential.

Free amino acids in the trichomonad cell have been quantitated,[99] with alanine found to be dominant. This is also the only amino acid released in detectable quantities as a metabolic end product.[50]

Of the enzymes of amino acid metabolism, aminotransferases are present.[100-102] Pyruvate, α-ketoglutarate, oxalacetate, and phenylpyruvate can accept the amino group from a number of amino acids; however, several amino acids tested did not serve as donors.[102] Separation and partial purification of the activities revealed the presence of at least four separate enzymes.[102] A single enzyme of unusual specificity catalyzes the transamination of ornithine and lysine; a second enzyme functions as a branched chain amino acid transferase; the third is an unstable alanine aminotransferase with lower specific activity than the other transferases; and the fourth, aspartate:α-ketoglutarate aminotransferase,[101] has been purified to homogeneity. In addition to aspartate, the last enzyme can utilize aromatic amino acids (phenylalanine, tyrosine, and tryptophan) as amino donors or phenylpyruvate as acceptor. Such specificity is unusual among eukaryotic aspartate aminotransferases. The significance of the aminotransferases for *T. vaginalis* remains unknown; for example, complete inhibition of the aspartate aminotransferase with gostatin does not interfere with the growth of the organism in a complex medium.[99]

The degradation of only a few amino acids has been studied. Respiration of *T. vaginalis* is not stimulated by amino acids with the exception of arginine,[90] which is metabolized to ornithine, ammonia, and carbon dioxide accompanied by substrate level phosphorylation by the sequential action of the enzymes of the arginine dihydrolase pathway:[46] arginine deiminase, ornithine carbamyltransferase, and carbamate kinase. While the activities of these enzymes have been demonstrated in *T. vaginalis*, those of arginase, urease, citrulline hydrolase,[46] and arginine decarboxylase[103] have not.

The polyamines putrescine, spermidine, and spermine are present, in quantities of about one regulatory functions in the living cell. Inhibition of polyamine biosynthesis in certain other parasitic protozoa suggested that this process can serve as a chemotherapeutic target. In *T. vaginalis* cells, putrescine is the predominant polyamine [100 to 200 nmol (mg protein)$^{-1}$], released into the medium in significant quantities.[103] In addition, both spermidine and spermine are present, in quantities of about one order of magnitude lower.[103,104] These amounts and proportions set *T. vaginalis* apart from other eukaryotes (i.e., protozoa and mammals) and make it resemble *Escherichia coli*. The evolutionary implications of this finding remain obscure.

Ornithine decarboxylase, a key enzyme of polyamine biosynthesis that produces putrescine, and S-adenosyl-L-methionine decarboxylase, which provides aminopropyl groups for the conversion of putrescine to spermidine and spermine, are present in *T. vaginalis*.[59,103,104] Ornithine decarboxylase is inhibited by α-difluoromethyl ornithine,[103] a compound with chemotherapeutic potential for another protozoan parasite, *Trypanosoma brucei*. Preliminary data showed no inhibition by this compound of in vitro growth of *T. vaginalis*, possibly because of relatively high levels of polyamines in the medium.[104]

Role and Metabolism of Sulfur-Containing Compounds

Sulfur-containing compounds are of major significance for living organisms. In view of the functional significance of iron-sulfur proteins in trichomonad flagellates, a detailed understanding of their sulfur metabolism would be of great interest. Unfortunately, our knowledge of this topic is very limited.

Sulfur-containing amino acids are present in *T. vaginalis*.[99] Cysteine is a component of ferredoxin[23] and of many other proteins. In the absence of folates, methionine might be the only compound in this organism that can serve as methyl donor.[43,105]

Although the pathways of sulfur metabolism have not been clarified as yet, some enzymes of *T. vaginalis* acting on sulfur-containing compounds have been characterized recently.[105] These include homocysteine desulfurase,[106] S-adenosylhomocysteine hydrolase,[107] and serine sulfhydrase.[108]

Free thiol groups exposed on the outer surface of the cell might be of special importance for *T. vaginalis*. Cells treated with *p*-chloromercuribenzene sulfonate, a thiol blocking agent impermeable to membranes, rapidly lose their viability.[109] This effect can be reversed by adding a compound with free thiols to the medium. *Trichomonas vaginalis* shares this high susceptibility with *Entamoeba histolytica* and *Giardia lamblia*, other anaerobic protozoa. The aerobic protozoa *Crithidia fasciculata* and *Paramecium tetraurelia* are much less susceptible to this compound. Free thiol groups have been detected on the surface of *G. lamblia*[34] and their existence, while not tested by similar methods, is likely to be found on the surface of *T. vaginalis*. It remains to be seen whether one of the functions of cysteine and other reducing components present in *T. vaginalis* growth media is the stabilization of the reactive thiol groups on the cell surface.

Thiol compounds, especially glutathione, together with enzymes able to reduce it, play a critical role in the defense of cells against free radicals. Recent studies indicate the existence of a marked heterogeneity of thiol compounds in various organisms.[110] *Entamoeba histolytica* does not contain glutathione.[111] As mentioned above, nothing is known of the presence of glutathione or related compounds in *T. vaginalis*. Trypanothione, a glutathione derivative probably restricted to kinetoplastid flagellates, has not been detected in *T. vaginalis*.[112]

Lipid Metabolism

Trichomonas vaginalis requires fatty acids and cholesterol as essential nutrients. All evidence indicates that these components are incorporated unchanged into neutral lipids and phospholipids and other derivatives.[35] Fatty acids do not stimulate respiration[90] and experiments with radiolabeled fatty acids showed no evidence of any modification of these compounds, for example, α-, β-, or ω-oxidation, chain elongation, and desaturation or saturation of internal double bonds. Radiolabeled cholesterol likewise remained unaltered.[35] De novo biosynthesis of fatty acids or cholesterol does not occur in *T. vaginalis*.[8,35]

Hydrolases

A number of hydrolytic enzymes acting on a multitude of substrates are found in *T. vaginalis*. The information available is spotty and provides little insight into the functional significance of these enzymes. It is likely that a subgroup of the hydrolases is lysosomal, being involved in intracellular digestion. Certain hydrolases might also be released by the cell and participate in pathogenesis.

Proteinases received the most detailed scrutiny. By electrophoretic separation of homogenates in hemoglobin-containing gels at least seven cysteine proteases (A to H), active at pH 4, have been detected.[113] Proteinases D and H have been partially purified and characterized.[114] Both are active on proteins (hide powder azure and azocasein) and on a number of synthetic substrates (peptide nitraminides containing arginine) with optimum activity in the pH range of 5 to 7. Proteinase D has a mass of 18 kDa and proteinase H of 64 kDa. Activation by dithiothreitol and action of a number of inhibitors confirm that both are cysteine proteases. The available data do not indicate the presence of noncysteine (i.e., thiol-independent) proteinases in *T. vaginalis*.

Glycosidases and other hydrolases of *T. vaginalis* involved in carbohydrate breakdown have not been studied to any extent. β-N-Acetylglucosaminidase with a pH optimum of 6 is present and is used in cell fractionation studies as a marker for hydrolase-containing large particles.[69] The presence of additional glycosidases is expected, in view of the large number of such enzymes detected in *T. foetus*, some of which have been characterized.[98,115-118]

RNAses and DNAses are also present in *T. vaginalis* but did not receive any attention, except for being held responsible for the difficulties they cause in nucleic acid purification.

Nonspecific phosphatases of *T. vaginalis*, partly localized in membrane-bounded organelles, have pH optima of around 5.[119] It has been shown that these enzymes liberate phosphate from sn-glycerol 2-phosphate, sn-glycerol 3-phosphate, glucose 1-phosphate, glucose 6-phosphate, and fructose 1,6-bisphosphate. No attempts were made to purify the enzymes; therefore, the number of enzymes responsible for the observed activities cannot be ascertained. An acid phosphatase activity hydrolyzing *p*-nitrophenyl phosphate, primarily localized in the small particle fraction, is used as a marker enzyme in cell fractionation studies.[69] A Mg^{2+} activated sn-glycerol 3-phosphatase, participating in glycerol production, has been mentioned previously in this chapter. No such activity was detected earlier,[119] a discrepancy that needs further studies. Measurement of acid phosphatase activity of cultures in microwell plates with chromogenic substrates is suggested as a rapid method to monitor growth in drug susceptibility assays.[120]

Trichomonas vaginalis produces β-hemolysis when grown on agar plates and the intensity of this activity has been shown to exhibit positive correlation with the pathogenicity for seven isolates tested.[121] The biochemical nature of the hemolytic factor and its potential role in pathogenesis remains to be established.

Cell Organelles

Cell Surface

Trichomonas vaginalis is surrounded by a single cell membrane (see Chap. 3, Figs. 3.4a, 3.5c), which is expected to be similar in its biochemical composition and function to membranes of other cells living an independent existence. Although it plays a major role in the interaction of the parasite with the host, and with its environment in general, information on it began accumulating only recently. To date, the various transport mechanisms functioning in this membrane received very little biochemical attention.

Subcellular fractions significantly enriched in the surface membrane of trichomonad cells have not been isolated yet. The protein composition of this membrane has been studied by electrophoretic separation of polypeptides that can be labeled with ^{125}I on the surface of intact organisms.[14] Two-dimensional electrophoresis reveals at least 30 such proteins.[15] A significant number of polypeptides recognized by immune sera from infected humans or experimental animals have been shown to be localized on the cell surface by immunofluorescence of intact organisms[14,20,122] or by immunoprecipitation of proteins of surface-labeled cells.[14,20] Monoclonal antibodies recognizing various surface proteins of *T. vaginalis*,[122] including a 115-kDa[18] and a 267-kDa[19] polypeptide, have been developed recently. *Trichomonas vaginalis* isolates reveal heterogeneity in regard to the presence or absence of certain surface proteins, especially of higher molecular weight.[15,19,20,122] Cells of certain isolates show an intraisolate phenotypic variation in the surface expression of the 267-kDa protein.[123] Studies with fluorescein-labeled lectins revealed the existence of a significant number of glycosyl groups on the surface,[16,17] and several surface polypeptides have been shown to be glycosylated.[18,19]

Cultured *T. vaginalis* binds various host macromolecules on its surface more or less tightly. Blood plasma components not easily removed from living organisms include plasminogen, fibrinogen, immunoglobulin G, lipoproteins A and B, transferrin, α_1-antitrypsin, and albumin.[124] In addition, binding of lactoferrin has been demonstrated.[44] Kinetic properties of the binding and its inhibition by competing molecules revealed the existence of receptors for at least α_1-chymotrypsin, lactoferrin, and lipoproteins.[44,124,125] The lipoprotein receptor is specific for apolipoprotein CIII and corresponds to a surface protein larger than 250 kDa. No data are available on the properties of the other receptors.

[125]I-labeled human low-density lipoprotein added to living cells is bound rapidly to its receptor at 4°C and 37°C and is subsequently internalized and degraded intracellularly at 37°C.[124] Lipoprotein uptake by this process is likely to provide necessary lipids for the growth of T. vaginalis. Human low-density lipoprotein as a supplement to serum-free medium can replace serum without decreased growth.[124] Since all routine media supporting good growth of T. vaginalis contain serum, it is likely that receptor-mediated uptake of lipoproteins is a major mechanism of satisfying the lipid requirements of this organism.

Receptors for iron-transport proteins also received some scrutiny.[44] Lactoferrin binds to the surface of T. vaginalis showing saturation kinetics, but is not internalized. Iron-free lactoferrin and lactoferrin saturated with iron bind equally. Two trichomonad proteins (178 kDa and 75 kDa) bind specifically to a lactoferrin-Sepharose affinity column and could correspond to the lactoferrin receptors. By comparing the fate of the apolactoferrin and ^{59}Fe bound to it, it was demonstrated that Fe^{3+} is removed by the cell from its bound form and is incorporated into intracellular iron-containing compounds. Exposure of cells to Fe^{3+}-saturated lactoferrin in nutrient medium results in a marked increase of pyruvate:ferredoxin oxidoreductase activity,[44] indicating increased production of iron-sulfur proteins by the cell. These results indicate that T. vaginalis is able to satisfy its iron requirement with the help of iron-transport proteins present in host secretions. Uptake of Fe^{2+} from the nutrient medium apparently does not require an exogenous protein.[10] Transferrin, the iron-transport protein of serum, also binds to T. vaginalis. This binding is not saturable, however, and transferrin cannot replace lactoferrin as an iron carrier.

The binding of certain host macromolecules to the surface has been suggested to be a component of adaptive mechanisms enabling T. vaginalis to survive in the host. Although this might well be the case for low-density lipoprotein and lactoferrin, for other components these considerations are not based on direct evidence.

Adherence of T. vaginalis to host cells is a major factor in host–parasite relationships,[126] possibly providing a surface as support in cell division and being involved in cytotoxicity.[127,128] Adherence depends on normal metabolism of T. vaginalis as shown by the inability of iodoacetate- or metronidazole-treated cells to adhere.[129] The responsible ligands on the cell surface are probably proteins, since they are sensitive to trypsin but not to treatment with periodate or neuraminidase. The regeneration of the ligands removed by trypsin is dependent on active protein synthesis and can be inhibited by cycloheximide. There is no experimentally documented biochemical information available on the mechanism of contact-dependent cytotoxicity of T. vaginalis.[127] Some aspects of this topic are discussed in Chapter 9.

Hydrogenosomes

Trichomonas vaginalis lacks morphologically recognizable mitochondria but contains hydrogenosomes (see Chap. 3, Figs. 3.2m,n, 3.4c,d).[74,130] These characteristic organelles are predominantly spherical or somewhat elongated structures, about 0.5 µm in diameter,[131] and occupy about 6% of the cell volume.[131] They are bounded by an envelope consisting of two closely apposed unit membranes,[132,133] and contain a granular matrix and often an electron-dense core. A flattened, membrane-bounded sac located between the two membranes might represent a special compartment within the organelle.

Hydrogenosomes had been recognized by light microscopists for a long time, as paraxostylar and paracostal granules, but only biochemical studies revealed their functional significance. They were named hydrogenosomes because they produce molecular hydrogen as a metabolic end product.[78] Originally this function was attributed to them on the basis of the presence of enzymes implicated in hydrogen production,[69,78] but recently it has been demonstrated directly.[81] According to present knowledge, certain characteristic steps of carbohydrate metabolism linked to substrate level phosphorylation occur in hydrogenosomes. These represent an integral part of the energy metabolism of *T. vaginalis* (see Fig. 5.2) and have been discussed in some detail earlier in the chapter. No additional functional role has been attributed to hydrogenosomes thus far.

Biochemical information on trichomonad hydrogenosomes has been obtained primarily from studies of *T. foetus*, but data available on *T. vaginalis* hydrogenosomes indicate that the main properties of these organelles in the two species are, at least qualitatively, identical. This fact needs to be contrasted to the significant differences in the cytosolic steps of carbohydrate metabolism of these two species.[49] The properties and functions of trichomonad hydrogenosomes are summarized in Table 5.5 and are described in more detail below.

Current views on the composition and functional capacities of trichomonad hydrogenosomes come primarily from studies on organelle-enriched fractions obtained by subcellular fractionation of *T. vaginalis* and other trichomonads homogenized in isotonic sucrose. Large-particle fractions separated by differential centrifugation are three- to five-fold enriched in hydrogenosomes.[49,69] Subsequent isopyknic centrifugation of such fractions in sucrose gradients yields a hydrogenosomal fraction with a final enrichment 10- to 15-fold relative to the initial homogenate,[69] close to the theoretic maximum corresponding to the 6% contribution of these organelles to the cell volume of *T. vaginalis*.[131] The equilibrium density of the hydrogenosomes in a sucrose gradient is about 1.25 g ml^{-1}. It can be shown by electron microscopy that hydrogenosomes are the dominant constituents of such fractions.

The composition of trichomonad hydrogenosomes is not well-known. They are characterized by the presence of a set of enzymes and electron transport components[22,69] participating in the metabolic processes leading from pyruvate to acetate, malate, CO_2, and H_2 as well as in the phosphorylation of ADP to ATP. Several of these constituents have been shown or are likely to be iron-sulfur proteins. According to EPR spectroscopic data iron-sulfur clusters are localized predominantly in the hydrogenosomes.[22] A key characteristic of the main hydrogenosomal oxidation-reduction processes is a low midpoint potential, indicating that the intraorganellar milieu is highly reducing. In addition, adenylate kinase[134] and a pyridine nucleotide-independent glycerol 3-phosphate dehydrogenase[49,98] have been demonstrated. The presence of an NADH oxidase activity and an enigmatic terminal oxidase was shown in *T. foetus* hydrogenosomes,[94] which contain also a part of the superoxide dismutase activity.[96] The presence of phospholipids, but not of cardiolipin,[36] has been demonstrated, contradicting an earlier report.[135] Ultrastructural-histochemical studies indicate the absence of cholesterol[136] from hydrogenosomal membranes of *T. foetus*. Glycogen is absent from the matrix of the organelle.[137] X-ray microanalysis disclosed high levels of Mg, Ca, and P_i in the intrahydrogeno-

TABLE 5.5. Biochemical properties (compounds and enzymes) present in and absent from trichomonad hydrogenosomes.

Present	Absent
Pyruvate as substrate (M)*	
Pyruvate:ferredoxin oxidoreductase (B)†	Pyruvate dehydrogenase complex (M)
Ferredoxin (M)	Cytochromes (M)
Hydrogenase (B)	
Terminal oxidase of unknown nature (cyanide-insensitive)	Cytochrome oxidase (M)
	Peroxisomal direct oxidases (P)‡
	Catalase (P)
Acetate:succinate CoA transferase	Acetyl phosphate and enzymes of its metabolism (B)
Succinate thiokinase (M)	Other enzymes of the tricarboxylic acid cycle (M)
	β-Oxidation of fatty acids (M, P)
Substrate level phosphorylation (M)	Oxidative phosphorylation (M)
Adenylate kinase (M)	ATPase (M)

*Characteristic of mitochondria.
†Characteristic of bacteria—listed only if not present also in mitochondria.
‡Characteristic of peroxisomes.

somal flattened sacs in *T. vaginalis*[138] confirming their demonstration by histochemical methods in *T. foetus*.[139] Hydrogenosomes most probably contain no DNA.[11,31]

Hydrogenosome-enriched fractions of *T. vaginalis* incubated in isotonic media under anaerobic conditions utilize pyruvate, converting it to acetate, malate, CO_2, and H_2 (Table 5.6).[81] The major product is acetate; malate is produced at a much lower rate. In the presence of 5% CO_2 the overall hydrogenosomal metabolism increases somewhat with a concomitant small increase of malate formation and decrease of H_2 production. Utilization of other substrates by *T. vaginalis* hydrogenosomes has not been reported. Data on the aerobic metabolism of *T. foetus* organelles[94] indicate that a few additional compounds also might serve as anaerobic substrates. Hydrogenosomes of this latter species exhibit a similar anaerobic metabolism, but exogenous CO_2 causes a shift to malate formation.[81] The rate of acetate production by isolated *T. vaginalis* hydrogenosomes [approximately 50 to 60 nmol min^{-1} (mg protein)$^{-1}$, corrected for the enrichment of the fraction] exceeds the rates observed in living *T. vaginalis* (see Table 5.2).

Under aerobic conditions, hydrogenosomes can use molecular oxygen as a terminal electron acceptor; thus they are respiratory organelles. No published data are available on *T. vaginalis* hydrogenosomes, but the aerobic metabolism of these organelles in *T. foetus* has been analyzed in some detail.[94,95,140] Respiration of these organelles is supported by only a few substrates: NADH, sn-glycerol 3-phosphate, pyruvate, and malate plus NAD. Intermediates of the tricarboxylic acid cycle (citrate, isocitrate, α-ketoglutarate, succinate, malate in the absence of NAD, and oxalacetate) and other organic acids (glyoxylate, glycolate, acetate, formate, and oxalate) are without effect.[94] The nature of the hydrogenosomal terminal oxidase remains to be elucidated. Respiration by *T. foetus* hydrogenosomes is not inhibited by cyanide or rotenone,[94] standard inhibitors of mitochondrial respiration.

The two unit membranes surrounding the hydrogenosome represent a permeability barrier, as indicated by the structure-bound latency of several enzymes.[78,81,94,134] It is likely that a number of transport systems are involved in the metabolism of the organelle in situ, but these have not been studied to date. A putative nucleotide transport system has very low sensitivity to atractyloside.[94]

Hydrogenosomes function in intact *T. vaginalis* under both anaerobic and aerobic conditions, as reflected by the production of acetate and CO_2.[32,51] In fact, the carbon flow through these organelles is not significantly affected by the absence or presence of O_2, even though the fate of the electrons liberated through pyruvate oxidation is quite different. In the first case protons and in the second, O_2 serve as the terminal acceptor.

The significance of hydrogenosomes for trichomonads is not entirely clear. It has been suggested that the extension of glycolysis by one oxidative step, resulting in the formation of one mole of ATP for each mole of acetate produced, increases the energy yield of carbohydrate catabolism.[74] Acetate production in *T. vaginalis* is accompanied, however, by glycerol production in nearly equimolar amounts. This process removes a triosephosphate from the pathway without the possibility of regaining the energy invested in the phosphorylation of hexoses; therefore, the potential energy gain in the hydrogenosome is compensated for by a loss in the glycolysis itself. Some key aspect of this question appears to be still unknown. These

TABLE 5.6. Anaerobic pyruvate metabolism of undisrupted hydrogenosomes of *T. vaginalis*.*

Gas phase	Utilized	Formed products			
	Pyruvate	Acetate	Malate	H_2	CO_2
	nmol min^{-1} (mg protein)$^{-1}$				
N_2	295	245	38	195	165
95% N_2 + 5% CO_2	315	255	46	170	—†

Recalculated from Steinbüchel A, Müller M: Anaerobic pyruvate metabolism of *Tritrichomonas foetus* and *Trichomonas vaginalis* hydrogenosomes. *Mol Biochem Parasitol* 20:57-65, 1986.
*Large-particle fractin enriched fourfold in hydrogenosomes, incubated at 30 °C in isotonic medium containing pyruvate, ADP, P_i, Mg^{2+}, and catalytic amounts of succinate.
†Production cannot be determined in the presence of exogenous CO_2.

considerations might not apply to *T. foetus* hydrogenosomes, because of the differences in the cytosolic metabolism of the two species.[49]

Trichomonas vaginalis[54] and *T. foetus*[141] strains with major deficiencies in hydrogenosomal metabolism have been developed by prolonged exposure (more than a year) to increasing concentrations of the antitrichomonal drug, metronidazole. These organisms contain hydrogenosomes, albeit with altered morphology, but lack pyruvate:ferredoxin oxidoreductase and hydrogenase and do not produce acetate or hydrogen as metabolic end products.[54,142] Such *T. vaginalis* strains are practically homolactic fermentors, whereas their *T. foetus* counterparts produce ethanol, both by entirely cytosolic pathways. The situation observed in these two species demonstrates that hydrogenosomal pyruvate oxidation is not indispensable for trichomonads. However, flagellates with major hydrogenosomal deficiencies grow more slowly than their parent strains. Evidently functional hydrogenosomes confer benefits on the organisms.

Biologic affinities and the evolutionary origin of hydrogenosomes are interesting but unresolved problems. These organelles are characterized primarily by their function, hydrogen production from pyruvate, and by the participation of several documented or putative iron-sulfur proteins in this process: pyruvate:ferredoxin oxidoreductase, hydrogenase, and ferredoxin (see Table 5.5). Such a combination of properties is found elsewhere among living organisms only in certain anaerobic bacteria. It is of interest that the most distinctive enzyme of hydrogenosomes, pyruvate:ferredoxin oxidoreductase can be found also in certain anaerobic protozoa that do not contain hydrogenosomes, namely, *Entamoeba* spp.[71,72] and *G. lamblia*.[70] These organisms, however, do not produce hydrogen, and their pyruvate:ferredoxin oxidoreductase is not localized in specialized organelles but in the cytosol. Thus the ability of a eukaryotic organism to produce hydrogen might be linked in an as yet unknown way to the existence of a membrane-limited organelle.

The functional and biochemical similarities of hydrogenosomes to certain groups of anaerobic bacteria suggest their origin from such bacteria.[74] This hypothesis has many attractive features, and is supported by a number of data. Certain components of hydrogenosomes and anerobic bacteria differ markedly, however.

Two of these deserve special mention, First, trichomonads contain [2Fe-2S] ferredoxins, which participate in H_2 production. This process in bacteria is usually linked to 2[4Fe-4S] ferrodoxins.[143-145] Ferrodoxins are found in mitochondria[144] and also in bacteria,[145,146] but those of bacteria do not participate in H_2 production. Second, hydrogenosomal conservation of the energy of the thioester bond of acetyl-CoA in trichomonads differs from the process found in bacteria, in which acetyl phosphate is an intermediate product. Hydrogenosomes are likely to remain a puzzle until detailed studies of their molecular composition shed more light on their affinities and origin.

Of all eukaryotic cells studied, hydrogenosomes were detected only in a few groups of anaerobic protozoa and fungi. These organelles, first described in 1973 in *T. foetus*,[78] were soon demonstrated also in other trichomonad species, *T. vaginalis*[69] and *Monocercomonas* sp.[96] The presence of morphologically similar structures in all members of the orders Trichomonadida and Hypermastigida, and limited metabolic data on additional trichomonad species strongly suggest that all these organisms contain hydrogenosomes. In fact, the presence of hydrogenosomes has been included among the taxonomic characters of Trichomonadida.[147] These organelles are also present in a number of anaerobic rumen ciliates[73,148-150] and very likely in certain ciliates[151] living in the sapropel, the anaerobic sediment of fresh and marine bodies of water. Recently, hydrogenosomes were detected in *Neocallimastix patriciarum*,[152] an anaerobic fungus of sheep rumen. The occurrence of an organelle with similar morphology and metabolic function in several protozoan and fungal groups all of which live in anaerobic habitats, although some are parasitic, some mutualistic, and some free-living, is of great interest. Its systematic distribution presents, however, a difficult question since the groups listed definitely belong to separate evolutionary lines. The extent of homologies of hydrogenosomes in these groups remains to be established. The types of enzymes involved in pyruvate oxidation and hydrogen formation seem to be the same. Some biochemical data indicate, however, that they might differ in certain other aspects; for example, energy conservation by hydrogenosomes of the rumen ciliate, *Dasytricha ruminantium* was reported to occur by a prokaryotic mechanism involving acetyl-phosphate,[73] a process not detected in trichomonad hydrogenosomes.[82]

Ribosomes and Endoplasmic Reticulum

Ultrastructural study shows that *T. vaginalis* contains ribosomes, both free and associated with membranes of the endoplasmic reticulum (see Chap. 3).

Golgi Apparatus

Like all members of Trichomonadida, *T. vaginalis* contains a Golgi apparatus (see Chap. 3, Figs. 3.2l, 3.3) (often referred to as parabasal apparatus), prominent in light microscopic preparations (see Chap. 3, Fig. 3.1f,g). Biochemical properties and functions of this organelle have not been studied in this group. Its large size indicates a significant role, no doubt analogous to that in other cells, that is, acting as an intermediate sorting and processing site for most polypeptides synthesized in the rough endoplasmic reticulum.

Lysosomes and Phagosomes

Both morphologic and cell fractionation studies show the presence in *T. vaginalis* of hydrolase-containing structures,[47,68] analogous to lysosomes. These have not been subjected to extensive studies. Intracellular digestion of bacteria ingested by *T. vaginalis* occurs in phagosomes,[47,48] likely to be lysosomal in nature.

Cytoskeleton

Like other trichomonad flagellates, *T. vaginalis* contains a highly evolved cytoskeleton (see Chap. 3), consisting of transitory structures involved in protoplasmic motility and adhesion as well as of permanent microtubular ones, the axostyle and the pelta in the body of the cell (see Fig. 3.3), and the flagellar microtubules (see Fig. 3.2i). The costa associated with the recurrent flagellum is a unique organelle present in species equipped with an undulating membrane (see Fig. 3.5a-c).

Areas containing organized microfilaments are often observed, especially when the organism is attached to a substratum,[126] and around developing food vacuoles.[48] Cytochalasin B, an inhibitor of microfilament formation, interferes with attachment[153] and phagocytosis.[48] The biochemical nature of *T. vaginalis* microfilaments is similar to those found in other organisms.[154] Actin constitutes about 10% of the total protein of *T. vaginalis*. It has been purified and was found to differ in certain details from actins of other organisms.[154]

The microtubular structures consist primarily of tubulins, as do microtubules from other organisms.[154] Low concentrations of the benzimidazole anthelmintics, mebendazole, and flunidazole inhibit the multiplication of *T. vaginalis* by affecting its microtubules.[155] Monoclonal and polyclonal antibodies against sheep brain tubulins react with *T. vaginalis* tubulin in immunoblots,[156] and they also demonstrate the microtubular structures microscopically by indirect immunofluorescence.[157] The use of a specific monoclonal antibody in the latter study indicated the presence of at least two α-tubulin isotypes, differing in topographic distribution in the cytoskeleton.[157]

The costa of *T. vaginalis* has not been isolated or subjected to biochemical analysis. In some Trichomonadidae; for example, *Trichomitopsis termopsidis,* this organelle was found to be protein in nature, with ATPase activity and high motility.[158] Although, as in most Trichomonadidae, *T. vaginalis* costa is nonmotile, it likely consists mainly of proteins. A report on the allegedly almost exclusively carbohydrate constitution of the costa in *T. foetus*[159] is most likely erroneous. In the procedure used, a large amount of the reserve glycogen, expected to cosediment with the costae, evidently masked the true composition of this organelle.

Absent Organelles

As mentioned in the introduction, the distinctive biochemical characteristics of *T. vaginalis* include several constituents and processes not commonly found in eukaryotic organisms. A no less distinctive feature is the lack of several of those common in other eukaryotes. The absence of mitochondria, peroxisomes, and glycosomes has already been mentioned. It will become apparent from the presentation that follows that although in all instances indirect evidence for the absence of such processes and organelles is quite strong, several major statements are based on extrapolations rather than on direct evidence obtained from *T. vaginalis*.

Absence of mitochondria is based on morphologic and biochemical data. The ultrastructure of the human urogenital trichomonad is well-known (see Chap. 3), and in no published micrograph can mitochondria be recognized. Earlier workers designated certain membrane-bound structures as mitochondria, usually because they felt that, as in most other cells, some organelles must correspond to mitochondria. Such designations were proposed, even at times when an objective analysis of early biochemical data should have led to the opposite conclusion. The obvious candidates for mitochondria were the hydrogenosomes. Although these latter structures share a few biochemical (Table 5.5) and even morphologic (see Chap. 3, Fig. 3.2n,m) properties with mitochondria, these are insufficient for identifying the two distinct kinds of organelles with each other.

Several constituents, characteristic alone or in combination, of mitochondria were found absent from trichomonads. Lack of the tricarboxylic acid cycle has been primarily inferred from the absence of stimulation of respiration by citrate, aconitate, isocitrate, and α-ketoglutarate.[87,90] Furthermore, the failure to demonstrate in homogenates the dehydrogenases of the cycle, with the exception of malate dehydrogenase,[87] spoke against the presence of mitochondria-linked metabolism. When extracts of cells were incubated with [2-^{14}C]pyruvate, no label was found in any of the tricarboxylic acid cycle intermediates.[87] No cytochromes were detected with a hand-held spectroscope in several trichomonad species, including *T. vaginalis*,[75] and recently with more sensitive equipment in *T. foetus*.[160] The susceptibility of respiration of *T. vaginalis* to various inhibitors discussed above also supports the idea of absence of cytochrome-mediated respiration. No energy-conserving ATPase typically associated with mitochrondria has been detected in *T. foetus*,[161] and it is not likely to be found in *T. vaginalis*. Inhibitor studies do not support the existence of electron-transport coupled phosphorylation. Of other mitochondrial constituents, no extranuclear DNA[11,31] or cardiolipin[36] has been found in *T. vaginalis* or in hydrogenosome-enriched fractions of this organism. These results cast doubt on an earlier report about the presence of DNA and cardiolipin in *T. foetus* hydrogenosomes.[135]

Absence of peroxisomes is also shown both by morphologic and biochemical data. The organism does not contain microbodies surrounded by a single unit membrane. Although hydrogenosomes were originally described as microbody-like organelles,[98] more detailed morphologic studies, especially the demonstration of two unit membranes composing the envelope of the organelle,[132,133] revealed sufficient differences between microbodies and hydrogenosomes to justify abandoning the "microbody-like" designation. With the demonstration that trichomonad hydrogenosomes are not peroxisomes, no morphologic candidates remain for these organelles.

Characteristic biochemical constituents of peroxisomes were not detected. *Trichomonas vaginalis* does not contain catalase[75,86] and is unable to perform the β-oxidation of lipids.[35] Although not tested directly, other peroxisomal direct oxidases are likely to be absent, as in the case of *T. foetus*.[94,98] Respiration of *T. vaginalis* is not stimulated by several substrates of peroxisomal respiration, for example, glycolate and lactate.[90] Catalase, when present in other trichomonads, for example, *T. foetus*,[75,98] is not particle-bound.[98]

The absence of glycosomes, characteristic microbody-like organelles of kinetoplastid flagellates,[162] is shown in trichomonads by the absence of microbodies and by the localization of all glycolytic enzymes in the nonsedimentable portion of the cytoplasm.[49,63,67]

Drug Metabolism

Human trichomoniasis is treated almost exclusively with a single class of compounds, 5-nitroimidazole derivatives (see Chap. 18), primarily metronidazole (1-hydroxyethyl, 2-methyl, 5-nitroimidazole). These compounds have extensive clinical use because of their high selectivity against anaerobes and minimal toxicity for aerobic organisms. The biochemical basis of the action of 5-nitroimidazoles is described below, a topic reviewed repeatedly in recent years.[163-168]

The action of metronidazole on *T. vaginalis* comprises four successive steps (Fig. 5.3):[163,168] (1) Entry of the drug into the cell, (2) its reductive activation to toxic products, (3) the action of these on intracellular targets, and (4) the release of inactive degradation products of the drug.

The entry of the drug into cells is by diffu-

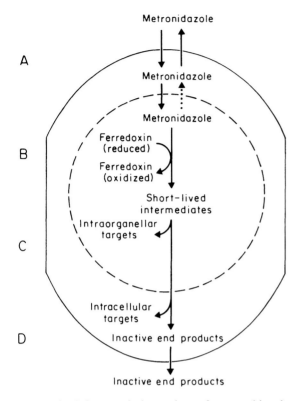

Fig. 5.3. Scheme of the action of metronidazole on *Trichomonas vaginalis*. *A,* Passage through the membranes of the cell and the hydrogenosome; *B,* reductive activation in the hydrogenosome; *C,* interaction with intracellular targets; *D,* release of inactive end products. (Modified with permission of the C.V. Mosby Co., from Müller M; Mode of action of metronidazole on anaerobic bacteria and protozoa. *Surgery* 93:165-171, 1983.)

sion, without the participation of active transport.[169] In cells with anaerobic metabolism, as discussed below, the drug, once inside the cell, undergoes metabolic modification. This modification decreases the intracellular concentration of the unchanged drug, thus increasing the gradient across the cell membrane. The gradient supports further entry of the drug into the cell, and continued metabolism leads to a marked intracellular accumulation of various derivatives. This process can be monitored with the aid of labeled 5-nitroimidazoles.[169]

Metronidazole and other 5-nitroimidazoles must be activated by the reduction of the nitro group to become cytotoxic.[164] The reduction occurs stepwise. Transfer of one electron to the nitro group results in the formation of the nitro-free radical ($R-NO_2^-$), the initial step of the overall reduction process that corresponds, on the average, to the transfer of four electrons.[170,171] The transfer of the additional electrons is assumed to occur by dismutation of the free radical ($2R-NO_2^- \rightarrow R-NO_2 + R-NO$). The further reduction products, probably the nitroso compound (R-NO), nitroso-free radical ($R-NO^-$), and hydroxylamine derivative (R-NHOH) have not been detected directly. The hydroxylamine derivative and possible other intermediates are short-lived, and after the reduction is completed, only fragments of the original molecule can be detected, which have no biologic activity.[165] The major end products of metronidazole reduction are acetamide and 2-hydroxyethyl oxamic acid.[165,172] The formation of these compounds by *T. vaginalis* has been demonstrated and used to measure the rate of reduction.[173]

The one-electron adduct free radical is easily reoxidized to the original compound by O_2 with the formation of superoxide radical. 5-Nitroimidazoles, however, have a stronger cytotoxic activity in the absence of O_2[169,174,175] indicating that the superoxide formed does not contribute significantly to this activity. These findings led to the suggestion that one or several of the short-lived intermediates of the reduction process represent the active cytotoxic products.

The intracellular targets of the toxic intermediates have not been identified definitively. The intermediates are most likely highly reactive, thus a multiplicity of the targets is possible. Present views favor DNA as the primary target. Model experiments demonstrated a marked interaction of chemically or polarographically[171,176,177] reduced metronidazole with purified DNA that occurs only if the reduction is performed in the presence of the target. If the drug is reduced before DNA is added, no interaction can be observed, reflecting the short life span of the reactive intermediates. The nature of the effect of reduced metronidazole on DNA is debated. Several authors noted a binding of the compound preferentially to guanine and cytosine residues,[171,177] but without causing major breaks in the DNA chain.[171,178] Others described a preferential effect on adenosine residues and multiple strand breaks.[166] Similar model experiments revealed less significant interaction with RNA and practically none with selected proteins.[177] Synthesis of nucleic acids and proteins is inhibited in metronidazole-treated *T. vaginalis*[179,180] and *Bacteroides fragilis*,[181] with DNA synthesis being affected first, suggesting DNA as the primary target in vivo.

A striking metabolic effect is the immediate cessation of H_2 production after a 5-nitroimidazole compound is added to a suspension of live *T. vaginalis*.[182-184] The inhibition of H_2 evolution was suggested as the primary mechanism of trichomonacidal action of the drug.[182] This explanation is no longer in favor, since the inhibition is reversible;[184] also suppression of H_2 production by aerobiosis or by certain metabolic inhibitors (2,4-dinitrophenol)[185] does not diminish the viability of the organisms. As indicated below, H_2 formation by *T. vaginalis* exposed to metronidazole or other 5-nitroimidazoles is inhibited because the nitro groups of the compounds compete with protons for the electrons. The hydrogen-producing system can also be damaged by the toxic intermediate, since H_2 production is not restored completely.[52] The activity of several other enzymes, however, did not decrease in homogenates of metronidazole-treated *T. foetus*.[186]

The selectivity of metronidazole and other 5-nitroimidazoles against anaerobic microorganisms is correlated with the selectivity of reductive activation, which is efficient in anaerobic organisms and for all practical purposes does not occur in aerobes. The underlying biochemical difference between susceptible and insusceptible organisms is in the nature of the electron-transfer components available as reductants for the nitro group. The electron affinity of the nitro group of 5-nitroimidazoles is low. In clinically used compounds, it corresponds to a one-electron midpoint potential of -450 to -480 mV;[187] therefore, its efficient reduction requires electron donors of low midpoint potential. It was recognized earlier that in the energy metabolism of *T. vaginalis* and other organisms susceptible to 5-nitroimidazoles low midpoint potential reactions, usually linked to ferredoxin, played an important role.[182,183]

Transfer of electrons to 5-nitroimidazoles has been demonstrated in living trichomonads and in hydrogenosome-enriched fractions. The first product of reduction of metronidazole, the one-electron adduct nitro-free radical, has been detected by EPR spectroscopy in *T. vaginalis*[188] and *T. foetus*.[189] Reduction by hydrogenosome-enriched fractions[190,191] shows the same requirements as hydrogenosomal metabolism. Cessation of H_2 evolution in protozoa exposed to 5-nitroimidazoles[182-184] suggests that reduced ferredoxin is reoxidized by these compounds and not by protons through the action of hydrogenase. H_2 evolution starts again when the drug is exhausted.[184,188] In fact, the detection of this effect gave the first clue in deciphering the previously enigmatic mode of action of 5-nitroimidazoles.

The source of electrons for the reduction of the nitro group in *T. vaginalis* and other organisms susceptible to metronidazole is primarily pyruvate oxidized by pyruvate:ferredoxin oxidoreductase. Activation is diminished when glycolysis is inhibited.[192] Strains of *T. vaginalis*[193] and *T. foetus*[142] in which a loss of pyruvate:ferredoxin oxidoreductase activity is induced are no longer susceptible to 5-nitroimidazoles.

It was also proposed earlier that ferredoxins, as electron transport proteins of low midpoint potential, are the direct reductants of the nitro group.[170,182,183,194] Recent work corroborated this assumption and indicated the privileged role of ferredoxin in the process. 5-Nitroimidazole derivatives are easily reduced by chemically reduced bacterial ferredoxins.[170] Efficient reduction is observed also when various ferredoxins in the presence of metronidazole are reduced by ferredoxin-linked oxidoreductases.[23,195-197] In the absence of ferredoxin, pyruvate:ferredoxin oxidoreductase and hydrogenase are rather inefficient electron donors for 5-nitroimidazoles.[23,197,198] Addition of ferredoxin from any source dramatically enhances the reduction. Methyl viologen can replace ferredoxin in this process, as it does in other ferredoxin-linked reactions. These observations indicate that enzymes catalyzing low midpoint potential reactions are insufficient for effective reduction without an appropriate electron-transport component, and also that ferredoxin itself is a critical factor in the process.

The rate of reduction of various nitroimidazoles and other nitro compounds depends on their midpoint potential if the electron donating system is a flavin, a flavoprotein, or even a ferredoxin-linked oxidoreductase in the absence of ferredoxin.[198,199] Compounds of more negative potential are reduced more slowly. It was, however, observed that ferredoxin eliminated this dependency for a group of 15 nitroimidazoles (with midpoint potentials in the range of -560 to -260 mV) that were all reduced at the same rate.[198] This observation complements the earlier assumptions on the selectivity of 5-

nitroimidazole drugs. Organisms in which ferredoxin-linked metabolic pathways play a major role are able to activate reductively nitro derivatives with low midpoint potential. Organisms in which ferredoxin plays no or only a minor metabolic role cannot effectively reduce these compounds, which show high selectivity. The difference between ferredoxin-mediated with ferredoxin-independent reduction decreases with decreasing negativity of the compounds; thus their selectivity for anaerobic organisms also decreases.

Lowered 5-nitroimidazole susceptibility in *T. vaginalis*[193] and *T. foetus*[141] can be induced experimentally. It can also be observed in a number of clinical isolates from patients in whom therapy with these drugs has failed (see Chapter 17).

Metronidazole resistance in *T. vaginalis* was induced by many investigators by growing the organisms in the presence of gradually increasing concentrations of the drug. Until recently only moderate (less than one order of magnitude) decreases in susceptibility have been reported.[200] In the past few years, however, in experiments lasting many months, *T. vaginalis* lines were obtained with several-hundred-fold decreased susceptibility, that is, with almost absolute resistance.[54] These organisms lack pyruvate:ferredoxin oxidoreductase and hydrogenase activities. In fact, they can be regarded as organisms without functional hydrogenosomes.[54] Similar results were reported earlier for *T. foetus*.[141,142] The high metronidazole resistance of trichomonads with deficient hydrogenosomal metabolism provides a strong argument for the essential role of hydrogenosomal enzyme systems in drug activation.

In contrast, isolates from treatment-refractory cases of trichomoniasis do not lack any of the major hydrogenosomal enzymes and do not differ in their overall carbohydrate metabolism from susceptible *T. vaginalis* isolates.[51] In fact, under anaerobic conditions they are highly susceptible to metronidazole. Lowered susceptibility is observed, however, when the susceptibility tests are performed under aerobic conditions.[201-204] This indicates that the observed lower clinical susceptibility of these isolates is due to an increased O_2 inhibition of reductive drug activation. Increased inhibition was demonstrated by determining the intracellular accumulation of label from radioactive metronidazole by the two groups of organisms[51] and also by observing an increased quenching by low O_2 concentrations of the EPR signal of metronidazole-free radical in resistant organisms.[188] The important role of O_2 is further emphasized by the observations that resistant organisms have a decreased tolerance for aerobiosis[193] and low affinity for O_2.[89] The molecular basis of these effects remains unknown.

Environments in the Host

As emphasized in the introduction, all information on the biochemistry of *T. vaginalis* has been obtained on organisms grown in culture and studied under artificial conditions. The data show that this species is able to multiply under a variety of environmental conditions and to survive under even more extreme ones. They also suggest that its metabolism and other activities depend on the environment. The biochemistry of *T. vaginalis* at the natural sites of infestation depends on the characteristics of these sites, which differ between female and male and also from site to site within the genitourinary tract. Little is known of the conditions in these environments with the exception of the vagina.

The predominance of symptomatic disease in women as contrasted to predominantly asymptomatic trichomoniasis in men suggests that either the vagina provides a more hospitable environment for *T. vaginalis* or reacts more markedly to the presence of the pathogen. The most luxuriant multiplication of *T. vaginalis* occurs in the vagina. Marked changes in the degree of symptoms and, more important, in the number of flagellates present are known to occur, however, in a female patient during the course of the infection. This fact indicates that changing vaginal factors are important in determining the multiplication of *T. vaginalis*. In the case of acute trichomoniasis, even within the vagina there may be significant biochemical differences in the environment. Trichomonads adhering to the vaginal epithelium encounter different conditions than those swimming freely in the discharge, at some distance from the wall. No specific data are available to define those environmental conditions that support maximal growth of the parasite.

The mutual interaction of *T. vaginalis* with the vaginal environment is rather poorly understood. The parasite depends on this environment for nutrients and possibly for mechanical substrates. Its presence leads to pathologic changes, which themselves may significantly modify the environment.

The vaginal fluid of healthy women contains nutrients needed by *T. vaginalis,* including glucose and other carbohydrates and free amino acids.[205-207] The relative importance of these nutrients in the survival and multiplication of the parasite remains to be analyzed. It is likely that sufficient amounts of carbohydrates are present as an energy source for in situ multiplication, although the possibility of amino acid–supported energy metabolism has been raised.[46] *T. vaginalis* engulfs and digests bacteria present in the vagina,[47,48] but the degree of contribution of this process to its nutrition is unknown.

The availability of O_2 in the vaginal lumen in trichomoniasis is of special interest, in view of the inhibitory effect of O_2 observed on in vitro growth.[83,85] It implies also the relationship of clinical metronidazole resistance to aerobic susceptibility levels observed for the organisms isolated in such cases.[201,203,204] Oxygen is not or is barely detectable in the lumen of the normal vagina.[208] Increase in the vaginal blood flow during sexual arousal[209] leads to a significant increase of pO_2 in the lumen. It is likely that a similar situation prevails in acute vaginal trichomoniasis when organisms attached to the inflamed vaginal mucosa probably encounter significant pO_2, while those swimming freely in the discharge away from the vaginal wall might be under almost complete anaerobiosis.

The typical discharge in trichomonal vaginitis is described as "purulent and frothy," although there can be major differences in its appearance. There is little biochemical information about its composition, and even if such information were available, the specific role of *T. vaginalis* would be difficult to determine. The frothiness of the discharge is possibly due, at least in part, to the copious gas production of *T. vaginalis*. The elimination of acidic end products of carbohydrate metabolism seems to have no major influence on the pH of the discharge.[210,211] Recent data show that diamines and polyamines possibly released by *T. vaginalis* can be detected in the discharge.[212,213]

As discussed above, the biochemical mechanisms of pathogenesis by *T. vaginalis* remain to be elucidated. Neither the role of the process of contact cytotoxicity nor the participation of factors possibly released by living or dead organisms are known. The presence of *T. vaginalis* elicits the appearance of specific antibodies and inflammatory cells in the vaginal fluid (see Chap. 4). Elucidation with the aid of biochemical methods of these intriguing interactions of the parasite and its host remains a challenging task for the future.

References

1. Shorb MS: The physiology of trichomonads. In Hutner SH (ed): *Biochemistry and Physiology of Protozoa*, vol 3, pp 383-457. New York: Academic Press, 1964.
2. Danforth WF: Respiratory metabolism. In Chen T-T (ed): *Research in Protozoology*, vol 1, pp 201-306. Oxford: Pergamon Press, 1967.
3. Ryley JF: Carbohydrates and respiration. In Florkin M, Scheer BT (eds): *Chemical Zoology*, vol 1, pp 55-92. New York: Academic Press, 1967.
4. Fulton JD: Metabolism and pathogenic mechanisms of parasitic protozoa. In Chen T-T (ed): *Research in Protozoology*, vol 3, pp 389-504. Oxford: Pergamon Press, 1969.
5. Honigberg BM: Trichomonads of importance in human medicine. In Kreier JP (ed): *Parasitic Protozoa*, vol. 2, pp 275–454. New York: Academic Press, 1978.
6. Honigberg BM: Trichomonads of veterinary importance. In Kreier JP (ed): *Parasitic Protozoa*, vol 2, pp 163-273. New York: Academic Press, 1978.
7. Michaels RM, Treick RW: The mode of action of certain 3- and 5-nitropyridines and pyrimidines. III. Biochemical lesions in *Trichomonas vaginalis*. *Exp Parasitol* 12:401-417, 1962.
8. Roitman I, Heyworth PG, Gutteridge WE: Lipid synthesis by *Trichomonas vaginalis*. *Ann Trop Med Parasitol* 72:583-585, 1978.
9. Mandel M, Honigberg BM: Isolation and characterization of deoxyribonucleic acid of two species of *Trichomonas* Donné. *J Protozool* 11:114-116, 1964.
10. Gorrell TE: Effect of culture medium iron content on the biochemical composition and metabolism of *Trichomonas vaginalis*. *J Bacteriol* 161:1228-1230, 1985.
11. Wang AL, Wang CC: Isolation and characterization of DNA from *Tritrichomonas foetus* and *Trichomonas vaginalis*. *Mol Biochem Parasitol* 14:323-335, 1985.
12. Kupferberg AB: Metabolic studies and chemotherapy of *Trichomonas vaginalis*. *Gynecologia* 149(suppl):114-121, 1960.
13. Alderete JF: Antigen analysis of several pathogenic strains of *Trichomonas vaginalis*. *Infect Immun* 39:1041-1047, 1983.
14. Alderete JF: Identification of immunogenic and antibody-binding membrane proteins of

pathogenic *Trichomonas vaginalis*. *Infect Immun* 40:284-291, 1983.
15. Alderete JF, Garza G, Smith J, Spence M: *Trichomonas vaginalis:* Electrophoretic analysis and heterogeneity among isolates due to high-molecular-weight trichomonad proteins. *Exp Parasitol* 61:244-251, 1986.
16. Wartoń A, Honigberg BM: Analysis of surface saccharides in *Trichomonas vaginalis* strains with various pathogenicity levels by fluorescein-conjugated plant lectins. *Z Parasitenkd* 69:149-159, 1983.
17. Choromański L, Beat DA, Nordin JH, Pan AA, Honigberg BM: Further studies on the surface saccharides in *Trichomonas vaginalis* strains by fluorescein-conjugated lectins. *Z Parasitenkd* 71:443-458, 1985.
18. Connelly RJ, Torian BE, Stibbs HH: Identification of a surface antigen of *Trichomonas vaginalis*. *Infect Immun* 49:270-274, 1985.
19. Alderete JF, Suprun-Brown L, Kasmala L: Monoclonal antibody to a major surface glycoprotein immunogen differentiates isolates and subpopulations of *Trichomonas vaginalis*. *Infect Immun* 52:70-75, 1986.
20. Alderete JF, Suprun-Brown L, Kasmala L, Smith J, Spence M: Heterogeneity of *Trichomonas vaginalis* and discrimination among trichomonal isolates and subpopulations with sera of patients and experimentally infected mice. *Infect Immun* 49:463-468, 1985.
21. Garber GE, Proctor EM, Bowie WR: Immunogenic proteins of *Trichomonas vaginalis* as demonstrated by the immunoblot technique. *Infect Immun* 51:250-253, 1986.
22. Chapman A, Cammack R, Linstead DJ, LLoyd D: Respiration of *Trichomonas vaginalis:* Components detected by electron paramagnetic resonance spectroscopy. *Eur J Biochem* 156:193-198, 1986.
23. Gorrell TE, Yarlett N, Müller M: Isolation and characterization of *Trichomonas vaginalis* ferredoxin. *Carlsberg Res Commun* 49:259-268, 1984.
24. Soliman MA, Ackers JP, Catterall RD: Isoenzyme characterization of *Trichomonas vaginalis*. *Br J Vener Dis* 58:250-256, 1982.
25. Gradus MS, Mathews HM: Electrophoretic analysis of soluble proteins and esterase, superoxide dismutase and acid phosphatase isoenzymes of members of the protozoan family Trichomonadidae. *Comp Biochem Physiol (B)* 81:229-233, 1985.
26. Ackers JP: Immunologic and biochemical procedures for identification of trichomonad genera, species, and strains. In Kulda J, Čerkasov J (eds): Proc Int Symp Trichomonads & Trichomoniasis, Prague, July 1985 (Post-Symp Publ Pt 1). *Acta Univ Carolinae (Prague) Biol* 30(3,4):229-234, (1986) 1987.
27. Brugerolle G, Méténier G: Localisation intracellulaire et caracterisation de deux types de malate déshydrogenase chez *Trichomonas vaginalis* Donné, 1836. *J Protozool* 20:320-327, 1973.
28. Tanaka K: Isoenzymes of malate dehydrogenase in *Trichomonas vaginalis*. *Kiseichugaku Zasshi* 20:391-398, 1971.
29. Proctor EM, Naaykens W, Wong Q, Bowie WR: Isoenzyme patterns of isolates of *Trichomonas vaginalis* from Vancouver. *Sex Transm Dis* 15:181-185, 1988.
30. Honigberg BM, Mohn FA: An improved method for the isolation of highly polymerized native deoxyribonucleic acid from certain protozoa. *J Protozool* 20:146-150, 1973.
31. Turner G, Müller M: Failure to detect extranuclear DNA in *Trichomonas vaginalis* and *Tritrichomonas foetus*. *J Parasitol* 69:234-236, 1983.
32. Mack SR, Müller M: End products of carbohydrate metabolism in *Trichomonas vaginalis*. *Comp Biochem Physiol (B)* 67:213-216, 1980.
33. Manners DJ, Ryley JF: Studies on the metabolism of protozoa. 6. The glycogens of the parasitic flagellates *Trichomonas vaginalis* and *Trichomonas gallinae*. *Biochem J* 59:369-372, 1955.
34. Etinger H, Halevy S: Lipid biochemistry of *Trichomonas vaginalis*. *Ann Trop Med Parasitol* 58:409-413, 1964.
35. Holz GG Jr, Lindmark DG, Beach DH, Neale KA, Singh BH: Lipids and lipid metabolism in trichomonads. In Kulda J, Čerkasov J, (eds): Proc Int Symp Trichomonads & Trichomoniasis, Prague, July 1985 (Post-Symp Publ Pt 1). *Acta Univ Carolinae (Prague) Biol* 30(3-4):299-311, (1986) 1987.
36. Paltauf G, Meingassner JG: The absence of cardiolipin in hydrogenosomes of *Trichomonas vaginalis* and *Tritrichomonas foetus*. *J Parasitol* 68:949-950, 1982.
37. Linstead D: New defined and semi-defined media for cultivation of the flagellate *Trichomonas vaginalis*. *Parasitology* 83:125-137, 1981.
38. Wang CC, Wang AL, Rice A: *Tritrichomonas foetus:* Partly defined cultivation medium for study of the purine and pyrimidine metabolism. *Exp Parasitol* 57:68-75, 1984.
39. Read CP: Comparative studies on the physiology of trichomonad protozoa. *J Parasitol* 43:385-394, 1957.
40. Raeburn S, Rabinowitz JD: Pyruvate:ferredoxin oxidoreductase. I. The pyruvate-CO_2 exchange reaction. *Arch Biochem Biophys* 146:9-20, 1971.
41. Čerkasovová A: Energetic metabolism of *Tritrichomonas foetus*. II. Accumulation of pyruvic acid and thiamine-deficient cultures. *Folia Parasitol (Prague)* 16:297-301, 1969.

42. Hollander DH, Leggett NC: Vitamin B_{12} requirement for the growth of *Trichomonas vaginalis in vitro*. *J Parasitol* 71:683-684, 1985.
43. Wang, CC, Cheng H-W: Salvage of pyrimidine nucleosides by *Trichomonas vaginalis*. *Mol Biochem Parasitol* 10:171-184, 1984.
44. Peterson KM, Alderete JF: Iron uptake and increased intracellular enzyme activity follow host lactoferrin binding by *Trichomonas vaginalis* receptors. *J Exp Med* 160:398-410, 1984.
45. Read CP, Rothman AH: Preliminary notes on the metabolism of *Trichomonas vaginalis*. *Am J Hyg* 61:249-260, 1955.
46. Linstead D, Cranshaw MA: The pathway of arginine catabolism in the parastic flagellate *Trichomonas vaginalis*. *Mol Biochem Parasitol* 8:241-252, 1983.
47. Brugerolle G: Mise en évidence du processus d'endocytose et des structures lysosomiques chez *Trichomonas vaginalis*. *C R Acad Sci Paris* 272:2558-2560, 1971.
48. Francioli P, Shio H, Roberts RB, Müller M: Phagocytosis and killing of *Neisseria gonorrhoeae* by *Trichomonas vaginalis*. *J Infect Dis* 147:87-94, 1983.
49. Steinbüchel A, Müller M: Glycerol, a metabolic end product of *Trichomonas vaginalis* and *Tritrichomonas foetus*. *Mol Biochem Parasitol* 20:45-55, 1986.
50. Chapman A, Linstead DJ, Lloyd D, Williams J: ^{13}C-NMR reveals glycerol as an unexpected major metabolite of the protozoan parasite *Trichomonas vaginalis*. *FEBS Lett* 191:287-292, 1985.
51. Müller M, Gorrell TE: Metabolism and metronidazole uptake in *Trichomonas vaginalis* isolates with different metronidazole susceptibilities. *Antimicrob Agents Chemother* 24:667-673, 1983.
52. Lloyd D, Kristensen B: Metronidazole inhibition of hydrogen production *in vivo* in drug-sensitive and resistant strains of *Trichomonas vaginalis*. *J Gen Microbiol* 131:849-853, 1985.
53. Gobert N, Chaigneau M, Savel J: Etude des gaz liberés au cours de la culture en anaerobiose de *Trichomonas vaginalis*. *C R Soc Biol (Paris)* 165:276-282, 1971.
54. Čerkasovová A, Novák J, Čerkasov J, Kulda J, Tachézy J: Metabolic properties of *Trichomonas vaginalis* resistant to metronidazole under anaerobic conditions. In Kulda J, Čerkasov J (eds): Proc Int Symp Trichomonads & Trichomoniasis, Prague, July 1985 (Post-Symp Publ Pt 2). *Acta Univ Carolinae (Prague) Biol*, 30(5,6):505-512, (1986) 1988.
55. Mason PR: Sensitivity testing of *Trichomonas vaginalis* using bromocresol purple indicator. *J Parasitol* 71:128-129, 1985.
56. Escario JA, Llera JLG, Fernández ARM: A new method for the "*in vitro*" screening of trichomonacides. In Kulda J, Čerkasov J (eds): Proc int Symp Trichomonads & Trichomoniasis, Prague, July 1985 (Post-Symp Publ Pt 2). *Acta Univ Carolinae (Prague) Biol,* 30(5,6):557-562, (1986) 1988.
57. Wellerson R, Doscher G, Kupferberg AB: Metabolic studies on *Trichomonas vaginalis*. *Ann NY Acad Sci* 83:253-258, 1959.
58. Ishiguro T: Gas chromatographic-mass spectrometric analysis of gases produced by *Trichomonas vaginalis in vitro*. *Sanka Fujinka Gakkai Zasshi* 37:1097-1102, 1985 (in Japanese).
59. White E, Hart D, Sanderson BE: Polyamines in *Trichomonas vaginalis*. *Mol Biochem Parasitol* 9:309-318, 1983.
60. Alderete JF, Garza GE: Soluble *Trichomonas vaginalis* antigens in cell-free culture supernatants. *Mol Biochem Parasitol* 13:147-158, 1984.
61. Mason PR, Patterson BA: Proliferative response of human lymphocytes to secretory and cellular antigens of *Trichomonas vaginalis*. *J Parasitol* 71:265-268, 1985.
62. Wellerson R, Kupferberg AB: On glycolysis in *Trichomonas vaginalis*. *J Protozool* 9:418-424, 1962.
63. Arese P, Cappuccinelli P: Glycolysis and pentose phosphate cycle in *Trichomonas vaginalis*. I. Enzyme activity pattern and the constant proportion quintet. *Int J Biochem* 5:859-865, 1974.
64. Wirtschafter S, Jahn TL: The metabolism of *Trichomonas vaginalis*. The glycolytic pathway. *J Protozool* 3:83-85, 1956.
65. Baernstein HD: Aldolase in *Trichomonas vaginalis*. *Exp Parasitol* 4:323-334, 1955.
66. Baernstein HD: Lactic dehydrogenase in *Trichomonas vaginalis*. *J Parasitol* 45:491-498, 1959.
67. Taylor MB, Berghausen H, Heyworth P, Messenger N, Rees LJ, Gutteridge WE: Subcellular localization of some glycolytic enzymes in parasitic flagellated protozoa. *Int J Biochem* 11:117-120, 1980.
68. Baernstein HD: Malic dehydrogenase in *Trichomonas vaginalis*. *J Parasitol* 47:279-284, 1961.
69. Lindmark DG, Müller M, Shio H: Hydrogenosomes in *Trichomonas vaginalis*. *J Parasitol* 61:552-554, 1975.
70. Lindmark DG: Energy metabolism of the anaerobic protozoon *Giardia lamblia*. *Mol Biochem Parasitol* 1:1-12, 1980.
71. Reeves RE, Warren LG, Susskind B, Lo H-S: An energy-conserving pyruvate-to-acetate pathway in *Entamoeba histolytica*. Pyruvate synthase and a new acetate thiokinase. *J Biol Chem* 252:726-731, 1977.
72. Lindmark DG: Certain enzymes of the energy metabolism of *Entamoeba invadens* and their

subcellular localization. In Sepulveda B, Diamond LS (eds): Proc Int Conf Amebiasis, pp 185-189. Mexico: Instituto Mexicano de Siguro Social, 1976.
73. Yarlett N, Hann AC, Lloyd D, Williams A: Hydrogenosomes in the rumen protozoon *Dasytricha ruminantium* Schuberg. *Biochem J* 200:365-372, 1981.
74. Müller M: The hydrogenosome. *Symp Soc Gen Microbiol* 30:127-142, 1980.
75. Ryley JF: Studies on the metabolism of the protozoa. 5. Metabolism of the parasitic flagellate *Trichomonas foetus*. *Biochem J* 59:361-369, 1955.
76. Bauchop T: Mechanism of hydrogen formation in *Trichomonas foetus*. *J Gen Microbiol* 68:27-33, 1971.
77. Kerscher L, Oesterhelt D: Pyruvate:ferredoxin oxidoreductase—new findings on an ancient enzyme. *Trends Biochem Sci* 7:371-374, 1982.
78. Lindmark DG, Müller M: Hydrogenosome, a cytoplasmic organelle of the anaerobic flagellate, *Tritrichomonas foetus*, and its role in pyruvate metabolism. *J Biol Chem* 248:7724-7728, 1973.
79. Lindmark DG, Müller M: Biochemical cytology of trichomonad flagellates. II. Subcellular distribution of oxidoreductases and hydrolases in *Monocercomonas* sp. *J Protozool* 21:374-378, 1974.
80. Adams MWW, Mortenson LE, Chen J-S: Hydrogenase. *Biochim Biophys Acta* 594:105-176, 1981.
81. Steinbüchel A, Müller M: Anaerobic pyruvate metabolism of *Tritrichomonas foetus* and *Trichomonas vaginalis* hydrogenosomes. *Mol Biochem Parasitol* 20:57-65, 1986.
82. Lindmark DG: Acetate production by *Tritrichomonas foetus*. In Van den Bossche H (ed): *Biochemistry of Parasites and Host-Parasite Relationships*, pp 15-21. Amsterdam: Elsevier, 1976.
83. Mack SR, Müller M: Effect of oxygen and carbon dioxide on the growth of *Trichomonas vaginalis* and *Tritrichomonas foetus*. *J Parasitol* 64:927-929, 1978.
84. Wellerson R, Doscher GE, Kupferberg AB: Carbon dioxide fixation in *Trichomonas vaginalis*. *Biochem J* 75:562-565, 1960.
85. Johnson G: Physiology of a bacteria-free culture of *Trichomonas vaginalis*. IV. Effect of hydrogen ion concentration and oxygen tension on population. *J Parasitol* 28:369-379, 1942.
86. Ninomiya H, Suzuoki-Ziro: The metabolism of *Trichomonas vaginalis*, with comparative aspects of trichomonads. *J Biochem* 39:321-331, 1952.
87. Wirtschafter S, Saltman P, Jahn TL: The metabolism of *Trichomonas vaginalis:* The oxidative pathway. *J Protozool* 3:86-88, 1956.
88. Zeuthen E: Oxygen uptake as related to body size in organisms. *Q Rev Biol* 28:1-12, 1953.
89. Yarlett N, Yarlett NC, Lloyd D: Metronidazole resistant clinical isolates of *Trichomonas vaginalis* have lowered oxygen affinities. *Mol Biochem Parasitol* 19:111-116, 1986.
90. Tsukahara T: Respiratory metabolism of *Trichomonas vaginalis*. *Jpn J Microbiol* 5:157-169, 1961.
91. Tanabe M: *Trichomonas vaginalis:* NADH oxidase activity. *Exp Parasitol* 48:135-143, 1979.
92. Linstead DJ, Bradley S: The purification and properties of two soluble reduced nicotinamide:acceptor oxidoreductases from *Trichomonas vaginalis*. *Mol Biochem Parasitol* 27:125-133, 1988.
93. Linstead D: Oxygen sensitivity and nitroimidazole action in *T. vaginalis*. *Wiad Parazytol* 29:21-31, 1983.
94. Čerkasov J. Čerkasovová A, Kulda J. Vilhelmová D: Respiration of hydrogenosomes of *Tritrichomonas foetus*. I. ADP-dependent oxidation of malate and pyruvate. *J Biol Chem* 253:1207-1214. 1978.
95. Lloyd D, Ohnishi T, Lindmark DG, Müller M: Respiration of *Tritrichomonas foetus*. *Wiad Parazytol* 29:37-39, 1983.
96. Lindmark DG, Müller M: Superoxide dismutase in the anaerobic flagellates, *Tritrichomonas foetus* and *Monocercomonas* sp. *J Biol Chem* 249:4634-4637, 1974.
97. Kitchener KR, Meshnick SR, Fairfield AS, Wang CC: An iron-containing superoxide dismutase in *Tritrichomonas foetus*. *Mol Biochem Parasitol* 12:95-99, 1984.
98. Müller M: Biochemical cytology of trichomonad flagellates. I. Subcellular localization of hydrolases, dehydrogenases, and catalase in *Tritrichomonas foetus*. *J Cell Biol* 57:453-474, 1973.
99. Rowe AF, Lowe PN: Modulation of amino acid and 2-oxo acid pools in *Trichomonas vaginalis* by aspartate aminotransferase inhibitors. *Mol Biochem Parasitol* 21:17-24, 1986.
100. Jaroszewicz L, Malyszko E: Aminotransferase activity of strains of the genus *Trichomonas*. *Exp Med Microbiol* 17:48-51, 1965.
101. Lowe PN, Rowe AF: Aspartate: 2-oxoglutarate aminotransferase from *Trichomonas vaginalis:* Identity of aspartate aminotransferase and aromatic aminotransferase. *Biochem J* 232:689-695, 1985.
102. Lowe PN, Rowe AF: Aminotransferase activities in *Trichomonas vaginalis*. *Mol Biochem Parasitol* 21:65-74, 1986.
103. North MH, Lockwood BC, Bremner AF, Cooms GH: Polyamine biosynthesis in trichomonads. *Mol Biochem Parasitol* 19:241-249, 1986.
104. Gillin FD, Reiner DS, McCann PP: Inhibition

105. Thong K-W, Coombs GH, Sanderson BE: S-Adenosylmethionine metabolism and transsulphuration reactions in trichomonads. In Kulda J, Čerkasov J (eds): Proc Int Symp Trichomonads & Trichomoniasis, Prague, July 1985 (Post-Symp Publ Pt 1). *Acta Univ Carolinae (Prague) Biol* 30(3-4):293-298, (1986) 1987.
106. Thong KW, Coombs GH: Homocysteine desulphurase activity in trichomonads. *IRCS Med Sci Libr Compend* 13:493-494, 1985.
107. Thong K-W, Coombs GH, Sanderson BE: S-Adenosylhomocysteine hydrolase activity in *Trichomonas vaginalis* and other trichomonads. *Mol Biochem Parasitol* 17:35-44, 1985.
108. Thong K-W, Coombs GH: L-Serine sulphydrase activity in trichomonads. *IRCS Med Sci Libr Compend* 13:495-496, 1985.
109. Gillin FD, Reiner DS, Levy RB, Henkart PA: Thiol groups on the surface of anaerobic parasitic protozoa. *Mol Biochem Parasitol* 13:1-12, 1984.
110. Fahey RC, Newton GL: Occurrence of low molecular weight thiols in biological systems. In Larsson A, Orrenius S, Holmgren A, Mannervik B (eds): *Functions of Glutathione: Biochemical, Physiological, Toxicological, and Clinical Aspects,* pp 251-260. New York: Raven Press, 1983.
111. Fahey RC, Newton GL, Arrick B, Overdank-Bogart T, Aley SB: *Entamoeba histolytica*: A eukaryote without glutathione metabolism. *Science* 224:70-72, 1984.
112. Fairlamb AH, Cerami A: Identification of a novel, thiol-containing co-factor essential for glutathione reductase enzyme activity in trypanosomatids. *Mol Biochem Parasitol* 14:187-198, 1985.
113. Coombs GH, North MJ: An analysis of the proteinases of *Trichomonas vaginalis* by polyacrylamide gel electrophoresis. *Parasitology* 86:1-6, 1983.
114. Lockwood BC, North MJ, Coombs GH: Proteolysis in trichomonads. In Kulda J, Čerkasov J (eds): Proc Int Symp Trichomonads & Trichomoniasis, Prague, July 1985 (Post-Symp Publ Pt 1). *Acta Univ Carolinae (Prague) Biol* 30(3-4):313-318, (1986) 1987.
115. Watkins WM: Enzymes of *Trichomonas foetus*: The action of cell-free extracts on blood-group substances and low-molecular-weight glycosides. *Biochem J* 71:261-274, 1959.
116. Edwards RG, Thomas P, Westood JH: The purification and properties of a β-N-acetylhexosaminidase from *Tritrichomonas foetus*. *Biochem J* 151:145-148, 1975.
117. Crampen M, Von Nicolai H, Zilliken F: Properties and substrate specifications of two neuraminidases from *Trichomonas foetus*. *Hoppe-Seyler's Z Physiol Chem* 360:1703-1712, 1979.
118. Yates AD, Morgan WTJ, Watkins WM: Linkage-specific β-D-galactosidases from *Trichomonas foetus*: Characterization of the blood-group B-destroying enzyme as a 1,3-β-galactosidase and the blood-group P_1-destroying enzyme as a 1,4-β-galactosidase. *FEBS Lett* 60:281-285, 1975.
119. Takeuchi T, Ohashi O, Asami K: Biochemical characterization of acid phosphatase in *Trichomonas vaginalis*. *Jpn J Parasitol* 21:159-167, 1972.
120. Latter VS, Walters MA: In vitro studies on the chemotherapy of *Trichomonas vaginalis*. In Kulda J, Čerkasov J (eds): Proc Int Symp Trichomonads & Trichomoniasis, Prague, July 1985 (Post-Symp Publ Pt 2). *Acta Univ Carolinae (Prague) Biol,* 30(5,6):553-556, (1986) 1988.
121. Krieger JN, Poisson MA, Rein MF: Beta-hemolytic activity of *Trichomonas vaginalis* correlates with virulence. *Infect Immun* 41:1291-1295, 1983.
122. Torian BE, Connelly RJ, Stephens RS, Stibbs HH: Specific and common antigens of *Trichomonas vaginalis* detected by monoclonal antibodies. *Infect Immun* 43:270-275, 1984.
123. Alderete JF, Kasmala L, Metcalfe E, Garza GE: Phenotypic variation and diversity among *Trichomonas vaginalis* isolates and correlation of phenotype with trichomonal virulence determinants. *Infect Immun* 53:285-293, 1986.
124. Peterson KM, Alderete JF: Selective acquisition of plasma proteins by *Trichomonas vaginalis* and human lipoproteins as a growth requirement for this species. *Mol Biochem Parasitol* 12:37-48, 1984.
125. Peterson KM, Alderete JF: Acquisition of α_1-antitrypsin by a pathogenic strain of *Trichomonas vaginalis*. *Infect Immun* 40:640-646, 1983.
126. Nielsen MH, Nielsen R: Electron microscopy of *Trichomonas vaginalis* Donné: Interaction with vaginal epithelium in human trichomoniasis. *Acta Pathol Microbiol Scand (B)* 83:305-320, 1975.
127. Honigberg BM: Biological and physiological factors affecting pathogenicity of trichomonads. In Levandowsky M, Hutner SH (eds): *Biochemistry and Physiology of Protozoa,* vol 2, pp 409-427. New York: Academic Press, 1979.
128. Krieger JN, Ravdin JI, Rein MF: Contact-dependent cytopathogenic mechanisms of *Trichomonas vaginalis*. *Infect Immun* 50:778-786, 1985.
129. Alderete JF, Garza GE: Specific nature of *Trichomonas vaginalis* parasitism of host cell surfaces. *Infect Immun* 50:701-708, 1985.
130. Müller M: Hydrogenosomes of trichomonad flagellates. In Kulda J, Čerkasov J (eds): Proc

Int Symp Trichomonads & Trichomoniasis, Prague, July 1985 (Post-Symp Publ Pt 1). *Acta Univ Carolinae (Prague) Biol* 30(3-4):249-260, (1986) 1987.
131. Nielsen MN, Diemer NH: The size, density, and relative area of chromatic granules ("hydrogenosomes") in *Trichomonas vaginalis* Donné from cultures in logarithmic and stationary growth. *Cell Tissue Res* 167:461-465, 1976.
132. Honigberg BM, Volkmann D, Entzeroth R, Scholtyseck E: A freeze-fracture electron microscope study of *Trichomonas vaginalis* Donné and *Tritrichomonas foetus* (Riedmuller). *J Protozool* 31:116-131, 1984.
133. Benchimol M, de Souza W: Fine structure and cytochemistry of the hydrogenosomes of *Tritrichomonas foetus*. *J Protozool* 30:422-425, 1983.
134. Declerck PJ, Müller M: Hydrogenosomal ATP:AMP phosphotransferase of *Trichomonas vaginalis*. *Comp Biochem Physiol (B)* 88:575-580, 1987.
135. Čerkasovová A, Čerkasov J, Kulda J, Reischig J: Circular DNA and cardiolipin in hydrogenosomes, microbody-like organelles of trichomonads. *Folia Parasitol (Prague)* 23:33-37, 1976.
136. Benchimol M, de Souza W: *Tritrichomonas foetus*: Localization of filipin-sterol complexes in cell membranes. *Exp Parasitol* 58:356-364, 1984.
137. Benchimol M, Elias CA, de Souza W: *Tritrichomonas foetus*: Ultrastructural localization of basic proteins and carbohydrates. *Exp Parasitol* 54:135-144, 1982.
138. Chapman A, Hann AC, Linstead D, Lloyd D: Energy-dispersive x-ray microanalysis of membrane-associated inclusions in hydrogenosomes isolated from *Trichomonas vaginalis*. *J Gen Microbiol* 131:2933-2939, 1985.
139. Benchimol M, Elias CA, de Souza W: *Tritrichomonas foetus*: Ultrastructural localization of calcium in the plasma membrane and in the hydrogenosome. *Exp Parasitol* 54:277-284, 1982.
140. Müller M, Lindmark DG: Respiration of hydrogenosomes of *Tritrichomonas foetus*. II. Effect of CoA on pyruvate oxidation. *J Biol Chem* 253:1215-1218, 1978.
141. Kulda J, Čerkasov J, Demeš P, Čerkasovová A: *Tritrichomonas foetus*: Stable anaerobic resistance to metronidazole *in vitro*. *Exp Parasitol* 57:93-103, 1984.
142. Čerkasovová A, Čerkasov J, Kulda J: Metabolic differences between metronidazole resistant and susceptible strains of *Tritrichomonas foetus*. *Mol Biochem Parasitol* 11:105-118, 1984.
143. Mortenson LE, Nakos G: Bacterial ferredoxins and/or iron-sulfur proteins as electron carriers. In Lowenberg W (ed): *Iron-Sulfur Proteins: Biological Properties*, vol 1, pp 37-64. New York: Academic Press, 1973.
144. Yasunobu KT, Tanaka M: The types, distribution in nature, structure-function, and evolutionary data of the iron-sulfur proteins. In Lowenberg W (ed): *Iron-Sulfur Proteins: Molecular Properties*, vol 2, pp 27-130. New York: Academic Press, 1973.
145. Yoch DC, Carithers RP: Bacterial iron-sulfur proteins. *Microbiol Rev* 43:384-421, 1979.
146. Knoell H-E, Knappe J: *Escherichia coli* ferredoxin, an iron-sulfur protein of the adrenodoxin type. *Eur J Biochem* 50:245-252, 1974.
147. Levine N et al (The Committee on Systematics and Evolution of the Society of Protozoologists): A newly revised classification of the Protozoa. *J Protozool* 27:37-58, 1980.
148. Yarlett N, Hann AC, Lloyd D, Williams AG: Hydrogenosomes in a mixed isolate of *Isotricha prostoma* and *Isotricha intestinalis* from bovine rumen contents. *Comp Biochem Physiol (B)* 74:357-364, 1983.
149. Yarlett N, Coleman GS, Williams AG, Lloyd D: Hydrogenosomes in known species of rumen entodiniomorphid protozoa. *FEMS Microbiol Lett* 21:15-19, 1984.
150. Snyers L, Hellings P, Bovy-Kesler C, Thines-Sempoux D: Occurrence of hydrogenosomes in the rumen ciliates Ophyroscolecidae. *FEBS Lett* 137:35-39, 1982.
151. Van Bruggen JJA, Zwart KD, van Assema RM, Stumm CK, Vogels GD: *Methanobacterium formicicum*, an endosymbiont of the anaerobic ciliate *Metopus striatus* McMurrich. *Arch Microbiol* 139:1-7, 1984.
152. Yarlett N, Orpin CG, Munn EA, Yarlett NC, Greenwood CA: Hydrogenosomes in the rumen fungus *Neocallimastix patriciarum*. *Biochem J* 236:729-739, 1986.
153. Cappuccinelli P, Varesio L: The effect of cytochalasin B, colchicine and vinblastine on the adhesion of *Trichomonas vaginalis* to glass surfaces. *Int J Parasitol* 5:57-61, 1975.
154. Cappuccinelli P, Sellitto C, Ziccioni D, Juliano C: Structural and molecular organization of *Trichomonas vaginalis* cytoskeleton. In Kulda J, Čerkasov J (eds): Proc Int Symp Trichomonads & Trichomoniasis, Prague, July 1985 (Post-Symp Publ Pt 1). *Acta Univ Carolinae (Prague) Biol* 30(3-4):211-217, (1986) 1987.
155. Juliano C, Martinotti MG, Cappuccinelli P: In vitro effect of microtubule inhibitors on *Trichomonas vaginalis*. *Microbiologica* 8:31-42, 1985.
156. Draber P, Rubino S, Draberová E, Viklicky V, Cappuccinelli P: A broad-spectrum antibody to alpha-tubulin does not recognize all protozoan tubulins. *Protoplasma* 128:201-207, 1985.
157. Juliano C, Rubino S, Zicconi D, Cappuccinelli

P: An immunofluorescent study of the microtubule organization in *Trichomonas vaginalis* using antitubulin antibodies. *J Protozool* 33:56-59, 1986.
158. Amos WB, Grimstone AV, Rothschild LJ, Allen RD: Structure, protein composition and birefringence of the costa, a motile flagellar root fibre in the flagellate *Trichomonas*. *J Cell Sci* 35:139-164, 1979.
159. Sledge WE, Larson AD, Hart LT: Costae of *Tritrichomonas foetus:* Purification and chemical composition. *Science* 199:186-188, 1978.
160. Lloyd D, Lindmark DG, Müller M: Respiration of *Tritrichomonas foetus:* Absence of detectable cytochromes. *J Parasitol* 65:466-469, 1979.
161. Lloyd D, Lindmark DG, Müller M: Adenosine triphosphatase activity of *Tritrichomonas foetus*. *J Gen Microbiol* 115:301-307, 1979.
162. Opperdoes FR, Misset O, Hart DT: Metabolic pathways associated with the glycosomes (microbodies) of the trypanosomatidae. In August JT (ed): *Molecular Parasitology,* pp 63-75. Orlando, Fla: Academic Press, 1984.
163. Müller M: Action of clinically utilized 5-nitroimidazoles on microorganisms. *Scand J Infect Dis Suppl* 26:31-41, 1981.
164. Müller M: Reductive activation of nitroimidazoles in anaerobic microorganisms. *Biochem Pharmacol* 35:37-41, 1986.
165. Goldman P, Koch RL, Yeung T-C, Chrystal EJT, Beaulieu BB Jr, McLafferty MA, Sudlow G: Comparing the reduction of nitroimidazoles in bacteria and mammalian tissues and relating it to biological activity. *Biochem Pharmacol* 35:43-51, 1986.
166. Edwards DI: Reduction of nitroimidazoles *in vitro* and DNA damage. *Biochem Pharmacol* 35:53-58, 1986.
167. Moreno SNJ, Docampo R: Mechanisms of toxicity of nitro compounds used in the chemotherapy of trichomoniasis. *Environ Health Perspect* 64:199-208, 1985.
168. Müller M: Mode of action of metronidazole on anaerobic bacteria and protozoa. *Surgery* 93:165-171, 1983.
169. Müller M, Lindmark DG: Uptake of metronidazole and its effect on viability in trichomonads and *Entamoeba invadens* under anaerobic and aerobic conditions. *Antimicrob Agents Chemother* 9:696-700, 1976.
170. Lindmark DG, Müller M: Antitrichomonad action, mutagenicity, and reduction of metronidazole and other nitroimidazoles. *Antimicrob Agents Chemother* 10:476-482, 1976.
171. Declerck PJ, de Ranter CJ: *In vitro* reductive activation of nitroimidazoles. *Biochem Pharmacol* 35:59-61, 1986.
172. Koch RL, Goldman P: The anaerobic metabolism of metronidazole forms N-(2-hydroxyethyl)-oxamic acid. *J Pharmacol Exp Therap* 208:406-410, 1979.
173. Beaulieu BB Jr, McLafferty MA, Koch RL, Goldman P: Metronidazole metabolism in cultures of *Entamoeba histolytica* and *Trichomonas vaginalis*. *Antimicrob Agents Chemother* 20:410-414, 1981.
174. Füzi M: Oxygen-dependent metronidazole-resistance of *Clostridium histolyticum*. Zentralbl Bakteriol Mikrobiol I Orig (A) 249:99-103, 1981.
175. Milne SE, Stokes EJ, Waterworth PM: Incomplete anaerobiosis as a cause of metronidazole 'resistance'. *J Clin Pathol* 31:933-935, 1978.
176. Edwards DI: The action of metronidazole on DNA. *J Antimicrob Chemother* 3:43-48, 1977.
177. LaRusso NF, Tomasz M, Müller M, Lipman R: *In vitro* interaction of metronidazole with nucleic acids. *Mol Pharmacol* 13:872-882, 1977.
178. LaRusso NF, Tomasz M, Kaplan D, Müller M: Absence of strand breaks in deoxyribonucleic acid treated with metronidazole. *Antimicrob Agents Chemother* 13:19-24, 1978.
179. Mihara M: Effects of antitrichomonas agents on the metabolism of nucleic acids in *Trichomonas vaginalis*. *Nippon Kagaku Ryohogakukai Zasshi* 17:1537-1544, 1969.
180. Ings RMJ, McFadzean JA, Ormerod WE: The mode of action of metronidazole in *Trichomonas vaginalis* and other microorganisms. *Biochem Pharmacol* 23:1421-1429, 1974.
181. Sigeti JS, Guiney DG Jr, Davis CE: Mechanism of action of metronidazole on *Bacteroides fragilis*. *J Infect Dis* 148:1083-1089, 1983.
182. Edwards DI, Mathison GE: The mode of action of metronidazole against *Trichomonas vaginalis*. *J Gen Microbiol* 63:297-302, 1970.
183. Edwards DI, Dye M, Carne H: The selective toxicity of antimicrobial nitroheteroxyclic drugs. *J Gen Microbiol* 76:135-145, 1973.
184. Müller M, Lindmark DG, Mack SR: *Trichomonas vaginalis:* Metabolism and the mode of action of antitrichomonad nitroimidazoles. *Bull Mem Soc Med Paris* 181:141-145, 1978.
185. Müller M, Nseka V, Mack SR, Lindmark DG: Effects of 2,4-dinitrophenol on trichomonads and *Entamoeba invadens*. *Comp Biochem Physiol (B)* 64:97-100, 1979.
186. Müller M, Lindmark DG, McLaughlin J: Mode of action of nitroimidazoles on trichomonads. In Van den Bossche H (ed): *Biochemistry of Parasites and Host-Parasite Relationships,* pp 537-544. Amsterdam: Elsevier, 1976.
187. Wardman P, Clarke ED: One-electron reduction potentials of substituted nitroimidazoles measured by pulse radiolysis. *J Chem Soc, Faraday Trans(I)* 72:1377-1390, 1976.
188. Lloyd D, Pedersen JZ: Metronidazole radical anion generation *in vivo* in *Trichomonas vaginalis*. Oxygen quenching is enhanced in a drug-

resistant strain. *J Gen Microbiol* 131:87-92, 1985.
189. Moreno SNJ, Mason RP, Muniz RPA, Cruz FS, Docampo R: Generation of free radicals from metronidazole and other nitroimidazoles by *Tritrichomonas foetus*. *J Biol Chem* 258:4051-4054, 1983.
190. Moreno SNJ, Mason RP, Docampo R: Distinct reduction of nitrofurans and metroindazole to free radical metabolites by *Tritrichomonas foetus* hydrogenosomal and cytosolic enzymes. *J Biol Chem* 259:8252-8259, 1984.
191. Chapman A, Cammack R, Linstead D, Lloyd D: The generation of metronidazole radicals in hydrogenosomes isolated from *Trichomonas vaginalis*. *J Gen Microbiol* 131:2141-2144, 1985.
192. Nseka K, Müller M: L'action des inhibiteurs de la glycolyse sur l'absorption du metronidazole par les protozoaires *Tritrichomonas foetus* et *Entamoeba invadens*. *C R Soc Biol (Paris)* 172:1094-1098, 1979.
193. Čerkasovová A, Čerkasov J, Kulda J: Resistance of trichomonads to metronidazole. In Kulda J, Čerkasov J (eds): Proc Int Symp Trichomonads & Trichomoniasis, Prague, July 1985 (Post-Symp Publ Pt 2). *Acta Univ Carolinae (Prague) Biol,* 30(5,6):485-503, (1986) 1988.
194. Coombs GH: Studies on the activity of nitroimidazoles. In Van den Bossche H (ed): *Biochemistry of Parasites and Host-Parasite Relationships,* pp 545-552. Amsterdam: Elsevier, 1976.
195. Chen J-S, Blanchard DK: A simple hydrogenase-linked assay for ferredoxin and flavodoxin. *Anal Biochem* 93:216-222, 1979.
196. Blusson H, Petitdemange H, Gay R: A new, fast, and sensitive assay for NADH-ferredoxin oxidoreductase detection in clostridia. *Anal Biochem* 110:176-181, 1981.
197. Marczak R, Gorrell TE, Müller M: Hydrogenosomal ferredoxin of the anaerobic protozoon, *Tritrichomonas foetus*. *J Biol Chem* 258:12427-12433, 1983.
198. Yarlett N, Gorrell TE, Marczak R, Müller M: Reduction of nitroimidazole derivatives by hydrogenosomal extracts of *Trichomonas vaginalis*. *Mol Biochem Parasitol* 14:29-40, 1985.
199. Clarke ED, Goudling KH, Wardman P: Nitroimidazoles as anaerobic electron acceptors for xanthine oxidase. *Biochem Pharmacol* 31:3237-3242, 1982.
200. de Carneri I, Trane F: *In vivo* resistance to metronidazole induced on four recently isolated strains of *Trichomonas vaginalis*. *Arzneim-Forsch* 21:377-381, 1971.
201. Meingassner JG, Thurner J: Strain of *Trichomonas vaginalis* resistant to metronidazole and other 5-nitroimidazoles. *Antimicrob Agents Chemother* 15:254-257, 1979.
202. Meingassner JG, Stockinger K: Untersuchungen zur Identifikation Metronidazole-resistenter *Trichomonas vaginalis*-Stämme *in vitro*. *Z Hautkr* 56:7-15, 1981.
203. Müller M, Meingassner JG, Miller WA, Ledger WJ: Three metronidazole resistant strains of *Trichomonas vaginalis* from the United States. *Am J Obstet Gynecol* 138:808-812, 1980.
204. Lossick JG, Müller M, Gorrell TE: In vitro drug susceptibility and doses of metronidazole required for cure in cases of refractory vaginal trichomoniasis. *J Infect Dis* 153:948-955, 1986.
205. Huggins GR, Preti G: Vaginal odors and secretions. *Clin Obstet Gynecol* 24:355-377, 1981.
206. Wagner G, Levin RJ: Vaginal fluid. In Hafez ESE, Evans TN (eds): *The Human Vagina,* pp 121-137. Amsterdam: North-Holland, 1978.
207. Moghissi KS: Vaginal fluid constituents. In Beller FK, Schumacher GFB (eds): *The Biology of the Fluids of the Female Genital Tract,* pp 13-23. New York: Elsevier, 1979.
208. Wagner G, Bohr L, Wagner P, Petersen LN: Tampon-induced changes in vaginal oxygen and carbon dioxide tensions. *Am J Obstet Gynecol* 148:147-150, 1983.
209. Wagner G, Levin R: Oxygen tension of the vaginal surface during sexual stimulation in the human. *Fertil Steril* 30:50-53, 1978.
210. Cohen L: Influence of pH on vaginal discharges. *Br J Vener Dis* 45:241-246, 1969.
211. Omer EF, el-Naeem HA, Ali MH, Catterall RD, Erwa HH: Effects of *Trichomonas vaginalis* on the pH and glycogen content of the vagina. *Ethiop Med J* 23:173-177, 1985.
212. Chen KCS, Amsel R, Eschenbach DA, Holmes KK: Biochemical diagnosis of vaginitis: Determination of diamines in vaginal fluid. *J Infect Dis* 145:337-345, 1982.
213. Sanderson BE, White E, Balsdon MJ: Amine content of vaginal fluid from patients with trichomoniasis and gardnerella associated nonspecific vaginitis. *Br J Vener Dis* 59:302-305, 1983.

6

Nucleic Acid Metabolism in *Trichomonas vaginalis*

Ching C. Wang

Introduction

It has been noted that all parasitic protozoa examined to date are unable to synthesize purine rings de novo. The studies on *Trypanosoma cruzi*,[1] *Leishmania braziliensis*,[2] *Plasmodium lophurae*,[3] *Trypanosoma mega*,[4] and *Eimeria tenella*[5] have provided well-documented cases of the lack of de novo purine nucleotide synthesis and hence the dependence of these organisms on purine salvage for survival. Some of the purine salvage enzymes in these parasites have also acquired unique properties differing from those of the host. For instance, allopurinol, a hypoxanthine analog unimportant in the metabolism of mammalian cells, is apparently converted by hypoxanthine phosphoribosyltransferase (HPRT) in the parasites to the corresponding nucleotide and is eventually incorporated into the parasite RNA.[6] These biochemical events seem to have rendered allopurinol active against *Leishmania*,[2] African trypanosomes,[7] *T. cruzi*,[8] and *E. tenella*.[5] Similar conversions of the purine nucleoside analogs, such as allopurinol riboside,[9] formycin B,[10] and 9-deazainosine,[11] to their nucleotides have also made these compounds effective antileishmanial and antitrypanosomal agents. The purine salvage enzymes in the parasitic protozoa thus not only serve essential biologic functions, but also possess sufficiently broad substrate specificities to become potential targets for antiparasitic chemotherapy.

Nucleic acid metabolism in anaerobic parasitic protozoa, *Trichomonas vaginalis*, *Tritrichomonas foetus*, *Giardia lamblia*, and *Entamoeba histolytica*, has been the subject of intensive studies in recent years. Being inhabitants of the mucosal surface of the host urogenital or intestinal tract, these parasites are surrounded by the host epithelial cells, which may be subject to rapid turnover caused by inflammatory reactions to parasitic infection. The turnover may provide the parasites with a rich source of host nucleic acids, which they could effectively utilize in view of their powerful phagocytic capability and high lysosomal activity.[12] Thus, the parasites could very well simplify their schemes of nucleic acid metabolism and survive on salvaging preformed nucleosides and bases. Indeed, recent investigations in several laboratories have demonstrated that *Trichomonas vaginalis*,[13,14] *Tritrichomonas foetus*,[15,16] *G. lamblia*,[17,18] and *E. histolytica*[19,20] are all incapable of de novo synthesis of purine or pyrimidine nucleotides. These parasites are apparently metabolically more deficient than the hemoflagellate and sporozoan parasites, known to depend primarily on de novo synthesis of pyrimidine nucleotides. The total lack of de novo synthesis of nucleotides, which has been reflected in the absence of dihydrofolate reductase and thymidylate synthetase in *Trichomonas vaginalis*, *Tritrichomonas foetus*, and *G. lamblia*,[13–18] is supplemented with very simple salvage pathways for nucleosides or bases of purine and pyrimidine. These pathways, unique for each of the four parasites, have become the focal points of further investigations. I shall present here only the purine and pyrimidine salvage in *T. vaginalis*.

Characterization of *T. vaginalis* DNA and estimation of its genomic size will be described. A highly specific double-stranded RNA (dsRNA) virus has been identified and isolated from many different strains of *T. vaginalis*. I shall look briefly into each of these interesting findings to introduce the reader to the current understanding of nucleic acid metabolism in *T. vaginalis*.

Purine Metabolism

Absence of De Novo Synthesis of Purine Nucleotides

In vitro cultivation of *T. vaginalis* can be carried out in a semidefined or defined medium.[21] The medium requires the presence of adenosine and guanosine, suggesting the inability of forming adenine and guanine rings by the parasite. Heyworth et al,[13] who incubated *T. vaginalis* for 6 hours with four radiolabeled potential purine ring precursors, glycine, bicarbonate, formate, and serine, were unable to detect radioactivity in the trichloroacetic acid (TCA)–precipitated nucleic acid fraction. Wang et al[22] pulse-labeled *T. vaginalis* with [^{14}C] glycine and [^{14}C]formate for 2 hours and analyzed the nucleotide pool by high performance liquid chromatography (HPLC); they found no detectable radioactivity associated with the purine nucleotides. Thus, all the evidence suggests strongly that *T. vaginalis* is incapable of de novo synthesis of purine nucleotides.

Purine Salvage Pathways

Heyworth et al[13] detected significant incorporation of radiolabeled adenine, guanine, adenosine, deoxyadenosine, and guanosine into the TCA-precipitable nucleic acid fraction of *T. vaginalis*. The radioactivity from labeled adenine appeared exclusively in the adenine of chromatographed nucleic acid hydrolysates, whereas the guanine precursor was limited to the guanine fraction. No significant incorporation of hypoxanthine or inosine into nucleic acids could be detected. Similar observations were made by Wang et al,[22] who found no significant labeling of purine nucleotides by radioactive hypoxanthine or inosine. Adenine and adenosine were incorporated only into adenine nucleotides, and guanine and guanosine only into guanine nucleotides. There seems to be no appreciable interconversion between adenine and guanine nucleotides.

The data from precursor incorporations appear to be supported by those on enzyme profiles. There is no detectable purine phosphoribosyltransferase activity in *T. vaginalis*. Adenosine kinase, guanosine kinase, adenosine phosphorylase, and purine nucleoside phosphorylase are the only major purine salvage enzymes active in *T. vaginalis*.[13] These results, supported by a study by Miller and Linstead,[23] point to adenosine and guanosine as the two immediate precursors of adenine and guanine nucleotides through the actions of adenosine and guanosine kinases. The major purine salvage pathways in *T. vaginalis* can be summarized in a very simple scheme presented below:

$$\text{Adenine} \leftrightarrow \text{Adenosine} \rightarrow \text{AMP} \rightarrow \text{ADP} \rightarrow \text{ATP}$$

$$\text{Guanine} \leftrightarrow \text{Guanosine} \rightarrow \text{GMP} \rightarrow \text{GDP} \rightarrow \text{GTP} \quad (1)$$

This scheme found further support in pharmacologic studies that demonstrated highly effective inhibition of the in vitro growth of *T. vaginalis* by well-known inhibitors of tumor cell adenosine kinase such as tubercidin, adenine arabinoside, toyocamycin, and sangivamycin.[22] Toyocamycin and sangivamycin were particularly active, having estimated 50% inhibitory concentrations (IC$_{50}$) of about 3 μm.[22] Mycophenolic acid, a ubiquitous cytotoxic agent exerting its strong specific inhibition of IMP dehydrogenase in eukaryotes and prokaryotes, had no effect on *T. vaginalis*. This lack of effect of mycophenolic acid could have been anticipated from the absence of incorporation of hypoxanthine or inosine into the nucleotide pool of the urogenital trichomonad. Since there is no need for IMP as the precursor of either AMP or GMP in this trichomonad, the presence of functional IMP dehydrogenase is nonessential for its survival, and there is indeed very little IMP dehydrogenase activity identifiable in the extracts of *T. vaginalis* (unpublished observation).

It can be concluded that *T. vaginalis* has an extraordinarily simple scheme of salvaging purine nucleosides. The adenosine and guanosine kinases in the parasite can be regarded as potential targets for anti–*T. vaginalis* chemotherapy.

Pyrimidine Metabolism

Absence of De Novo Synthesis of Pyrimidine Nucleotides

No activities of dihydrofolate reductase, thymidylate synthetase, aspartate transcarbamylase, dihydroorotase, orotate phosphoribosyltransferase, or orotidine-5′-phosphate decarboxylase could be detected in extracts of *T. vaginalis*.[14,24] It is therefore very likely that this protozoon is incapable of de novo synthesis of pyrimidine rings. In our subsequent studies, we were able to demonstrate a total lack of incorporation of potential pyrimidine ring precursors, such as bicarbonate, aspartate, or orotate into the pyrimidine nucleotide pool of *T. vaginalis*.[14] Nor were we able to detect the radiolabeled precursors in the DNA or RNA fractions of this parasite following a 2-hour incubation at 37 °C.[14] In the semi-defined culture medium, it is necessary to include uridine, cytidine, and thymidine to obtain growth of *T. vaginalis*.[21] There is little doubt that this parasite cannot synthesize its own pyrimidines.

Pyrimidine Salvage Pathways

Further investigations of pyrimidine metabolism in *T. vaginalis* indicated that exogenous cytidine, uridine, thymidine, and uracil could be taken up by the organisms in vitro and converted to nucleotides, whereas cytosine and thymine were excluded.[14] Cytidine and uridine were the most actively incorporated substrates; their incorporations competed with each other but were hardly affected by uracil or thymidine. Incorporation of uracil, strongly inhibited by cytidine or uridine, was unaffected by thymidine; whereas incorporation of thymidine was not affected by any other substrate. Extraordinarily high activities of uridine phosphorylase and cytidine deaminase were found in extracts of *T. vaginalis*.[14] These observations suggest efficient conversions of exogenous uracil and cytidine to uridine before their incorporations into the nucleotide pool. Uridine must thus be the most dominant and immediate precursor of some of the pyrimidine nucleotides in *T. vaginalis*. This conclusion is further supported by the finding in the parasite extracts of high levels of uridine phosphotransferase[14] and uridine kinase[23] activities, which could be involved in converting uridine to UMP in living organisms. There was, however, little conversion of uracil nucleotides to cytosine nucleotides in *T. vaginalis* even after prolonged pulse and chase with radiolabeled precursors.[14] These findings suggest the necessity of converting exogenous cytidine to CMP by the cytidine phosphotransferase activity detectable in the trichomonad extracts. The major pyrimidine salvage pathways in *T. vaginalis* may be presented in the following scheme:

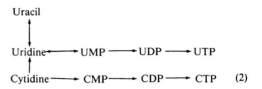

(2)

These pathways represent a simplified scheme with little interconversion between the two types of nucleotides. The enzymes responsible for converting uridine and cytidine to the corresponding nucleotides could be regarded as potential chemotherapeutic targets.

Deoxyribonucleotide Metabolism

Absence of Conversion of Ribonucleotides to Deoxyribonucleotides

Incorporation of exogenous thymidine into TMP, TDP, and TTP of *T. vaginalis* is not interfered with by other pyrimidine bases or ribonucleosides.[14] The process, initiated by membrane-bound thymidine phosphotransferase activity, led eventually to an exclusive incorporation of thymidine into the flagellate's DNA.[14] Exogenous uridine was found incorporated exclusively into *T. vaginalis* RNA, as would have been expected from the absence of dihydrofolate reductase and thymidylate synthetase from this organism. However, when a mixture of radiolabeled adenosine, guanosine, uridine, and cytidine was incubated with *T. vaginalis* at 37 °C for hours, all the radioactivity was found in the RNA fraction.[25] This unusual observation, coupled with the finding that deoxyadenosine, deoxyguanosine, deoxycytidine, and thymidine were all incorporated into the DNA fraction, suggested strongly that there is little conversion of ribonucleotides to deoxyribonucleotides in *T. vaginalis*.[25] This suggestion was further supported by the observation that no ribonucleoside diphosphate reductase or ribonucleoside triphosphate reductase activities could be detected in fresh extracts of the trichohomonad,[25] even though the same assay revealed significant ribonucleoside diphosphate reductase activity in extracts of *T. foetus*. The absence of ribonucleotide reductase in *T. vaginalis* further confirms the lack of conversion of ribonucleotides to deoxyribonucleotides.

Salvage of Deoxyribonucleosides

The negative findings described above raised the question as to possible sources of dATP, dGTP and dCTP needed for DNA synthesis in *T. vaginalis*. Therefore, the extracts and membrane fractions of *T. vaginalis* were reexamined for other possible enzyme activities. The membrane fraction, which sedimented at 100,000 g, was found to contain, in addition to thymidine phophotransferase, deoxyadenosine, deoxyguanosine, and deoxycytidine phosphotransferases.[25] The activities of these four enzymes were present at similar specific levels, i.e. about 1.0 nmol min^{-1} (mg protein)$^{-1}$, which represented an activity level capable of converting 1.7 pmol of each of the four deoxyribonucleosides to the corresponding 5′-monophosphates by 10^6 *T. vaginalis* in 1 minute. This conversion rate is more than adequate to fulfill the requirement for DNA synthesis in log-phase of the protozoan (see below).

Trichomonas vaginalis membranes were solubilized in Triton X-100, and approximately 20% of each of the four phosphotransferase activities were recovered in the solubilized fraction.[25] These activities, associated with the two major isozymes I and II, had a pH optimum of 5.0 to 6.0 and recognized TMP, dAMP, dGMP, dCMP, dUMP, FdUMP and *p*-nitrophenylphosphate as the phosphate donors. Enzyme I, purified ten-fold by DEAE-Sepharose chromatography and Sephacryl 200 filtration, was totally freed of the acid phosphatase of *T. vaginalis:* its molecular weight was estimated at 200 kDa. It was found to have strict substrate specificity for the four deoxyribonucleosides found in DNA and K_m values of 2 to 3 mM for these deoxyribonucleosides, which act on each other as competitive inhibitors. Enzyme I is capable of hydrolyzing *p*-nitrophenylphosphate with a Michaelis constant of 0.74 mM, but the rate of hydrolysis is enhanced by thymidine, suggesting that thymidine may be an acceptor with a stronger preference for phosphate than for water. This enzyme, designated as deoxyribonucleoside phosphotransferase, has little effect on ribonucleosides and could be another interesting target for antitrichomonal chemotherapy.

Isolation and Characterization of DNA

DNA Isolation

Isolation of DNA from *T. vaginalis* has been relatively difficult because of the presence of highly active DNase in the homogenates.[26] A phenol extraction method was applied for this purpose,[26] but there was no indication of the DNA size thus isolated to see if it could be useful for genetic engineering. A routine procedure was recently established in our laboratory by lysing *T. vaginalis* in 4 M guanidinium thiocyanate, centrifuging in CsCl buoyant density gradient, and then purifying the DNA band by NACS-37 column chromatography.[27] The DNA purified by this procedure acted as a single component in ion-exchange chromatography, agarose gel electrophoresis, CsCl buoyant density gradient centrifugation, and thermal denaturation.[27] It has an estimated size larger than 23.5 kilobase (kb), and exhibits highly reproducible restriction patterns in agarose gel electrophoresis suitable for cloning purposes.

DNA Characterization

Trichomonas vaginalis DNA has a melting temperature of 84 °C suggesting a GC content of 36%; the DNA has 35% to 42% hyperchromicity when fully melted. Cot analysis revealed the presence of 13.3% highly repetitive sequences, 53.3% moderately repetitive sequences, and 33.3% unique sequences. The complexities of unique sequences in *T. vaginalis* DNA are estimated to be 2.5 × 10^7 base pairs,[27] which is about the same as that of *Trypanosoma brucei*[28] but about six times the size of *Escherichia coli* chromosomal DNA.[29] From the genomic size and the percentage of repetitive sequences, the content of DNA in each *T. vaginalis* cell is calculated to be about 81 fg.[27]

Discovery of a Double-Stranded RNA Virus with *T. vaginalis* as Its Specific Host

The Double-Stranded RNA

During isolation of DNA from *T. vaginalis*, a linear double-stranded RNA (dsRNA), with an estimated size of 5.5 kb, was identified and isolated by CsCl buoyant density gradient centrifugation, NACS-37 column chromatography, and agarose gel electrophoresis.[27] Electron microscopic examination indicated a linear double-stranded structure 1.5 μm in length, with no apparent hairpins or loops. Boiling in 30% dimethyl sulfoxide denatured the dsRNA into single strands of 1.5 μm and shorter fragments.[30] HPLC analysis of dsRNA hydrolysate revealed that it consisted of 23.4% G, 23.4% C, 23.0% A, and 30.3% U; it melted at a transition temperature of 81.7 °C in 75 mM NaCl and 7.5 mM sodium citrate, pH 7.0, with 7% to 15% hyperchromicity. Forty different *T. vaginalis* strains from all over the world contained this dsRNA. However, three isolates resistant to metronidazole, IR78, CDC85, and MRP-2MT, and strain RU375, sensitive to metronidazole, lacked the dsRNA. It is not clear at this time what factor(s) common to the four *T. vaginalis* strains could account for this absence.

The Virus

It became apparent during recent studies that dsRNA remained intact in crude homogenates of *T. vaginalis* kept for hours at room temperature, while the rest of the DNA and RNA rapidly degraded.[31] This remarkable stability of dsRNA held true even in the presence of added pancreatic ribonuclease A; however, it disappeared from crude homogenates upon addition of sodium dodecyl sulfate (SDS) or proteinase K. Differential centrifugations of the homogenates sedimented the dsRNA mainly in the pellet obtained at 12,500 g. Upon further homogenizations, most of the dsRNA was released into the 105,000-g pellets. When these pellets were analyzed by CsCl buoyant density gradient centrifugation, a UV-absorbing band with a ρ value of 1.468 g ml^{-1} was identified and isolated This band, extracted with phenol, precipitated with ethanol, and when analyzed by agarose gel electrophoresis, it was found to contain only the dsRNA. When the band fraction was fixed, negatively stained, and examined in an electron microscope, it was found to consist of a uniform population of icosahedral viruslike particles with a diameter of 33 nm (Fig. 6.1). Examination of these viruslike particles by SDS–polyacrylamide gel electrophoresis revealed one major protein band with an estimated molecular weight of 85 kDa; it could be the capsid protein of this viruslike particle. This virus, found only in *T. vaginalis* among the many different species of protozoan parasites examined in our laboratory, may not be directly correlated with susceptibility of the virus-infected *T. vaginalis* strains to metronidazole, as was mentioned previously. However, the possibility of the virus being eliminated from the trichomonad by repeated metronidazole treatments cannot be ruled out at this time.

Conclusion

Recent studies of the nucleic acid metabolism in *T. vaginalis* have revealed many interesting aspects worthy of further explorations for antitrichomonal chemotherapy. One remarkable aspect of the metabolism is reflected in its multiple deficiencies. There is in the parasite no detectable de novo synthesis of purine nucleotide, de novo synthesis of pyrimidine rings, and no dihydrofolate reductase, thymidylate synthetase, ribonucleotide reductase, or purine or pyrimidine phosphoribosyltransferase activities. This extensive deficiency in nucleic acid metabolism, not reported from the prokaryotes and eukaryotes studied thus far, represents clearly a major, unique vulnerability of *T. vaginalis*. Its adenosine kinase, guanosine kinase, uridine phosphotransferase, and/or uridine kinase and deoxyribonucleoside phosphotransferase are most definitely indispensable enzymes for the survival and growth of this protozoon.

Further investigations of these enzymes are warranted for possible designs of specific inhibitors as potential targets for therapy. Ironically, however, the multiple metabolic deficiencies in *T. vaginalis* have not yet led to discoveries of new enzyme inhibitors capable of acting as chromotherapeutic agents. On the contrary, the absence of dihydrofolate reductase and IMP dehydrogenase in *T. vaginalis* has enabled the organism to become totally insensitive to spe-

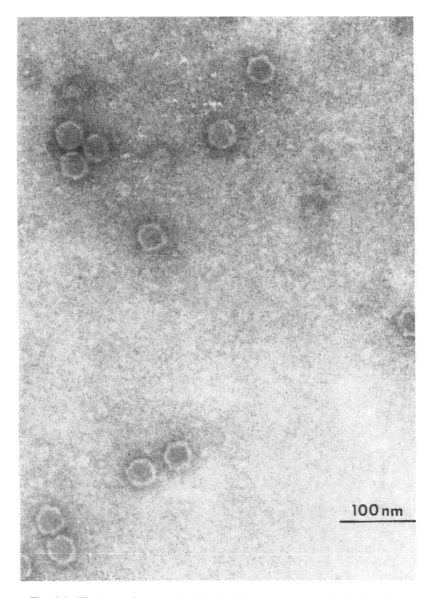

Fig. 6.1. Electron micrograph of the *Trichomonas vaginalis* dsRNA virus.

cific enzyme inhibitors such as methotrexate[14] and mycophenolic acid,[22] both common cytotoxic agents.

The successful isolation of relatively intact *T. vaginalis* DNA has indicated the feasibility of studies of the parasite by genetic engineering techniques. Discovery of a dsRNA virus specifically associated with *T. vaginalis* represented the first finding of such a virus in a protozoan. This finding may suggest the likelihood of using this virus as a transforming vector for *T. vaginalis*. Manipulations of RNA may be difficult, but the viral dsRNA may be reverse-transcribed to its cDNA, which can be recombined with foreign gene inserts or genes lethal to the parasite. Since the cDNA of Q β RNA[32] and poliovirus RNA[33] can be used to infect host cells and to produce the original RNA viruses, it is conceivable that the cDNA of *T. vaginalis* dsRNA virus could be made infective for *T. vaginalis*.

The cDNA with a potentially lethal gene insert could then be tested for its possible lethal effect on *T. vaginalis*.

References

1. Berens RL, Marr JJ, LaFon SW, Nelson DJ: Purine metabolism in *Trypanosma cruzi*. Mol Biochem Parasitol 3:187-196, 1981.
2. Marr JJ, Berens RL: Antileishmanial effect of allopurinol. II. Relationship of adenine metabolism in *Leishmania* spp. to the action of allopurinol. J Infect Dis 136:724-732, 1977.
3. Walsh CJ, Sherman IW: Purine and pyrimidine synthesis by the avian malaria parasite, *Plasmodium lophurae*. J Protozool 15:763-770, 1968.
4. Bone GJ, Steinert M: Isotopes incorporated in the nucleic acids of *Trypanosoma mega*. Nature 178:308-309, 1956.
5. Wang CC, Simashkevich PM: Purine metabolism in a protozoan parasite *Eimeria tenella*. Proc Natl Acad Sci USA 78:6618-6622, 1981.
6. Marr JJ, Berens RL, Nelson DJ: Antitrypanosomal effects of allopurinol: Conversion *in vitro* to aminopyrazolopyrimidine nucleotides by *Trypanosoma cruzi*. Science 201:1018-1020, 1978.
7. Berens RL, Marr JJ, Brun R: Pyrazolopyrimidine metabolism in African trypanosomes: Metabolism similarities to *Trypanosoma cruzi*. Mol Biochem Parasitol 3:187-196, 1981.
8. Avila JL, Avila A: *Trypanosoma cruzi*: Allopurinol in the treatment of experimental acute Chagas' disease. Exp Parasitol 51:204-208, 1980.
9. Nelson DJ, LaFon SW, Tuttle JV, Miller WH, Miller RL, Krenitsky TA, Elion GB, Berens RL, Marr JJ: Allopurinol ribonucleoside as an antileishmanial agent. J Biol Chem 254:11544-11549, 1979.
10. Carson DA, Chang K-P: Phosphorylation and antileishmanial activity of formycin B. Biochem Biophys Res Commun 100:1377-1383, 1981.
11. Marr JJ, Berens RL, Cohn NK, Nelson DJ, Klein RS: Biological action of inosine analogs in *Leishmania* and *Trypanosoma* spp. Antimicrob Agents Chemother 25:292-295, 1984.
12. Honigberg BM: Trichomonads of veterinary importance. In Kreier JP (ed): *Parasitic Protozoa*, vol 2, pp 164-275. New York: Academic Press, 1978.
13. Heyworth PG, Gutteridge WE, Ginger CD: Purine metabolism in *Trichomonas vaginalis*. FEBS Lett 141:106-110, 1982.
14. Wang CC, Cheng H-W: Salvage of pyrimidine nucleosides by *Trichomonas vaginalis*. Mol Biochem Parasitol 10:171-184, 1984.
15. Wang CC, Verham R, Rice A, Tzeng S-F: Purine salvage by *Tritrichomonas foetus*. Mol Biochem Parasitol 8:325-337, 1983.
16. Wang CC, Verham R, Tzeng S-F, Aldritt SM, Cheng H-W: Pyrimidine metabolism in *Tritrichomonas foetus*. Proc Natl Acad Sci USA 80:2564-2568, 1983.
17. Wang CC, Aldritt SM: Purine salvage networks in *Giardia lamblia*. J Exp Med 158:1703-1712, 1983.
18. Aldritt SM, Tien P, Wang CC: Pyrimidine salvage in *Giardia lamblia*. J Exp Med 161:437-445, 1985.
19. Lo H-S, Wang CC: Purine salvage in *Entamoeba histolytica*. J Parasitol 71:662-669, 1985.
20. Albach RA, Booden T, Boonlayangoor P: *Entamoeba histolytica*: Autoradiographic analysis of nuclear sites of RNA synthesis. Exp Parasitol 42:248-259, 1977.
21. Linstead D: New defined and semi-defined media for cultivation of the flagellate *Trichomonas vaginalis*. Parasitology 83:125-137, 1981.
22. Wang CC, Verham R, Cheng H-W, Rice A, Wang AL: Differential effects of inhibitors of purine metabolism in two trichomonad species. Biochem Pharmacol 33:1323-1329, 1984.
23. Miller RL, Linstead D: Purine and pyrimidine metabolizing activities in *Trichomonas vaginalis* extracts. Mol Biochem Parasitol 7:41-51, 1983.
24. Hill B, Kilsby J, Rogerson GW, McIntosh RT, Ginger CD: The enzymes of pyrimidine biosynthesis in a range of parasitic protozoa and helminths. Mol Biochem Parasitol 2:123-134, 1981.
25. Wang CC, Cheng H-W: The deoxyribonucleoside phosphotransferase of *Trichomonas vaginalis*: A potential target for anti-trichomonal chemotherapy. J Exp Med 160:987-1000, 1984.
26. Honigberg BM, Mohn FA: An improved method for the isolation of highly polymerized native deoxyribonucleic acid from certain protozoa. J Protozool 20:146-150, 1973.
27. Wang AL, Wang CC: Isolation and characterization of DNA from *Tritrichomonas foetus* and *Trichomonas vaginalis*. Mol Biochem Parasitol 14:323-336, 1985.
28. Borst P, Fase-Fowler F, Frasch ACC, Hoeijmakers JHJ, Weijers PJ: Characterization of DNA from *Trypanosoma brucei* and related trypanosomes by restriction endonuclease digestion. Mol Biochem Parasitol 1:221-246, 1980.
29. Britten RJ, Graham DE, Neufeld BR: Analysis of repeating DNA sequences by reassociation. Methods Enzymol 29E:363-418, 1974.
30. Wang AL, Wang CC: A linear double-stranded RNA in *Trichomonas vaginalis*. J Biol Chem 260:3697-3702, 1985.
31. Wang AL, Wang CC: The double-stranded RNA in *Trichomonas vaginalis* may be originated from virus-like particles. Proc Natl Acad Sci USA 83:7956-7960, 1986.
32. Racaniello VR, Baltimore D: Cloned polio virus complementary DNA is infectious in mammalian cells. Science 214:916-919, 1981.
33. Taniguchi T, Palmieri M. Weissmann C: Qβ phage formation in the bacterial host. Nature 274:223-228, 1978.

7

Cultivation

David Linstead

Introduction

The history of our understanding of human urogenital trichomoniasis and the organism *Trichomonas vaginalis,* which is the causative agent of this disease, is intimately bound up with the history of efforts to cultivate the organism in vitro. Only with axenic cultivation did it become possible to show conclusively that *T. vaginalis* was the sole originator of the rather diverse signs and symptoms seen in human infections. The disease has been produced in parasite-free female subjects by inoculation of axenic cultures on a number of occasions, producing some of the pathologic changes characteristic of the naturally acquired infection. Early experiments of this type are described in Trussell's classic book,[1] while later experiments are noted by Honigberg.[2]

The ability to grow the organism axenically has made possible the many advances in the understanding of the physiology, biochemistry, and immunology of the organism that have occurred in recent years. Axenic cultivation has also meant that in vitro screening procedures for potential antitrichomonal agents could readily be developed and this has contributed to the sustained interest that has been shown in advancing the available chemotherapy of the disease. Experimental manipulation of the media used for the cultivation of *T. vaginalis* has made a contribution in its own right to our understanding of the biosynthetic capabilities of the organism, and with the recently acquired ability to grow the organism in defined and semidefined media, the possibility to greatly extend this facet is raised.

Cultivation of *Trichomonas vaginalis* Donné

Historical Aspects

As with other parasitic protozoa such as *Entamoeba histolytica,* there was a lengthy phase in which *T. vaginalis* could only be propagated in xenic cultures. This phase of cultivation of the organism is now largely of historical interest. The first apparently successful reports of multiplication of the urogenital trichomonad outside a host date from the early 1920s. The importance of serum in supporting growth was noted even in the earliest studies,[3] and the adoption of the serum- and starch-containing media, which had proved of value in cultivation of the intestinal amebae, illustrated an affinity between these organisms and the trichomonads that has been given a rational biochemical explanation in recent years. There are reports of monoxenic cultivation of *T. vaginalis* including the apparently successful use of *Candida albicans*[4] as the growth-promoting partner organism. This observation is surprising because some of the modern reports indicate that the two organisms tend not to be associated in vivo[5] and are antagonistic when cultivated together.[6]

The Achievement of Axenic Cultivation—The Use of Antibiotics

The modern era of research on *T. vaginalis* dates from the first successful experiments on the axenic cultivation of the organism. This in turn was dependent on the availability of antibiotics to eliminate selectively the bacterial contaminants that inevitably accompany the organism when it is transferred from the patient into a primary culture. Suitable antibiotics became available in the mid-1940s, and the establishment of axenic cultures, as a relatively straightforward procedure, dates from that period.

Penicillin was employed by Adler and Pulvertaft[7] and Johnson and Trussell,[8] both of whom reported the value of this agent in 1944. There have been many subsequent reports using penicillin alone and in combination with a variety of other antibiotics effective against both bacteria and fungi. The latter can be a troublesome contaminant of primary cultures and may appear at any time during routine cultivation if there is a momentary lapse of sterile procedures. Streptomycin,[9] neomycin,[10] and chloramphenicol[11] have been recommended for use either alone or as an adjunct to penicillin for the elimination of bacteria. Mycoplasmas may sometimes be present as contaminants of primary clinical isolates[12,13] and, unless procedures such as electron microscopy or specific fluorescent stains are used, they may easily be overlooked. *Mycoplasma hominis* and *Ureaplasma urealyticum* are quite common in the vagina[14] and the possibility of their presence in cultures of material of recent clinical origin should always be borne in mind. The elimination of mycoplasmas can be ensured by the use of a suitable antibiotic. Kanamycin is probably the preferred compound for this purpose,[12,13] although tylosin (Tylocine), which has found wide applicability in ridding tissue cultures of contaminating mycoplasmas[15] and has very little activity against eukaryotes, may also be of value. Fungi, particularly yeast species, are more difficult to eliminate from contaminated cultures by the use of antibiotics, and the time-honored V-shaped tube,[16] which takes advantage of trichomonad motility, is at its most effective with a monoxenic contamination with dense immotile yeast cells. Antifungal antibiotics can be employed with due caution, however, and both nystatin (Mycostatin)[12] and amphotericin B (Fungizone)[17] have been used. In the author's opinion, which is shared by many other investigators, the former is to be preferred, despite its lack of solubility that renders the formation of solutions of accurately known concentration difficult. Amphotericin B is rather toxic to *T. vaginalis* and its overenthusiastic use against a refractory fungal contamination can lead to loss of the culture.

Recommended antibiotics for use in axenization of *T. vaginalis*, and the rescue of contaminated cultures are given in Table 7.1, together with suggestions for appropriate concentrations.

For the initiation of cultures directly from clinical specimens for diagnostic purposes, penicillin plus streptomycin in the quantities indicated is generally adequate. When isolating from experimentally infected animals in which the probability of fungal contamination is high, the addition of nystatin is necessary. Kanamycin may be added when it is desired to eliminate the possibility of mycoplasma contamination.

The author, and many other investigators, believe strongly that antibiotics should not be used during routine propagation of *T. vaginalis* for experimental purposes, once axenization has been achieved. There can be no guarantee that the presence of an antibiotic, however seemingly inocuous, does not subtly alter the physiology and biochemistry of the organism. The finding that cultivation in the presence of antibiotics reduces pathogenicity in *Trichomonas gallinae*[18] adds strength to this argument.

The Refinement of Complex Media and Their Contribution to Diagnosis

Once axenic growth of *T. vaginalis* had been obtained, a major stimulus to the development of better culture media was the desire to use cultivation of clinical specimens as an adjunct to the diagnosis of trichomoniasis. There is no doubt that it is of the greatest value in this respect. The complex media that have been developed for the cultivation of *T. vaginalis* are variations of the traditional nutrient broths that have served both the microbiologist and the parasitologist well. Good growth requires attention to certain specific requirements of the organism, which are mainly attributable to its basically anaerobic character. Media capable of supporting growth of *T. vaginalis* can be compounded from the following basic components.

TABLE 7.1. Treatments generally recommended for eliminating organisms contaminating primary cultures of *T. vaginalis*.

	Types of contaminant								
	Bacteria				Fungi				Mycoplasmas
	A. First choice		B. For resistant organisms		A. First choice		B. For resistant organisms		
Name	Dosage per ml	Name	Dosage per ml	Name	Dosage per ml	Name	Dosage per ml	Name	Dosage per ml
Benzyl penicillin	1000 IU	Gentamycin	Up to 200 µg	Nystatin (Mycostatin)	50 U	Amphotericin B (Fungizone)	15 µg	Kanamycin	100 µg
Streptomycin sulfate	100 µg	Floxacillin (penicillinase-resistant)	500 µg			Miconazole nitrate	50–100 µg	Tylosin (Tylocine)	60–120 µg
		Neomycin	500 µg						
		Chloramphenicol	250 µg						

Buffered Salts Solution

Trichomonas vaginalis is not highly demanding in its salt requirements; however, many media contain NaCl and others include additional "physiologic" cations such as Ca^{2+} and Mg^{2+}. Since the hydrogenosome of *T. vaginalis* contains large amounts of nonheme iron-sulfur proteins associated with anaerobic electron transport processes, the organism has a definite iron requirement and additional inorganic iron in the form of ferrous sulfate ($FeSO_4$) can be included in the medium with benefit[19]; indeed the content of hydrogenosomal iron-containing proteins correlates closely with the quantity of iron in the medium.[20]

The urogenital trichomonad produces lactic and acetic acids as a result of glucose catabolism,[21] and as a consequence it tends to acidify the media in which it grows. While it appears tolerant of low pH, the inclusion of a suitable buffer, usually in the form of a mixture of monobasic and dibasic potassium phosphates, is highly desirable in promoting a lengthy period of logarithmic growth.

Protein Hydrolysate

A mixture of amino acids and small peptides supplied by various undefined tryptic or peptic digests of animal protein ("peptones") are components of many published media.

Liver Digest or Yeast Extract

Liver digests and yeast extracts (sometimes both are specified) serve as sources of vitamins, purines and pyrimidines, and many other nutrients.

Reducing Agent(s)

Trichomonas vaginalis thrives best when oxygen is either entirely absent or much reduced in quantity. This can be achieved by extensively deoxygenating media and equilibrating them with inert gases such as pure nitrogen or argon, using techniques borrowed from the handling of obligate anaerobic bacteria. The shortest doubling times and most luxuriant growth are achieved by these extreme means. However, for most practical purposes less elaborate precautions to avoid oxygen suffices to allow satisfactory growth. The normal procedure is to add one or more reducing agents, which serve to lower the redox potential of the medium. The two most popular are cysteine and ascorbic acid, although others such as thioglycollic acid and thiomalic acid have been employed. Both cysteine and ascorbic acid work well and can be used with advantage together as in Diamond's excellent trypticase-yeast-maltose (TYM) medium.[22] The composition of this medium is given in Table 7.2. Ascorbic acid, however, has the advantage of being a stable, freely soluble substance that does not contribute to making the culture malodorous. Cysteine is metabo-

TABLE 7.2. Diamond's TYM (trypticase-yeast-maltose) medium.[22]

Component	Amount (in g except where indicated)
Trypticase (BBL)*	20.00
Yeast extract (BBL)	10.00
Maltose	5.00
L-Cysteine HCl	1.00
L-Ascorbic acid	0.20
K_2HPO_4	0.80
KH_2PO_4	0.80
Agar (Difco Bacto)	0.50
Distilled H_2O	900 ml

*Baltimore Biological Laboratory.

Dissolve buffer salts in 600 ml of water. Add and dissolve all remaining ingredients except agar in the order in which they occur. Adjust pH to 6.0 with 1N HCl and add agar. (For species other than *T. vagianlis* pH is adjusted to 6.8 to 7.0 with 1H NaOH.) Heat to dissolve agar, then distribute in required volumes and autoclave. Diamond[32] recommends the use of 10% sheep or bovine serum as a supplement; the present author has also had good results using 8% to 10% vol/vol horse serum. The complete medium should be stored at 4 °C and Diamond recommends use within 10 days. See the footnote to Table 7.4 for a comment on the use of agar in medium for the cultivation of *T. vaginalis*.

lized by *T. vaginalis* in part to ethanethiol, which contributes a characteristic and unpleasant odor to growing cultures. The modification of TYM, given in Table 7.3, which eliminates cysteine, may therefore be considered an improvement.

Serum

The urogenital trichomonad does not grow in an otherwise adequate medium if serum is not present. Many different sera have been tried; horse, calf, and human sera are the most useful because of their availability. Of these, horse serum is in some respects the most satisfactory, and when economically priced, it is to be recommended. The experiments reported by a number of authors[23-25] leave no doubt that the role of serum is to supply lipids, in particular fatty acids and cholesterol, in a form that can be assimilated by the organism without toxicity. There is apparently no special nutrient or high molecular weight growth factor supplied by serum that cannot be accounted for in terms of our current understanding of the nutrition of *T. vaginalis*.

Choice of Complex Media for Diagnosis and Routine Cultivation

Several authors have discussed the relative merits of the alternative methods of diagnosis available, and there is general agreement that cultivation achieves the maximum number of positive findings (see reference no. 5 for a recent study involving a large patient group). In Chapter 16 of this book McMillan discusses the use of culture media for diagnosis and concludes that "when laboratory facilities are adequate, cultivation should be employed routinely in the diagnosis of trichomoniasis." An ideal medium for diagnostic purposes should be available either off the shelf as a single premixed powder that can be reconstituted and autoclaved before use, or as readily available, consistent, dry ingredients that can be processed and formulated. For this reason media are to be preferred that do not employ ill-defined fresh liquid extracts of, for example, liver.

Tables 7.2 to 7.6 contain detailed recipes and instructions for the preparation of the original TYM medium,[22] TYM modified by Hollander[26] and Kulda et al,[13] CACH medium,[27] CPLM medium,[28] and STS medium.[29] These media all fulfill the criteria that their components are commercially available from large supply houses as dry ingredients. There is no absolute evidence that one is greatly superior to the others for diagnostic purposes, although various comparative trials have been performed[10,17,30,31]; see also Chapter 16 in this book). Consistently good results have been obtained with CPLM medium and Rayner's advocacy of this medium for diagnosis[31] is persuasive.

The present author's laboratory has used all

TABLE 7.3. Culture medium for *T. vaginalis* modified after Hollander[26] and Kulda et al.[13]

Component	Amount (in g except where indicated)
Trypticase (BBL)	20.00
Yeast extract (Difco)	10.00
Maltose (Sigma)	5.00
L-Ascorbic acid (analytic grade)	1.00
KCl	1.00
KHCO$_3$	1.00
KH$_2$PO$_4$	1.00
K$_2$HPO$_4$	1.00
FeSO$_4$·7H$_2$O	0.18
Glass distilled H$_2$O	900 ml

After dissolving ingredients by heating and stirring, cool, check pH, which should be about pH 6.8, and adjust if necessary with KOH and HCl. Dispense in appropriate volumes as required and sterilize at 15 lb pressure for 15 minutes. Sterilized medium can be stored at 4 °C for at least 1 month.

Immediately before use add sterile, heat-inactivated horse serum (mycoplasma-free grade) to 10% final concentration together with appropriate antibiotics if indicated (see Table 7.1).

It is the author's experience that a lower initial pH (pH 6.0 to 6.2) can provide conditions more immediately favorable to the outgrowth of small inocula (e.g. approximately 10^2 organisms/ml). This may not be true for all strains, but may be helpful in certain difficult cases of primary isolation. The higher pH medium is recommended for general purposes.

TABLE 7.4. CACH medium modified after Müller and Gottschalk.[27]

Component	Amount (in g except where indicated)
Trypticase (BBL)	20.00
Yeast extract (Difco)	5.00
Neutralized liver digest (e.g., Oxoid L27)	10.00
Cysteine HCL (chromatographically homogenous)	1.00
L-Ascorbic acid (analytic grade)	1.00
Maltose (Sigma)	1.00
Glucose (analytic grade)	10.00
NaCl	6.00
KH_2PO_4	3.00
$MgSO_4 \cdot 7H_2O$	0.20
$CaCl_2$	0.10
Agar (Difco Bacto)	0.5–2.00*
Glass distilled H_2O	900 ml

Dissolve all the components by heating and stirring, and filter through two thicknesses of Whatman No. 2 filter paper on a Buchner funnel. When the solution has cooled adjust it to pH 6.4 to 6.5 with NaOH. Dispense in appropriate volumes, and sterilize under 15 lb pressure for 15 minutes. Store at 4 °C for no longer than 1 month. Add heat-inactivated horse serum to a final concentration of 10% just before use. Also add antibiotics if required. This medium is richer than the one described in Table 7-2 and has general utility for isolating *T. vaginalis* from clinical cases and experimentally infected animals.[65] It has the potential merit of also supporting the growth of *Candida albicans* and other trichomonad species.

*Many authors recommend the addition of small amounts of agar (0.05% to 0.2%) to culture media to increase viscosity and thus slow down oxygen diffusion into the medium. This may be of considerable value when small inocula are used (e.g., in clinical diagnosis) to reduce the degree of exposure of the trichomonads to oxygen. It is unsatisfactory, however, in those instances (e.g., preparation of cells for biochemical or immunologic experiments) in which harvesting or concentration of the organisms is accomplished by centrifugation, as obviously the viscosity interferes with this procedure.

TABLE 7.5. CPLM (Cysteine-Peptone-Liver-Maltose) medium.[10,17,28]

Component	Amount (in g except where indicated)
Peptone (Difco Bacto)	32.00
Agar (Difco Bacto)	1.60
Cysteine HCl	2.40
Maltose	1.60
Liver infusion (Difco Bacto)	320 ml*
Ringer's solution	960 ml†

*Liver infusion: Place 330 ml of distilled water in a beaker, add 20 g Difco Bacto liver infusion powder and infuse for 1 hour at 50 °C. Raise temperature to 80 °C for 5 minutes and cool. Filter off coagulated protein through a coarse filter on a Buchner funnel.

†Ringer's solution consists of NaCl 0.6% wt/vol; $NaHCO_3$, KCL, and $CaCl_2$, 0.01% wt/vol of each.

Mix Ringer's solution and prepared liver infusion. Add solid ingredients and dissolve by heating and stirring. Continue heating until agar melts. Add 0.7 ml 0.5% wt/vol aqueous methylene blue. Adjust pH to 5.8 to 6.0 with NaOH or HCl. Sterilize at 15 lb pressure for 15 minutes. Distribute in required volumes and add 20% vol/vol serum before use. Methylene blue provides an indicator of anaerobiosis, and tubes may be used while a yellowish reduced zone is present. The original authors recommended human serum and room temperature storage. The present author has found horse serum at 10% vol/vol an acceptable substitute. See footnote to Table 7.4 regarding the inclusion of agar in culture media for *T. vaginalis*.

TABLE 7.6. STS (Simplified Trypticase Serum) medium.[29]

Component	Amount (in g except where indicated)
Trypticase (BBL)	20.00
Cysteine HCl	1.50
Maltose	1.00
Agar (Difco Bacto)	1.00
Distilled H_2O	950 ml

Mix the above ingredients and adjust to pH 6.0 with NaOH or HCl, heat to dissolve the agar. If desired add 0.6 ml of 0.5% wt/vol aqueous methylene blue as an indicator of reduction. Adjust to volume and dispense into required volumes while still warm. Sterilize by autoclaving at 15 lb pressure for 15 minutes. Supplement with sterile inactivated serum immediately before use. The original authors recommended 5% vol/vol human serum. The present author has used 8% to 10% bovine or horse serum as a satisfactory substitute. The agar component in this medium is essential; the medium is unsatisfactory when it is omitted.

the media described in Tables 7.2 to 7.7 for routine propagation of many different *T. vaginalis* strains. All perform well, with Diamond's TYM and its modification being perhaps the most consistently reliable. For diagnostic purposes all these media should be made as freshly as possible and exposure of the sample to air between being taken from a patient and inoculation into diagnostic medium minimized. The possibility of positive cultures appearing as late as 96 hours after inoculation should be borne in mind.[31]

Many authors recommend the addition of small amounts of agar (0.05% to 0.2%) to culture media to increase viscosity and therefore discourage oxygen diffusion into the medium. This is of undoubted value when small inocula are used to reduce the degree of oxygen exposure that the organisms receive, and therefore it is appropriate for media destined for diagnostic purposes. For routine subculture and bulk cultivation it is undesirable because it prevents concentration of the organisms by centrifugation.

In the author's laboratory *T. vaginalis* is routinely propagated either in 100 × 8 mm screw-cap borosilicate glass tubes of 8-ml nominal capacity, or in disposable plastic tubes. A volume of 6.5 ml of medium plus serum is dispensed into each glass tube and an inoculum of up to 0.5 ml added. When disposable plastic tubes are economical, the 100 × 13 mm screw-cap tube holding 7 ml supplied by Falcon has been found to be excellent, and volumes should be scaled accordingly. The residual gas phase in the tube is air, and tubes are tightly capped. Cultures are incubated at 37 °C and examined daily. Subcultures are best made before stationary phase is reached (i.e., approximately 10^6 organisms/ml in most complex media) because loss of viability may set in rapidly at this stage. Subcultures should normally have an initial cell density of 5 × 10^3 to 1 × 10^4 organisms/ml. *Trichomonas vaginalis* is conveniently counted using an improved Neubauer counting chamber and approximate magnification of × 200. Duplicate counts of duplicate samples are recommended for accurate quantitation.

When larger-scale cultures are desired, the normal progression is to 100-ml (4-oz) medical flats and then to 1-L Winchester bottles. Ten liters of well-grown culture gives 2 × 10^{10} cells, which is adequate for most biochemical purposes.

TABLE 7.7. Modified CMRL (Connaught Medical Research Laboratory) 1066 medium.[44]

Component	Amount (in ml for a total of 1l)
CMRL 1066 (Gibco) 10 × concentrate	100
HEPES/1M KOH, pH 7.4	20
Glucose 9% wt/vol, ascorbic acid 1% wt/vol	100
Ammonium ferric citrate 3 g/l	10
Glass distilled H_2O	770

Adjust pH to 6.0 with 1N HCl or 1N NaOH as necessary. Filter-sterilize (0.22 μm membrane); osmolarity, 248 mOsm. This medium does not store well and should be prepared immediately before use, at which time fetal calf serum is added to a final concentration of 8% vol/vol.

Growth of *T. vaginalis* on Solid Media and Its Use in Cloning

Trichomonas vaginalis can be successfully grown on solid media, although the technique is not easy, for under these circumstances the organism becomes very susceptible to desiccation and oxygen stress. The following method is modified from Samuels[35] and uses the BBL Gas Pak anaerobic container for incubation. An alternative variant using desiccators with alkaline pyrogallol for incubation is given by Diamond.[32]

TYM medium, or its modification, and CPLM are suitable as basic media. Begin with 20 ml of the appropriate warm, serum-free medium, containing 1.6% wt/vol of agar (Difco Bacto or Oxoid Ionagar) and pour into a sterile 100-mm plastic petri dish. If desired, 0.2 ml of an antibiotic mixture can be added immediately before pouring. Each plate is then placed into an airtight box containing a 5-g piece of solid CO_2 and left for at least 15 minutes to harden and absorb the CO_2. For the overlay melt 10 ml of the chosen medium containing 0.8% wt/vol of agar and hold in a water bath at 40 °C. Quickly add 1 ml horse serum and 0.1 ml antibiotic mixture if desired. Cool to 37 °C, inoculate with 0.5 ml of *T. vaginalis* suspension and immediately pour over base layer. Set the dish in the CO_2 chamber to harden and absorb gas.

As soon as the plates are hard, transfer them to a Gas-Pak anaerobic jar in which a 2 to 3 g piece of solid CO_2 has been allowed to evaporate. Activate the appropriate number of hydrogen-generating envelopes dependent upon the volume of the jar, and place these inside. An anaerobic indicator strip (BBL) may also be added if desired. The jar is then quickly sealed and incubated at 37 °C for 3 to 5 days. Depending on inoculum size this method can be used to give well isolated colonies or confluent growth within the overlay.

An alternative approach to the cultivation of *T. vaginalis* in a solid medium was adopted by Hollander.[26] This method uses a "sloppy" agar (0.36 wt/vol) medium in a single layer in a petri dish. The medium employed is a modification of Diamond's TYM medium similar to that given in Table 7.3. Plates are incubated in an anaerobic jar at 37 °C for 5 days. Hollander reported a good correlation between numbers of colonies and numbers of flagellates in the inoculum, and the method could therefore be used for estimation of viability as colony-forming ability in a manner analogous to that used for bacteria.

Both these methods are potentially useful for obtaining clones of *T. vaginalis*. The use of clone-derived cultures is of great value in studies on the immunology, experimental pathology, and biochemistry of *T. vaginalis* because it eliminates the uncertainties associated with nonhomogeneous populations. A word of caution should, however, be expressed on the use of single-step dilution and colony formation in agar as a dependable method of cloning. *T. vaginalis* cells, particularly of some strains, have an inherent tendency to aggregate and associate with one another.[2] Because of this the possibility that a colony has arisen from more than one cell must seriously be considered.

This problem can partly be overcome by recovering the cells from a colony by cutting out the agar region containing it and gentle macerating it in liquid culture medium. The organisms can then be grown up and the process of single colony selection repeated several times. The alternative is to use a cloning method in which the presence of a single cell as progenitor is unambiguously established, for example, by microscopic examination of single drops of diluted culture followed by growth of drops containing a single organism in the wells of a microtitration plate in an anaerobic jar.

A useful method for isolation of single organisms was described by Kulda and Serbus.[36] This technique utilizes an apparatus devised for isolation of microspores. It was mentioned by Kulda and Honigberg[37] in connection with isolation of *Tritrichomonas foetus* clones. The reader is referred to their report for details of this method.

Establishing the Requirements—Toward a Defined Medium

The seeds that have borne fruit in recent years in the form of major advances in our understanding of the nutrition of *T. vaginalis* were first sown by Shorb, Samuels, and other workers in the 1950s and 1960s. The author was stimulated by reading Shorb's 1964 review[38] to begin his own efforts to produce a defined medium for the cultivation of the organism. Shorb was optimistic in her review about the imminence of a defined medium for the cultivation of trichomonad species. This was based on work that, considering the difficulties at the time of securing pure samples of some of the nutrients involved, was remarkably prescient. The work carried out in this period established that trichomonads had definite nutritional requirements in three main areas: vitamins, nucleic acid precursors, and lipids.

The study of the requirements of T. vaginalis for water-soluble and fat-soluble vitamins is still in its infancy, and recent work has not addressed this question to the same degree as the other two areas; therefore, pointers from early work may still be of value. There are reports[39,40] that Trichomonas gallinae requires nicotinamide, choline, pyridoxamine or pyridoxine, pantothenate, folic acid, and biotin. Modern studies would discount folic acid or p-aminobenzoate as likely requirements for Trichomonas vaginalis. The only evidence that bears on the requirement of trichomonads for fat-soluble vitamins is the report by Shorb[38] that Trichomonas gallinae requires tocopherol.

Several early studies[38] pointed to a requirement by trichomonads for nucleic acid precursors, and Sprince et al[41] suggested that Trichomonas vaginalis required at least three bases, adenine, guanine, and uracil, for growth. This is in remarkable agreement with recent studies, which indicate that adenine, guanine, uracil, and thymidine are required as base or nucleoside. The requirement for thymidine is notoriously difficult to demonstrate by omission experiments because quantitatively it is so small.

Early studies on the lipid requirements of trichomonads have also proved to be quite reliable. Several authors[42,43] investigated the role of cholesterol in replacing serum for the growth of trichomonads and concluded that indeed there was a specific requirement for this sterol, which serum normally furnishes. This observation has subsequently been fully confirmed for T. vaginalis by the author's own studies using pure cholesterol in a defined medium.[44] In the same period Lund and Shorb[40] investigated the role of serum in supplying fatty acids for the growth of trichomonads. Using several species, although not T. vaginalis, they found a confusing picture. However, their studies did point to fatty acids being important as growth factors normally supplied by serum. In one series of experiments these authors were able to show that Tritrichomonas foetus and Tritrichomonas suis required both a saturated and an unsaturated fatty acid as well as cholesterol for the replacement of serum. Recent studies[44-46] have supported a specific requirement for fatty acids and emphasized the inability of T. vaginalis to modify the fatty acids it receives from the medium.

Growth of T. vaginalis in Defined and Semidefined Media

Although the many studies referred to above laid a sound basis for an advance to a defined or semidefined medium for growth of T. vaginalis, this advance was not made until 1981 when the present author reported the development of successful media for this purpose.[44] The strategy adopted in evolving a defined medium was similar to that adopted by Cross and Manning[47] in developing a defined medium for growth of the procyclic trypomastigate forms of Trypanosoma brucei. This was to take a suitable mammalian tissue culture medium and adapt it by adding particular factors known to be of importance in the cultivation of trichomonads. The tissue culture medium CMRL 1066 was found to allow quite reasonable growth when modified in this way and the recipe for the modification is given in Table 7.7.

The factors key to the use of this medium for the successful growth of T. vaginalis are:

1. A tissue culture medium was chosen that supplied purines and pyrimidines.
2. The buffering capacity of the medium and the quantity of glucose supplied was increased to take account of the generation of energy by glycolysis and the production of acid end products of T. vaginalis.
3. The iron content of the medium was increased in response to the known requirement of trichomonads for this element.
4. The redox potential of the medium was lowered by the addition of ascorbic acid.

The growth of the Bushby strain in this medium during the first 14 subcultures after initiation from complex medium is shown in Fig. 7.1. The average doubling in this medium was 11.6 ± 3.08 hours ($n = 15$), compared with 5.91 ± 0.60 hours ($n = 10$) for the same strain identically maintained in modified Diamond's TYM medium with 10% horse serum. The strain was maintained in modified CMRL 1066 medium for over 50 consecutive subcultures, indicating the suitability of the medium for long-term cul-

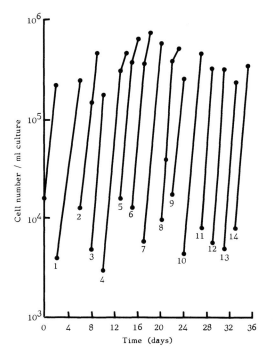

Fig. 7.1. Growth in modified CMRL 1066 medium. Organisms from modified Diamond's TYM medium (Table 7.3) were inoculated into modified CMRL 1066 medium supplemented with 8% inactivated fetal calf serum. Cell densities were counted daily for the first 36 days (14 subcultures). (All figures show results obtained with *T. vaginalis* Bushby strain.)

tivation. Organisms could also be cyropreserved in this medium using a standard technique[44] and reinoculated directly into it.

Using modified CMRL 1066 medium, a substitution for serum was developed based on a synthetic triglyceride, cholesterol, and bovine serum albumin. It was prepared in the following manner. Stock solutions of cholesterol (chromatographic grade, Sigma) and DL-glyceryl-1-palmitate-2-oleate-3-stearate (GPOS, Calbiochem) were prepared at 20 mg/ml in analytic grade absolute ethanol by warming to 37°C. A stock solution of 5 % wt/vol of bovine albumin (various sources, see text and figures) in distilled water was prepared, the pH of the solution was adjusted to 7.1 with 0.1 M NaOH, and the solution was sterilized by filtration. To prepare 5 ml of serum substitute, 50 μl of stock GPOS and 50 μl of stock cholesterol solutions were added to a sterile screw-cap 20-ml bottle. A 5-ml volume of the albumin was added to a second similar bottle and both bottles were warmed to 37 °C. The contents of the albumin-containing bottle were then rapidly added under sterile conditions to the bottle containing GPOS and cholesterol and the contents were immediately and thoroughly mixed with a "whirlimixer." It is important to follow these instructions carefully to generate a stable, slightly opalescent emulsion. If this does not occur discard the preparation and repeat the process. A successful emulsion kept at 4 °C remains usable for several months.

This serum substitute was added to culture media at a level of 10% vol/vol giving a final albumin concentration of 0.5% wt/vol and cholesterol and GPOS concentrations of 20 μg/ml. Serum substitutes were successfully prepared according to this method using Calbiochem "fatty-acid-poor" albumin and Armour "Cohn Fraction V" albumin.

When the serum substitute was combined with modified CMRL 1066 it was possible for the first time to obtain a reproducible repeated subculture of *T. vaginalis* in a serum-free medium. The progress of a subculture series initiated in this medium is illustrated in Figure 7.2. Up to 15 consecutive subcultures have been performed in this medium, indicating that it does indeed provide all the nutritional requirements of *T. vaginalis;* therefore, the possibility of carryover from the initiating culture can be discounted. The doubling time in this medium was about 14 hours.

To avoid the limitations of a medium that was purchased as a preformulated concentrate, the author has devised two media, DL7 and DL8 (first reported in reference 44), which can be modified at will to test the requirements for its components. The composition of these two media is detailed in Tables 7.8 and 7.9.

Both DL7 and DL8 media support good growth of *T. vaginalis* when supplemented with 8% inactivated fetal calf serum, whether this is in the natural state or has been extensively dialyzed. The growth characteristics of the Bushby strain of *T. vaganlis* in the two media supplemented with 8% fetal calf serum are shown in Figures 7.3 and 7.4. The doubling times were 8.73 ± 2.72 hours ($n = 11$) for DL7 medium, and 7.99 ± 1.44 hours ($n = 14$) for DL8 medium.

Both media also support growth when a serum substitute is used in place of serum. Figure 7.5 shows the progress of a culture series initi-

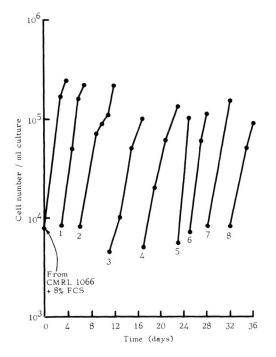

Fig. 7.2. Growth in modified CMRL 1066 medium with serum substitute consisting of 0.5% wt/vol of fatty-acid-poor bovine serum albumin (Calbiochem), 20 μg/ml cholesterol, and 20 μg/ml GPOS. Growth was followed over the next 36 days (8 subcultures).

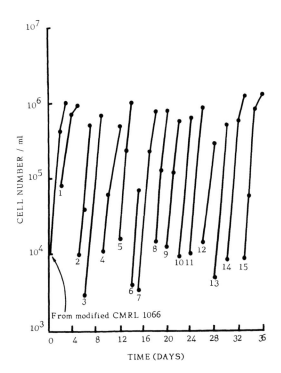

Fig. 7.3. Growth in DL7 medium. Organisms from CMRL 1066 medium were inoculated into DL7 medium with 8% inactivated fetal calf serum. Growth was followed over the next 36 days (15 subcultures).

ated in DL7 medium in which serum was replaced by 0.5 % wt/vol of Calbiochem "fatty-acid-poor" albumin with 20 μg/ml GPOS and 20 μg/ml cholesterol. Figure 7.6 shows the progress of a similar culture series using DL8 medium in which the serum substitute was 0.5% wt/vol of Armour "Cohn Fraction V" albumin with 20 μg/ml GPOS and 20 μg/ml cholesterol.

The opportunity afforded by the good growth of *T. vaginalis* in DL7 medium with serum substitute has been taken to observe more rigorously the lipid requirements of *T. vaginalis*. Using DL7, serum substitutes were tried that employed "essentially fat-free" albumin (Sigma), pretreated with activated charcoal at low pH and extensively dialyzed to remove fatty acids, together with cholesterol and various combinations of pure fatty acids. Repeated subculture was successful only when palmitic, stearic, and oleic acids were supplied as well as cholesterol (Fig. 7.7), providing strong nutritional evidence for the dependence of *T. vaginalis* on exogenous fatty acid sources, as well as for its uniquely limited capacity to modify those fatty acids with which it is supplied.

Both DL7 and DL8 media can be used to examine the detailed nutritional requirements of *T. vaginalis*, and neither in any sense represents a minimal medium for the cultivation of the organism. There is real scope for the systematic elimination of components, particularly vitamins, many of which are probably superfluous. Evidence bearing on the requirement or lack of it of *T. vaginalis* for a number of the vitamins in these media is available from other recent sources and this is discussed subsequently. A modified version of DL8 medium, called HUT (Hypoxanthine-Uracil-Thymidine) medium, has been used by Wang et al[48] to study purine and pyrimidine metabolism in *Tritrichomonas foetus*, and it is probable that with suitable adjustments the medium could also be used for the cultivation of many other trichomonad species.

TABLE 7.8. DL7 culture medium.[44]

Component	Concentration (mg/l except where indicated)
Salts	
NaCl	3000
K_2HPO_4	500
$MgSO_4 \cdot 7H_2O$	750
$CaCl_2 \cdot 6H_2O$	350
Trace elements*	5 ml
Amino acids	
Cysteine HCl	500
MEM amino acid solution 50 × (Gibco)	30 ml
MEM nonessential amino acids 100 × (Gibco)	15 ml
Nucleic acid precursors	
Adenosine	20
Cytidine	20
Uridine	20
Guanosine	20
Thymidine	10
Carbohydrates	
D(+) Glucose	2000
DL-α-Glycerophosphate diNa·$6H_2O$	1000
Ascorbic acid	1000
D(−) Ribose	100
D(+) Glucosamine HCl	100
Vitamins	
Stock solution A (100 × concentrate)†	10 ml
Stock solution B (1,000 × concentrate)‡	1 ml
Miscellaneous	
Tween 80 (tissue culture grade)	50

*Trace elements solution (200 × concentrate)
†Vitamin stock solution A (100 × concentrate)
‡Vitamin stock solution B (1000 × concentrate)

	mg/100 ml
$FeSO_4 \cdot 7H_2O$	100.0
$ZnSO_4 \cdot 7H_2O$	100.0
$CoSO_4 \cdot 7H_2O$	50.0
$CuSO_4 \cdot 5H_2O$	20.0
$MnSO_4 \cdot 4H_2O$	10.0
$Na_2MoO_4 \cdot 2H_2O$	0.6

	mg/100 ml
d-Biotin	2
D-Pantothenic acid hemi-calcium salt	80
Nicotinamide	20
Pyridoxal HCl	40
Pyridoxal-5'-phosphate	5
Riboflavin	20
Thiamine HCl	20
p-Aminobenzoic acid	10
Choline chloride	10
i-Inositol	20
Folic acid (dissolve with NaOH)	100
Vitamin B_{12}	10
D-pantethine	10

Vitamin solution A is adjusted to pH 7.4 with 0.1N NaOH or 0.1N HCl distributed in 10-ml aliquots and stored frozen at −20 °C.

TABLE 7.8. Continued

	mg/100 ml
Coenzyme A	50
Nicotinamide adenine dinucleotide	100
Flavin mononucleotide	100
Flavin adenine dinucleotide	50
Putrescine diHCL	20
DL-carnitine HCl	100
Nicotinamide adenine dinucleotide phosphate	50
p-Hydroxybenzoic acid	20
Deoxyribose	400
Dithiothreitol	100

Vitamin solution B is dissolved in 10 mM HEPES buffer pH 7.4 and stored in 1-ml aliquots at $-70\,°C$.

Methods of preparation: Single strength DL7 medium is prepared by mixing the individual components and adjusting to pH 7.1 with 1.0 M NaOH; the osmolarity should be about 233 mOsm. It can be stored single strength at $-20\,°C$ for up to 3 months before use.

When required, single strength DL7 medium is thawed, warmed up to 37 °C and stirred thoroughly to redissolve any precipitate. It may then be filter-sterilized (0.22 μm membrane) and combined with the desired serum or serum substitute before inoculation. The reader's attention is directed to the composition of the trace element solution, which was incorrectly reproduced in the original publication. Cu appeared twice rather than Co and Cu. Literal adherence to the original published recipe yields a medium with levels of Cu that are possibly toxic to some strains of *T. vaginalis*.

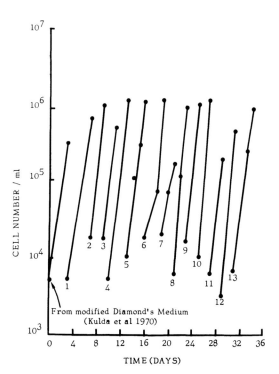

Fig. 7.4. Growth in DL8 medium. Organisms from modified Diamond's TYM medium were inoculated into DL8 medium containing 8% inactivated fetal calf serum. Growth was followed for the next 36 days (13 subcultures).

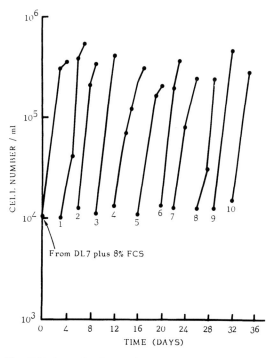

Fig. 7.5. Growth of organisms transferred from DL7 medium supplemented with 8% fetal calf serum to DL7 medium with serum substitute consisting of 0.5% wt/vol of "fatty-acid-poor" bovine serum albumin (Calbiochem) with 20 μg/ml cholesterol, and 20 μg/ml GPOS. Growth is shown over the next 36 days (10 subcultures).

TABLE 7.9. DL8 culture medium.[44]

Component	Concentration (mg/l except where indicated)
Salts	
NaCl	2000
KCl	2000
KH_2PO_4	1000
K_2HPO_4	500
Fe^{2+} gluconate (Serva)	55
$MgSo_4 \cdot 7H_2O$	750
$CaCl_2 \cdot 6H_2O$	350
Trace elements*	5 ml
Amino acids	
L-Glutamine	440
MEM essential amino acid solution 50 × (Gibco)	30 ml
MEM nonessential amino acid solution 100 × (Gibco)	15 ml
Nucleic acid precursors	
Adenine sulfate	70
Guanosine	30
Uridine	50
Thymidine	15
Carbohydrates	
Maltose	5000
DL-α-Glycerophosphate $diNa \cdot 6H_2O$	100
Ascorbic acid	1000
D(+) Glucosamine HCl	100
D(+) Galactosamine HCl	50
N-Acetyl-D-glucosamine	10
N-Acetyl neuraminic (Sigma type VI)	10
Vitamins	
Vitamin stock solution C†	10 ml

*Same as in DL7 culture medium, see Table 7.8
†Vitamin stock solution C (100 × concentrate)

	mg/100 ml
d-Biotin	20
D-Pantothenic acid hemi-calcium salt	100
Nicotinamide	50
Pyridoxamine HCl	50
Riboflavin	10
Thiamine HCl	10
p-Aminobenzoic acid	5
Coenzyme Q_{10} (dissolved in ethanol)	1
Choline chloride	10
i-Inositol	100
Reduced glutathione	100
Folic acid (dissolved with NaOH)	20
Vitamin B_{12}	10
2-Deoxy-D-ribose	50
Dithiothreitol	100
Tween 80	100

Vitamin solution C is adjusted to pH 7.4 with 0.1 M NaOH and stored frozen at −20 °C.

Method of preparation: Single strength medium DL8 is prepared by mixing the necessary components and adjusting the pH to 7.1 with 1.0 M NaOH; osmolarity is approximately 220 mOsm. This medium stores less well than DL7 medium, but can be kept frozen at −20 °C for short periods. The medium is thawed and sterilized before use as described for DL7 medium. The medium should be very pale greenish-yellow. The appearance of a brown color, or brown precipitate, is an indication that the medium should be discarded.

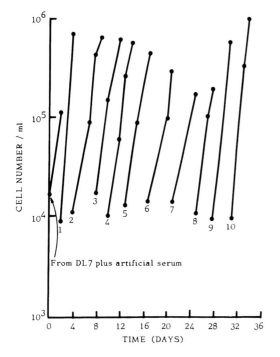
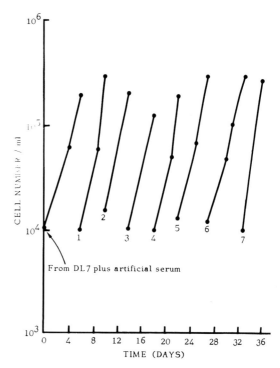

Fig. 7.6. Growth of organisms transferred from DL7 medium with serum substitute to DL8 medium with serum substitute consisting of 0.5% wt/vol of Armour "Cohn Fraction V" albumin with 20 µg/ml cholesterol and 20 µg/ml GPOS. This was prepared as described in the text. Progress of this culture series was followed for the next 36 days (10 subcultures).

Fig. 7.7. Growth with a fully defined serum replacement. Organisms from a DL7 culture with serum substitute were inoculated into DL7 medium with a serum replacement (see reference no. 44) consisting of 0.5% wt/vol of "essentially fat-free" bovine albumin (Sigma), 20 µg/ml cholesterol, 6.6 µg/ml oleic acid, 6.6 µg/ml stearic acid, and 6.6 µg/ml palmitic acid. The progress of this culture series was followed for the next 36 days (7 subcultures).

Current Understanding of the Nutrition of *T. vaginalis*

Early studies on the specific nutritional requirements of *T. vaginalis* are summarized in Shorb's review.[38] This section concentrates on recent advances in four areas: purines and pyrimidines, lipids, vitamins, and iron.

Purines and Pyrimidines

The biochemical study by Heyworth et al[49] indicated that *T. vaginalis* was unable to synthesize purines de novo, requiring them, therefore, to be salvaged from the growth medium. Moreover, this study demonstrated that *T. vaginalis* lacked the ability to carry out normal interconversion of purine nucleotides, thus preventing it from using hypoxanthine or inosine as purine sources. This inability to interconvert purines is highly unusual, even among parasitic protozoa, in which the sporozoa, for instance, utilize hypoxanthine as a preferred purine source.[50,51]

In the author's laboratory (unpublished observations) the nutritional evidence for purine and pyrimidine requirements has been checked using DL7 medium. To enable repeated subculture, the purine requirements noted were adenine or adenosine plus guanine or guanosine, confirming the lack of interconversion of purines and acceptability of these purines in the form of either bases or nucleosides. The pyrimidine requirements could be satisfied by cytidine plus thymidine. The utilization of the bases cytosine and thymine was not checked. The results with regard to pyrimidine utilization are in accord with the observed pattern of pyrimidine metabolizing enzymes in extracts,[52,53] which would permit interconversion of cytidine and uridine, but not de novo synthesis of the pyrimidine ring. Thymidine is required specifically because of the lack of dihydrofolate reductase

Lipids

The nutritionally identified requirement for cholesterol has been reported by several workers and is referred to earlier in the sections on establishing a defined medium. The requirement for saturated and unsaturated fatty acids, also identified nutritionally, is discussed there also. Additional biochemical evidence is available that bears on the nature of this requirement and the manner in which it may be satisfied when the parasite is in its natural environment, the vagina. Roitman and co-workers[45] observed that *T. vaginalis* could not synthesize de novo any major class of lipids from acetate or glucose. This observation has been confirmed in the author's laboratory using defined medium DL7.[44] Holz et al[54] have recently reported that *T. vaginalis* cannot oxidize (α, β, ω), chain elongate, desaturate, or saturate internal double bonds, or introduce branches or rings in the fatty acids it receives from the growth medium. Using radiolabeled cholesterol, no evidence was found for ring or side chain desaturation or saturation, alkylation or dealkylation, or isomerization of this sterol.

The manner in which *T. vaginalis* normally obtains its lipid requirements is suggested by the recent work of Peterson and Alderete,[55] who showed that this species avidly and specifically bound human lipoproteins, and that these were subsequently internalized. Purified lipoproteins could be used instead of serum in a complex medium; this supported high multiplication rates, indicating that both sterol and fatty acid requirements were satisfied from this source.

Vitamins and Cofactors

There is direct or indirect evidence of the importance of several of the vitamins included in the DL7 and DL8 media that could be valuable to workers seeking a more minimal mixture. *p*-Aminobenzoic acid and folic acid are unlikely to be required by *T. vaginalis* because of the absence of one carbon metabolism employing folate cofactors.[53] Choline is not required by the urogenital trichomonad, since this base cannot be taken up by the intact organism.[56] Coenzyme Q_{10} is also not required; there is no evidence for any quinone involvement in electron transport in this parasite. On the positive side the recent detection of pyridoxal phosphate–dependent amino transferases in *T. vaginalis*[57] suggests that this species requires this vitamin in one form or another.

Experiments in the author's laboratory (with S. Bradley, unpublished) have shown that exogenous inositol is freely incorporated into inositol phospholipids by *T. vaginalis*. In the absence of exogenous phospholipid sources, this substance may well be a nutritional requirement.

Metals

Intriguing experiments have been performed by Peterson and Alderete[58] on the mechanism of iron uptake by *T. vaginalis*. These authors showed that this organism possessed specific surface receptors for human lactoferrin, an iron-sequestering protein naturally found in the secretion of mucosal surfaces such as the vagina. These receptors permitted the internalization by the trichomonad of radiolabeled iron bound to lactoferrin. Exogenous iron bound to lactoferrin was found to be highly effective in stimulating the activity of pyruvate: ferredoxin oxidoreductase, one of the iron-sulfur proteins of the hydrogenosome, when added to the growth medium. These results suggest strongly that *T. vaginalis* in the human host obtains its vital iron requirements by scavenging iron-loaded lactoferrin with a high affinity receptor.

Cryopreservation of *T. vaginalis*

Cryopreservation is an important backup for any laboratory interested in the long-term maintenance of *T. vaginalis* strains. Liquid nitrogen storage provides an insurance against loss caused by contamination or accident, and eliminates possible problems of changing pathogenicity and antigenic character during repeated in vitro subculture. When large numbers of strains are maintained it also reduces the risk of cross-contamination and misidentification of individual isolates. A number of methods have been reported as successful for cryopreservation, and it is clear that the urogenital trichomonad is not particularly exacting in its requirements in this respect. However, attention to the rate of freezing, the maintenance of refrigeration, and the precise manner of thawing is important to securing high viability and rapid reestablishment of cultures.

Trichomonads should be stored either immersed in liquid nitrogen ($-196\,°C$), or in the vapor phase above liquid nitrogen ($-170\,°C$), in suitable ampules sealed to prevent entry of the liquid phase. A consensus has not emerged on either the optimal rate of cooling during the initial freezing process or a particular favored suspension medium, although dimethyl sulfoxide (DMSO) is clearly the preferred cyroprotectant.[59] Ivey and Hall[60] studied the effect of the cooling rate on survival and concluded that 2 °C or 5 °C per minute was satisfactory, while Diamond[61] had good results with 8 °C per minute. Diamond[61] and other authors such as Honigberg et al.[62] note the importance of traversing the region of release of latent heat of fusion as quickly as possible by rapid supercooling. A description of the cryopreservation method favored by Honigberg's group[62] using Linde freezing equipment is given in Chapter 16. A simpler method, originally developed for the cryopreservation of *Entamoeba* trophozoites,[63] is offered here as an alternative. It is adaptable to any means of slow cooling that provides for cooling rates of 1 °C per minute, and still achieves high viability despite its simplicity. The procedure used is as follows. Organisms from a well growing, but *not* stationary-phase culture are spun down and suspended at a density of approximately 10^7/ml in an appropriate culture medium in which they have been growing. DMSO is added with continual shaking to the suspension to give a final concentration of 7.5% vol/vol. Aliquots of 1 ml of the suspension are transferred to 3-ml screw-cap sterile plastic freezing ampules, several samples normally being frozen from each culture. All manipulations must be carried out aseptically. The ampules are transferred to an automatic liquid nitrogen–cooled freezing apparatus or other device to ensure uniform slow cooling and are cooled, at a rate of 1 °C per minute, from ambient temperature to $-30\,°C$. Once at this temperature ampules can be transferred directly to liquid nitrogen or to liquid nitrogen vapor for subsequent storage. The transfer should be made as rapidly as possible to avoid any rewarming. Viability with increasing length of storage has been discussed by several authors; however, provided the liquid nitrogen refrigeration is maintained without fault, cultures that are viable on thawing immediately after freezing by this procedure can be expected to maintain this state for many years.

Care during thawing and dilution of the DMSO is repaid by more rapid reestablishment of cultures. Ampules taken from liquid nitrogen are allowed to thaw completely at room temperature and the contents are aseptically transferred to a normal culture tube. An equal volume of fresh complete medium at room temperature is then added with gentle mixing and the cell suspension is allowed to stand for 5 to 10 minutes. Further fresh medium is then added, 1 ml at a time, over a period of about 20 min, until the tube is almost full. A normal subculture (5% to 10% of volume) is then made from this tube and both tubes are incubated at 37 °C. Tubes should be examined daily thereafter until further subculture is possible.

Cultivation of Other Trichomonad Species Occurring in Humans

Trichomonas tenax (O. F. Müller)

Although their inherent pathogenicity is debatable, two other species of thrichomonad are found in man: *Trichomonas tenax,* the oral trichomonad, and *Pentatrichomonas hominis* from the large intestine. *Trichomonas tenax* is significantly more difficult to axenize and maintain in culture than *Trichomonas vaginalis,* and the extent of knowledge of this species is correspondingly scanty. In general, the complex serum-containing media that are suitable for *T. vaginalis* will support growth of *Trichomonas tenax* after adjustment, if necessary, of the pH to pH 7.0 to 7.5, which is reported to be optimal for this organism.[1] *Trichomonas tenax* was first obtained in axenic culture by Diamond,[64] who used a complete medium containing TTY (tryptose-trypticase-yeast) broth supplemented with horse serum, chick embryo extract, and antibiotics. The composition of this medium is shown in Table 7.10. There are no reports known to me of the direct axenization of *Trichomonas tenax* from host material or polyxenic cultures. The sole success in axenization is that of Diamond,[32,64] in which he employed a monoxenic culture with *Trypanosoma cruzi* as a basis for transfer to

TABLE 7.10. Medium for the culture and axenization of *T. tenax* according to Diamond.[32, 64]

Basal Broth TTY*	
Component	Amount (in g except where indicated)
Tryptose (Difco)	10.0
Trypticase (BBL)	10.0
Yeast extract (BBL)	10.0
Glucose	5.0
L-Cysteine HCl	1.0
Ascorbic acid	0.2
NaCl	5.0
K_2HPO_4	0.8
HK_2PO_4	0.8
Distilled H_2O	To 1600 ml

*Dissolve salts in 600 ml of distilled water. Add tryptose and dissolve by heat (50 °C). Add remaining ingredients by stirring. Adjust pH to 7.0 with 1 M NaOH. Bring total volume to 1600 ml. Distribute in 160-ml amounts. Add 0.1 g Bacto agar (Difco) to each. Autoclave and cool to 45 °C.

Crude Chick Embryo Extract, 25% ($CEEC_{25}$)†	
Components	Amount (in g except where indicated)
A. *Buffered Saline:*	
NaCl	5.0
KH_2PO_4	1.6
Na_2HPO_4	1.6
Distilled H_2O	1000 ml
B. *Chick Embryos* (11–12 day)	1 g tissue/2 ml sterile buffered saline

†To prepare the chick embryo extract indicated by Diamond[32,64] $CEEC_{25}$, aseptically harvest 11- to 12-day-old chick embryos. Remove eyes and beak. Weigh and place in a sterile blender (Waring). Add chilled buffered saline 2 ml/g tissue and blend for 2 minutes. Refrigerate for 1 hour, remove liquid from beneath froth, and transfer to sterile screw-cap centrifuge tubes. Centrifuge 20 minutes, 850 g, 4 °C. Collect supernatant fluid, pool and distribute in 10-ml amounts. Rapid-freeze and store at −20 °C.

Method of preparation of the complete medium: To each 160 ml of TTY broth and liquified agar at 45 °C add 20 ml horse or bovine serum and 10 ml vitamin mixture NCTC107 (K.C. Biologicals Inc., Lenexa, Kansas 66215, USA; see also reference no. 32). Distribute in 9.5-ml amounts. Store at 4 °C. Use within 10 days. Immediately before use warm to room temperature and add 0.5 ml $CEEC_{25}$.

axenic culture. To establish axenic cultures, 2 to 4 × 10^5 trichomonads from a 72-hour monoxenic culture were transferred to tubes containing TTY–serum–chick embryo extract (Table 7.10). These tubes were incubated at 35.5 °C and examined daily for the multiplication of the trichomonad. Subculture occurred once multiplication was established. The *Trypansoma cruzi* associate died out by the second or third subculture. Large inocula were used initially. *Trichomonas tenax,* like some strains of *Trichomonas vaginalis,* has a tendency to grow attached to the wall of the culture tube, but despite this Diamond[32] reported yields of 1 to 3 × 10^6 organisms/ml. Once the cultures were well established (more than 10 subcultures), chick embryo extract could be eliminated.

Cryopreservation of *T. tenax* has also proved difficult, with only partial success reported, and several observations unconfirmed. Use of DMSO as a cryoprotectant would appear to offer the best chance of success, together with a cooling protocol devised for *T. vaginalis*. Numbers of duplicate frozen stabilates should be prepared as Honigberg[2] reports difficulty in reestablishing cultures from thawed material.

Pentatrichomonas hominis (Davaine)

Although generally more amenable to cultivation and axenization than *T. tenax*, *Pentatrichonmonas hominis* can be troublesome to axenize directly, because of the extensive intestinal bacterial flora that accompanies it. Xenic cultivation is readily achievable, however, using one of the complex media suitable for *T. vaginalis*. With care and judicious use of combinations of antibiotics from Table 7.1, axenic cultures can be obtained. Once stable growth of axenic cultures is obtained (see reference 22), the organism is no more fastidious in its growth requirements than the urogenital trichomonad. The present author has used the medium in Table 7.2 supplemented with 10% horse serum for the successful cultivation of *P. hominis*. For this purpose the pH of the medium should be adjusted to pH 7.0 rather than pH 6.8.

Pentatrichomonas hominis can be cryopreserved without undue difficulty; either Honigberg's method described by McMillan (see Chapter 16), or the abbreviated method described earlier in the section on nutrition of *T. vaginalis* may be employed successfully. Both of these methods use DMSO as cryoprotectant.

Unfortunately rather few strains of *P. hominis* are in axenic cultivation. The isolation of additional strains would stimulate an assessment of the biochemical and immunologic attributes of this interesting organism and of the extent to which it differs from the much better known *T. vaginalis*.

References

1. Trussell RE: Trichomonas vaginalis *and Trichomoniasis*. Springfield, Ill: Charles C Thomas, 1947.
2. Honigberg BM: Trichomonads of importance in human medicine. In Kreier JP (ed): *Parasitic Protozoa*, vol 2, pp 296-454. New York: Academic Press, 1978.
3. Lynch KM: Cultivation of *Trichomonas* from the human mouth, vagina, and urine. *Am J Trop Med Hyg* 2:521-538, 1922.
4. Sorel D: Trois techniques de recherche du *Trichomonas vaginalis:* leur valeurs comparés. *Presse Még* 62:602-604, 1954.
5. Fouts AC, Kraus SJ: *Trichomonas vaginalis:* Reevaluation of its clinical presentation and laboratory diagnosis. *J Infect Dis* 141:137-143, 1980.
6. Smith RF, Horen P: Inhibition of *Trichomonas vaginalis* by fungi during associative growth. *Sex Transm Dis* 7:172-174, 1980.
7. Adler S, Pulvertaft RJV: The use of penicillin for obtaining bacteria-free cultures of Trichomonas vaginalis Donné 1837. *Ann Trop Med Parasitol* 38:188-189, 1944.
8. Johnson G, Trussell MH: Physiology of bacteria-free *Trichomonas vaginalis*. VII. Temperature in relation to survival and generation. *Proc Soc Exp Biol Med* 57:242-254, 1944.
9. Quisno RA, Foter MJ: The use of streptomycin in the purification of cultures of *Trichomonas vaginalis*. *J Bacteriol* 51:404, 1946.
10. Lowe GH: A comparison of current laboratory methods and a new semi-solid culture medium for the detection of *Trichomonas vaginalis*. *J Clin Pathol* 18:432-434, 1965.
11. McEntegart MG: The application of a haemagglutination technique to the study of *Trichomonas vaginalis* infections. *J Clin Pathol* 5:275-280, 1952.
12. Honigberg BM, Livingstone MC, Frost JK: Pathogenicity of fresh isolates of *Trichomonas vaginalis:* "The mouse assay" versus clinical and pathologic findings. *Acta Cytol* 10:353-361, 1966.
13. Kulda J, Honigberg BM, Frost JK, Hollander DH: Pathogenicity of *Trichomonas vaginalis*. *Am J Obstet Gynecol* 108:908-918, 1970.
14. Osborne NG, Grubin L, Pratson L: Vaginitis in sexually active women: Relationship to 9 sexually transmitted organisms. *Am J Obstet Gynecol* 142:962-967, 1982.
15. Periman D, Rahman SB, Semar JB: Antibiotic control of mycoplasmas in tissue culture. *Appl Microbiol* 15:82-85, 1967.
16. Glaser RW, Coria NA: Purification and culture of *Trichomonas foetus* (Riedmüller) from cows. *Am J Hyg* 22:221-226, 1935.
17. Lowe GH: A comparison of culture media for the isolation of *Trichomonas vaginalis*. *Med Lab Technol* 29:389-391, 1972.
18. Stabler RM, Honigberg BM, King VM: Effect of certain laboratory procedures on virulence of the Jones' Barn strain of *Trichomonas gallinae* for pigeons. *J Parasitol* 50:36-41, 1964.
19. Gorrell TE: Iron enhances H_2 production by *Trichomonas vaginalis*. *J Protozool* 27:17A, 1980.
20. Gorrell TE: Effect of culture medium iron content on the biochemical composition and metabolism of *Trichomonas vaginalis*. *J Bacteriol* 161:1228-1230, 1985.
21. Mack SR, Müller M: End products of carbohydrate metabolism in *Trichomonas vaginalis*. *Comp Biochem Physiol (B)* 67:213-216, 1980.
22. Diamond LS: The establishment of various trichomonads of animals and man in axenic cultures. *J. Parasitol* 43:488-490, 1957.

23. Samuels R: Growth of axenic trichomonads in a serum-free medium. In Prog Protozool, Proc 2nd Int Conf Protozool, 1965. *Excerpta Med Found Int Congr Ser* 91:200, 1965.
24. Samuels R, Beil E: Serum-free medium for axenic cultures of trichomonads. *J Protozool* 9 (suppl):19, 1962.
25. Nakabayashi T, Miyata A: Examination of a milk medium in the cultivation of *Trichomonas vaginalis*. *Trop Med* 10:39-49, 1968.
26. Hollander DH: Colonial morphology of *Trichomonas vaginalis* in agar. *J Parasitol* 62:826-828, 1976.
27. Müller WA, Gottschalk C: Standardisierung der parasitologischen Nachweisverfahren für Trichomononadeninfektionen. *Angew Parasitol* 11: 170-176, 1970.
28. Johnson G, Trussell RE: Experimental basis for the chemotherapy of *Trichomonas vaginalis* infections. I. *Proc Soc Exp Biol Med* 54:245-249, 1943.
29. Kupferberg AB, Johnson G, Sprince H: Nutritional requirements of *Trichomonas vaginalis*. *Proc Soc Exp Biol Med* 67:304-308, 1948.
30. Cox PJ, Nicol CS: Growth studies of various strains of *T. vaginalis* and possible improvements in the laboratory diagnosis of trichomoniasis. *Br J Vener Dis* 49:536-539, 1973.
31. Rayner CFA: Comparison of culture media for the growth of *Trichomonas vaginalis*. *Br J Vener Dis* 44:63-66, 1968.
32. Diamond LS: Lumen dwelling protozoa: *Entamoeba*, trichomonads, and *Giardia*. In Jensen JP (ed): In Vitro *Cultivation of Protozoan Parasites*, pp 65-109. Boca Raton, Fla: CRC Press, 1983.
33. Meerovitch E: *Entamoeba, Giardia,* and *Trichomonas*. In Taylor AER, Baker JR (eds): *Methods of Cultivating Parasites* In Vitro, pp 19-37. London: Academic Press, 1978.
34. Jírovec O, Petrù M: *Trichomonas vaginalis* and trichomoniasis. *Adv Parasitol* 6:117-188, 1968.
35. Samuels R: Agar techniques for colonizing and cloning trichomonads. *J Protozool* 9:103-107, 1962.
36. Kulda J, Serbus C: Isolation of trichomonad clones by means of a "monosporic isolation apparatus." *Folia Parasitol (Prague)* 15:163-167, 1969.
37. Kulda J, Honigberg B: Behavior and pathogenicity of *Tritrichomonas foetus* in chick liver cell cultures. *J Protozool* 16:479-495, 1969.
38. Shorb MS: The physiology of trichomonads. In Hutner SH (ed): *Biochemistry and Physiology of Protozoa*, vol 3, pp 383-457. New York: Academic Press, 1964.
39. Jones L, Smith BF: Certain B complex vitamins as growth promoting factors for *Trichomonas gallinae*. *Exp Parasitol* 8:509-514, 1959.
40. Shorb MS, Lund PG: Requirement of trichomonads for unidentified growth factors, saturated and unsaturated fatty acids. *J Protozool* 6:122-130, 1959.
41. Sprince H, Goldberg ER, Kucker G, Lowy RS: The effect of ribonucleic acid and its nitrogenous constituents on the growth of *Trichomonas vaginalis*. *Ann NY Acad Sci* 56:1016-1027, 1953.
42. Mandel M, Honigberg BM: The response of *Trichomonas gallinae* to cholesterol and dihydrocholesterol. *Anat Rec* 128:586, 1957.
43. Lund PG, Shorb MS: Steroid requirements of trichomonads. *J Protozool* 9:151-154, 1962.
44. Linstead D: New defined and semi-defined media for cultivation of the flagellate *Trichomonas vaginalis*. *Parasitology* 83:125-137, 1981.
45. Roitman I, Heyworth PG, Gutteridge WE: Lipid synthesis by *Trichomonas vaginalis*. *Ann Trop Med Parasitol* 72:538-585, 1978.
46. Linstead D: Further studies on the cultivation of *Trichomonas vaginalis* in a defined medium. *Parasitology* 81:18-19, 1980.
47. Cross GAM, Manning JC: Cultivation of *Trypanosoma brucei* spp. in semi-defined and defined media. *Parasitology* 67:315-331, 1973.
48. Wang CC, Wang AL, Rice A: *Tritrichomonas foetus*: Partly defined cultivation medium for study of the purine and pyrimidine metabolism. *Exp Parasitol* 57:68-75, 1984.
49. Heyworth PG, Gutteridge WE, Ginger CD: Purine metabolism in *Trichomonas vaginalis*. *FEBS Lett* 141:106-109, 1982.
50. Irvine AD, Young ER, Purnell RE: The in vitro uptake of tritiated nucleic acid precursors by *Babesia* spp. of cattle and mice. *Int J Parasitol* 8:19-24, 1978.
51. Schmidt G, Walter RD, Konigk E: Adenosine kinase from normal mouse erythrocytes and from *Plasmodium chabaudi:* Partial purification and characterisation. *Tropenmed Parasitol* 25:301-308, 1974.
52. Miller RL, Linstead D: Purine and pyrimidine metabolising activities in *Trichomonas vaginalis* extracts. *Mol Biochem Parasitol* 7:41-51, 1983.
53. Wang CC, Chang HW: Salvage of pyrimidine nucleosides by *Trichomonas vaginalis*. *Mol Biochem Parasitol* 10:171-184, 1984.
54. Holz GG Jr, Lindmark DG, Beach DH, Neale KA, Singh BN: Lipids and lipid metabolism of trichomonads. In Kulda J, Cerkasov J (eds): Proceedings of the International Symposium on Trichomonads & Trichmoniasis, Prague, July 1985 (Post-Symp Publ Pt 1), *Acta Univ Carolinae* (Prague) *Biol* 30 (3,4):299-311, (1986) 1987.
55. Peterson KM, Alderete JF: Selective acquisition of plasma proteins by *Trichomonas vaginalis* and human lipoproteins as a growth requirement for this species. *Mol Biochem Parasitol* 12:37-48, 1984.
56. Linstead D, Bradley S: Phosphatidylcholine and

phosphatidylethanolamine biosynthesis in *Trichomonas vaginalis*. Abstracts Spring Meet Br Soc Parasitol, Nottingham, 1985, p 36, 1985.
57. Lowe PN, Rowe AF: Aspartate: 2-Oxoglutarate aminotransferase from *Trichomonas vaginalis*. *Biochem J* 232:689-695, 1985
58. Peterson KM, Alderete JF: Iron uptake and increased intracellular enzyme activity follow host lactoferrin binding by *Trichomonas vaginalis* receptors. *J Exp Med* 160:398-410, 1984.
59. Lumsden WHR, Robertson DHH, McNeillage GJC: Isolation, cultivation, low temperature preservation and infectivity titration of *Trichomonas vaginalis*. *Br J Vener Dis* 42:145-154, 1966.
60. Ivey MH, Hall DG: Use of solid medium techniques to evaluate factors affecting the ability of *Trichomonas vaginalis* to survive freezing. *J Parasitol* 61:1101-1103, 1975.
61. Diamond LS: Freeze-preservation of protozoa. *Cryobiology* 1:95-102, 1964.
62. Honigberg BM, Farris VK, Livingston MC: Preservation of *Trichomonas vaginalis* and *Trichomonas gallinae* in liquid nitrogen. In Prog Protozool, 2nd Int Conf Protozool, 1965. *Excerpta Med Found Int Congr Ser* 91:199, 1965.
63. Neal RA, Latter VS, Richards WHG: Survival of *Entamoeba* and related amoebae at low temperature. II. Viability of amoebae and cysts stored in liquid nitrogen. *Int J Parasitol* 4:353-360, 1974.
64. Diamond LS: Axenic cultivation of *Trichomonas tenax*, the oral flagellate of man. I. Establishment of cultures. *J Protozool* 9:442-444, 1962.
65. Meingassner JG, Georgopoulos A, Patoshka M: Intravaginale Infektionen der Ratte mit *Trichomonas vaginalis* und *Candida albicans:* Ein Modell zur experimentellen Chemotherapie. *Tropenmed Parasitol* 26:395-398, 1975.

8

Employment of Experimental Animals in Studies of *Trichomonas vaginalis* Infection

Jaroslav Kulda

Introduction

Susceptibility of nonspecific laboratory hosts to experimental infections with pathogenic trichomonads was recognized by early investigators, who established disease-producing infections in mice by parenteral inoculation of bacteria-free parasites of pigeons, *Trichomonas gallinae*,[1] and of cattle, *Tritrichomonas foetus*.[2] Surprisingly, early attempts to infect small mammals with *Trichomonas vaginalis* were mostly unsuccessful.[3] The first account of lasting experimental infections of the laboratory mouse obtained with the human urogenital trichomonad was published by Schnitzer et al.[4] in 1950. These authors provided also the first descriptions of lesions in mice caused by *T. vaginalis* after intraperitoneal, intramuscular, and subcutaneous inoculation. Thereafter, numerous investigators used laboratory rodents in studies of host-parasite interactions, virulence assays, and drug tests. The laboratory mouse has been the animal most widely used, although rats, guinea pigs, and hamsters have also been employed. The inoculation routes most frequently used were intraperitoneal, subcutaneous, and intravaginal. The literature on the subject is voluminous, and exhaustive coverage is not attempted in this chapter. Early data on animal experiments with *T. vaginalis* were listed by Trussell.[3] Later the subject was repeatedly reviewed,[5-8] usually with emphasis on the use of laboratory animals in antitrichomonal drug assays.[5,6,8] The most complete survey of the literature until 1977 was published by Honigberg.[9]

Although numerous data are available on animal experiments with trichomonads, not all contribute significantly to an understanding of the disease and the parasite. The complex environment of the human vagina, changing in response to the hormonal cycle and challenging the parasite with defense mechanisms specific to the site, can hardly be paralleled in experimental animals, except perhaps in some nonhuman primates. Therefore, the finding of a suitable animal model of human trichomoniasis is a difficult task that has not yet been accomplished satisfactorily. Intravaginal infections of animals may be expected to simulate most reliably the human disease, but even here the parasite meets conditions that differ considerably from those occurring in the human genitourinary tract. The ectopic infections (i.e., unnatural tissue infections) expose the parasite to host responses that do not occur at the natural site. Therefore, their results have to be viewed with caution to avoid erroneous interpretations. Still, useful assays can be performed and valuable information obtained if the animal models are used sensibly. An understanding of the specific features of the host-parasite interactions at a given site of a particular host and knowledge of the limits of the model used are prerequisites for meaningful employment of laboratory animals in studies of trichomoniasis.

Intravaginal Infections

Introduction

Intravaginal infections of experimental animals with *T. vaginalis* are likely to provide manifestations that are most similar to those found in human urogenital trichomoniasis. However, this route of inoculation presents numerous technical difficulties and has met with limited success. Successful intravaginal infections were reported for monkey, guinea pig, hamster, rat, and mouse but repeated attempts to infect these animals frequently yielded inconsistent results or failed.

Susceptibility of laboratory animals to intravaginal infection is related to their endocrine status and varies with changing phases of their sexual cycle. Failure to recognize this relationship could be the main source of difficulties in infecting the animals. Basic data on the reproductive physiology of laboratory animals that may be useful in establishing *T. vaginalis* intravaginally can be found in other works.[10,10a,10b]

Sexual Cycles

Of the laboratory animals used in trichomoniasis research, only the Old World monkeys (e.g., *Macaca*) have a menstrual cycle similar to that in humans. Sexually mature female laboratory rodents have a different type of sexual cycling, separated by short periods of heat (estrus), the estrous cycle. Essentially, the menstrual and the estrous cycles are similar. Both are governed by complex endocrine relationships among hypothalamus (releasing factors), pituitary (gonadotropins), and ovary (estrogens, progesterone). Both cycles include the preovulatory follicular period dominated by ovarian estrogens and the postovulary luteal phase controlled by progesterone secreted by the corpus luteum. There are, however, substantial differences between the two cycles (Fig. 8.1).

1. The estrous cycle is characterized by a specific period of sexual receptivity closely correlated with ovulation. In the menstrual cycle the receptivity is fairly unrestricted and the ovulation occurs in the midcycle.
2. In the estrous cycle, the response of uterine endometrium to the stimulatory effect of progesterone is mild and the phase of its development is followed by a resorptive process without sloughing off and hemorrhage. In the menstrual cycle, an extensive development of endometrium during the phase of hormonal stimulation is followed by sloughing off of endometrium and bleeding after the fall of progesterone levels.
3. In estrous animals, the estrogen levels drop sharply after ovulation, while in menstruous primates the residual levels of estrogen remain relatively high and, in synergism with progesterone, influence the development of endometrium.
4. A transient preovulatory peak of progesterone occurs in the estrous cycle, exerting, in synergism with estrogens, a positive feedback effect on secretion of gonadotropins needed to start ovulation. This does not happen in the menstrual cycle, in which estrogen alone induces the preovulatory surge of gonadotropins.
5. Finally the estrous cycle of most laboratory animals is considerably shorter than the menstrual cycle of monkeys and humans.

Estrous Cycle and Infection

The *estrous cycle* of laboratory animals includes four major phases: *proestrus, estrus, metestrus,* and *diestrus.* During the proestrus and estrus, active proliferation of epithelium occurs in the reproductive organs culminating in ovulation at the end of estrus. Degenerative epithelial changes occur in metestrus, which follows estrus shortly after ovulation. Diestrus, the longest phase of the cycle, is the period of quiescence or moderate cell growth. The levels of estrogen start to increase in diestrus reaching a peak in the midproestrus and decreasing sharply before ovulation. Preovulatory increase in progesterone occurs in midproestrus; the hormone levels drop in estrus and rise again after ovulation, decreasing gradually in diestrus.

Optimal conditions for intravaginal growth of *T. vaginalis* are apparently achieved under the influence of increased levels of estrogens occurring during proestrus and preovulatory estrus. Change of the phase to metestrus and diestrus could terminate the infection or induce its latent phase. Therefore, a supportive treatment with estrogens is usually employed to keep the animals in protracted estrus, thus aiding the establishment of persisting infection. In certain animals (rat) ovariectomy is recommended to stabilize the effect of estradiol treatment and to prevent interference of naturally produced ovarian steroids.

Monitoring the Estrous Cycle

The stage of the estrous cycle should be monitored to determine the optimum time for trichomonad inoculation. It can be assessed by cytologic examination of vaginal smears[10] or by

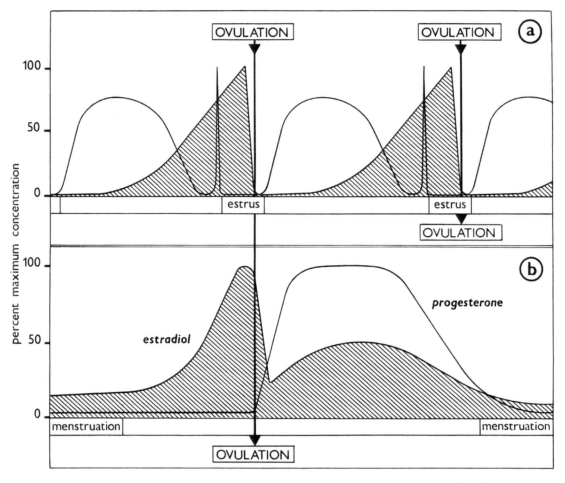

Fig. 8.1. Diagrammatic comparison of estrous (**a**) and menstrual (**b**) cycles. The former cycle is relatively short, with estrus, the period of heat, being synchronized with ovulation. After ovulation, the estradiol levels (*shaded areas*) drop abruptly to zero. The menstrual cycle, significantly longer than the estrous cycle, is unrelated to sexual receptivity of the female. The fall of the progesterone (*solid line*) level at the end of the luteal phase induces periodic bleeding; ovulation occurs around the midcycle. There is a marked decrease of the estradiol level following ovulation; however, relatively high residual amounts of this hormone are maintained throughout the entire cycle. A rise in the progesterone level occurs in the luteal phases of both cycles; however, the pronounced progesterone peak preceding ovulation during the estrous cycle is not evident in the course of the menstrual cycle.

determining changes in plasma hormone levels with the aid of radioimmunoassay.[11]

Mouse and Rat

Characterization of the cycle by means of vaginal smears is fast, simple, and useful in routine examination of the mouse and rat. It is based on a detection of changes in vaginal epithelium during the cycle. Briefly, the epithelium becomes cornified, the cornified layer is eventually shed, and an influx of leukocytes follows. The exfoliative cytology at estrus (Fig. 8.2b) is characterized by the presence of anucleate cornified epithelial cells and the absence of leukocytes. In metestrus, numerous leukocytes are present together with sparse cornified cells and squamous nucleate cells. Diestrus (Fig. 8.2c) is characterized by a predominance of leukocytes, while in proestrus (Fig. 8.2a) the leukocytes are sparse, and rounded nucleate cells predominate. The absence of leukocytes in vaginal

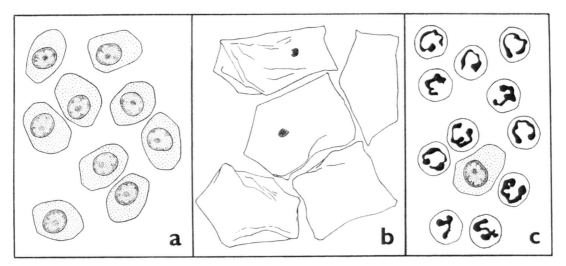

Fig. 8.2. Cytologic picture typical of vaginal smears taken from rats and mice in the three main phases of the estrous cycle. **a.** Proestrus, characterized by predominance of nucleated epithelial cells. **b.** Estrus, characterized by the presence of typically anucleated, cornified epithelial cells and by the absence of leukocytes. **c.** Diestrus. Some epithelial cells are seen among numerous leukocytes, but no anucleated cornified epithelial cells are evident.

smears during preovulatory estrus and their influx in the postovulatory period are common to all laboratory rodents. However, the patterns of exfoliative cytology occurring in the hamster and the guinea pig are different.

Hamster

In the hamster three different pictures are seen during the short period of estrus.[10] Therefore, the vaginal smears are of little value for determination of estrus. The end of estrus, however, is marked by a copious postovulatory discharge. An opaque, creamy white, very viscous discharge filling the vagina extends to the outside and can be easily recognized. Because the estrous cycle of the hamster is very regular, the appearance of the discharge can be used for predicting the time of the next estrus. Animals showing the discharge at the morning examination are likely to enter the next estrus 4 days later in the evening.

Guinea Pig

In the guinea pig sampling of vaginal secretions is restricted by the presence of an epithelial vaginal membrane that covers the vaginal orifice during most of the cycle.[12] The membrane opens periodically, usually 1 day before the estrus, and remains open for 2 to 3 days. The cytologic pictures seen during this short period are characteristic of proestrus, estrus, and the late phase of estrus around the time of ovulation, with an early influx of leukocytes.[12,13]

Effect of Estrogens on Infection

Factors directly responsible for enhancing the growth of trichomonads in estrogenized animals are not known. There is some evidence that the effect of estrogens is indirect and site-specific, since estradiol was found to be injurious to *T. vaginalis* in vitro[14] and did not exert any stimulatory effect on subcutaneous infection of the mouse with this trichomonad.[9] Increased glycogen content was observed in the vaginal mucosa of spayed rats after treatment with estradiol, suggesting a possible association between an increased susceptibility to infection and accumulation of a substrate that could be used as a nutrient by the parasite.[15]

The physiologic background of differential susceptibility of laboratory animals to intravaginal infection with trichomonads, and mechanisms by which estrogens stimulate the infection, require further studies. Such investigations would make intravaginal models far more amenable to routine use and would also provide for a better understanding of the role played by ovarian hormones in modifying sus-

Monkey

Intravaginal infections of nonhuman primates with *T. vaginalis* may yield results that are most relevant to human trichomoniasis. This possibility has not yet been fully exploited and baseline studies are needed to develop adequate methods and to define optimum conditions for the establishment of lasting infections.

Experimental Hosts

At least four species of monkeys appear to be susceptible to the infection: the rhesus monkey, *Macaca mulatta*[16-20]; the crab-eating macaque (cynomolgus monkey), *Macaca irus*[20]; the stump-tailed macaque, *Macaca arctoides*[21]; and the squirrel monkey, *Saimiri sciureus*.[22,23] Attempts to infect chimpanzees, marmosets, and tamarins were unsuccessful,[22] but the number of inoculated animals was too small to provide for valid conclusions of their susceptibility. A short-lived infection (6 days) was induced in one of the two inoculated owl monkeys, *Aotus trivirgatus*.[22]

The conditions prevailing in the rhesus monkey seem to approach more closely those found in the human vagina than those obtained in the New World monkeys. Sexually mature *M. mulatta* females have a regular menstrual cycle[24] and their sexual receptivity is not limited to a distinct breeding season; the vaginal pH is acid, pH 4.4 to 6.9.

The squirrel monkey is a polyestrous animal with a restricted breeding season and has no vaginal bleeding. Unlike the acid pH of the human vagina, that of the squirrel monkey is neutral, and its vaginal flora does not include lactobacilli.[22]

Inoculation and Susceptibility

Inoculation Procedures

Before inoculation, the monkeys have to be examined for the presence of indigenous trichomonads by microscopy and culture of vaginal and rectal contents. If trichomonads are present either in the vagina or the intestine, the animals must be treated with one of the 5-nitroimidazole derivatives. Metronidazole (25 mg/kg/day for 5 days), administered by a stomach tube to animals slightly sedated with ketamine, was used to treat the squirrel monkey.[23]

To inoculate the animals, about 10^6 or 5×10^5 trichomonads from culture, suspended in fresh media, were administered in 1.0- or 0.5-ml volumes, to the rhesus and the squirrel monkeys, respectively. The inocula were administered in single doses with a tuberculin syringe,[18] catheter,[22] or sterile feeding tube inserted into the vagina.[23] To prevent leakage of inoculum, the vaginal orifice was plugged with a gelatinous sponge[21,22] or the labia were held closed for 1 minute following the inoculation.[23]

Usually, an experimental infection of monkeys with *T. vaginalis* was accomplished without supportive treatment. The only report on the use of estrogen involved females of *Macaca arctoides*. The animals were inoculated intramuscularly with estradiol valerate (2 mg/kg) on days 7 and 1 before inoculation and on postinoculation day 7.[21] No attempt was made to investigate the effect of hormonal treatment on the infection rate.* Exceptionally high inocula (25 to 30×10^6) used for infecting the stump-tailed macaques[21] do not seem to have any advantage in establishing the infection.

Horizontal transmission of the infection among the squirrel monkeys was accomplished by inserting a cotton-tip swab first into the vagina of the infected and then into the vagina of the uninfected females.[23] Attempts to transmit the infection sexually have not been reported.

Susceptibility

The proportion of infected monkeys varied in different studies. Usually, successfully infected animals, all sexually mature females, outnumbered those in which the infection failed. The young rhesus monkeys seemed to be refractory to infection until the first menses.[18] In the squirrel monkey, susceptibility depended apparently on the estrous cycle phase. Trichomonads were successfully established in monkeys inoculated during estrus, while attempts to infect the animals before the onset of estrus failed.[23] The infection once established was not eliminated at the completion of the cycle but persisted with intermittent latency over several months.[22,23]

---

*Infection rate is defined as the percentage of successfully infected animals of the total number of animals inoculated.

Course of Infection and Pathologic Manifestations

Course of Infection

Trichomonas vaginalis infection reported in both rhesus and squirrel monkeys appeared to be asymptomatic and in most instances self-limiting, lasting from several days to 4 months.[17, 21-23] The self-limiting infection need not be the rule in *M. mulatta*.[17,18] Several rhesus monkeys were reported to maintain the initial infection over 1 year[18] and an infection persisting for 3 years was noted in three other rhesus monkeys used for drug testing.[19]

Great variations in the density of the trichomonad population were observed in both squirrel monkey[23] and rhesus monkey.[18] In *M. mulatta* these fluctuations showed some periodicity related to the hormonal cycle of the host, with the largest populations being observed shortly before and after the menses. The trichomonad numbers were low in the midcycle; detection of parasites during this period was often difficult.

Reappearance of trichomonads after a latent phase of infection was observed in two squirrel monkeys following their transfer to a common cage shared by six noninfected animals, including a male.[22] It was suggested that the stressful situation caused by group caging provoked the conversion from a latent infection to a detectable one; and it is known that certain changes (e.g., hormonal) in the environment cause exacerbation of the disease in female patients harboring *T. vaginalis*.

Pathologic Manifestations

Monkeys infected with *T. vaginalis* did not show clinical manifestations of disease.[17,18,22] An occasional discharge containing few inflammatory cells developed after inoculation of trichomonads into the vagina of squirrel monkeys; however, persistence of the discharge was not closely associated with the presence of parasites.[22] In *M. mulatta* the amount of the vaginal discharge appeared not to be correlated with the *T. vaginalis* infection; colposcopic examination revealed no signs of vaginitis or cervicitis.[18]

A histopathologic lesion similar to that occurring in humans was observed in the surgically removed cervix of a squirrel monkey infected with the virulent Balt 42 strain of *T. vaginalis*. The changes, localized mainly in the squamocolumnar region, included focal degeneration of the epithelium with intracellular edema and vacuolization, presence of interepithelial polymorphonuclear leukocytes, and pathologic vascularity with dense infiltration of mononuclear cells in subepithelial tissue.

Immune Response

Antibody response to experimental infection with *T. vaginalis* was observed in four squirrel monkeys.[22] All animals developed IgG serum antibody response after the primary or booster inoculation, with peak titers in the range 1:100 to 1:500. The secretory antibodies, both IgG and IgA, appeared in the vaginal mucus after repeated inoculation (second or third) sometimes in the absence of detectable trichomonads. Their maximum titers ranged from 1:800 to 1:1600 and 1:100 to 1:1600 for IgG and IgA, respectively. Both humoral and local responses were transient and disappeared or showed lower titers within 6 to 33 weeks. Previous infection provided little, if any, protection against subsequent challenge. Although the small number of animals does not allow generalization, the brief duration of antibodies and lack of protection against reinfection in monkeys is compatible with findings reported from naturally infected humans.

Indigenous Trichomonads

Monkeys can be naturally infected with several trichomonad species inhabiting their oral cavity, intestine, and even vagina. The vaginal trichomonad *Tetratrichomonas macacovaginae* was reported from the rhesus monkey.[25] This inadequately characterized species, not observed since its original description, might have been *Pentatrichomonas*. In addition, some intestinal trichomonads can occasionally become established in the vagina or contaminate vaginal samples. Natural or experimental infections with the common intestinal trichomonad, *Pentatrichomonas hominis* were reported for both the rhesus and the squirrel monkey.[26-28] Recently, a new trichomonad, *Tritrichomonas mobilensis*, was described from the intestine of the squirrel monkey.[29] The organism appears the possess a certain pathologic potential,[30,31] is widespread in laboratory colonies of squirrel monkeys, and occurs apparently also in other species of New World monkeys.[32] Cultures of *T. mobilensis* were obtained from vaginal samples, but the actual source of the organism was most probably a fecal contamination of the swabs.[23]

Apart from the oral species, *Trichomonas tenax* (Fig. 8.3c), all trichomonads of the monkey can easily be distinguished from *Trichomonas vaginalis* by their structure (Fig. 8.3a,b). The main differentiating feature common to all is the presence of a free posterior flagellum that continues beyond the posterior end of the undulating membrane; this flagellum is absent from *T. vaginalis* (Fig. 8.3d).

Evidently, indigenous trichomonads of monkeys can interfere with experimental infections as potential contaminants; they have to be eradicated before these primates are used in experiments.

Guinea Pig

That an intravaginal infection with *T. vaginalis* can be established in the adult guinea pig is evident from a series of reports published by a group of Polish investigators[33-42] (for review article see reference no. 9). The guinea pig model appears to be of potential use in studies on the progression of pathologic changes in trichomoniasis. The infections were symptomatic and the histopathologic and cytopathologic changes found in the infected animals were similar to those reported from women. Published accounts, focused on the description of patho-

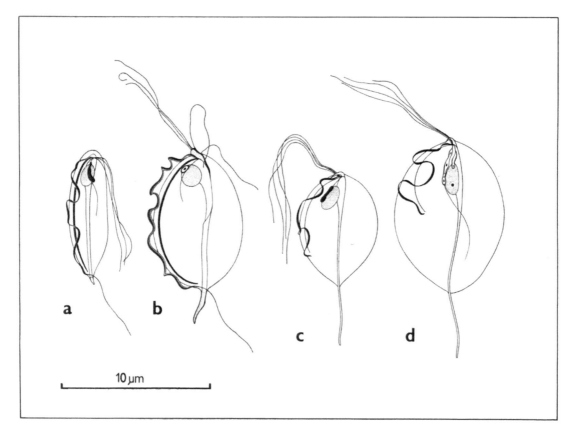

Fig. 8.3. Composite diagrams of trichomonad species indigenous to monkeys (**a-c**) and of *Trichomonas vaginalis* (**d**) from the urogenital tract of humans. **a.** *Trichomonas mobilensis*, from the colon. **b.** *Pentatrichomonas hominis*, an inhabitant of the large intestine which can establish itself in the monkey's vagina. **c.** *Trichomonas tenax*, from the oral cavity. (**a**: Reproduced with permission of the Society of Protozoologists, from Culberson DE et al. *Trichomonas mobilensis* n. sp. (Zoomastigophorea: Trichomonadida) from Bolivian squirrel monkey *Saimiri boliviensis boliviensis*. *J Protozool* 33:301-304, 1986. **b-d**, Original diagrams of protargol-stained organisms, E. Nohýnková del.)

logic changes, gave little information about the behavior of parasites in the infected host, and failed to provide a standardized protocol for infection procedures. No attempts were made to ascertain relative virulence of different *T. vaginalis* strains for the guinea pig.

Inoculation, Infectivity, and Susceptibility

According to Kazanowska et al[42] guinea pig females were successfully infected without a supportive hormonal treatment during any phase of the sexual cycle and also during pregnancy. Animals were inoculated with 1 ml of axenic cultures of the trichomonad (0.8 to 2.0×10^5 cells in cysteine-peptone-liver-maltose [CPLM] medium) isolated from patients with acute trichomoniasis. Fairly large series of animals were used in the individual experiments, but the percentage of successful infections was not specified. Apparently, lasting infections were established, but exact data on their duration and on recovery of trichomonads in the course of the infection have not been published. Occasional remarks indicated microscopic findings of trichomonads on postinoculation days 2 and 12; the presence of "motionless forms," of uncertain identity, was reported up to 1 month postinoculation.

Experimental vaginal infection of guinea pigs with *T. vaginalis* ought to be characterized more precisely to develop a standard procedure. It is especially important to ascertain whether, and to what extent, the infection is influenced by the sexual cycle of the animals and by estrogen treatment. A procedure that might be applicable to *T. vaginalis* was developed for intravaginal infection of guinea pigs with *Tritrichomonas foetus*.[43]

Pathologic Manifestations

The acute phase of experimental trichomoniasis in the guinea pig was manifested by abundant purulent discharge from the vagina, reddening and swelling of the vulva, loss of appetite, and general weakness. Inflammation appeared 24 to 75 hours after inoculation. Alterations of squamous epithelium were detectable on cytopathologic examination of vaginal smears from postinoculation day 4; histopathologic changes in the vaginal portion of the cervix developed about 8 days later. The acute form of infection eventually transformed into a chronic one. Microscopically, the transition from the acute to the chronic inflammatory phase was noted between postinoculation days 20 and 24.[33,34,40,42]

Spread of the infection to the upper parts of the genital tract was observed, resulting in endometritis and inflammation of the oviduct mucosa.[34,41] Abortion of macerated fetuses commonly occurred if pregnant animals were inoculated.[34] Inflammatory involvement of the lower urinary tract was also reported and lesions were examined by histologic[37] and histochemical[38] techniques (reviewed in reference 9).

Cytopathology

Cytopathologic examination of vaginal smears revealed marked inflammation with masses of polymorphonuclear neutrophils and some mononuclear cells. The leukocytes frequently formed characteristic clusters on degenerating epithelial cells[34,35] corresponding to those described in human infections with *T. vaginalis*.[44] Exfoliated squamous epithelial cells showed marked alterations of nuclei that often were enlarged and hyperchromatic, with granular clumping of chromatin. Some nuclei were irregular in shape, showing signs of karyorrhexis as well as fragmentation and protrusion of chromatin particles into the cytoplasm.[33-35] Nuclei were often located eccentrically, and some cells were multinucleate. A narrow perinuclear halo, another characteristic feature of cells seen in human infections, was frequently present.[33,34]

Histopathology

Histopathologic examination of early lesions showed acute inflammation involving increased epithelial and subepithelial vascularity and infiltration of epithelium and underlying stroma with polymorphonuclear leukocytes. Focal erosions of epithelium were also observed. The germinative layer of the epithelium displayed characteristic cellular lesions with vacuolization, perinuclear halos, and hyperchromasia of nuclei. Cytochemical examination indicated increased levels of cytoplasmic RNA, decreased levels of nuclear DNA, and accumulation of acid mucopolysaccharides in the surface layers of the vaginal mucosa.[33,34,39,41]

In the phase of transition from acute to chronic inflammation an increased frequency of mast cells was observed in the vaginal portion of the cervix. The mast cells, belonging predominantly to the mature granular population, were located perivascularly and throughout the subepithelial connective tissue.[40] Association of the mast cells with vaginal trichomoniasis was observed in endocervical smears of humans, and the possibility of their involvement in the immunologic or immunopathologic response to the infection was suggested.[45] The guinea pig model might be useful in experimental studies aimed at elucidating these relationships.

Evidently, the cytopathologic and histopathologic changes observed in the vagina of guinea pigs infected with *T. vaginalis* closely parallel those occurring in humans. Kazanowska[41] stressed on many occasions the similarity of these changes to those occurring in precancerous states and demonstrated analogous lesions in guinea pigs treated with chemical carcinogens (methylcholanthrene). The guinea pig might represent an appropriate model for examining the reaction of host tissues to trichomonad infection. This model could be used for the investigation of putative relationships between trichomoniasis and the development of neoplastic changes in the hosts.

Hamster

Infection with trichomonads can be established intravaginally in the golden hamster, *Mesocricetus auratus* without the use of supportive hormonal treatment. The model has been used mainly for assaying drugs against *T. foetus*,[46-48] but an intravaginal infection of the hamster with *T. vaginalis* has also been accomplished.[49-52] Infections with *T. vaginalis* were asymptomatic, persisting from several weeks to several months, and trichomonads could be serially passaged over long periods. However, the susceptibility of the hamster to spontaneous intravaginal infection with indigenous intestinal trichomonads[49,50,52] renders questionable many results obtained with this model.

Inoculation Procedures

To infect hamsters, cultures concentrated by centrifugation were administered in small volumes of medium (0.05 to 0.1 ml), and the inoculations were repeated at least once at 24- or 48-hour intervals. When specified, the inocula contained 3×10 or 2×10^6 organisms.[50,51] The infection rates varied from 20% to 33% and from 59% to 90% after a single and multiple dose, respectively. In an experiment employing a small series of spayed and estrogenized animals[51] there was no improvement of the infection rate (three of nine animals inoculated with a single dose were infected). Results with *T. foetus* based on a large series of inoculated animals[48] also revealed large random variations in the number of infected animals, and the extent of variation was not reduced by estrogen treatment.

Indigenous Intestinal Trichomonads and Their Effect on Experimental Intravaginal Infections with T. vaginalis

The colon and cecum of laboratory rodents, and of hamsters in particular, are inhabited by a variety of flagellates, including several species of trichomonads.[53] These occasionally contaminate samples collected in the course of experiments and in certain instances become established intravaginally. Therefore, experimental infections should be monitored by reisolation of inoculated organisms and ascertaining their actual identity.

Five morphologic species of trichomonads occur in laboratory rodents (Fig. 8.4a-e). All are nonpathogenic for their hosts, inhabit the cecum and colon, and are excreted with feces. The largest and most common species, *Tritrichomonas muris*, can be readily recognized in wet mounts or in Giemsa-strain smears; for determination of other species, staining with protargol is recommended. Morphologic characteristics of the aforementioned trichomonad species, as well as their taxonomy, are discussed in several reports.[54-61] Distinguishing these species from *T. vaginalis* is easy—all of them have a free posterior flagellum extending beyond the end of the undulating membrane, while no such free flagellum is present in *T. vaginalis*.

The most troublesome contaminant is *P. hominis*, a parasite found in various mammals including humans. This species is relatively resistant to unfavorable conditions outside the body, being able to survive in moist feces or water at room temperature for several days.[62] It is also the only trichomonad of rodents that grows in standard culture media, under both axenic and xenic conditions. Thus it can cause false-positive results if samples for monitoring experimental infections by culture are contaminated with feces.

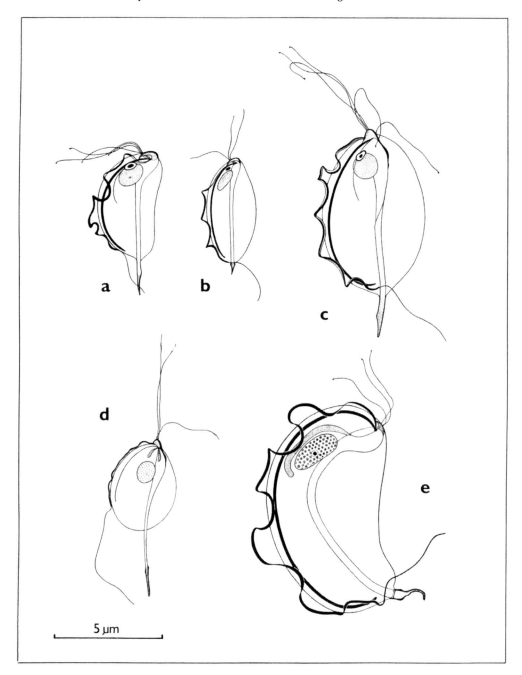

Fig. 8.4. Composite diagrams of trichomonad species found in the cecum and colon of laboratory rodents. **a.** *Tetratrichomonas microti.* **b.** *Tritrichomonas minuta.* **c.** *Pentatrichomonas hominis.* **d.** *Trichomitus wenyoni.* **e.** *Tritrichomonas muris.* **a,b,d,e**: Specimens from the cecum of a Syrian hamster; **c** was obtained from organisms in culture. (All figures are original diagrams of protargol-stained flagellates, E. Nohýnková del.)

Actually, spontaneous infections of the vagina of the hamster by intestinal trichomonads or infections induced by contamination with fecal material in the course of manipulating the animals pose a serious problem. Uhlenhuth and Shoenherr[49] found an intravaginal infection with *P. hominis* in 13 of 130 hamsters examined. Overgrowth by an intestinal trichomonad, most probably *P. hominis,* was observed also in a *T. vaginalis* line serially passaged in hamsters.[52] Vershinsky[50] noted in two hamsters an intravaginal infection with a small trichomonad having three anterior flagella that might have been *Trichomitus wenyoni*. Susceptibility of hamsters to intravaginal infection with the abundant intestinal trichomonads renders the model susceptible to contamination and vitiates its usefulness for studies of experimental infection with *T. vaginalis*.

Rat

Intravaginal infection with *T. vaginalis* can be established in spayed rats treated with estrogens. The model is suitable for drug testing because the infection is sufficiently persistent to allow post-treatment observations. It should prove useful for studying hormonal modifications of the *T. vaginalis* infection, and mechanisms favoring establishment of infection in the estrogenized host. The major disadvantage of the model is the need for surgical removal of ovaries before the animals are used in experiments. The rat model does not seem to be useful in pathogenicity studies. Symptomatic infections are infrequent,[63] and the responses of the animals are apparently unrelated to the virulence of the inoculated *T. vaginalis* strains.

Inoculation, Infectivity, and Susceptibility

Inoculation Procedures

The technique, introduced by Cavier and Mossion,[64,65] was subsequently improved by investigators who modified the estrogen treatment and the exact inoculation method (Table 8.1). A brief description of a current technique used in the author's laboratory is given below. It is largely based on recommendations of J. G. Meingassner (reference 66 and personal communication).

Recommended inoculation protocol (see also Table 8.1) is as follows. Adult virgin female rats (Wistar, 180 to 200 g), anesthetized with pentobarbital (40 mg/kg, administered intraperitoneally), are spayed. After a 2-week rest period, the rats are given a single subcutaneous inoculation of a long-lasting estradiol derivative (e.g., Progynon-Depot; 10 mg/kg of estradiol undecylate) to induce lasting estrus. On the fourth day following estradiol treatment, vaginal smears are examined to confirm the presence of estrus. Then the rats, anesthetized with pentobarbital, are inoculated with 3 to 5 × 10^5 logarithmic-growth phase *T. vaginalis* in 0.2 ml of Diamond's trypticase-yeast-maltose (TYM) medium. Using a blunt cannula, the inoculum is placed at the bottom of the vagina, and the vaginal orifice is sealed with a gelatin sponge. The vaginal contents are examined microscopically and by cultivation for the presence of trichomonads on postinoculation day 7 by using washings of the vagina with the medium. Earlier examinations tend to disturb the establishment of the infection.

Infectivity and Susceptibility

The infection rate reported in successful attempts varied between 50% and 100% (Table 8.1), and the infection persisted for 14 to 76 days, depending on the duration of induced estrus. A marked increase in the infectivity rate was observed in animals infected intravaginally with *Candida albicans* (10^7 cells in 0.05 ml) 1 day before inoculating trichomonads (Table 8.2).[66]

Failures to infect estrogen-treated rats were also reported,[48,51,67] and it is generally believed that the establishment of intravaginal infection in rats is difficult.[9] The results of Michaels et al[67] suggest that the success of inoculation depends both on the strain of the parasite employed and on the strain of the rat. These authors were unable to infect rats with a laboratory strain of *T. vaginalis* maintained in vitro for a prolonged period. Using a fresh isolate they succeeded in infecting one of the two rat strains employed (Long Evans); the other strain (Sprague-Dawley) proved insusceptible. Successful intravaginal infection of the Sprague-Dawley rats, however, was reported by others.[68] In my experience, lack of success in infecting the rat usually is caused by failure to induce an estrus of sufficient duration or by leakage of inoculum. Furthermore, some animals tend to remove the inoculum from the vagina. Proper estrogen treatment and monitoring of the estrus, inoculation of the rats

Estrogen treatment			Infection protocol			Results			References
Form of hormone and method of administration	Dosage	Rat weight (g)	Period after estrogen treatment	Kind of inoculum and no. of inoculations	Total no. of animals	Percent infected animals	Period of infection (days)		
Estradiol benzoate (sc inoculation)	1.2–5.0 µg/animal/day	100–200[†]	4–14 days	One drop of 48-h culture on 3 successive days	NG[§]	90	40–70		64, 65, 69, 72, 73
	100 µg/animal/day	120–200[‡]	6–8 days	NG	NG	70	NG		69
Estradiol pellet (sc inoculation)	10 mg/animal	150–200[†]	2 wk	4×10^5 organisms from 48-h culture in 0.2 ml (single dose)	53	79	NG		15
	20 mg/animal 30–40 mg/animal	120–200[†,‡] 180–220[*]	11 days 7 days	NG 0.1–0.2 ml of 48-h culture sediment (single dose)	NG 31	100 60	NG 30		69 70
Estradiol benzoate microcrystalline suspension (Agofollin-Depot) (sc inoculation)	5 mg/animal every 8 days	100–120[†]	2 wk	NG	NG	NG	60		65
	10 mg/animal every 2 wk	100–120[†]	2 wk	One drop of 48-h culture on 3 successive days	100	80–100	NG		65
	17 mg/kg	180–200[†]	19 days	$3–5 \times 10^5$ organisms from log-phase culture	60	49	50		Kalašová and Kulda (unpublished data)
Estradiol cyclopentyl propionate (sc inoculation)	0.1 mg/animal	NG[§]	4 days	One drop of 48-h culture on 3 successive days	NG	NG (one of 2 strains used infective for some of the rats)	NG		67
Estradiol undecylate (Progynon-Depot) (sc inoculation)	10 mg/kg	200–250[†]	4 days	$4–5 \times 10^5$ organisms from 48-h culture in 0.2 ml (single dose) + *Candida albicans*	62	92	NG		66

[*]sc, Subcutaneous. [†]Spayed rats. [‡]Intact rats. [§]Data not given (NG) in the original reports.

TABLE 8.2. Effect of *Candida albicans* on the establishment of *T. vaginalis* in vaginas of spayed rats subjected to estrogen treatment.

Strain	Without *Candida*		With *Candida**	
	No. of rats	Percent infection	No. of rats	Percent infection
E (Vienna)[†]	46	29	62	92
TV 7-37 (Prague)[‡]	34	32	12	75

*Candida albicans, 10^7 cells in 0.05 ml per rat, inoculated intravaginally 1 day before infection with trichomonads.
[†]Data from Meingassner et al.[66]
[‡]Unpublished data of Nedvědová and Kulda.

anesthetized by barbituate, and prevention of inoculum loss by insertion of a gelatin sponge into the vagina should improve the success of infection experiments.

Effect of Hormonal Treatment

Dependence of the *T. vaginalis* infection on the treatment of rats with estrogens was clearly demonstrated,[15,69,70] and cytologic monitoring proved that the spontaneous disappearance of parasites was correlated with the termination of estrus[70] (also Nedvědová and Kulda, unpublished observations). Interruption of estrus induced by treating estrogenized rats with progesterone,[71] testosterone,[72] or desoxycorticosterone[73] also eliminated the infection.

Further studies are needed to determine factors directly responsible for rendering the vaginal milieu of the estrogen-treated host favorable for establishment of *T. vaginalis*.

Spayed rats treated with estrogen showed no shift in vaginal pH.[74] Regardless of the hormonal treatment, the vaginal pH of the spayed rats was neutral, ranging between 6.6 and 7.5. The same range was observed in the intact rats irrespective of the phase of their sexual cycle.

Quantitative changes were noted in the vaginal bacterial flora following the treatment of spayed rats with estrogen, marked by proliferation of *Escherichia* and enterococci. Lactobacilli have not been found either in normal or in spayed and estrogenized rats.[75]

Nakamura[15] found the glycogen content to increase markedly in the vagina of estrogen-treated rats. Distribution of glycogen was uneven throughout the vagina with the highest glycogen content being present in the upper part, near the portio vaginalis cervicis. The Japanese worker[15] proved this area to be the site preferentially colonized by *T. vaginalis*. In subsequent experiments, he[15] used rats with alloxan-induced diabetes, possessing heavy deposits of glycogen in the vaginal tissues, and successfully established *T. vaginalis* infection in these animals. On the basis of these results Nakamura concluded that the most important factor governing establishment of *T. vaginalis* in the estrogen-treated rat was the increase of glycogen deposits in the vaginal mucosa. These conclusions are compatible with clinical observations emphasizing the importance of hormonally induced glycogen deposits in the vaginal mucosa of humans that correlate with susceptibility to *Trichomonas* infection in adult women and newborns and with its lack in adolescent girls.[7]

Pathogenicity

The rat model has been employed mostly for drug testing,[67,68,70,76] little attention having been paid to the pathogenic effects of *T. vaginalis* in the genitourinary tract of the infected rat. Results obtained in the author's laboratory[63,77] showed that *T. vaginalis* can induce inflammatory discharge and histopathologic lesions in some of the infected animals. With two *T. vaginalis* strains employed, the percentage of symptomatic infections was rather low (20%) and the severity of lesions did not reflect virulence levels estimated on the basis of histopathologic findings in patients and of the subcutaneous mouse assays.

Microscopic examination of symptomatic animals in an early phase of the infection (3 days after inoculation) showed foci of ameboid trichomonads adhering to the vaginal epithelium (Fig. 8.5b,c) and acute inflammation of the vaginal mucosa with acanthosis of the epithelial layer, interepithelial leukocytes, and dense infiltration of subepithelial connective tissue with polymorphonuclear leukocytes.

The immune response of the rat to intravaginal infection has not been studied.

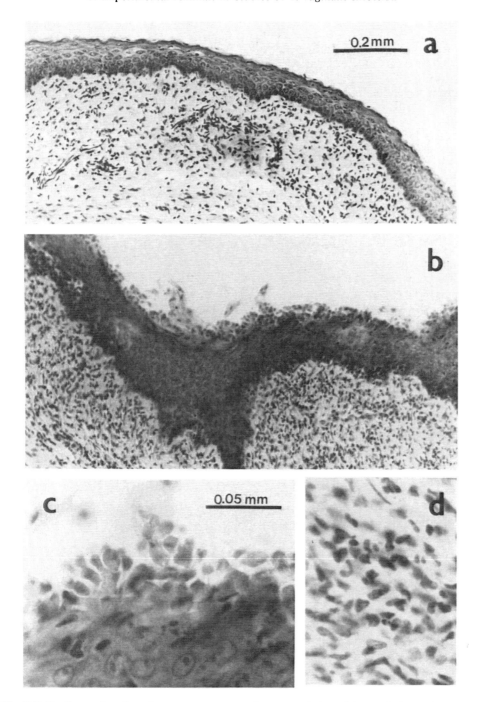

Fig. 8.5. Sections of vaginae from estrogen-treated rats, noninfected (control) (a) or infected with an avirulent strain of *T. vaginalis* (TV 17-48), 3 days after inoculation (b-d). b. Note the acanthotic thickening of the epithelial layer to which are attached numerous trichomonads. c,d. At a higher magnification, the trichomonads can be seen in greater detail (c), and infiltration with polymorphonuclear leukocytes is evident in the subepithelial stroma (d). Baker's formaldehyde fixative (BF), Mayer's hematoxylin & eosin stain (MH & E). Scale in c applies also to d. (Original figures illustrating data summarized briefly in reference no. 63.)

Mouse

Establishment of intravaginal infection with *T. vaginalis* in sexually mature mice treated with estrogens was successfully accomplished[51,78-80] using different modes of inoculation:

1. Mice were inoculated with parasites from axenic cultures.
2. Trichomonads were serially passaged from mouse to mouse using a primary culture from vaginal washings as the inoculum.[80]
3. Infection was transmitted sexually by successive mating of the male with infected and noninfected females pretreated with estrogen.[78]

The model has not yet been sufficiently explored and has to be developed further. It should prove convenient for assaying antitrichomonal drugs and for investigating certain aspects of host-parasite interactions (e.g., infectivity). No data have been published so far on the host response to the established intravaginal infection with *T. vaginalis*.

Inoculation, Infectivity, and Susceptibility

Inoculation Procedures

Pretreatment of mice with estrogens is essential to produce a lasting infection, but spaying is not necessary. According to Cappuccinelli et al,[78] the estrogen treatment is needed only to establish the parasites; thereafter the infection is maintained without the exogenous hormone. A lasting infection (1 to 3 months), however, becomes established in some animals only, while in others the infection is short-lived (1 week). The phase of the sexual cycle and, accordingly, the time interval between the administration of estrogen and the inoculation of the parasites is of importance for a successful establishment of the parasites. The highest infection rate was observed in early proestrus.[80] Unlike the British group,[80] the Italian workers[78] succeeded also in infecting a small number of mice not treated with estrogen.

Some investigators[78] recommended repeated inoculation; others[51] suggested that a single inoculation is sufficient if the number of parasites in the inoculum is increased to 5×10^5. Plugging the vagina with a gelatin sponge[79,80] to avoid the loss of inoculum apparently increases significantly the infection rate.

The following inoculation protocol is based largely on the results of Coombs et al.[80] Adult female mice (25 to 30 g) are injected subcutaneously with a long-lasting estradiol derivative (40 mg/kg). On the third day after administration of estrogen, the mice are inoculated with 1×10^5 logarithmic-phase trichomonads suspended in 0.02 ml of culture medium (e.g., TYM) with about 0.3% agar. The vagina is then sealed with a small piece of gelatin sponge. Mice can be inoculated with a tuberculin syringe equipped with a blunt cannula. The procedure is facilitated by keeping the animals anesthetized with barbiturate (e.g., 200 mg/kg of sodium hexobarbital administered intraperitoneally).[51] To examine the inoculated mice, the vagina is washed with about 0.02 ml of a culture of medium and the sample transferred to a small volume of the medium supplemented with antibiotics.

Susceptibility and Infectivity

Susceptibility of mice to infection does not seem to be strain-specific, because at least seven strains of these rodents proved to be susceptible.[78-80] The infection rate of different *T. vaginalis* strains varied from 15% to 100% (Table 8.3). However, the number of infected mice in each group decreased with time, infection persisting for 1 to 3 months in a few animals only. Occasionally, after repeated inoculation of large doses of parasites (1×10^6), trichomonads spread through the upper genitalia to the abdominal cavity, eventually causing a fatal abdominal infection.

Preliminary comparative experiments with ten fresh isolates of *T. vaginalis*[80] are difficult to interpret because of the small number of mice employed (5 or 6 per group). Again, the infection rate and the persistence of the infection varied among the isolates (17% to 100% and 3 to 65 days, respectively); at least one strain appeared to be noninfective. As is apparent from Table 8.4, a marked infectivity decrease was observed in a fresh isolate in the course of prolonged axenic in vitro cultivation. Maintenance of the same strain by subpassaging in the vagina of mice largely preserved the original infectivity level.[80]

Support of Infection with *Candida*

To improve the intravaginal infection of mice with trichomonads, Meingassner[79] injected the

TABLE 8.3. Procedures resulting in the establishment of lasting intravaginal *T. vaginalis* infections in laboratory mice.

Estrogen treatment			Infection protocol		Results		
Form of hormone and method of administration	Dosage	Mouse wt. or age	Period after estrogen treatment (days)	No. trichomonads and vol. inoculation	Total no. of mice	Percent infected animals	Period of infection (days)	Reference
Estradiol valerate (sc* inoculation in sesame oil)	0.5 mg/animal every 10 days	30 g	3	One inoculation of 5×10^3 or 2×10^6 in 0.05 ml	48	52 (37–75)	>40	78
	0.5 mg/animal in a single dose	30 g	3	Two inoculations of 5×10^5 at 24-h intervals in 0.05 ml	76	99	50	78
Estradiol undecylate (Progynon-Depot) (SC and intraperitoneally in sesame oil)	40 mg/kg in 2 equal doses	25–30 g	3	One inoculation of 10^5 trichomonads + *Candida albicans*† in 0.05 ml	27	94	NG‡	79
Estradiol benzoate microcrystalline suspension (SC inoculation)	0.5 mg/animal weekly	18–20 g	7	One inoculation of 1.5×10^5 (4 different strains) in 0.05 ml	77	26 (15–40)	14–21	51
				One inoculation of 5×10^5 or two inoculations of 5×10^5 and 3×10^5 at 24- or 48-h intervals all in 0.05 ml	48	62 (50–65)	42	51
Estradiol cypionate (Sigma) (SC in corn oil)	40 mg/kg in a single dose	2–3 mo	2	One inoculation of 10^5 in 0.02 ml	180	52	92	80

*sc, Subcutaneous. †Trichomonads and *Candida* are administered in a mixed inoculum. ‡Data not given (NG) in the original report.

TABLE 8.4. Infectivity decrease of *T. vaginalis* strain 39 after prolonged axenic cultivation.*

Maintenance			Number of mice infected after days in vivo or in vitro				Maximum length of infection (in days)
Method	No. of months	No. mice per group	3	7	14	28	
In vivo (control)[†]	7	15	8	4	2	1	45
In vitro[‡]	1	14	7	6	4	3	65
	3	6	3	3	2	1	37
	7	6	1	1	0	0	7
	10	6	0	0	0	0	<3

*Data from Coombs et al.,[80] reproduced with permission of Charles University Publishing House, Prague, Czechoslovakia.
[†]Trichomonads were maintained by serial mouse-to-mouse passages, using primary cultures obtained from vaginal washings as the primary inocula.
[‡]Trichomonads were maintained by serial passages in axenic cultures.

vaginas of these experimental hosts with inoculum consisting of a mixture of *T. vaginalis* and *Candida*. The yeast cells from a stabilate containing 3.6×10^7 cells/ml were added at a ratio of 1:80 (vol/vol) to a standard trichomonad dose of 10^5 per mouse immediately before injection. The inoculum included also 1000 units of penicillin and 1000 μg streptomycin/ml. When the mixed yeast–trichomonad suspension was employed, the infection rate equaled 95% on postinfection day 4.[79] No information was given about the persistence of infection.

Effect of Immunosuppression

Results of immunosuppressive treatment on susceptibility of mice to infection are inconclusive. An increase in the infection rate up to 100% was observed after pretreatment of mice with dexamethasone[51] (0.1 mg/kg of dexamethasone-21-isonicotine given for 6 days preceding infection). However, the number of mice used in the experiment was too small (six) to allow evaluation of the effect. Irradiation of mice apparently did not affect their susceptibility to infection,[78] and cyclophosphamide was said actually to decrease the infection rate.[80]

Pathologic Manifestations

The pathogenic effect of an established *T. vaginalis* infection on the genitourinary organs of female mice has not been investigated. However, Patten et al.[81] followed the effect of repeated intravaginal injections of *T. vaginalis* on the epithelium of mouse cervix. Trichomonads from axenic cultures isolated from patients with cytologic evidence of dysplasia were inoculated weekly into young adult C₃H mice (about 2×10^3 trichomonads per mouse) for 12 weeks. At weekly intervals cytopathologic specimens (Papanicolaou smears) were examined and compared with those of controls treated with sterile culture medium. In a subsequent experiment, the animals were examined histologically.

It was demonstrated that the exposure to *T. vaginalis* resulted in dysplastic epithelial changes in about 22% of the animals. The changes were reversible after treatment was discontinued. The reaction appeared 3 weeks after the initial exposure to the parasites and disappeared completely within 7 weeks of the last injection. The cellular changes included enlargement and hypochromasia of nuclei, clumped chromatin pattern, and binucleation. Histopathologic examination showed delayed maturation of the epithelium with immature cells occupying the whole epithelial layer. The changes were comparable qualitatively and quantitatively to those observed in dysplasia induced in mouse by chemical carcinogens.[82]

Infection of the Male Urogenital System

Published data on the use of laboratory animals in studies of the pathogenic effects of *T. vaginalis* on the male urogenital system are sparse[83-85] and provide little information relevant to natural infection of men.

Infection of Urinary Passages

Experimental infection of the urethra and the bladder with *T. vaginalis* was attempted in male guinea pigs but no significant pathologic changes were noted. The inoculation resulted in a short-lived increase in the numbers of leukocytes and desquamated epithelial cells in the urine sediment. Surgical placement of *T. vaginalis* in the bladder or in the renal pelvis did not cause any pathologic manifestations. No data were presented with regard to the behavior and persistence of parasites in the urinary tract of inoculated animals.

Intratesticular Infection

Intratesticular infection with *T. vaginalis* was established in rats and guinea pigs and the progression of pathologic changes was followed to ascertain whether there were any morphologic changes pathognomonic of trichomoniasis.[84]

The study was performed on 122 rats and supplemented by observations on 9 guinea pigs. Animals were inoculated with seven axenic isolates (4×10^6 in 0.3 ml of medium per animal) from both men and women.

The inoculated parasites penetrated between the seminiferous tubules causing rapid development of lesions. Migration of leukocytes into the inoculation area could be observed as early as 4 hours after inoculation. Within 24 hours necrotic foci were formed around the damaged seminiferous tubules. The germinative epithelial cells were destroyed first; this was followed by total dispersal of the tubules throughout the structureless necrotic areas. Dense infiltrate of polymorphonuclear neutrophils lined the periphery of the lesion. Occasionally, groups of eosinophils appeared in the infiltrate. A granulating tissue began to form in the adjacent area on postinfection days 3 and 4, proliferating toward the necrotic foci. More organized connective tissue formed at the periphery. Several weeks after inoculation plasma cells appeared in the connective tissue. By the end of the first month after inoculation and later, small foci of epithelioid cells and elements resembling giant cells appeared in the connective tissue proliferating in the necrotic foci. Impregnation of the necrotic area with calcium, localized mainly in the area of destroyed seminiferous tubules, could be observed from the third week after inoculation. On the basis of these observations the authors concluded that the experimental orchiditis caused by *T. vaginalis* was accompanied by pathologic changes resembling those occurring in tuberculosis or brucellosis.[84]

Ectopic Infections

Experimental infections produced by inoculation of *T. vaginalis* into various extravaginal sites were extensively studied in mice. Occasional use of other animals is of marginal importance. Subcutaneous and intraperitoneal infections were reported for the guinea pig,[85,86] the rat was infected subcutaneously,[87] and trichomonads were established in the anterior chamber of the rabbit eye.[88,89] In the mouse, the animal most susceptible to extravaginal infection with *T. vaginalis,* this parasite can be established in various sites including the peritoneal and thoracic cavities,[90] as well as in the muscular and subcutaneous tissues and in the scrotum.[91] In light of this, the following section deals exclusively with infections of mice. Intraperitoneal and subcutaneous infections, which can be produced consistently and without technical difficulties, have been generally used in studies of pathogenicity,[7,9] immunology,[9,92] and chemotherapy[5,6,8] of *T. vaginalis.*

Intraperitoneal Infection

Intraperitoneal inoculation of *T. vaginalis* into mice causes a fibropurulent peritonitis with abscesses and necrotic foci in abdominal organs and production of ascitic fluid. The severity of the infection, although dependent on the responsiveness of the individual animal, reflects the inherent virulence of the inoculated strain. A pattern of gross-pathologic changes and mortality rate of infected mice are generally reproducible under standard experimental conditions. However, the correlation of a strain's virulence for mice with the severity of disease in human patients need not be perfect (see section on relative virulence, below). In contrast to the intravaginal infection, intraperitoneal infection of mice with *T. vaginalis* does not present technical problems. The latter method has been in use over 3 decades. Its main features have been described [5,86,93-99] and published data reviewed.[7,9] Intraperitoneal infection of mice has been used in various arrangements for assessing virulence of *T. vaginalis* strains and for drug assays.

Inoculation, Infectivity, and Susceptibility

To inoculate the animals, logarithmic–growth phase parasites from axenic cultures resuspended in fresh medium are administered in a single dose. In my experience, the optimum inoculum contains 10^6 trichomonads in 0.5 ml of medium. This dose results in an infection rate close to 100% for most *T. vaginalis* strains, still allowing the observer to differentiate among the pathogenic effects of the isolates. *T. vaginalis* can also be maintained by serial mouse-to-mouse passages. For such transfers, primary cultures of trichomonads recovered from the abdominal cavity on postinoculation days 9 or 10 are used as the inocula.[100-102]

All fresh isolates appear to be infective when inoculated intraperitoneally into mice.[86,94,97,100] Differences in their virulence become evident in the subsequent phases of infection.

Certain differences have been found in the susceptibility of different inbred strains of mice to *T. vaginalis*. Landolfo et al[103] found BALB/c mice to be the most susceptible, while mice with the A-background genes were more resistant. In the latter strains the progression of pathologic changes was slow and mortality 60% to 80% lower than that of the BALB/c mice (Fig. 8.6). A genetic comparison of strains suggested that susceptibility of mice to *T. vaginalis* infection may be controlled by genes located mainly outside the major histocompatibility complex.

Progression and Pathologic Manifestations

Most of the inoculated parasites are destroyed by nonspecific cytotoxic activity of peritoneal cells shortly after they are introduced into the

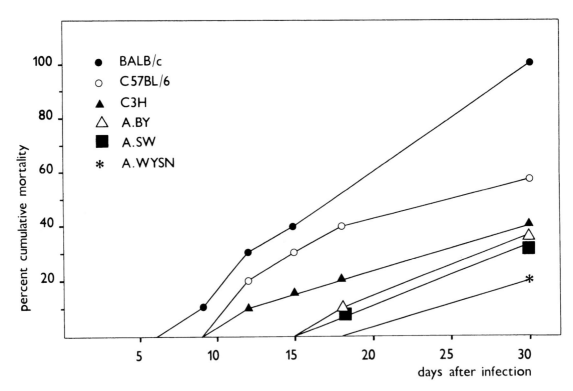

Fig. 8.6. Graphic representation of relative susceptibility, as expressed by percentage of mortality, of various inbred strains of mice to intraperitoneal infection with 8×10^5 cells of *T. vaginalis* strain isolated from a patient suffering acute trichomoniasis. Twenty mice were used for each strain. (Reproduced with permission of Istituto Sieroterapico Milanese, from Landolfo S. et al: Genetic control of *Trichomonas vaginalis* infection. I. Resistance or susceptibility among different mouse strains. *Boll Ist Sieroterap Milan* 58: 48-51, 1979.)

abdominal cavity. In mice inoculated with 10^6 trichomonads, only a few intact parasites can be found 15 minutes after inoculation; within 6 hours, the parasites are cleared from the peritoneal exudate (Nedvědová and Kulda, unpublished data). In vitro studies indicate that natural cytotoxicity against *T. vaginalis* is mediated primarily by macrophages.[104]

The surviving trichomonads retreat to abdominal organs (pancreas, liver, mesentery lymph nodes) establishing isolated foci in which the parasites multiply. Subsequently, the infection spreads from these foci by development and progressive extension of tissue lesions. Viable parasites are absent from the purulent ascites during the early infection stages. In mice infected with virulent strains, trichomonads reappear in ascitic fluid between postinoculation days 5 and 6 and become abundant during the terminal phase of infection.

Gross Pathologic Changes

The predilection site for development of early lesions is the pancreas, where macroscopic abscesses form as early as 12 hours after inoculation. Depending on the virulence of the strains employed, the lesions either remain small and localized, or grow progressively, resulting in large necrotic foci spreading to adjacent organs.

Early invasions are noted also in the liver, where they start typically in the area of the hilus. With virulent strains, the infection spreads distally causing necrosis that primarily affects the lower part of the central lobe and the base of the liver (Fig. 8.7b,c). Despite extensive necrosis, the shape and size of the liver remains unchanged. After some time, lesions appear also in the mesentery lymph nodes and occasionally in the suprarenals and genitalia. The spleen and kidney are seldom invaded, but splenomegaly is frequent.

A characteristic feature of the more advanced infection is the penetration by necrotic caseous matter of the liver, pancreas, stomach, and spleen, located in the upper part of the abdominal cavity, which often adhere to the adjacent thickened peritoneum; the viscera are covered also with fibropurulent exudate (Fig. 8.7c). Necrotic caseous matter may also be seen on the mesentery and in the lower part of the abdominal cavity.

As the infection progresses, the volume of the ascitic fluid increases. Shortly before the animal's death, yellowish-gray ascites of the fibrinopurulent type may be present in quantities exceeding 3 ml in each mouse. The final phase of infection is a generalized fibrinopurulent peritonitis. Progression of gross pathologic changes in experimental trichomoniasis of mice is shown in Fig. 8.7a–d.

The course of infection may vary with the virulence of the parasite strains employed and with susceptibility of the mice. Some animals may die before the lesions are fully developed. Sometimes the infection is not fatal—the lesions are limited in size and heal (Fig. 8.7e,f).

Histopathologic Changes

Histopathologic changes accompanying intraperitoneal infection of mice with *T. vaginalis* may be divided into three basic types, reflecting the virulence of the strains used.[87,94,95,98]

1. In mice infected with *highly virulent* strains, exudative inflammatory changes predominate. Invasion of abdominal organs results in necrosis of the liver and abscesses or phlegmonous inflammatory changes in other organs. The demarcation of lesions is poor and parasites form extensive invasive zones in contact with host tissue. Although some proliferative changes appear, reparatory processes are disorganized, and trichomonads, present in large numbers, cause persistent inflammation.

2. Pathologic changes caused by strains of *moderate virulence* are of chronic type; they are associated with productive changes. There is a marked organization of fibrinopurulent exudate, and the connective tissue containing numerous plasma cells permeates the inflammatory foci.[98] Abdominal organs, including the liver, usually are invaded but the extent of lesions is limited. Characteristic of this type of infection are foci of chronic peritonitis, with an intensive growth of connective tissue and adhesions. Peritonitis may cause death.

3. Infections with strains of *low virulence* are characterized by minimum exudation and early onset of productive changes. Tissue invasions, if any, are minimal; the liver is not invaded. Lesions are circumscribed and the parasites are prevented from contact with

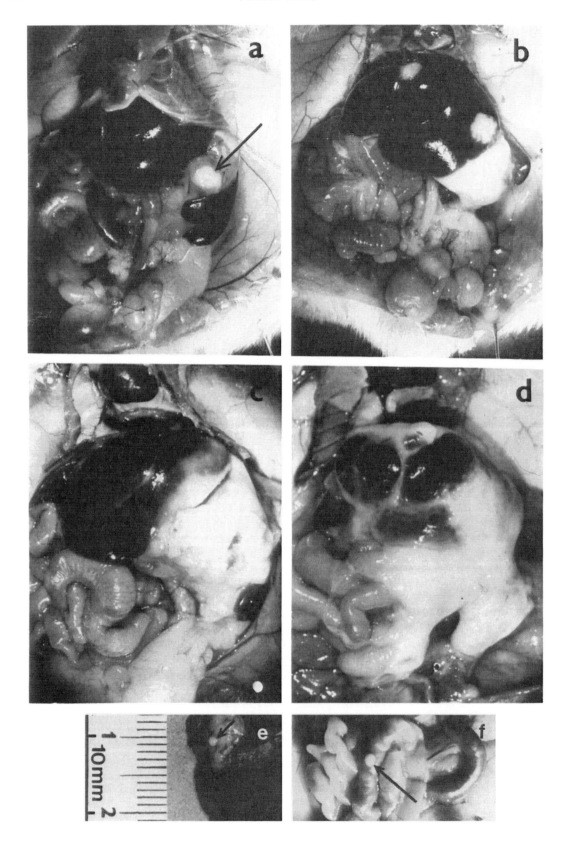

normal host tissue. Trichomonads, present in low numbers, may eventually be eliminated and the lesions filled with granulation tissue containing epithelioid and giant cells.[98] Frequently, the parasites persist in encapsulated lesions for many weeks.

Lesions in the pancreas and liver, important in early development of infection, have been studied most extensively.[94,96,98-100,105] Invasion of the pancreas occurs irrespective of strain virulence. Early lesions are small abscesses, surrounded by a delicate fibrous layer, with a center infiltrated by polymorphonuclear neutrophils and a wide peripheral zone in which parasites intermingle with leukocytes (Fig. 8.8). Kalašová-Nedvědová[77] observed, in addition to early abscesses, congested capillary vessels filled with trichomonads and leukocytes, suggestive of an intravascular origin of the lesions. Other investigators assumed that the pancreas was invaded from the outside.[100] In mice infected with virulent strains, the abscesses spread rapidly. Multiplying trichomonads palisade along the periphery of the lesions, the trichomonad mantles being followed by a zone of leukocytes and then a core central necrosis. The inflammation causes complete disorganization of invaded tissue, atrophy of glandular follicles, and edema and fibrous proliferation in the surrounding area. The glandular parenchyma is not invaded directly, being protected by an envelope of connective tissue.[97] Eventually, the inflammation becomes subchronic; it is characterized by increasing participation of mononuclear cells and extensive proliferation of granulation tissue.

In mice infected with low-virulence strains, the early pancreatic lesions are similar, but enlargement of the abscess is limited. Trichomonads fail to develop a distinct zone at the periphery of the lesions, being separated from the abscess walls by leukocytes. The more prominent fibrous walls and granulation tissue surrounding the abscess limit and thus localize the infection foci.

The liver can be invaded by more virulent strains. Invasion takes place from the outside, beginning in areas covered with fibrinopurulent membranes that contain numerous trichomonads. The parasites penetrate perivascularly and pass through the liver in dense zones leaving areas of necrosis behind (Fig. 8.9a,b). These zones are separated from the parenchyma by narrow mantles of leukocytes and macrophages that do not prevent direct contact of trichomonads with hepatocytes. Behind the advancing parasites there are prominent zones of degenerating leukocytes followed by necrotic tissue.

Results of histochemical studies[100] indicated depletion of glycogen from the parenchyma of invaded liver, accumulation of lipids and proteinaceous substances in the parasitized areas of liver and pancreas, and presence of neutral mucopolysaccharides in the fibrinous substances.

Electron microscopic observation of the infected liver[105] revealed disorganization of liver parenchyma caused by disruption of desmosomes and degradation of collagen into a fibrinoid substance. Hepatocytes in the invaded area were devoid of glycogen, contained numerous lipid inclusions, and showed alteration of mitochondria. Like those in human infections,[106,107] ameboid trichomonads attached to the host cells by means of cytoplasmic extensions of agranular cytoplasm rich in microfilaments[108]. Although in vitro experiments indicated contact-dependent cytotoxicity of *T. vaginalis*,[109] no signs of contact cytolysis could be demonstrated in vivo by electron microscopy. Cell membranes of hepatocytes were intact at

---

Fig. 8.7. Photographs of viscera in the peritoneal cavities of mice necropsied at various times after intraperitoneal inoculation with axenic cultures of high virulence (TV 7-37) (**a–d**) or low-virulence (**e,f**) strains of *T. vaginalis*. **a**. Postinoculation day 4. In early infection, note a pancreatic abcess (*arrow*). **b,c**. Postinoculation days 8 and 11. The infection involves progressive hepatic necrosis (**b,c**) leading to confluent necrosis (**c**). **d**. Postinoculation day 14. Generalized fibro-purulent peritonitis is seen at this infection stage. **e,f**. Only small localized abscesses (*arrow*) are evident in the abdominal cavity. The scale in millimeters, to the left of **e**, is applicable to all photographs. (From Nedvědová and Kulda, unpublished.)

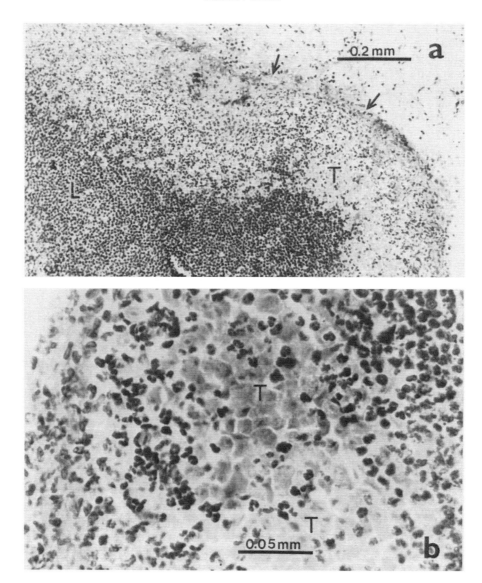

Fig. 8.8. Section of pancreatic abscess produced by a virulent *T. vaginalis* strain at 12 h postinoculation. **a.** Dense core of leukocytes (*L*) surrounded by delicate fibrous layer (*arrows*). Trichomonads intermingled with leukocytes are seen the wide peripheral zone (*T*) of the abscess. **b.** Interrelationships between the parasites and leukocytes are shown in greater detail at a higher magnification. BF, Masson's Trichrome. (From Nedvědová and Kulda, unpublished.)

the site of parasite attachment. A prominent Golgi apparatus and numerous transport vesicles suggested high secretory activity of trichomonads. However, the exocytic and endocytic processes occurred on the parasite surface away from the area of direct contact with the host cell.

Intraperitoneal Mouse Assays for Virulence

Mortality

A majority (80% to 95%) of random isolates of *T. vaginalis* can kill a mouse after intraperitoneal inoculation (Table 8.5), and this prop-

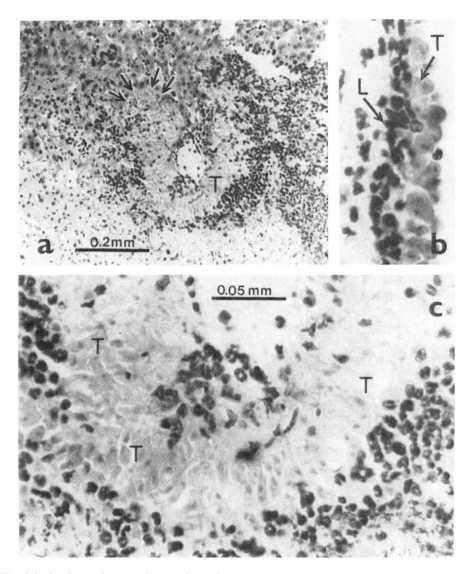

Fig. 8.9. Sections of mouse liver 7 days after intraperitoneal inoculation with a virulent *T. vaginalis* strain TV 7-37. **a.** In an overview photomicrograph, a zone rich in parasites is seen in the perivascular area. In the upper part of this area (*arrows*) the trichomonads (*T*) are in direct contact with hepatocytes. **c.** Zone of invading trichomonads is seen in greater detail at a higher magnification. **b.** Contact between the layer of the parasites (*T*) and that of the leukocytes (*L*) located at the liver surface is clearly evident. BF, MH & E. Scale in **c** applies also to **b**. (From Nedvědová and Kulda, unpublished.)

erty can be employed in simple laboratory assays.

The number of inoculated trichomonads necessary to cause lethal infection may differ for different trichomonad strains, but the dose response is rather variable, not allowing calculation of LD_{50} endpoints.*

---

*Number of inoculated parasites needed to kill 50% of inoculated animals.

TABLE 8.5. Percentages of random *T. vaginalis* isolates from geographically widely separated areas capable of killing laboratory mice on intraperitoneal inoculation.

Mice				Trichomonads			Strains		
							Capable of killing mice		
Strain or source	Wt. or age	No. in group	Period of observation (days)	Inoculum size	Period of in vitro cultivation	Total No.	No.	%	Reference
A/LN (NIH Anim. Prod. Sect.)	32–38 days	18–40	60	10^7	3–6 mo	12	10	85	96
Albino unspecified	18–20 g	10	42	4×10^6	Unspecified	80	76	95	95
Albino unspecified random mating (Nat. Labs. St. Louis, Missouri)	20–30 g	10	14	10^6	3–4 wk	30	24	80	97
Albino "H" A-background random mating (SEVAC, Prague)	18–20 g	20–30	21	10^6	1–3 wk	25	23	92	Kulda (unpublished data)

The *mortality rate,* that is, the percentage of mice that die within a given time after administration of a standard inoculum, can provide a meaningful assay endpoint. The *mean day of death* may be useful as an additional criterion in assays involving strains that cause high mortality of mice.[97,110] With an inoculum of 10^6 trichomonads, the recommended time of observation in a mortality test is 3 weeks. A minimum of 30 mice per group should be inoculated in two independent experiments if the mortality rate is to be used for evaluation of virulence. Among the 23 mouse-killing strains examined by the author (Table 8.5), the mortality rate varied between 16% and 95%. Inoculated mice died between postinoculation days 5 and 21, with the largest numbers of mice dying between days 9 and 11 after infection.

Parasite strains causing high (90% to 100%) mortality rates can be used for assaying antitrichomonal drugs or other factors affecting the trichomonads. The effects are assayed by inoculating mice with a standard dose of a virulent strain; changes in the mortality rate and/or in the time of death are compared with those of untreated controls.

Virulence Indices

Estimation of gross-pathologic changes in the abdominal cavity of mice provides another criterion for evaluation of *T. vaginalis* virulence. Two assays, have been proposed, both based on semiquantitative assessment of pathologic manifestations that allow statistical treatment of the results. In these assays, mice infected with a standard inoculum are observed for a given time, then the survivors are killed. Both the mice that succumb to the infection and those killed are examined at necropsy for the presence of trichomonads and for gross-pathologic changes in the abdominal cavity. The pathologic changes in each mouse are evaluated and rated on different bases. Experimental conditions and rating criteria employed in the two assays, compared in Table 8.6, are described below.

(1). *The virulence index*[98] ranges from 0 (maximum virulence) to 10 (lack of virulence). The authors published a table (Table 2 in reference 98) that facilitates rating of different pathologic manifestations. The criteria were selected on the basis of histopathologic studies of the progression of intraperitoneal infection of mice.[87,94,95,98] They include the death of an inoculated animal (maximum rating) and emphasize injury to the liver, but volumes of ascites are not considered. High inocula of parasites (4 × 10^6) and the relatively short observation period (10 days) might decrease the sensitivity of the assay.

(2). *The mean infestation (infection) index*[111] ["l'indice moyen d'infestation"(IMI)], ranging from 0 (lack of virulence) to 16 (maximum virulence) is based on rating lesions in the liver, pancreas, and remaining sites of the peritoneal cavity, and on the quantity of ascites. The extent of lesions in each area and the volume of ascites are rated from 0 (absent) to 4 (maximum). If parasites are undetected at an examined site, the rating of a particular lesion is divided by 2. The partial rating is then totaled to obtain the "infestation (infection) index" for an individual mouse. The arithmetic mean of the indices of all mice infected with a particular strain is the mean infestation (infection) index. The assay is based on studies of experimental trichomoniasis in mice by Gobert et al,[99,100] which provide support for inclusion of the pancreatic lesions as a criterion qualifying for independent evaluation. The extended period of observation (3 weeks) provides for better differentiation of the pathogenic effects of different trichomonad strains and allows for evaluation of mortality. Actually, failure to include mortality among the criteria may distort the rating of highly virulent strains that kill mice before the gross-pathologic changes are fully developed.

Careful employment of the two assays should provide comparable results. Cavier's assay[64,65] is simpler but more subjective than that based on the elaborate table of criteria provided by the Estonian workers.[98]

It is evident from the data published by Teras and Rõigas[98] that low-virulence strains are the least frequent, representing about 10% of all of those examined. Incidence of strains with high and moderate virulence is about the same (45% each) (Table 8.7).

Changes of Virulence

Virulence of *T. vaginalis* decreases after prolonged maintenance in axenic cultures. This is evident from the decrease of their virulent indices and loss of their ability to kill intraperitoneally inoculated mice. Using the mortality assay, a significant decrease of virulence was detected in 8 of 12 strains investigated after 3

TABLE 8.6. Comparison of intraperitoneal mouse assays employing gross-pathologic criteria for virulence evaluation.

Virulence index*		Mean infestation (infection) index†	
Mice		Mice	
Weight (g)	18–20	Weight (g)	18–20
No. in group	10–12	No. in group	10–20
Inoculum		Inoculum	
No. of trichomonads	4×10^6	No. of trichomonads	10^6
Vol. of medium	0.5 ml	Vol. of medium	0.5 ml
Length of observation (days)	10	Length of observation (days)	21
Criteria	Death	Criteria	Hepatic lesions
	Hepatic necrosis		Pancreatic lesions
	Lesions in stomach and spleen region		Peritoneal lesions
	Lesions in intestinal region		Ascitic fluid
	Fibropurulent coating of abdominal organs		Presence of trichomonads
	Presence of trichomonads		
Rating	If infection fatal within 10 days: virulence index = 0	Rating	Rating of each individual lesion and ascitic fluid: 0 (none) to 4 (maximum no.)
	In survivors until postinoculation day 10: rating of intensity (+ to +++) of each individual gross-pathologic criterion or absence thereof		If trichomonads absent, rating divided by 2
	Virulence index assessed on the basis of partial rating given in Table 2, reference no. 98		Total of partial rating
Index: Range	0–10 per mouse	Index: Range	0–16 per mouse
No virulence	0	Maximum virulence	16
Maximum virulence	10	No virulence	0

*Method of Teras and Röigas.[98] †Method of Cavier et al.[111]

TABLE 8.7. Percentages of high, moderate, and low virulence strains among random isolates of *T. vaginalis* as estimated by an intraperitoneal mouse assay.

Sex of patients	No. of strains examined	No. of mice per strain	Virulence for mice*						References
			High		Moderate		Low		
			No.	%	No.	%	No.	%	
Women	50	7	28	56	13	26	9	18	87, 94
Women	24	10	7	29	16	66	1	4	95
Men	56	10	22	39	30	54	4	7	95
Women	48	10–12	26	54	18	38	4	8	102
Women	98	10–11	46	47	40	41	12	12	98
Men	73	10–11	28	38	36	49	9	12	98
Women	25	20–30	11	44	11	44	3	12	Kulda (unpublished data)
TOTAL:	374		168	45	164	44	42	11	

*Virulence indices correspond to those suggested by Teras and Rõigas.[98]

to 4 months of cultivation.[97,112] Examination of 15 *T. vaginalis* strains with the aid of the intraperitoneal mouse assay[101,102] showed substantial virulence attenuation of stocks maintained in vitro for 8 to 32 months. Testing of the strains cultivated for shorter periods (e.g., 4 months) did not reveal major changes in the initial virulence. In general, virulence attenuation was more pronounced in highly virulent strains. Prolonged treatment with antibiotics[9,112] and induced resistance to osarzol[101] appeared to enhance attenuation.

An increase in virulence was observed in strains with low or moderate virulence after two to three intraperitoneal passages in mice.[101,102]

In studying the virulence of laboratory-maintained strains it is essential to use fresh isolates or their cryostabilates. It has been established unequivocally that the initial virulence can be retained by cryopreservation of the parasite in the presence of dimethylsulfoxide or glycerol as the cryoprotectants. By this method the original virulence levels of the organisms can be preserved for many years.[9,112,113]

Subcutaneous Infection

Inoculation, Infectivity, and Susceptibility

Subcutaneous inoculation of *T. vaginalis* into mice results in the development of a localized abscess at the injection site. Development of the lesion depends on the ability of the parasite to multiply in the inoculation site, and on a process involving the influx and death of leukocytes, lysis of the abscess wall, and renewed multiplication of parasites accompanied by the spreading of the lesion. Under standard conditions, the rate of growth of the lesion and duration of the progressive phase of its development is characteristic of a given strain, reflecting its virulence potential. Consequently, the lesion volume can be used to estimate the virulence level of a given strain.[114,115] The subcutaneous infection of mice with trichomonads, easy to establish and producing lesions amenable to direct measurement, offers a convenient model for laboratory studies. It has been successfully applied to virulence and drug assays[5-9] as well as to generating antitrichomonal sera or primed spleen cells for producing monoclonal antibodies against trichomonad surface antigens.[116]

Inoculation Procedures

Since Schnitzer et al[4] found the mouse susceptible to subcutaneous infection with trichomonads, numerous investigators employed this model for drug testing.[5,6] Subcutaneous infections with *T. vaginalis* were produced by single or multiple inoculations into the ventral,[4,117] dorsal,[114,118,119] or hind leg[92] areas, with parasites suspended in a variety of media and the inocula ranging from 7×10^3 to 1.4×10^7 cells. Honigberg[114,115] applied subcutaneous infection of mice to studies of trichomonad pathogenicity and virulence, and developed a standard procedure providing generally reproducible results (see section, Subcutanous Mouse Assay for Virulence, below).

According to Honigberg's method, the subcutaneous lesions can be consistently produced by inoculation of mice with 8 to 9×10^5 trichomonads in 0.5 ml of medium containing agar (0.05% to 0.075%). This dose ensures high infection rates and allows differential growth of abscesses characteristic of an individual trichomonad strain. Variations of inoculation doses ranging between 6×10^5 and 10^6 cells do not affect substantially the size of the resulting abscesses; however, the employment of much higher parasite numbers ($\geq 2 \times 10^6$ cells) tends to impair the sensitivity and usefulness of the virulence assays.

The presence of agar in the inoculation medium favors establishment of infection.[114,120] Consequently agar-free media are not recommended for suspending the parasites. Because the concentration of agar in the inoculum may influence the size of lesions, it is necessary to employ a standard concentration in all instances.[114] Immunosuppressive treatment is unnecessary, since agar-containing inocula readily establish the infection, although a minor increase of susceptibility to inocula without agar was observed in cortisone-treated mice.[120]

Shaved mice have to be used in assays that involve measurement of lesions. One person should hold the mouse without distorting the shape of the lesion, while measurements ought to be taken by another person. The length and width of the lesion can be measured with calipers, and the height with the aid of a metric ruler. Some experience is required to obtain reproducible measurements.

For immunization, large doses of parasites (5

× 10⁶) should be injected into both flanks of a mouse. Before inoculation, the trichomonads have to be thoroughly washed with phosphate-buffered saline (PBS) to remove contaminating medium components. The flagellates are inoculated in PBS without agar. High IgG response to exposed surface proteins of trichomonads can be obtained by administration of the parasites in two injections given at 14-day intervals. The second inoculation is then followed by a booster injection 4 weeks later. Primed spleen cells for production of monoclonal antibodies should be harvested 3 days after the final inoculation.[116]

Infectivity

Under standard conditions, the infection rate of fresh isolates of *T. vaginalis* approaches 100%, irrespective of strain virulence. Occurrence of noninfective strains is exceptional. Assays with large numbers of *T. vaginalis* strains[114,121-125] showed that the strains differed in the sizes of lesions they produced, and not in their infection rates. Even strains with very low virulence established transient infections and produced visible, albeit small, abscesses in most of the inoculated mice.

In contrast to the foregoing findings were those published by Krieger et al.[126] Among seven clinical isolates of *T. vaginalis* they examined, two were found noninfective for C_3H mice. The other isolates reportedly caused abscesses only in a certain percentage of these hosts (5% to 71%), and the authors used the proportion of lesion-producing infections to assess the virulence of their strains. In light of our extensive experience, Krieger's results seem improbable. They might be explained by considerable differences from the standard inoculation procedure (e.g., the employment of unshaven mice, or the use of agar-free inocula).

Susceptibility

Different strains of mice respond differently to subcutaneous infection with trichomonads. As shown with *T. gallinae*,[127] the mean volumes of 5-day lesions produced by the standard inoculum can differ significantly for some strains. Moreover, the tendency of abscesses to rupture and drain is more pronounced in certain strains of mice (e.g., DBA/2) than in others. Of the eight strains examined, the BALB/c was found most satisfactory. Since the mean volumes of lesions vary with the mouse strains employed, the absolute measurements obtained with different strains of mice cannot be compared directly. However, relative differences among trichomonads of different virulence are comparable, as has been confirmed for BALB/c, CBA, and C57B1/6 mice (Kulda, unpublished data).

Subcutaneous infection with *T. vaginalis* was successfully established in hairless hr-C_3H mice.[21] Although the convenience of a nude mouse for monitoring subcutaneous lesions cannot be doubted, the limited accessibility and cost of the athymic animals, owing to their poor breeding capacity, would hardly favor their extensive use in assaying drug sensitivity or virulence of trichomonads.

Most of the comparative data on *T. vaginalis* virulence have been obtained with the aid of C57B1/6 mice.[114,115,121-125] Although the response of this strain to trichomonad infection is less pronounced than that of several other strains examined (e.g., the mean volumes of abscesses are about 30% lower than in BALB/c mice), there are no major drawbacks to their use. Employment of this strain for virulence assays with *T. vaginalis* is recommended because it renders feasible direct comparison of the data with the results obtained by various workers on many previous occasions.

Progression and Pathologic Manifestations of Infection

Gross Pathology

The lesion caused by subcutaneous inoculation of *T. vaginalis* is a hard abscess protruding above the body surface (Fig. 8.10). The abscess contains compact purulent matter and many trichomonads. Typically, the abscess grows during the first 2 weeks after inoculation and eventually ruptures and drains. Abscesses caused by virulent strains start to open by the end of the first week after inoculation; small abscesses produced by strains with low virulence do not necessarily rupture. After draining, the lesion usually heals with formation of a scar. Occasionally, the ruptured lesion may close and produce a secondary abscess; rarely, the abscess fails to heal for a prolonged period.[114]

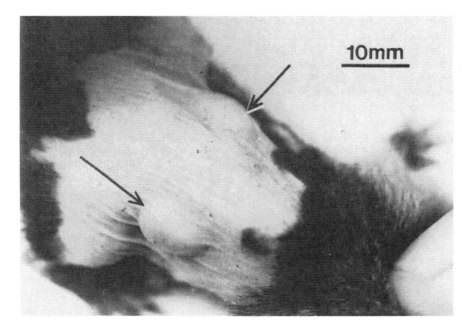

Fig. 8.10. Subcutaneous abscesses seen on the flanks of a C57BL/6 mouse 6 days after inoculation with 8×10^5 *T. vaginalis*. (Original photograph by J. Chalupsky.)

Behavior of Parasites

In contrast to the situation observed in intraperitoneal infection, there is no immediate host response suppressing the parasites during the earliest phases of a subcutaneous infection. Trichomonads start to multiply shortly after inoculation, their numbers increasing progressively.[115,119] The first phase of multiplication occurs throughout the injection pocket; subsequently, the organisms palisade against the wall of the lesion.

Histopathology

Histopathologic changes are shown in Fig. 8.11a,b. The infection induces migration of polymorphonuclear neutrophils, which enter the injection pocket and line the margin of the lesion. The influx of leukocytes continues along with intensive multiplication of parasites. Although some trichomonads are phagocytosed, the leukocytes are ineffective in suppressing the infection, trichomonads increasing in number and intermingling with leukocytes in the peripheral zone. Eventually, the tightly packed parasites form a distinct mantle against the collagen margin of the injection pocket pushing the leukocytic mantle inward. Continual influx of leukocytes into the lesion results in the formation of a second leukocytic mantle that, again, is moved inward by the advancing zone of multiplying trichomonads, the process being cyclically repeated. The innermost zone of leukocytes degenerates and the center of the abscess becomes necrotic.

In the adjacent tissues, marked fibroblastic activity occurs starting on postinoculation day 2 and increases in the course of the subsequent infection phases. There is a rise in the number of mononuclear cells and, eventually, granulation tissue is formed. Death of leukocytes within the lesion apparently contributes to cytolytic processes injurious to host tissue.[115] The lesion spreads by lysis of the wall and releases parasites that penetrate the interstitial spaces of the surrounding edematous tissue. The collagen lining of the lesion is replaced by a fibrous wall. Edema is more pronounced and fibrosis is less effective in infections with highly virulent strains; repeated breakouts and spreading of the parasites result. Finally, the lesion is walled off by a thick layer of granulation tissue that trichomonads cannot penetrate.

This process of lesion development is very similar in different trichomonad strains,[115] the differences among the strains being quantita-

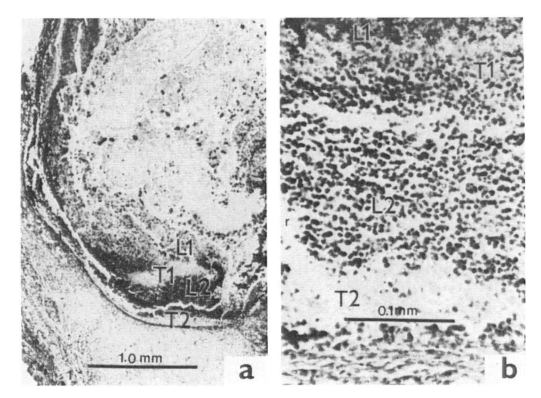

Fig. 8.11. Section of subcutaneous lesions examined on the 5th day after subcutaneous infection of C57BL/6 mice with 8×10^5 *T. vaginalis* belonging to a virulent C-1:NIH strain. **a.** The 2nd trichomonad (*T2*), 2nd leukocytic (*L2*), 1st trichomonad (*T1*), and 1st leukocytic (*L1*) mantles are evident on the periphery of the abscesses, the mantles proceeding from the periphery toward the center of the lesion whose central core is necrotic. **b.** At a higher magnification, signs of degeneration are evident in the earlier trichomonad (*T1*) and leukocytic (*L1*) mantles (cf. *T2* and *L1*). Verhoef van Gieson stain. (Reproduced with permission of the American Society of Parasitologists, from Frost JK, Honigberg BM: Comparative pathogenicity of *Trichomonas vaginalis* and *Trichomonas gallinae* for mice. II. Histopathology of subcutaneous lesions. *J Parasitol* 48:898-918, 1962.)

tive rather than qualitative. The main variables are the duration of the cycle, the duration of the period of progression, and the severity of the response. Because of these differences, the size of the lesion measured at a given time after inoculation is characteristic of the strain and reflects its virulence.

Subcutaneous Mouse Assay for Virulence

Honigberg developed[114] and standardized[121] a quantitative assay for trichomonad virulence, based on measurements of 6-day abscesses produced by subcutaneous administration of standard inocula of trichomonads to inbred mouse strains.

Assay Procedure

Shaved inbred mice C57BL/6 of either sex, weighing 18 to 20 g, are inoculated into both flanks with 8 to 9×10^6 trichomonads suspended in 0.5 ml of fluid thioglycollate medium* supplemented with 5% inactivated horse serum. To prepare inocula, late logarithmic-growth phase organisms from the second transfer in agar-free medium (e.g., TYM, CPLM) are centrifuged, washed in PBS, and resuspended to the desired concentration in the thioglycollate medium. On the sixth postinocula-

---

*BBL Fluid Thioglycollate Medium, USP, (11260) Div. Becton, Dickinson and Co:, Cockeysville, MD 21030.

tion day, the length (L), width (W), and height (H) of each lesion are measured and the volume of the protruding portion of the abscess, corresponding to about one-half of a spheroid, is calculated according to the following formula[114]:

$$\text{vol} = 0.5236 \times L \times W \times H$$

To obtain reliable data, assessment of the mean volumes should be based on measurements of at least 30 intact abscesses. Inoculation of at least 20 mice per group is recommended, because some lesions may open before being measured.

The rupture of the lesion before or on postinoculation day 6 may reflect a virulence-associated property of a trichomonad strain. However, abscesses can also rupture for reasons unrelated to virulence; thus correlation between percentage of opened lesions and virulence must be made with caution.

Reliability

Reliability of the assay was questioned by Delachambre,[128] who made a statistical analysis of results and found an excessive variability in responses of the individual mice (BALB/c) to inoculation with two T. vaginalis clones. In contradistinction, measurements obtained for several T. vaginalis strains and clones by Honigberg and his associates[114,124] showed satisfactory homogeneity when subjected to analysis of variance. Lack of statistical homogeneity of results may reflect faulty technique or the presence of interfering factors that must be identified and eliminated. Failure to distinguish virulence levels of different T. vaginalis strains mentioned by Alderete[129,130] was undoubtedly caused by the use of excessively high numbers of parasites in the inocula.

Successful performance of the assay demands skill, patience, and strict adherence to the standard protocol. If these conditions are fulfilled, highly reproducible results can be obtained (see Table 8.8).

Size of Lesions

Mean volumes of 6-day abscesses produced in C57BL/6 mice by 77 fresh, random isolates of T. vaginalis examined by Honigberg et al[125] and Kulda et al (reference no. 123 and unpublished results) formed a continuum ranging from 60 mm^3 to 230 mm^3. There was a single strain (Balt 53) distinguished from others by very low virulence for mice, producing abscesses of about 25 mm^3 volume. No cytopathologic changes occurred in the patient infected with this strain. Four strains producing the largest abscesses, with mean volumes above 200 mm^3 (Balt 42, Balt 44, TV 7-37, TV 84-00) were isolated from patients with severe histopathologic changes present in the cervix uteri. The strain-specific size of subcutaneous lesions in mice indicates apparently an intrinsic virulence potential of the strains examined. However, as in other virulence assays, a correlation between the results obtained in these laboratory animals and pathogenic effects of trichomonads in patients need not necessarily be absolute (see section, Relative Virulence of T. vaginalis for Mouse and Humans, below).

Intramuscular Infection of Mouse

Intramuscular inoculation of axenic T. vaginalis cultures into mice produces localized abscesses at the inoculation site.[4,97,134] Inocula of 1.5×10^5 to 2.5×10^6 trichomonads in 0.2-ml agar-containing media (CPLM, TYM) with or without serum[4,97] or in 0.1 ml of saline[134] produced lesions, the volume and persistence of which were influenced by the infection dose.[4] Typically, palpable swelling of inoculated tissue developed within 3 days after infection; then the lesion gradually increased to form a large abscess after 14 to 18 days. Fully developed lesions measured 500 to 2200 mm^3 and contained greenish material full of parasites and leukocytes. The abscess diminished after 8 to 12 weeks, but still contained viable trichomonads.

The possibility of using intramuscular infection for assaying trichomonad virulence was suggested by Gavrilescu et al,[134] according to whom differences in the histopathology of lesions (acute or chronic abscesses) could be correlated with the clinical form of disease in patients.[134] However, other investigators[97] found intramuscular infection unsuitable for evaluation of virulence because of considerable variations in the size of the lesions and poor accessibility to the infection site.

Intramuscular infection of mice was successfully employed in immunologic studies[135-137]; it was found to provide effective protection (80% to 100%) against repeated challenge by the intramuscular or the intraperitoneal route. The protection was unaffected by splenectomy or

TABLE 8.8. Comparison of virulence estimates of five *T. vaginalis* strains obtained by the standardized subcutaneous mouse assay[121] at various times.*

Parasite strain	Mean volumes, in mm³, of 6-day subcutaneous lesions									
	Ref. 121 (1966)[†]		Ref. 122 (1970)[†]		Ref. 132 (1980)[†]		Ref. 133 (1983)[‡]		Ref. 131 (1983)[‡]	
	n[§]	Mean ± S.E.[∥]	n	Mean ± S.E.	n	Mean ± S.E.	n	Mean ± S.E.	n	Mean ± S.E.
JH30A	42	151.25 ±6.98	74	151.45 ±6.25	—	—	30	139.84 ±15.96	18	145.83 ±9.59
JH31A	21	71.29 ±5.15	49	78.60 ±3.98	—	—	22	72.73 ±10.21	37	75.18 ±2.49
JH32A	45	78.99 ±5.47		82.37 ±3.88	—	—	—	—	30	77.44 ±3.88
JH34A	43	108.66 ±5.76	71	117.05 ±4.64	44	117.98 ±6.10	44	116.07 ±5.14	29	104.60 ±3.90
JH162A	—	—	70	75.64 ±2.84	37	83.87 ±5.09	41	72.94 ±5.58	22	78.80 ±5.80
JH384A	—	—	—	—	—	—	42	137.34 ±11.40	41	138.63 ±14.50

*Most data from Wartón and Honigberg,[131] reprinted with permission of Springer-Verlag International.
[†]Noncloned populations.
[‡]Clones.
[§]Number of determinations.
[∥]Sample standard error of the mean.

blockage of spleen by Thorotrast; however, it could not be correlated with blood serum level of agglutinating antibodies. No recent reports are available on the use of this intramuscular inoculation route.

Relative Virulence of *T. vaginalis* for Mouse and Humans

Attempts to correlate results of mouse assays with pathogenic effects of *T. vaginalis* in humans have met with certain difficulties. Clinical findings are difficult to quantify and the assessment of the severity of disease relies mostly on subjective evaluation. Moreover, clinical manifestations are affected by responsiveness of the individual host. There is some evidence that changes in the cervix uteri, revealed on cytopathologic examinations of vaginal smears or histopathologic studies of cervical biopsies reflect most accurately the virulence potential of a given *T. vaginalis* strain.[121,122,125] It should be pointed out, however, that even the pathologic findings reflect the response of a single patient to a parasite strain. In light of the foregoing considerations a trend for an agreement rather than perfect correlation can be expected between the results of laboratory tests and clinical findings.

Intraperitoneal Assay

Estonian workers[87,94,98] compared virulence estimated on the basis of the intraperitoneal mouse assay with clinical findings recorded from patients. One hundred fifty-one isolates from women and 73 from men were employed in their studies. The results of these studies are summarized in Table 8.9. Although the correlation was far from perfect, the occurrence of strains with higher virulence for mice was considerably lower in patients with latent and chronic disease than in those with acute or subacute trichomoniasis.

Mortality Assay

By evaluating mortality of mice after intraperitoneal inoculation, Reardon et al[96] found differences between two *T. vaginalis* strains, one isolated from a patient with severe vaginitis and the other from a person with a mild infection. However, the authors were unable to correlate mortality of the mice with clinical manifestations associated with 11 strains obtained from patients presenting moderate symptoms. Also, Ivey and Hall[97] were unable to distinguish by the mortality assay virulence of 30 fresh isolates of the urogenital trichomonad isolated from either symptomatic or asymptomatic patients; however, they were able to demonstrate virulence difference in strains tested by the subcutaneous assay.

Subcutaneous Assay

Most satisfactory results were obtained in studies using the subcutaneous assay[121,122] with a total of 11 *T. vaginalis* strains documented by complete records of clinical and pathologic findings. Eight of the strains showed a high degree of positive correlation between the virulence for mice and various parameters of the vaginal and cervical disease in patients; minor differences were found in two strains. The single nonconforming strain, highly pathogenic for mice, was isolated from a patient with moderate disease. In a subsequent study[125] including a nonselected sample of 52 *T. vaginalis* strains, the results of the subcutaneous assay were compared with clinical findings and with the results of a cytopathologic examination based on Papanicolaou smears from the vaginal pool and pancervical scrapings; when available, cervical biopsies were studied histopathologically. Surprisingly, cellular changes in the cervical epithelium were the only parameter showing a statistically significant relationship with the results of the mouse assay. Clinical symptoms and signs of vaginitis or inflammation of the vagina and cervix detected by cytopathologic examination did not correlate with the mean volumes of the subcutaneous lesions.

Comparison of Mouse Assays

To date, the subcutaneous mouse assay is regarded as superior to other tests[9,138] because it allows more precise quantification of the results and appears to be less affected by various factors that may modify the course of an experimental infection. However, data allowing objective comparison of different assays still remain sparse. The only report on direct comparison of intraperitoneal and subcutaneous virulence assays[97] denied the existence of relationships between the values obtained by the two methods, and dismissed the usefulness of the intraperitoneal assay for virulence evalua-

TABLE 8.9. Proportions of *T. vaginalis* strains, judged by the intraperitoneal mouse assay to have high, moderate, and low virulence levels, among isolates from patients presenting acute or subacute or else chronic or latent urogenital trichomoniasis.

Patients		Strains from cases of acute or subacute trichomoniasis							Strains from cases of chronic or latent trichomoniasis							References
			Virulence for mice							Virulence for mice						
			High		Moderate		Low			High		Moderate		Low		
Sex	No.	No.	No.	%	No.	%	No.	%	No.	No.	%	No.	%	No.	%	
Women	53	37	25	68	9	24	3	8	16	3	19	4	25	9	56	87, 94
Women	98	56	34	61	19	34	3	5	42	12	29	21	50	9	21	98
Men	73	42	20	47	21	50	1	2	31	8	26	15	48	8	25	98

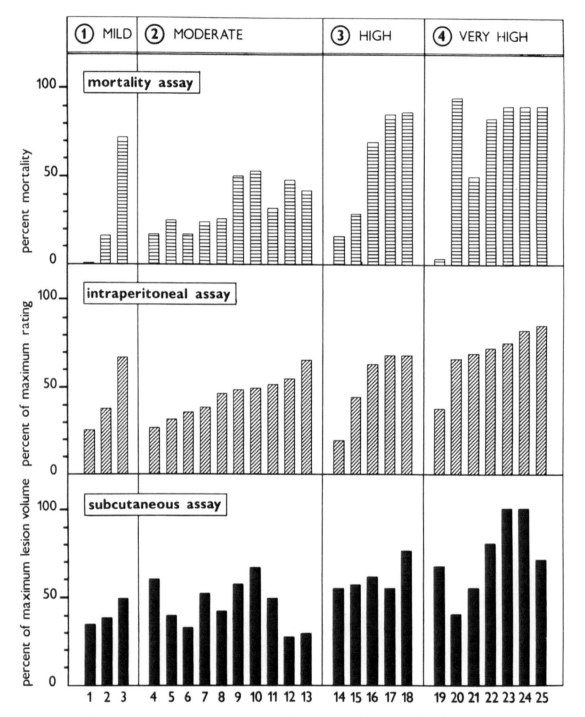

Fig. 8.12. Comparison of virulence estimates for a group of 25 fresh *T. vaginalis* isolates by the "mortality" (20 to 30 mice per strain), "intraperitoneal" (Cavier's index) (20 to 30 mice per strain) and "subcutaneous" (measurements of 30 to 60 intact abscesses per strain) mouse assays. The 25 strains were divided into four groups according to their virulence levels on the basis of histopathologic evaluation of the changes found in cervical biopsies obtained from the donors of the strains. The four groups were characterized as follows: **1.** *Mild*—normal epithelium; no or only very minor inflammation. **2.** *Moderately virulent*—acanthosis of the

tion. This extreme view is not supported by observations made in our laboratory.

To provide some comparative data, hitherto unpublished results of a cooperative study, partly reported in an abstract[123] are presented in Figure 8.12. The diagram shows a comparison of virulence of 25 fresh *T. vaginalis* isolates as estimated by the subcutaneous, intraperitoneal (Cavier's modification), and mortality assays. Pathogenic effects of the strains on patients were evaluated on the basis of histopathologic examinations of cervical biopsies. The strains were divided into four groups according to severity of the lesions, as defined in the figure legend.

Irrespective of the assay employed, the strains with the highest virulence for mice fitted into the group of isolates that caused the most severe lesions in humans. None of the methods was clearly superior to the others. A substantial disagreement in the virulence rating based on the subcutaneous and intraperitoneal mouse assays was observed in five strains. Despite these differences, the combined results of the three mouse assays were complementary rather than contradictory in correlating with histopathologic findings in patients. There was, however, one strain, isolated from a patient with no apparent disease, that exhibited rather high virulence in all assays.

Discrepancies between the results of various virulence assays can be explained by sensitivity differences of the assays to the possibly multiple virulence-controlling factors present in various trichomonad strains. It is apparent from studies on the effect of cultivation on virulence that the intraperitoneal assay, particularly the mortality test, are more sensitive for detecting virulence attenuation. Indeed, loss or considerable decrease in the ability to kill mice after intraperitoneal inoculation was observed after in vitro cultivation of *T. vaginalis* strains for 3 to 4 months.[110] With the aid of the intraperitoneal assay[101] based on rating of pathologic changes,[98] marked reduction in virulence was observed after 8 months of cultivation. Finally, significant changes in virulence were detected by the subcutaneous assay after cultivation for over 2 years.[114] These findings have been confirmed by observation of a virulent *T. vaginalis* strain, maintained in vitro for 1 year (Stejska and Kulda, unpublished data) in which no attenuation was detected by the subcutaneous assay (mean volume of lesions about 180 mm^3), while decrease in the mean "infestation" (infection) index[111] of about 20% was demonstrated by the intraperitoneal assay. Furthermore, the mortality dropped from 80% to 23%, and the mean time to death increased from 9.8 to 17.6 days. It is tempting to infer that these differences reflect interaction with factors of cell-mediated cytotoxicity[104] that exert a powerful antitrichomonal effect in early phases of the intraperitoneal infection, but do not play an important role in subcutaneous infections of mice. The relevance of these findings to virulence determinants operating in humans remains to be elucidated.

Available data suggest that the mouse assays for virulence reflect intrinsic biologic properties of trichomonad strains that might be of importance in considering potential virulence of the parasite for humans. All three aforementioned assays appear to yield reproducible results provided that they (1) are performed according to established standard procedures; (2) employ sufficiently large groups of suitable mouse strains; and (3) use viable parasites derived from axenic cultures of freshly isolated strains or their cryostabilates. In the final analysis, the choice of the assay should depend on the requirements of a particular experiment.

References

1. Bos A: Ueber Trichomoniasis bei Tauben. IV. Mitteilung. Pathogenität von *Trichomonas columbae* für Mäuse. *Zentralbl Bakteriol Parasitenkd Infectionskr Hyg (A)* 132:453-458, 1934.
2. Morgan BB: Inoculations of *Trichomonas foe-*

---

cervical epithelium; moderate cellular changes (e.g., vacuolization, perinuclear halos, moderate leukocytic infiltration of the subepithelial stroma). **3.** *Highly virulent*—acanthosis, pathologic vascularity, cellular changes typical of mild dysplasia, and marked inflammation of the subepithelial tissue. **4.** *Very highly virulent*—focal erosions of the cervical epithelium and proliferation of granulation tissue; cellular changes characteristic of marked to severe dysplasia. (From Kulda et al, unpublished data.)

tus (Protozoa) in white rats and mice. *Proc Helminthol Soc Wash* 9:68-71, 1942.
3. Trussel RE: *Trichomonas vaginalis and Trichomoniasis.* Springfield, Ill: Charles C Thomas, 1947.
4. Schnitzer RJ, Kelly DR, Leiwant B: Experimental studies on trichomoniasis. 1. The pathogenicity of trichomonad species for mice. *J Parasitol* 36:343-349, 1950.
5. Schnitzer RJ: Chemotherapy of trichomonad infection. In Schnitzer RJ, Hawkin F (eds): *Experimental Chemotherapy,* vol 1, pp 289-331. New York: Academic Press, 1963.
6. Michaels RM: Chemotherapy of trichomoniasis. In Goldin A, Hawkin F, Schnitzer RJ (eds): *Advances in Chemotherapy,* vol 3, pp 39-108. New York: Academic Press, 1968.
7. Jírovec O, Petrů M: *Trichomonas vaginalis* and trichomoniasis. In Dawes B (ed): *Advances in Parasitology,* vol 6, pp 117-188. London: Academic Press, 1968.
8. Meingassner JG, Heyworth PG: Intestinal and urogenital flagellates. In Schonfeld H (ed): *Antiparasitic Chemotherapy. Antibiotic Chemotherapy,* vol 30, pp 163-202. Basel: S. Karger, 1981.
9. Honigberg BM: Trichomonads of importance in human medicine. In Kreier JP (ed): *Parasitic Protozoa,* vol 2, pp 275-454. New York: Academic Press, 1978.
10. Hafez ESE (ed): *Reproduction and Breeding Techniques for Laboratory Animals.* Philadelphia: Lea & Febiger, 1970.
10a. UFAW: *Handbook on the Care and Management of Laboratory Animals,* 5th ed. Edinburgh: Churchill Livingstone, 1976.
10b. Melby EC Jr, Altman NH (eds): *Handbook of Laboratory Animal Science.* Cleveland, CRC Press, 1974.
11. Butcher RL: Changes in gonadotropins and steroids associated with unilateral ovariectomy of the rat. *Endocrinology* 101:830-840, 1977.
12. Wagner JE, Manning PJ (ed): *The Biology of the Guinea Pig.* New York: Academic Press, 1976.
13. Young WC: The vaginal smear picture, sexual receptivity and the time of ovulation in the guinea pig. *Anat Rec* 67:305-325, 1937.
14. Głebski J: The influence of some hormones on the morphological and biological features of *Trichomonas vaginalis* in culture conditions. *Wiad Parazytol* 15:261-262, 1969 (In Polish, English summary).
15. Nakamura N: Studies on experimental vaginal trichmoniasis in rats, with special reference to factors influencing the establishment of infection. *Jpn J Parasitol* 20:7-23, 1971 (In Japanese, English summary).
16. Kessel JF, Gafford JA Jr: Observations on the pathology of trichomonas vaginitis and on vaginal implants with *Trichomonas vaginalis* and *Trichomonas intestinalis. Am J Obstet Gynecol* 39:1005-1014, 1940.
17. Trussel RE, McNutt SA: Animal inoculations with pure cultures of *Trichomonas vaginalis* and *Trichomonas foetus. J Infect Dis* 69:18-28, 1941.
18. Johnson G, Kupferberg AB, Hartman CG: Cyclic changes in vaginal populations of experimentally induced *Trichomonas vaginalis* infections in rhesus monkeys. *Am J Obstet Gynecol* 59:689-692, 1950.
19. Cuckler AC, Kupferberg AB, Millman N: Chemotherapeutic and tolerance studies on amino-nitrothiazoles. *Antibiot Chemother* 5:540-555, 1955.
20. Eugere E, Lynch V, Thoms RK: Observations on vaginal trichomoniasis in monkeys. *J Parasitol* 42 (suppl):22, 1956.
21. Raether W, Seidenath H: The activity of fexinidazole (HOE 239) against experimental infections with *Trypanosoma cruzi* trichomonads and *Entamoeba histolytica. Ann Trop Med Parasitol* 77:13-26, 1983.
22. Street DA, Taylor-Robinson D, Hetherington CM: Infection of female squirrel monkeys (*Saimiri sciureus*) with *Trichomonas vaginalis* as a model of trichomoniasis in women. *Br J Vener Dis* 59:249-254, 1983.
23. Gardner WA, Culberson DE, Scimeca JM, Brady AG, Pindak FF, Abee CR: Experimental genital trichomoniasis in the squirrel monkey (*Saimiri sciureus*) model. *Genitourin Med* 63:188-191, 1987.
24. Williams RF, Hodgen GD: The reproductive cycle in female macaques. *Am J Primatol* 1(suppl):181-192, 1982.
25. Hegner R, Ratcliffe H: Trichomonads from the vagina of the monkey, from the mouth of the cat and man and from the intestine of the monkey, opossum and prairie dog. *J Parasitol* 14:27-35, 1927.
26. Hegner R: Experimental transmission of trichomonads from intestine and vagina of monkeys (*Macacus rhesus*). *J Parasitol* 14:261-264, 1928.
27. Dobell C: Researches on the intestinal protozoa of monkeys and man. VI. Experiments with the trichomonads of man and the macaques. *Parasitology* 26:531-577, 1934.
28. Hollander DH, Gonder JD: Indigenous intravaginal pentatrichomonads vitiate the usefulness of squirrel monkeys (*Saimiri sciureus*) as models for trichomoniasis in man. *Genitourin Med* 61:212-213, 1985.
29. Culberson DE, Pindak FF, Gardner WA, Honigberg BM: *Tritrichomonas mobilensis* n. sp. (Zoomastigophorea: Trichomonadida) from the Bolivian squirrel monkey *Saimiri boliviensis boliviensis. J Protozool* 33:301-304, 1986.
30. Pindak FF, Mora de Pindak M, Abee CR, Gar-

dner WA Jr: Detection and cultivation of intestinal trichomonads of squirrel monkeys (*Saimiri sciureus*). *Am J Primatol* 9:197-205, 1985.
31. Culberson DE, Scimeca JM Jr, Gardner WA Jr, Abee CR: Pathogenicity of *Tritrichomonas mobilensis*: Subcutaneous inoculation in mice. *J Parasitol* 74: 774-780, 1988.
32. Bunton TE, Lowenstine LJ, Leininger R: Invasive trichomoniasis in a *Callicebus moloch*. *Vet Pathol* 20:491-494, 1983.
33. Soszka S, Kazanowska W, Kuczyńska K: On injury of the epithelium of the vagina caused by *Trichomonas vaginalis* in experimental animals. *Wiad Parazytol* 8:209-215, 1962.
34. Ginel W: On changes in the vaginal mucous membrane of pregnant animals caused by *Trichomonas vaginalis* Donné. *Wiad Parazytol* 8:218-221, 1962.
35. Sztykiel Z: On the behavior of leukocytes in the cause of experimental infection of guinea pig with *Trichomonas vaginalis*. *Wiad Parazytol* 8:227-233, 1962.
36. Kazanowska W: The mucous membrane of the uterus (of experimental animals) following intravaginal infection with *Trichomonas vaginalis*. *Wiad Parazytol* 8:223-228, 1962.
37. Soszka S, Kazanowska, W, Kuczyńska K, Łotocki W, Klimowicz L: Experimental infection of the urinary tract of guinea pigs by *T. vaginalis*. *Wiad Parazytol* 19:447-452, 1973 (In Polish, English summary).
38. Kazanowska W, Dubiel C, Knapp P, Racz R: Histochemical studies of the urinary tract in guinea pigs infected by *T. vaginalis*. *Wiad Parazytol* 19:453-460, 1973 (In Polish, English summary).
39. Skrzypiec R: Biomorphological changes in the vaginal part of the uterus in a guinea pig as a result of infection with *Trichomonas vaginalis*. *Wiad Parazytol* 21:377-388, 1975 (In Polish, English summary).
40. Skrzypiec R: Behaviour of the common part of uterus of guinea pigs after infection with *Trichomonas vaginalis* II. The appearing mast cells. *Wiad Parazytol* 25, 51-61, 1979 (In Polish, English summary).
41. Kazanowska W: Morphological and histochemical phenomena in the mucous membrane of genital organs in the course of trichomonadosis. *Wiad Parazytol* 12:139-150, 1966 (In Polish, English summary).
42. Kazanowska W, Kuczyńska K, Skrzypiec R: Pathology of *T. vaginalis* infection in experimental animals. *Wiad Parazytol* 29:63-66, 1983.
43. Maestrone G, Semar R: Experimental intravaginal infection with *Trichomonas foetus* in guinea pigs. *Chemotherapia* 12:137-145, 1967.
44. Hypes RA, Ladewig PP: Leucocytic clusters on epithelial cells in cervico-vaginal smears: A presumptive test for *Trichomonas* infection. *Am J Clin Pathol* 26:94-97, 1956.
45. Kobayashi TK, Fujimoto T, Okamoto H, Yuasa M, Sawaragi I: Association of mast cells with vaginal trichomoniasis in endocervical smears. *Acta Cytol* 27:133-137, 1983.
46. Rubin R, Cordray DR: A new approach in chemotherapeutic trials against *Trichomonas* infections. *Am J Vet Res* 19:249-251, 1958.
47. Michaels RM, Strube RE: Antitrichomonal agents 5-nitrothiazoles, 5-nitropyridines and 5-nitropyrimidines. *J Pharm Pharmacol* 13:601-610, 1961.
48. Ryley JF, Stacey GT: Experimental approaches to the chemotherapy of trichomoniasis. *Parasitology* 53:303-320, 1963.
49. Uhlenhuth P, Shoenherr KE: Untersuchungen über die Übertragugnsmöglichkeiten vershiedener Trichomonadearten auf kleine Versuchstiere. *Z Immunitaetsforsch Exper Ther* 112:48-56, 1955.
50. Vershinsky BV: Experiment on the infection of *Mesocricetus aureatus* with *Trichomonas vaginalis* Donné, 1836. *Zool Zh* 36:1774-1776, 1957 (In Russian).
51. Wildfever A: Die Chemotherapie der vaginalen Trichomoniasis und Candidosis der Maus. *Arzneim-Forsch* 24:937-943, 1974.
52. Latter VS, Linstead DJ, Beesley JE, Walters MA, Bradley S, Steward TW, Ackers JP: Intravaginal infection of golden hamster by *Trichomonas vaginalis*. Abstracts Int Symp Trichomonads & Trichomoniasis, Prague, July 1985, p 56. Prague: Local Organizing Committee, Dept Parasitol Faculty Sci, Charles Univ, 1985.
53. Flynn RJ (ed): *Parasites of Laboratory Animals*. Ames: Iowa State University Press, 1973.
54. Wenrich DH: Intestinal flagellates of rats. In Hegner BH, Andrews J (eds): *Problems and Methods of Research in Protozoology*, pp 124-142. New York, Macmillan, 1930.
55. Wenrich DH: Morphology of the intestinal trichomonad flagellates in man and of similar forms in monkeys, cats, dogs and rats. *J Morphol* 74:189-211, 1944.
56. Kirby H: The structure of the common intestinal trichomonad of man. *J Parasitol* 31:163-175, 1945.
57. Kirby H, Honigberg BM: Flagellates of the caecum of ground squirrels. *Univ Calif, Berkeley, Publ Zool* 53:315-366, 1949.
58. Gabel JR: The morphology and taxonomy of the intestinal protozoa of the American woodchuck *Marmota monax* Linnaeus. *J Morphol* 94:473-549, 1954.
59. Wenrich DH, Nie D: The morphology of *Trichomonas wenyoni* (Protozoa, Mastigophora). *J Morphol* 85:519-529, 1949.
60. Wenrich DH, Saxe LH: *Trichomonas microti* n. sp. (Protozoa, Mastigophora). *J Parasitol* 36:261-269, 1950.

61. Daniel WA, Mattern CFT, Honigberg BM: Fine structure of mastigont system in *Tritrichomonas muris* (Grassi). *J Protozool* 18:575-586, 1971.
62. Hegner R: Experimental studies on the viability and transmission of *Trichomonas hominis*. *Am J Hyg* 8:16-34, 1928.
63. Nedvědová H, Kulda J: Pathogenicity of a virulent *Trichomonas vaginalis* strain for the laboratory rat after intravaginal inoculation. *J Protozool* 25:39A, 1978.
64. Cavier R, Mossion X: Essais d'infestation expérimentale de la ratte par *Trichomonas vaginalis* (Donné, 1837). *C R Acad Sci (Paris)* 242:2412-2414, 1956.
65. Cavier R, Mossion X: Nouveaux essais d'infestation expérimentale de la ratte par *Trichomonas vaginalis* (Donné, 1837). *C R Acad Sci (Paris)* 243:1807-1809, 1956.
66. Meingassner JG, Georgopolous A, Patoschka M: Intravaginale Infektionen der Ratte mit *Trichomonas vaginalis* und *Candida albicans*: Ein Modell zur experimentellen Chemotherapie. *Tropenmed Parasitol* 26:395-398, 1975.
67. Michaels RM, Peterson LJ, Stahl GL: The activity of substituted thiosemicarbazones against *Trichomonas vaginalis* and *Trichomonas foetus in vitro* and in experimental animals. *J Parasitol* 48:891-897, 1962.
68. Meingassner JG, Mieth H: Die Wirksamkeit bekannter systemisch wirksamer Trichomonazide im Infektionsmodell *Trichomonas vaginalis*/Ratte. *Arzneim-Forsch* 27:638-639, 1977.
69. Combescot CH, Pestre M, Domenech A, Verain A: Rôle du terrain endocrinien au cour de l'infestation expérimentale de la ratte albinos par *Trichomonas vaginalis* (Donné, 1837): Etude de la flore vaginale et du pH vaginal. In Hewer HR, Riley ND (eds): *Proc 15th Int Cong Zool, 1958.* Sect 8 Pap 2, pp 1-3 published by 15th Int Cong Zool, Linnean Soc. of London 1959.
70. Asami K: Effects of metronidazole on *Trichomonas vaginalis* in culture and in an experimental host. *Am J Trop Med Hyg* 12:535-538, 1963.
71. Combescot CH, Pestre M, Domenech A: Action de la progestérone sur l'infestation expérimentale de la ratte albinos par *Trichomonas vaginalis*. *C R Soc Biol (Paris)* 151:332-334, 1957.
72. Combescot CH, Pestre M, Domenech A: Action de la testostérone sur l'infestation expérimentale de la ratte albinos par *Trichomonas vaginalis*. *C R Soc Biol (Paris)* 151:953-954, 1957.
73. Combescot CH, Domenech A, Pestre M: Action de la désoxycorticostérone sur l'infestation expérimentale de la ratte albinos par *Trichomonas vaginalis*. *C R Soc Biol (Paris)* 151:702-703, 1957.
74. Combescot CH, Pestre M, Domenech A: pH vaginal et infestation expérimentale a *Trichomonas vaginalis* chez la ratte albinos. *C R Soc Biol (Paris)* 151:549-551, 1957.
75. Combescot CH, Verain A, Pestre M, Domenech A: Etude de la flore vaginale chez la ratte albinos. Effets de la castration et des oestrogènes. *C R Soc Biol (Paris)* 151:551-553, 1957.
76. Cavier R, Savel J, Quemerais MJ: L'essai pharmacologique des medicaments trichomonacides sur la ratte expérimentalement infestée par *Trichomonas vaginalis* Donné, 1837. *Therapie* 15:361-367, 1960.
77. Kalašová-Nedvědová H: *Experimental Infections with* Trichomonas vaginalis *in the Laboratory Rat and Mouse* (in Czech), M.Sc. thesis. Faculty of Science, Charles University, Prague, 1977.
78. Cappuccinelli P, Lattes C, Cagliani I, Ponzi AN: Features of intravaginal *Trichomonas vaginalis* infection in the mouse and the effect of estrogen treatment and immunodepression. *G Batteriol Virol Immunol Ann Osp Maria Vittoria Torino* (Parte I: Ser Microbiol) 67:31-40, 1974.
79. Meingassner JG: Comparative studies on the trichomonacidal activity of 5-nitroimidazole derivatives in mice infected s.c. or intravaginally with *T. vaginalis*. *Experientia* 33:1160-1161, 1977.
80. Coombs GH, Bremmer AF, Markam DJ, Latter VS, Walters MA, North MJ: Intravaginal growth of *Trichomonas vaginalis* in mice. In Kulda J, Čerkasov J (eds): Proc Int Symp Trichomonads & Trichomoniasis, Prague, July 1985 (Post-Symp Publ Pt 1). *Acta Univ Carolinae (Prague) Biol* 30 (3-4):387-392, (1986) 1987.
81. Patten SF Jr, Hughes CP, Reagan JW: An experimental study of the relationship between *Trichomonas vaginalis* and dysplasia in the uterine cervix. *Acta Cytol* 7:187-190, 1962.
82. Reagan JW, Wentz WB: Changes in the mouse cervix antedating induced cancer. *Cancer* 12:389-395, 1959.
83. Soszka S, Kazanowska W, Kuczyńska K, Łotocki W, Klimowicz L: Experimental infection of the urinary tract of guinea pigs by *T. vaginalis*. *Wiad Parazytol* 19:447-452, 1973 (In Polish, English summary).
84. Rõigas E, Teras J, Ridala V, Tompel H: Pathomorphological reaction of testicles of white rats and guinea pigs infected with *Trichomonas vaginalis* Donné. *Wiad Parazytol* 15:315-317, 1969.
85. Jitriphai P: Pathogenicity of *Trichomonas vaginalis* in genital tract of experimental male animals. *J Med Assoc Thailand* 54:714, 1971.
86. Newton WL, Reardon LV, De Leva AM: A comparative study of the subcutaneous inoculation of germ-free and conventional guinea

pigs with two strains of *Trichomonas vaginalis. Am J Trop Med Hyg* 9:56-61, 1960.
87. Teras J: Experimental investigation of pathogenicity of *Trichomonas vaginalis* (in Russian), thesis. State University of Tartu, 1954.
88. De Carneri I: Variation of the sensitivity of a strain of *Trichomonas vaginalis* to metronidazole after culturing in the presence or absence of the drug. In Corradetti (ed): *Proc Int Cong Parasitol Rome, 1964*, vol 1, pp 366-367. Oxford: Pergamon Press, 1966.
89. Kean BH, Weldt JT: Transmission of *Trichomonas vaginalis* in eye of animals. *Proc Soc Exper Biol Med* 89:218-219, 1955.
90. Weldt JT, Kean BH: Experimental ocular trichomoniasis: Pathologic observations. *Am J Pathol* 32:1135-1145, 1956.
91. Iwai S: Experimental inoculation of pure cultures of *Trichomonas vaginalis* into small laboratory animals. *Jpn J Parasitol* 6:136-144, 1957.
92. Tsai YH, Price KE: Experimental models for the evaluation of systemic trichomoniacides. *Chemotherapy* 18:348-357, 1973.
93. Honigberg BM: Immunology of trichomonads, with emphasis on *Trichomonas vaginalis:* a review. In Kulda J, Čerkasov J (eds): Proc Int Symp Trichomonads & Trichomoniasis, Prague, July 1985 (Post-Symp Publ Pt 1). *Acta Univ Carolinae (Prague) Biol* 30 (3-4):321-336, (1986) 1987.
94. Bogovsky PA, Teras J: Pathologico-anatomical changes in white mice in intraperitoneal infection by cultures of *Trichomonas vaginalis. Med Parazitol Parazit Bolezni* 27:194-199, 1958. (In Russian, English sumamry).
95. Rõigas EM: On trichomonad etiology of inflammations of the genito-urinary tract of men (in Russian), thesis. Academy of Science Estonian SSR, Tallinn, 1961.
96. Reardon LV, Ashburn LL, Jacobs L: Differences in strains of *Trichomonas vaginalis* as revealed by intrapertoneal injections into mice. *J Parasitol* 47:527-532, 1961.
97. Ivey MH, Hall DG: Virulence of different strains of *Trichomonas vaginalis* in the mouse. *Am J Trop Med Hyg* 13:16-19, 1964.
98. Teras J, Rõigas E: Characteristics of the pathomorphological reaction in cases of experimental infection with *Trichomonas vaginalis. Wiad Parazytol* 12:161-172, 1966.
99. Gobert JG, Georges P, Savel J, Genet P, Piette M: Etude de l'endoparasitisme expérimentale de *Trichomonas vaginalis* chez la souris. II. Etude cytologique et histologique. *Ann Parasitol Hum Comp* 44:687-696, 1969.
100. Gobert JG, Truchet M, Savel J: Etude de l'endoparasitisme expérimentale de *Trichomonas vaginalis* chez la souris. IV. Etude histochimique des lésions chez les animaux parasites. *Ann Parasitol Hum Comp* 46:511-522, 1971.
101. Laan IA: On changeability of pathogenicity, agglutinability and fermentative activity of *Trichomonas vaginalis* (in Russian), thesis. Academy of Science Estonian SSR, Tallinn, 1965.
102. Laan IA: On the effect of passages in vitro and in vivo on the pathogenicity, agglutinative ability and fermentative activity of *Trichomonas vaginalis. Wiad Parazytol* 12:173-182, 1966.
103. Landolfo S, Martinotti MG, Martinetto P, Forni G: Genetic control of *Trichomonas vaginalis* infection. I. Resistance or susceptibility among different mouse strains. *Boll Ist Sieroter Milan* 58:48-51, 1979.
104. Landolfo S, Martinotti MG, Martinetto P, Forni G: Natural cell mediated cytotoxicity against *Trichomonas vaginalis*. I. Tissue, strain, age distribution, and some characteristics of the effector cells. *J Immunol* 124:508-514, 1980.
105. Brugerolle G, Goberg JG, Savel J: Etude ultrastructurale des lésions viscérales provoquées par l'injection intrapéritoneale de *Trichomonas vaginalis* chez la souris. *Ann Parasitol Hum Comp* 49:301-318, 1974.
106. Nielsen MH, Nielsen R: Electron microscopy of *Trichomonas vaginalis* Donné: Interaction with vaginal epithelium in human trichomoniasis. *Acta Pathol Microbiol Scand (B)* 83:305-320, 1975.
107. Ovčinnikov NM, Delectorskij VV, Turanova EN, Yashkova GN: Further studies of *Trichomonas vaginalis* with transmission and scanning electron microscopy. *Br J Vener Dis* 51:357-375, 1975.
108. Kulda J, Nohýnková E, Ludvík J: Basic structure and function of the trichomonad cell. In Kulda J, Čerkasov J (eds): Proc Int Symp Trichomonads & Trichomoniasis, Prague, July 1985 (Post-Symp Publ Pt 1). *Act Univ Carolinae (Prague) Biol* 30 (3-4):180-198, (1986) 1987.
109. Krieger JN, Ravdin JI, Rein MF: Contact-dependent cytopathogenic mechanisms of *Trichomonas vaginalis. Infect Immun* 50:778-786, 1986.
110. Lindgren RD, Ivey MH: The effect of cultivation and freezing on the virulence of *Trichomonas vaginalis* for mice. *J Parasitol* 50:226-229, 1964.
111. Cavier RE, Gobert JG, Savel J: Application d'une methode d'infestation intrapéritonéale de la souris par *Trichomonas vaginalis* a l'étude pharmacologique des trichomonacides. *Ann Pharm Fr* 30:637-642, 1972.
112. Ivey MH: Virulence preservation of recent isolates of *T. vaginalis. J Parasitol* 61:550-552, 1975.
113. Diamond LS, Bartgis IL, Reardon LV: Virulence of *Trichomonas vaginalis* after freeze-

preservation for 2 years in liquid nitrogen vapor. *Cryobiology* 1:295-297, 1965.
114. Honigberg BM: Comparative pathogenicity of *Trichomonas vaginalis* and *Trichomonas gallinae* to mice. I. Gross pathology, quantitative evaluation of virulence, and some factors affecting pathogenicity. *J Parasitol* 47:545-571, 1961.
115. Frost JK, Honigberg BM: Comparative pathogenicity of *Trichomonas vaginalis* and *Trichomonas gallinae* for mice. II. Histopathology of subcutaneous lesions. *J Parasitol* 48:898-918, 1962.
116. Alderete JF, Suprun-Brown L, Kasmala L: Monoclonal antibody to a major surface glycoprotein immunogen differentiates isolates and subpopulations of *Trichomonas vaginalis*. *Infect Immun* 52:70-75, 1986.
117. Barr FS, Brent BJ: The activity of various drugs against *Trichomonas vaginalis* in vitro. *Antibiot Chemother* 10:637-639, 1960.
118. Lynch JE, Holley EC, Margison JE: Studies on the use of the mouse as a laboratory animal for the evaluation of antitrichomonal agents. *Antibiot Chemother* 9:508-514, 1955.
119. Paronikjan GM: Obtaining of experimental model of trichomonad infection. *Izv Akad Nauk Arm SSR Biol Nauki* 11:51-57, 1958 (In Russian).
120. Jeffries L, Harris M: Observations on the maintenance of *Trichomonas vaginalis* and *Trichomonas foetus:* The effects of cortisone and agar on enhancement of severity of subcutaneous lesions in mice. *Parasitology* 57:321-334, 1967.
121. Honigberg BM, Livingstone MC, Frost JK: Pathogenicity of fresh isolates of *Trichomonas vaginalis*. "The mouse assay" versus clinical and pathologic findings. *Acta Cytol* 10:353-361, 1966.
122. Kulda J, Honigberg BM, Frost JK, Hollander DH: Pathogenicity of *Trichomonas vaginalis:* A clinical and biologic study. *Am J Obstet Gynecol* 108:908-918, 1970.
123. Kulda J, Zavadil M, Vojtěchovská M, Dohnalová M, Karásková J, Kunzová E: Comparison of the pathogenicity of *Trichomonas vaginalis* for natural and experimental host. Abstracts Cong Int Particip Trichomoniasis, Bratislava, Sept. 1977, p 58. Bratislava: Org Comm Cong Trichomoniasis, Faculty Med, Comenius Univ, 1977.
124. Kuczyńska K, Choromański L, Honigberg BM: Comparison of virulence of clones of two *Trichomonas vaginalis* strains by the subcutaneous mouse assay. *A Parasitenkd* 70:141-146, 1984.
125. Honigberg BM, Gupta PK, Spence MR, Frost JK, Kuczyńska K, Choromański L, Wartoń A: Pathogenicity of *Trichomonas vaginalis:* Cytolpathologic and histopathologic changes of the cervical epithelium. *Obstet Gynecol* 64:179-184, 1984.
126. Krieger JN, Poisson MA, Rein MF: Beta-hemolytic activity of *Trichomonas vaginalis* correlates with virulence. *Infect Immun* 41:1291-1295, 1983.
127. Kulda J: Notes on the applicability of different strains of laboratory mice for the evaluation of virulence of trichomonads by the method of Honigberg. *Acta Parasitol Pol* 13:93-102, 1965.
128. Delachambre D: Difficultés d'obtention d'un test fiable pour déterminer la virulence d'un flagellé parasite (*Trichomonas vaginalis*). *C R Acad Sci (Paris)* 292:613-618, 1981.
129. Alderete JF: Antigen analysis of several pathogenic strains of *Trichomonas vaginalis*. *Infect Immun* 39:1041-1047, 1983.
130. Alderete JF: Enzyme linked immunosorbent assay for detecting antibody to *Trichomonas vaginalis:* Use of whole cells and aqueous extracts as antigens. *Br J Vener Dis* 60:164-170, 1984.
131. Wartoń A, Honigberg BM: Analysis of surface saccharides in *Trichomonas vaginalis* strains with various pathogenicity levels by fluorescein-conjugated plant lectins. *Z Parasitenkd* 69:149-159, 1983.
132. Wartoń A, Honigberg BM: Lectin analysis of surface saccharides in two *Trichomonas vaginalis* strains differing in pathogenicity. *J Protozool* 27:410-419, 1980.
133. Su-Lin K-E, Honigberg BM: Antigenic analysis of *Trichomonas vaginalis* strains by quantitative fluorescent antibody methods. *Z Parasitenkd* 169:162-181, 1983.
134. Gavrilescu M: Results of experimental intramuscular inoculation of *Trichomonas vaginalis*. *Microbiol Parasitol Epidemiol* 7:349-356, 1962 (In Romanian, English summary).
135. Schnitzer RJ, Kelly DR: Short persistence of *Trichomonas vaginalis* in reinfected immune mice. *Proc Soc Exp Biol Med* 82:404-406, 1953.
136. Kelly DR, Schnitzer RJ: Experimental studies on trichomoniasis. II. Immunity to reinfection in *T. vaginalis* infection in mice. *J Immunol* 69:337-342, 1952.
137. Kelly DR, Schumaker A, Schnitzer RJ: Experimental studies in trichomoniasis. III. Influence of the site of the immunizing infection with *Trichomonas vaginalis* on the immunity of mice to homologous infection by different routes. *J Immunol* 73:40-43, 1954.
138. Honigberg BM: Biological and physiological factors affecting pathogenicity of trichomonads. In Lewandowsky M, Hutner SH (eds): *Biochemistry and Physiology of Protozoa*, 2nd ed, vol 2, pp 409-427. New York: Academic Press, 1979.

9

Host Cell–Trichomonad Interactions and Virulence Assays in In Vitro Systems

B. M. Honigberg

Introduction

As in other instances, most information on trichomonad–cell (or, rarely, tissue) culture systems involves *Trichomonas vaginalis*. The data on in vitro interactions between lymphocytes, monocyte-macrophage lineage cells, and polymorphonuclear leukocytes, on the one hand, and trichomonads, on the other, are limited to the urogenital species. Therefore, this chapter deals primarily with experimental results obtained by using a variety of cell (and some tissue) cultures exposed to the urogenital trichomonad. The very limited information about the effects of *Trichomonas tenax* and *Pentatrichomonas hominis* on vertebrate cells in vitro is summarized at the end of the chapter.

Trichomonas vaginalis Donné, 1836

Effects of Different Parasite Strains on Various Vertebrate Cells

In this section, based on numerous articles and some abstracts,* interactions between parasites and cell cultures are described which ultimately result in at least some abnormal changes caused by the former in the latter components of the in vitro systems.†

In most instances little attention has been given by the investigators to the cytopathologic changes caused by the parasite; often, the time required for complete destruction of the vertebrate cell layers or the amount of radiolabels released by the vertebrate cells was considered the important criterion in estimating virulence of *T. vaginalis* strains. Since knowledge of the nature and extent of the pathologic manifestations is helpful in virulence assays, I shall at-

---

The data obtained in the author's laboratory were the result of investigations supported by Grants and Fellowships (the latter awarded to B. M. H. and his associates) from various sources, but primarily by Research Grants R01 AI00742 and R01 AI16176, Training Grants T01 AI00226 and T32 AI07109, all from the National Institute of Allergy and Infectious Diseases [awarded to B. M. H.], and Biomedical Research Support Grant RR07048, from the National Research Resources Institute [awarded to the University of Massachusetts], U. S. Public Health Service.

---

*Abstracts are used in those instances in which they constitute the only record, that is, where full reports dealing with a given experiment are not available.
†Diametrically opposite results of such interactions will be discussed in the section, "Interactions Between *T. vaginalis* and Lymphocytes, Monocyte-Macrophage Lineage Cells, and Polymorphonuclear Neutrophils, pp. 192-205.

tempt to discuss all aspects of the parasite-vertebrate cell interactions in in vitro systems.

Is Cytotoxicity Exclusively Contact-Dependent?

All the experiments discussed below are summarized in Table 9.1, to which the reader should refer for many of the details.

As far as can be ascertained, Hogue[1] was the first to report the injurious effects of *T. vaginalis* on tissue cultures. It is evident from Table 9.1 that she employed a rather simple method and that her experiments involved an undetermined strain of the parasite. In light of Hogue's cultivation method,[1] the conclusion that the cytopathic effects (CPE) were caused by toxins produced by the trichomonads without participation of mechanical action was not justified by her limited results. Of significance, however, was the fact that her findings led many U.S. parasitologists to the conclusion that the urogenital flagellate of humans ought to be considered at least a potential pathogen. Because of the paucity of data concerning the identity and virulence of the parasite strain and the methods of cultivation and examination, the statements of Kotcher and Hoogasian[2] that the damage caused by *T. vaginalis* to cell line and primary tissue cultures depended exclusively on direct mechanical action of the flagellates must be viewed with caution. However, as will become evident from the discussion that follows, mechanical action, or at least injury resulting from direct contact between the parasites and the culture cells, plays a very important part in pathogenesis associated with urogenital trichomoniasis.

Although, like most of the other reports, that of Christian et al[3] lacks data dealing with virulence for women and/or experimental animals, of the several *T. vaginalis* strains they employed, it includes more data about the methods and experimental procedures used. These workers also made careful cinemicrographic records of their observations. It is of interest that incubation of the cell cultures with inocula of approximately 5,000 organisms resulted in a typical progression of events noted for *T. vaginalis* infections by subsequent investigators. This progression involved the following phases: (1) Flagellates were moving freely over cell sheets without any CPE evident in the cell cultures. (2) Organisms were no longer visible, perhaps because they were motionless or remained in very close association with culture cells. (Some parasites may have become intracellular, see references 4,5; see also Chap. 14.) (3) There appeared small cell-free plaques (lesions) lined with the parasites. In the centers of the plaques the trichomonads were seen to tumble the no longer attached culture cells. Areas several cells deep surrounding the lesions were infiltrated by ameboid flagellates (as shown in Figs. 9.3e, f; 9.4a; and 9.8a) which dislodged the culture cells and caused the enlargement of the existing plaques and formation of new ones. (4) The plaques became confluent and most of the culture cells were destroyed. Some trichomonads attacked the few remaining cells; others collected into large active balls (behavior referred to as "swarming"). On the other hand, inoculation of small parasite numbers (i.e., 500) resulted in "stationary phase" in which the flagellates could remain for a long time—in some instances no CPE could be seen for over 1 month after infection.

As far as the primary mechanism of damage inflicted by the trichomonads on healthy culture cells is concerned, Christian et al,[3] who could not demonstrate any toxicity of filtrate of mixtures of trichomonads and damaged HeLa cells, suggested that while these results did not completely rule out the possibility of very labile toxic substances closely associated with the parasite, observations of the infected HeLa cell cultures ". . . reveal such strenuous trichomonal activity that . . . mechanical action could possibly be a major factor in cell destruction." This degree of caution does not characterize many of the reports discussed in this section of the chapter.

The brief report of Honigberg and Ewalt[6] included a suggestion of the ability of a *T. vaginalis* strain, originally isolated from a case of severe vaginitis and maintained for a prolonged period in axenic culture, to invade cells other than phagocytes and to be actively phagocytized by macrophages. On the basis of their limited observations, these authors suggested also that "the parasites seem to have both direct and indirect effects [caused by toxic substances] upon the [culture] cells" (chick liver cell monolayers and HeLa cells). However, in light of the preliminary nature of the report and because the parasite strain (C-1:NIH) had been culti-

TABLE 9.1. Employment of cell-culture systems in studies of interactions between *Trichomonas vaginalis* and host cells and for virulence evaluation.[a,b]

					Trichomonas vaginalis strains				
		Cell culture exposed (for time) at 37°C to				Virulence evaluation			
						Laboratory			
						Cell culture assay			
Kinds of cultures	Media used in experiments	Living parasites [Nos.]	Parasite-derived substances [cytopathogen. + or −]	Nos. and designations of strains	Medical[c]	Cytologic exam.	Release or binding of radioactive isotopes (also staining [10,16])	Mouse assay[d]	Agreement between lab. assays and/or between lab. assays and med. data[e]	References
Hang.-drop human fetal intest., lung muscle; chick embryo heart, intest., leg muscle	Tyrode-Locke-Lewis; chick or hum., EE[f]; human plasma	Yes (2–4; up to 24 h)	Yes (~5 h) [+]	1	N.G.[g]	Yes (detailed)	No	No	N.A.[h]	Hogue (1943)[1]
Primary chick embryo explant; human synovial cells	N.G.	Yes	Yes [−]	1	N.G.	Yes	No	No	N.A.	Kotcher, Hoogasian (1957)[2]
HeLa cell monolay.	BME[i] in Hanks' BSS[j], 20% calf S,[k] P&S[l]	Yes [500–5000]	Yes [−]	Several	N.G.	Yes (detailed)	No	No	N.A.	Christian et al (1963)[3]
Chick liver cell monolay. (primary); HeLa cells	1 pt. CPLM NA[m] + 2 pts. modif. Huff et al med.[n] (chick liver) or BME w/inact. horse S (HeLa)	Yes (2–12 h)	Yes (20 h) [+]	1 (C-1:NIH)	Severe vaginitis	Yes (detailed)	No	Yes (s.c.)	Fair (lab. assays & med. data)	Honigberg, Ewalt (1963)[6]
Monkey kidney cell monolay. (1st passage)	M199 modif., 2% calf S; 1% delipidized fract. dried milk (vol.?) + 0.2 ml TYM[o] pH6	Yes (18–60 h, depend. on strain) [4 × 10⁵]	Yes [−]	2 (101, 108)	No	Yes (detailed)	No	Yes (s.c.)	Yes, (some between lab. assays)	Kulda (1967)[7]
Chick liver cell monolay. (primary)	1 pt. CPLM NA + 2 pts. modif. Huff et al med.	Yes [10⁵]	Yes [+, reversible]	2 (JH30A, JH32A)	Yes	Yes (detailed)	No	Yes (s.c.)	Yes (high between lab. assays & med. data)	Farris, Honigberg (1970)[5]
Chick embryo fibroblast monolay.	Hanks' BSS, lactalbum. hydrolysate, 20% horse S	Yes (usually 24 h, up to 4 d) [5 × 10³, 10⁴, 10⁵]	No	8	Yes[p]	Yes (detailed)	No	No	Yes (some between cell cult. & med. data)	Toś-Luty et al (1973)[12]
HEp-2 cell monolay. in most experim. (5 cell lines tested)	N.G.	Yes w/ or w/o avirulent *Staph.* (6–12 h) [*T.v.*: cult. cell ratio = 1:100 cells; bact. — 10⁷]	No	16	N.G.	Yes	No	No	N.A.	Zhordaniia et al (1974)[33]
Rabbit kidney tubul. epithel. cell (RK13) monolay.	1 pt. M199, 10% fetal bov. S, 25 mM HEPES, P&S + 2 pts. *T.v.* med.[s1]	Yes (up to 48 h) [2 × 10⁶]	No	1 (LUMP 889)	Yes (vaginal disch.)	Yes (detailed; light & EM)	No	No	N.A.	Heath (1981)[13]
McCoy strain cell monolay.	2 ml Eagle's MEM[r], 10% fet. bov. S + 0.2 ml TYM	Yes (3–24 h) [3.5 × 10⁵]	Yes [−]	25 (non-cloned) 16 (clones)	N.G.	Yes (detailed; standard indices)	No	Yes (i.p.) (standard indices)[14]	Yes (high between lab. assays)	Brasseur, Savel (1982)[14]

157

TABLE 9.1. *Continued*

Kinds of cultures	Media used in experiments	Cell culture exposed (for time) at 37 °C to		Nos. and designations of strains	*Trichomonas vaginalis* strains Virulence evaluation					References
		Living parasites [Nos.]	Parasite-derived substances [cytopathogen. + or −]		Medical[c]	Laboratory			Agreement between lab. assays and/or between lab. assays and med. data[f]	
						Cell culture assay		Mouse assay[d]		
						Cytologic exam.	Release or binding of radioactive isotopes[10,16] (also staining[10,16])			
Chinese hamster ovary (CHO-K1) cell monolay.	1 ml F-W[u] med. for *T.v.* + 0.1 ml same med. w/ *T.v.* cultiv. for 48 h	Yes (max. difference in CPE[f] after 4 h) [10⁶ = ~6/CHO cell]	No	6 (4 from symptomatic, 2 from asympt. patients)	Yes[x]	Yes (detailed; standardized)	No	Yes (s.c., modif.[11]) [also hemolys.]	Yes [high between lab. assays (cell cult., hemolys., s.c., mouse) & med. data]	Krieger et al (1983)[11]
4 kinds of cell cult.[v] (HeLa most sensitive)	2 pts. Eagle's MEM, 10% fet. bov. S, P&S, + 1 pt. TYM, 10% inact. horse S	*Monolay.* [up to 5×10⁴ and 2×10⁵ (18 h or 8 h maxima) pellets (for [³H] thymidine exper.)]	Yes (20 h) [Filtrates of cult. supernat. of log-phase *T.v.*: −]	1 (NYH 286) from long-term cultiv. [−]	N.G.	Yes [staining crystal violet (colorimetry); phase-contrast microsc. (cytol.)]	Yes [³H] thymidine-labeled cult. cells	Yes (s.c.) (modif.[85])	N.A.	Alderete, Pearlman (1984)[16]
HEp-2, HeLa monolay. cells primarily [3 other cell lines tried[w]]	2 pts. Dulbecco's modif. MEM, 10% fet. bov. S, P&S + 1 pt. TYM, inact. horse S	Yes for adher.: (30 min) [2 × 10⁶ radiolab. *T.v.* + 4 × 10⁵ cells; for cytotox. (18 h) [5 × 10⁴ *T.v.*: 5 × 10⁴ cells]	No	4[x] (most exper. w/ radiolabeled NYH 286)[y]	N.G.	Yes *Monolay.* For adher.: dk.-field microsc.	Yes For adher.: from [³H]thymidine-labeled *T.v.*; for cytotox.: crystal violet release[16]	Yes[y] (s.c.) (modif.[82])	N.A.	Alderete, Garza (1985, 1987)[18,83]
BHK (baby hamster kidney) cell monolay.	0.2 ml Eagle's MEM, 5% fet. bov. S, 2 mM glutamine (also nystatin, Fungizone, gentamycin)	Yes (up to 52 h) [1.4 × 10⁴/well and four 10-fold dilutions of this conc.]	Yes [lysates; +, esp. in absence of serum]	2 (*T.v.* 1, *T.v.* 2)	Yes vaginitis	Yes [w/ or w/o serum; w/ equal orig. inoc. (1.4 × 10⁴) *T.v.* 1 had higher final CPE titers]	No	No	N.A.	Gentry et al (1985)[17]
CHO-K1 (Chinese hamster ovary) cells	1 ml serum-free F-W med.	Yes (monolay.,[aa] also pellets[bb]) Also effect of certain cmpds. on cell killing by *T.v.*[cc]	Yes (filtrates & sonic extracts) [−]	3 (CHAR-1, CHAR-2, CHAR-3)	Yes symptomatic vaginitis	Yes (detailed; standardized scale for 0–100% monolay. remaining intact)	Yes (good agreement between ¹¹¹InOx from CHO cells & cell deaths)	No	N.A.	Krieger et al (1985)[20]
HeLa cell monolay.	As in reference 10	As in reference 10	No	5[dd] (NYH 286; IR78; JH31A; RU 375, JHH)	N.G.	No	Yes [staining: crystal violet (colorimetry, FACS[ee]]	N.A.	N.A.	Alderete et al (1986)[43]

158

Cell system	Medium	Monolayer	Cytopathic effect (CPE)	No. of T.v. strains	Clinical source	Microscopy	Radiolabel assay	In vivo	Reference
WISH (human amnion epith.-like cell line) monolay.	RPMI 1640 w/o serum or antibodies	Yes (1 h) [5 × 10⁵] w/gentle agitation	No	1	(D) fresh isolate	Yes "acute trichomoniasis"	No	Yes (from [³H]thymidine bound by adherent T.v.)[f]	No	N.A.	Martinotti et al (1986)[35]
16 types of cell monolay. cultures (6[gg] most suscept. to T.v.; 6th most sensit. to CDF[h])	GMP[ii]	Yes (up to 6 days) [10³, 10⁴, 10⁵]	Yes CDF [+]	1		Yes	No	N.A.	Pindak et al (1986)[19]
Vaginal epith. cell monolay. ("primary")	2 ml Eagle's MEM, 10% fet. bov. S, S & neomycin + 100 μl TYM w/o agar	Yes (mostly 24 h) [10⁵–10⁶]	No	3	[1711 old; 130354 from severe vaginitis; 10 fresh (symptoms?)]	Yes (from 2)	Yes (detailed; light and EM; standardized)	No	Yes (high between cell cult. & med. data)	Rasmussen et al (1986)[15]
MDCK[44] cell monolay.	M199 medium, 10% bov. S	Yes (72 h: pH 5–6) [parasite to MDCK ratio 5:1; effects of various substances[d] on T.v. monolay. adhesion & lysis	No	1 (J1)	N.G.	Yes (lt. and EM)	No	No	N.A.	Silva Filho, De Sousa (1986)[23]
4 cell cult. systems[mmm]	Appropriate cell cult. med.[nn] (2 ml. + 0.5 ml MDM[oo]/Leighton tube; 0.5 ml + 0.2 ml MDM/tissue cult. well]	Yes (up to 65 h) [2.5 × 10⁵ T.v./tube; 10³/well]	No	? (G8 clone lab. strain IR78; "recent clinical isolates[x]")	N.G.	Yes	No	Yes (s.c.) (also intra-vag[84])	Yes (between lab. assays)	Bremner et al (1987)[31]
CLID (Mouse fibroblasts) monolay.	RPMI 1640, 10% fet. bov. S, 50 μM β-mercaptoethanol	Yes (2, 3, or 6 h) [2 × 10⁵/well]	No	1 (TO-37) fresh isolate	"Acute vaginal trichomoniasis"	Yes (detailed) (l.t. microsc. of living T.v. & Giemsa stained preps.; also cell counts; scanning EM)[pp]	No	No	N.A.	Juliano et al (1987)[32]
Baby hamster kidney (BHK) cell monolay.	N.G.	Yes (No data)	Yes (CLS[qq]) [+] (Also protease action)	N.G.	N.G.	Yes	No	No	N.A.	Ortega et al (1987)[18]
5 cell lines[r]; (HeLa most sensitive)	3 pts. PBS (+)[ss] 1 pt. V-bouillon.[tt] 2 pts. Eagle's MEM, 10% inact. calf S	Yes Pellet [4 × 10⁴ to 2.5 × 10⁶] w/ 2.5 × 10⁵ cult. cells	No	1 (4FM) laboratory strain	N.G.	No	Yes (⁵¹Cr from labeled cells)	No	N.A.	Fukui, Takamori (1988)[21]

[a]Special characteristics and significance of the various systems are described in text.
[b]With two exceptions,[14,31] no indication of using cloned T. vaginalis populations was found in the reports in this table.
[c]Detailed pathologic data are not given in most instances. Given in reference 5; to a lesser degree in reference 13.

[d] Either subcutaneous (s.c.) or intraperitoneal (i.p.) assay, unmodified or modified (modif.).
[e] Agreement between medical evaluation and either one or both lab. assays, or between lab. assays.
[f] Embryo extract.
[g] Not given.
[h] Not applicable.
[i] Basal Medium Eagle.
[j] Balanced Salt Solution.
[k] Serum.
[l] Penicillin and streptomycin in typically low conc. (~100 U/ml; ~100 μg/ml).
[m] Cysteine-peptone-liver-maltose medium[85] for trichomonads, without agar.
[n] Huff et al tissue culture medium[86] modified by Honigberg and Ewalt[6]; 5% chick (C) EE_{50}; 15% heat-inactivated chicken serum, 80% Gay's BSS, pH 7.0–7.2.
[o] Diamond's medium[22] for trichomonads.
[p] Inflammation of the vagina; inflammation of the urinary bladder; cervical cancer; myoma; adnexal tumor.
[q] *Trichomonas vaginalis*.
[r] Minimum Essential Medium
[s] Feinberg-Whittington medium[87] for *T. vaginalis*.
[t] Cytopathic effect.
[u] Symptomatic (4 patients) with the following symptoms and signs—*symptoms*: change in character of vaginal discharge (4/4), vulvo-vaginal irritation (4/4), vaginal tenderness during exam (4/4), dyspareunia (2/4), dysuria (2/4); *signs*: excessive purulent vaginal discharge (4/4), erythema of vaginal walls (4/4), granular vaginitis with punctate hemorrhages (2/4). **Asymptomatic** (2 patients)—no obvious symptoms or signs.
[v] HEp-2, HeLa, African green monkey kidney cells (Vero); primary cultures of normal baboon testis (NBT) cells.
[w] Normal human skin fibroblast cells, strains: CCD-27SK, CCD-50SK [CCD-57SK, not used?]; human foreskin fibroblast cells. (All these cells were grown in Eagle's MEM with nonessential amino acids and 20% fet. bov. S).
[x] NYH286, JH31A, RU 375, JHHR. All four strains would appear to have been cultivated for prolonged periods in axenic cultures. Except for the actual period of in vitro cultivation after 1977 (12 transfers up to that time; see reference 88), and the resulting effects, the medical history and the results of the s.c. mouse assay for virulence of strain JH31A are well documented.[23] Since apparently no relevant data were provided by Dr. M.R. Spence to Dr. Alderete's group about JHHR, the history of this strain cannot be ascertained.
[y] The results of the standard s.c. mouse assay (for references see Wartón and Honigberg[3]) are at variance with those of the "modified" procedure introduced by Alderete.[82]
[z] Various observations were reported by Alderete and his coworkers[10,16,38a,42,43] on the basis of experiments using NYH 286, a nonamoeboid *T. vaginalis* strain that has been cultivated by serial passages for a long time, and primarily HeLa cell line.
[aa] The first sign of monolayer destruction noted after 5-hr exposure of 2×10^5 CHO-K1 cells to 2×10^5 trichomonads; also with constant CHO cell numbers and inoculum size increasing from 1.6×10^4 to 1.6×10^6 there was a decrease in time needed for a complete destruction of the monolayers.
[bb] The results of the pellet experiments[20] [4h interaction of the parasites (10^6) and CHO-K1 cells (2×10^5), ratio 5:1; also experiments with varying parasite numbers and constant CHO cell numbers] support the "one hit" hypothesis of host cell killing by *T. vaginalis*. For discussion of the application of this hypothesis to the urogenital trichomonads, see the relevant part of the section Is Cytotoxicity Exclusively Contact Dependent? which closely follows Table 9.1.
[cc] Cytochalasin D, a microfilament inhibitor, reduced target cell killing in monolayer and pellet cell systems. Colchicine and vinblastine, microtubule function inhibitors, did not.
[dd] NYH286, IR78, and JH31A are designated as "long-term" grown isolates, and RU 375 and JHH as "fresh isolates." In the absence of the history of strain RU 375, which came from Dr. D. M. Müller (The Rockefeller Univ.), there are no data on the length of cultivation since the time of its in vitro isolation. The remarks with regard to the history of strain JHH made in footnote *x* probably pertain also to JHH trichomonads.
[ee] Fluorescence-activated cell-staining. This method provided the basis for differentiation of the strains used by Alderete et al[43] into positive and negative phenotypes with regard to their reaction with the monoclonal antibody C20A3. For discussion of this problem, see section Putative Immunologic Factors.
[ff] In a study of effects of various factors on cytadherence (for details, see text).
[gg] HeLa, HeLa 229, HEp-2, HamLu, McCoy, Rk-13.
[hh] HeLa 229, McCoy, HeLa, HEp-2, L929, BHLu.
[ii] Cell Detachment Factor, e.g., McCoy cells grown in 75 cm² flask, then refed with 50 ml GMP[19] (see jj) and inoc. with 5 ml of *T. vaginalis* in GMP. After 72 h at 37°C supernatant fluid was collected by centrif., adj. to pH 7.3, and filtered (0.45 μm pore diameter). Polyethylene glycol was added, 10 g/100 ml fluid. After overnight storage at 4°C, the fluid was spun at 2800 g and the sediment suspended in M199 in 1/10 of the original cult. volume, then filter-sterilized.
[jj] Empirical Growth Medium[19]: MEM 900 ml, heat-inact. fet. bov. S. 100 ml; 1 M HEPES buffer, 30 ml; glucose, 20 g; ascorbic acid, 100 mg; pantothenic acid, 20 mg; kanamycin sulfate, 200 mg; chloramphenicol, 20 mg; amphotericin B, 2.5 mg; pH 6.8.
[kk] Madin Darby Canine Kidney cell line.
[ll] Trypsin, cytochalasin B, neuraminidase, several protease inhibitors.
[mm,nn] Murine-macrophage-like cell line P388D (HOMEM medium[89]); mouse myeloma cell line P3-X63-Ag8-653 (HOMEM medium); primary chick-embryo cells (EBM with Hanks' BSS); Chinese hamster ovary cells CHO-K1 [Ham's F12 medium (see *ATCC Catalogue of Cell Lines & Hybridomas*, 6th ed. Rockville, MD: Am Type Cult Collection, 1988)]. All media supplemented with FBS + P&S + Amphotericin B).
[oo] Modified Diamond's Medium[24] + inactivated horse S.
[pp] In a study of effects of antiskeletal compounds on cytopathic effects and cytadherence of *T. vaginalis* to cell culture elements. The study of phagocytosis reported in the same paper did not involve the use of cell cultures.
[qq] Crude Lysate Supernatants.
[rr] HeLa, Vero, LLC-MK 2, FL, L929.
[ss] Dulbecco's phosphate buffered saline with 0.9 mM $CaCl_2$ and 0.5 mM $MgCl_2$.
[tt] 2% liver digest; 2% polypeptone; 0.2% L-cysteine HCl; 0.5% NaCl; 1% glucose (pH 5.8); 10% heat-inact. horse serum.

vated for a long time before being employed in these experiments, the results of Honigberg and Ewalt[6] must be viewed with no less caution than those of Hogue[1] or Kotcher and Hoogasian.[2]

Of interest are the observations of Kulda,[7] who attempted to evaluate virulence of two *T. vaginalis* strains for mice, as estimated by the original subcutaneous assay[8] and by their CPE in monkey kidney monolayer cell cultures (for certain details of Kulda's method, see Table 9.1; for the CPE, see his report[7]). Although clear virulence differences could be demonstrated between these parasite strains for mice, according to Kulda[7] "no essential difference was . . . found in the final effect of *Trichomonas vaginalis* strains 101 [less virulent] and 108 [more virulent]." However, in infections with the latter, the cell culture destruction was complete by 48 hours after inoculation, while in those involving the former, the cell cultures were not destroyed completely until 60 or more hours after infection. Evidently, Kulda's statement notwithstanding,[7] differences in virulence between two *T. vaginalis* strains were revealed by the cell-culture assay. In light of the aforementioned results and some of the subsequent data published by others, it would appear that Kulda was overpessimistic in his concluding statement ". . . that it will be hardly possible to use tissue culture for the . . . [development] of a simple virulence test." The only accurate part of this prediction probably is the word "simple"—it seems that to be truly meaningful a cell-culture assay could not be very simple.

Kulda[7] indicated also that according to Honigberg and Ewalt[6], suppression of cell division and the appearance of abnormal nuclear and cytoplasmic changes evoked by exposure of cell culture to parasite-free filtrates of trichomonad-rich cultures constituted a response to some toxic substances presumably eliminated by the parasites. He asserted, however, that similar alterations could be seen in uninfected aging monkey kidney cell cultures; the addition of cell-free filtrates or supernatant fluids from active trichomonad cultures merely increased the frequency of the pathologic changes. This statement confirms the generally recognized fact that various harmful factors (e.g. metabolic products of target and effector cells) cause similar CPE in vertebrate cell cultures.

At this point we come to the report of Farris and Honigberg[5] which, although often subject to misinterpretation, appears to constitute in many respects a turning point in the use of cell cultures in studies of *Trichomonas vaginalis*-"host" cell interactions. These workers employed chick liver monolayer cultures, consisting of macrophages, fibroblast-like, and epithelial cells, which proved to be very sensitive to *T. vaginalis* in preliminary experiments. Admittedly, the assay described by Farris and Honigberg[5] was rather difficult. (1) It included careful evaluation of the clinical and pathologic manifestations accompanying the infection in the human donors, a condition met only very rarely by the investigators working with cell culture assays (see Table 9.1). (2) It employed freshly isolated strains or at least strains that were stabilated in liquid nitrogen soon after their isolation in axenic cultures. (3) It took into account the potential effects of the generation times (G) of the strains examined on the results by using a virulent strain with an atypically short G that was comparable to the G of the mild isolate, short G values being characteristic of such strains. In other instances, more complex manipulations may have to be used to equalize the G values of the strains tested. [Since, on the basis of several closely reproducible assays, the populations isolated from human hosts are quite homogeneous with regard to their virulence,[9] there appears to be no need to employ cloned populations of the strains to be examined.] (4) The cell cultures included several cell types, for example, macrophages, fibroblast-like, and epithelial cells. (5) The conditions (i.e. growth phase, cytologic appearance, etc.) of the cell cultures were monitored. (6) Care was exercised with regard to the size of the inocula. (7) The composition of the medium, its total volume, and physico-chemical properties were closely controlled as were the conditions of incubation, for example, the composition of the atmosphere and temperature. (8) The experimental preparations were accompanied by many control-experimental ones. (9) All preparations were subjected to careful cytologic examinations, supported by photomicrographs.

Clearly, the system employed by Farris and Honigberg[5] cannot be considered suitable for a routine virulence assay. However, it can contribute a basis for various assays and should aid in the establishment of correlations between medical evaluation and cell culture assay and/or various other laboratory assays.

A brief synopsis of the Farris-Honigberg[5] report, illustrated by photomicrographs selected from among those used for the original paper, is included here to recapitulate the observations that may be important for the understanding of some facets of interactions between *T. vaginalis* and vertebrate cell cultures.

The relatively virulent JH30A strain trichomonads, which tended to be ameboid in cell cultures, exhibited a tendency to adhere to all three elements of these cultures, i.e. macrophages, fibroblast-like ("fibroblast"), and epithelial cells. As pointed out by a few authors (e.g. references 5, 10), the parasites typically applied themselves to the cells by their ventral surfaces, that is, surfaces opposite to those invested with the undulating membranes. This situation was especially clear with regard to the macrophages (Fig. 9.1a–d)*; in some instances more than one parasite was applied to a single phagocyte (Fig. 9.1b–d). The abnormal changes evident in the cytoplasm and nuclei of the culture cells indicated that the parasites caused widespread injuries to the cells to which they were closely applied. Of special interest is the condition evident in Fig. 9.1e—it appears that

---

*Evidently the chick macrophages present in the liver monolayer cultures had low natural cytotoxicity for virulent, but evidently not mild, trichomonads—in this system they acted as target and not as effector cells.

Figs. 9.1 to 9.4. Photomicrographs of trypsin-dispersed chick liver cell cultures. The typically monolayer cultures are spread over the surface of the coverglass facing away from the Leighton tube wall (for the cultivation method, see reference 5). The experimental cultures were infected with the relatively virulent *T. vaginalis* strain JH30A or with the very mild (nearly avirulent) strain JH32A of this species (for culture media and experimental procedures, see the data for reference 5 in Table 9.1). The term "h" refers to the length of time in hours the cell cultures were exposed to the trichomonads, to cell-free filtrates of active trichomonad cultures, or to media, that is, 1 part CPLM to 2 parts cell culture medium (CPLM control), or to tissue culture medium alone (system control). (All figures reproduced with permission of the American Society of Parasitologists, from Farris VK, Honigberg BM: Behavior and pathogenicity of *Trichomonas vaginalis* Donné in chick liver cell cultures. *J Parasitol* 56: 849-882, 1970.)

Fig 9.1. All cultures show interactions between the trichomonads (effectors) and chick liver cells (targets). Except for those in **g** and **k**, the parasites belong to strain JH30A. **a.** Twelve h. Ameboid flagellate (*T*) applied to a macrophage (*M*); a trichomonad (*T*) showing many organelles, seen to the left of the foregoing phagocyte, is likely adhering to chick fibroblast. **b.** Twenty h. Macrophage (*M*) with ameboid trichomonads (*T*) adhering by their entire ventral surfaces (i.e. surfaces at 180° from the undulating membrane and costa) to its opposite ends. Note also the parasite (*T*) near the middle of the right edge of the figure; it may adhere to a chick liver cell. Flagellates, especially numerous in the upper right corner, are distributed over the chick cell area included in the micrograph. **c.** Twenty h. Three parasites (*T*) applied to a macrophage (*M*). **d.** Twenty h. Two trichomonads (*T*), the lower probably dividing, are applied to a macrophage (*M*). **e.** Eight h. Part of an ameboid trichomonad (*T*) near the right margin of the figure is adhering to a macrophage (*M*). An ameboid flagellate (*T*) seen in the center of the upper half of the micrograph appears to be pulling and perhaps engulfing a strand of debris (*d*) whose consistency resembles that of the degenerated cytoplasm (*c*) of a destroyed phagocyte or fibroblast. Other trichomonads (*T*) seem to be applied to fibroblasts (*F*); although the cytoplasm of the fibroblast near the upper left margin of the micrograph is not clearly discernible or may be degenerated (*c*), its nucleus (*F.nu.*) is clearly visible. **f.** Two h. Healthy-looking parasite, with all the organelles except for the anterior flagella and the costa clearly visible, is lodged within a macrophage. The cytoplasm of the macrophage is stained only very lightly, but its two nuclei are clearly evident. **g.** Twenty h. JH32A. A parasite, showing signs of only very slight degeneration, is enclosed in a food vacuole of an apparently normal macrophage. **h.** Two h. One apparently healthy ameboid parasite (*T*) is seen within a food vacuole of a macrophage, in the lower center of the figure. Another, perhaps somewhat degenerated, rounded parasite (*T*) is lodged in a second vacuole of this phagocyte, above and to the left of the first. A third, slightly ameboid healthy-looking flagellate (somewhat anterior to the previous two) appears to adhere to the fibroblast layer. The remaining structures in the figure are fibroblast nuclei, some of which appear unhealthy. **i.** Eight h. More or less distinctly ameboid trichomonads in an elongate macrophage. The nu-

9. Host Cell–Trichomonad Interactions and Virulence Assays in In Vitro Systems 163

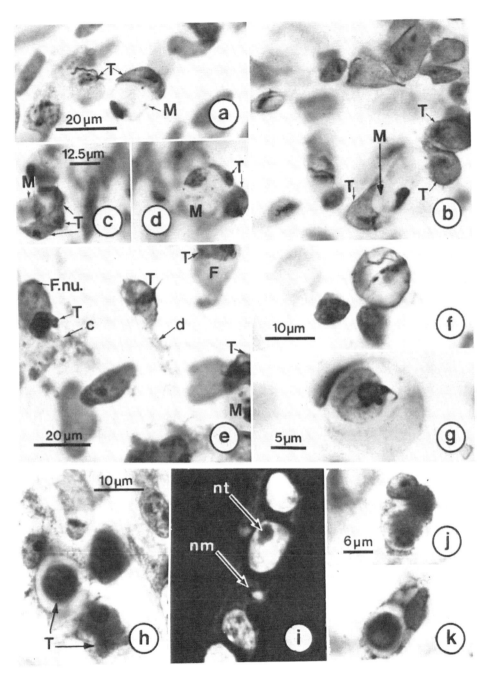

cleus of the largest parasite (*nt*), which appears black in this micrograph, actually stained green for DNA. The nucleus of the phagocyte contains a nucleolus (white in this figure) which actually took a brick-red RNA stain. The cytoplasm of the macrophage was faintly reddish, while that of the trichomonads had an intense brick-red color (high RNA level). (Acridine orange fluorescence; exciter filter Corning No. 5840.) **j**. Two h. Swollen degenerating parasite within a macrophage. The trichomonad cytoplasm stains lightly and its nuclear chromatin shows margination. **k**. Two h. JH32A. Abnormal flagellate, with a conspicuous cytoplasmic vacuole, is seen within a healthy-looking macrophage. The same scales should be used, respectively, for **a** and **b**; **c** and **d**; **h** and **i**; and **j** and **k**. 9.1 **a–g**: Bouin's fixative (B), protargol stain (P); **h–k**: May-Grünwald-Giemsa stain (MGG).

the trichomonad (T), seen in the center of this figure, pulled away from a cell carrying with it a fragment of the phagocyte or perhaps fibroblast cytoplasm (d).

The typically nonameboid and smaller flagellates belonging to the mild JH32A strain had much less tendency to settle on the culture cells or to establish intimate contact with those upon which they came to rest. It seems, therefore, that cytadherence may play a less important role in degeneration of the vertebrate cell cultures infected with the mild than with the virulent strains. The relatively large numbers of macrophages containing one or more trichomonads in their cytoplasm (Fig. 9.1f–k) in early infection indicated that the highest phagocytic activity occurred at that time. Furthermore, phagocytosis was much more frequent in cultures infected with the mild JH32A strain. Phagocytic activity declined in cultures exposed to both parasite strains, but this decline was more precipitous in cell cultures inoculated with the more virulent JH30A trichomonads, in which typically only little phagocytosis was seen during infection. Many of the intracellular parasites of the latter strain appeared normal [Fig. 9.1f, h (in the latter figure, note the ameboid organisms near the center of the lower margin)], although some of them looked unhealthy (Fig. 9.1j). On the other hand, most of the mild strain JH32A flagellates lodged in the food vacuoles of the macrophages showed signs of degeneration (Fig. 9.1k). Although, as stated by Krieger et al,[20] "... T. vaginalis kills target cells without phagocytosis," the often abnormal appearance of the macrophages containing JH30A trichomonads suggests that the engulfment process plays a role in destruction of normally phagocytic cells by virulent parasite strains. Relatively few fibroblasts and epithelial cells harbored trichomonads of either strain within their cytoplasm, but the "invasion" of these culture cells was significantly more frequent in cultures exposed to the virulent trichomonads. It seems that the latter parasites were able in some way to enter relatively healthy fibroblasts which then showed signs of cytoplasmic and nuclear degeneration (Fig. 9.2a, b). The same situation was observed with regard to the epithelial cells; some of them that contained healthy-looking parasites within their cytoplasms appeared otherwise nearly normal (Fig. 9.2c); others, although harboring evidently normal parasites, showed extensive degenerative changes (Fig. 9.2d). Although the mechanisms whereby the parasites enter presumably nonphagocytic cells in vitro or in vivo are not understood, on occasion T. vaginalis can be seen within the cytoplasm of nonphagocytic cells[4,5,12] (Chaps. 14 and 15), a fact that has been disregarded by the majority of parasitologists and medical men.

It was stated by many investigators that cytopathogenicity of T. vaginalis for vertebrate cells in in vitro systems is in very large measure [13–15] or even exclusively[16,20] contact-dependent, involving cytadherence followed by cytotoxicity. Also, Farris and Honigberg[5] noted that CPE was greatly dependent upon direct contact between the parasite and culture cells and that, although "the general degenerative changes similar to those described [from cell cultures infected with living pathogenic-strain trichomonads] were seen in fibroblasts exposed to filtrates of these parasites . . . the level of such changes . . . [was usually] far lower." Furthermore, according to these investigators,[5] many changes described for fibroblasts in cultures infected with the virulent JH30A strain were noted in those inoculated with the mild JH32A trichomonads; however, they were much less extensive. Filtrates obtained from cultures of the mild strain appeared to cause only minimal injury to the fibroblasts. For details of the pathologic changes caused by the parasites in fibroblasts, the reader is referred to Fig. 9.3a, b, e; 9.4b, c in this chapter. With regard to the epithelial cells, pathologic manifestations, such as vacuolization of cytoplasm and nuclear pyknosis, were noted in cultures infected with both the mild and the virulent strains (Fig. 9.3c); however, they appeared much later in the course of infection and were less extensive in the cell cultures exposed to the mild strain. Only minor abnormal changes were observed in the epithelial cells in the presence of filtrates from cultures of the virulent strain, and virtually none were seen in the presence of such filtrates from cultures of the mild parasites.

Gentry et al[17] demonstrated pathologic changes in baby hamster kidney (BHK) cells infected with TV 1 and TV 2 T. vaginalis strains, and observed differences in the time required for the more virulent, TV 1, and the less virulent, TV 2 parasites to cause development of such changes. They also reported the appearance of comparable cytopathic effect in BHK cells exposed to cell-free lysates of the urogeni-

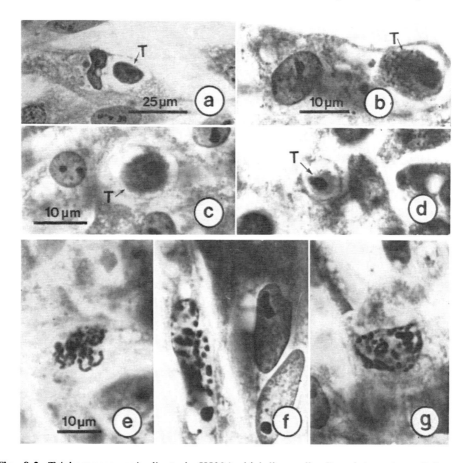

Fig. 9.2. *Trichomonas vaginalis* strain JH30A-chick liver cell culture interactions. Only one preparation (**d**) shows an infection with strain JH32A. **a, b.** One h. Single infections of fibroblasts by the parasites (*T*). The intracellular parasites appear healthy, but the fibroblasts show various degrees of degenerative changes manifested in all instances by more or less pronounced cytoplasmic vacuolization and in some cases by coarsely granular appearance (**a**), smudging (**b**), and margination (not shown) of the chromatin in often distorted nuclei (**a**). **c.** Twelve h. Healthy-appearing trichomonad (*T*) within an epithelial cell cytoplasm. Beyond some cytoplasmic vacuolization, the parasitized cell has only a few degenerative changes. The scale is applicable also to **d. d.** Sixteen h. JH32A. Trichomonad, with lightly staining cytoplasm and an abnormal-looking nucleus, in a degenerated liver epithelial cell, the cytoplasm of which is highly vacuolated and partly retracted. The epithelial cells around the one harboring *T. vaginalis* show various cytoplasmic and nuclear changes. **e.** Twelve h. Fibroblast in normal prophase; the nuclear envelope has been dissolved. The scale is applicable also to **f** and **g. f, g.** Twenty h. Abnormal prophase stages. The chromosomes are fragmented, the fragments of various sizes being scattered throughout the nucleoplasm. In some instances chromatin margination against the still typically present nuclear envelope can be seen (**g**). Vacuolization of the cytoplasm (**f**) or its retraction toward the nucleus are also often associated with arrested prophases. All figures: MGG.

tal trichomonads. For the first 40 hours the cytopathologic changes were produced by the lysates only when serum-free medium was employed. The time required for the appearance of CPE was inversely proportional to the amount of lysate protein used. The authors suggested that proteases released from hydrogenosomes during homogenization might be responsible for such changes. They also considered the possibility that "the beta hemolysin of patho-

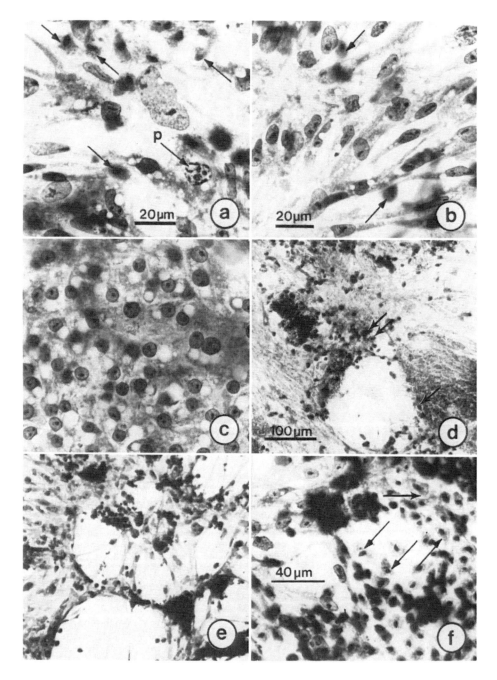

Fig. 9.3. All micrographs are of infections with strain JH30A, except for **c**. **a**. Twenty h. Fibroblasts with many adherent trichomonads, several of which are indicated by arrows. Some of the protozoa are applied to areas of the cover glass that are already free of fibroblasts. The remaining fibroblasts show pronounced vacuolization and retraction of the cytoplasm. The fibroblast nuclei often have abnormal shapes and range in size from quite narrow to giant ones (e.g. the nucleus near the center of the figure). The chromatin tends to be quite condensed in many nuclei and appears smudged in some. Frankly pyknotic nuclei, and others showing margination of the chromatin material, are also evident. Note the abnormal prophase (*p*) near the lower right corner. **b**. Twenty h. Fibroblasts showing cytoplasmic and nuclear changes described above in the legend for **a**. However, in this field, there are many

genic *T. vaginalis* might be a cytotoxic cysteine protease similar to the one . . . observed in *E. histolytica*," and thus could be responsible for the cytopathogenic activity of the trichomonad lysates. Gentry et al[17] admitted, however, that they failed to demonstrate the ability of the parasites to release a CPE-producing factor under "growth conditions." They suggested that a possible reason for the difficulty in demonstrating toxins in the supernatant fluids of *T. vaginalis* cultures was that the toxic substances were ". . . bound to serum proteins; the cytotoxic cathepsin of *E. histolytica* has been shown to be inactivated thus."

The relative paucity of data included in the generally interesting report renders difficult a critical evaluation of all the results and assumptions presented by Gentry et al.[17]

Cytopathic effects caused in BHK cell cultures by supernatant fluids of crude lysates of *T. vaginalis* were reported also by Ortega et al.[18] The CPE was inhibited by fetal bovine serum and phenyl methyl sulfonyl fluoride (PMSF); it was enhanced by cysteine hydrochloride. The protease activity of the "crude lysate supernatants" was evaluated by using azocoll as a substrate. Cysteine hydrochloride and dithiothreitol increased this activity, while iodacetamide, PMSF, and Ep495 inhibited it; pepstatin had no effect. All the results suggested the presence of a thiol protease in the lysate.

Pindak et al[19] reported that pathologic changes similar to those observed in the presence of trichomonads were noted in mammalian cell cultures exposed to the cell detaching factor (CDF) (see Effects on Attachment of Cell Culture Monolayers, below) in the absence of the parasites.

Throughout the chapter thus far we have been dealing with various aspects of the parasite-target cell interactions and trichomonad strain virulence evaluation in target cell monolayer cultures. However, virulence has also been assessed by some investigators with the aid of "pellet experiments."[16,20,21] In this method the effector (trichomonad), radiolabeled or unlabeled, and target (vertebrate culture) cells are mixed in specified ratios; then the mixture is pelleted by centrifugation. The pellets are incubated under given sets of conditions for given time periods. After incubation, they are resuspended in appropriate media and the percentages of surviving target cells are estimated by microscopic techniques or by release of radiolabel from the lysed target cells.

Krieger et al[20] used the pellet method, in addition to the one involving cell monolayers, to demonstrate, with the aid of ^{111}Indium oxine label, that trichomonads kill the target cells in culture by an extracellular process. These workers also employed the pellet technique in studies of the mechanism of cytotoxicity of trichomonads for Chinese hamster ovary (CHO) cells. Further to ascertain if the cytopathic effect of *T. vaginalis* on the target cells fitted the "one-hit hypothesis," developed originally for evaluation of the contact-dependent cytopathogenicity of lymphocytes (for references, see 20) and

---

◁ fewer trichomonads (*arrows*) attached to the chick liver cells; furthermore, the cytoplasm of more fibroblasts is evident. The scale is applicable also to c. **c.** Sixteen h. Strain JH32A. Cytoplasm of the epithelial cells contains vacuoles of various sizes; these vacuoles tend to be relatively large in this field. The nuclei vary in size. Some of them appear quite normal; others show loss of chromatin or its condensation into relatively coarse granules. Smudging of chromatin material is found in some cells, as are also pyknotic nuclei. The effect is very similar but more pronounced in infections involving the relatively virulent JH30A trichomonads. The field with JH32A was chosen because of its relative clarity and because it showed cytopathologic changes in the near absence of parasites. **d.** Two h. Early lesion in an island of epithelial cells. Trichomonads (*single arrow*) are applied to the inner margin of the lesion; they are numerous also (*double arrow*) in the regions adjacent to the cell-free area. **e.** Sixteen h. The area riddled with lesions is a network of cell strands remaining from the chick cell monolayer. A field of fibroblasts with adherent parasites is seen on the left of the figure. Many of the cell strands are covered with trichomonads, some of which are seen also in the cell-free areas For scale, see **d. f.** Twenty h. Typical area of a chick cell culture in a final stage of destruction. Relatively few degenerated chick cells are interspersed among the typically ameboid virulent trichomonads (some of which are indicated by *arrows*). The flagellates, which coat the cover glass, now nearly denuded of chick cells, tend to adhere to one another (see the upper middle and upper right areas of the figure). All figures: MGG.

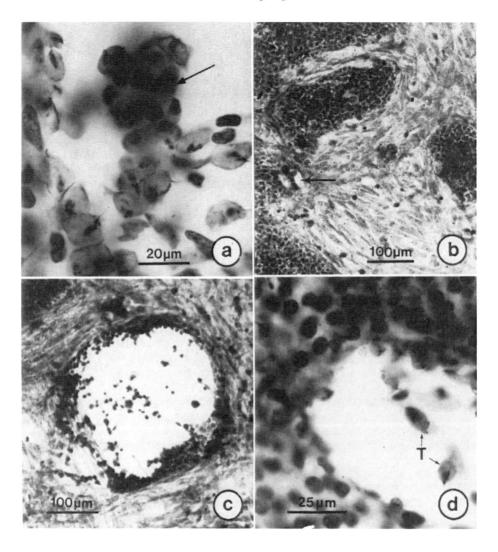

Fig. 9.4. All photomicrographs are of infections with the mild JH32A strain, except for **a**. **a**. Twenty h. Strain JH30A. Margin of a lesion is shown. In the left part of the figure, the ameboid trichomonads palisade against the mostly degenerating cells forming the lesion's margin (cf. Fig. 9.8a). BP. **b**. Four h. Only very few parasites rest on the surface of the virtually normal chick cell monolayer. Very small cell-free areas (one indicated by an *arrow*), which possible represent early lesions, are seen near the left margin of the figure. MGG. **c**. Sixteen h. Cell-free area in an epithelial cell island of a monolayer. Trichomonads and some macrophages are evident in the lumen of the lesion. Some of the flagellates are arranged along its margin. Only a few parasites are seen outside the area free of vertebrate cells. Most of the chick culture cells around the lesion appear quite healthy. MGG. **d**. Sixteen h. Part of a typical lesion in an epithelial cell area. Nonameboid trichomonads (*T*) are seen in the lumen of the lesion; some tend to attach themselves to the margin of the cell-free area. BP.

was therefore contact-dependent, Krieger et al[20] used the appropriate calculations to elucidate the quantitative aspects of CHO cell killing by the urogenital trichomonads. The results indicated that one trichomonad killed one mammalian cell, and only on direct contact. Evidently, the killing process was contact-dependent.

Not all investigators find the pellet method entirely satisfactory for many reasons, e.g. precipitous pH changes, which may be responsible for artifactual damage to the cells; also specific and nonspecific damage caused by other factors. Having had experience with the pellet method in other connections, some of us find that experiments involving this technique must be very carefully controlled and that all results must be considered with caution.

It was reported some time ago that a comparison of cultures infected with virulent strain trichomonads and those exposed to their cell-free filtrates indicated that the living flagellates played a crucial role in the formation and enlargement of the lesions.[5] Mild strains (e.g. JH32A) caused no lesion formation in vertebrate cell cultures during the early stages of infection,[5] and no lesions or any other CPE were noted in chick liver cultures exposed to cell-free supernatant fluids of active cultures of mild strain trichomonads.[5,14] However, in considering the question of *T. vaginalis*-dependent cytotoxicity, one cannot disregard the results obtained by various investigators with regard to the pathologic changes caused by cell-free filtrates of active trichomonad cultures, of lysates of these parasites, and of other *Trichomonas*-derived materials (for references, see the preceding paragraphs of this section). Also, according to Brasseur and Savel,[14] nonconcentrated or 50× concentrated supernatant fluids from *T. vaginalis* cultures maintained in Diamond's medium[22] for 12–72 hours caused slight CPE in McCoy, but not in HeLa, cell cultures.

The effect of cell-free filtrates from *T. vaginalis* cultures on fibroblast mitosis (Fig. 9.5), typically arrested in prophase (Fig. 9.2f, g; Fig. 9.3a; cf. Fig. 9.2e of normal prophase), has been demonstrated.[5] The filtrates from cultures of the virulent strain inhibited cell division more strongly than the living organisms of the mild strain. It was also evident that reversal of mitotic inhibition occurred in cell cultures exposed for 16 h to filtrates of cultures of both strains (Fig. 9.5). Clearly, the inhibiting capacity of the filtrates appears to be labile under the experimental conditions, for example, incubation at 37°C.

The probability of elimination of cytotoxic substances by *T. vaginalis* into the medium is reinforced by the reports on inhibition of destruction of Darby canine cell monolayer by *T. vaginalis* in the presence of various protease inhibitors (e.g. leupeptin, thymostatin, antipapain) in the "interactive medium" (M199 with bovine serum)[23] and by those dealing with release of hydrolases.[24] However, according to Alderete and Pearlman,[16] although proteolytic activity was still present in supernatant fluids of trichomonad cultures and in concentrated fractions of such fluids, they were not cytotoxic for mammalian cell cultures. These workers suggested several possible explanations for the lack of cytotoxicity: very low levels of toxic substances in the filtrates, masking of such substances by the culture medium components, the types of vertebrate cell cultures employed, and the kinds of assays used in evaluation of the cytopathic effects. Evidently, Alderete and his co-workers, who, as is apparent from their various reports, strongly supported the idea of contact-dependence of *T. vaginalis* cytotoxicity, admitted that CPE of cell-free supernatant fluids from trichomonad cultures should perhaps be demonstrated under suitable conditions. In addition, Rasmussen et al,[15] who thought that cytopathogenicity of the urogenital trichomonad was associated with parasites adhering to the target cells, questioned whether cytotoxicity depended on mechanical action of the flagellates or on the activity of lytic factors bound to the parasite membranes.

In light of the foregoing considerations, the suggestion of Krieger et al[20] that the presence of pathologic changes in the absence of living parasites in some areas of cell cultures (see references cited in this chapter) or of vaginal epithelium in *T. vaginalis*-infected women[25] is due to the fact that the flagellates have moved away from the tissues they have previously damaged, need not be the only possible explanation of this condition.

With regard to the mechanisms of pathogenicity of *T. vaginalis*, perhaps the conclusion arrived at by Heath[13] on the basis of his observations, to be discussed later in this chapter, and those of others describes accurately the actual situation. Behavior of *T. vaginalis* in cell cultures suggested that both "mechanical and chemical components operated" in human trichomoniasis. Their adherence to the epithelial layers of the vagina and ectocervix as well as their "protrusive activity during ameboid crawling" may be the causes of disruption of the squamous epithelium and, to some extent, of necrosis and increased levels of "desquamation of the vaginal epithelium." However, the markedly increased "subepithelial vascularity" seen in the walls of the vagina and the cervix

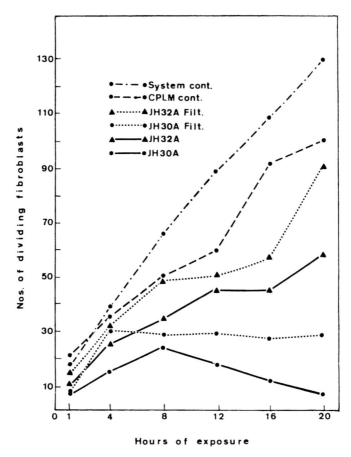

Fig. 9.5. Mean numbers of fibroblast mitoses in experimental and control-experimental chick liver cell cultures. The numbers were based on average counts of cells in division in a total of 100 random microscope object fields (each with a surface of 0.056 mm^2) delimited by an ocular net micrometer. For each experimental and control experimental system, 50 fields were examined in each of two MGG-stained preparations. (Reproduced with permission of the American Society of Parasitologists, from Farris VK, Honigberg BM: Behavior and pathogenicity of *Trichomonas vaginalis* Donné in chick liver cell cultures. *J Parasitol* 56: 849–882, 1970.)

"is often out of proportion" to the numbers of flagellates attached to the epithelium. Thus, diffusible chemical factors produced by the parasites were probably involved in causing the pathologic changes characteristic of infections with the urogenital trichomonad.

Progression of the Pathologic Process in Cell Cultures

The sequence of events accompanying infection of HeLa cells as described by Christian et al[3] was outlined in the earlier part of this chapter. To justify a generalized description of the progression of the pathogenic process, it seems necessary to provide accounts of this process as detailed by several investigators employing various kinds of cell cultures and different strains of *T. vaginalis*. (For all the accounts given below, the parasite strains, types of cell cultures, and the important methods employed are detailed in Table 9.1.)

Farris and Honigberg.[5] Strain JH30A. The first cell-free areas, formed most frequently in the epithelial cell islands, were seen at 2 hours after inoculation with the virulent parasites (Fig. 9.3d). Trichomonads palisaded against the internal margins of the lesions (arrow) and were abundant around the cell-free spaces (double arrow). By 4 hours, the lesions enlarged by retraction of the cytoplasm of the peripheral epithelial cells and of that of the adjacent fibroblasts. As the infection progressed, the lesions increased in number and diameter (Fig. 9.3e). The cells lining the lesions were covered with parasites which extended their pseudopodia over the cell surfaces. [Undoubtedly the interrelationship, to be discussed later in this chapter, between the cell culture and trichomonads corresponded to those shown by transmission electron microscopy between the vaginal epithelial cells and *T. vaginalis* in women (Fig. 9.6a) and between cultures of these cells and the same parasite species (Fig. 9.6b).] A few usually unhealthy-appearing macrophages, often with the parasites closely apposed to their external surfaces or containing trichomonads within food vacuoles, were present in or around the cell-free areas; some of the debris collecting in the vicinity of the lesions consisted of fragments of degenerated phagocytes. By 16 or 24 hours after inoculation many areas of the cell culture monolayers were riddled with lesions (Fig. 9.3e). The remaining culture cells tended to slough off the cover glasses; usually, only small fragments of the cell sheets remained attached after 24 hours of infection. The parasites adhered to the typically degenerated cells in and around lesions (Figs. 9.3f, 9.4a). In the final infection stages, the cell-free spaces (spreading over nearly the entire coverglass surfaces) were covered with trichomonads (Fig. 9.3f).

Strain JH32A. Cultures exposed for even as long as 4 hours to the mild trichomonads contained either no or only occasional very small lesions (arrow) (Fig. 9.4b). By 8 hours, distinct, but still small cell-free areas were seen in the cell monolayers (Fig. 9.4c). The cells located at the margins of lesions showed signs of cytoplasmic and nuclear degeneration. Some usually nonameboid parasites appeared to adhere to the lesion margins and were found also within the cell-free areas. The intimate association between the parasites and the culture cells characteristic of infections with the virulent strain was seen infrequently (if, indeed, ever) in cultures exposed to the JH32A organisms. Numerous active macrophages, often with degenerating parasites in their cytoplasm, were evident around the small lesions. Although with time the cell-free areas became larger, they never reached the size or number of those caused by the JH30A trichomonads, at least for the duration of the experiments. The larger lesions, some of which were noted 16 hours after infection, were located in the epithelial cell islands resembling those noted in much earlier infection of cultures exposed to the virulent strain. Active macrophages, typically containing degenerated parasites, were still abundant around the cell-free areas. Some of the typically nonameboid trichomonads appeared to be applied to the internal margins of the cell-free areas (Fig. 9.4d); some nonameboid ellipsoidal flagellates were noted also in the lumina of the lesions. Even at 20-24 hours after infection, most degenerated cells were seen near the lesions. No cultures containing the JH32A trichomonads that were examined within the time limits of the experiments were riddled with cell-free areas.

Of major significance in the foregoing account is the fact that *T. vaginalis* strains differing in virulence appear to evoke cell-culture reactions that vary not only in the time periods required to destroy the vertebrate cells, but also in certain qualitative and quantitative characteristics of the interactions between the effector and target cells.

As far as can be ascertained, the only cytochemical investigations of the effects *T. vaginalis* exerted on vertebrate cell cultures were conducted by Sharma and Honigberg,[26-29] who employed the relatively virulent JH30A strain of the parasite and chick liver cell monolayers in their experiments. In many instances, especially those involving the macrophages, the cytochemical aspects of direct trichomonad-chick liver cell culture interactions were demonstrated. The changes caused by *T. vaginalis* in the levels of nucleic acids, glycogen, lipids and proteins, as well as in the activities of many enzymes, revealed with the aid of various cytochemical methods, are summarized in Table 9.2.

Tos-Luty et al.[12] The differences in the postinfection times of appearance of the degenerative changes in chick fibroblasts and in the intensity of these changes depended on the virulence levels of the strains employed.

After 4 hours of infection, numerous small cytoplasmic vacuoles were noted, and single trichomonads appeared within the cells. Intensification of pathologic changes progressed

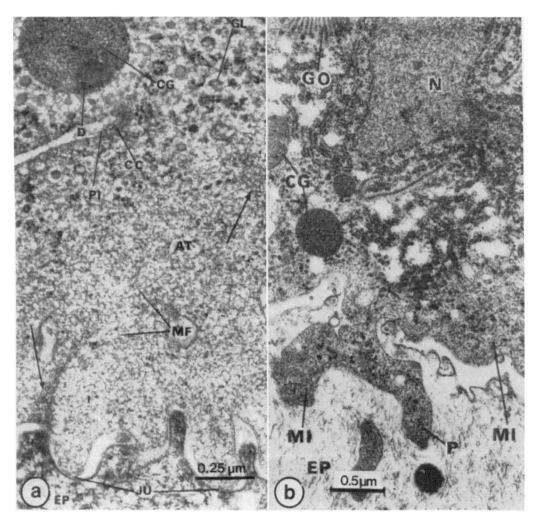

Fig. 9.6. Transmission electron micrographs showing the areas of adhesion of *T. vaginalis* to human vaginal epithelial cells in vivo and in vitro.

a. Ultrathin section from a human biopsy passing through an area in which *T. vaginalis* (*AT*) is apposed by its pseudopod (cf. Fig. 3.7a) to an epithelial cell (*EP*) from the vaginal lining. The cytoplasm of the epithelial cell shows signs of degeneration. The parasite "ectoplasm" contains microfilaments (*MF*) which in some regions form well defined bundles (*arrows*). The surfaces of the trichomonad and the epithelial cell interdigitate, and in certain areas their segments come into direct contact with each other (*JU*). A typical hydrogenosome (*CG*) with a darker area (*D*) is seen in the upper left corner of the figure. Part of a pinocytotic invagination (*PI*) is covered by the cell coat (*CC*). Glycogen granules (*GL*), often arranged in groups that resemble the rosette-like aggregates characteristic of trichomonads (cf. Fig. 3.5d), "are surrounded by an . . . [electron-lucent] zone." (This latter area is clearer in Fig. 3.7a.) [Reproduced with permission of Munksgaard, Denmark, from Nielsen MH, Nielsen R: Electron microscopy of *Trichomonas vaginalis* Donné: Interaction with vaginal epithelium in human trichomoniasis. *Acta Pathol Microbiol Scand* (B) 83:305-320, 1975.]

b. Ultrathin section from a *T. vaginalis*-human vaginal epithelial cell culture system 24 h after infection with strain 130354, isolated from a case of trichomoniasis accompanied by severe symptoms. The section passes through an area in which pseudopods (*P*) of a trichomonad penetrate rather deeply into the cytoplasm of a partially degenerated epithelial cell (*EP*). The parasite appears to be apposed very closely to the mammalian cell by its pseudopods, which contain clearly evident microfilaments (*MI*) (cf. Figs. 3.7a and 9.6a). (Reproduced with permission of the British Medical Association, from Rasmussen SE et al: Morphological studies of the cytotoxicity of *Trichomonas vaginalis* to normal human vaginal epithelial cells in vitro. *Genitourin Med* 62:240-246, 1986.)

TABLE 9.2. Cytochemical analysis of chick liver cell cultures infected with a relatively pathogenic strain of *T. vaginalis*.*

Compound or enzyme	Localization in cytoplasm		Present in nucleus	Present in type of cells[†]	Postinfection levels	
	In granules or droplets	Diffuse			In parasite-free macrophages[‡]	In other cell types
DNA	−	+	+	F,E,M	Unchanged	Unchanged
RNA	+	+	+	F,E,M	Unchanged	Decreased
Glycogen	+	−	−	E,M	Unchanged	Decreased
Lipids[§]	+	−	−	F,E,M	Increased	Increased
Protein						
Tyrosine-containing	+	±	−	(F)[∥],E,M	Unchanged	Increased in E
Protein-bound amino acids	+	+	+	F,E,M	Increased	Decreased
Alkaline phosphatase	+	−	−	F,E,M	Increased	Decreased
Acid phosphatase	+	+	−	F,E,M	Increased	Increased
Adenosine triphosphatase	+	+	+	F,E,M	Unchanged	Decreased
5-Nucleotidase	+	+	+	F,E,M	Increased	Decreased
Malate dehydrogenase	+	−	−	F,E,M	Increased (granular and diffuse)	Increased (granular and diffuse)
Lipase	+	+	−	E,M	Unchanged	Decreased
Nonspecific esterase	+	±	−	F,E,M	Increased	Increased
Monoamine oxidase	+	−	−	F,E,M	Increased	Decreased
Glucose-6-phosphatase	+	+	+	E,M	Increased (somewhat)	Decreased
Glucosan phosphorylase	+	+	−	E,M	Unchanged (or somewhat increased)	Decreased
Glucose-6-phosphate dehydrogenase	+	−	−	E,M	Increased	Decreased
α-Glycerophosphate dehydrogenase	+	−	−	F,E,M	Unchanged	Increased

Reprinted from Honigberg BM: Trichomonads of importance in human medicine. In Kreier JP (ed): *Parasitic Protozoa*, vol 2, pp 275–454. New York: Academic Press, 1978, reproduced with permission of Academic Press.
*For the cytochemical methods and discussion of results, see the reports of Sharma and Honigberg.[26–29]
[†]F, fibroblasts; E, epithelial cells; M, macrophages.
[‡]In all instances decrease of the amounts of given compound and in enzymatic activity was noted in macrophages to which the parasites were applied or which harbored trichomonads within their cytoplasm.
[§]Most of the lipids were neutral fats; however, some unsaturated lipids were also found in all three cell types.
[∥]The level of tyrosine-containing proteins, as demonstrated by Millon's tests, was very low in the fibroblasts, and no change could be seen in the reaction level among these cells in experimental cultures.

during the next 24 hours. In fibroblast cultures infected with *T. vaginalis* strains from patients with inflammation of the vagina or of the urinary bladder the majority of the culture cells showed a high degree of cytoplasmic vacuolization. Nuclear changes were manifested by pyknosis and chromatin margination. Although the intercellular areas tended to be enlarged as a result of cytoplasmic retraction, the tissue cultures retained their monolayer appearance. Strains isolated from women with cervical cancer or uterine myoma caused complete destruction of the cell monolayers after 24 hours. The majority of the fibroblasts sloughed off the glass; those still adhering had small often deformed nuclei. Vacuolated or condensed cytoplasm formed a narrow band around the nucleus. Trichomonads were observed within the culture cells. Although lacking in detail, the foregoing account is among the few referring to intracellular trichomonads in either in vitro or in vivo systems.

Heath.[13] No medical record of trichomoniasis or results of laboratory virulence assays of the *T. vaginalis* strain were given.

It was observed by light and scanning electron microscopy (SEM) that 30 minutes after infection most parasites were swimming actively over the cell monolayers. By 60 minutes, they formed large aggregates (of up to 200 organisms) in a process referred to as "swarming" or "rosetting." After 1-2 hours, the aggregates and surrounding individual organisms lay motionless on the monolayer sheets, the flagella and undulating membranes of the parasites remaining motile. Since the parasites could not be detached by gentle shaking, they must have been adhering to the culture cells. Between 2 and 3 hours after inoculation, small cell-free lesions were noted underneath or to one side of the trichomonad aggregates attached to the monolayers; no lesions were found in the cell culture areas free of the parasites. With time, progressively more lesions developed in association with the parasite aggregates, the flagellates often settling in these lesions.

At 6-10 hours, about 10% of the cell culture monolayers were disrupted, some of the lesions reaching up to 1 mm in diameter. When examined by SEM, the structure of the cells seen in and around the lesions was altered (Fig. 9.7a, b). The trichomonads palisaded against the cells at the internal margins of lesions (Fig. 9.8a) or else adhered to the few cells remaining in the lesions' lumina (Figs. 9.7b; 9.8a). Most of the flagellates appeared to be ameboid (Fig. 9.7b). In some organisms, large pseudopods were evident at the posterior poles around the projecting segment of the axostyles (Fig. 9.7d). The ability of *T. vaginalis* to lift the cell culture monolayers and to migrate under them suggested that "the protrusive activity during ameboid crawling" was responsible, in part at least, for detachment of the mammalian cells from the substrate and thus for enlargement of the lesions. A cell with a large smooth pseudopod protruding underneath an epithelial cell, and probably acting like a shovel, is shown in Fig. 9.7c. In addition, smaller filopods (white arrows) are evident in the posterior part of the parasite.

A view obtained by transmission electron microscopy of a transverse section through a lesion (Fig. 9.8b, from reference 15), similar to the one shown by scanning electron microscopy (Fig. 9.8a), is seen in the subadjacent micrograph. In this latter section, one can visualize the spatial relationships among the trichomonads and the human vaginal epithelial cells in an in vitro system. These relationships provide additional insight into the mechanisms whereby the parasite can cause detachment and destruction of the host cells.

From 10 hours onward, the lesions continued to enlarge and tended to coalesce; at about 36 hours after infection, most of the cell culture monolayer was completely destroyed. The medium was replete with debris from lysed cells; also $1-2 \times 10^6$ trichomonads were seen swimming in the medium and crawling on the coverglasses.

Brasseur and Savel.[14] Description of the progression of the pathogenic process by these French investigators is quite similar to those given previously by others. However, Brasseur and his colleagues,[30] who examined 25 noncloned and 16 cloned isolates in McCoy monolayer cell cultures (see Table 9.1), were able to differentiate "quantitatively" among virulence levels of the strains they examined by employing the "In Vitro Virulence Index" (Table 9.3). Brasseur and his coworkers[30] used these indices in their later experiments, for example, in those which compared virulence of *T. vaginalis*

Fig. 9.7. Scanning electron micrographs of *T. vaginalis* strain derived from LUMP 889 stabilate (London Sch. Hyg. Trop. Med.) in RK_{13} epithelial cell monolayer culture. (For the available history of the parasite strain, see Heath[13] in Table 9.1.) **a.** Lesion in the cell monolayer 6 hours after infection with the trichomonads. **b.** Higher magnification micrograph of the lesion shown in **a**. Isolated epithelial cells (*ec*), some of which have retraction fibers (*black arrows*), are evident in the lesion. Parasites (*tv*) adhere to a number of the culture cells. **c.** Trichomonad with a large pseudopod that protrudes beneath the epithelial cell seen in the lower part of the figure. Toward the posterior end of the parasite note the small filopodial extensions (*small white arrows*). The structure (marked by the *black arrow*) on the trichomonad surface is a pinocytotic pit; small vesicles from lysed epithelial cells are distributed over the cell membrane of the flagellate. **d.** Bizarre ameboid trichomonad crawling over an epithelial cell. Two large pseudopods (*p*), adjacent to the projecting posterior segment of the axostyle (*ax*), remain in close contact with the epithelial cell surface. The apparent absence of the periflagellar canal mentioned by Heath[13] is undoubtedly an artifact. The flagellum (*arrow*) emerging from the body "a short distance from the others" is probably the recurrent flagellum (cf. Fig. 3.2d). (Reproduced with permission of the British Medical Association, from Heath JP: Behaviour and pathogenciity of *Trichomonas vaginalis* in epithelial cell cultures. *Br J Vener Dis* 57:106-117, 1981.)

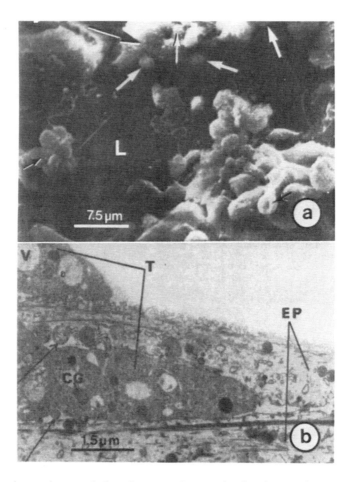

Fig. 9.8 Scanning and transmission electron micrographs showing spatial relationships between axenic isolates of *Trichomonas vaginalis* and epithelial cell cultures.

a. Scanning electron micrograph of a lesion (*L*) in an RK_{13} epithelial cell monolayer seen at 6 h after infection with *t. vaginalis* LUMP 889 stabilate-derived strain. Groups of trichomonads (*small black arrows with narrow white edges*) adhere to the epithelial cells lining the lesion. Flagellates (indicated by *large white arrows*) are applied to the sides and lower surface of the part of the cell monolayer that is lifted off the substrate (*large black arrow with a narrow white edge,* seen near the left part of the upper margin of the figure). (Reproduced with permission of the British Medical Association, from Heath JP: Behaviour and pathogenicity of *Trichomonas vaginalis* in epithelial cell cultures. *Br J Vener Dis* 57:106-117, 1981.)

b. Transmission electron micrograph of a section through a culture of human vaginal epithelim 24 h after infection with *T. vaginalis* strain 130354 freshly isolated from a patient presenting with severe symptoms of urogenital trichomoniasis. Flattened trichomonads (*T*) are seen on the surface as well as between the epithelial cells (*EP*). Since the epithelial cell membrane was broken in the course of infection, cytoplasmic organelles (indicated by the *long fine arrows* in the far left of the figure) are lodged in intercellular spaces. Hydrogenosomes (*CG*) and food vacuoles (*V*) are seen in the trichomonad cytoplasm. (Reproduced with permission of the British Medical Association, from Rasmussen SE et al: Morphological studies of the cytotoxicity of *Trichomonas vaginalis* to normal human vaginal epithelial cells in vitro. *Genitourin Med* 62:240-246, 1986.)

TABLE 9.3. "Virulence Index" of *T. vaginalis* in vitro (McCoy cell monolayers).

Category	Subcategory			Percent or no.	Weighted values*
Cytoplasmic retraction	NA†			≤20%	1
				21–50%	2
				>50%	5
Conditions of monolayer	Lesions	Size		≤20 cells	1
				>20 cells	2
		No.		≤1/field‡	1
				1–5 /field	2
	Destruction of monolayer			Partial: 10–50%	10
				Total: 51–100%	20

Reprinted from Brasseur P, Savel J: Evaluation de la virulence des souches de *Trichomonas vaginalis* per l'étude de l'effet cytopathogène sur culture des cellules. *C R Soc Biol (Paris)* 176:849–860, 1982, with permission of the Societé de Biologie, Paris.
*In arbitrary units.
†Not applicable.
‡Diameter of the object field is not given.

strains with the chemotactic effects exerted by supernatant fluids of the flagellates on polymorphonuclear leukocytes.

Unfortunately Brasseur and Savel[14] failed to correlate the virulence indices with the severity of the vaginal and cervical diseases as reflected in clinical and pathologic manifestations in women from whom the *T. vaginalis* strains were isolated. Actually, the authors attempted to correlate the in vitro results with the results they obtained for these strains by the intraperitoneal mouse assay (see Chap. 8). On the basis of the latter correlations, they stated: "All the indices presented a very significant correlation with the tests performed in mice, with the exception of six cases in which the Swiss mice showed resistance to inoculation with the parasites . . . Furthermore, the in vitro test for virulence [using] McCoy cells appears to be more sensitive [and] the index values are relatively higher and have a better statistical distribution . . . [It also has] the advantage of being easy and quick to perform."

Krieger et al.[20] The observations of Krieger and his coworkers on an in vitro system involving *T. vaginalis* strains from cases of "symptomatic vaginitis" and Chinese hamster ovary (CHO) cell cultures were in general agreement with those detailed above for the accounts of previous workers. As indicated earlier in this chapter, according to Krieger et al[20] "trichomonads kill cells only by direct contact."

Pindak et al.[19] Using the empirical growth medium (GMP), which provided for slower development of the cytopathic manifestations caused by *T. vaginalis* strains, these authors noted that in the presence of the parasites the usual progression of events in a variety of mammalian cell cultures (see discussion of *T. vaginalis* preference for certain types of cells and also Table 9.1) extended for a longer period than that noted by other investigators using richer media. In GMP-cultivated systems the adherent organisms produced small focal cell-free lesions (plaques) in the cell culture monolayers within 1 day, and a nearly total destruction of such cultures by the enlarging plaques was achieved 3 days after infection. The cell detaching factor (CDF) (see section entitled Effects on Attachment of Cell Culture Monolayers, p. 191) was said to cause CPE similar to that observed in the presence of the parasites. Thus, Pindak et al[19] observed in a "cell-free (filter sterilized) product" an effect usually found in the presence of live organisms. However, CDF production could not be correlated with the virulence levels of the trichomonad strains as evaluated by clinical and pathologic manifestations found in the human donors of the isolates or by the subcutaneous mouse assay.

Rasmussen et al.[15] Strain 130354 (freshly isolated from a case of severe vaginitis) and 10 (designated as a freshly isolated strain) trichomonads moved actively over the CHO cell monolayer for about 1 hour after inoculation (Fig. 9.9a). Thereafter they settled on the culture cells in small aggregates or singly and became ameboid (Fig. 9.9b). Forty-eight hours after inoculation, the parasites formed sheets that partly covered the epithelial cell monolayers (Fig. 9.9c). Adherence of the fresh isolates 130354 and 10 appeared to be associated with their cytotoxic effects, all cell destruction being originally confined to the areas with adherent trichomonads. In the presence of these strains cytopathic effects were noted in monolayers exposed to as few as 10^3 parasites per well; larger inocula caused total destruction of the mammalian cell cultures.

A laboratory trichomonad strain, 1711, cultivated for a long time, remained ovoidal after having been added to the vaginal epithelium monolayers. The flagellates moved over the cell cultures singly or more typically in large aggregates consisting of up to several hundred flagellates (Fig. 9.9d). Adherence of single organisms or of aggregates to the mammalian cell cultures was noted rarely after 3½ hours (Fig. 9.9e) or even at 24 hours after inoculation (Fig. 9.9f). Furthermore, with strain 1711, cytopathic effects were noted only in the cell cultures exposed to a minimum of 10^5 parasites per well.

In several respects the account of Rasmussen,[15] illustrated by clear photomicrographs, is more similar to that of Farris and Honigberg,[5] with which it should be compared, than to those of the progression of *T. vaginalis* infection in vitro published by some other workers. The properties of the strains and/or the culture conditions might have been responsible for the aforementioned similarities and differences. It seems also that the formation of large aggregates ("swarming") is not an indispensable prerequisite for the attachment of the trichomonads to vertebrate cell cultures and that it may occur in strains which have only a very limited capacity for cytadherence, for example, strain 1711.

Bremner et al.[31] By using several vertebrate cell lines (see Table 9.1) and old strain(s) as well as recent clinical isolates of *T. vaginalis*, these investigators were able to quantify virulence of the parasite strains in vitro when the "initial density of trichomonads and the ratio of trichomonads to [vertebrate] . . . cells were within certain ranges." "In the mixed culture the mammalian cell numbers decreased almost immediately . . . [and] detached and died by 65 hours." At lower parasite densities, the culture cells were unaffected and the flagellates died rapidly; at higher densities, the trichomonads overgrew even when no vertebrate cells were present. Although the progression of the pathologic process is not described, the applicability of the in vitro system to virulence evaluation of the human urogenital trichomonad is emphasized.

Juliano et al.[32] The findings of these workers about the events occurring in CLID mouse fibroblasts infected with a *T. vaginalis* strain from a patient presenting with "acute vaginal trichomoniasis"[32] conformed closely to those

---

Fig. 9.9. Progression of infection in human vaginal epithelial cell monolayers by *T. vaginalis* strain 130354 freshly isolated from a patient with severe symptoms of trichomoniasis, and by an established laboratory strain, 1711, originally described in 1975. Inocula of 10^3 organisms per well of tissue culture plate were used with both strains. Phase-contrast photomicrographs of living organisms. The scale in a is applicable to all the figures. (Reproduced with permission of the British Medical Association, from Rasmussen SE et al: Morphological studies of the cytotoxicity of *Trichomonas vaginalis* to normal human vaginal epithelial cells in vitro. Genitourin Med 62:240-246, 1986.)

a-c. Strain 130354. a. Immediately after inoculation with the trichomonads, a few individuals or small groups of typically ameboid (*A*) parasites adhere to intact mammalian cell monolayer. b. After 3 1/2 hours most parasites, all of them ameboid (*A*), adhere to the vaginal epithelial cells (*EP*) which, however, remain largely undamaged. c. At 22 hours after infection, sheets of the trichomonads cover large cell-free areas (lesions) formed in the epithelial cell monolayers.

d-f. Strain 1711. d. Immediately; e. 3-1/2 h; and f. 22 h after inoculation most of the trichomonads (*T*) have formed aggregates including many typically nonameboid organisms, but none of them adhere to the epithelial cell (*EP*) monolayers, which show no pathologic changes.

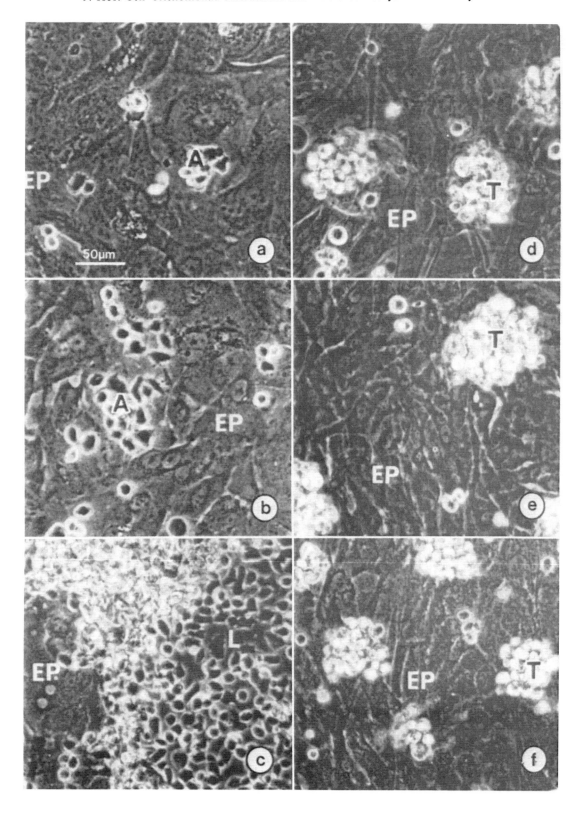

found in the more detailed reports mentioned in the text of this chapter and in Table 9.1. For example, at about 30 minutes after inoculation, preparations examined by scanning electron microscopy revealed that the parasites adhered to the surface of the mammalian cells by pseudopods, "especially by thin filipodial tips." As early as the first postinfection hour, some of the culture cells showed retracted cytoplasm. By 6 hours, pathologic changes affected almost all cells in the monolayers, many of them sloughing off the glass and thus causing the formation of cell-free lesions.

The accuracy of the base-line observations of Juliano et al[32] imparts credibility to those they reported for the effects exerted by certain compounds on the cytopathologic process occurring in the CLID cell-trichomonad system. (These findings are discussed later in the present chapter.)

Summary of Results of Investigations

In light of the foregoing, the progression of the pathologic process caused by *T. vaginalis* in cell cultures can be summarized as follows:

1. The parasites swim freely over the vertebrate cell monolayers.
2. Many strains form aggregates consisting of numerous cells ("swarming").
3. The more virulent strains tend to settle on and adhere to the vertebrate culture cells; the less inherently virulent (or cultivated for prolonged periods) strains have less tendency for cytadherence.
4. The typically ameboid virulent trichomonads apply themselves very closely to the cell culture elements.
5. Cell-free areas (lesions) form by cytoplasm retraction and cytolysis underneath or to the side of the adherent flagellates.
6. The lesions enlarge, the ameboid activities of the parasites playing an important role in the formation of the cell-free areas.
7. Certain of the cytopathic changes in the vertebrate cell cultures may be caused by substances given off by the parasites into the environment (some of these changes appear to be reversible).
8. With time, the lesions enlarge and the culture cells slough off the substrate. It must be kept in mind that the times of onset of the various changes depend on the parasite strain, inoculum size, parasite:target cell (P/C) ratio, and other known and as yet unknown factors operating in the trichomonad-cell culture systems.

Spatial Relationships, at the Fine-Structural Level, Between the Effector (T. vaginalis) and Target Cells in In Vivo and In Vitro Systems

Figure 9.6a is an electron micrograph of an ultrathin section passing through the wide (1.4 μm, on average) peripheral zone of an ameboid *T. vaginalis (AT)* and through the area of its contact with a very narrow band of a vaginal epithelial cell *(EP)*, showing signs of degeneration. In the parasite cytoplasm one can see a dense network of about 2-5 nm microfilaments *(MF)*. The MFs usually have a random distribution; however, bundles of these organelles (arrows) situated at nearly right angles to the surface membrane are evident in many organisms. The bundles are especially conspicuous opposite the remnants of desmosomal attachment plaques (or tonofilaments) in the adjacent epithelium. The surface membrane of a parasite, closely applied to the adjacent epithelial cell, frequently interdigitates with its membrane. The areas in which the parasite and the host cell membranes are closely apposed appear as tight junctions *(JU)*. A similar kind of spatial relationship between a degenerating vaginal epithelial cell and a trichomonad of a strain isolated from a patient with severe symptoms of trichomoniasis (Fig. 9.6b) was reported from an in vitro system by Rasmussen et al.[15] The pseudopod (P) formed by the parasite projects deeply into the cytoplasm of a vaginal epithelial cell *(EP)* which shows signs of degeneration. At the contact site of a trichomonad and an epithelial cell, the cytoplasm of the parasite "consisted almost exclusively of a network of microfilaments [MI] . . . Sometimes endocytic vacuoles with dense material were [also] noted".[15]

Does T. vaginalis have a Preference for Certain Types of Cells in In Vitro Systems?*

It is evident from Table 9.1 that various kinds of cell cultures have been employed in the studies of *T. vaginalis*-vertebrate cell culture interactions. The best comparisons of suitability of

---

*In all instances comparison of the media used in the experiments with the various kinds of cell cultures and *T. vaginalis* is given in Table 9.1.

cell culture types for the aforementioned studies are the investigations in which several such types were used with one trichomonad strain by single groups of workers. Among those who reported this kind of investigation were: Zhordaniia et al,[33] Alderete and Pearlman,[16] Alderete and Garza,[10] Pindak et al,[19] Bremner et al,[31] and Fukui and Takamori.[21]

Zhordaniia et al.[33] These workers employed cultures of the following cell lines: HEp-2 (epithelial cells from epidermoid carcinoma of the human larynx), A-1 (human amnion cells), L (fibroblast[†] cells from subcutaneous tissue of mice), Ch.L. (diploid cells from human lung), and MA (monkey kidney cells). The cell culture types most sensitive to *T. vaginalis* were HEp-2 and A-1; furthermore, since all subsequent experiments were carried out with HEp-2 cells, it appears that they provided the system of choice.

Alderete and Pearlman.[16] These investigators used HEp-2, HeLa (cells from epithelioid carcinoma of human cervix uteri), Vero (African Green Monkey kidney fibroblasts), and NBT (normal baboon testicular cells, primary cultures). After having performed all their experiments using Müller's *T. vaginalis* strain, NYH 286, with HeLa cell monolayers, they compared the responses of the other cell culture types to the same strain of the parasite.[16] The results of the study indicated that: "After incubation for 20 hours with a low parasite inoculum [12.5 × 10^3 and 25 × 10^3] (giving a parasite:cell [P/C] ratio of 1:4), Vero cells appeared most refractory, while HEp-2 cells were highly susceptible to trichomonad killing." At higher parasite densities (5 × 10^4 and 10^5), however, cultures of all four types of mammalian cells appeared to be equally susceptible to the damage by the urogenital trichomonad.

Alderete and Garza.[10] These workers* employed the epithelial HEp-2 and HeLa cell lines; several human skin fibroblast lines, CCD-275K, CCD-505K, and presumably CCD-575K (results with this line are not shown); as well as human foreskin fibroblast cells. As in the previous study,[16] interactions of *T. vaginalis* strain NYH 286 with the several cell lines were compared. Two parameters, that is, the P/C ratios, using [^{3}H]thymidine-labeled trichomonads, and cytotoxicity indices, using the crystal violet staining technique (for the method of cytotoxicity determination used to ascertain the suitability of the mammalian cell cultures for the parasite-target cell investigations, see references 10 and 16). On the basis of their study, the authors concluded that ". . . surface parasitism of HeLa and HEp-2 epithelial cells [expressed by the P/C] was always higher than the levels of attachment observed with . . . [the] fibroblast cell lines, including human skin and foreskin cells."

As far as cytotoxicity was concerned, the epithelial HeLa and HEp-2 cells were readily destroyed by strain NYH 286 parasites. The extent of destruction of the fibroblasts, reflected in the cytotoxicity indices, did not exceed 30% of the damage inflicted by these trichomonads on the aforementioned epithelial cells. According to Alderete and Garza,[10] a similar low level of cell killing was noted with mouse skin fibroblasts exposed to NYH 286 strain and with fibroblasts from all sources infected with other isolates of the urogenital trichomonad.

It should be noted that in all three aforementioned sets of experiments[10,16,33] epithelial cells were found to be more susceptible than fibroblasts to infection and destruction by *T. vaginalis*. Farris and Honigberg[5] also observed that in chick liver cell monolayers infected with both a moderately virulent and a mild strain of the urogenital trichomonad, the lesions formed first in the areas of the epithelial cell islands (Fig. 9.3d and Fig. 9.4c).

Pindak et al.[19] This group investigated the effect of a *T. vaginalis* strain isolated from a patient suspected of having genital herpes on 15 kinds of established lines and primary isolates of cell cultures. The established cell lines were: HeLa, HeLa 299, HEp-2, RD, McCoy, L_{925} (clone of L cell strain), RK_{13}, MDCK, Vero, and BS-C-1 (see *ATCC Catalogue of Cell Lines and Hybridomas,* 6th ed. Rockville, MD: Am Type Cult Collect, 1988); primary or low passage human foreskin fibroblast (FSK) cells; two lines of baby hamster lung fibroblasts, HAMLu and BHLu; and apparently continuous lines, Bamas and CnK, developed from primary cultures of cynomolgus monkey kidney cells obtained from commercial sources.

[†]Here and elsewhere below epithelial cells (= epithelial-like cells) and fibroblasts (= fibroblast-like cells).
*In certain instances the results of Alderete and his collaborators differed from those of other investigators; some of these are mentioned in the section Effects of Prolonged In Vitro Cultivation.

According to Pindak et al,[19] trichomonads inoculated into the cell cultures attached to the monolayers and multiplied, producing focal cell-free lesions. These lesions expanded gradually until the entire cell culture sheet was destroyed. With minor variation, the progression of the pathologic changes occurred in 14 of the 15 types of cell culture. However, in HeLa, HeLa 299, HEp-2, HamLu, McCoy, RK_{13}, and L_{929} cell cultures inocula of 10^3, 10^4, and 10^5 trichomonads gave rise to dense populations of these parasites; only the two higher inocula could initiate such populations in the presence of MDCK, RD, and Vero cells. The remaining cell cultures supported less growth, while FSK cells "may have actually exerted some deleterious effect [on *T. vaginalis*]."

It ought to be noted that the cell types belonging to the first and second groups, which were able to support excellent growth of the parasites in cultures initiated with the smallest- or intermediate-size inocula, included some fibroblast lines. It appears, therefore, that the *T. vaginalis* isolate employed by Pindak et al[19] did about as well in fibroblasts as in epithelial cell cultures. Only information gained from experiments using the various cell culture systems and a variety of *T. vaginalis* isolates can aid in resolving the apparent discrepancies between the results of the latter investigators and other workers. It appears, however, that Pindak et al[19] also favored the RK_{13} (epithelial) cells which, "because of their hardiness and greater tolerance for the low pH associated with the proliferation of *T. vaginalis* . . . can be maintained for longer periods without spontaneous degeneration of the monolayer."

Bremner et al.[31] One cloned laboratory strain and recent clinical isolates with a murine macrophage-like cell line (P338D), mouse myeloma cell line (P3-X63-Ag8-653), primary chick embryo cell cultures, and Chinese hamster ovary cells (CHO-K1) were employed by this group of Scottish experts.

Evidently, the four vertebrate cell culture types were affected in a similar manner by the urogenital trichomonads in all experiments. These kinds of cultures, two of which (P3-X63-Ag8-653, CHO-K1) consisted of epithelial cells, did not affect the progression of infection in terms of the numbers of parasites resulting from comparable inocula of different trichomonad strains and in the extent of destruction inflicted by these parasites on the cell cultures. However, "the upper and lower limits of the effective range [trichomonads overgrew only if they affected the target cells] were found to be dependent upon the trichomonad [strain] used."

Fukui and Takamori.[21] The studies were limited to experiments involving the pellet method. A single laboratory strain (4FM) of *T. vaginalis* and five cell lines: HeLa, Vero, LLC-MK$_2$ (Rhesus monkey kidney epithelial cells), FL (human amnion epithelial cells), and L_{929} were used. All cells were labeled with ^{51}Cr. In the trichomonad-mammalian cell pellets (most of which included HeLa cells), the ^{51}Cr release was time-dependent, temperature-dependent, and parasite dose-dependent, the radiolabel release reaching its maximum in 3-hour incubation with 3.2×10^5 and 2.5×10^6 organisms at 37°C and 30°C, respectively. The mammalian cell lines were ranked as follows with regard to their sensitivity to the trichomonads: HeLa (epithelial), FL (epithelial), L_{929} (fibroblasts), Vero (fibroblasts), and LLC-MK$_2$ (epithelial). As reported by some previous workers,[10,16,33] epithelial cells appeared to be the most susceptible to injury by *T. vaginalis*. Indeed, the susceptibility would be anticipated, since the urogenital trichomonads remain in close contact with the epithelial cells in the urogenital system of their human hosts. Still, in other instances[19,31] non-epithelial cell types were also found to be sensitive to the cytopathic effects of *T. vaginalis*. Only additional experiments using different strains of the urogenital trichomonad and a variety of vertebrate cell cultures can help to establish unequivocally the superiority of epithelial cells in studies of parasite-culture cell interactions in vitro.

Effects of Various Factors on Adherence to, and Cytopathogenicity of T. vaginalis for Cell Cultures and on Attachment of Cell Cultures to Substrates

Effects on Adherence and Cytopathogenicity

*Inherent Virulence.** An extensive account of the pathologic changes caused in chick liver

---

*The examples cited in this section have been reviewed in more detail, albeit with different emphasis, in some of the preceding sections.

monolayer cell cultures by a relatively virulent and a very mild (nearly avirulent) strain, as estimated by clinical and pathologic findings and confirmed by the subcutaneous mouse assay[34] was already given[5] (see Figs. 9.1 to 9.4). It is evident from this account that the effect of the two strains on cell cultures was quite different. The pathologic changes caused by the virulent parasites in all cell types present in the chick liver monolayers appeared much sooner and were far more extensive than those observed after inoculation of the mild-strain trichomonads. Distinct differences in abnormal changes caused in chick embryo fibroblast cultures by *T. vaginalis* strains isolated from women with diseases of the urogenital tract associated with pathologic manifestations differing in gravity were reported subsequently by Toś-Luty et al.[12] Brasseur and Savel,[14] who used 25 noncloned strains and 16 cloned substrains derived from these strains with McCoy cell line monolayers, were able to differentiate "quantitatively" (using a "virulence index" devised by them) among the virulence levels of their isolates. There was a statistically significant correlation between pathogenicity of 19 of the 25 strains and 4 clones as evaluated by the intraperitoneal mouse assay (see Chap. 8), modified by Brasseur and Savel,[14] and as assessed with the aid of cell cultures.

Krieger et al[11] studied pathologic effects on CHO-K1 epithelial cell cultures of four *T. vaginalis* strains isolated from symptomatic patients (see footnote u to Table 9.1) and of two strains from asymptomatic women. Although all the parasite strains employed by these investigators eventually destroyed the monolayers, the kinetics of destruction were different for the several strains. The largest differences were observed at four hours after inoculation at which time the mean percent of monolayer destruction by the trichomonad strains isolated from patients with "clinical vaginitis" equalled 65.8 ± 17.0 (SD), while the mean for the strains from asymptomatic women was 26.7 ± 5.8, the two mean values being statistically different at the 2% level ($P \leq 0.02$). The results of the cell culture test correlated very closely with those obtained with the aid of the quantitative β-hemolysis assay ($r = 0.94$) and to a lesser degree with that of the subcutaneous mouse assay as modified by Krieger et al[11] (for remarks on this modification, see Chap. 8).

The authors did not rank the virulent strains by any of the laboratory assays, because they felt that "objective standards for amount and character of discharge and for inflammation of vaginal mucosa are not available [and that] a patient's perception of the severity of her symptoms is also modified by highly subjective, nonquantifiable, psychological factors. . . ." Therefore, clinical severity is difficult to evaluate and its "correlation with any test has to be considered with caution." Evidently, for the study of Krieger et al[11] the subjects were chosen because they presented with "extremes of the spectrum of clinical disease and could be easily placed into the two groups." They suggested also that "correlation with disease of intermediate severity would be more difficult" to demonstrate by a laboratory assay. The foregoing assertions are supported by the results reported by some investigators, but not by others; certain groups of workers consider pathologic manifestations more meaningful than clinical ones (see preceding parts of this Chap.; also Chaps. 8 and 14).

Bremner et al[31] were able to assess virulence of *T. vaginalis* strains on the basis of their effects on four types of vertebrate cell cultures (see footnote mm to Table 9.1). Evidently "the upper and lower limits of the effective range were found to be dependent upon the trichomonad line used." Two parameters were considered to be of special importance: the parasite:vertebrate culture cells ratio and the length of time required for destruction of the cell cultures.

Prolonged In Vitro Cultivation. Sixteen *T. vaginalis* strains isolated from patients suffering from "urogenital diseases" and HEp-2 cell cultures were employed by Zhordaniia et al[33] in a series of experiments. Strains grown in culture for 1 or 6 months caused destruction of 30% (of 1,000) HEp-2 cells; those cultivated for 1 year or longer had a detrimental effect on only 10% (of 1,000) mammalian cells in culture. A correlation between the length of time in vitro of *T. vaginalis* strains and attenuation of their virulence for cell cultures was observed also on many occasions by Honigberg and his coworkers (unpublished data).

According to Rasmussen et al,[15] freshly isolated *T. vaginalis* strains, one from a case of "severe vaginitis" (Fig. 9.9a–c) and another of unknown origin, had a different appearance from an established laboratory strain (cultivated axenically for many years) (Fig. 9.9d–f) on being inoculated into first-transfer cultures of human vaginal epithelium. The fresh isolate from a case of "severe vaginitis" also caused much more damage to the human cell cultures than the laboratory strain. On a scale of 0–4 units reflecting the extent of cell culture destruction (0–100%), in the presence of 10^4 trichomonads per tissue culture well, the freshly isolated trichomonads exerted cytotoxic effects rated at 3–4 units, while the destruction caused by the established laboratory strain was estimated at no more than 1 unit. Also Bremner et al[31] demonstrated that prolonged in vitro cultivation of the urogenital trichomonads resulted in attenuation of their virulence.

In assessment with the aid of mammalian cell cultures of the relative virulence of *T. vaginalis* strains and in differentiation between the effects of freshly isolated and established laboratory lines using such cultures, the results of Alderete and his coworkers[10,16] appear to be at variance with those of many other investigators. Alderete and Pearlman[16] stated: "Furthermore isolates from infected women but grown for less than one week in our laboratory were similarly evaluated for their cytopathogenic potential using the [mammalian cell culture] assay system. Results identical with those observed for strain [NYH] 286 were obtained." In a subsequent publication Alderete and Garza[10] indicated that, when tested for their ability to attach to HeLa cells, NYH 286 parasites typically gave a P/C ratio of 0.74 and isolate JH31A was characterized by a higher P/C value (1.0). "Two fresh isolates, JHHR and RU 375, were less efficient in their ability to parasitize cell surfaces, yielding P/C ratios of 0.64 and 0.46, respectively." According to these workers,[10] *T. vaginalis* strain NYH 286 parasites were not seen to exhibit "ameboid motility" and did not penetrate the cell culture layers or become trapped under the monolayers.

It should be noted that strain NYH 286 is a laboratory strain which had been cultivated for a long time in Dr. Müller's laboratory at The Rockefeller University, until it was cryopreserved (Müller, personal communication). Certain investigators who tested this strain in cell culture systems consider it rather atypical in that it does not tend to become ameboid and, indeed, does not crawl under the cell monolayers, an activity typical of most, especially more virulent, strains (see Figs. 9.7c; 9.8a, b). The behavior described for NYH 286 cells appears similar to that characteristic of originally mild strains[5] (Fig. 9.4b–d) or of established laboratory strains grown in culture for many years, for example, 1711[15] (Fig. 9.9d–f). The large initial inoculum of 2×10^5 organisms used to cause monolayer destruction suggests that the NYH 286 parasites were rather mild when used by Alderete and his collaborators. This is not surprising since, as indicated by Rasmussen et al,[15] destruction of culture cells appears to be predicated on very close contact between the membranes of the ameboid parasites and of the epithelial cells (Fig. 9.6b). Perhaps less damage is inflicted on the host cells by strains which, even when employed in large numbers, tend to adhere to the epithelial cells on relatively few occasions, and, when they do, their ameboid shape is not obvious (e.g. Figs. 9.4c, d; 9.9d–f). It is not clear also why infections of mammalian cell cultures with JH31A strain trichomonads, which are among the mildest stocks, as well documented by the medical records of the patients from whom they were isolated and demonstrated by the subcutaneous mouse assay,[34] were characterized by a rather high (1.0) P/C ratio.[10] [As correctly asserted by Dr. Alderete, there are no truly avirulent strains of *T. vaginalis*.]

Nothing is known about the medical history of the woman donor of strain JHHR, nor is there any record of the number of serial in vitro passages before it was used in the experiments reported by Alderete and Garza.[10] We know, however, that RU 375 is a laboratory strain that was cryopreserved after having been maintained in culture for a long time in Dr. Müller's laboratory. Virulence of this strain could have been, and probably was, attenuated by the prolonged in vitro cultivation. The foregoing information notwithstanding, it can be assumed on the basis of the statements published by Alderete and his coworkers[10,16] that the "fresh isolates" grown in their laboratory for less than a week were RU 375 and JHHR.

Activity of Living Parasites. As might have been expected, only living parasites adhered to cell cultures and caused pathologic changes in such cultures. Glutaraldehyde- or formalin-fixed trichomonads exhibited neither of these activities.[10,16,23,35] Furthermore the extent of adherence was dependent on the parasite concentration and on temperature.[10,16,20,23,35] Alderete and Garza[10] noted only 22% of adherence to HeLa cells at 4°C when compared to that at 37°C, while according to Silva Filho and Souza[23] no trichomonads adhered to MDCK cells at 4°C. After HeLa cells with trichomonads attached at 37°C were placed at 4°C, many of the flagellates detached, the number of parasites that continued to adhere to the target cells being reduced by about 80%.[10] These results suggested to the authors[10] "that trichomonad metabolism and membrane fluidity are important for *T. vaginalis* adherence to host cells."

Apparent Specific Receptor Sites. On the basis of their results which demonstrated "stoichiometric competition of attachment of ... [³H]thymidine-labeled *T. vaginalis* NYH 286 by unlabeled homologous and RU 375 heterologous isolates," Alderete and Garza[10] suggested that adherence of the trichomonads to HeLa cells occurred at specific receptor sites, being "mediated by specific receptor-ligand reactions." Some time earlier, Brasseur and Savel[14], taking into account the situation observed in *Entamoeba histolytica* and *Giardia lamblia,* hypothesized that adherence of *T. vaginalis* to host cells, which constituted an important step in the process whereby this parasite expressed its virulence potential, depended on the presence of glycoprotein membrane receptors. Some of the observations of Martinotti et al[35] on trichomonad adherence inhibition by bacteria (discussed in the next section) appear to be of interest in this connection.

Bacteria. According to Zhordaniia et al,[33] in the presence of the accompanying microflora ("principally staphylococci") at a 10^7 concentration, *T. vaginalis* strains (isolated from cases of "urogenital diseases") and cultivated for up to 1 or 6 months before being inoculated into HEp-2 cell cultures at a concentration of about one parasite to 100 mammalian cells, destroyed 80% or 50% of these cells in 6 to 12 hours after infection. In the absence of bacteria, the same trichomonad strain, maintained for the same periods in culture prior to inoculation in the mammalian cell monolayers, destroyed only about 30% of the monolayers. No pathologic changes were noted in cell cultures incubated with the staphylococci alone.

Quite different results were obtained by Martinotti et al[35] who used *T. vaginalis* strain D isolated from a case of "acute trichomoniasis" and cultivated for about six or seven serial transfers in Diamond's[22] medium without agar before being inoculated into human amnion (WISH) cell monolayers. Adherence of the parasites to WISH cells was inhibited by preincubation with *Lactobacillus fermentum* or *Streptococcus agalactiae* (group B streptococci), both obtained from vaginal isolates. The lactobacilli appeared to be more inhibitory than the streptococci, and with both bacterial species the inhibition was concentration-dependent. Also, with both bacteria, adherence inhibition was lower when they were preincubated together with the trichomonads during the adherence assay. In the case of streptococci, pronounced inhibition was noted only with very high concentrations of the bacteria.

The discrepancies between the results obtained by Zhordaniia et al[33] and Martinotti et al[35] can perhaps be explained by those between the bacteria they employed in their trichomonad adherence experiments. The latter workers[35] interpreted their observations in terms of competition between the bacteria and trichomonads for receptor sites on WISH cell surfaces. Since maximum exclusion of *T. vaginalis* was achieved by preincubation with the lactobacilli or streptococci rather than by coincubation of the flagellates and bacteria with the mammalian cell cultures, Martinotti and coworkers[35] felt that "a priority of colonization may be a crucial factor." Furthermore, on the basis of the data obtained by others with bacteria and mycotic agents (for relevant studies, see reference 35), they suggested that size might also play an important role in competition of different organisms for a particular site.

Anticytoskeleton Compounds. Rather extensive studies were conducted on the effects of anticytoskeleton compounds (anti-microtubule and anti-microfilament) on cytadherence of *T. vaginalis* to mammalian cell cultures,[10] on cytotoxicity of this parasite, ultimately resulting in destruction of the cell culture monolayers,[20] and on both cytadherence and mammalian cell destruction.[32] Among these were four microtubule inhibitors: colchicine,[10,20,32] vinblastine,[20] as well as mebendazole and griseofulvin[32]; one microtubule stabilizing agent, taxol[32]; and two microfilament inhibitors, cytochalasin (cyto) B[10,32] and cytochalasin (cyto) D[20]. As in many other instances dealing with *T. vaginalis*-vertebrate cell culture interactions, there are some conflicting results in the accounts published by different investigators.

According to Alderete and Garza,[10] pretreatment of the urogenital trichomonad for 30 min with 0.2 μg/ml of colchicine reduced cytadherence by 40%; pretreatment with 0.1 μg/ml of cyto B resulted in 20% reduction of parasite adhesion. The authors suggested that the effect of the cytoskeleton inhibitors could cause either "inefficient or improper" localization of ligands or nonspecific "perturbations and toxicity" of the parasite membrane unrelated to the events occurring in the process of adhesion. Also, according to Silva Filho and Souza,[23] cyto B inhibited adherence of *T. vaginalis* to MDCK cell cultures.

Krieger et al[20] reported that 1 or 10 μg of cyto D in Feinberg-Whittington (FW) medium (see footnote s in Table 9.1) without serum (NS) inhibited destruction of CHO cell monolayers. However, in the pellet system, only 80% inhibition was noted with 10 μg of cyto D, and none with 1 μg. The urogenital trichomonads (10^6/ml) preincubated for 2 hours in the same culture medium with 1 μM colchicine, then inoculated in this medium containing colchicine into CHO cell cultures failed to exert any cytopathic effect on the hamster cell monolayers after a 5-hour incubation. The parasites (10^6/ml) suspended in FW-NS medium containing 1 μM vinblastine caused only minor (16%) inhibition of cytopathic effects in CHO cell monolayers after 4- and 5-hour exposure, but no such effect on mammalian cell killing was noted in pellets. Krieger and his collaborators,[20] who presented a long and informative discussion of the entire problem, appeared to be uncertain as to whether the microfilaments played an important role in adherence of trichomonads to the mammalian cells or in the subsequent pathologic changes leading to destruction of these cells in monolayers and the pellets. These workers[20] cited electron microscope studies[25] of *T. vaginalis*-epithelial cell interactions in explanation of the need for intact microfilaments in trichomonad adhesion (see Fig. 9.6a). They entertained also several possibilities that might explain the effects of vinblastine on CHO cells, for example, reduced motility of the trichomonads caused by this microtubule inhibitor.

Juliano et al[32] used a freshly isolated *T. vaginalis* strain TO-37 and CLID mouse fibroblasts in experiments involving, in addition to the commonly used compounds, i.e. colchicine (250 μM, first dissolved in distilled water) and cyto B (210 μM)*, other pharmacologic substances known to affect the cytoskeletal components. Among these were mebendazole (34 μM) and griseofulvin (140 μM), found previously to be effective against *T. vaginalis* microtubules,[36] as was also taxol (10 μM), a microtubule "stabilizing agent."

As estimated on the basis of cytoplasm retraction seen with the aid of light microscopy and scanning electron microscopy, as well as on that of the mean numbers of fibroblasts per light-microscope field (Fig. 9.10), various compounds inhibited to differing degrees the pathologic changes in CLID cell monolayers, especially during early infection, for example, 2 hours after inoculation. Similar results were obtained with regard to cytadherence, the P/C ratios being always lower after exposure of the mammalian cells to trichomonads pretreated for 24 hours with the antimicrofilament or antimicrotubule compounds.

Cell cultures exposed to mebendazole- or griseofulvin-pretreated parasites had no obvious damage during the first 60 minutes after inoculation—no cell-free lesions could be seen in such cultures. The CPE-inhibitory effects of all the compounds, except for cyto B, decreased with time, for example, after a 6-hour exposure to mebendazole-treated trichomonads; the pathologic changes in CLID cells resembled those noted after exposure to untreated (control) flagellates. Evidently at 2 hours after infection mebendazole protected the mammalian cells most and taxol protected them least. It seems that the most lasting and progressively increasing CPE inhibition was exerted by cyto B (Fig. 9.10). However, during the initial 2 hours of infection this antimicrofilament compound had the least inhibitory effect on the mammalian cell culture destruction.

Although, according to Juliano et al,[32] it is possible that ". . . inhibition of [the] cytopathic effect . . . [by] anticytoskeletal compounds could be due to action on other cell targets," mebendazole and griseofulvin, the drugs most effective in CPE inhibition in the in vitro system, also had the most pronounced effect on growth and viability of *T. vaginalis*.[36] Further-

---

*All compounds, except for colchicine, were first dissolved in 0.1% concentration of dimethyl sulfoxide (DMSO). At this concentration, DMSO did not appear to affect viability of the parasites.

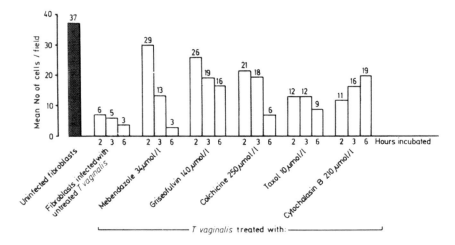

Fig. 9.10. Mean numbers of CLID mouse fibroblasts per microscope object field at × 200 magnification in cultures of these fibroblasts incubated for 2, 3, or 6 h with *T. vaginalis* strain TO-37 (originally isolated from a patient with "acute vaginal trichomoniasis") untreated or pretreated with the anticytoskeleton compounds indicated in the histogram. (Reproduced with permission of the British Medical Association, from Juliano C et al: Action of anticytoskeletal compounds on in vitro cytopathic effect, phagocytosis, and adhesiveness of *Trichomonas vaginalis*. Genitourin Med 63:256-263, 1987.)

more, none of the compounds employed in the CPE inhibition experiments affected protein or nucleic acid (DNA and RNA) synthesis.[32] Finally, since the inhibition of CPE in the in vitro system by the compounds acting on the cytoskeleton was found to be reversible and their effect on the cytoskeleton of the trichomonads was not,[36] "pathogenicity of *T. vaginalis* . . . could have various components, some of which might depend on the cytoskeleton, and others, which . . ." start acting later in the progression of the pathologic changes and depend on other mechanisms.[32]

More work needs to be done with the anticytoskeleton compounds before their role in pathogenicity of *T. vaginalis* is better understood. It seems that, as indicated by Krieger et al,[20] cytochalasins, especially cyto D, have strong inhibitory effects on the CPE caused by the urogenital trichomonads in in vitro systems. This conclusion is reinforced by the finding that microfilaments, with which become associated adhesive proteins, for example, fibronectin, in the process of cell adhesion[37] (for trichomonads, see reference 25), play an important role in cytadherence and cytopathogenicity of these parasites. However, in light of the results of other workers (e.g. reference 36), the importance of microtubules in these processes cannot be dismissed.

Proteases. To test whether surface proteins of *T. vaginalis* were involved in the apparently "specific receptor-ligand reactions" (see section on *Apparent Specific Receptor Sites*, p. 185), Alderete and Garza[10]* treated the parasites (10^7 organisms), especially of strain NYH 286, with either trypsin or pronase before incubating them primarily with HeLa cell monolayers. Treatment with trypsin reduced the trichomonad adherence by 40–60%; pronase reduced the attachment by over 90%, as estimated by [^{3}H]thymidine radioactivity release. Of interest was the observation that protease treatments did not impair motility of the flagellates. The trypsin effect was reversible, the ability to attach to HeLa cells being restored to the trichomonads which, after exposure to this enzyme, were incubated in TYM medium for 1 or 2 hours—during this period the flagel-

---

*In all experiments detailed in this section Alderete and Garza[10] employed HeLa cell cultures and probably the NYH 286 strain of *T. vaginalis*, which is a nonameboid laboratory strain that has been maintained in culture for prolonged time by serial transfers.

lates evidently resynthesized the proteins. Cycloheximide, which by itself had but little effect on adherence, when used in combination with trypsin, eliminated the reversibility of inhibition by this enzyme of parasite adhesion to the cell monolayers.

The foregoing findings, in combination with the experimental results showing "that the rapid, saturable trichomonad adherence [to HeLa cells]" was competitive and specific (see section on *Apparent Specific Receptor Sites,* p. 185.), suggested that the antigenic surface proteins were the parasite adhesins participating in "adhesin-receptor type reactions," involved in attachment of *T. vaginalis* to cell cultures[10]. Four such putative protein adhesins were identified in strain NYH 286 using the parasite-HeLa cell culture system.[38,38a] These proteins, removable by trypsin treatment of living trichomonads, "were not synthesized by a non-adherent (Adh⁻)" subpopulation that was isolated,[38a] but were synthesized by "revertent adherent (Adh⁺)" parasites derived from the Adh⁻ population.[38a]* An antibody to one of these four proteins inhibited adhesion of the flagellates to HeLa cells.

According to Silva Filho and Souza,[23] the addition to *T. vaginalis* strain Jt-MDCK cell line (see footnote kk to Table 9.1) culture system of protease inhibitors, such as TLCK (tosyl-lysine chloromethyl ketone), PMSF (phenyl methyl sulfonyl fluoride), TPCK (tosyl-phenol alanyl chloromethyl ketone), iodoacetic acid, leupeptin, chymostatin, and antipapain, caused a significant reduction in the CPE exerted by the parasite on the mammalian cells. Inhibition of pathogenicity of *T. vaginalis* G3 clone for the P3-X63-Ag8-653 cell line (see footnotes mm, nn to Table 9.1) by leupeptin, an inhibitor of cysteine proteases, was reported also by Bremner et al.[31] Since leupeptin did not inhibit *T. vaginalis* grown in Diamond's[22] medium, these authors emphasized the usefulness of mammalian cell cultures in screening drugs against the urogenital trichomonads.

Protease activity of "crude lysate supernatants" (CLS), which "parallels the production of *Trichomonas vaginalis* CPE," was investigated by Ortega et al,[18] who used BHK cell line (see Table 9.1) in their experiments. They found that: (1) with azocoll as the substrate, cysteine hydrochloride dithiothreitol enhanced this activity; (2) iodoacetamide, PMSF, and Ep 495 inhibited it; and (3) pepstatin exerted no effect on the trichomonad protease. Their results suggested to Ortega et al[18] that the enzyme in question was a thiol protease.

Metabolism Inhibitors. Iodoacetate (0.5 mM), a glycolysis inhibitor (see also reference 23), and metronidazole (10 µg/ml) (for action of this drug, see Müller[39] and various references in his review) were used for treatment of the trichomonads (2×10^6) at 37°C for 30 minutes.[10] "Cellular integrity" was ascertained by the trypan-blue exclusion test and by radioactivity released by the [³H]thymidine-labeled parasites. The radiolabeled flagellates, suspended in the parasite-cell culture medium, were added to HeLa cell monolayers. In other instances, the urogenital trichomonads were incubated with the human cell cultures for 30 minutes. Thereafter, the monolayers were washed and covered with the homologous medium containing either of the metabolism inhibitors. Both compounds either inhibited the trichomonad adherence by over 95% (in the first kind of experiments) or caused detachment of nearly all parasites from the HeLa cells (in the second kind of experiments). To prove that the mammalian cells were unaffected by iodoacetate or metronidazole, live trichomonads were added to the monolayers from which previously the flagellates were released by treatment with either of the two inhibitory compounds—normal adherence resulted. Evidently, attachment of the parasites to mammalian cells in vitro was dependent on active metabolism of the trichomonads.

Carbohydrates. Only negative results with regard to adherence inhibition of *T. vaginalis* to HeLa cells were obtained by Alderete and Garza[10] in experiments using the following carbohydrates (singly or in combination): *N*-acetylglucosamine, α-methylmannopyranoside, glucose, fructose, sucrose, and maltose (at 250 mM final concentrations). The trichomonads, suspended in the medium (see Table 9.1) containing these sugars, were added to the cell monolayers. Furthermore, the urogenital parasites [2×10^6 in phosphate-buffered saline (PBS)] exposed to neuraminidase (10 µg) for 30 minutes or to periodate (10 mM) for 10 minutes were not

---

*Reference 38a, the complete report containing further details about the proteins thought to act as adhesins, was added in proof. For details, the reader should consult this reference.

affected in their capacity to adhere to HeLa cells treated in the same manner with the two compounds. According to the authors,[10] all the results provided further support for the lack of involvement of carbohydrates in attachment of the trichomonads to mammalian cell cultures. However, Silva Filho and Souza[23] found neuraminidase from *Vibrio cholerae* to render *T. vaginalis* strain Jt "more adhesive" to MDCK cell monolayers (see also The Question of Lectins, below).

Female Sex Hormones. Of interest were the results reported by Martinotti et al[35] in the WISH cell-trichomonad system in which the epithelial monolayers had been pretreated with female hormones, that is, α- and β-estradiol, estrone, estriol, and progesterone. Long incubation (18 hours) of WISH cells with 0.1 ng/ml of either estradiol was needed to enhance adherence of the parasites to the cell surfaces, the maximum level of this enhancement being recorded at the 100 ng/ml concentration of the hormones. An initial rise in the numbers of *T. vaginalis* attached to WISH cells was observed at 1 ng/ml concentration of estrone and progesterone and these numbers peaked at a 100 ng/ml concentration. No effect on parasite adhesion was exerted by estriol. According to the authors,[35] the results support the hypothesis that changes in the levels of female sex hormones influence adherence of *T. vaginalis* to the epithelial cells in vivo, "thereby affecting colonization and infection" of the urogenital system epithelium.

Putative Immunologic Factors. All work on immunologic factors, said to constitute one of the pathogenicity components of *T. vaginalis*, was done by Alderete and his coworkers.[41-44] The activity of the factors, presumably responsible for the presence or absence of virulence among strains of the urogenital trichomonad, could be estimated by the length of time required for destruction of HeLa cell cultures.[43,44] It depended on the presence in the cell culture-parasite system of an antibody to a specific surface immunogen. This antibody had the attributes of the IgG2a-type monoclonal antibody (MAb), C20A3, prepared by Alderete et al,[42] to the surface immunogen. The reaction between the MAb and the parasites, resulting in lysis of the latter, was found to be complement-independent, but temperature-dependent and time-dependent.[41]

The antigen was demonstrated by various immunologic techniques using MAb C20A3 and, in some instances, a monospecific antiserum.[43] Its apparent molecular mass *(m)* was estimated at about 267 kDa.[42] Furthermore, certain chemical properties of this high molecular-weight antigen were ascertained with the aid of trypsinization as well as of radiolabeled carbohydrates and lectins (Con A and wheat germ agglutinin). The results indicated the glycoprotein nature of the immunogen.

Of special interest to the present discussion are the enzyme-linked immunosorbent assays (ELISA) using whole cells and detergent extracts of trichomonads (for relevant references, see reference 42) and, above all, the indirect fluorescent antibody test (IFAT), performed with living flagellates at 4°C and, therefore, specific for surface antigens. Also, fluorometry and fluorescence-activated cell sorting (FACS) (as given in reference 43) were employed in the investigations to be summarized briefly below.

Alderete and his coworkers[41-44] reported the following results:

(1) It was demonstrated by ELISA using detergent extracts that trichomonads which were resistant to lysis even in the presence of MAb C20A3 concentrations exceeding many times that found usually effective[41] and in which the 267 kDa antigen could not be demonstrated by the whole-cell ELISA, still possessed this immunogen.[42] In such instances, the large glycoprotein probably was not located on the surface or else was present in only very small quantities.[43] (2) Populations of *T. vaginalis* strains, including fresh isolates,[44] consisted mainly of two phenotypes (negative and positive). One phenotype, resistant to MAb C20A3-mediated lysis and nonreacting with this MAb (as shown by IFAT), was referred to as neg (MAR$^-$); it could constitute the entire population or a part thereof, that is MAR$^-$ subpopulation. The other phenotype, sensitive to MAb C20A3-mediated lysis and reacting with the MAb, could constitute only a part of a population; it was styled as pos (MAS$^+$). There were also subpopulations that, although resistant to the antibody-mediated lysis, reacted with MAb C20A3; they were the pos (MAR$^+$) phenotypes. (3) Established laboratory strains used in the study[43] and many fresh isolates[44] homogeneous for the neg phenotype did not undergo phenotypic variation even on prolonged (e.g. a 4-year period for strain RU 375) in vitro cultivation; however, no stable pos-phenotype populations

were found among the trichomonad strains and no clones of this phenotype could be established.[43,44] (4) Heterogeneous strains, noncloned and cloned, when analyzed over extended periods (several weeks, but typically not for a few days[44]), showed pos-to-neg and neg-to-pos phenotype variations. Alderete et al[43] stated that "... the rate of ... disruption of cells in monolayer cultures ... was influenced by the phenotype of the trichomonads. All pos phenotype clones and subpopulations (MAR$^+$) yielded diminished levels of host cytotoxicity. On the other hand, trichomonal populations with neg phenotype ... were effective in host cell killing ... Finally, in all cases, a change in phenotype from pos to neg paralleled the enhanced ability of trichomonads to ... disrupt cell monolayers." Actually, it was shown that HeLa cell monolayers were destroyed by neg-phenotype parasites by 16 or 18 hours of incubation. However, "pos phenotype clones and subpopulations of *T. vaginalis* NYH 286 could ... [kill HeLa cells] after extended incubation times [i.e. 24 hours—see Fig. 9 in reference 43] ... No changes in phenotype occurred during this prolonged incubation..."[43] It seems that a difference of 6 hours incubation need not be significant. The process of phenotype-controlled cytopathogenicity is not entirely clear. Much work needs to be done before this process, which showed no correlation with the symptoms found in the patients from whom the parasites were isolated,[44] can be adequately explained.

Of general interest are the recent findings of Wang et al[45] based on the examination of 28 *T. vaginalis* isolates, which included established laboratory strains as well as fresh isolates. The strains were analyzed for their phenotype as determined by their resistance to and reactivity with the C20A3 MAb as well as for the presence of viral double-stranded RNA (dsRNA). Only the heterogeneous trichomonad populations containing both the pos and neg phenotypes (as determined by fluorescence) and able to undergo phenotypic variations were found to contain dsRNA. The isolates and subpopulations of the heterogeneous phenotypes which retained dsRNA continued as heterogeneous populations; however, loss of this ribonucleic acid on prolonged cultivation was paralleled by disappearance of the surface disposition of the 267 kDa antigen. These events were construed to implicate the viral dsRNA in the sequestration of the immunogen into the trichomonad membrane.[45] The observation of the correlation between the loss of the dsRNA and of the expression of the 267 kDa antigen as the result of prolonged cultivation would lead to an unexpected conclusion—the parasite virulence would be enhanced in the process of cultivation, an occurrence that is at variance with much of the available information on *T. vaginalis*–host relationship.

Other Factors. Of interest with regard to *T. vaginalis* surface glycoproteins is the abstract of Silva Filho and Souza.[40] Evidently, specific receptors for laminin (LAM), a molecular mass *(m)* 1000 kDa glycoprotein known to be involved in cell adhesion, are widely distributed on the surface of *T. vaginalis*. These receptors were demonstrated by immunologic (polyclonal and monoclonal antibodies), immunochemical [sodium dodecyl sulfate-polyacrylamide gel electrophoresis (SDS-PAGE) followed by immunoblotting], and immunoelectron microscopic (colloidal gold-LAM complexes) methods.

Difluoromethylornithine (DFMO), which inhibits ornithine decarboxylase, was reported by Bremner et al[31] to prevent *T. vaginalis* clone G3 from causing pathologic changes in a mouse myeloma cell line (see footnotes mm, nn to Table 9.1). The cytopathogenicity-suppressing activity of DFMO could not be assayed in axenic cultures of the urogenital trichomonad; thus the value of the cell culture system in drug testing was emphasized by the authors.

Using a parasite-cell culture system involving a fresh isolate of *T. vaginalis* strain D radiolabeled with [^{3}H]thymidine and WISH cell monolayers, Martinotti et al[35] tested the effect of fetal bovine serum (FBS) and concanavalin A (Con A) on adhesion of the trichomonads to the mammalian cell monolayer. In some instances the presence of FBS in the RPMI 1640 medium used for experiments appeared to increase the numbers of parasites adhering to WISH cells (see also reference 23). Some WISH cell cultures were pretreated with Con A (10 mg/ml) or with a mixture of this lectin and α-D-methylmannoside (30 mM). Con A alone inhibited trichomonad adhesion to the mammalian cells by about 10.4%, but no changes in the levels of trichomonad adherence were noted when cells pretreated with a mixture of the lectin and sugar were used.

Effects on Attachment of Cell Culture Monolayers

Pindak et al[19] reported activity of a "cell detaching factor" (CDF) in polyethylene glycol-concentrated filtrates of the fluids from *T. vaginalis*-cell culture (involving primarily McCoy, but also HEp-2 and RK-13 lines) systems in GMP (see footnote jj to Table 9.1). The CDF was prepared according to the method summarized in Table 9.1 (footnote ii). Control suspensions were obtained from cell cultures inoculated with heat-killed trichomonads or from noninfected cell cultures by the method used for the experimental preparations. Cell culture-parasite systems involving six *T. vaginalis* strains known to differ in virulence for women and experimental animals (mice), as well as for mammalian cell cultures, were found to produce CDF. The experimental suspension at 1:10 dilutions inhibited monolayer formation on incubation for 24 hours at 37°C in cultures of six cell lines (HeLa 229, McCoy, HeLa, HEp-2, L_{929}, BHL_4). At 1:20 dilution, only HeLa 229 cells were completely blocked; of the remaining five cell lines, only BHL_4 was inhibited significantly less than the others. The other 10 cell lines used by Pindak et al[19] (see section on preference of *T. vaginalis* for certain cell culture types, p. 181; also LLC-MK) were unaffected by the CDF; they formed normal monolayers in the presence of this factor. Freshly prepared HeLa 229 cell suspensions added to supernatant fluids obtained from cultures maintained with CDF for 1 week failed to attach, but those added to supernatant fluids from cultures grown in the absence of CDF formed normal monolayers. These results suggested to the authors[19] that ". . . the observed effect was not due to the depletion of nutrients or to some metabolic product of normal cells." It is evident also from the data that the effect of the CDF on mammalian cells is reversible.

According to Pindak et al,[19] CDF activity: (1) could not be demonstrated in fluids from cell culture-trichomonad systems containing heat-killed parasites; (2) was not observed in preparations obtained from systems including heat-killed or ethanol-fixed McCoy cells in which the flagellates merely survived; (3) could not be found in freeze-thawed supernatant fluids from cell culture-trichomonad systems; (4) appeared not to be correlated with virulence levels of the parasite strains; (5) was associated with moieties having molecular mass greater than 100 kDa, as estimated by selective ultrafiltration; (6) was associated with thermolabile (60°C for 1 hour) molecules that were sensitive to pH below 5; (7) for heretofore obscure reasons, was produced by the trichomonads in the presence of only some types of mammalian cell lines.

One must agree with Pindak and his collaborators that far more study is needed before the nature and significance of CDF can be explained and related to human trichomoniasis.

The Question of Lectins

A comparison between the progression of cytopathologic changes caused by *T. vaginalis* in vertebrate cell cultures, as described at the beginning of this chapter, and those reported for *Entamoeba histolytica* in such cultures (CHO cell monolayers),[46] reveals many similarities between the CPE exerted by the flagellates and by the amebae. In light of these similarities, one would expect to find some lectin or lectins on the surfaces of the urogenital trichomonads similar to those that have been proved to mediate adherence of *E. histolytica* to mammalian culture cells in vitro and thus to contribute to lysis of these cells (e.g. references 46–49).

No lectin-like compounds have yet been reported from *T. vaginalis,* although two such substances were identified in *Tritrichomonas mobilensis* Culberson, Pindak, Gardner and Honigberg, 1986, an intestinal parasite of squirrel monkeys. One of them was a hemagglutinin (especially effective with monkey and human red cells) found by Pindak et al[50,51] in supernatant fluids of early subpassages of active trichomonad cultures; it was thought to be "a metabolic product of the organism rather than an integral cellular component."[50] The agglutinating capacity of this compound was not inhibited by ten different sugars commonly bound by lectins. On the basis of a 50-fold activity increase in retentate obtained by passing the supernatant fluid through a membrane with a cutoff of 100 kDa, the molecular mass *(m)* of the lectin-like substance was estimated to be at least 100 kDa. It was inactivated completely by incubation at 70°C, but only partially by an 18-hour exposure to 56°C; also, it remained stable when kept for 4 hours at room temperature in an environment with pH ranging from 4.0 to 9.0. The hemagglutinin was degraded by prolonged treatment with protease K, but was affected much less by trypsin or collagenase; exposure to papain increased the hemagglutination titer of the lectin-like substance. Supernatant fluids from *Trichomonas vaginalis* and *P. hominis* cultures did not contain any substance resembling the hemagglutinin reported from *Tritrichomonas mobilensis*.

More recently, Demeš et al[51a] described an adhesin with lectin properties from *Tritrichomonas mobilensis*. CHO cell suspensions and hemagglutination were employed in assays of the kinetics and nature of cytadherence of the intestinal trichomonad. Attachment of the parasites to CHO cells was concentration- and time-dependent, and attachment and hemagglutination were inhibited by sialic acid and sialyllactose; a number of other sugars had no effect. Exposure of the parasites to low (4°C) temperature or trypsin treatment did not reduce the ability of the flagellates to attach to the mammalian cells. Thus there is on the surface of *Tritrichomonas mobilensis* a lectin with specificity for sialic acid. In this respect the intestinal trichomonad resembles *E. histolytica*, which, however, was shown to have galactose- or *N*-acetylgalactosamine-binding lectin.[47,52]

It seems that the marked glycosylation of *Trichomonas vaginalis*[53,54] and *Tritrichomonas mobilensis*[51a], as demonstrated with the aid of fluorescein-conjugated lectins in the quantitative fluorescent antibody method and by agglutination of the parasites using various lectins, need not be related to the presence of lectins on the surfaces of these species. It is likely that with regard to the urogenital trichomonad, at least, the carbohydrate residues of the surface glycoproteins react with lectins and that some differences in the reaction levels depend on those in the relative amounts of these residues.[42]

Although on the basis of the results they obtained with several sugars, neuroaminidase, and periodate (see Effects of Prolonged In Vitro Cultivation, above), Alderete et al[55] concluded that carbohydrates were not involved in pathogenicity of *T. vaginalis* for mammalian cells, they did not dismiss completely such a possibility. Indeed, we should continue our search for participation of surface sugars in the parasite-vertebrate cell culture interactions. It will be remembered, for example, that pretreatment of WISH cell monolayers with Con A inhibited trichomonad adhesion, but that no significant changes in the adherence levels were observed when these mammalian cells were pretreated with a mixture of Con A and α-D-methylmannoside.[35] As far as the role played in virulence of *T. vaginalis* by sugar residues present on the surfaces of the flagellates is concerned, Choromański et al[54] reported that "more binding sites for soybean agglutinin (SBA) were found on the virulent than on avirulent strains." The sugar residues apparently associated with virulence were D-lactose residues.

Interactions Between *T. vaginalis* and Lymphocytes, Monocyte-Macrophage Lineage Cells, and Polymorphonuclear Neutrophils

Numerous investigators reported immune responses of a variety of white blood cells and tissue macrophages to *T. vaginalis*. Yet, as pointed out by Mason and Patterson,[56] "Despite these responses, chronic infection with the parasite is common, and immunity to re-infection is poor."

Lymphocytes

Mitogenic stimulation of T-lymphocyte population suppressor cell subsets by extracts of certain protozoa, for example, *E. histolytica*[57] and *Naegleria fowleri*,[58] has been reported to cause weak immune host responses. Furthermore, there is evidence for cell-mediated responses in *T. vaginalis* infections of humans (for references, see Mason and Paterson[56]; Honigberg[59]). It has been implied also that factors suggested for other protozoan infections might be responsible for the weak immune response in human trichomoniasis as well.[56] However, in light of the results of Yano et al,[60] stimulation of suppressor cell populations of T lymphocytes seems unlikely. These workers added to cultures containing peripheral blood lymphocytes (PBL) in RPMI 1640 medium (with fetal bovine serum, kanamycin, and 7-mercaptoethanol) equal volumes of soluble *T. vaginalis* strain C-1:NIH antigen or purified protein derivative of tuberculin (PPD) antigen. The cultures were incubated in 5% CO_2 in air for 3 to 6 days. At 16 to 24 hours before harvesting, [³H]methylthymidine was added to the cultures. The PBL from patients showed a high response to stimulation by *T. vaginalis* antigen as estimated on the basis of [³H]methylthymidine incorporation. However, the results obtained by Yano et al[60] also indicated that the soluble *T. vaginalis* antigen did not contain factors stimulating PBL mitogenesis; lymphocytes from uninfected persons responded to the PPD antigen but not to that of *T. vaginalis*. The responses of the PBL from persons infected with trichomonads to the soluble antigen from *T. vaginalis* or to the PPD antigen were nearly completely eliminated after treatment of these lymphocytes with anti-Leu 1 monoclonal antibody (MAb) or by anti-Leu 3a MAB; both MAbs were used with complement. However, the PBL responses were unaffected by treatment with anti-Leu 2a MAbs (also used with complement). It is evident from the foregoing results "that *T. vaginalis*-induced proliferation responses of PBL are mainly mediated by Leu 1 and Leu 3a-positive cells (helper/inducer T cells) in the PBL population."[60]

As far as can be ascertained, there is a single report (Martinotti et al[61]) preceding that of Yano et al[60]; it deals with cellular response to *T. vaginalis* as evaluated by stimulation of peripheral lymph-node cells and spleen lymphocytes. The soluble *T. vaginalis* antigen was prepared from organisms (strain AI) grown in TYM medium[22] for 12 hours; upon being washed and resuspended in Earle's Balanced Salt Solution (BSS) the trichomonads were sonicated, stirred overnight, and centrifuged. The supernatant fluid, which was dialyzed against Tris-HCl and had its protein content adjusted to 5 mg/ml, was freeze-dried and stored at $-40°C$ until used.

Female syngeneic Swiss/T mice received: (1) three 15-mg intraperitoneal inoculations (i.p.) of the parasite antigen; (2) a 15-mg i.p. inoculation plus one intravaginally (i.vag.) introduced ovule (each containing 15 mg of the antigen in paste) eight times on alternate days (a double dose of the antigen was administered 8 days later); (3) one i.vag. ovule 8 times on alternate days and two ovules eight days later.

Significant stimulation was noted in the PBL from mice inoculated via the i.p. route only; however, strong stimulation was noted also among spleen lymphocytes from mice receiving i.p. and i.vag. (ovules) for immunization. The authors[61] attempted to provide an explanation for the limitations of the lymphocyte reactivity, but they offered it quite tentatively.

The last full paper dealing with interactions between PBL and *T. vaginalis* was that of Mason and Patterson,[56] who set out to determine "whether human peripheral lymphocytes were responsive to secretory and cellular proteins of pathogenic and non-pathogenic strains of this protozoan." Cultures of one virulent and one mild *T. vaginalis* strain were initiated by placing vaginal swabs from women infected with these parasites into Diamond's[22] medium without agar. Virulence of the strains was determined on the basis of clinical manifestations in women infected with each, and by the employment of the subcutaneous mouse assay as modified by Mason and Forman.[62]

Overnight *T. vaginalis* cultures, washed with PBS (pH 7.3), were resuspended in this solution at a concentration of 2×10^7 organisms/ml and incubated at 37°C for 3 to 4 hours. The supernatant fluid obtained by centrifugation was passed through a 0.45 µm-pore diameter filter to remove all particulate material, then dialyzed overnight against distilled water at 4°C. The sedimented flagellates were resuspended in distilled water and disrupted by three consecutive freeze-thawings. The suspension was centrifuged, filtered, and dialyzed. After determination of the protein content of the secretory (TSP) and cellular (TCP) products, both preparations were freeze-dried and stored at $-20°C$ until used.

Heparinized blood samples from women harboring *T. vaginalis* and from noninfected persons found negative for antitrichomonad antibodies as demonstrated by IFAT, were layered onto Hypaque Ficoll and centrifuged. Thereafter, the mononuclear cell (MC) layer was extracted and washed with PBS before lysing the erythrocytes with ammonium chloride. The MC layer was washed again and suspended in RPMI 1640 medium supplemented with 20% fetal bovine serum and 10 mM Hepes buffer (pH 7.3). Viability of the mononuclear cells was established by the trypan blue exclusion test; 95% of cells remained viable. Stimulated MCs were tested in microtiter plates, using 10^5 MCs in 0.2 ml of medium. Upon addition in 50 µl of 12.5-100 µg/ml of TSP or TCP protein, the plates were incubated at 37°C. Control wells contained 10^5 MCs alone in 0.2 ml medium or in medium with phytohemagglutinin. For the final 18 hours of incubation [^{3}H]thymidine was added to each well. After incubation, cells were collected on glass-fiber discs, washed with PBS, and subjected to radioactivity (cpm) counting.

No increase of [^{3}H]thymidine uptake on incubation with TSP or TCP was recorded for lymphocytes from persons having no trichomonad infections. However, in the presence of either TSP or TCP, lymphocytes obtained from individuals infected with the urogenital trichomonads had a markedly increased thymidine uptake.

Increased [³H]thymidine uptake was noted after 3 to 4 days of incubating lymphocytes with TCP or TSP from virulent and mild parasites, maximum stimulation being achieved at 5 to 6 days of incubation. However, no significant differences in this uptake were noted between lymphocytes exposed to either of the antigens prepared from the virulent or mild strains of the urogenital parasite (Table 9.4).

The attempts of Mason and Patterson[56] to explain the lack of differences in reactions of the lymphocytes to *T. vaginalis* strains differing in virulence have been tentative, at best, and this question must be the subject of further studies. It is known, however, that stimulated lymphocytes can secrete substances enhancing motility and killing capacity of phagocytic cells. It has been demonstrated, for example, that monocyte-macrophage lineage cells[63] and polymorphonuclear leukocytes[64] are capable of killing the urogenital trichomonads in in vitro systems. Thus, stimulation of lymphocytes by *T. vaginalis* may well constitute an important factor in modulating the inflammatory reaction in infections involving this parasite.

Subsequently, Mason[65] observed that, in vitro, spleen cells obtained from mice infected with various *T. vaginalis* strains were stimulated with trichomonad antigens, the stimulation, that is, lymphoproliferative response, being measured by incorporation of [³H]thymidine. He reported that: (1) A marked lymphoproliferative response was obtained in spleen cells from mice infected with a virulent strain when these cells were exposed to the antigen prepared from the same strain. (2) A less marked response of the aforementioned spleen cells occurred on their exposure to an antigen derived from an avirulent strain. (3) Spleen cells from mice infected with an avirulent *T. vaginalis* strain had about equally strong responses to antigens prepared from both virulent and avirulent stocks. The aforementioned results suggested that while cell-mediated responses were evoked by both virulent and avirulent strains, some special antigens associated with virulent trichomonads appeared to enhance these responses.

A brief report dealing with the role of immune T lymphocytes in induction of macrophage activation for killing of *T. vaginalis* was published by Martinotti et al.[66] Resident peritoneal macrophages from unstimulated Balb/c mice were seeded in the wells of a microtiter plate. Next, T lymphocytes obtained from spleen cells of mice immunized with live *T. vaginalis* were added to the wells to yield in each a T lymphocyte:macrophage ratio of 10:1 or 100:1. Bovine serum albumin (BSA)-specific T lymphocytes were obtained in a similar manner for use in control experiments. [³H]Thymidine-labeled *T. vaginalis* were then added to the T lymphocyte mixture-containing wells, and the macrophage-mediated cytotoxicity was estimated 48 hours later by the release of the radiolabel (mean cpm) from the flagellates. In all instances, after 24 hours, a strong proliferative response of T lymphocytes from *T. vaginalis*-immunized mice was noted in the presence of the parasites.

It was evident from the results that T lymphocytes from animals immunized with the urogenital trichomonad induced a high level of parasite-killing activity in the resident macrophages. Significantly lower killing capacity for *T. vaginalis* was induced in the macrophages by T lymphocytes derived from mice inoculated with BSA. T lymphocytes themselves, obtained from mice immunized with *T. vaginalis* or BSA, did not lyse the parasites.

A similar experimental system was employed by Martinotti et al[66] to test the ability of human PBL to activate human monocytes for killing of *T. vaginalis*.

Macrophages and Monocytes

A report on natural cell-mediated toxicity (CMC) of macrophages or macrophage-like cells directed in mice against the primarily ex-

TABLE 9.4. Lymphocyte response to antigens of two *T. vaginalis* strains.*

Antigen	cpm (mean ± SE)	
	Virulent	Avirulent (mild)
Cellular	7817 ± 1040	7640 ± 1266
Secretory	6421 ± 672	7017 ± 1311

Reproduced from Mason PR, Patterson BA: Proliferative response of human lymphocytes to secretory and cellular antigens of *Trichomonas vaginalis*. *J Parasitol* 71:265–268, 1985, with permission of the American Society of Parasitologists.

*The results shown were obtained from experiments in which 10^5 lymphocytes (from four patients showing *T. vaginalis* infection) were incubated with antigens at a concentration of 50 µg/ml for 5 days.

tracellular parasite *T. vaginalis* was published by Landolfo et al.[67] A 48-hour microcytotoxicity in vitro assay using [^{3}H]thymidine-labeled trichomonad strain (D) isolated from "an acute human vaginal infection" was employed. Strain and age distribution of the mice from which the cells responsible for the cytotoxicity (effector cells) have been isolated were described, as was also the tissue distribution of these cells in their donors. Sixteen mouse strains served as donors of the effector cells, but many experiments were performed using BALB/c strain animals. Because of space limitations, the methods used for obtaining the effector cells, for aiding in their adherence to plastic surfaces, for their fractionation on nylon wool columns, for removal of the cells by feeding them with carbonyl iron powder and subjecting them to magnets, for irradiation, as well as for isolation and [^{3}H]thymidine-labeling of the trichomonads will not be summarized here. (For details of these methods see reference 67). On the other hand, the steps involved in the cytotoxicity assay will be briefly recapitulated. The cytotoxic activity was tested in microtiter plates using cells suspended (in progressively increasing concentrations) in RPMI 1640 medium with 2-mercaptoethanol, gentamycin, mycostatin, and heat-inactivated fetal bovine serum (FBS). The cell suspensions (0.1 µl) were added to 0.1 ml volumes of protozoan suspensions (2×10^5/ml). Spontaneous release of radioactivity (control) was determined by adding 0.1 ml of medium alone; it never exceeded 15% to 20% of the maximum release after 48 hours. Maximum release was estimated by adding 10% sodium dodecyl sulfate (SDS) in water; it equalled about $80 \pm 5\%$ of the total radioactivity incorporated, as estimated by solubilization of an equal number of radiolabeled flagellates with NCS directly in the scintillation vials. After 48-hour incubation at 37°C in 5% CO_2 in air atmosphere, the suspension was centrifuged and the radioactivity of the supernatant fluid was measured in a liquid scintillation spectrometer. The percentage of cytotoxicity was calculated with the aid of the following formula:

$$\% \text{ cytotoxicity} = \frac{\text{cpm of experimental release} - \text{cpm of spontaneous release}}{\text{cpm of maximum release} - \text{cpm of spontaneous release}}$$

The following results were listed by Landolfo et al:[67] (1) Normal BALB/c lymphoid cell preparations were cytotoxic at effector:target (E:T) ratios of 50:1, 25:1, and 12:1. (2) The tissue distribution of the natural cell-mediated cytotoxicity (CMC) revealed that the resident peritoneal cells (PC) without any "deliberate" stimulation or activation were the most cytotoxic. With regard to their cytotoxicity level they were followed, in sequence, by the cells from the lungs, spleen, and bone marrow. Peripheral blood, lymph node, and thymus cells had little, if any, detectable cytotoxicity. It should be noted that the peritoneal exudate cells obtained 4 days after i.p. inoculation of light mineral oil (LMO) caused a very high CMC. (3) Since it has been shown that serum factors can modulate CMC in some in vitro systems, PC cytotoxicity was tested in a medium supplemented with 10% heat-inactivated FBS, screened by the *Limulus* test for bacterial endotoxin contamination, or in 1% fresh adult BALB/c mouse serum. Since comparable PC cytotoxicity levels were noted, the idea that cytotoxicity for *T. vaginalis* was modulated by local environment in vitro could be eliminated. (4) The use of the erythrocyte-lysing buffer did not affect the toxicity levels of the effector cells. (5) Since it has been shown that natural CMC varied among mouse strains and that these differences were influenced by the genes found within and outside the H-2 complex, a comparison was made of the levels of natural CMC for *T. vaginalis* in BALB/c mice with its levels in mice differing in the H-2 complex and background genes. The natural CMC of normal PC from 8- to 10-week-old mice belonging to several strains showed large differences; however, completely negative results were never observed. Irrespective of their haplotype, the BALB background animals had high CMC (mean values ranging from 18 to 53%) at all the E:T ratios employed (Table 9.5). At E:T ratios of 25:1, the highest CMC levels were observed in BALB/c nu/nu mice (mean value = 49%). (For additional discussion of this entire question, see reference 67). (6) Effector cells from BALB/c mice ranging in age from 1 to 42 weeks did not show any obvious differences in the CMC levels. Evidently, the age of the donor mice of these cells had "little, if any, detectable effect on CMC . . . [of] *T. vaginalis*." (7) The natural CMC for the urogenital trichomonad was found to be the property of the cells adherent to nylon wool fibers. The lack of complete recovery of cytotoxicity from the adherent population of cells might have depended on the injury caused to them by the mechanical manipulations involved in their

TABLE 9.5. Natural cell-mediated cytotoxicity (CMC) for various *T. vaginalis* strains of normal peritoneal cells (PC) from different strains of mice.

Mouse strain	H-2 haplotype	No. of experiments	Cytotoxicity*		
			25:1[†]	12:1	6:1
BALB/c	d	18	28 (18–46)	30 (15–52)	18 (13–31)
BALB/c nu/nu	d	3	49 (41–53)	39 (23–61)	36 (17–49)
BALB.B	b	3	31 (20–41)	43 (36–52)	20 (14–28)
BALB.K	k	3	30 (24–33)	37 (18–46)	34 (31–40)
DBA/2	d	4	27 (18–37)	24 (19–30)	17 (11–28)
A.WySn	a	4	25 (14–38)	27 (17–36)	18 (10–30)
A.TL	t1	4	9[‡] (7–11)	9[‡] (5–13)	8[‡] (6–10)
A.BY	b	3	17 (13–27)	18 (13–24)	14 (11–19)
C57BL/6CR	b	3	27 (19–36)	30 (18–41)	16 (10–24)
B10	b	3	31 (22–37)	22 (14–29)	20 (16–25)
B10.M	f	4	22 (20–25)	18 (13–22)	19 (15–23)
B.PL	u	4	19 (14–22)	18 (11–23)	17 (14–21)
B10.RIII	r	3	14 (10–20)	15 (12–19)	13 (11–17)
B10.S	s	3	19 (16–23)	17 (11–27)	19 (14–24)
CBA/HeN	k	3	7[‡] (2–12)	9[‡] (2–17)	8[‡] (4–15)
CBA/J	k	1	8	10	NT[§]
C3H/CR	k	3	24 (16–37)	22 (14–32)	16 (12–21)
C3H/HeJ	k	3	29 (21–35)	23 (17–29)	21 (14–30)
N:NIH/Dp	q	4	8[‡] (6–10)	10[‡] (7–14)	8[‡] (5–11)

Reproduced from Landolfo S et al: Natural cell-mediated cytotoxicity against *Trichomonas vaginalis* in the mouse: I. *J Immunol* 124: 508–514, 1980, with permission of Williams & Wilkins, Baltimore, MD.
*Mean percent cytotoxicity (3 replicates/experiment) followed by the range in parentheses.
[†]Effector:target cell ratio.
[‡]Statistically significant difference ($P \leq 0.05$) between CMC of BALB/c strain at the corresponding E:T ratio.
[§]Not tested.

separation. (8) Treatment with carbonyl iron powder and magnet, used to remove phagocytic cells, reduced the natural CMC by about 60%. After two treatments, the CMC appeared to be completely lost; yet 90% of the cells remained still viable as demonstrated by the trypan-blue exclusion test; they gave also a high mitogenetic response to phytohemagglutinin and lipopolysaccharide from bacteria. (9) One or two treatments of the PCs with anti-Thy 1, 2 serum and complement (C) had no effect on the CMC of these cells. Similar results were obtained when the PCs were subjected once or twice to rabbit anti-mouse Ig and C. However, this treatment eliminated the ability of these cells to respond mitogenically to phytohemagglutinin and lipopolysaccharide. (10) One or two consecutive treatments of the macrophage-like cells with anti-Ia.8 serum had no effect on their CMC; thus these cells belonged to an Ia$^-$ population.

Although the foregoing results[67] indicated that the mouse cells with natural toxicity for *T. vaginalis* belonged to the monocyte-macrophage lineage, Mantovani et al[63] thought it necessary to confirm these findings with human cells.

Mononuclear cells were obtained from peripheral blood of *Trichomonas*-free male and female human volunteers. Monocytes among these cells were identified by their structure, uptake of neutral red, and staining with nonspecific esterase. Aliquots of mononuclear cell mixture suspended in RPMI 1640 supplemented with FBS or human AB-type serum were enriched for monocytes by being attached to plastic petri dishes. After washing off the nonadherent cells, the attached ones, recovered by incubation with EDTA in PBS, were almost exclusively (over 90%) identifiable monocytes. [^{3}H]-labeling of the trichomonads and the cytotoxicity assay were similar to those employed previously by Landolfo et al[67] for the mouse system.

The following results were reported by Mantovani et al[63] for the human monocytes: (1) Human peripheral mononuclear blood cells that adhered to plastic exhibited high cytotoxicity for *T. vaginalis*. Depending on the donors of the monocytes, cytotoxicity (percentage of parasite lysis) of the effector cells at E:T ratios as low as 3:1 ranged from 3.7 ± 0.9* to 39.9 ±

*Sample standard deviation

1.6% after 48-hour incubation. (2) On the basis of structural and functional criteria (in a separate study), the authors noted that most cells in monocyte populations had receptors for sheep red blood cells, suggesting that they belonged to the T-cell lineage. (3) Nonadherent cell preparations containing less than 2% monocytes were only very weakly cytotoxic and preparations with only 10–25% monocytes had cytotoxic activity falling between that of nearly pure monocytes and that of nonadherent cells. (4) In vitro exposure to silica, known to inhibit the monocytes, reduced greatly cytotoxicity mediated by the host cells. (5) As in the mouse system, in the human-cell system several lines of evidence suggested that CMC for *T. vaginalis* was associated with the monocyte-macrophage lineage. (6) It seemed that "the spontaneous cytotoxicity of monocytes against protozoa such as *Toxoplasma gondii* or *T. vaginalis* could represent one line of in vitro resistance to these infectious agents."

There are two additional articles published by the Italian workers, Martinotti et al,[68,69] who have concerned themselves with CMC for *T. vaginalis* involving the monocyte-macrophage lineage. These reports deal also with the effects of cytoskeleton-disrupting and other compounds on cytotoxicity.[68,69] Since the second report[69] is more detailed, it will be discussed here.

All experiments were performed with the monocyte-macrophage lineage cells derived from 8- to 10-week-old male and female F_1 mice obtained by crossing C57BL/6 and DBA strains. The *T. vaginalis* strain D previously employed by the Italian group was cultivated in modified Diamond's[22] medium in a manner identical to that used in the earlier experiments.[63,67] As before,[63] the trichomonads were subsequently suspended in RPMI 1640 medium supplemented with the several ingredients, for example, 10% FBS, and labeled with [^{3}H]thymidine. Resident peritoneal macrophages were obtained from mice by lavage of the peritoneal cavity. This procedure yielded 90% of small mononuclear cells and macrophages. The cells were washed in Earle's BSS containing heparin, then pelleted by centrifugation. The pellets were resuspended in RPMI 1640 medium without serum. Differential counts were performed on samples of these suspensions.

Cytotoxicity assays were carried out in microtiter plates as described in the previous reports (e.g. reference 67). RPMI medium with 2-mercaptoethanol, gentamycin, and mycostatin was used. A suspension of 2×10^6 peritoneal macrophages/ml was placed in each of the microtiter plate wells. After 24-hour incubation, the adherent cells were washed and 0.1 ml aliquots of a suspension of [^{3}H]thymidine-labeled *T. vaginalis* (2×10^4/ml) were added to each well. The plates with the experimental and control-experimental (maximum and spontaneous release of the radiolabel) wells were then incubated for 48 hours at 37°C. Thereafter, the suspensions were centrifuged, and radioactivity of 0.1 ml aliquots of the supernatant fluids was determined in a scintillation spectrometer. The formula employed for the calculation of percent cytotoxicity was given previously in this chapter for the experiments of Landolfo et al.[67]

In addition to the foregoing experiments using intact macrophages, these cells were employed for the preparation of supernatant fluids.[69] For this purpose, the macrophages were diluted to a final density of 2×10^6 cells/ml and plated in 16-mm wells of microtiter plates. After 2 hours, the cultures consisted of highly purified macrophages, as determined by their structure, phagocytic activity, and staining with nonspecific esterase. Subsequently 2×10^5 trichomonads/ml were added and the macrophage-trichomonad mixtures were incubated for an additional 24 hours. The cultures appeared structurally normal and healthy and no extensive release of lactic dehydrogenase was recorded. The supernatant fluids collected from the macrophage-trichomonad mixtures were transferred to siliconized tubes on ice. The tubes were immediately centrifuged at 10,000 g for 10 min at 4°C, and the supernatant fluids were either tested directly or stored at $-20°C$ for later use.

The investigation of Martinotti et al[69] included also testing the effects on CMC of cytoskeleton-disrupting agents. Colchicine, vinblastine, and deuterium oxide (D_2O) were employed to disrupt microtubules, leading to the disappearance of the polarity of the cells and inhibiting cell secretion (see reference 7 in Fouts and Kraus[70]). Cytochalasin (cyto) B was used to investigate the role of actin-containing filaments, membrane transport, and the role played by cell motility in macrophage-mediated cell toxicity for *T. vaginalis*. In most respects the aforementioned experiments of Martinotti et al[69] resembled those reported earlier by the same group.[68] The possible toxicity of the cytoskeleton-affecting compounds was tested as follows: [^{3}H]thymidine-labeled effector (macrophages) and target (trichomonads) cells were incubated separately from each other with the highest experimental concentration of each of the drugs. After 48-hour incubation, the amounts of [^{3}H]thymidine released into the supernatant fluid were measured. The specific release of radioactivity, as estimated by cpm, never exceeded 2% with any of the aforementioned compounds when compared with the control values obtained after incubation of the macrophage-trichomonad systems in the absence of a given compound. In all instances, viability of the effector cells after 48-hour incubation was monitored by the trypan-blue exclusion test—over 95% of these cells remained viable.

The following results were obtained in the experiments performed by Martinotti et al.[69] (1) Various dilutions of the supernatant fluids obtained after incubation of the effector cells together with the trichomonads lysed the parasites in a dose-dependent fashion (Fig. 9.11). Lysis of *T. vaginalis,* although at a lower level, occurred also in the presence of supernatant fluids from macrophages alone; such fluids obtained from cultures of trichomonads alone had no cytolytic activity. (2) To find out whether the factors causing cytolysis of the parasites that were produced by the effector cells required metabolically active parasites, lysates of *T. vaginalis* were mixed with the macrophages. Only relatively little cytotoxicity was observed in the

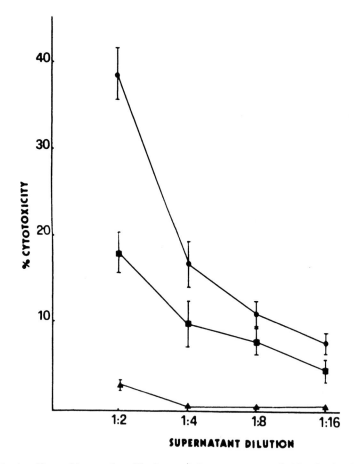

Fig. 9.11. Lytic effect of increasing dilutions of the supernatant fluids obtained from mixed cultures of uninduced resident peritoneal mouse macrophages and *T. vaginalis* strain D (●——●); from macrophage cultures (■——■); and from axenic cultures of *T. vaginalis* (▲——▲). For experimental details, see section Monocytes and Macrophages in text and also Ref. 69. (Reproduced with permission of the American Society of Microbiologists, from Martinotti MG et al: Natural macrophage cytoxicity against *Trichomonas vaginalis* is mediated by soluble lytic factors. *Infect Immun* 41:1144-1149, 1983.)

supernatant fluids of these mixtures in comparison with the lytic effects of such fluids recorded from systems containing living target and effector cells. Furthermore, no lysis of trichomonads was caused by supernatant fluids from systems using macrophages and living unrelated B77 line cells as inducing target cells. Lysates from resident macrophages exhibited no cytolytic activity. (3) In the supernatant fluids obtained from mixtures of viable macrophages and trichomonads, macrophage-mediated lysis of the flagellates was observed after a 4-hour contact. Cytotoxic activity increased constantly after 8 to 14 hours, reaching a plateau at 18 to 24 hours of incubation (for kinetics of production of cytotoxic supernatant fluids from macrophages cultivated with *T. vaginalis,* see Fig. 9.12). Similar cytotoxicity, albeit at a much lower level, was noted with supernatant fluids from cultures of macrophages maintained in the absence of parasites. (4) Pretreatment of macrophages with colchicine or vinblastine increased cytotoxicity levels against *T. vaginalis*; DO_2 enhanced this activity in only one of four experiments. Treatment of the macrophage-parasite system with cyto B, which because of the reversible mechanism of its action had to be maintained throughout the experiment, reduced greatly the macrophage-mediated toxicity directed against the urogeni-

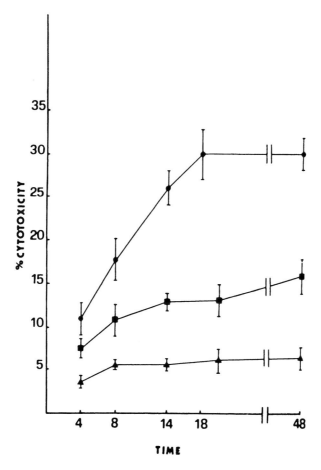

Fig. 9.12. Kinetics of cytotoxicity production in supernatant fluids obtained from mixed cultures of uninduced resident peritoneal mouse macrophages and *T. vaginalis* strain D (●——●); from macrophage cultures (■——■); and from axenic cultures of *T. vaginalis* (▲——▲). For experimental details, see section Monocytes and Macrophages in text and also Ref. 69. (Reproduced with permission of the American Society of Microbiologists, from Martinotti MG et al: Natural macrophage cytoxicity against *Trichomonas vaginalis* is mediated by soluble lytic factors. *Infect Immun* 41:1144–1149, 1983.)

tal trichomonads, in comparison with the cytotoxicity levels recorded for control experiments. (5) Active protein synthesis and metabolism are known to be needed for Bacillus Calmette Guérin-induced cytolytic activity of macrophages against tumors (Ref. 2 in Martinotti et al[69]). In light of this, experiments were carried out to learn whether macrophage-mediated cytotoxicity for T. vaginalis also required active protein synthesis and metabolism of the effector cells. It became evident from the results of these experiments that all cytotoxicity was eliminated from mixtures of T. vaginalis and macrophages cultivated together for 30 minutes with cyclohexamide or puromycin, both potent metabolic inhibitors. In addition, control experiments were performed to establish if the reduction of cytotoxic effects by puromycin and cyclohexamide depended on their effects on macrophages alone. It was established that: (a) Neither enhancement nor reduction of spontaneous release of [^{3}H]thymidine by macrophages occurred in the presence of either of the metabolic inhibitors. (b) Viability of T. vaginalis was not affected by these compounds, and very similar growth curves of the parasites were obtained with untreated cells and with those derived from puromycin- or cyclohexamide-treated cultures after the inhibitors were completely removed by washing. (6) The cytotoxic factors present in the supernatant fluids from the macrophage-T. vaginalis mixtures were found to have the following attributes: (a) Extensive dialysis of these supernatant fluids against the culture medium caused no loss of their cytotoxic capacity. (b) The cytotoxic activity was not affected by treatment with 10% FBS. (c) This activity was unaltered by exposing the macrophage-parasite systems to 56°C for 30 minutes; however, it was reduced after these systems were maintained for 1 hour at this temperature. The activity was eliminated completely when they were kept at 90°C for 10 min.

Some important points were brought out by Martinotti et al[69] with regard to CMC in human urogenital trichomoniasis. (1) The understanding that cytotoxicity for T. vaginalis was mediated by the monocyte-macrophage lineage cells was extended by the finding that lysis of the target organism was due to the release of certain soluble factors by normal macrophages. (2) Production of these factors by the macrophages was found to be significantly enhanced by the presence of T. vaginalis. This latter point was supported by the finding that supernatant fluids obtained from cultures of uninduced effector cells maintained alone for 24 hours exhibited a 10-20% (basal level) of cytotoxicity for trichomonads as reflected in the amount of [^{3}H]thymidine-dependent radioactivity, while higher cytotoxicity was caused by supernatant fluids from suspension of uninduced macrophages incubated for 24 hours with live urogenital trichomonads. (3) In light of the results observed with the B77 cells, the effector cells appeared to be able to distinguish between mammalian cells and protozoa. (4) The evident increase of CMC following treatment of the effector cells with colchicine and vinblastine might depend on "enhanced release of lysosomal content[s] in the extracellular compartment."[69] In light of the reported results, the effect of D_2O could not be considered as conclusive. On the other hand, decrease of CMC after cyto B treatment of the macrophages, a drug known to disrupt microfilaments, might have been attributable to blocking "of transport across the cytoplasmic membrane[s] of the secretion products or the block[ing] of macrophage migration and impairment of the close contact between effector and target cells."[69] (5) The results obtained by treatment of the macrophages with the protein synthesis inhibitors, cyclohexamide and puromycin, suggested that interference with protein synthesis inhibited the production of the cytotoxins secreted into the surrounding medium and that the macrophages synthesized the cytotoxic factors on stimulation by T. vaginalis. (6) The cytotoxic factors appeared not to be dialyzable and, on the basis of their response to a temperature of 56°C, seemed to belong in two groups—one more thermolabile and the other more thermostable. The former retained its cytolytic activity after a 30-minute exposure to 56°C but was reduced thereafter; the latter retained a part of its cytotoxic capacity when kept for 60 minutes at this elevated temperature—its cytotoxic effects became about equal to those exerted by the supernatant fluids from suspensions of macrophages alone. (7) It has been demonstrated previously in various tumor cell systems that CMC depended on intimate contact of cells followed by release of soluble lytic substances such as hydrogen peroxide (H_2O_2) or superoxides that contain the highly reactive O_2^-

ion, proteases, arginase, and lysozyme (for relevant references, see reference 69). Martinotti et al[69] extended these studies to the protozoan parasite *T. vaginalis*. Evidently natural toxicity for *T. vaginalis* is mediated by normal resident macrophages which release "at least two soluble factors with heterogeneous mechanisms of production and molecular nature."[69]

Interactions between T. vaginalis and Polymorphonuclear Neutrophils

Although the trichomonad-polymorphonuclear neutrophil (PMN) interactions constitute an integral part of those between *T. vaginalis* and the host system, they have been singled out by some investigators as a separate entity.

Trichomonas vaginalis infections in women range from virtually asymptomatic, without any evidence of inflammation, to ones characterized by severe inflammation and spontaneous acute leukorrhea.[59,70] Although often considered typical of urogenital trichomoniasis, vaginal discharge is found only in approximately 60% of patients harboring *T. vaginalis*[70]; still PMNs can be found in the lower parts of the genital tract in the absence of leukorrhea.[64] According to Fouts and Kraus,[70] "the concept that *T. vaginalis* is not found in women entirely free of leukorrhea . . . should be discarded." They conceded, however, a positive correlation between large amounts of discharge and the success of positive diagnosis, because of the greater numbers of parasites present. A similar view was expressed by Atia and Thin[71] who felt that epithelial changes constituted the most important criteria in diagnosing inflammatory states in urogenital trichomoniasis [and some other sexually transmitted diseases (STDs) e.g. gonorrhea] and that in light of variations among individuals and those related to the menstrual cycle, the numbers of leukocytes could not be considered a reliable basis for rating the inflammatory changes. Yet they stated that a marked excess of leukocytes was not to be ignored in correlating inflammatory reactions. The experimental results discussed later in this section, according to which secretions of virulent *T. vaginalis* strains stimulate stronger chemotactic responses of PMNs than those of avirulent strains, are consistent with the observations of large numbers of PMNs (and of trichomonads) in the vaginal discharges of patients presenting with high-grade inflammations.

Although it is well known that PMNs can kill a variety of bacterial, mycotic, and even helminth parasites, little information has been gathered about their ability to attack trichomonads and about the process whereby they can destroy these protozoa. The killing process of *T. vaginalis* by PMNs, described by Rein et al[64] is preceded by and dependent on the attraction of PMNs, by chemotaxis, to the sites of parasite invasion. The process of chemotaxis of PMNs in bacterial infections has been well documented (e.g. reference 72). As far as urogenital trichomoniasis is concerned, it has not been clearly understood whether the PMN infiltration is the consequence of epithelial cell destruction, of invasion by pyogenic bacteria, or if it is stimulated by some substances given off by the trichomonads themselves.

It is known that strains of *T. vaginalis* differ in their inherent virulence for humans (for relevant references, see reference 59), experimental animals (see Chap. 8), and cell cultures (see Progression of the Pathologic Process in Cell Cultures). In light of this, isolates differing in virulence were examined for the effects they, and especially the products given off by them into the environment (solutions in which they are suspended), exert on PMN chemotaxis.

Mason and Forman[73] isolated *T. vaginalis* strains from patients attending an antenatal clinic. No information was provided about either the clinical or pathologic aspects of the infections found in the donors of the isolates, nor were there any data given on the results of laboratory animal assays for virulence. The authors[73] described chemotactic responses of PMNs (obtained from *Trichomonas*-negative nonimmune donors) to secretions of the urogenital trichomonads found in the supernatant fluids of the parasite suspensions after various incubation periods in Hanks' Balanced Salt Solution (HBSS) with 10% human serum at 37°C. The chemotactic responses of PMNs were studied with the aid of modified Boyden chambers,[74] the leukocyte suspensions being placed in the upper chamber and the test material (trichomonad-free supernatant fluid, containing the parasite secretions) in the lower chambers. Maximum chemotactic responses were noted using supernatant fluids from suspensions of 1.2×10^5 trichomonads incubated in

HBSS with human serum at 37°C for 20 minutes. Longer incubation reduced chemotactic activity which appeared to be lost after 2-hour incubation. Dilution (1:2) with HBSS of the supernatant fluid obtained after a 60-minute incubation increased the chemotactic activity. The question of chemokinesis (increased activity of cells due to the presence of a chemical substance, without any reference to a concentration gradient) was also investigated using the Zigmond-Hirsch[75] system to distinguish chemotaxis and chemokinesis. The results suggested that both processes were involved in PMN responses to trichomonad products present in the supernatant fluids.

The results indicated that: (1) Formalinized trichomonads failed to release chemotactic factors. (2) No chemotactic activity remained in a supernatant fluid obtained after a 20-minute incubation and then heated at 60°C for 60 minutes. (3) A 2-hour dialysis of a supernatant fluid resulting from a 40-minute incubation of trichomonads in HBSS with human serum did not affect significantly the ability of the secretion to stimulate chemotaxis—the PMN/HPF (high power field) ratio of the dialyzed secretion equalling $58.2 \pm 7.1^*$ ($n = 6$) in comparison with 10.6 ± 2.0 ($n = 6$) for HBSS control and 70.2 ± 8.0 ($n = 10$?) obtained with supernatant fluid of trichomonads incubated for 20 minutes in HBSS with serum.

Acute trichomoniasis in women is typically associated with changes in the normal bacterial flora (possibly pyogenic bacteria multiply); epithelial cell lysis also occurs. Both of these factors are known to cause PMN chemotaxis[72] and, therefore, can be responsible for PMN infiltration of the sites invaded by the trichomonads. It is evident, however, from the results reported by Mason and Forman[73] that some factor or factors secreted by the urogenital trichomonads attract PMNs in vitro. These factors are released quite rapidly, are effective at low rather than at high concentrations, and stimulate both true chemotaxis and chemokinesis. Heat lability and failure to cross dialysis membranes suggest a proteinaceous nature of the chemotactic substances.

The practical significance of the data reported by Mason and Forman[73] would be enhanced if they demonstrated differences among the chemotactic secretions produced by *T. vaginalis* strains with various virulence levels and if PMNs were shown to be instrumental in killing the parasites.

With regard to the first problem, Mason and Forman[62] extended their studies of chemotactic PMN responses to secretion of *T. vaginalis* by using parasite strains that differed in virulence. The inherent virulence differences of the parasite strains were estimated by a modification of the subcutaneous mouse assay introduced by Mason and Forman.[62] No meaningful information was provided about the clinical or pathologic manifestations the several strains caused in their human donors. Fourteen isolates came from 36 asymptomatic patients attending an antenatal clinic; 16 were obtained from 27 women seen at an STD clinic. All the latter subjects presented with symptoms of urogenital infection, although "trichomoniasis was not necessarily the infection diagnosed clinically." The methods of preparation of the trichomonads and of the PMNs were similar to those described previously.[73]

PMNs were obtained from: (1) male and female laboratory workers without "clinical evidence of trichomoniasis" and negative for anti-trichomonad antibodies ($n = 10$); (2) nonpregnant women from an STD clinic with signs of urogenital disease ($n = 27$) ("trichomoniasis was not necessarily diagnosed clinically" in these patients); (3) pregnant women who attended an antenatal clinic and presented with no symptoms ($n = 36$). In all cases, suspensions of washed leukocytes in HBSS containing 10% human serum (negative for anti-*T. vaginalis* antibodies) or bovine serum albumin were employed.

An additional 5 ml of blood were allowed to clot. The resulting serum was used for IFAT in examinations for antibodies to *T. vaginalis* and for preparation of *Escherichia coli* endotoxin-activated serum (EAS).

As in their previous study,[73] Mason and Forman[62] evaluated chemotaxis using the Boyden chamber technique with the associated PMN staining and counting procedures. To confirm that the PMNs had normal chemotactic responses, EAS controls were used. Marked chemotactic response of PMNs from the laboratory workers was noted with secretions of

---

*Sample standard deviation

both virulent and mild *T. vaginalis* strains. This activity was about the same at dilutions giving the equivalent of 20-min or longer incubations of the parasites in HBSS. At higher dilutions, the secretions of the virulent strains stimulated a stronger chemotactic PMN response than those obtained from suspensions of the mild ones. No differences were noted between responses of PMNs from *T. vaginalis*-infected and trichomonad-free persons. However, the leukocytes from pregnant women had significantly greater chemotactic responses to secretions from virulent strains than those from women attending the STD clinic; no such differences were noted with secretions from avirulent trichomonads.

It appears that increased sexual activity or discontinuation of use of contraceptives may increase the frequency of trichomoniasis in pregnancy.[75a] It is possible also that intensified chemotaxis of PMNs in response to *T. vaginalis* secretions may contribute to "greater awareness of infection in this group."[62]

In a brief abstract, Chikunguwo et al[76] reported that although, on biochemical analysis, secretions from cultivated virulent and mild strains were found to contain lactate, protein (of low molecular weight), carbohydrate, and lipid, those from the virulent strain included less protein and lactate. Gel filtration, paper and ion-exchange chromatography, and paper electrophoresis were used to "characterize the chemotactic factor in active fractions." All the results suggested that this factor was likely to be a peptide of molecular mass of about 900 Da "with a negative charge at alkaline pH." The peptide was chymotrypsin-sensitive, but trypsin-resistant.[76]

Results pertaining to the chemotactic factors similar to those published by Mason and Forman[62,73] were reported by Brasseur et al.[30] These workers used 15 *T. vaginalis* strains isolated from patients attending an STD clinic in a hospital in Rouen (France). The parasites, cultivated in TYM medium without agar and with 10% serum for 24 hours, were washed in HBSS, then suspended in this solution at a concentration of 2×10^6/ml and incubated at 37°C in an atmosphere of 5% CO_2 in air for various times. Thereafter they were centrifuged and the supernatant fluid was kept frozen until used. PMNs were obtained from human volunteers without a history of urogenital trichomoniasis and negative for anti-*T. vaginalis* antibodies.

Chemotaxis and chemokinesis of PMNs were assessed by a modification of the "under agarose" method of Nelson et al.[77] It was noted that although chemotactic activity of PMNs was observed in the presence and absence of human serum, it was higher in its presence. The highest chemotactic activity was recorded in 1:5 dilution of cell-free supernatant fluids (with HBSS) obtained after 30-min incubation—the mean chemotactic index (the ratio of chemotaxis:chemokinesis) of supernatant fluids equalled 1.37 ± 0.15 ($n^* = 17$?) [cf. the negative HBSS control index of 1.02 ± 0.03 ($n = $?)].

There was a positive correlation between the chemotactic indices and virulence of strains, as assayed by the Brasseur and Savel[14] cell culture method (for details of the method, see Progression of the Pathologic Process in Cell Cultures, above). For example, perhaps the most virulent strains examined by these workers had a virulence index of about 87 and a chemotactic index of approximately 1.55. On the other hand, a virtually avirulent strain, with a virulence index of about 12.5, had a chemotactic index around 1.15.

Storage of the supernatant fluids at $-20°C$ reduced significantly their chemotaxis-stimulating potency, e.g. after a 4-month storage, the chemotactic index of one strain was found to be reduced from 1.41 ± 0.1 to 1.09 ± 0.07 (cf. "negative" control index 1.02 ± 0.03).

It was indicated by Mason and Forman[62] that most of the chemotactic activity was retained in the nondialyzable fraction of the supernatant fluid of trichomonad suspension. However, according to Brasseur et al[30]: "In contrast with [the] previous findings (Mason and Forman, 1980)[73] we found that chemotactic activity was at least partially dialyzable." For example, the chemotactic index before dialysis equalled 1.53 ± 0.1; it was 1.49 ± 0.15 in the nondialyzable and 1.29 ± 0.01 in the dialyzable fraction (dialyzate).

No statistically significant differences were noted in the chemotactic responses of PMNs from patients infected with *T. vaginalis* and

---

*In many instances Brasseur et al[30] do not provide the n value that indicates the number of determinations.

from uninfected persons ($P > 0.5$). This situation obtained not only in those instances in which supernatant fluids from *T. vaginalis* suspensions were employed, but also in those utilizing ECF (*E. coli* culture filtrates) as chemoattractants.

Antitrichomonal drugs, such as 5-nitroimidazoles (metronidazole or ornidazole) did not appear to interfere with chemotactic activity. However, polyamines, e.g. putrescine and spermidine, eliminated by *T. vaginalis* into the medium, were found to act as chemoattractants for human PMNs. These polyamines, at concentrations of 5 µg/ml and 50 µg/ml, respectively, had chemotactic indices equalling 1.35 ± 0.01 and 1.21 ± 0.06.

In general, the conclusions drawn by Brasseur et al[30] were quite similar to those of Mason and Forman.[62,73]

The results of Rasmussen and Rhodes[78] dealing with the chemotactic responses of human PMNs to *T. vaginalis,* which were performed with suspensions of living parasites rather than with supernatant fluids obtained from such suspensions, differed in some respects from those reported previously.[30,62,73] The method employed by Rasmussen and Rhodes[78] for the analysis of chemotaxis was a modification of the Nelson et al[77] procedure. Living trichomonads of five strains at 10^2, 10^3, or 10^5 concentrations were placed in 10 µl of Eagle's Minimum Essential Medium (MEM) in the outer wells. In all instances, the flagellates were obtained from 24-hour (logarithmic phase of growth) populations cultivated in TYM medium. Each suspension contained also either 50% fresh human plasma or 50% heat-inactivated human serum. Since the results were the same with either of these additives, the authors concluded that complement activation did not affect chemotactic activity. The PMNs, placed in the center well in a concentration of 2.5×10^5 in 10 µl of Eagle's MEM, were obtained from adult female laboratory workers. Although not indicated, the incubation time can perhaps be assumed from the report of Nelson et al[77] to have been about 2 hours.

No differences were noted among the chemotactic indices of the five *T. vaginalis* isolates obtained from the Neisseria Department, Statens Seruminstitut, Denmark, although these strains differed in cytotoxicity (thus in virulence for their hosts). On the basis of these results Rasmussen and Rhodes[78] concluded that "although chemotaxis is undoubtedly an important factor in the inflammatory response, it is unlikely that the different manifestations of infection seen in the clinic can be solely explained by a chemotactic response." Furthermore, they believed that "the onset of infection is due to the adherence of cytotoxic *T. vaginalis* and a subsequent accumulation of PMN[s] leads to the . . . inflammatory response." Since there are differences in the level of inflammation (and the accompanying leukorrhea) in urogenital trichomoniasis caused by various strains, and assuming that in many instances the inflammatory process is dependent upon the presence of trichomonads and not of accompanying nontrichomonad microorganisms, the findings of Rasmussen and Rhodes appear to be at variance with those of the previous investigators. The results of Mason and Forman[62,73] and Brasseur et al[30] suggested that the stronger PMN chemoattracting capacity of the more virulent strains would result in a higher degree of inflammation accompanied by leukorrhea.

The last contribution to be considered with regard to *T. vaginalis*-PMN interactions in an in vitro system is that of Rein et al[64] (for details of the methods, see reference 64). The conditions involving the phagocytosis and killing of the parasites conformed to those commonly observed with microorganisms (e.g. reference 79). Lack of parasite killing on incubation of the trichomonad-leukocyte system with bovine albumin rather than with fresh 10% noninactivated human serum indicated that the attachment of the trichomonads to the phagocytes depended on the activation of complement. Additional tests, including the use of trypan blue, suggested that activation of complement was by the alternative pathway; it would then lead to the attachment of the C3a component to the parasites. The thus opsonized trichomonads could bind to the specific C3a receptors on the leukocytes. The killing of *T. vaginalis* by PMNs occurred in aerobic environment, in "room air"; under anaerobic conditions ("Eh of less than -42 mV with no measurable oxygen"), 88% of the flagellates survived. The O_2-dependent nature of the killing system was confirmed by the findings that catalase and superoxide dismutase, the enzymes inactivating H_2O_2 and the superoxide anion, which are characteristically involved in the O_2-dependent killing mechanism of microorganisms by phagocytes, inhibited destruction of *T. vaginalis* by PMNs. The residual trichomonacidal activity of the phagocytes in the apparently strictly anaerobic environment was explained as follows: "Indeed even the small amount of oxygen dissolved in plastic tissue culture plates was sufficient to permit . . . killing of trichomonads in an oxygen-free atmosphere" by the PMNs "that seemed to be adept at extracting even traces of oxygen from the environment." There is in the vaginal environment a certain amount of O_2, whose level may vary at different times. In any event, the levels of this gas appear to be high enough to allow for the killing of *T. vaginalis* by PMNs.

Because of their likely relevance to the occurrences in natural *T. vaginalis* infections, the findings of Rein et al,[64] the only ones providing a detailed account of destruction of *T. vaginalis* by PMNs, are included in this chapter.

Of special interest are the light microscopic findings involving cinemicrography, that represent the only detailed sequence of events in the process of destruction of trichomonads by phagocytic host cells. Some of these events are shown in Figure 9.13. Under the experimental conditions used by Rein et al,[64] PMNs were attracted to the parasites which they pursued (chemotaxis). Upon coming in contact with the trichomonads, the leukocytes attempted to engulf them, but because of the relative dimensions of the flagellates and the PMNs, the latter were usually unsuccessful. However, when several phagocytes attacked a trichomonad, the parasite typically became immobile (Fig. 9.13a). The PMNs extended pseudopodia over the surface of the flagellate; ultimately the trichomonad membrane ruptured, the cytoplasmic inclusions spilling out of the cell (Fig. 9.13b). Subsequently, the parasites were broken up into smaller fragments (Fig. 9.13c-f) that could be phagocytized. In no instance was agglutination, killing, or lysis of *T. vaginalis* caused by the serum alone.

On the basis of the findings detailed above and some additional observations, Rein et al[64] presented a discussion of the killing phenomenon to which the reader is referred for further details. This discussion includes also consideration of certain unexpected results: (1) the lack of inhibition by superoxide dismutase of reduction of nitroblue tetrazolium (NBT) at the interface between PMNs and the parasites; and (2) effective killing of the trichomonads by the PMNs from patients with chronic granulatomatous disease (these leukocytes do not produce either H_2O_2 or the O_2^- radical). In light of all the information, the authors concluded that: (1) The trichomonads attract large numbers of PMNs. (2) The parasites activate complement via the alternative pathway, the C3a component being then involved in their attachment to the phagocytes. (3) The PMNs spread over the surface of the flagellates and activate, at the interfaces, oxidative trichomonacidal systems. These systems destroy the membranes of the parasites which are then fragmented. (4) The fragments are phagocytized by the PMNs.

As indicated previously in this chapter, there is no agreement as to the role played by phagocytosis in destruction of *T. vaginalis*. Although the relevance to natural infection of destruction of the urogenital parasites by PMNs in the in vitro system[64] may be questioned by some workers (e.g. Krieger et al[11]), the process described by Rein et al is of general interest with regard to the mechanisms whereby phagocytic cells can attack parasites.

Conclusion

It is evident from the preceding sections that a relatively large body of information has been accumulated about the effects of *T. vaginalis* on the host cells and about those of lymphocytes, monocyte-macrophage lineage cells, and polymorphonuclear leukocytes (neutrophils) on this parasite. Still, much more research is needed before the basic phenomena involved in the host cell-trichomonad interactions are adequately understood. Such an understanding would be of major importance not only with regard to pathogenesis in trichomoniasis, a common sexually transmitted disease, but would contribute also to the comprehension of many facets of diseases caused by protozoan and nonprotozoan parasites.

Trichomonas tenax (O. F. Müller, 1773) and *Pentatrichomonas hominis* (Davaine, 1860)

There are relatively few accounts dealing with the effects of the oral and intestinal trichomon-

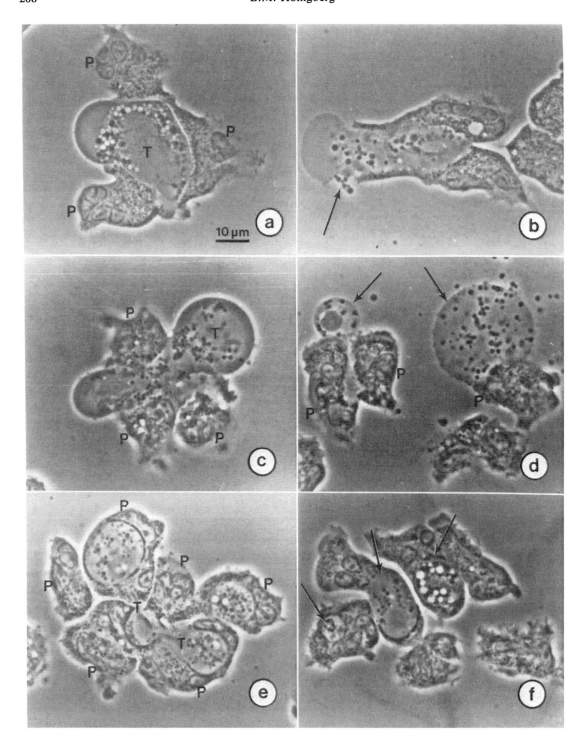

ads of man, *T. tenax* and *P. hominis,* on vertebrate cell cultures, and even these accounts are scanty. In light of its relative paucity, the information pertaining to both species will be dealt with together. In considering pathogenicity of *T. tenax* it should be remembered that obtaining axenic cultures of this flagellate takes a long time (see reference 59 for the cultivation of this species) and that therefore to date one cannot assay virulence of truly fresh isolates.

As far as can be ascertained, among the earliest (if not the earliest) reports on the effects of *T. tenax* and *P. hominis* on vertebrate cell cultures was that of Kulda.[7] Except for the parasite species, the pH of 7.0 of TYM medium[22] used for the intestinal trichomonad, and tryptose-trypticase-yeast (TTY) medium[80] employed for the oral flagellate, all the conditions in experiments involving these parasites were essentially the same as those listed for *T. vaginalis* (see Table 9.1). However, unlike the urogenital trichomonad, *T. tenax* and *P. hominis* had no apparent cytopathic effect on monkey kidney cell monolayers and did not multiply in their presence, merely surviving up to 48 hours of incubation.

The effects of the intestinal trichomonad were investigated subsequently by Zhordaniia et al.[33] The parasite-cell culture systems were the same as those used by the Russian investigators for *T. vaginalis* (see Table 9.1). However, in place of *Staphylococcus*, the bacteria that accompanied *P. hominis* in some of the experiments, were "intestinal rods," said to be the "principal" microorganisms found in xenic cultures of the trichomonads isolated from persons suffering from "intestinal disorders and other diseases."[33] By themselves, these bacteria caused no pathologic changes in HEp-2 cell monolayers. Of a total of 12 strains, 3 were cultivated for 1 year or longer, 3 were grown in culture for 6 months, and 6 for only 1 month. In all instances the cell cultures were examined microscopically at 6 to 12 hours after inoculation with the trichomonads. In the presence of bacteria, 50% of HEp-2 cells in the monolayers were destroyed by the trichomonads cultivated for 1 month, 30% by those grown in vitro for 6 months, and 10% by those cultivated for at least 1 year. *Pentatrichomonas hominis* cultivated for the three corresponding time periods, i.e. 1 or 6 months or 1 year or more, but inoculated without bacteria into the HEp-2 cell cultures, destroyed 10% of the cells in the monolayers after the first and second period of in vitro growth; however, no HEp-2 cells were injured by trichomonads cultivated for one year or more. Evidently, by themselves, the intestinal flagellates were only slightly pathogenic. Furthermore, their virulence was attenuated by prolonged cultivation.

Alderete and his coworkers[10,16] published some information on the effects on HeLa cell monolayers of a *T. tenax* laboratory strain, HS-4:NIH (cultivated for a long time), which did not react with the C20A3 MAb to the 267 kDa surface glycoprotein (see *Putative Immunologic Factors,* p. 189) and did not appear to synthe-

---

Fig. 9.13. Interaction, in "room air" at 37°C in the presence of 10% human serum, between polymorphonuclear neutrophils (PMNs) and *T. vaginalis* isolated in axenic cultures from symptomatic women as revealed by phase-contrast microscopy. The scale in **a** applies to all figures.

a,b. Lysis of trichomonads. **a.** Several PMNs (*P*) are in contact with a trichomonad (*T*). **b.** PMN extends its pseudopod along the surface of a parasite whose membrane has ruptured, with the resulting spillage of the organism's inclusion-filled cytoplasm (*arrow*). (The time elapsed between **a** and **b** was less than 5 min.)

c-f. Effects of multiple PMNs on trichomonads. **c.** Three PMNs attacking a single trichomonad (*T*). **d.** Attack results in fragmentation of the parasite (*arrows*). (A 3-min period intervened between **c** and **d**.) **e.** Single trichomonad (*T*), attacked by six PMNs, is disrupted. **f.** Individual fragments of this parasite (*arrows*) are being phagocytized by the neutrophils. [Fig. 9.13a-f reproduced (using prints provided by MF Rein) with permission of the American Society of Microbiologists, from Rein MF et al: Trichomonacidal activity of human polymorphonuclear neutrophils: Killing by disruption and fragmentation. *J. Infect Dis* 142:575-585, 1980.]

size this antigen.⁴³* As estimated by the crystal violet-staining method, in comparison with *T. vaginalis* strains, little or no cytopathic effect was exerted by the oral trichomonad on the mammalian cell cultures. At twice the concentration used with *T. vaginalis* strain NYH 286, *T. tenax* caused no more than about 10% of CPE exerted by the former species.¹⁶ Furthermore, on the basis of the amount of radioactivity (cpm) recorded for the [³H]thymidine-labeled oral trichomonads adhering to HeLa cell cultures, these flagellates did not attach themselves to the mammalian cells.¹⁰

General Conclusions

After the interrelationships among chemical attributes, immunologic phenomena, and clinical and, more important, pathologic manifestions have been clarified, and certainly we are further from achieving any of these goals with regard to trichomonads other than *T. vaginalis,* the problems of host-parasite interactions will have to be analyzed with the tools of molecular biology. It should be emphasized, however, that in all instances, these techniques would be useful only if applied to appropriate material and to problems that, on the basis of results obtained with other methods, are ready to be analyzed by this more sophisticated approach. It seems that to date the approaches to the many problems of human trichomoniases have been primarily pragmatic and rather diffuse; thus they failed to contribute as well as they might have to our understanding of these cosmopolitan parasitic infections.

References

1. Hogue MJ: The effect of *Trichomonas vaginalis* on tissue-culture cells. *Am J Hyg* 37:145-152, 1943.
2. Kotcher E, Hoogasian AC: Cultivation of *Trichomonas vaginalis* Donné, 1837, in association with tissue cultures. *J Parasitol* 43(suppl):39, 1957.
3. Christian RT, Miller NF, Ludovici PP, Riley GM: A study of *Trichomonas vaginalis* in human cell culture. *Am J Obstet Gynecol* 85:947-954, 1963.
4. Frost JK, Honigberg BM, McLure MT: Intracellular *Trichomonas vaginalis* and *Trichomonas gallinae* in natural and experimental infections. *J Parasitol* 47:302-303, 1961.
5. Farris VK, Honigberg BM: Behavior and pathogenicity of *Trichomonas vaginalis* Donné in chick liver cell cultures. *J Parasitol* 56:849-882, 1970.
6. Honigberg BM, Ewalt AC: Preliminary observations on pathogenicity of *Trichomonas vaginalis* for cell cultures. In Ludvik J, Lom J, Vávra J (eds): *Prog Protozool, Proc 1st Int Congr Protozool,* Prague, August 1961, pp. 568-569, Prague: Publ House, Czechoslovak Acad Sci, 1963.
7. Kulda J: Effect of different species of trichomonads on monkey kidney cell cultures. *Folia Parasitol (Prague)* 14:295-310, 1967.
8. Honigberg BM: Comparative pathogenicity of *Trichomonas vaginalis* and *Trichomonas gallinae* to mice. I. Gross pathology, quantitative evaluation of virulence, and some factors affecting pathogenicity. *J Parasitol* 47:545-571, 1961.
9. Kuczyńska K: Comparison of virulence of clones of two *Trichomonas vaginalis* strains by the subcutaneous mouse assay. *Z Parasitenkd* 70:141-146, 1984.
10. Alderete JF, Garza GE: Specific nature of *Trichomonas vaginalis* parasitism of host cell surfaces. *Infect Immun* 50:701-708, 1985.
11. Krieger JN, Poisson MA, Rein MF: Beta-hemolytic activity of *Trichomonas vaginalis* correlates with virulence. *Infect Immun* 41:1291-1295, 1983.
12. Toś-Luty S, Latuszyńska J, Stępkowski S: Comparative studies on the pathogenicity of selected strains of *T. vaginalis* and *T. gallinae* for tissue cultures. *Wiad Parazytol* 19:419-427, 1973 (in Polish, English summary).
13. Heath JP: Behaviour and pathogenicity of *Trichomonas vaginalis* in epithelial cell cultures. *Br J Vener Dis* 57:106-117, 1981.
14. Brasseur P, Savel J: Evaluation de la virulence des souches de *Trichomonas vaginalis* par l'étude de l'effet cytopathogène sur culture des cellules. *C R Soc Biol (Paris)* 176:849-860, 1982.
15. Rasmussen SE, Nielsen MH, Lind J, Rhodes JM: Morphological studies of the cytotoxicity of *Trichomonas vaginalis* to normal human vaginal epithelial cells in vitro. *Genitourin Med* 62:240-246, 1986.
16. Alderete JF, Pearlman E: Pathogenic *Tricho-*

---

*It may be noted that if the absence of reactivity with the C20A3 MAb is to be compared with the situation described for *T. vaginalis*,⁴³ the oral trichomonad would belong to the MAR⁻ phenotype, thus being cytopathogenic.

monas vaginalis cytotoxicity to cell culture monolayers. *Br J Vener Dis* 60:99-105, 1984.
17. Gentry GA, Lawrence N, Lushbaugh W: Isolation and differentiation of herpes simplex virus and *Trichomonas vaginalis* in cell culture. *J Clin Microbiol* 22:199-204, 1985.
18. Ortega YR, Gilman RH, Miranda E, Yang J, Sterling C, Fedorka P, O'Hare J: Cytopathic effects induced *in vitro* by *Trichomonas vaginalis:* A comparison with *Entamoeba histolytica* and *Giardia lamblia*. *Abstracts 36th Annual Meet Am Soc Trop Med Hyg,* Los Angeles, Nov-Dec 1987, Abstract 109, p 117, Lawrence, Kansas: Am Soc Trop Med Hyg (Allen Press). 1987.
19. Pindak FF, Gardner WA Jr, Mora de Pindak M: Growth and cytopathogenicity of *Trichomonas vaginalis* in tissue cultures. *J Clin Microbiol* 23:672-678, 1986.
20. Krieger JN, Ravdin JI, Rein MF: Contact-dependent cytopathogenic mechanisms of *Trichomonas vaginalis*. *Infect Immun* 50: 778-786, 1985.
21. Fukui K, Takamori Y: Determination of cytopathogenic effects of *Trichomonas vaginalis* on cultured cells by assaying ^{51}Cr release. *Zentralbl Bakteriol Mikrobiol Hyg (A)* 267:450-456, 1988.
22. Diamond LS: The establishment of various trichomonads of animals and man in axenic cultures. *J Parasitol* 43:488-490, 1957.
23. Silva Filho FC, Souza W: Interaction of *Trichomonas vaginalis* and *Tritrichomonas foetus* with epithelial cells. 2nd Meet Brazil Soc Protozool, Canambu, Brazil, Nov 9, 1986. *Mem Inst Oswaldo Cruz* 81(suppl):30, 1986.
24. Lockwood BC, North MJ, Coombs GH: The release of hydrolases from *Trichomonas vaginalis* and *Tritrichomonas foetus*. *Mol Biochem Parasitol* 30:135-142, 1988.
25. Nielsen MH, Nielsen R: Electron microscopy of *Trichomonas vaginalis* Donné: Interaction with vaginal epithelium in human trichomoniasis. *Acta Pathol Microbiol Scand (B)*83:305-320, 1975.
26. Sharma NN, Honigberg BM: Cytochemical observations on chick liver cell cultures infected with *Trichomonas vaginalis*. I. Nucleic acids, polysaccharides, lipids, and proteins. *J Parasitol* 52:538-555, 1966.
27. Sharma NN, Honigberg BM: Cytochemical observations on proteins, alkaline and acid phosphatases, adenosine triphosphatase and 5'-nucleotidase in chick liver cell cultures infected with *Trichomonas vaginalis*. *J Protozool* 14:126-140, 1967.
28. Sharma NN, Honigberg BM: Cytochemical observations on malic dehydrogenase, lipase, nonspecific esterase, and monoamine oxidase in chick liver cell cultures infected with *Trichomonas vaginalis*. *J Protozool* 16:171-181, 1969.
29. Sharma NN, Honigberg BM: Cytochemical observations on glucose-6-phosphatase, glucosan phosphorylase, glucose-6-phosphate dehydrogenase, and α-glycerophosphate dehydrogenase in chick liver cell cultures infected with *Trichomonas vaginalis*. *Int J Parasitol* 1:67-83, 1971.
30. Brasseur P, Ballet JJ, Savel J: *Trichomonas vaginalis*–derived chemotactic activity for polymorphonuclear cells. In Kulda J, Čerkasov J (eds): Proc Int Symp Trichomonads & Trichomoniasis, Prague, July, 1985 (Post-Symp Publ, Pt 1). *Acta Univ Carolinae (Prague) Biol* 30(3,4):343-351, (1986) 1987.
31. Bremner AF, Coombs GH, North MJ: Studies on trichomonad pathogenicity: The effects of *Trichomonas vaginalis* on mammalian cells and mice. In Kulda J, Čerkasov J (eds): Proc Int Symp Trichomonads & Trichomoniasis, Prague, July, 1985 (Post-Symp Publ, Pt 1). *Acta Univ Carolinae (Prague) Biol* 30 (3,4): 381-386, (1986) 1987.
32. Juliano C, Monaco G, Bandiera P, Tedde G, Cappuccinelli P: Action of anticytoskeletal compounds on in vitro cytopathic effect, phagocytosis, and adhesiveness of *Trichomonas vaginalis*. *Genitourin Med* 63:256-263, 1987.
33. Zhordaniia TK, Bakhutashvili VI, Namoradze GI, Dzotsenidze LL: Studies on interaction between trichomonads of man (*Trichomonas hominis* Davaine, 1860 and *Trichomonas vaginalis* Donné, 1837) and cell cultures. *Med Parazytol Parazit Bolezni* 43:647-650, 1974 (in Russian, English summary).
34. Honigberg BM, Livingston MC, Frost JK: Pathogenicity of fresh isolates of *Trichomonas vaginalis:* "The mouse assay" versus clinical and pathologic findings. *Acta Cytol* 10:353-361, 1966.
35. Martinotti MG, Martinetto P, Savoia D: Adherence of *Trichomonas vaginalis* to cell culture monolayers. *Eur J Clin Microbiol* 5:320-323, 1986.
36. Juliano C, Martinotti MG, Cappuccinelli P: In vitro effect of microtubule inhibitors on *Trichomonas vaginalis*. *Microbiologica* 8:31-42, 1985.
37. Hynes RO, Destree AT, Wagner DD: Relationships between microfilaments, cell-substratum, adhesion and fibronectin. *Cold Spring Harbor Symp Quant Biol* 46:659-670, 1982.
38. Garza GE, Kasmala L, Alderete JF: Identification of putative *Trichomonas vaginalis* adhes-

ins and heterogeneity among trichomonads in a parent population for adherence phenotype. *Abstracts. Annual Meet Am Soc Microbiol,* 1987, Abstract B-233, p. 63. Washington DC: Am Soc Microbiol for American Society for Microbiology, 1987.

38a. Alderete JF, Garza GE: Identification and properties of Trichomonas vaginalis proteins involved in cytadherence. *Infect Immun* 56:28-33, 1988.

39. Müller M: Energy metabolism of protozoa without mitochondria. *Annu Rev Microbiol* 42:465-468, 1988.

40. Silva Filho FC, de Souza W: The occurrence of laminin receptors on the surface of Trichomonas vaginalis and Tritrichomonas foetus. Program and Abstract 40th Annual Meet Soc Protozool, Univ Illinois Urbana, July 1987, p 28, 1987.

41. Alderete JF, Kasmala L: Monoclonal antibody to a major glycoprotein immunogen mediates differential complement-independent lysis of Trichomonas vaginalis. *Infect Immun* 53:697-699, 1986.

42. Alderete JF, Suprun-Brown L, Kasmala L: Monoclonal antibody to a major surface glycoprotein immunogen differentiates isolates and subpopulations of Trichomonas vaginalis. *Infect Immun* 52:70-75, 1986.

43. Alderete JR, Kasmala L, Metcalfe E, Garza GE: Phenotypic variation and diversity among Trichomonas vaginalis isolates and correlation of phenotype with trichomonal virulence determinants. *Infect Immun* 53:285-293, 1986.

44. Alderete JF, Demĕs P, Gombošová A, Valent M, Yánoška A, Fabušová H, Kasmala L, Garza GE, Metcalfe EC: Phenotypes and protein-epitope phenotypic variation among fresh isolates of Trichomonas vaginalis. *Infect Immun* 55:1037-1041, 1987.

45. Wang A, Wang CC, Alderete JR: Trichomonas vaginalis phenotypic variation occurs only among trichomonads infected with the double-stranded RNA virus. *J Exp Med* 166:142-150, 1987.

46. Ravdin JI, Guerrant RL: Role of adherence in cytopathogenic mechanisms of Entamoeba histolytica: Study with mammalian tissue culture cells and human erythrocytes. *J Clin Invest* 68:1305-1313, 1981.

47. Petri WA Jr, Joyce MP, Broman J, Smith RD, Murphy CF, Ravdin JI: Recognition of the galactose- or N-acetylgalactosamine-binding lectin of Entamoeba histolytica by human immune sera. *Infect Immun* 55:2327-2331, 1987.

48. Chadee K, Petri WA Jr, Innes DJ, Ravdin JI: Rat and human colonic mucins bind to and inhibit adherence lectin of Entamoeba histolytica. *J Clin Invest* 80:1245-1254, 1987.

49. Salata RA, Cox JG, Ravdin JI: The interaction of human T-lymphocytes and Entamoeba histolytica: Killing of virulent amoebae by lectin-dependent lymphocytes. *Parasite Immunol* 9:249-261, 1987.

50. Pindak FF, Gardner WA Jr, Mora de Pindak M, Abee CR: Detection of hemagglutinins in cultures of squirrel monkey intestinal trichomonads. *J Clin Microbiol* 25:609-614, 1987.

51. Pindak FF, Mora de Pindak M, Gardner WA Jr, Abee CR: Basic properties of Tritrichomonas mobilensis hemagglutinin. *J Clin Microbiol* 26:1460-1463, 1988.

51a. Demeš P, Pindak FF, Wells DJ, Gardner WA Jr: Adherence and surface properties of Tritrichomonas mobilensis, an intestinal parasite of the squirrel monkey. *Parasitol Res* (in press), 1989.

52. Petri WA Jr, Smith RD, Schlesinger PH, Murphy CF, Ravdin JI: Isolation of the galactose-binding lectin that mediates the in vitro adherence of Entamoeba histolytica. *J Clin Invest* 80:1238-1244, 1987.

53. Wartoń A, Honigberg BM: Analysis of surface saccharides in Trichomonas vaginalis strains with various pathogenicity levels by fluorescein-conjugated plant lectins. *Z Parasitenkd* 69:149-159, 1983.

54. Choromański L, Beat DA, Nordin JH, Pan AA, Honigberg BM: Further studies on the surface saccharides in Trichomonas vaginalis strains by fluorescein-conjugated lectins. *Z Parasitenkd* 71:443-458, 1985.

55. Alderete JF, Suprun-Brown L, Kasmala L, Smith J, Spence M: Heterogeneity of Trichomonas vaginalis and discrimination among trichomonal isolates and subpopulations with sera of patients and experimentally infected mice. *Infect Immun* 49:463-468, 1985.

56. Mason PR, Patterson BA: Proliferative response of human lymphocytes to secretory and cellular antigens of Trichomonas vaginalis. *J Parasitol* 71:265-268, 1985.

57. Diamantstein TC, Tridl M, Klos M, Gold D, Hahn H: Mitogenicity of Entamoeba histolytica for murine lymphocytes. *Immunology* 41:347-352, 1980.

58. Ferrante A, Smythe C: Mitogenicity of Naegleria fowleri extract for murine T-lymphocytes. *Immunology* 51:461-468, 1984.

59. Honigberg BM: Trichomonads of importance in human medicine. In Kreier JB (ed): *Parasitic Protozoa,* vol 2, pp 275-454. New York: Academic Press, 1978.

60. Yano A, Yui K, Aosai F, Kojima S, Kawana T, Ovary Z: Immune response to *Trichomonas vaginalis*. IV. Immunochemical and immunobiological analyses of *T. vaginalis* antigen. *Int Arch Allergy Appl Immunol* 72:150-157, 1983.
61. Martinotti MG, Cagliani I, Lattes C, Cappuccinelli P: Immune response and degree of protection in mice immunized with *Trichomonas vaginalis* antigen. *G Batteriol Virol Immunol* 70:3-12, 1977.
62. Mason PR, Forman L: Polymorphonuclear cell chemotaxis to secretions of pathogenic and nonpathogenic *Trichomonas vaginalis*. *J Parasitol* 68:457-462, 1982.
63. Mantovani A, Polentarutti N, Peri G, Martinotti G. Landolfo S: Cytotoxicity of human peripheral blood monocytes against *Trichomonas vaginalis*. *Clin Exp Immunol* 46:391-396, 1981.
64. Rein MF, Sullivan JA, Mandell GL: Trichomonacidal activity of human polymorphonuclear neutrophils: Killing by disruption and fragmentation. *J Infect Dis* 142:575-585, 1980.
65. Mason PR: *Trichomonas vaginalis:* Response of mouse spleen cells to antigens prepared from pathogenic and non-pathogenic strains. *Abstracts Int Symp Trichomonads & Trichomoniasis*, Prague, July 1985, p 48. Prague: Local Organizing Committee, Dept Parasitol, Faculty Sci, Charles Univ, 1985.
66. Martinotti MG, Jemma C, Giovarelli M, Musso T: Induction of macrophage activation by immune T lymphocytes for *Trichomonas vaginalis* killing. *Abstracts Int Symp Trichomonads & Trichomonasis*, Prague, July 1985, p 49. Prague: Local Organizing Committee, Dept Parasitol, Faculty Sci, Charles Univ, 1985.
67. Landolfo S, Martinotti MG, Martinetto P, Forni G: Natural cell-mediated cytotoxicity against *Trichomonas vaginalis* in the mouse. I. Tissue, strain, age distribution, and some characteristics of the effector cells. *J Immunol* 124:508-514, 1980.
68. Martinotti MG, Gallione MA, Martinetto P, Landolfo S: Role of cytoskeleton in natural cell-mediated cytotoxicity against *Trichomonas vaginalis*. *Microbiologica* 5:389-391, 1982.
69. Martinotti MG, Cofano F, Martinetto P, Landolfo S: Natural macrophage cytotoxicity against *Trichomonas vaginalis* is mediated by soluble lytic factors. *Infect Immun* 41:1144-1149, 1983.
70. Fouts AC, Kraus SJ: *Trichomonas vaginalis:* Re-evaluation of its clinical presentation and laboratory diagnosis. *J Infect Dis* 141:137-143, 1980.
71. Atia WA, Thin RN: Cervical cytology in genital infection. *Br J Vener Dis* 51:331-332, 1975.
72. Snyderman R, Pike M: Biological aspects of leukocyte chemotaxis. In Day NK, Good RA (eds): *Comprehensive Immunology, vol 2: Biological Amplification Systems in Immunology*, pp 159-181. New York: Plenum Medical, 1977.
73. Mason PR, Forman L: In vitro attraction of polymorphonuclear leucocytes by *Trichomonas vaginalis*. *J Parasitol* 66:888-892, 1980.
74. Boyden S: The chemotactic effect of mixtures of antibody and antigen on polymorphonuclear leucocytes. *J Exp Med* 115:453-466, 1962.
75. Zigmond SH, Hirsch JG: Leukocyte locomotion and chemotaxis. New methods for evaluation and demonstration of a cell-derived chemotactic factor. *J Exp Med* 137:387-410, 1973.
75a. Brown MT: Trichomoniasis. *Practitioner* 209:639-644, 1972.
76. Chikunguwo SM, Mason PR, Read JS: Investigations of *Trichomonas vaginalis* and chemotactic factor. *Abstracts Int Symp Trichomonads & Trichomoniasis*, Prague, July 1985, p 57. Prague: Local Organizing Committee, Dept Parasitol, Faculty Sci, Charles Univ, 1985.
77. Nelson RD, Quie PG, Simmons RL: Chemotaxis under agarose: A new and simple method for measuring chemotaxis and spontaneous migration of human polymorphonuclear leukocytes and monocytes. *J Immunol* 115:1650-1656, 1975.
78. Rasmussen SE, Rhodes JM: Chemotactic effect of *Trichomonas vaginalis* on polymorphonuclear leucocytes in vitro. In Kulda J, Cerkasov J (eds): Proc Int Symp Trichomonads & Trichomoniasis, Prague, July, 1985 (PostSymp Publ, Pt 1). *Acta Univ Carolinae (Prague) Biol* 30(3, 4):353-355, (1986) 1987.
79. Roitt I, Brostoff J, Male D: *Immunology*. St. Louis. Toronto: C.V. Mosby, 1985.
80. Diamond LS: Axenic cultivation of *Trichomonas tenax*, the oral flagellate of man. I. Establishment of cultures. *J Protozool* 9:442-444, 1962.
81. Lumsden WHR, Robertson DHH, McNeillage GJC: Isolation, cultivation, low temperature preservation and infectivity titration of *Trichomonas vaginalis*. *Br J Vener Dis* 42:145-154, 1966.
82. Alderete JF: Antigen analysis of several pathogenic strains of *Trichomonas vaginalis*. *Infect Immun* 39:1041-1047, 1983.
83. Alderete JF, Garza GE: *Trichomonas vaginalis* attachment to host cell surfaces and role of cytadherence in cytotoxicity. In Kulda J, Čerkasov J (eds): Proc Int Symp Trichomonads & Tricho-

moniasis, Prague, July, 1985 (Post-Symp Publ, Pt 1). *Acta Univ Carolinae (Prague) Biol* 30(3,4):373-380, (1986) 1987.

84. Coombs GH, Bremner AF, Markham DJ, Latter VS, Walters MA, North MJ: Intravaginal growth of *Trichomonas vaginalis* in mice. In Kulda J, Čerkasov J (eds): Proc Int Symp Trichomonads & Trichomoniasis, Prague, July, 1985 (Post-Symp Publ, Pt 1). *Acta Univ Carolinae (Prague) Biol* 30(3,4):387-392, (1986) 1987.

85. Johnson G, Trussell RE: Experimental basis for the chemotherapy of *Trichomonas vaginalis* infections: I. *Proc Soc Exp Biol Med* 54:245-249, 1943.

86. Huff CG, Pipkin AC, Weathersby AB, Jensen DV: The morphology and behavior of living exoerythrocytic stages of *Plasmodium gallinaceum* and *P. fallax* and their host cells. *J Biophys Biochem Cytol* 7:93-101, 1960.

87. Feinberg JG, Whittington MJ: A culture medium for *Trichomonas vaginalis* Donné and species of *Candida*. *J Clin Pathol* 10:327-329, 1957.

88. Su-Lin K-E, Honigberg BM: Antigenic analysis of *Trichomonas vaginalis* strains by quantitative fluorescent antibody methods. *Z Parasitenkd* 69:161-181, 1983.

89. Berens RL, Brun R, Krassner SM: A simple monophasic medium for axenic culture of haemoflagellates. *J Parasitol* 62:360-365, 1976.

10

Microflora Associated with *Trichomonas vaginalis* and Vaccination Against Vaginal Trichomoniasis

Carol A. Spiegel

Introduction

The purpose of this chapter is to discuss the vaginal flora associated with *Trichomonas vaginalis* and to consider the use of a vaccine for vaginal trichomoniasis. I feel it is necessary first to discuss the ecology of the vagina, and thus changes in the vaginal environment over time and the effects of these changes on the endogenous vaginal flora are discussed. The chapter also includes a consideration of the bacterial flora present in trichomoniasis as well as in other vaginal infections.

Microbial Ecology: An Overview

Microbial ecology is a very complex subject. For example, at first glance an environment such as the mouth may look relatively large and homogeneous, but we find on further examination that the niches are quite varied so that different species colonize the tongue, buccal mucosa and pharynx, the occlusal portions of the teeth, the smooth surfaces of the teeth above and below the gingival margin, the gingival crevice in health, during gingivitis, periodontitis, and juvenile periodontosis, and so on. In fact, in discussing the econiche of an organism which is 1 μm in diameter it may be most accurate to consider only the sphere of space immediately surrounding it. This degree of variability occurs at all of the mucous membranes including the oral cavity, the gastrointestinal tract, and the female genitourinary tract.

In discussing the vaginal flora I will use the term "endogenous" rather than "normal" flora to avoid implying that the organisms may never be associated with pathology. Toxic shock syndrome is a fitting example. It is caused by overproduction of toxin in an altered environment by an organism not thought of as a pathogen at that site.

Some of the characteristics of an econiche that affect the organisms that will inhabit it are discussed below.

Access

Under normal circumstances the fetus has no bacterial flora; it is sterile. Animals can be kept germ-free by restricting their exposure to bacteria. A newborn becomes colonized by the organisms it encounters as it passes through the birth canal, by organisms on the hands and lips of those who care for the infant, and by the items the infant puts in his or her mouth including food. Organisms absent from the newborn's environment cannot colonize it.

Adherence

Organisms colonize mucous membrane surfaces but not saliva, discharge, or feces. To keep from being washed away they must have some mechanism of attachment such as the fimbriae (pili) found on many gram-negative organisms or the glycocalyx produced by staphylococci. We now know that microbial attachment is not haphazard but specific and that organisms compete for the finite number of sites available. Our endogenous flora prevent attachment of pathogens to these sites and overgrowth by them.

Temperature

Human mucous membranes have a temperature of about 37°C. Some organisms such as dermatophytes prefer the cooler temperature of the skin.

pH

Every organism has a pH range at which it can grow and an optimum pH. Under normal conditions the stomach is so acid that it has no resident flora. However, in achlorhydria there is overgrowth by a wide variety of organisms present in the foods we eat. Changes in vaginal pH have a profound effect on the flora as is discussed below.

Nutrients

An organism must be able to metabolize the nutrients present in an environment to survive there. These include breakdown products of metabolism by other organisms. For example, an organism that cannot break down glycogen but can utilize its subunits of maltose may be able to colonize the niche only if an organism with the extracellular enzyme for glycogen metabolism is also present.

Bacteriocins

To improve its own chances of survival an organism may produce a low molecular weight compound that inhibits other bacteria. Such compounds are called bacteriocins. The bacteriocins produced by an organism inhibit related species, the organisms most likely to have similar growth requirements and therefore be in the environment competing for the same nutrients and attachment sites. This is thought to be the biologic function of antibiotics produced by streptomycetes in the soil.

Moisture

Moisture is of course essential for growth. Gram-positive organisms are generally less susceptible to the ill effects of drying than are gram-negative ones. The skin, for instance, is colonized by gram-positive organisms such as staphylococci and corynebacteria.

Nonspecific Host Factors

The colonizing organisms must be able to withstand all of the various hostile factors mobilized by the human host. These include mucus, lysozyme, complement, shedding of epithelial cells, fluid flow, opsonins, and phagocytic cells among others.

Specific Host Factors

The main specific factor is antibody. Antibodies can make the organism "sticky" so that it is more readily removed by phagocytes or they can block the organism's receptors for attachment sites. This is the basis for vaccination. IgA is usually the antibody of greatest significance at mucous membranes.

Hormones

Hormonal changes caused by anxiety may effect a change in the fecal flora. Hormonal changes in pregnancy affect the gingival flora in part by serving as nutrients. Hormonal changes in the female have a tremendous impact on the vaginal flora, as described below.

The Dynamic Vaginal Environment and Its Microflora

The wall of the vagina is lined with modified squamous epithelial cells that are shed and replaced from below. The cells are carried out of the vagina in the transudate, which contains mucous. Glycogen, when present, is the main carbon source. The amount of glycogen and the vaginal fluid pH are inversely related, that is, when the glycogen content is high the vaginal pH is low. The pH and the oxidation-reduction potential are also inversely related. When the vaginal pH is low, the oxidation-reduction potential is high, rendering the environment less hospitable for anaerobic bacteria.

The flora of the human vagina is under the influence of all the variables described above. The hormonal influence is such a major factor affecting the composition of the vaginal flora that they cannot be discussed separately. These changes follow two cycles. The first is the "life cycle," which has up to five phases; neonatal, premenarcheal, postpubertal, pregnant, and postmenopausal. The second is the menstrual cycle.

Cruickshank and Sharman[1] performed careful studies to characterize the vaginal flora throughout the life cycle of the female. This work was done, however, before anyone appreciated that anaerobic bacteria may have been present and before reproducible anaerobic technology was available. Therefore it should be kept in mind that their data, while accurate, are incomplete.

Vaginal Ecology of the Neonate

The newborn female has no endogenous vaginal flora. The vaginal fluid pH, however, is low because of the presence of maternal estrogens. Under their influence the newborn's vaginal epithelium is "mature" and the cells glycogen-rich. By the time she is 3 days old, the newborn's vagina is colonized by lactobacilli, coagulase-negative staphylococci, and enterococci.

As the estrogens are metabolized, the glycogen content of the epithelial cells decreases and the pH rises. The flora becomes more fecal in origin, with *Escherichia coli* and *Proteus* spp. predominating.

Vaginal Ecology of the Premenarcheal Girl

Until puberty the estrogen and, therefore, the glycogen content remains low and the pH high. Scanty lactobacilli are accompanied by coagulase-negative staphylococci, coliforms, diphtheroids, and clostridia. Girls 11 or more years old have fewer coliforms and more lactobacilli than girls less than 2 years old.[2] *Bacteroides fragilis* is more common in this age group than it is after puberty.[3]

Vaginal Ecology after Puberty

When estrogen production begins, the vaginal epithelial cells become glycogen-rich, the pH falls, and lactobacilli predominate. Lactobacilli can utilize some sources of glycogen, and some workers have assumed that the lactic acid they produce is responsible for lowering the pH. This hypothesis, however, does not explain the acid pH in the noncolonized newborn.

Effect of the Menstrual Cycle on the Composition of the Vaginal Flora

This stage in the life cycle is complicated by the monthly cycle. Studies of the endogenous vaginal flora are fraught with potential problems. One must carefully choose the women to be studied. The number of lifetime sexual partners must be considered because it is directly related to the prevalence of *Mycoplasma*.[4] One must consider the contraceptive method being used since oral contraceptives predispose to candida vaginitis and the intrauterine contraceptive device to bacterial vaginosis,[5] formerly called nonspecified vaginitis. Subjects prone to recurrent urinary tract infections are more likely to be colonized by Enterobacteriaceae.[6,7] Women who are selected as normal controls because they have no symptoms may have an asymptomatic infection.[5] Douching will decrease the flora and may alter relative proportions of organisms for a time. One must also consider the phase of the menstrual cycle.

One of two study designs have been used in the characterization of the vaginal endogenous flora: qualitative or prevalence studies[8,9] and quantitative studies.[10,11] The former method assumes that for alterations in the flora to be considered significant there must be the appearance or disappearance of species. The latter method suggests that shifts in the proportions of organisms may be of significance. Relatively recent reviews of the vaginal flora are available,[12,13] and therefore the topic is not extensively reviewed here. However, to summarize, it is known that the vaginal flora is composed of multiple facultative bacteria including *Lactobacillus* spp., *Gardnerella vaginalis*, group B streptococcus, *Enterococcus*, viridans streptococci, corynebacteria and *Staphylococcus*, as well as multiple anaerobic species including species of *Bacteroides* other than *B. fragilis*, *Peptostreptococcus* spp., *Eubacterium* spp., *Bifidobacterium* spp., and *Lactobacillus* spp.

The presence and quantity of these organisms vary with the time of the menstrual cycle. This has been most fully examined by Sautter and Brown,[11] who characterized the vaginal flora in seven women by cultivation of this flora several times a week for 1 month. Of the seven women studied, four used oral contraceptives, two wore an intrauterine contraceptive device, and one used no contraceptive method. Facultative lactobacilli were the predominant organisms in

95% of the 65 samples collected, followed by *Corynebacterium* spp. (80%), *Mycoplasma hominis* and *Ureaplasma urealyticum* (78% each), *G. vaginalis* (60%), *Candida albicans* (50%), and *Bacteroides* (50%), including *B. melaninogenicus, B. asaccharolyticus, B. bivius,* and *B. disiens*. When one of these bacteria was found in a woman it usually was found in over half of her specimens. *Bacteroides* were more prevalent in the first half of the menstrual cycle, as had previously been noted.[14]

In another study that examined menstrual and postmenstrual flora in 35 women, none of whom used oral contraceptives or wore an intrauterine device, only *Staphylococcus epidermidis,* diphtheroids, and *Peptostreptococcus anaerobius* were more prevalent during menses.[15]

Vaginal Ecology after Menopause

When estrogen production ends, the glycogen content decreases and the pH rises. The flora, including staphylococci, diphtheroids, and enterococcus, resembles that found before menarche. The mucous membrane is dry. Women often complain of itching, which can be alleviated with estrogen therapy.

The Vaginal Flora Associated with Vaginitis

In 1921 Schroder[16] described three grades of "vaginal cleanliness" to differentiate endogenous from abnormal vaginal flora. In an effort to provide sufficient grades for all known vaginal infections Jírovec and Petrů[17] created seven grades:

Grade 0. Normal in a premenarcheal girl—no lactobacilli, scanty other flora
Grade I. Normal in premenopausal women—many epithelial cells and lactobacilli, few leukocytes
Grade II. Nonpurulent discharge—many epithelial cells, no lactobacilli, but many different bacteria, few leukocytes
Grade III. Purulent discharge—as in II, plus many leukocytes
Grade IV. Gonorrheal discharge—intracellular gram-negative diplococci
Grade V. *Trichomonas* discharge
Grade VI. Yeast vaginitis

There are three well accepted alterations of the vaginal flora that indicate vaginitis: yeast vaginitis, bacterial vaginosis, and *T. vaginalis* vaginitis.

Yeast Vaginitis

Yeast vaginitis is most often caused by *C. albicans* and is represented by Grade VI.

Bacterial Vaginosis (Nonspecific Vaginitis)

Bacterial vaginosis does not appear to be caused by any single agent. The syndrome is characterized by the presence of a thin homogeneous discharge adherent to the vaginal walls, an elevated vaginal fluid pH (greater than 4.5), clue cells, and a "fishy" amine odor detectable on addition of potassium hydroxide.[5] Some as yet unknown alteration in the environment leads to a decrease in the number of lactobacilli and an increase in the number and variety of anaerobes including *B. bivius, B. disiens,* black-pigmented *Bacteroides, Peptostreptococcus* spp.[18-21], and *Mobiluncus* spp.[22,23] as well as the facultative organism *G. vaginalis*.[24] *Mycoplasma hominis* is also associated with this syndrome,[25,26] but its presence may only reflect prior sexual experience. Note that except perhaps for *Mobiluncus* spp. all these organisms belong to the endogenous vaginal flora. In bacterial vaginosis, however, these organisms increase in number with the elimination of lactobacilli.[19,21] This change in flora is readily apparent when a direct smear of vaginal fluid is examined[27] and is represented by Grade II. A large number of leukocytes is not usually seen in bacterial vaginosis (Grade III) and may represent contamination by cells produced in response to an undiagnosed cervicitis. In a student health population 50% of the clinically diagnosed cases of bacterial vaginosis may be asymptomatic.[5] The drug of choice for treating this syndrome is metronidazole.[28,29]

Trichomonas vaginalis Vaginitis

Culture Studies

Bacteria

There are relatively few studies that have discussed or examined the bacterial flora associated with trichomoniasis and yet there is a great deal of variability in the results reported. Nicoli et al[30] compared the prevalence of organisms in women with and without trichomoniasis. There was an increase in the prevalence of *Edwardsiella tarda, Citrobacter intermedium, Enterobacter cloacae, Acinetobacter calcoaceticus, Moraxella liquefaciens, Moraxella lwoffi, Streptococcus mitis, S. milleri, Staphylococcus aureus, S. epidermidis, Bacillus megaterium, Neisseria gonorrhoeae,* and *G. vaginalis.* There was a decrease in the prevalence of *Lactobacillus* spp., *E. coli, Proteus mirabilis, Pseudomonas aeruginosa,* and *Streptococcus faecalis.* The prevalence of *Klebsiella pneumoniae, Staphylococcus saprophyticus,* and anaerobic bacteria was unchanged. They also stated that *Leptothrix vaginalis* frequently accompanies *T. vaginalis.* This finding has not been supported by other studies.[31] *Leptothrix* are sheathed environmental organisms. The name *Leptothrix vaginalis* is not taxonomically correct. An oral anaerobic organism, *Leptotrichia buccalis* has a similar morphology and may be the large filamentous morphotype seen on some direct smears.

Mason et al[32] examined the flora of 104 women with and 232 without trichomoniasis and found no difference in the prevalence of *Mycoplasma*, yeast, lactobacilli, group B or group D streptococci, or diphtheroids. Anaerobic bacteria were not sought. Levison et al[33] found an association between *T. vaginalis* and *Bacteroides.* Of 11 women who had *Bacteroides* 8 had trichomonads, whereas of 16 without *Bacteroides* only 2 had trichomonads. There is an increased risk of endometritis in pregnant women who have trichomoniasis at term and most of these infections are caused by anaerobic bacteria. DeLouvois et al[34] studied 280 pregnant women and found that the isolation rate of *T. vaginalis, Bacteroides* spp., *M. hominis, N. gonorrhoeae,* and anaerobic gram-positive cocci was increased when lactobacilli were absent. Brockman and Höhne[35] (summarized in reference 36) found that metronidazole-susceptible *Bacteroides* spp. and peptostreptococci predominated in trichomoniasis. Hite et al[37] also found an increase in the prevalence of peptostreptococci, black-pigmented *Bacteroides*, other *Bacteroides* spp., and facultatively anaerobic streptococci, and a decrease in lactobacilli. Von Eicher[38] drew a picture showing trichomonads in a mixed bacterial flora containing curved rods resembling *Mobiluncus* and lacking lactobacilli. Robinson and Mirchandani[39] noted an almost complete association between symptomatic trichomoniasis and streptococci, which were usually hemolytic. It is possible that these represent *G. vaginalis.*

Several authors have noted an increased prevalence of gonorrhea in patients with trichomoniasis.[30,40,41] Judson[42] showed very clearly that the prevalence rates for the two infections varied independently in patients attending the Denver Metro Health Clinic.

Fungi

Several authors have noted a lower prevalence of *C. albicans* in patients with trichomoniasis,[30,41,43] while some have seen no difference.[32] If the vaginal pH were elevated in trichomoniasis, one would predict a decrease in the presence of acid-loving (acidophilic) and acid-requiring (aciduric) organisms such as yeast and lactobacilli. An elevation of the pH has been reported.[39,41,44]

Exacerbations During Menses and Pregnancy

Exacerbations of trichomoniasis reportedly occur during menses[44-46] and pregnancy.[44] A self-limited infection has been induced in rhesus monkeys inoculated intravaginally with an axenic culture of *T. vaginalis.*[45] An increase in the number of trichomonads was seen during menses in all five monkeys tested but not during every menstrual cycle. There were also instances of increased trichomonads at other times in the cycle.

The reasons for these observations are unclear. Menstrual blood might serve as a stimulant or nutrient. The elevation in the vaginal pH with a concomitant decrease in the oxidation-reduction potential would enhance the growth of anaerobic organisms including *T. vaginalis.*

In Vitro Studies

In an in vitro study *T. vaginalis* was shown to survive longer when grown with living cultures of *Micrococcus* spp., *Staphylococcus aureus,* or *Streptococcus faecalis.* Survival time was shortened when *Proteus vulgaris* or *Pseudomonas aeruginosa* were added.[47] Killed organisms had no significant effect.

The presence of atypical or "involutional" forms of lactobacilli have been described in trichomoniasis.[48,49] When a high concentration of *T. vaginalis* was incubated with lactobacilli the latter became involutional. In the presence of a high concentration of lactobacilli *T. vaginalis* did not grow well.[48]

Phagocytosis of Vaginal and Cervical Flora

Mycoplasma are commonly seen in patients with trichomoniasis.[50] In electron microscopic studies of vaginal biopsies the predominant organism observed in phagolysosomes was mycoplasma. Only rarely were bacteria seen.[51] Trichomonads collected by cervical scraping have been examined by electron microscopy and phagocytosis of bacteria has been documented.[52] Trichomonads also ingest and kill *N. gonorrhoeae* and *M. hominis*. Ingestion of *Chlamydia trachomatis* could not be demonstrated but its survival was decreased when mixed with *T. vaginalis*.[53]

Animal Models

Street et al[54] attempted to induce trichomoniasis in six squirrel monkeys. Bacterial cultures were obtained from four of them, one of which was culture-positive for *T. vaginalis*. No differences were noted between the flora of the culture-positive and the culture-negative animals. No uninoculated control animals were examined.

Honigberg et al[55] used a subcutaneous mouse assay to examine 52 strains of *T. vaginalis* for differences in pathogenicity. Vaginal cultures for *N. gonorrhoeae*, *G. vaginalis*, *Torulopsis glabrata* (syn. of *Candida glabrata*), *M. hominis*, *U. urealyticum*, group B streptococci, and lactobacilli were also obtained. Anaerobic bacteria were not sought. The size of the mouse lesion was proportional to the severity of the epithelial abnormality in the patient. There was no relationship between the size of the lesion produced in the mouse and the bacteria isolated from the patient.

Summary and Conclusions

Jírovec and Petrů[17] commented that some physicians doubted the significance of *T. vaginalis* because it was found in women with no signs or symptoms of vaginitis. They further described phases in the development of trichomoniasis (Grade V) that would explain these various clinical pictures. These phases would also explain variations in the bacterial flora that accompany *T. vaginalis* infection.

V/A. Trichomoniasis acuta—Occurs shortly after contact with the organism. The number of epithelial cells and lactobacilli is decreased, while the number of trichomonads and leukocytes is increased. This is a short-lived phase rarely seen by the clinician.

V/B. Culminating trichomoniasis—Patients most often go to their clinician during this phase. Many trichomonads, leukocytes, and bacteria of various morphologies are present. Lactobacilli are absent and epithelial cells are present in smaller numbers.

V/C. Chronic trichomoniasis—There are many epithelial cells, few leukocytes, and a variety of bacteria other than lactobacilli. Trichomonads are present in varying numbers.

I/V (a combination of Grades I and V). Latent stage—There are normal numbers of lactobacilli and epithelial cells and few leukocytes and trichomonads. Culture is usually required to detect the trichomonads.

V/VI (a combination of Grades VI and VII). Trichomonads are found with yeast and yeastlike organisms.

The natural history of trichomoniasis is not understood. It is not known if patients actually progress through these stages. If they do, it is not known whether the progression is controlled by virulence factors possessed by the infecting strain of *T. vaginalis* or by the host response. Some of the virulence factors have been identified and discussed by Honigberg.[44] It is known that 25% to 50% of women with trichomoniasis are asymptomatic. The definitive studies to characterize the bacterial flora associated with the various phases of trichomoniasis have not yet been done. Trichomonads ingest bacteria presumably because they provide a source of nutrients. It is not unreasonable to speculate that some bacteria would be preferred to others.

It is interesting to note that from what is known about the bacterial flora associated with "culminating" and "chronic" *T. vaginalis* infection, it appears to be similar to that found in bacterial vaginosis. Microscopically they are very similar (Grade II and Grade V). There are many bacteria of a variety of morphotypes present, but lactobacilli are absent. Various anaerobic organisms appear to be associated with both syndromes and both are treated with metronidazole. *Trichomonas vaginalis* is associated with involutional forms of lactobacilli that morphologically and biochemically resemble *G. vaginalis*. Large numbers of epithelial cells are present, and both are characterized by a foul odor.[5,36] The main difference is the presence of leukocytes in trichomoniasis and their usual absence in bacterial vaginosis. In "latent" trichomoniasis lactobacilli are the predominant bacteria. A large proportion of these patients are probably colonized also by *G. vaginalis*.[24] Lactobacilli produce lactic acid and *G. vaginalis* produces acetic acid as metabolic byproducts. These two compounds inhibit the activity of metronidazole in an aerobic environment,[56] which would exist in the vagina under these circumstances. This may help explain the induction of the latent stage by therapy and the difficulty in eradicating *T. vaginalis* from such patients.

It would be naive to ignore the interactions between the trichomonad and the endogenous vaginal flora. However, because both trichomonads and the anaerobic bacteria associated with bacterial vaginosis respond to treatment with metronidazole, it will be very difficult to separate the roles played by each. The availability of an agent with activity against either the trichomonad or the bacteria, but not both, would be very useful in defining the roles of each. Such investigation is essential to our understanding of the pathogenesis of vaginal trichomoniasis.

Vaccination Against Vaginal Trichomoniasis

Rationale

Vaccination can be a very effective method of prevention. A killed or attenuated strain of the offending organisms is used to stimulate an immune response. Another substance that carries a cross-reacting antigen and so will stimulate a cross-reacting immune response can serve the same function. The latter principle was used in the development of a vaccine for trichomoniasis called SolcoTrichovac.[36]

Immunization can effect either humoral or cell-mediated immunity. Antibody can prevent infection by blocking binding sites and preventing adherence of the organism, binding to and inactivating a toxin, allowing binding of complement and cell lysis, or enhancement of phagocytosis by host cells. Secretory IgA is the immunoglobulin of greatest significance at mucous membranes. Immunization can also create antigenic memory in lymphocytes and enhance cell-mediated immunity.

In the development of SolcoTrichovac aberrant strains of organisms described as *Lactobacillus acidophilus* were isolated from women with vaginitis. These strains are described as "shortened even coccoidal forms of 'lactobacilli' which have largely lost the ability to produce lactic acid."[49] To be included in the genus *Lactobacillus* an organism must produce lactic acid as its predominant by-product; therefore, the identity of the isolates used in the development of this vaccine is in question. The description is consistent with that of *G. vaginalis*. The vaccine is composed of 7×10^9 inactivated cells per 0.5-ml dose.

Antibody Titers Following Vaccination with SolcoTrichovac

Milovanović and co-workers[49] diagnosed 97 women with trichomoniasis by wet mount and culture. All were given three intramuscular injections of 0.5 ml of vaccine at 2-week intervals and a booster dose 12 months after the first dose.[49] Serum agglutinating titers against the organisms in the vaccine were determined 6 weeks and 4, 7, and 12 months after the first dose of vaccine and 2 weeks after the booster injection. The geometric mean titer rose from 1:56 before vaccination to a high of 1:320 at 6 weeks. This measures serum antibody, which gives no information about local antibody or protection against infection. There was a broad range of titers before vaccination; 60 patients had a titer of 1:80 or greater, suggesting that many of the women had prior exposure to this antigen. The authors do not report how long these women had trichomoniasis or whether they had had it in the past. If they did have antibody from a prior infection, it would appear that it was not protective antibody. Because there was no control group of sham-vaccinated patients with trichomoniasis, it is not known if the rise in titer might be a response to the infection rather than the vaccination.

Using the same protocol as above but without the booster vaccination at 12 months Milovanović et al[57] documented an increase in total vaginal IgA after vaccination in 12 women, but again the authors did not report prior episodes of trichomoniasis and there was no control group. It is difficult to separate the effects of the vaccination from those of the infection. Three of the 12 patients showed a decrease in titer and had trichomoniasis 2 weeks after the third dose of vaccine. It is not clear whether these were persistent or recurrent infections.

Effect of SolcoTrichovac on Vaginal Bacterial Flora

The accompanying bacterial flora was studied in a group of 36 patients vaccinated on days 0, 14, and 28.[58] Anaerobic bacteria were not considered pathogens and so were not studied. The prevalence of coliform bacteria other than *E. coli* apparently decreased after vaccination. Since these organisms are not vaginal pathogens and therefore reflect colonization rather than infection, the significance of this observation is not clear. The increase in the prevalence of *S. aureus* from 14% to 35% is notable. The isolates were not examined for toxin production. The prevalence of *Lactobacillus* spp. increased from 10% before vaccination to a high of 72% on day 42 and had decreased to 50% at the 4 month follow-up. Because the natural history of trichomoniasis is not understood, the absence of a control group of unvaccinated women makes these data difficult to interpret.

Clinical Trials with SolcoTrichovac

The effect of vaccination on the vaginal flora of 19 women with trichomoniasis diagnosed by wet mount and 46 women with vaginitis caused by other agents has been studied by Goisis et al.[59] The methods used to make a diagnosis in the latter group were not given. It is not reported if these patients had signs or only symptoms. Only five of them had an organism previously associated with vaginitis, *C. albicans*. Since up to 50% of women without vaginitis are colonized by *C. albicans,* its significance in these patients is unclear. Anaerobic bacteria were not sought. The women were treated only with SolcoTrichovac, but the sexual partners of the women with trichomoniasis were treated with metronidazole.

At the follow-up examinations the vaginal pH decreased to normal and lactobacilli were seen in increased numbers. The serum antibody titer against the vaccine strains was increased in both the patients with and without trichomoniasis. Again there was no control group of unvaccinated patients with trichomoniasis to monitor the natural history of the flora and antibody titers.

In another clinical trial[60] 102 women were vaccinated. On interview, 32 (31%) reported a previous history of trichomoniasis, but the length of time since the last infection was not documented. Symptomatic women ($n = 52$) were also treated with intravaginal metronidazole. Asymptomatic women were either untreated ($n = 28$) or swabbed intravaginally with an Albothyl (the active ingredient is metacrasolsulfonic acid) solution of unstated concentration. Data collected at the postvaccination examinations were not separated by treatment group. The response was determined by alleviation of symptoms and absence of trichomonads in wet mount or cytologic smear, but not by cultivation.

Before vaccination all patients had Grade V flora.[17] Six months after completing the vaccination regimen only 4 (5.2%) of 77 patients still had trichomoniasis. However, only 28 (36%) had Grade I flora, while 39 (51%) had Grade II or Grade III flora. These patients may have had chronic trichomoniasis (Grade V/C) in which culture may be required to detect *T. vaginalis*. Again, this picture may in part represent the natural history of untreated trichomoniasis.

In the only blind placebo-controlled study, reported by Litschgi,[61] SolcoTrichovac was examined for its ability to protect patients against recurrent trichomoniasis. One hundred fourteen women with vaginal trichomoniasis diagnosed by wet preparation were treated with an oral or systemic nitroimidazole, as were their partners, and then randomly assigned to either a vaccine or placebo group. Trichomonads were sought by wet preparation, and culture and serum antibody levels were determined. The episodes of trichomoniasis in the two groups of patients are given in Table 10.1. At the 4- and 12-month follow-up visits the vaccine appeared to confer considerable protection against trichomoniasis. Litschge[61] does not differentiate recurrent from persistent infections; therefore, the number of patients involved is not known. The number of patients with recurrent trichomoniasis prior to vaccination did not differ between the treatment and placebo groups. Unfortunately no data on serum antibody levels were provided.

The results of an open multicentered study have also been reported.[62] No unvaccinated control group was studied. All 87 patients re-

TABLE 10.1. Vaginal trichomoniasis in patients vaccinated with SolcoTrichovac and placebo.

Treatment group	Prevalence of vaginal trichomoniasis on follow-up (No. positive/No. examined)		
	0–6 wk	6 wk–4 mo	4–12 mo
Vaccine			
Symptomatic*	0	1/55	0
Asymptomatic	3/61	0	0
TOTAL	3/61	1/55	0/55
Placebo			
Symptomatic*	3/53	10/48	8/45
Asymptomatic	0	9/48	7/45
TOTAL	3/53	19/48	15/45

Adapted from Litschgi M: SolcoTrichovac in the prophylaxis of trichomonad reinfection. Gynäkol Rundsch 23 (suppl 2): 72-76, 1983.
*Patients with symptomatic disease were treated with a nitroimidazole and returned to the study population.

ceived vaccine. Twenty-six symptomatic patients also received local therapy such as douches or creams. The remaining 39 symptomatic patients received antitrichomonal therapy. The 26 patients who had mild or no symptoms received no additional treatment. In 70% of the patients symptoms had been present for at least 2 weeks and as long as a year. The number of motile trichomonads decreased after vaccination but the number of organisms did not differ between study groups. Three and 12 months after therapy 9% and 5% of patients had trichomoniasis as determined by examination of wet preparations.

Another clinical trial was performed on a group of 94 women with trichomoniasis who had had *T. vaginalis* infection in the preceding year.[63] Patients were treated only with three doses of vaccine given intramuscularly at 2-week intervals. No control group was included. Four weeks after the first dose of vaccine 70 (77%) of 91 patients were found to be free of trichomonads on microscopic examination. There was no significant change at subsequent visits. Forty-six (49%) of the patients dropped out of the study. Seventy percent of the dropouts were trichomonad-free at their last visit.

Conclusions

The efficacy of SolcoTrichovac has not yet been adequately determined. The data from the blind placebo-controlled study were encouraging,[61] but the apparent alteration of the vaginal flora[58] is worrisome. Further studies of both the clinical and ecologic effects of this vaccine appear to be warranted.

In the coming years vaccine development for trichomoniasis may take a different direction, that is, use of the trichomonad. It has recently been shown that there is an antibody response in humans to trichomoniasis. An enzyme-linked immunosorbent assay using whole cells[64,65] or an aqueous protein extract[65] was used to detect the presence of trichomonad-specific IgM, IgG, and IgA in the sera from infected women. IgG and IgA were found in vaginal fluid but the presence of serum antibody could better differentiate current from past infection.[64]

Trichomonas vaginalis is antigenically heterogeneous.[66,67] However, all of 88 strains from geographically diverse sites reacted with one or both of two of nine monoclonal antibodies tested.[67] It is not yet known if any of the antibodies studied to date are protective. Because *T. vaginalis* was reported to conceal some of its surface antigens,[66] antibodies raised in vitro after disruption of the cell's integrity may not be the ones raised in vivo.

A great deal more work must be done before a vaccine for vaginal trichomoniasis, whether trichomonad-free or trichomonad-containing, is commercially available.

References

1. Cruickshank R, Sharman A: The biology of the vagina in the human subject. *J Obstet Gynecol Br Emp* 41:208-226, 1934.
2. Hammerschlag MR, Alpert S, Rosner I, Thurston, P, Semine D, McComb D, McCormack WM: Microbiology of the vagina in children: Normal and potentially pathogenic organisms. *Pediatrics* 62:57-62, 1978.
3. Hammerschlag MR, Alpert S, Onderdonk AB, Thurston P, Drude E, McCormack WM, Bartlett JG: Anaerobic microflora of the vagina in children. *Am J Obstet Gynecol* 131:853-856, 1978.
4. McCormack WM, Almeida, PC, Bailey PE, Grady EM, Lee Y-H: Sexual activity and vaginal colonization with genital mycoplasmas. *JAMA* 221:1375-1377, 1972.
5. Amsel R, Totten PA, Spiegel CA, Chen KCS, Eschenbach D, Holmes KK: Nonspecific vaginitis: Diagnostic criteria and microbial and epidemiologic associations. *Am J Med* 74:14-22, 1983.
6. Stamey TA, Sexton CC: The role of vaginal colonization with Enterobacteriaceae in recurrent urinary infections. *J Urol* 113:214-217, 1975.
7. Pfau A, Sacks T: The bacterial flora of the vaginal vestibule, urethra and vagina in premenopausal women with recurrent urinary tract infections. *J Urol* 126:630-634, 1981.
8. Corbishley CM: Microbial flora of the vagina and cervix. *J Clin Pathol* 30:745-748, 1977.
9. Osborne NG, Wright RC, Grubin L: Genital bacteriology: A comparative study of premenopausal women with postmenopausal women. *Am J Obstet Gynecol* 135:195-198, 1979.
10. Bartlett JG, Onderdonk AB, Drude E, Goldstein C, Anderka M, Alpert S, McCormack WM: Quantitative bacteriology of the vaginal flora. *J Infect Dis* 136:271-278, 1977.
11. Sautter RL, Brown WJ: Sequential vaginal cultures from normal young women. *J Clin Microbiol* 11:479-484, 1980.
12. Brown WJ: Microbial ecology of the normal vagina. In Hafez ESE, Evans TN (eds): *In The Human Vagina*. New York: Elsevier/North Holland Biomedical Press, 1978.
13. Larsen B, Galask RP: Vaginal microbial flora: Practical and theoretical relevance. *Obstet Gynecol* 55(suppl):100s-112s, 1980.
14. Neary MP, Allen J, Okubadejo OA, Payne DJH: Preoperative vaginal bacteria and postoperative infections in gynecological patients. *Lancet* 2:1291-1294, 1973.
15. Johnson SR, Petzold CR, Galask RP: Qualitative and quantitative changes of the vaginal microflora during the menstrual cycle. *Am J Reprod Immunol Microbiol* 9:1-5, 1985.
16. Schröder R: Zur Pathogenese und Klinik des vaginalen Fluors. *Zentralbl Gynäkol* 45:1350-1361, 1921.
17. Jírovec O, Petrů M: *Trichomonas vaginalis* and trichomoniasis. *Adv Parasitol* 6:117-188, 1968.
18. Goldacre MJ, Watt B, Loudon N, Milne LJR, Vessey MP: Vaginal microbial flora in normal young women. *Br Med J* 1:1450-1453, 1979.
19. Spiegel CA, Amsel R, Eschenbach D, Schoenknecht F, Holmes KK: Anaerobic bacteria in nonspecific vaginitis. *N Engl J Med* 303:601-607, 1980.
20. Taylor E, Blackwell AL, Barlow D, Phillips I: *Gardnerella vaginalis*, anaerobes, and vaginal discharge. *Lancet* 1:1376-1379, 1982.
21. Blackwell AL, Fox AR, Phillips I, Barlow D: Anaerobic vaginosis (non-specific vaginitis): Clinical, microbiological, and therapeutic findings. *Lancet* 2:1379-1382, 1983.
22. Spiegel CA, Eschenbach DA, Amsel R, Holmes KK: Curved anaerobic bacteria in bacterial (nonspecific) vaginosis and their response to antimicrobial therapy. *J Infect Dis* 148:817-822, 1983.
23. Thomason JL, Schreckenberger PC, Spellacy WN, Riff LJ, LeBeau LJ: Clinical and microbiological characterization of patients with nonspecific vaginosis associated with motile, curved anaerobic rods. *J Infect Dis* 149:801-808, 1984.
24. Totten PA, Amsel R, Hale J, Piot P, Holmes KK: Selective differential human blood bilayer media for isolation of *Gardnerella (Haemophilus) vaginalis*. *J Clin Microbiol* 15:141-147, 1982.
25. Pheifer TA, Forsyth PS, Durfee MA, Pollock HM, Holmes KK: Nonspecific vaginitis: Role of *Haemophilus vaginalis* and treatment with metronidazole. *N Engl J Med* 298:1429-1434, 1978.
26. Paavonen J, Miettinen A, Stevens CE, Chen KCS, Holmes KK: *Mycoplasma hominis* in nonspecific vaginitis. *Sex Transm Dis* 10(suppl):271-275, 1983.
27. Spiegel CA, Amsel R, Holmes KK: Diagnosis of bacterial vaginosis by direct Gram stain of vaginal fluid. *J Clin Microbiol* 18:170-177, 1983.
28. Amsel R, Critchlow CW, Spiegel CA, Chen KCS, Eschenbach D, Smith K, Holmes KK: Comparison of metronidazole, ampicillin, and amoxicillin for treatment of bacterial vaginosis (nonspecific vaginitis): Possible explanation for the greater efficacy of metronidazole. In Finegold SM (ed): *US Metronidazole Conference, Proceedings from a Symposium*. New York: Biomedical Information Corporation, 1982.
29. Eschenbach DA, Critchlow CW, Watkins H, Smith K, Spiegel CA, Chen KCS, Holmes KK: A dose-duration study of metronidazole for the treatment of nonspecific vaginosis. *Scand J Infect Dis* 40(suppl):73-80, 1983.
30. Nicoli R-M, Nourrit J, Muniglia A, Michel-Nguyen A: Le flagellé *Trichomonas vaginalis* et son

environnement bactérien en milieu vaginal. *Ann Parasitol* 56:23-31, 1981.
31. McLellan R, Spence MR, Brockman M, Raffel L, Smith JL: The clinical diagnosis of trichomoniasis. *Obstet Gynecol* 60:30-34, 1982.
32. Mason PR, MacCallum M-J, Poyntner B: Association of *Trichomonas vaginalis* with other microorganisms. *Lancet* 1:1067, 1982.
33. Levison ME, Trestman I, Quach R, Sladowski C, Floro CN: Quantitative bacteriology of the vaginal flora in vaginitis. *Am J Obstet Gynecol* 133:139-144, 1979.
34. deLouvois J, Hurley R, Stanley V: Microbiol flora of the lower genital tract during pregnancy: Relationship to morbidity. *J Clin Pathol* 28:731-735, 1975.
35. Brockman J, Höhne C: Bakteriologische Aspekte der Trichomonadenkolpitis. *Zentralbl Gynäkol* 101:722-726, 1979.
36. Pavić R, Stojković L: Vaccination with Solco-Trichovac: Immunological aspects of a new approach for therapy and prophylaxis of trichomoniasis in women. *Gynäkol Rundsch* 23(suppl 2):27-38, 1983.
37. Hite KE, Hesseltine HC, Goldstein L: A study of the bacterial flora of the normal and pathologic vagina and uterus. *Am J Obstet Gynecol* 53:233-240, 1947.
38. von Eicher W: *Selenomonas* in der Scheide. *Zentralbl Gynäkol* 52:1775-1778, 1968.
39. Robinson SC, Mirchandani G: Observations on vaginal trichomoniasis. IV. Significance of vaginal flora under various conditions. *Am J Obstet Gynecol* 91:1005-1012, 1965.
40. Tsao W: Trichomoniasis and gonorrhoeae. *Br Med J* 1:642-643, 1969.
41. Fouts AC, Kraus SJ: *Trichomonas vaginalis:* Reevaluation of its clinical presentation and laboratory diagnosis. *J Infect Dis* 141:137-143, 1980.
42. Judson F: The importance of coexisting syphilitic, chlamydial, mycoplasmal, and trichomonal infections in the treatment of gonorrhea. *Sex Transm Dis* 6(suppl 2):112-119, 1979.
43. Hurley R, Leask BGS, Faktor JA, deFonseka CI: Incidence and distribution of yeast species and of *Trichomonas vaginalis* in the vagina of pregnant women. *J Obstet Gynaecol Br Commonw* 80:252-257, 1973.
44. Honigberg BM: Trichomonads of importance in human medicine. In Kreier JP (ed): *Parasitic Protozoa,* vol 2, pp. 275-454. New York, Academic Press, 1978.
45. Johnson G, Kupferberg AB, Hartman CG: Cyclic changes in vaginal populations of experimentally induced *Trichomonas vaginalis* infections in rhesus monkeys. *Am J Obstet Gynecol* 59:689-692, 1950.
46. Rein MF, Chapel TA: Trichomoniasis, candidiasis, and the minor venereal diseases. *Clin Obstet Gynecol* 18:73-88, 1975.
47. Szreter H: Influence of microorganisms on the survival rate of *Trichomonas vaginalis* in physiological salt solution. *Wiad Parazytol* 25:409-415, 1979.
48. Soszka S, Kuczyńska K: Influence of *T. vaginalis* on the physiological flora of the vagina. *Wiad Parazytol* 23:519-523, 1977.
49. Milovanović R, Grčić R, Stojković L: Serological study with SolcoTrichovac, a vaccine against *Trichomonas vaginalis* infection in women. *Gynäkol Rundsch* 23(suppl 2):39-45, 1983.
50. Mardh P-A, Stormby N, Westrom L: Mycoplasma and vaginal cytology. *Acta Cytol* 15:310-315, 1971.
51. Nieslen MH, Nielsen R: Electron microscopy of *Trichomonas vaginalis* Donné: Interaction with vaginal epithelium in human vaginal trichomoniasis. *Acta Pathol Microbiol Scand* 83:305-320, 1975.
52. Garcia-Tamayo J, Nuñez-Montiel JT, deGarcia HP: An electron microscopic investigation on the pathogenesis of human vaginal trichomoniasis. *Acta Cytol* 22:447-455, 1978.
53. Street DA, Wells C, Taylor-Robinson D, Ackers JP: Interaction between *Trichomonas vaginalis* and other pathogenic micro-organisms of the human genital tract. *Br J Vener Dis* 60:31-38, 1984.
54. Street DA, Taylor-Robinson D, Hetherington CM: Infection of female squirrel monkeys (*Saimiri scuireus*) with *Trichomonas vaginalis* as a model of trichomoniasis in women. *Br J Vener Dis* 59:249-254, 1983.
55. Honigberg BM, Gupta PK, Spence MR, Frost JK, Kuczyńska K, Choromański L, Wartoń A: Pathogenicity of *Trichomonas vaginalis:* Cytopathologic and histopathologic changes of the cervical epithelium. *Obstet Gynecol* 64:179-184, 1984.
56. Clarkson TE, Coombs GH: The antagonistic effects of acetate and lactate upon the trichomonacidal activity of metronidazole. *J Antimicrob Chemother* 11:401-406, 1983.
57. Milovanović R, Grčić R, Stojković L: IgA antibodies in the vaginal secretion after vaccination with SolcoTrichovac. *Gynäkol Rundsch* 23(suppl 2):46-49, 1983.
58. Milovanović R, Grčić R, Stojković L: Changes in the vaginal flora of trichomoniasis patients after vaccination with SolcoTrichovac. *Gynäkol Rundsch* 23(suppl 2):50-55, 1983.
59. Goisis M, Magliano E, Goisis F: Effects of vaccination with SolcoTrichovac on the vaginal flora and the morphology of the Doderlein bacilli. *Gynäkol Rundsch* 23(suppl 2):56-63, 1983.
60. Lorenz V, Ruttgers H: Clinical experience using SolcoTrichovac in the treatment of trichomonas infections in women. *Gynäkol Rundsch* 23(suppl 2):64-71, 1983.
61. Litschgi M: SolcoTrichovac in the prophylaxis

of trichomonad reinfection. *Gynäkol Rundsch* 23(suppl 2):72-76, 1983.
62. Rippman ET: SolcoTrichovac in medical practice: An open, multicenter study to investigate the antitrichomonal vaccine SolcoTrichovac. *Gynäkol Rundsch* 23(suppl 2):77-84, 1983.
63. Elokda HH, Andrial M: The therapeutic and prophylactic efficacy of SolcoTrichovac in women with trichomoniasis. *Gynäkol Rundsch* 23(suppl 2):85-88, 1983.
64. Street DA, Taylor-Robinson D, Ackers JP, Hanna NF, McMillan A: Evaluation of an enzyme-linked immunosorbent assay for the detection of antibody to *Trichomonas vaginalis* in sera and vaginal secretions. *Br J Vener Dis* 58:330-333, 1982.
65. Alderete JF: Enzyme linked immunosorbent assay for detection antibody to *Trichomonas vaginalis:* Use of whole cells and aqueous extract as antigen. *Br J Vener Dis* 60:164-170, 1984.
66. Alderete JF, Suprun-Brown L, Kasmala L, Smith J, Spence M: Heterogeneity of *Trichomonas vaginalis* and discrimination among trichomonal isolates and subpopulations with sera of patients and experimentally infected mice. *Infect Immun* 49:463-468, 1985.
67. Krieger JN, Holmes KK, Spence MR, Rein MF, McCormack WM, Tam MR: Geographic variation among isolates of *Trichomonas vaginalis:* Demonstration of antigenic heterogeneity by using monoclonal antibodies and indirect immunofluorescence technique. *J Infect Dis* 152:979-984, 1985.

Reference Added in Proof

Alderete JF: Does lactobacillus vaccine for trichomoniasis, SolcoTrichovac, induce antibody reaction with *Trichomonas vaginalis? Genitorurin Med* 64:118-123, 1988.

11

Clinical Manifestations of Urogenital Trichomoniasis in Women

Michael F. Rein

Introduction

Vaginitis is a common clinical syndrome said to account for half of patient visits to private gynecologists[1] and for one quarter of visits by women to clinics for sexually transmitted diseases (STD).[2] Trichomoniasis accounts for about a third of vaginal infections diagnosed in public clinics. Although certain commercial topical vaginal therapeutics claim to be broad spectrum and appropriate for all common forms of vaginitis, their effectiveness for trichomoniasis is in fact quite limited,[3] and optimal patient management depends on specific etiologic diagnosis.[4,5] Differential diagnosis of acute vaginitis cannot be based reliably on clinical features alone but depends as well on some rapid laboratory evaluations.[6]

Trichomonas vaginalis is highly site specific. It is capable of producing infection in the human urogenital tract and not, for example, in the mouth or colon,[7] which may, however, be inhabited by other species of flagellates.[7] The clinical features of trichomoniasis in women are therefore limited to the urogenital tract.

Interpretation of Clinical Information

Reviews of clinical features of disease often provide information on frequency of specific signs and symptoms in series of patients. Such data must be regarded with suspicion, because selection biases profoundly influence the prevalence of certain clinical features. Thus, for example, women presenting to gynecologists are very likely to be symptomatic and to have sought medical attention because of symptoms. On the other hand, many women, including those with relatively high levels of sexual activity, will go to STD clinics because they have been referred by symptomatic male partners or because they have been identified through contact tracing. The percentage of such women with asymptomatic genital infections is therefore increased. Thus the sensitivity of a clinical finding, that is, the percentage of patients with infection who manifest the clinical feature, may be considerably lower among women seen in STD clinics than among women presenting to vaginitis clinics or private gynecologists.

The predictive value of a clinical feature is even more profoundly influenced by the nature of the population being studied. The predictive value of a finding (PVpos) is the percentage of patients with the finding who actually have the disease being sought.[8] PVpos is highly dependent on the prevalence of disease in the population, and in a low-prevalence population even excellent tests yield a high percentage of false-positive results. If, for example, a particular clinical feature were found in 90% of patients with trichomoniasis (sensitivity = 0.90) and in no more than 10% of patients without trichomoniasis or who in fact had other conditions (specificity = 0.90), the presence of the feature would be strongly correlated with the diagnosis of trichomoniasis. However, if one were working with a population in which the true prevalence of disease were only 0.10, that is, only 10% of patients actually had trichomoniasis, then even this excellent diagnostic feature would have a poor predictive value. Indeed, in this setting only about 50% of patients with the feature would actually have trichomoniasis.

Thus, the clinician is cautioned not to make the diagnosis of trichomoniasis on the basis of any single clinical feature but rather to collect as much information as practical and then to evaluate the patient in the light of all clinical and laboratory data.

Clinical Features of Infection

Sites of Infection

Intravaginal Infection

Trichomonas vaginalis principally infects squamous epithelium in the genital tract.[9] In the adult, the exocervix is thus susceptible to trichomonal attack, but the organisms are only rarely found in the endocervix.[10] *Trichomonas vaginalis* does not cause purulent endocervical discharge, and the clinician who finds a mucopurulent cervical discharge in a patient with trichomoniasis is best advised to consider a coincidental genital infection with *Neisseria gonorrhoeae, Chlamydia trachomatis,* or herpes simplex virus.[11]

Extravaginal Infection

Trichomonads can be recovered from the urethra and Skene's glands in 90% of infected women.[10,12,13] The organisms may be identified on wet mount of material recovered from the urethra in about half of infected women.[12] Infection at these sites may cause dysuria or discharge from the urethra or Skene's ducts.[14] Trichomonads have been identified in voided urine (isolated case reports) and in 15% to 30% of small series of infected women.[15-18] Organisms have also been recovered from urine specimens obtained by bladder catheterization,[16,18] but not from suprapubic aspiration.[19] It thus remains at least somewhat uncertain whether urinary trichomoniasis represents true bladder infection or contamination from the periurethral glands. Organisms have not been recovered from the ureters,[18] and their presence in the upper urinary tract, although reported,[20,21] must be very rare.

Trichomonads have been identified in the fallopian tubal exudate of a woman with acute salpingitis,[20] but this observation has not been repeated. Thus, *T. vaginalis* cannot be considered a major cause of salpingitis, and the clinician encountering a patient with trichomoniasis and marked adnexal tenderness is cautioned to entertain the possibility of simultaneous infection with other organisms such as *N. gonorrhoeae* or *C. trachomatis,* which are clearly identified as pathogens in the fallopian tubes. Some workers have speculated that these organisms might in fact be transported to the fallopian tubes by adhering to trichomonads, but there is no experimental evidence supporting such pathogenetic cooperation.

Perhaps the major significance of trichomonal infection outside the vagina rests in its explaining the relatively low cure rates obtained with topical vaginal therapy. Such treatments (see Chap. 18) may eradicate organisms from the vagina but leave them in the urinary tract resulting in relapse even in the absence of post-treatment sexual contact.

Symptoms

Asymptomatic Infection

About 9% to 56% of trichomonal infections are asymptomatic in various population groups.[22-29] The spectrum of clinical trichomoniasis in women thus ranges from the asymptomatic carrier state to flagrant vaginitis. Factors responsible for this variability are incompletely understood but include differences in the intrinsic virulence of individual strains (Chap. 9) and differences in individual host susceptibility. Clinical evidence for the latter includes the observation that even among several infected female sexual partners of an individual man, the spectrum of disease ranges from mild to severe. Since these epidemiologically linked women are most likely infected with the same strain of trichomonad, differences in clinical expression probably relate to host factors.

In addition, symptoms of trichomoniasis may vary over time in the same woman. About one-third of asymptomatically infected women have been observed to become symptomatic within 6 months.[22] Although one cannot discount the possibility that such women became superinfected with more virulent strains, a number of observers have noted the frequency with which symptoms of trichomoniasis either appear or exacerbate during or immediately following a menstrual period.[23-25] These observations, that asymptomatic infection can become symptomatic and that infection with an individual strain may produce symptomatic disease in some women but not in others, support the contention that women with asymptomatic trichomoniasis should be treated. The medical consequences of chronic asymptomatic trichomoniasis for the individual woman are not defined.

Symptomatic Infection

On the basis of experimental infections and limited clinical observations, the incubation period for trichomoniasis has been estimated to range from 3 to 28 days.[16,23,25,30]

The frequency with which various symptoms are encountered in women with trichomoniasis is presented in Table 11.1. The prevalence of individual symptoms varies widely in different studies and most likely reflects the degree of stoicism of the populations studied or the vigor with which a history of symptoms is sought. Many women do not spontaneously complain of particular symptoms, but indicate their presence if the proper questions are asked.

Vaginal discharge and vulvovaginal irritation are the most common presenting complaints. Discharge is described by 50% to 75% of symptomatic women presenting to STD clinics[22,26,27,29,31] and by at least that many women presenting to other facilities. The discharge is perceived as malodorous by only about 10% of infected patients.[14,17] Vulvovaginal pruritus or irritation is noted by one-fourth to four-fifths of infected women.[22,26,27,29,32-34] Pruritus is often severe and may awaken the patient at night. As one might therefore expect, up to one-half of infected women note some degree of dyspareunia.[31] Dysuria is described by about one-third of patients.[22,34,35] It may be perceived as internal suggesting urethral origin, or more commonly as external, probably produced by urine washing over the inflamed vulva.[36,37]

Pain in the lower abdomen is noted by only 5% to 12% of infected women[26,28] and is more likely to be localized to the right than to the left lower quadrant.

There is no evidence that trichomoniasis is more symptomatic in pregnancy or among users of oral contraceptives.[34,38]

Signs

On examination, the vulva manifests diffuse erythema or excoriation in less than 20% of patients.[26,27] Among different women, the labia range from normal-appearing to markedly erythematous and edematous.

Vaginal discharge is usually loose and not adherent, and it may run out onto the vulva.

Insertion of the speculum sometimes causes considerable discomfort. On many occasions the physician is unable to complete the speculum examination on the first visit of a patient with severe vaginal trichomoniasis. In this situation, a swab may be blindly inserted into the vagina, and microscopic examination of material recovered thereupon is usually adequate to make a diagnosis of trichomoniasis. The patient is treated, and the examination is completed at a return visit, 72 hours later.

On insertion of the speculum (Fig. 11.1), excessive discharge is observed in 50% to 75% of patients.[22,26,27,32-34] The discharge is yellow or green in only 5% to 20% of cases.[14,22,32,39] Classically described as frothy, the discharge is in fact seen to contain bubbles in only about 10% of infected women in most series,[22,29,30,40] although froth has been described in up to 50% of women in other studies.[27,31]

It is clinically difficult to assess the color of genital epithelia, because natural individual col-

TABLE 11.1. Prevalence of clinical features among women with trichomoniasis.

Clinical features and laboratory results	Prevalence
Symptoms	
None	9%–56%
Discharge	50%–75%
Malodorous	~10%
Irritating, pruritic	23%–82%
Dyspareunia	10%–50%
Dysuria	30%–50%
Lower abdominal discomfort	5%–12%
Signs	
None	~15%
Diffuse vulvar erythema	10%–20%
Excessive discharge	50%–75%
Yellow, green	5%–20%
Frothy	10%–50%
Vaginal wall inflammation	40%–75%
Strawberry cervix	~2%
Laboratory findings	
pH > 4.5	66%–91%
Positive whiff test	~75%
Wet mount	
Excess PMNs	~75%
Trichomonads	66%–80%
Fluorescent antibody	~90%
Gram stain	<1%
Acridine orange	~60%
Giemsa's stain	~50%
Pap smear	~70%

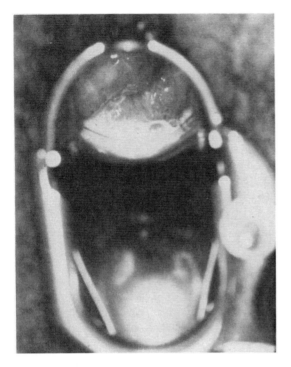

Fig. 11.1. Excessive frothy discharge is revealed on insertion of the vaginal speculum. Note also inflammation of the exocervix. (Reproduced with permission of McGraw-Hill, from Holmes KK, Mårdh PA, Sparling PF, Weisner PG [eds]: *Sexually Transmitted Diseases*. New York: McGraw-Hill, 1984.)

oration varies tremendously. The vaginal walls are recognized as erythematous in perhaps one-third to two-thirds of patients.[14,34] In severe cases, the vaginal walls may present a granular appearance (Fig. 11.2) caused by capillary proliferation and tiny hemorrhages.[40,41] When viewed through the colposcope, the vaginal walls reveal a characteristic double cresting of capillaries.[41] The external surface of the cervix generally shares in the inflammatory process. In perhaps 2% of cases, punctate hemorrhages result in the finding termed "strawberry cervix."[29] If inflammation is particularly severe, inflammatory cells and debris may collect on the vaginal walls yielding a pseudomembrane.[14] Although vaginal discharge sometimes pools in the cervical os, careful examination should reveal no evidence of endocervical discharge in pure trichomonal infection.

Perhaps 15% of infected women in some series show no vaginal inflammation, and about 4% show no evidence of disease even by colposcopy.[25,41]

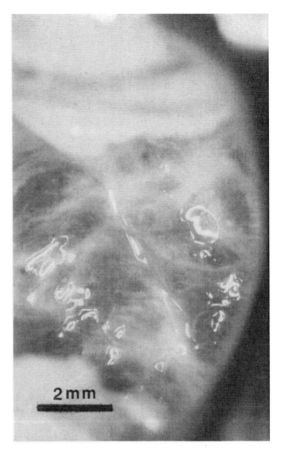

Fig. 11.2. Closeup view of the vagina and exocervix showing the granular appearance associated with severe disease. (Original photograph.)

Bimanual examination often produces discomfort. Much of this is related to vaginal wall tenderness, which must be differentiated from true adnexal tenderness. Frank inflammation of the fallopian tubes is not encountered in pure trichomonal infection, and its presence should prompt evaluation for additional sexually transmitted diseases.

Accuracy of the Clinical Diagnosis

None of the clinical features of vulvovaginal trichomoniasis is sufficiently specific to permit accurate etiologic diagnosis. The patient's perception of symptoms is highly unreliable and subject to considerable psychological overlay.[42] Discharge of either cervical or vaginal origin exits through the introitus and is perceived by patients as "vaginal discharge." Among young, sexually active women, differential diagnosis of vaginal symptoms usually includes trichomoniasis, candidiasis, and bacterial vaginosis (formerly referred to as nonspecific vaginitis, clue cell vaginitis, and bacterial leukorrhea).[6] There is considerable overlap among the clinical features of these three infections. Patients with vulvovaginal candidiasis are perhaps more likely to complain of vulvar irritation and pruritus and rather less discharge, whereas patients with bacterial vaginosis complain of odor, mild to moderate amounts of discharge, and relatively little vulvar irritation.[6] There is, however, so much overlap that differential diagnosis always involves rapidly performed laboratory studies. The clinician is strongly cautioned against attempting to make an etiologic diagnosis on purely clinical grounds.

Clinical Approach to the Patient with Vulvovaginitis

History

Although the history is, by itself, unreliable in differential diagnosis, certain features can provide helpful preliminary clues. The sexually transmitted nature of trichomoniasis is not disputed (see Chap. 17), and bacterial vaginosis, too, is apparently a sexually transmitted disease.[6] Thus a history of contact with a new sexual partner in the 3 to 4 weeks preceding the onset of symptoms may provide some evidence in favor of one of these diagnoses. On the other hand, vulvovaginal candidiasis does not depend on sexual transmission. Risk factors for candidiasis include recent use of antibiotics,[43,44] oral contraceptives,[26,45,46] pregnancy,[47-49] diabetes mellitus, and even tight and insulating clothing.[50] None of these is defined as a risk factor for trichomoniasis, so their presence provides clues to alternative diagnoses.

Symptoms of vulvar irritation, common in trichomoniasis, are also common in patients with vulvovaginal candidiasis. In patients with bacterial vaginosis, on the other hand, inflammation is minimal, and associated irritative symptoms are relatively uncommon.[6,51,52]

Physical Examination

The patient is examined in the lithotomy position. A good light source is essential. The vulva should be examined for evidence of erythema, excoriation, and lesions. Grouped vesicles or ulcerations indicate herpes simplex infection.[53] A diagnosis of candidiasis is suggested by areas of diffuse erythema surrounded by satellite lesions, which are tiny erythematous papules or papulopustules. The examiner should also search for the verrucous lesions of condyloma acuminata[54] or the umbilicated papules of molluscum contagiosum.[55]

The speculum is then inserted. It should be lubricated only with warm water, because commercial lubricants containing antimicrobial substances can affect the sensitivity of subsequent cultures. Erythema of the vaginal walls is assessed. Bacterial vaginosis is usually associated with only minimal if any inflammation.[6,51]

Next, one assesses the nature of the vaginal discharge. Normal, physiologic vaginal discharge has a finely granular appearance. Completely homogeneous, creamy discharge should be regarded as abnormal. Bubbles are usually not present in normal discharge. Observing bubbles, even in an asymptomatic woman, should prompt further investigations. The bubbles may be seen in either trichomoniasis or bacterial vaginosis and have been said to be somewhat larger in the former condition.[6] Material for wet mount is collected by sweeping a cotton swab through the vaginal fornices. The material can then be agitated in 1 ml of saline in a test tube. Some clinicians prefer to place a drop of saline on a microscope slide and transfer a loop of vaginal material directly onto the slide to which a cover slip is then applied. A second drop of vaginal material may be combined with a drop of 10% potassium hydroxide and treated in the same way.

Once adequate vaginal material has been obtained, the exocervix is wiped off with a large cotton swab. Endocervical material can then be recovered on swabs inserted into the cervix. The presence of grossly purulent cervical discharge, the appearance of green or yellow color in the endocervical material recovered on a swab, or the observation of excess numbers of polymorphonuclear neutrophils (PMN) on microscopic examination of endocervical material supports the diagnosis of mucopurulent cervicitis.[11]

On bimanual examination, many patients with trichomoniasis are found to have palpable hypertrophy of the vaginal rugae suggesting active vaginitis but not specifically trichomoniasis. In addition, patients with trichomoniasis sometimes have vaginal wall tenderness that must be differentiated from true adnexal tenderness.

Rapid Laboratory Evaluation

Determination of pH

After the speculum is withdrawn from the vagina, the pH of vaginal secretions can be determined easily through the use of indicator papers. I have found nitrazine paper to be useful because it has a pH range of 4.5 to 7.0, so that any observed color change indicates an elevated vaginal pH. The pH of normal vaginal secretions that have not been contaminated with menstrual blood, semen, or cervical discharge, is 4.5 or less.[56-61] The vaginal pH tends to remain normal in patients with vulvovaginal candidiasis.[56-58] On the other hand, 66% to 91% of patients with trichomoniasis were found, in old studies, to have vaginal pH elevated beyond 4.5 with the majority having pH levels above 5.0.[16,62] The vaginal pH is, however, also elevated in 90% of women with bacterial vaginosis.[6,51,52,61] Thus while an elevated vaginal pH in a symptomatic woman argues against the diagnosis of vulvovaginal candidiasis, it has little value in differentiating trichomoniasis from bacterial vaginosis.

Whiff Test

After the pH of secretions has been determined, two drops of 10% potassium hydroxide solution can be added to the material collected in the lower lip of the speculum. The elaboration of a pungent, aminelike, "fishy" odor constitutes a positive whiff test. Like the vaginal pH determination, the whiff test is negative in patients with candidiasis but is positive in more than 50% of patients with trichomoniasis.[6,63]

Wet Mount

Final differentiation of trichomoniasis from bacterial vaginosis depends principally on the results of microscopic evaluation. A drop of the suspension of vaginal material prepared during physical examination is transferred to a microscope slide, a cover glass is applied, and the preparation is examined as a wet mount. One can use a regular light microscope with the substage condensor racked down or the substage diaphragm closed to increase contrast. Phase-contrast microscopy is becoming available and increases the ease with which the wet mount may be evaluated. In the patient with trichomoniasis, squamous vaginal epithelial cells are normal. Their borders are sharply defined, their nuclei are easily discerned, and they have a relatively clean appearance (Fig. 11.3a). In a normal wet mount, one sees small numbers of PMNs, with a ratio of PMN/epithelial cells of 1 or less in 90% of uninfected women. On the other hand, PMN/epithelial cell ratios greater than 1 were observed in 47 of 65 (72%) women seen in our clinic with trichomoniasis (Fig. 11.3a,b) and did not provide clinical or laboratory evidence for other genital infection. Indeed, 15% of these women had PMN/epithelial cell ratios greater than 10 to 1. Characteristically, the number of PMNs is normal in patients with bacterial vaginosis.[6] Thus in the absence of cervicitis in a symptomatic woman, a finding in vaginal discharge of excessive numbers of PMNs with an elevated vaginal pH suggests a diagnosis of trichomoniasis.

In florid infection, vaginal epithelial cell turnover is rapid, and one may observe increased numbers of parabasilar cells.[3,14] These are ovoid and about two-thirds the size of typical squamous cells. They indicate vaginal wall inflammation and are not specific for trichomoniasis.

The diagnosis of trichomoniasis is best confirmed by observing the parasite (Fig. 11.3a,b). The overall sensitivity of the wet mount has been analyzed in a large number of series and ranges from 50% to 90%. Recent experience, however, defines an overall sensitivity of approximately 66% to 80%.[6,13,22,29,34,64-78] The wet mount can certainly be negative in patients with trichomoniasis, and the failure to observe protozoa does not rule out the diagnosis.

Stained Smears

The Gram stain of vaginal material is not useful for identifying trichomonads[22,79] and is not recommended as part of the routine workup for trichomoniasis. By contrast, candidiasis may be diagnosed by Gram stain, and the technique is as good as the wet mount in making a diagnosis of bacterial vaginosis.[80]

Although hematologic stains, such as Giemsa's stain, can be applied to dried specimens of vaginal discharge, in most studies,[22,81] its sensitivity was found to be lower than that of the

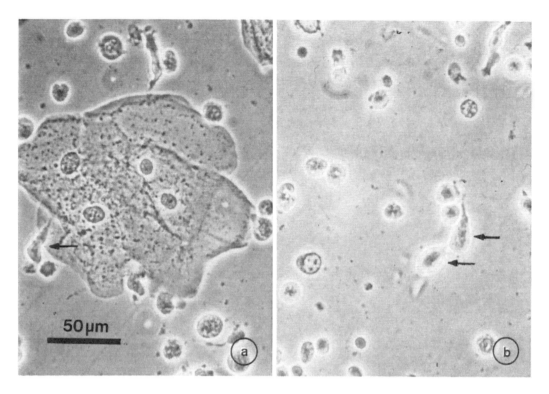

Fig. 11.3a,b. Wet mounts of vaginal material from patients with trichomoniasis. a. Squamous epithelial cells are clean and have sharp borders. Numerous polymorphonuclear neutrophils (*PMN*) are evident. A single elongate trichomonad (*arrow*) is seen near the left border of the figure. b. Numerous PMNs are distributed throughout the field. Two trichomonads (*arrows*) are seen to the right of the center of the figure. (a,b. Phase contrast; same magnification.) (Reproduced with permission of J.B. Lippincott, from Rein MF: Trichomoniasis. In Spittell JA (ed): *Clinical Medicine*. Philadelphia: J.B. Lippincott, 1986.)

wet mount, but it was somewhat higher in others.[82] Dry preparations may also be stained with acridine orange and examined with a fluorescence microscope. The technique detects trichomonads in about 60% of specimens,[69,72] and the diagnostic value does not exceed significantly that of the standard wet mount.

Fluorescent monoclonal antibodies have recently become commercially available for the identification of trichomonads in fixed vaginal smears.[83] The test has only been evaluated in small numbers of women and only compared to the wet mount rather than the culture. The fluorescent antibody (FA) test appears to be more sensitive than the wet mount, and assuming that the wet mount can detect 75% of cases, one can estimate that the sensitivity of the FA test is approximately 90% among high-risk women.

Trichomonads are often reported from cervical cytologic preparations stained with the Papanicolaou method, but some workers consider this an unreliable method for diagnosis.[14,84] In our hands, the specificity of the technique has been high, with the diagnosis confirmed by culture in all of 28 women referred because of positive Pap smears. The sensitivity of the Pap smear is, however, more problematic and varies from 30% to 90% in the published literature, with most series suggesting a sensitivity of about 70%, roughly equivalent to that of the wet mount.[67,70-74,76,78,82,84]

The wet mount remains the preferred microscopic approach to the differential diagnosis of vaginitis in women. Of the techniques named, it alone is reasonably sensitive for trichomoniasis and, importantly, for other forms of vaginitis as well. It yields results in minutes.

Conclusions

Using data from the history and physical examination, the vaginal pH determination, the whiff test, and direct microscopy, an acceptable differential diagnosis can usually be made on the patient's first presentation with vaginal symptoms. In difficult or recurrent cases, cultures for *Trichomonas* may be very helpful (see Chap. 7). Whenever possible, a specific etiologic diagnosis should be made, and therapy should be narrowly directed.

Trichomoniasis clearly serves as a marker for other sexually transmitted diseases. The patient with trichomoniasis should therefore be assiduously studied for the presence of other infections such as gonorrhea or chlamydial disease, which may be clinically silent but eventually of greater medical significance. Treatment of a patient with sexually transmitted disease always involves management of that patient's sexual partners.

References

1. Fleury FJ: Adult vaginitis. *Clin Obstet Gynecol* 24:407-438, 1981.
2. Centers for Disease Control: Nonreported sexually transmitted disease. *MMWR* 28:61-63, 1979.
3. Merrell-National Laboratories: *Review of clinical studies with AVC cream and its individual ingredients for the treatment of vaginitis.* Presented to the Fertility and Maternal Health Drugs Advisory Committee and the Anti-infective and Topical Drugs Advisory Committee, Food and Drug Administration, Washington, May 1979.
4. Sweet RL: Importance of differential diagnosis in acute vaginitis. *Am J Obstet Gynecol* 152:921-923, 1985.
5. Lossick JG: Treatment of *Trichomonas vaginalis* infection. *Rev Infect Dis* 4:S801-S813, 1982.
6. Rein MF, Holmes KK: "Nonspecific vaginitis," vulvovaginal candidiasis, and trichomoniasis: Clinical features, diagnosis, and management. In Remington JS, Swartz MN (eds): *Current Clinical Topics in Infectious Diseases* vol. 4, pp 281-315. New York: McGraw-Hill, 1983.
7. Honigberg BM: Trichomonads of importance in human medicine. In Kreier JP (ed): *Parasitic Protozoa,* vol 2, pp 275-454. New York: Academic Press, 1978.
8. Griner PF, Mayewski RJ, Mushlin AI, Greenland P: Selection and interpretation of diagnostic tests and procedures: Principles and application. *Ann Intern Med* 94:559-592, 1981.
9. Nielsen MH, Nielsen R: Electron microscopy of *Trichomonas vaginalis* Donné: Interaction with vaginal epithelium in human trichomoniasis. *Acta Pathol Microbiol Scand (B)* 83:305-320, 1975.
10. Grys E: Topografia rzęsistkowicy w narządzie rodnym kobiety. *Wiad Parazytol* 10:122, 1964.
11. Brunham RC, Paavonen J, Stevens CE, Kiviat N, Kuo C-C, Critchlow CW, Holmes KK: Mucopurulent cervicitis—the ignored counterpart in women of urethritis in men. *N Engl J Med* 311:1-6, 1984.
12. Wallin JE, Thompson SE, Zaidi A, Wong K-H: Urethritis in women attending an STD clinic. *Br J Vener Dis* 57:50-54, 1981.
13. Whittington MJ: Epidemiology of infections with *Trichomonas vaginalis* in the light of improved diagnostic methods. *Br J Vener Dis* 3:80-91, 1957.
14. Gardner HL: Trichomoniasis. In Gardner HL, Kaufman RH (eds): *Benign Diseases of the Vulva and Vagina,* 2nd ed, pp. 243-272. Boston: G. K. Hall, 1981.
15. Peterson WF, Stauch JE, Ryder CD: Metronidazole in pregnancy. *Am J Obstet Gynecol* 94:343-353, 1966.
16. Trussell RE: Trichomonas vaginalis *and Trichomoniasis,* 1st ed. Springfield, Ill: Charles C Thomas, 1947.
17. Grys E: Localization of *T. vaginalis* in the genitalia and urinary tract of women. *Wiad Parazytol* 19:371-373, 1973.
18. Pattyson RA: *Trichomonas vaginalis* vaginitis: A laboratory and clinical study. *NY State J Med* 37:41-51, 1937.
19. Stamm WE, Wagner KF, Amsel R, Alexander ER, Turck M, Counts GW, Holmes KK: Causes of the acute urethral syndrome in women. *N Engl J Med* 303:409-415, 1980.
20. Gallai Z, Sylvestre L: The present status of urogenital trichomoniasis: A general review of the literature. *Appl Ther* 8:773-778, 1966.
21. Suriyanon V, Nelson KE, Ayudhya VCN: *Trichomonas vaginalis* in a perinephric abscess: A case report. *Am J Trop Med Hyg* 24:776-780, 1975.
22. Rein MF: Trichomoniasis in VD clinic women. Paper presented at the Ann Meet, Am Public Health Assoc, Washington, D.C., November 1, 1977.
23. Catterall RD: Trichomonal infection of the genital tract. *Med Clin North Am* 56:1203-1209, 1972.
24. Brown MT: Trichomoniasis. *Practitioner* 209:639-644, 1972.
25. Jírovec O, Petrů M: *Trichomonas vaginalis* and trichomoniasis. *Adv Parasitol* 6:117-188, 1968.
26. Oriel JD, Patridge BM, Denny MJ, Coleman JC: Genital yeast infections. *Br Med J* 4:761-764, 1976.

27. Wisdom AR, Dunlop EMC: Trichomoniasis: Study of the disease and its treatment. *Br J Vener Dis* 41:90-96, 1965.
28. Zigas V: An evaluation of trichomoniasis in two ethnic groups in Papua, New Guinea. *Sex Transm Dis* 4:63-65, 1977.
29. Fouts AC, Kraus SJ: *Trichomonas vaginalis:* Reevaluation of its clinical presentation and laboratory diagnosis. *J Infect Dis* 141:137-143, 1980.
30. Catterall RD, Nicol CS: Is trichomonal infestation a venereal disease? *Br Med J* 1:1177-1179, 1960.
31. Hughes HE, Gordon AM, Barr GTD: A clinical and laboratory study of trichomoniasis of the female genital tract. *J Obstet Gynaecol Br Commonw* 73:821-827, 1966.
32. Shaw HN, Henriksen E, Kessel JF, Thompson CF: Clinical and laboratory evaluation of "vagisol" in the treatment of *Trichomonas vaginalis* vaginitis. *West J Surg Obstet Gynecol* 60:563-570, 1952.
33. Hager WD, Brown ST, Kraus SJ, Kleris GS, Perkins GJ, Henderson M: Metronidazole for vaginal trichomoniasis: Seven day vs single dose regimens. *JAMA* 244:1219-1220, 1980.
34. Clark DH, Solomons E: An evaluation of routine culture examination for *Trichomonas vaginalis* and *Candida. Am J Obstet Gynecol* 78:1314-1319, 1959.
35. Mason PR, Patterson B: Epidemiology and clinical diagnosis of *Trichomonas vaginalis* infection in Zimbabwe. *Cent Afr J Med* 29:53-56, 1983.
36. Komaroff AL, Pass TM, McCue JD, Cohen AB, Hendricks TM, Friedland G: Management strategies for urinary and vaginal infections. *Arch Intern Med* 138:1069-1073, 1978.
37. Komaroff AL, Friedland G: The dysuria-sterile pyuria syndrome. *N Engl J Med* 303:452-453, 1980.
38. Bramley M: Study of female babies of women entering confinement with vaginal trichomoniasis. *Br J Vener Dis* 52:58-62, 1976.
39. Ronnike F: Discharge symptomatology in 1000 gynaecological cases examined with a view to *Trichomonas vaginalis:* Results of metronidazole (Flagyl) treatment. *Acta Obstet Gynecol Scand* 42(suppl 6):63-70, 1963.
40. Ganju D, Anjneyulu R: Leucorrhoea due to *Trichomonas vaginalis*—clinical trial with hamycin. *Hindustan Antibiot Bull* 9:69-73, 1966.
41. Kolstad P: The colposcopical picture of *Trichomonas vaginalis. Acta Obstet Gynecol Scand* 43:388-398, 1965.
42. Dodson MG, Friedrich EG: Psychosomatic vulvovaginitis. *Obstet Gynecol* 51(suppl):23-25, 1978.
43. Caruso LJ: Vaginal moniliasis after tetracycline therapy. *Am J Obstet Gynecol* 90:374-378, 1964.
44. Oriel JD, Waterworth PM: Effects of minocycline and tetracycline on the vaginal yeast flora. *J Clin Pathol* 28:403-406, 1975.
45. Apisarnthanarax P, Self ST, Ramer G, Monks LF: Oral contraceptives and candidiasis. *Cutis* 77-82, 1974.
46. Diddle AW, Gardner WH, Williamson PJ, O'Connor KA: Oral contraceptive medications and vulvovaginal candidiasis. *Obstet Gynecol* 34: 373-377, 1969.
47. Hopsu-Havu VK, Gronroos M, Punnonen R: Vaginal yeasts in parturients and infestation of the newborns. *Acta Obstet Gynecol Scand* 59:73-77, 1980.
48. Morton RS, Rashid S: Candidal vaginitis: Natural history, predisposing factors and prevention. *Proc R Soc Med* 70(suppl 4):3-6, 1977.
49. Bland PB, Rakoff AE, Pincus IJ: Experimental vaginal and cutaneous moniliasis: A clinical and laboratory study of certain monilias associated with vaginal, oral and cutaneous thrush. *Arch Dermatol Syphilol* 36:760-780, 1937.
50. Elegbe IA, Botu M: A preliminary study on dressing patterns and incidence of candidiasis. *Am J Public Health* 72:176-178, 1982.
51. Vontver LA, Eschenbach DA: The role of *Gardnerella vaginalis* in nonspecific vaginitis. *Clin Obstet Gynecol* 24:439-460, 1981.
52. Pheifer TA, Forsyth PS, Durfee MA, Pollock HM, Holmes KK: Nonspecific vaginitis: Role of *Haemophilus vaginalis* and treatment with metronidazole. *N Engl J Med* 298:1429-1434, 1978.
53. Corey L, Holmes KK: Genital herpes simplex virus infections: Current concepts in diagnosis, therapy, and prevention. *Ann Intern Med* 98:973-983, 1983.
54. Oriel JD: Genital warts. *Sex Transm Dis* 4:153-159, 1977.
55. Brown ST, Nalley JF, Kraus SJ: Molluscum contagiosum. *Sex Transm Dis* 8:227, 1981.
56. Drake SM, Evans BA, Gerken A: Vaginal pH and microflora related to yeast infections and treatment. *Br J Vener Dis* 56:107-110, 1980.
57. Peeters F, Snauwaert R, Segers J, van Cutsem J, Amery W: Observations on candidal vaginitis: Vaginal pH, microbiology, and cytology. *Am J Obstet Gynecol* 112:80-86, 1972.
58. Baldson MJ: Comparison of miconazole-coated tampons with clotrimazole vaginal tablets in the treatment of vaginal candidosis. *Br J Vener Dis* 57:275-278, 1981.
59. Baldson MJ, Taylor GE, Pead L, Maskell R: *Corynebacterium vaginale* and vaginitis: A controlled trial of treatment. *Lancet* 1:501-504, 1980.
60. Chen KCS, Forsyth PS, Buchanan TM, Holmes KK: Amine content of vaginal fluid from untreated and treated patients with nonspecific vaginitis. *J Clin Invest* 63:828-835, 1979.
61. Gardner HL, Dukes CD: *Haemophilus vaginalis*

vaginitis: A newly defined specific infection previously classified "nonspecific" vaginitis. *Am J Obstet Gynecol* 69:962-976, 1955.
62. Elden CA: An evaluation of a particular mode of therapy of *Trichomonas vaginalis*. *Am J Obstet Gynecol* 43:1054-1056, 1942.
63. Chen KC, Amsel R, Eschenbach DA, Holmes KK: Biochemical diagnosis of vaginitis: Determination of diamines in vaginal fluid. *J Infect Dis* 145:337-345, 1982.
64. Teokharov BA: Nongonococcal infections of the female genitalia. *Br J Vener Dis* 45:334-340, 1969.
65. Rothenberg RB, Simon R, Chipperfield E, Catterall RD: Efficacy of selected diagnostic tests for sexually transmitted diseases. *JAMA* 235:49-51, 1962.
66. Tsao W: Trichomoniasis and gonorrhea. *Br Med J* 1:642, 1969.
67. Spence MR, Hollander DH, Smith J, McCaig L, Seiodl D, Brockman M: The clinical and laboratory diagnosis of *Trichomonas vaginalis* infection. *Sex Transm Dis* 7:168-171, 1980.
68. Ackers JP, Lumsden WHR, Catterall RD, Coyle R: Antitrichomonal antibody in the vaginal secretions of women infected with *T. vaginalis*. *Br J Vener Dis* 51:319-323, 1975.
69. Martin RD, Kaufman RH, Burns M: *Trichomonas vaginalis* vaginitis: A statistical evaluation of diagnostic methods. *Am J Obstet Gynecol* 87:1024-1027, 1963.
70. Chintana T, Sucharit P, Chongsuphajaisiddhi T, Tongprasoeth N, Suphadtanaphongs W: A study of the diagnostic methods for *Trichomonas vaginalis* infection. *Southeast Asian J Trop Med Public Health* 10:81-86, 1979.
71. Hayes BS, Kotcher E: Evaluation of techniques for the demonstration of *Trichomonas vaginalis*. *J Parasitol* 46(suppl):45, 1960.
72. Hipp SS, Kirkwood MW, Gaafar HA: Screening for *Trichomonas vaginalis* infection by use of acridine orange fluorescent microscopy. *Sex Transm Dis* 6:235-238, 1979.
73. Nagesha CN, Anathrakrishna NC, Sulochana, P: Clinical and laboratory studies on vaginal trichomoniasis. *Am J Obstet Gynecol* 106:933-935, 1970.
74. Oller LZ: Routine exfoliative cytology for cancer and trichomonas detection at a clinic for venereal diseases. *Br J Vener Dis* 41:304-308, 1965.
75. Robertson DHH, Lumsden WHR, Fraser KF, Hosie DD, Moore DM: Simultaneous isolation of *Trichomonas vaginalis* and collection of vaginal exudate. *Br J Vener Dis* 45:42-43, 1969.
76. Thin RNT, Atia W, Parker JDJ, Nicol CS, Canti G: Value of Papanicolaou-stained smears in the diagnosis of trichomonas, candidiasis and cervical herpes simplex virus infection in women. *Br J Vener Dis* 51:116-119, 1975.
77. Mansukhani S, Gadgil RK, Irani SB: Laboratory aid in the diagnosis of trichomonal vaginitis. *J Indian Med Assoc* 36:447-449, 1961.
78. Wolinska WH: The value of a prepared culture medium in the diagnosis of *Trichomonas vaginalis*. *Am J Obstet Gynecol* 77:306-308, 1959.
79. Cree GE: *Trichomonas vaginalis* in gram-stained smears. *Br J Vener Dis* 44:226-227, 1968.
80. Spiegel CA, Amsel R, Holmes KK: Diagnosis of bacterial vaginosis by direct Gram stain of vaginal fluid. *J Clin Microbiol* 18:170-177, 1983.
81. Cox PJ, Nicol CS: Growth studies of various strains of *Trichomonas vaginalis* and possible improvements in the laboratory diagnosis of trichomoniasis. *Br J Vener Dis* 49:536-539, 1973.
82. Mason PR, Super H, Fripp PJ: Comparison of four techniques for the routine diagnosis of *Trichomonas vaginalis* infection. *J Clin Pathol* 29:154-157, 1976.
83. Smith RF: Detection of *Trichomonas vaginalis* in vaginal specimens by direct immunofluorescence assay. J Clin Microbiol 24:1107-1108, 1986.
84. Perl G: Errors in the diagnosis of *Trichomonas vaginalis* infection as observed among 1199 patients. *Obstet Gynecol* 39:7-9, 1972.
85. Holmes KK, Mardh PA, Sparling PF, Weisner PG (eds): *Sexually Transmitted Diseases*. New York: McGraw-Hill, 1984.
86. Rein MF: Trichomoniasis. In Spittell JA (ed): *Clinical Medicine*. Philadelphia: JB Lippincott, 1986.

12

Epidemiology and Clinical Manifestations of Urogenital Trichomoniasis in Men

John N. Krieger

Introduction

At least three species of trichomonads may be important in human medicine: *Trichomonas tenax* (O.F. Müller), *Pentatrichomonas hominis* (Davaine), and *Trichomonas vaginalis* Donné[1]; each appears to be limited to a specific site. *Trichomonas tenax* inhabits the oral cavity and *P. hominis* inhabits the colon, which is the primary site of Trichomonadida in other animal hosts. Infestation with either *T. tenax* or *P. hominis* has seldom been associated with morbidity. In contrast to the other human trichomonads, *T. vaginalis* infects the urogenital tract and has been implicated as a common cause of urogenital disease.[2-7]

The first clinical description of *T. vaginalis* was published in 1836 when Donné identified the protozoon in unstained preparations of vaginal discharge mixed with saline.[8] This procedure, the "wet mount" examination, remains the primary test for diagnosis of trichomoniasis in routine medical practice.

Trichomonas vaginalis was long considered to be a common but harmless organism,[9-11] an opinion unchanged either by Kunstler's description of the trichomonads in the female urinary tract in 1883, or by Marchand's identification of these organisms in the male urinary tract in 1884.[12,13] Over the next 6 decades many reports described the clinical features of trichomoniasis, but practicing physicians were primarily interested in the diagnosis and treatment of gonorrhea and syphilis. With the availability of effective antibiotics, it became apparent to clinicians that there was a significant incidence of other sexually transmitted diseases.[14] Attention was again directed toward *T. vaginalis* as a common cause of urogenital morbidity.

Despite the widely accepted concept that trichomoniasis is transmitted almost exclusively by sexual contact, clinical reports generally do not consider male urogenital trichomoniasis. For example, Short et al[15] recently compared the prevalence of diseases among heterosexual and homosexual men and heterosexual women attending a clinic for sexually transmitted diseases. *T. vaginalis* was identified by wet mount examination of 331 (14%) of 2357 women attending the clinic. In contrast, no cases of trichomoniasis were identified among 6071 men. From the design of the clinical protocol it is apparent that *T. vaginalis* was not considered to be a pathogen in men. This viewpoint is widespread among venereologists, specialists in infectious diseases, and urologists in the United States. Thus, it is not surprising that most general physicians also consider *T. vaginalis* to be an unusual but harmless commensal organism in the male genitourinary tract.

This chapter reviews clinical studies of *T. vaginalis* infections in men. First we consider why most studies have been limited to investigation of trichomoniasis in women. We then discuss the epidemiology and clinical manifestations of *T. vaginalis* infections in the male. Finally we review suggestions for clinical classification of male urogenital trichomoniasis.

Research that yielded the author's data was supported in part by Research Grants 2 RO1 AI121092 and AI21422, from NIAID, U.S. Public Health Service.

Why the Epidemiology and Clinical Manifestations of Trichomoniasis in Men Have Been Incompletely Defined

Although thousands of studies have been completed on the epidemiology and clinical features of human urogenital trichomoniasis, most of these studies concern various aspects of the disease in women.[1,9,10] There are several possible explanations for the relative paucity of studies investigating trichomoniasis in men (Table 12.1). Trichomoniasis in women is associated with well-described clinical syndromes, considerable morbidity, and standard diagnostic algorithms.[2,3,6] In contrast, trichomoniasis in men is associated with poorly delineated clinical syndromes, uncertain morbidity, and there is no familiar algorithm available for diagnosis.[4] Although the presentation of trichomoniasis in women is carefully considered in clinical textbooks and teaching, the manifestations of trichomoniasis in men have been omitted from standard texts. Consequently, most practicing American physicians do not consider trichomoniasis in the differential diagnosis of genitourinary symptoms in men.

Epidemiology and Clinical Manifestations of Trichomoniasis in Men

Evidence for Heterosexual Transmission of *T. vaginalis*

Much of our information on the epidemiology of trichomoniasis in men is indirect and has been inferred from the large number of studies carried out in women. Because the epidemiology of trichomoniasis in women is the subject of Chap. 17 in this book, we shall review only aspects of these studies pertinent to the disease in men.

In the past, there was considerable debate on the mode of transmission of urogenital trichomoniasis.[29-35] It was demonstrated that trichomonads could survive for short periods on toilet seats, washcloths, clothing, and in bathwater.[26] Although nonvenereal transmission by such contaminated fomites may explain reports of trichomoniasis in a few patients, such as sexually mature virgins,[15] it is now accepted that transmission of trichomoniasis by such nonvenereal mechanisms is most unusual.[2,6,9,15]

Several types of evidence support the consensus that *T. vaginalis* is transmitted by direct genital contact in almost all cases. First, trichomoniasis is a urogenital tract disease. The highest prevalence of trichomoniasis occurs during the years of peak sexual activity, in patients attending clinics for sexually transmitted diseases, and in prostitutes.[4] Furthermore, trichomoniasis is commonly associated with other sexually transmitted diseases.[37-39]

The especially high prevalence of trichomoniasis among contacts of infected patients strongly supports the view that it is transmitted primarily by sexual contact (Table 12.2). In various studies trichomonads were isolated from 14% to 60% of male sexual partners of infected women. Conversely, trichomonads were isolated from 67% and 100% of female partners of infected men in two studies from the United Kingdom.[30,36]

Additional support for the importance of sexual contact in transmission of trichomonia-

TABLE 12.1. Comparison of clinical features of urogenital trichomoniasis in women and in men.

Aspect*	Women	Men
Literature (1)	Extensive, over 10,000 articles; many recent	Lmited, under 100 articles; few in last decade
Clinical syndromes	Well-described; familiar to clinicians	Poorly delineated; unfamiliar to clinicians
Morbidity (1,4)	Readily apparent; considerable	?
Diagnosis		
"Clinical cases"	Standard tests: "Hanging drop" "Wet mount"	? Value
Other cases (16-24)	Urinalysis, cytology, special stains	Urinalysis, cytology(?), special stains(?)
Cultures (1,4,25)	Useful	Useful
Serology (26-28)	Investigational	Investigational

*Reference nos. in parentheses.

TABLE 12.2. Epidemiologic evidence suggesting that trichomoniasis is transmitted by sexual contact.

Female or male contacts	Men or women with trichomoniasis	Prevalence (%)	No. patients	Country	Reference
Female	Men	100	56	U.K.	30
Female	Men	67	118	U.K.	36
Male	Women	60	30	U.K.	40
Male	Women	45	207	U.K.	13
Male	Women	42	31	U.S.	29
Male	Women	39	364	Yugoslavia	41
Male	Women	33	24	U.K.	36
Male	Women	30	50	Poland	42
Male	Women	28	94	U.S.	43
Male	Women	22	37	U.K.	44
Male	Women	20	641	Poland	45
Male	Women	14	21	U.K.	25

sis is derived from human inoculation studies performed between 1940 and 1955 (prior to the availability of effective antitrichomonal therapy). These studies fulfilled Koch's postulates in both men and women and clarified certain aspects of pathogenesis. Intravaginal inoculation of women resulted in establishment of infection in up to 75% of subjects, regardless of the bacterial flora initially present.[46] Symptoms and physical findings characteristic of vaginal trichomoniasis occurred after a 5- to 28-day incubation period. Lanceley and McEntegart[47] reported similar studies in ten male "volunteers." None of five control patients developed symptoms or clinical manifestations following intraurethral inoculation with sterile growth medium (Table 12.3). In contrast, urethritis occurred in all five experimental patients who were each inoculated with identical medium containing 4×10^6 *T. vaginalis*. Prostatitis was documented in two patients by microscopic examination of expressed prostatic secretions. Both patients with prostatitis had severe urethritis, with protozoa found on microscopic examination. One patient developed "moderately severe" urethral discharge containing trichomonads. The two remaining patients experienced transient, mild urethral discharge containing no demonstrable protozoa. Although all patients developed mild urethritis within 24 hours of inoculation, trichomonads could not be demonstrated in urethral scrapings, urine, or prostatic fluid until 6 to 9 days after inoculation.

Thus, clinical and experimental studies support the concept that in almost all cases tricho-

TABLE 12.3. Clinical findings in men following intraurethral inoculation of *T. vaginalis* in sterile medium.

Patient no.	Inoculum*	Clinical findings†	Protozoa observed‡ (days)
1	T.v.	Severe urethritis, prostatitis	6-9§
2	T.v.	Severe urethritis, prostatitis	6-44
3	T.v.	Moderately severe urethritis	9-13
4	T.v.	Mild urethritis	—
5	T.v.	Mild urethritis	—
6	M	—	—
7	M	—	—
8	M	—	—
9	M	—	—
10	M	—	—

Based on studies reported by F Lanceley and MG McEntegart.[47]
*T.v. intraurethral inoculation with 4×10^6 *T. vaginalis* organisms; M, intraurethral inoculation with sterile growth medium.
†The distinction between urethritis and prostatitis was based on examination of prostatic secretions.
‡Urethritis developed in all patients inoculated with *T. vaginalis* but protozoa were observed in only 3 of 5 cases.
§One patient was lost to follow-up after 50 days.

monads are transmitted on the male genitalia and/or in the male lower urogenital tract where they have the potential to produce disease.

Definition of the Issues

The major epidemiologic and clinical issues in male trichomoniasis may be considered in three categories: What is the relationship between the presence of *T. vaginalis* and nongonococcal urethritis? Is there any relationship between trichomoniasis and other genitourinary tract syndromes? What is the natural history of *T. vaginalis* infection in men? Each of these issues are considered sequentially.

Trichomonas vaginalis *as a Cause of Nongonococcal Urethritis*

Much of the literature on the epidemiology of trichomoniasis is from Eastern Europe; it has been summarized by Trussell,[9] Jírovec and Petrů,[10] and Honigberg.[1] Tables 12.4 and 12.5 compare several investigations from various geographic areas. Clearly, trichomoniasis is worldwide in distribution, although the reported prevalence in male populations is a function of many factors, such as diagnostic techniques, patient selection, and the orientation of the investigators.

As is the case with other sexually transmitted diseases, definition of appropriate control populations is difficult.[4] Several possibilities have been suggested, including asymptomatic men, civilian or military; patients with no obvious genitourinary tract abnormalities; and men with gonococcal urethritis (Table 12.4). In these studies *T. vaginalis* was seldom isolated from asymptomatic men with no apparent genitourinary abnormalities. In contrast, *T. vaginalis* was found in up to 18% of patients with gonococcal urethritis.

The importance of *T. vaginalis* as a cause of genitourinary syndromes in men is controversial.[4] Many investigators, particularly those who are primarily interested in sexually transmitted diseases in women, believe that asymptomatic transmission is the rule for men infected with *T. vaginalis*. Such asymptomatic male carriers are viewed as epidemiologically important because they transmit symptomatic disease to their female partners. This opinion is widespread among American workers who have identified *T. vaginalis* as an infrequent cause of nongonococcal urethritis or prostatitis.[52-57] Some studies used methods for isolation of multiple genitourinary tract pathogens and reasonable control groups. However, despite these comprehensive studies, the etiologies remain uncertain for 30% to 50% of cases of nogonococcal urethritis and the great majority of cases of prostatitis.[57] Review of the methodology demonstrates that direct examination of unstained specimens was used to exclude infection by *T. vaginalis* in many studies. In expert hands this may not detect 50% to 90% of infections in men. Occasional studies, primarily concerned with other urogenital pathogens, have included cultures for *T. vaginalis,* "for the sake of completeness." Thus, negative results for fastidious organisms, such as *T. vaginalis,* should be interpreted cautiously.

In contrast, many European studies have found that infection with *T. vaginalis* is a major cause of morbidity in men.[4] Clinical conditions that have been associated with trichomoniasis

TABLE 12.4. Prevalence of trichomoniasis in "control" male populations.

Population	Country	No. patients	Prevalence (%)	Reference
Gonococcal urethritis	U.S.	108	18	48
	U.S.	30	17	36
	U.S.*	141	10	49
	U.K.	40	8	40
	Poland	96	4	42
	U.K.	285	1	47
Criminals: (routine urine specimens)	Japan	191	4	50
	Japan	1303	3	50
	U.K.	227	0	25
No genitourinary abnormality	Poland	34	3	42
Asymptomatic				
Civilian	U.K.	22	0	36
Military	U.S.	27	0	51

*Includes patients with both gonococcal and nongonoccal urethritis.

TABLE 12.5. Prevalence of trichomoniasis in men with nongonoccal urethritis.

Country	No. patients	Prevalence (%)	Reference
Chile	2482	68	58,59
U.S.	75	41	48
	179	2	60
	118	1	54
Canada	324	11	14
U.K.	326	15	36
	2300	6	61
	1646	6	34
	310	6	31
France	288	12	62
Poland	128	11	42
	765	10	63
Japan	100	17	50

include nongonococcal urethritis, prostatitis, balanoposthitis, epididymitis, urethral stricture disease, and infertility. The strongest epidemiologic association is between *T. vaginalis* and nongonococcal urethritis.[52-57] Some of the studies linking trichomoniasis and nongonococcal urethritis are summarized in Table 12.5. The highest reported prevalence was 68% among men in Chile,[58,59] and the lowest was 1% among men in Seattle.[54] The median prevalence of *T. vaginalis* in men with nongonococcal urethritis was 11% in these studies. It was found in clinical studies[58,59] that nongonococcal urethritis occurred in 16% of male sexual partners of women with trichomonal vaginitis. In similar studies, Weston and Nicol[13] found that 93 (45%) of 206 sexual contacts of infected women subsequently developed nongonococcal urethritis. The favorable symptomatic response to metronidazole therapy in men with trichomonad-positive nongonococcal urethritis provides additional support for an etiologic role for *T. vaginalis* in this syndrome.[4,43] None of the studies discussed in this paragraph employed methods for isolation of urogenital pathogens other than gonococci.

On balance, studies described in this section support an etiologic role for *T. vaginalis* in some men with nongonococcal urethritis. However, the precise proportion of cases of nongonococcal urethritis attributable to trichomoniasis has not been reliably established. It is entirely possible that the significance of *T. vaginalis* as a cause of nongonococcal urethritis will vary in different geographic areas and different patient populations.

Trichomonas vaginalis *as a Cause of Other Urologic Syndromes*

Although many workers have described involvement of the prostate gland in men with trichomoniasis (Table 12.6), the role of *T. vaginalis* in prostatitis is not clear.[4] Identification of *T. vaginalis* in prostatic fluid was first reported by Drummond,[67] who found the protozoa in expressed prostatic secretions from four of five

TABLE 12.6. Identification of *T. vaginalis* in men with urologic conditions other than nongonoccal urethritis.*

Syndrome	Country	No. patients	Prevalence (%)	Reference
Prostatitis[†]	U.S.	26	100	60
	Japan	70	28	64
	France	178	22	65
	Japan	946	9	66
"Lower tract symptoms not responding to routine therapy"	U.S.	42	72	23
Infertility	Japan	609	4	67

*All patients had suspected *T. vaginalis* urethritis or prostatitis.
[†]As defined by the authors of each study, often including some men with nongonoccal urethritis.

husbands of infected women. Subsequent workers identified *T. vaginalis* in specimens obtained from 9% to 100% of patients with prostatitis.[23,60,63-66,68] Sylvestre et al[14] described prostatic involvement in 10 (29%) of 35 men with trichomonal urethritis. An especially high prevalence of trichomoniasis, up to 85%, has been reported in patients who do not respond to multiple courses of antibiotics or who have longstanding symptoms.[47,51,68,69] Unfortunately, studies of the role of *T. vaginalis* in prostatitis uniformly suffered from major methodologic deficiencies.[4] None of these investigators employed a quantitative or widely accepted definition of "prostatitis."[53,57,70] No procedures were employed for isolation of other common sexually transmitted urogenital pathogens, for example, *Chlamydia trachomatis, Ureaplasma urealyticum,* or herpesvirus. Furthermore, control groups were generally absent or inadequate.

Less rigorous observations suggest that infection with *T. vaginalis* may be associated with various other lower genitourinary tract conditions including balanoposthitis, other inflammatory processes affecting the external genitalia, urethral stricture disease, epididymitis, and infertility. Inflammation of the foreskin (posthitis) and glans penis (balanitis) has been described as one feature of trichomoniasis by many investigators.[1,4,9,10,14,33,40] Balanitis with or without posthitis occurred in 2 (11%) of 18 men with trichomoniasis described by Watt and Jennison[40] and in four (4%) of 91 men described by Wilson and Ackers.[25] It has been suggested that the presence of phimosis or "excessive length" of the prepuce are predisposing factors for development of trichomonal balanoposthitis.[14,59] *Trichomonas vaginalis* has also been isolated as the only pathogen in chronic draining sinuses on the median raphe of the penis[71] and has been suggested as one consideration in the differential diagnosis of penile ulcers.[69,72]

Urethral abnormalities have been associated with trichomoniasis in several studies. Catterall and Nicol[30] documented strictures in 5 (9%) of 56 infected men, while Weston and Nicol[13] found strictures in 11 (50%) of 22 men with persistent urethritis caused by *T. vaginalis*. Urethral strictures have been associated with longstanding and recurrent disease.[4,13,30,73] However, it is unclear whether presence of a preexisting anatomic obstruction favors development of persistent trichomonal urethritis or whether persistent urethral inflammation led to stricture formation. Presence of congenital urethral abnormalities, such as hypospadias and meatal stenosis, has also been suggested as a risk factor for development of trichomoniasis.[10] None of the studies described in this paragraph included control groups and little consideration was given to other sexually transmitted organisms, such as *C. trachomatis* or *Neisseria gonorrhoeae,* which may be associated with development of urethral strictures.

Epididymitis caused by infection with *T. vaginalis* was initially described by Liston and Lees in 1940.[74] There have been five more recent case descriptions.[75-77] Although conventional bacterial cultures were negative, no attempts were made to isolate other urogenital pathogens, for example, *Chlamydia,* which commonly causes epididymitis in populations of sexually active men. The significant prevalence of *T. vaginalis* in some studies of infertile men, as well as adverse effects of *T. vaginalis* on sperm motility in vitro, have led some workers to consider this parasite an unusual cause of male infertility.[78] This possibility would appear to be unlikely since other studies have shown that *T. vaginalis* is highly prevalent in many populations of pregnant women and there are no controlled studies supporting an association between trichomoniasis and infertility. Finally, some observations suggest that infection with *T. vaginalis* may be associated with both erectile dysfunction and premature ejaculation,[58,59] but there is no evidence for a causal relationship.

Data implicating *T. vaginalis* as the etiologic agent in urologic conditions other than nongonococcal urethritis are incomplete at best. The prostate gland appears to be involved in some patients with trichomoniasis, but *T. vaginalis* does not appear to be a common cause of prostatitis in the absence of urethritis. Balanoposthitis and other inflammatory conditions of the male genitalia may on occasion be related to infection with *T. vaginalis*. It appears that trichomoniasis may rarely involve genitourinary organs such as the epididymis. The data linking *T. vaginalis* with other urinary tract syndromes are weak.

The Natural History of T. vaginalis *Infections in Men*

There have been few studies of the natural history of untreated trichomoniasis in men. One of the major unresolved questions is what proportion of infected men develop symptoms. In their extensive review Jírovec and Petrů[10] state that "in the male the infection is mostly latent or persisting in subclinical form. . . ." Similar findings were reported by Willcox.[35] Watt and Jennison[40] investigated the husbands of 30 women undergoing therapy for chronic trichomonal vaginitis. Eighteen (60%) of the 30 men had positive cultures for *T. vaginalis* but 11 (61%) of the infected patients had no clinical abnormalities. Four of the seven remaining patients had evidence of urethritis, that is, mucous threads; one had both urethritis and prostatitis; and two developed trichomonal balanoposthitis. According to other workers[58,59] 16% of sexual contacts of women with trichomoniasis developed nongonococcal urethritis, while Weston and Nicol[13] reported that 23% of a similar population developed urethritis. In contrast, Catterall[60] found that 108 (86%) of 126 infected men were symptomatic. Thus, there is wide variation in the reported proportions of symptomatic disease among men infected with *T. vaginalis*.

The incubation period for development of symptomatic trichomonal urethritis was evaluated in several studies. It was found to equal from 3 to 8 days,[58,59] 5 days or less,[13] and from 6 to 9 days in experimental trichomonal urethritis.[47] However, much wider variability in the incubation period was reported by Czech investigators,[10,62] according to whom this period ranged from 1 to 5 days in approximately 20% of men, from 6 to 10 days in 40%, from 11 to 15 days in about 10%, from 2 to 4 weeks in 20%, and from 1 to 3 months in approximately 9%.

The relative proportions of different clinical presentations of trichomoniasis in men were evaluated in several studies. Two general designs have been employed for investigating the clinical syndromes related to *T. vaginalis* infection in men: evaluation of sexual partners of infected women and investigations of men presenting genitourinary tract complaints. As discussed above, most investigations have shown that the majority of sexual contacts of infected women are asymptomatic.[10,35,40] There are few studies of men with documented trichomoniasis. Weston and Nicol[13] evaluated 93 men with trichomoniasis—52% had obvious urethral discharge, 26% complained of dysuria, and 5% noted increased urinary frequency, while 21% had no genitourinary symptoms. Wisdom and Dunlop[34] described 92 men with symptomatic trichomoniasis. Urethral discharge was the most common problem (64%); other complaints included urethral irritation (52%), urethral pain (24%), increased urinary frequency (5%), lower abdominal pain ("probably due to cystitis") (4%), and balanoposthitis (3%). Only 6 (7%) of 92 cases had evidence of prostatitis. Catterall[60] described the clinical findings in the 108 symptomatic cases in his series of 126 men infected with *T. vaginalis*. Ninety men (83%) complained of urethral discharge, 26 (24%) noted pruritus, 15 (14%) had dysuria, 4 (4%) experienced increased urinary frequency, 3 (3%) noted bloody urethral discharge or hematuria, and 1 (1%) had epididymitis. Further evaluation demonstrated evidence of urethral strictures in ten cases, and "all ten showed a marked tendency to relapse." Evidently, urethritis was the major syndrome in all these clinical cases, while other syndromes were significantly less common. In contrast, Kuberski[59] found that 20 (77%) of 26 men with genitourinary trichomoniasis had clinical evidence of prostatitis.

Spontaneous resolution of trichomonal infection in men has been reported by several investigators. The most convincing study was that of Weston and Nicol,[13] who investigated the prevalence of trichomoniasis among male sexual partners as a function of the interval between last sexual exposure and examination (Table 12.7). When the interval between exposure and examination was 2 days, *T. vaginalis*

TABLE 12.7. Prevalence of urogenital trichomoniasis among sexual partners of infected women.

Interval between last sexual exposure and evaluation (days)	*T. vaginalis* identified/contacts examined	Prevalence (%)
2	16/23	70
4	16/29	55
6	7/27	26
8	10/25	40
10	8/21	38
14	9/30	30
20	8/15	53
30	3/13	23
40	1/4	—
50	2/2	—
60	1/4	—
60+	5/6	—
Unknown	7/8	—
TOTAL: 93/207		MEAN: 45

Based on data of Weston TE, Nicol CS: Natural history of trichomonal infection in males. *Br J Vener Dis* 39: 251-257, 1963, reproduced with permission of the British Medical Association, London, England.

was found in 70% of patients, but after 2 weeks infection could be identified in only 30% of patients. Beyond that period it became more difficult to evaluate the prevalence of *T. vaginalis* because of the relatively small numbers of patients examined. Other workers suggest that spontaneous resolution of untreated trichomoniasis is most unusual. Coutts et al[58] stated that "chronic infection of the male genitourinary tract is very common." Watt and Jennison[40] found that 13 (87%) of 15 men infected with *T. vaginalis* were still infected after 2 weeks, but the patients apparently were not instructed to refrain from intercourse with infected partners. Catterall[60] treated 12 men with oxytetracycline (now known to be totally ineffective) for 5 days. Follow-up after 3 months showed that 10 (83%) still harbored the protozoa. Similar findings were reported by Whittington[36] and in several studies from Eastern Europe.[9,10] Thus, the long-term consequences of untreated trichomoniasis in men have not been completely defined.

In summary, several tentative conclusions may be drawn from the limited studies of the natural history of urogenital trichomoniasis in the male:

1. The spectrum of disease ranges from totally asymptomatic to florid urethritis complicated by prostatitis.
2. The incubation period for development of symptomatic disease appears to be 10 days or less in most patients.
3. Nongonococcal urethritis is the most common clinical syndrome and other conditions appear to be unusual.
4. Significant questions remain concerning possible spontaneous resolution of infection and the long-term consequences of chronic infection by *T. vaginalis* in the male.

Classification of Male Urogenital Trichomoniasis

Several schemes have been proposed for classification of trichomoniasis in men. One of the most elaborate is that used by various investigators in Europe.[1,10,41,62] Many distinct stages of *T. vaginalis* infection are distinguished in this classification: primary acute, primary subchronic, primary latent, primary subacute, secondary subacute, secondary latent, secondary acute, secondary subchronic, and chronic. It is suggested that the primary acute stage occurs shortly after sexual contact with an infected woman and is characterized by profuse urethritis and, perhaps, prostatitis. The primary subchronic stage has a gradual onset and scant discharge. The primary latent phase is not associated with symptoms, but the infection remains contagious for subsequent sexual contacts. The primary acute stage may progress to a secondary subacute stage characterized by less severe symptoms. The primary subchronic phase may lead to an asymptomatic secondary latent phase. The primary latent phase may lead to florid symptoms, i.e., secondary acute stage, or mild exacerbations, i.e., secondary subacute phase. Any stage may eventually progress to a chronic phase. Although such classifications have a theoretic appeal and may be useful for explaining the course of trichomoniasis in particular patients, this elaborate scheme appears to be unnecessarily complex for routine clinical application.

Less complicated classifications have been employed by a number of other investigators.[13,14,34,40,49,59] In general these clinical classifications describe three presentations of male urogenital trichomoniasis: asymptomatic carrier state, acute symptomatic disease, and mild symptomatic disease. Asymptomatic carriers are usually identified by investigation of sexual partners of infected women and occasionally by routine urinalysis carried out for other indications. As discussed above, the relative proportion of asymptomatic male carriers is open to dispute, but the asymptomatic carrier state appears to be common. Acute disease, characterized by purulent urethritis, occasionally accompanied by prostatitis or other complications, appears to be an unusual presentation of trichomoniasis. Most symptomatic men with trichomoniasis present with mild urethritis. Usually such patients are indistinguishable from those with nongonococcal urethritis of other etiologies.[37,38,55,56,79] Presence of long-standing symptoms, failure of multiple courses of antibacterial therapy, or a history of sexual exposure to a woman with vaginitis should lead to strong consideration of trichomoniasis in the differential diagnosis.

Summary and Conclusions

Trichomonas vaginalis was first described 150 years ago and is now recognized as a major cause of morbidity in women. Although essentially all cases of trichomoniasis are transmitted by direct genital contact, *T. vaginalis* infection in male carriers is associated with poorly delineated clinical syndromes and uncertain morbidity. Most available studies have evaluated the prevalence of trichomoniasis in various male populations. *Trichomonas vaginalis* was seldom isolated from asymptomatic men or those with no obvious genitourinary tract abnormalities but was found in up to 18% of men with gonococcal urethritis.

The strongest epidemiologic association is between trichomoniasis and nongonococcal urethritis. Investigation of male sexual partners of infected women demonstrates that 16 to 45% develop nongonococcal urethritis and that many infected patients remain asymptomatic. Unfortunately, very few studies of *T. vaginalis* in men with nongonococcal urethritis have employed methods for isolation of urogenital pathogens other than gonococci. For this reason the precise proportion of cases of nongonococcal urethritis attributable to trichomoniasis has not been reliably established but it probably represents no more than 10% of patients with this syndrome. Because nongonococcal urethritis appears to be the most common sexually transmitted disease in men, this figure suggests that male trichomoniasis may be associated with considerable morbidity.

The prostate gland appears to be involved in some men with trichomoniasis, but *T. vaginalis* seems to be an unusual cause of prostatitis in the absence of urethritis. Balanoposthitis and other inflammations of the external genitalia and epididymitis may occasionally result from infection with *T. vaginalis*. There are weak data supporting associations between *T. vaginalis* and other genitourinary tract syndromes.

Limited investigations of the natural history of male urogenital trichomoniasis support the concept that the clinical spectrum ranges from totally asymptomatic carrier state to florid urethritis complicated by prostatitis. The incubation period for development of symptoms appears to be 10 days or less in most patients. Nongonococcal urethritis is clearly the most common clinical syndrome, while other conditions are unusual. Important questions remain concerning the issues of spontaneous resolution of infection and the long-term consequences of chronic asymptomatic trichomoniasis in the male.

For clinical purposes men with urogenital trichomoniasis may be classified in three groups. Asymptomatic carriers are usually identified by investigation of sexual contacts of infected women or occasionally by urinalysis obtained for other reasons. Acute trichomoniasis in men, characterized by profuse purulent urethritis and, occasionally, by prostatitis, appears to be unusual. Most symptomatic men with trichomoniasis present with symptoms and signs typical of nongonococcal urethritis.

References

1. Honigberg BM: Trichomonads of Importance in Human Medicine. In Kreier JP (ed): *Parasitic Protozoa*, vol 2, pp 275-454. New York: Academic Press, 1978.
2. Fouts AC, Kraus SJ: *Trichomonas vaginalis:* Reevaluation of its clinical presentation and laboratory diagnosis. *J Infect Dis* 141:137-143, 1980.
3. Fleury FJ: Adult vaginitis. *Clin Obstet Gynecol* 24:407-438, 1981.
4. Krieger JN: Urologic aspects of trichomoniasis. *Invest Urol* 18:411-417, 1981.
4. McLellan R, Spence MR, Brockman M, Raffel L, Smith JC: The clinical diagnosis of trichomoniasis. *Obstet Gynecol* 60:30-34, 1982.
6. Rein MF, Holmes, KK: "Nonspecific vaginitis," vulvovaginal candidiasis, and trichomoniasis: Clinical features, diagnosis and management. In Remington JS, Swartz MN (eds): *Current Clinical Topics in Infectious Diseases*, pp 281-315. New York: McGraw-Hill, 1983.
7. Altchek A: Pediatric vulvovaginitis. *J Reprod Med* 29:359-375, 1984.
8. Donné A: Animacules observés dans les matières purulentes et le produit des sécrétions des organes génitaux de l'homme et de la femme. *C R Hebd Acad Sci (Paris)* 3:385-386, 1836.
9. Trussell RE: *Trichomonas vaginalis and trichomoniasis*, 1st ed., Springfield, Ill., Charles C Thomas, 1947.
10. Jírovec O, Petrů M: *Trichomonas vaginalis* and trichomoniasis. *Adv Parasitol* 6:117-188, 1968.
11. de Leon E: Trichomoniasis. In Marcial-Rojas RH (ed): *Pathology of Protozoal and Helminthic Diseases*, pp 124-138. Baltimore: Williams & Wilkins, 1971.
12. Kunstler J: *Trichomonas vaginalis* Donné. *J Microg* 8:317-331, 1884.
13. Weston TE, Nicol CS: Natural history of trichomonal infection in males. *Br J Vener Dis* 39:251-257, 1963.
14. Sylvestre L, Belanger M, Gallai Z: Urogenital trichomoniasis in the male: Review of the literature and report on treatment of 37 patients by a new nitroimidazole derivative (Flagyl). *Can Med Assoc J* 83:1195-1199, 1960.

15. Short SL, Stockman DL, Wolinsky SM, Trupei MA, Moore J, Reichman RC: Comparative rates of sexually transmitted diseases among heterosexual men, homosexual men and heterosexual women. *Sex Transm Dis* 11:271-274, 1984.
16. Kiviat NB, Paavonen JA, Brockway J, Critchlow, CW, Brunham RC, Stevens CE, Stamm WE, Kuo CC, de Rouen T, Holmes KK: Cytologic manifestations of cervical and vaginal infections. I. Epithelial and inflammatory cellular changes. *JAMA* 253:989-995, 1985.
17. Greenwood JR, Kirk-Hillaire K: Evaluation of acridine orange stain for detection of *Trichomonas vaginalis* in vaginal secretions. *J Clin Microbiol* 14:699, 1981.
18. Krieger JN, Holmes KK, Spence MR, McCormack WM, Tam MR: Geographic variation among *Trichomonas vaginalis:* Demonstration of antigenic heterogeniety using monoclonal antibodies and the indirect immunofluorescence technique. *J Infect Dis* 152:979-984, 1985.
19. Papanicolaou GN, Wolinska WH: Vaginal cytology in trichomonas infestation. *Int Rec Med* 168:551-556, 1955.
20. Mason PR, Super H, Fripp PJ: Comparison of four techniques for the diagnosis of *Trichomonas vaginalis* infections. *J Clin Pathol* 29:154-157, 1976.
21. Wachtel EJ, Hudson EA: The usefulness of cytology. *Br J Hosp Med* 23:256-265, 1980.
22. Levett PN: A comparison of five methods for the detection of *Trichomonas vaginalis* in clinical specimens. *Med Lab Sci* 37:85-88, 1980.
23. Summers JL, Ford ML: The Papanicolaou smear as a diagnostic tool in male trichomoniasis. *J Urol* 107:840-842, 1972.
24. Hipp SS, Kirkwood MW, Gaafar HA: Screening for *Trichomonas vaginalis* infection by use of acridine orange fluorescent microscopy. *Sex Transm Dis* 6:235-238, 1979.
25. Wilson A, Ackers JP: Urine culture for the detection of *Trichomonas vaginalis* in men. *Br J Vener Dis* 56:46-48, 1980.
26. Su-Lin K-E: Antibody to *Trichomonas vaginalis* in human cervicovaginal secretions. *Infect Immun* 37:852-857, 1982.
27. Mathews HM, Healy GR: Evaluation of two serological tests for *Trichomonas vaginalis* infection. *J Clin Microbiol* 17:840-843, 1983.
28. Ackers JP, Lumsden WHR, Catterall RD, Coyle R: Antitrichomonal antibody in the vaginal secretions of women infected with *T. vaginalis*. *Br J Vener Dis* 51:319-323, 1975.
29. Burch TA, Rees CW, Reardon LV: Epidemiological studies on human trichomoniasis. *Am J Trop Med Hyg* 8:312-318, 1959.
30. Catterall RD, Nicol CS: Is trichomonal infestation a venereal disease? *Br Med J* 1:1177-1179, 1960.
31. Lancely F: *Trichomonas vaginalis* infections in the male. *Br J Vener Dis* 29:213-217, 1953.
32. Trussell RE, Plass ED: The pathogenicity and physiology of a pure culture of *Trichomonas vaginalis*. *Am J Obstet Gynecol* 40:883-890, 1940.
33. Wisdom AR, Dunlop EMC: Trichomoniasis: Study of the disease and its treatment. II. The disease and its treatment in men. *Br J Vener Dis* 41:93-96, 1965.
34. Wisdom AR, Dunlop EMC: Trichomoniasis: Study of its treatment in women. II. The disease and its treatment in women. *Br J Vener Dis* 41:90-93, 1965.
35. Willcox RR: Epidemiological aspects of human trichomoniasis. *Br J Vener Dis* 36:167-174, 1960.
36. Whittington MJ: Epidemiology of infections with *Trichomonas vaginalis* in the light of improved diagnostic methods. *Br J Vener Dis* 33:80-91, 1957.
37. Krieger JN: Evaluation and treatment of unconventional genitourinary tract infections. *Semin Urol* 3:193-199, 1985.
38. Rein MF: Urethritis. In Mandell GL, Douglas RG Jr, Bennett JE (eds): *Principles and Practice of Infectious Diseases,* pp 720-729. New York: John Wiley, 1985.
39. Rein MF: Vulvovaginitis and cervicitis. In Mandell GL, Douglas RG Jr, Bennett JE (eds): *Principles and Practice of Infectious Diseases,* pp 729-738. New York: John Wiley, 1985.
40. Watt L, Jennison RF: Incidence of *Trichomonas vaginalis* in marital partners. *Br J Vener Dis* 36:163-166, 1960.
41. Koštić, P: Importance de l'étude du trichomonas chez l'homme. *Urol Int* 9:171-177, 1959.
42. Hoffmann B, Kilczewski W, Małyszko E: Studies on trichomoniasis in males. *Br J Vener Dis* 37:172-175, 1961.
43. Scharpia HE: Studies on metronidazole (Flagyl) in the therapy of urogenital trichomoniasis in the male patient. *J Urol* 93:303-306, 1965.
44. Ackers JP, Catterall RD, Lumsden WHR, McMillan A: Absence of detectable local antibody in genitourinary tract secretions of male contacts of women infected with *Trichomonas vaginalis*. *Br J Vener Dis* 54:168-171, 1978.
45. Hoffman B, Małyszko E: Investigation of *Trichomonas vaginalis:* Incidence among inhabitants of Białystok county using different laboratory methods. *Wiad Parazytol* 15:355-356, 1969.
46. Hesseltine HC, Wolters SL, Campbell A: Experimental human vaginal trichomoniasis. *J Infect Dis* 71:127-130, 1942.
47. Lanceley F, McEntegart MG: *Trichomonas vaginalis* in the male: The experimental infections of a few volunteers. *Lancet* 1:668-671, 1953.
48. Feo LG, Varano NR, Fetter TR: *Trichomonas vaginalis* in urethritis of the male. *Br J Vener Dis* 32:233-235, 1956.
49. Butler WJ, Spence MR, Brockman MT: Trichomoniasis: A male urethritis, in preparation.
50. Kawamura N: Studies on *Trichomonas vaginalis*

in the urological field. Parts I-V. *Jpn J Urol* 60:15-49, 1969.
51. Kuberski T: Evaluation of the indirect technique for study of *Trichomonas vaginalis* infections, particularly in men. *Sex Transm Dis* 5:97-102, 1978.
52. Mardh PA, Colleen S: Search for uro-genital tract infections in patients with symptoms of prostatitis. *Scand J Urol Nephrol* 9:8-16, 1975.
53. Meares EJ Jr: Prostatitis syndromes: New perspectives about old woes. *J Urol* 123:141-147, 1980.
54. Holmes KK, Handsfield HH, Wang SP, Wentworth BB, Turck M, Anderson JB, Alexander ER: Etiology of nongonococcal urethritis. *N Engl J Med* 292:1199-1205, 1975.
55. Bowie WR: Urethritis in males. In Holmes KK, Mardh PA, Sparling PF, Wiesner PJ (eds): *Sexually Transmitted Diseases,* pp 638-649. New York: McGraw-Hill, 1984.
56. Krieger JN: Urethritis: Etiology, diagnosis, treatment and complications. In Gillenwater JY, Grayhack JT, Howards SS, Duckett JW (eds): *Adult and Pediatric Urology,* vol 2, pp 1343-1374. Chicago: Year Book Medical Publishers, 1987.
57. Krieger JN: Prostatitis syndromes: Pathophysiology, differential diagnosis and treatment. *Sex Transm Dis* 11:100-112, 1984.
58. Coutts WE, Silva-Inzunza B, Tallman B: Genitourinary complications of non-gonococcal urethritis and trichomoniasis in males. *Urol Int* 9:189-208, 1959.
59. Kuberski T: *Trichomonas vaginalis* associated with nongonococcal urethritis and prostatitis. *Sex Transm Dis* 7:135-136, 1980.
60. Catterall RD: Diagnosis and treatment of trichomonal urethritis in men. *Br Med J* 2:113-115, 1960.
61. Durel P, Roiron-Ratner V, Siboulet A, Sorel C: Nongonococcal urethritis. *Br J Vener Dis* 30:69-72, 1954.
62. Jíra J: Zur Kenntnis der männlichen trichomoniase. *Zentralbl Bakteriol Parasitenkd Infektionskr Hyg (A)* 172:310-329, 1958.
63. Kawamura N: Trichomoniasis of the prostate. *Jpn J Clin Urol* 27:335-344, 1973.
64. Verges J: Les prostatites à trichomonas formes rondes (T.F.R.). *J Urol Nephrol* 85:357-361, 1979.
65. Kimura S: Studies on *Trichomonas vaginalis* in urological patients. *Jpn J Urol* 56:455-475, 1966.
66. Ohmura K: Studies on the *T. vaginalis* infection in male genitourinary tracts. *Jpn J Parasitol* 9:510-514, 1960.
67. Drummond AC: Trichomonas infestation of the prostate gland. *Am J Surg* 31:98-103, 1936.
68. Gallai Z, Sylvestre L: The present status of urogenital trichomoniasis: A general review of the literature. *Appl Ther* 8:773-778, 1966.
69. Fullilove RE Jr: *Trichomonas vaginalis* in men. *J Med Soc NJ* 80:94-96, 1983.
70. Meares EM Jr, Stamey TA: Bacteriologic localization patterns in bacterial prostatitis and urethritis. *Invest Urol* 5:492-518, 1968.
71. Soendjojo A, Pindha S: *Trichomonas vaginalis* infection of the median raphe of the penis. *Sex Transm Dis* 8:255-257, 1981.
72. Sowmini N, Vijayalakshmi K, Chellmuthiah C, Sundaram M: Infections of the median raphe of the penis: Report of three cases. *Br J Vener Dis* 49:469-474, 1972.
73. Riba LW, Harrison RM: Strictures of the male urethra and *Trichomonas vaginalis. Surg Gynecol Obstet* 71:369-371, 1940.
74. Liston WG, Lees R: *Trichomonas vaginalis* infestation in male subjects. *Br J Vener Dis* 16:34-55, 1940.
75. Fisher I, Morton RS: Epididymitis due to *Trichomonas vaginalis. Br J Vener Dis* 45:252-253, 1969.
76. Amar AD: Probable *Trichomonas vaginalis* epididymitis. *JAMA* 206:417-418, 1967.
77. Krieger JN: Epididymitis, orchitis and related conditions. *Sex Transm Dis* 11:173-181, 1984.
78. Tuttle JP Jr, Holbrook TW, Derrick FC: Interference of human spermatozoal motility by *Trichomonas vaginalis. J Urol* 118:1024-1025, 1977.
79. Rein MF, Chapel TA: Trichomoniasis, candidiasis, and the minor venereal diseases. *Clin Obstet Gynecol* 18:73-88, 1975.

13

Urogenital Trichomoniasis in Children

Alicja Kurnatowska and Alina Komorowska

Introduction

A review of the literature reveals that much less attention has been given to problems of *Trichomonas vaginalis* infections of children than to urogenital trichomoniasis in adults. In this chapter we attempt to review *T. vaginalis* trichomoniasis in children using data from the works of others and much information, published and hitherto unpublished, collected by us in Poland.

Diagnosis

Anatomical and hormonal differences and those in the nature of the material collected from the urogenital organs of a child in its various developmental stages influence greatly the prevalence as well as certain properties of *T. vaginalis*. In light of this, diagnosis of trichomoniasis in children is more difficult than in adult women or men, despite the fact that, in adults, a definitive diagnosis is based on finding the parasite in material collected from the patients.

Collection of Material for Examination

In infections of young children, including newborns, infants, and preschool groups, the important diagnostic material is the morning urine. This urine, collected in sterile containers, must be delivered to the laboratory, in a constant temperature-maintaining vessel, within 30 to 40 minutes from the time of collection. Older children are encouraged to urinate (a few hours after their previous micturition) in the physician's office just before examination.

In girls in all stages of sexual development, one should collect, simultaneously with urine, vaginal contents from the vestibule and, whenever possible, through the aperture in the hymen, from deeper parts of the vagina. In girls during puberty, it is often possible to obtain vaginal contents through the use of a speculum and with it also to collect mucus from the canal and surface of the cervix uteri. Girls who started to menstruate should be examined in the early estrogenous or late luteal phases of the cycle. These periods have been selected because, according to earlier studies, the largest *T. vaginalis* populations were found at those times, irrespective of the clinical picture.[1] In some cases it is necessary to collect urine by a urethral catheter with or without cystoscopy. One can also obtain contents of the urethra by pressing its external orifice. Often, especially in infections of young girls, the clinician is unsuccessful in obtaining sufficient material for examination.

In boys at all stages of development, before collecting urine, the examiner obtains, by means of a sterile cotton swab, material for smears from the surface of the glans and from under the prepuce. Subsequently, urethral contents are collected by the means described above for girls. In addition to urine voided naturally, urine obtained during catheterization of the bladder should be examined.

Material from the vestibule of the vagina, the vagina, and the vaginal portion of the cervix

---

This chapter was translated from Polish by B.M. Honigberg.

uteri is collected on a cotton swab wound around a thin wooden or metal applicator. A 2- to 3-mm platinum loop is employed to obtain urethral contents from children of both sexes.

Examination of Microscopic Preparations and Cultivation

If there is enough material in a sample collected from a child, one fresh and two permanent stained preparations should be made. For fresh preparations either 0.9% NaCl or 0.1% safranin in 0.9% NaCl are used. Despite the fact that such preparations constitute an important method for examination of the urine sediment obtained by centrifugation, they can be used only for general orientation purposes. They ought to be examined within 15 minutes of collection; otherwise they should be kept in a moist chamber at 37°C until examined. The microscopic examination can be made by phase-contrast, dark-field, or bright-field microscopy, at magnifications of 100 × to 600 ×. Addition of safranin facilitates identification of the parasites, because, in contrast to those of epithelial cells, nuclei of white cells and bacteria do not stain with safranin.

Permanent preparations from various materials are made as films on slides. The films are fixed in 70% methanol and either Giemsa- or Wright-stained.[2] The stained preparations are examined in bright field under oil immersion (magnification of × 1000 to × 1500). In such preparations one finds pleomorphic protozoa with blue-staining cytoplasm and with oval or spindle-shaped cherry-red nuclei situated close to the anterior end of a parasite. The nuclei contain large chromatin granules. Above the nucleus, there is a group of kinetosomes from which originate the flagella. If the fixation process is successful, one can also see the axostyle and flagella, all of which stain dark blue. The second permanent preparation should be Gram-stained. It is employed for evaluation of the bacterial flora accompanying *T. vaginalis*. Fungi, primarily yeasts,* are recognized in Giemsa- or Wright-stained preparations. In infections with yeasts there are evident vegetative cells, blastospores, and pseudohyphae; in infections involving other fungi, one finds, for example, true hyphae, astrospores, or conidia.

The aforementioned methods of preparation

---

*For "Delimitation of the yeasts," see reference 62.

of permanent slides from material obtained from children suspected of infection with *T. vaginalis* are based on the successful results obtained by us in the course of comparative studies carried out on the vaginal contents of women infected with *T. vaginalis*. These methods have been chosen from several kinds of fixation and more than ten staining methods.[3] Many characteristics, arrived at by morphometric techniques, of *T. vaginalis* from children correspond to those obtained by these methods for this parasite found in infections of adults,[4] with the aid of light and scanning electron microscopy.[5,6] There exist, however, certain quantitative differences between the trichomonads obtained from young girls and from adult women. Among these differences are the higher percentages of rounded organisms and smaller variation coefficients for the shape of *T. vaginalis* found in girls. There are also some additional morphometric differences between populations of this parasite seen in young girls and from adult female patients.

Certain structural characteristics of *Trichomonas* in young girls facilitate diagnosis. For example, the presence in Giemsa-stained and Wright-stained smears of cells with nucleoplasmic ratios of about 10, characteristic of trichomonads, allows their differentiation from rounded cells from the deep layer of the squamous epithelium that lines the vagina. These rounded cells, always present in young girls, are characterized by a centrally located nucleus that accounts for one-third to one-fifth of the cell area. Therefore, in children in whom the parasite densities are low, as in newborns, infants, and girls up to 10 years of age, the chances of correct diagnosis are greatly enhanced by the use of Giemsa-stained or Wright-stained preparations (Table 13.1). However, this statement does not hold true for smears stained according to the Papanicolaou method.

In general, a higher percentage of *T. vaginalis* infections can be detected by cultivation of the material collected from patients suspected of harboring this parasite. Since the trichomonad populations are typically low in children, the employment of axenic cultivation is of special importance.

Epidemiology

The prevalence of infection with the urogenital trichomonad in children is inadequately under-

TABLE 13.1. Comparison (in percent) of the efficiency of various methods of diagnosing *T. vaginalis* in children, assuming that 100% of cases are detected by cultivation in Roiron-Ratner medium.[94]

Methods	Percent of positives ±SD*
Wet preparations	
0.9% NaCl	30.8 ± 2.3
0.1% safranin in 0.9% NaCl	34.5 ± 2.4
Fixed and stained preparations	
Papanicolaou's	25.5 ± 2.8
Gram's	35.3 ± 2.4
Giemsa's	77.0 ± 2.1
Wright's	84.8 ± 1.8
Cultivation	
Simić[93]	95.3 ± 1.1
Roiron-Ratner[94]	100.0 ± 0.0

*Sample standard deviation.

stood, especially in boys. This is caused by various difficulties encountered in screening for trichomoniasis; most reports deal with children suffering from a variety of diseases of these organs and are attended to by specialists not necessarily familiar with *T. vaginalis*.

Prevalence of *T. vaginalis* Infection in Girls during a 25-Year Period (1960–1985)

Analysis of our data derived from examination of 19,024 girls, 0 to 18 years of age, yielded the following results. Among the patients attending the Outpatient Pediatric Gynecology Clinic (henceforth referred to as "the Clinic") for various reasons, the mean percentage of patients infected with *T. vaginalis* equaled approximately 4.99 ± 0.16%,* while among those examined in the Center for Parasitic and Mycotic Diseases (henceforth referred to as "the Center"), it was approximately 7.0 ± 0.5%; the difference between these means is statistically significant ($P < 0.01$). In comparing the results obtained in sequential quinquennial periods there was found a nearly sixfold decrease in the frequency of trichomoniasis in young girls attending the Clinic, and a twofold decline among patients seen at the Center. The data reflecting the foregoing results are shown in Table 13.2.

Distribution of frequency of recognized trichomoniasis in biennial periods is shown in Figure 13.1. During the 1970s the percentage of infections noted among patients attending the Clinic (*curve a*) is significantly smaller ($P < 0.01$) than in patients examined at the Center (*curve b*). It is not possible directly to compare our results obtained in the quinquennial and biennial periods with the data published by others because there have been no other 25-year investigations reported.

The early publications from the 1950s and 1960s usually dealt with small numbers of cases of trichomoniasis in girls. However, some workers[7-10] obtained results similar to ours (6% to 12%), with regard to prevalence of *T. vaginalis* infections. In the 1970s, the frequency of trichomoniasis reported for girls varied widely. It was said to have ranged from 1.9%[11] to 14%.[12] The data obtained during the last quinquennial period were in the range given for the 1970s.[13-16] The comparison of the results of various authors is shown in Table 13.3; one must

*Sample standard deviation (SD).

TABLE 13.2. Frequency of *T. vaginalis* of genitourinary organs in girls attending the Outpatient Pediatric Gynecology Clinic (Group I) and the Center for Parasitic and Mycotic Diseases (Group II).

	Group I			Group II		
	Total no.	Positive for *T. vaginalis*		Total no.	Positive for *T. vaginalis*	
Years	examined	n*	% ± SD†	examined	n	% ± SD
1960–64	2512	246	9.79 ± 0.59	—‡	—	—
1965–69	3290	227	6.90 ± 0.44	376	33	8.78 ± 1.46
1970–74	4380	244	5.57 ± 0.35	663	64	9.65 ± 1.15
1975–79	2504	68	2.72 ± 0.32	665	42	6.32 ± 0.94
1980–84	3905	61	1.56 ± 0.20	808	39	4.83 ± 0.75
1985	730	18	2.47 ± 0.57	87	4	4.60 ± 2.25
TOTAL	17,325	864	4.99 ± 0.16	2599	182	7.00 ± 0.50

*Number of girls positive for *T. vaginalis*.
†Sample standard deviation.
‡Not determined.

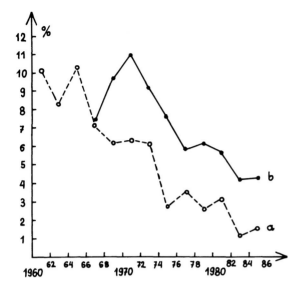

FIGURE 13.1. Prevalence of *T. vaginalis* infection among girls (0 to 18 years of age) attending the Outpatient Pediatric Gynecology Clinic (*curve a*) or the Center for Parasitic and Mycotic Diseases (*curve b*) in Łódź, Poland, between 1960 and 1985.

remember, however, that not all the authors used cultivation methods in demonstrating *T. vaginalis* in the vaginal contents of girls, and this method is known to provide for the detection of the highest percentage of positive results.

Frequency of Urogenital Trichomoniasis in Various Developmental Stages of Girls

In our investigations, similar to those reported by Peter and Vesely,[17] we distinguished three developmental stages, each characterized by different susceptibility to *T. vaginalis* infection. These stages are: (1) newborns and infants (up to 1 year of age); (2) childhood (up to 10 years of age); and (3) maturation or puberty (10 to 18 years of age).

During the first stage, there are sufficiently high levels of maternal estrogens to provide conditions favorable for the establishment of *T. vaginalis* infection. The second period, childhood, which starts after cessation of estrogen activity in infants, is characterized by inactivity of the genital organs related to changes referred to as *involutio postnatalis*. The conditions prevalent in the vagina, including pH of 7 to 8, do not favor the development of the urogenital trichomonad. Finally, the phases of prepuberty, puberty, and adolescence, especially the last, are accompanied by estrogenic activity as well as by anatomical and physiologic changes of the genital organs, including lowering of the pH to between 4 and 6.5, which provide an environment suitable for the establishment of *T. vaginalis*. In 2727 girls who were examined for various reasons we analyzed the prevalence of the *T. vaginalis* infections in all the aforementioned developmental periods. The data from these investigations are presented in Table 13.4. The statistical differences among the mean percentages were significant—in developmental periods I and II P was <0.05 and between II and III, P was <0.01. However, there was no such difference between the mean percentages of infection in developmental periods I and III ($P>0.05$). A histogram showing the frequency of *T. vaginalis* infections in girls at increasing ages is shown in Figure 13.2. In this figure, there are evident two peaks, II and III, of *T. vaginalis* infection in girls. It should be added that among six girls infected with *T. vaginalis* during childhood (period II), there were also signs of early sexual maturation; one of these girls actually started to menstruate. Peter and Vesely[17] also pointed out a relationship between precocious infections with *T. vaginalis* and pubertas praecox, and many other authors[18-22] emphasized the rarity of this parasite in girls in childhood (period II). In small groups of girls examined, trichomoniasis was not found until 13 years of age [23,24] or else there were reported only very few cases.[16,25-28] It appears that the factors favoring *T. vaginalis* infection during childhood may be foreign bodies introduced into the vagina or else female pinworms migrating into the genital organs.[20,29,30]

Infection Routes

Infection with *T. vaginalis* in newborns is acquired during delivery. Despite the well developed prophylactic measures against trichomoniasis in the period of pregnancy during which there are required examinations of the vaginal contents and treatment of all diagnosed *T. vaginalis* infections, in some instances there occurs infection of the child from the mother, especially in pelvic deliveries. In infancy, the child can acquire the urogenital trichomonad from the mother or from another person who takes care of him or her. In all children, before and after puberty, infection with *T. vaginalis* may

TABLE 13.3. Prevalence of urogenital trichomoniasis in girls as reported by various authors between the years 1956 and 1983.

	Patients examined				
			Positive for *T. vaginalis*		
Years	Age (yr)	Total no.	No.	% ± SD*	Reference
1956	1–14	84	3	3.5 ± 2.00	95
1958	0–12	58	3	5.2 ± 2.92	96
1959	9–12	110	4	3.6 ± 1.76	97
1959	1–11	35	0	—	23
1963	0–18	833	63	7.5 ± 0.91	32
1964	0–18	933	92	9.8 ± 1.06	7
1964	6–14	151	8	5.3 ± 1.83	98
1966	0–17	2727	165	6.0 ± 0.45	9
1969	0–18	619	74	12.0 ± 1.31	10
1969	0–13	50	1	2.0 ± 1.98	27
1970	0–18	2062	71	3.4 ± 0.40	99
1971	0–18	2479	190	7.6 ± 0.53	100
1971	10–18	362	14	3.9 ± 1.02	101
1972	0–18	350	49	14.0 ± 1.85	12
1972	0–18	2613	164	6.2 ± 0	102
1972	0–18	400	15	3.7 ± 0.94	103
1972	0–18	5554	59	1.07 ± 0.41	104
1972	8–14	70	6	8.6 ± 3.35	28
1973	0–9	936	18	1.9 ± 0.46	105
1973	10–18	936	78	8.3 ± 0.90	105
1974	0–18	1278	25	1.9 ± 0.38	11
1975	0–11	30	1	3.3 ± 3.30	26
1976	0–18	286	33	11.5 ± 1.89	106
1977	0–10	211	21	9.9 ± 2.06	107
1977	0–18	1090	50	4.6 ± 0.66	71
1978	0–18	1120	102	9.1 ± 0.86	108
1978	0–15	98	2	2.0 ± 1.41	25
1980	0–13	38	0	—	24
1980	0–18	1643	75	4.6 ± 0.52	13
1981	15–19	1413	37	2.6 ± 0.12	14
1983	10–12	119	4	3.4 ± 0.45	16
1983	0–18	1122	114	10.1 ± 0.90	15

*Sample standard deviation.

also be indirect, depending on the environment and especially the family environment. In this period, however, one has to consider that members of both sexes may serve as sources of infection. The routes of infection may be indirect as well as direct in girls who become sexually active.

Infections in Family Units

In the early stages of our investigations,[31] we examined 40 women from families in which young girls were infected with *T. vaginalis* and 20 women from families of children free of this parasite. The correlation coefficient value ($r = +0.31$) indicated statistically significant correlation ($P < 0.05$) between *T. vaginalis* infection in girls and in adult women from their immediate surroundings. The preliminary results encouraged us to undertake further investigations, which we carried out between 1967 and 1972 and between 1980 and 1985. These studies involved three groups of families: Group I—trichomoniasis in adult women; Group II—trichomoniasis in adult men; and Group III—trichomoniasis in children. From various families, 472 *T. vaginalis*-infected women, mothers or grandmothers and in some cases women who took care of the children, were examined. Urogenital trichomoniasis was diagnosed also in one-third of the sexual partners of those women as well as in 37 children (7.8 ± 1.2%), the majority of whom were girls. In 67 families in which *T. vaginalis* was diagnosed originally in men, infection with this parasite could be demonstrated in the majority of their female sexual partners (91 ± 3.9%) as

TABLE 13.4. Comparison of frequency of trichomoniasis diagnosed in a sample of 2727 girls in various stages of sexual development.

	Patients		
	Total no. examined	Positive for *T. vaginalis*	
Stage of development		No.	% ± SD*
I—Newborn and infancy	106	9	8.5 ± 2.70
II—Childhood	1001	6	0.6 ± 0.24
III—Puberty and adolescence	1620	150	9.3 ± 0.72
TOTAL	2727	165	6.1 ± 0.46

*Sample standard deviation.

well as in 9 children (13.4 ± 4.3%). In the last group, which consisted of 109 families, *T. vaginalis* was found first in children—it was demonstrated in 93 girls and 16 boys. The parasite was diagnosed also in the majority of the mothers (72.5 ± 4.3%), and in almost half of the fathers (43.1 ± 4.7%).

It should be pointed out that urogenital trichomoniasis is three times more frequent in those women who are sexual partners of men in whom *T. vaginalis* infection can be demonstrated; the difference between the percentages of infection in the males and females is statistically significant ($P < 0.001$). However, in Group III, in which *T. vaginalis* was found first in the children more frequently than in men, the infection appeared to have originated in women, especially in mothers. In the latter group, the difference between the percentages of infection between men and women was also statistically significant ($P < 0.05$). More recently there were found three cases of *T. vaginalis* and *Candida* infection of the prostate in men. One of these cases belonged to Group III, because trichomoniasis was found first in children.

Infections in Groups of Children Other Than Family Units

Studies on *T. vaginalis* infections in children were also conducted in groups other than family units. Examination of children (up to 10 years of age) from orphanages, preschool groups, and also school-age groups, led to identification of infection foci. For example, in the course of examination of 70 girls, between 2 and 5 years of age, in one of the preschool centers in Łódź, not a single case of trichomoniasis was found (Kurnatowska, unpublished data). Similar results were obtained by Gorzędowska,[32] who investigated 60 girls of the same age in an orphanage and a preschool center in Wrocław. It is necessary, however, once more to underscore the fact that in the period of childhood the sources of infection of girls may be persons from their immediate environment,[31] especially if these persons use the same beds or toiletries (e.g., sponges, towels) or live in one-room apartments. In young people, there are also possibilities of indirect infections (nonsexual route) in childrens' homes, in residential schools, and in student hostels,[3,32] as well as of direct infections (sexual contacts). Valent et al[33] described a group of girls examined in childrens' homes, among whom *T. vaginalis* was diagnosed in over one-third (34.6%) of the indi-

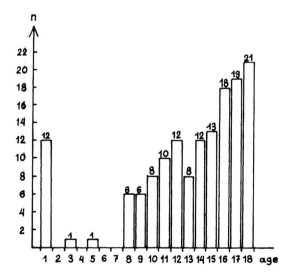

FIGURE 13.2. Histogram showing the frequency of *T. vaginalis* infection in 2727 girls of various ages.

viduals; however, a majority of them were not virgins. Gorzedowska[32] in her investigations of girls from high schools and one residential school found a similar percentage (30%) of urogenital trichomoniasis; she demonstrated that more *T. vaginalis* infections were diagnosed among girls living in common rooms that housed four people. Sagone[34] emphasized the possibility of spreading *T. vaginalis* infection by exchange of bathing suits among girls. Other authors[35] asserted that *T. vaginalis* could retain its viability in a drop of discharge on bathing supplies for 2 to 6 hours, especially at temperatures ranging between 25°C and 37°C. Our own investigations revealed prolonged survival of *T. vaginalis* in mudbaths and in bathing waters in one of the Polish spas.[36] Neistein et al[37] emphasized the possibility of survival of *T. vaginalis* in warm mineral waters and on moist bathing implements up to 2 to 3 hours.

The reader should be reminded that examination of nearly 20,000 girls who reported to our Clinic, and especially of those who complained about vaginal discharge, revealed a decline in the prevalence of urogenital trichomoniasis (see Fig. 13.2). This decline appears to be related to a threefold decrease in the prevalence of *T. vaginalis* infection among adult women in the area of Łódź during the past 11 years.[38] Undoubtedly, the decline has resulted in lowering the opportunities for infection of girls by direct contact with their mothers or other women in their immediate surroundings. We believe that the decline of urogenital trichomoniasis is the result of screening of adult women accompanied by prophylactic measures with regard to cancer of genital organs, sexually transmitted diseases (STDs), or infection that can be transmitted during delivery. However, data published by the World Health Organization (WHO)[39] do not warrant an optimistic outlook for decline of *T. vaginalis* trichomoniasis; in many countries its prevalence is rising.

Clinical Aspects

General Picture

The clinical picture of urogenital trichomoniasis in children depends on their sex, the stage of sexual development, and to some degree on the bacteria and mycotic agents accompanying *T. vaginalis* in the biocenosis of the urogenital organs. Our clinical experiences, and those of most other workers, with trichomoniasis of children included primarily girls, and only very rarely boys.

In interviews with our patients, we took into consideration, in addition to the customary data pertaining to childhood diseases, and especially to diseases of the urinary system, information dealing with the occurrence of trichomoniasis in the family as well as in other persons in the immediate surroundings of the child.

In girls, special attention was given to vaginal discharges as well as to pruritus and a burning sensation in the vulva and vagina. With regard to the urethra, we considered pain on micturition, polyuria, or other ailments of the urinary system. The clinical changes were defined objectively by describing the nature of the vaginal contents as well as the appearance of the vulva and of the vaginal vestibule. Whenever the insertion of the speculum was possible, the state of the vagina and of the cervix uteri was described, taking into consideration the following parameters: swelling and reddening, erosion, disseminated spots or lumps, and other small changes. In the course of palpation, there was noted the absence or presence of pain in the lower abdominal regions. In postmenarcheal girls attention was paid to the degree of development of the secondary and tertiary sexual characteristics, the character of the menstrual cycle, and also to the beginning of sexual activity.

In boys attention was paid to itching and burning of the urethra, pain at micturition, frequency of micturition, and other ailments of the urinary system. Among the objective clinical changes the appearance of the prepuce and the contents originating from under the prepuce were considered as well as swelling and reddening of the urethral orifice and pains during palpation. In the period of puberty we evaluated the development of secondary and tertiary sexual characteristics and determined the onset of sexual activity. In a certain proportion of cases it was also necessary to consider the results of x-ray and ultrasound examinations that in some children, especially those with recurrent trichomoniasis, helped to discover developmental abnormalities of the urinary system which favored persistence of *T. vaginalis* infection. Considering the results of evaluation by the various aforementioned methods of the material taken from a patient, we attempted to recognize unifocal trichomoniasis in an organ or multifocal trichomoniasis, most frequently including infection of the sexual and excretory organs. *Trichomonas vaginalis* infections complicated by mycoses were classified separately; these also could be unifocal or multifocal. For obvious reasons, in addition to the results of laboratory examinations, we considered all the data brought out during interviews and objective clinical signs before making the final diagnosis. In cases of family-group investigations, we took into account also data pertaining to all the family members of the *T. vaginalis*-infected child. It should be added that in various forms of *T. vaginalis* infections, there ought to be considered also the results of bacteriologic examinations, which aid in prescribing the regimen for treatment of the child and of members of his or her family, or in other group infections, of the persons from the child's immediate surroundings.

Accompanying Bacterial Flora

In a group of 756 patients who harbored *T. vaginalis* and were examined for the presence of aerobic and anaerobic bacteria, the most frequently isolated were gram-positive cocci. Among them, there were enterococci and streptococci; the second most frequently observed group of bacteria was represented by gram-negative rods, that is, *Escherichia coli*. Among gram-positive rods, there was isolated also *Gardnerella vaginalis*. The numerical data pertaining to the prevalence of various bacteria are presented in Table 13.5. It should be added that we were unsuccessful in isolating *Lactobacillus* strains, which according to certain authors quite frequently accompany infection with *T. vaginalis*.[40] In the analysis of the bacterial strains isolated from material obtained by a vaginoscope from 67 girls (3 months to 16 years of age), Gerstner et al isolated *Lactobacillus* rather frequently[41]—this bacterium was isolated from 36 cases of vulvovaginitis (11.1%) and from 31 girls without inflammation (38.7%). In both groups, as in girls harboring *T. vaginalis*, the most frequent bacteria were gram-positive cocci as well as gram-negative rods, genera characteristic of biocenosis of the intestinal tract. McCormack et al,[42] who examined the frequency of various etiologic factors of STDs among college students, found anaerobic and microaerophilic strains of bacteria in 466 persons, mycotic agents in 309, and *T. vaginalis* in 14.

TABLE 13.5. Groups of genera and species of bacteria isolated from 756 juvenile patients infected with *T. vaginalis*.

Bacteria	Isolates No.	% ± SD*
Gram-positive rods		
Gardnerella vaginalis	84	11.1 ± 1.14
Gram-negative rods		
Escherichia coli	105	13.9 ± 1.26
Proteus mirabilis	38	5.0 ± 0.79
Proteus morgani	26	3.4 ± 0.66
Klebsiella pneumoniae	32	4.2 ± 0.54
Klebsiella sp.	19	2.5 ± 0.57
Enterobacter sp.	8	1.1 ± 0.38
Gram-positive cocci		
Enterococci	135	17.9 ± 1.39
Streptococcus viridans	72	9.6 ± 1.07
Group B *Streptococcus*	82	10.8 ± 1.13
Group D *Streptococcus*	56	7.4 ± 0.95
Staphylococcus epidermidis	18	2.4 ± 0.55
Gram-negative cocci		
Neisseria gonorrhoeae	6	0.8 ± 0.32

*Sample standard deviation.

Trichomoniasis

Unifocal Trichomoniasis

In newborns and infants, *T. vaginalis* infections are most frequently acute; they are only rarely chronic. In girls the sex organs are usually involved, the signs being swelling and reddening of the vulva and perineum. Sometimes there are spots and erosions in the vaginal vestibule, because the vaginal discharge through the opening of the hymen causes continual irritation and maceration of the vulva. In some cases one can observe "swellings" involving the entire perineum and groin. If one does not find *T. vaginalis* in the urethral contents, but only in the urine, which rinses out the parasites from the urinary passages and from the vaginal vestibule, there may be serious difficulties in recognizing the infection, especially when *T. vaginalis* appears in the vaginal contents in very small amounts. On the other hand, when one finds the parasite as the predominant cellular component of the vaginal contents usually diagnosis of a unifocal infection of the sex organs is correct. This diagnosis is confirmed by successful results of local treatment of vaginal trichomoniasis in girls.

An illustration of the infections acquired during birth and involving *T. vaginalis* in newborns and infants can be found in the cases described below.

Case No. 1

ZF, girl, 6 days old, born after 32 weeks of pregnancy, weighing 1900 g, showing respiratory insufficiency. The baby had to be kept in an incubator. There was evidence of swelling and reddening of the vulva and of the vaginal vestibule, in which there were small spots. There was purulent copious vaginal discharge. In permanent Wright-stained preparations numerous trichomonads and white blood cells were seen. In Gram-stained preparations mixed bacterial flora was found. In preparations of the urinary sediment taken with the aid of a microcatheter no *T. vaginalis* was noted. An axenic culture of the protozoon was isolated in Roiron-Ratner medium from the vaginal contents. Inoculation of urine into the same medium and also into Sabouraud's medium gave negative results for fungi. The mother stated that she was treated for trichomoniasis and cured before her pregnancy and that no *T. vaginalis* was discovered in the course of pregnancy. After diagnosis of this parasite in the infant, an extensive examination of the mother was conducted; despite the absence of clinical symptoms, *T. vaginalis* was found on cultivation. On the other hand, fungi could not be cultivated in Sabouraud's medium. It should be added that on microscopic examination of fresh preparations, as well as of permanent preparations stained by various methods, the parasite was not found in urine or in the vaginal contents of the mother.

Diagnosis: Trichomoniasis vulvovaginalis. Infection was acquired from the mother (family infection). Local treatment of the child with ornidazole and local and oral administration of this drug to the mother resulted in cure.

Case No. 2

JG, girl, 8 days of age, weighing 3200 g, brought to term. The baby was taken home. There appeared signs of inflammation of the vulva on the sixth day of life. The infant was restless and crying. Swelling and reddening of the vulva and vaginal vestibule were observed on examination. There was also a copious yellow-greenish vaginal discharge and yellow-brownish spots were noted on the diapers. In fresh and in permanent Giemsa-stained preparations there were found numerous *T. vaginalis* and occasional red and white blood cells. Gram-stained preparations revealed occasional gram-positive rods (*Lactobacillus* and *Gardnerella* sp.) and numerous gram-negative rods (*E. coli*). In cultures isolated on Simić's medium there were numerous *T. vaginalis*. No trichomonads were found during the first examination of the mother by various methods. Only 14 days after parturition, a *T. vaginalis* strain was isolated in Roiron-Ratner medium. No fungi could be grown in Sabouraud's medium from the material obtained either from the mother or from the child.

Diagnosis: Trichomoniasis vulvovaginalis. Infection was acquired from the mother (family infection). After local treatment of the child with metronidazole and after employment of this drug orally and locally in the child's mother, a cure was achieved.

Case No. 3

OW, girl, 11 weeks old, weighing 4600 g. The baby was carried to term; she was taken home.

The baby was restless and tended to cry. There was acute inflammation of the vulva and vagina, with erosions of the vagina, as well as copious yellowish purulent discharge from the vagina into the vaginal vestibule. Numerous trichomonads were seen on routine examinations of the vaginal contents and of the urine in stained preparations; positive cultures were obtained from these materials. In Wright-stained preparations there were found numerous epithelial cells of the superficial and intermediate layers; these observations suggested estrogen-dependent maintenance of the vaginal epithelium. In Gram-stained preparations, there were observed gram-negative rods of the genus *Proteus,* which could be cultivated also from the bladder urine obtained with the aid of a microcatheter. On the other hand, no growth of *T. vaginalis* was obtained in Roiron-Ratner medium or of mycotic species in Sabouraud's medium. Despite the absence of clinical symptoms and signs of trichomoniasis in the mother, it was possible to cultivate *T. vaginalis* from her vaginal contents in Roiron-Ratner medium; however, no fungi grew in Sabouraud's medium.

Diagnosis: Trichomoniasis vulvovaginalis. Infection was acquired from the mother (family infection). After treatment with tinidazole administered both locally and orally to the child and by the same routes to her mother, complete cure was achieved. However, the infant required additional treatment with antibiotics effective against *Proteus* sp. isolated from the urine.

It should be pointed out that in the first two cases, there was symptomless trichomoniasis in the mother. The infection was discovered only after taking several samples of the vaginal contents and inoculating them into Roiron-Ratner medium. Two cases of trichomoniasis in newborns were described also from a city hospital ward in Łódź.[43] Infection with *T. vaginalis* was accompanied by a generally very poor state of health. One case was a premature female baby who weighed 1800 g and whose delivery involved manual manipulation after 7 months of pregnancy, and was necessitated by premature outflow of the waters and a falling out of the umbilical cord. The second case was a 3000-g baby girl carried to term and delivered under normal circumstances. From the third week of life, she has circulatory insufficiency and breathing difficulties caused by coarctation of the aorta. After treatment with tinidazole, the state of both infants improved significantly, and they achieved a complete cure of the trichomoniasis. However, for neither patient, was it possible to establish the source of the infection, that is, to find the person from whom *T. vaginalis* was acquired by both babies.

Schwarz et al[44] commenting on the rare occurrence of *T. vaginalis* in newborns (they found only two cases of trichomoniasis among 948 subjects examined) described the course of this infection in two female babies.

Case A. The child, weighing 2100 g, was born in the 37th week of pregnancy. On the 14th day of life there was noted inflammation of the vulva and vagina, with copious and foamy discharge in the vaginal vestibule; the infection entailed also changes in the urine (leukocyturia). Numerous *T. vaginalis* were observed. Treatment was achieved by oral administration of metronidazole.

Case B. The child was born in the 30th week of pregnancy; it weighed 1330 g. *Escherichia coli* and *Aerobacter aerogenes* (10^6 bacteria per milliliter) as well as significant numbers of leukocytes were found in the urine. *Trichomonas vaginalis* could not be cultivated. Cephalosporin was administered. The recurrence of signs and symptoms of infection of the urinary tract persisted for up to 6 weeks, when *T. vaginalis* infection was discovered and the child was treated orally with metronidazole.

In the mothers of the aforementioned premature babies, the results of culture attempts were negative on many occasions. The attending physicians were successful in cultivating *T. vaginalis* from the vaginal contents of both mothers only 5 weeks after parturition. However, in both instances the possibility existed of infections during delivery.

Despite the numerous difficulties in differentiating unifocal from multifocal infections in girls, we attempted to separate these two types, because they require different courses of treatment. Therefore, in addition to the data obtained from a group of 228 patients, which showed a high degree of correlation between unifocal *T. vaginalis* infections and clinical manifestations, we analyzed similar results from a group of 263 girls examined in the course of a mass survey. The results of these examinations are presented in Table 13.6.

Among the girls examined in the survey, the majority ranged from 8 to 18 years, because, as pointed out previously (see the Epidemiology section), *T. vaginalis* infections appear to be very rare during the period defined as "quiescent." However, it is in this period that one may better analyze vulvovaginitis caused by bacteria, especially those infecting the nasopharyngeal region or other segments of the upper respiratory tract,[45] as well as vulvovaginitis dependent on the effects of fungi, the presence of foreign bodies in the vagina, the effects of masturbation, or the effects of chemical agents, for example, detergents used in washing of underwear made of artificial fiber. Finally poor hygienic conditions should also be considered.[46-48]

Depending on the age of the girls and the degree of their sexual maturity, as estimated in "cytohormonal" preparations (staining according to Shorr's method and by Harris' hematoxylin), it is possible to evaluate the estrogenic reactions of the vaginal epithelium. In a sample of 623 girls between the ages of 10 and 17 the correlation between the appearance of *T. vaginalis* in the vaginal material and values of the karyopyknosis index (KI) was analyzed by chi-square test. The results obtained in these investigations are shown in Figure 13.3. There was a significant statistical correlation ($P < 0.01$) between *T. vaginalis* infections and low, intermediate, or high estrogen reaction (KI = 10% to 80%). Trichomoniasis was rarely found in girls with low (KI < 10%) or very high (KI > 80%) estrogen levels in the vaginal epithelium.[49] In this time of sexual maturation in girls as well as in adult women with unifocal trichomoniasis of the vagina, the disease may be asymptomatic, acute, or chronic. Asymptomatic infections in girls can be discovered in the course of screening investigations or during studies of family groups, usually only by culture methods. Girls with acute or chronic trichomoniasis seek advice in the Clinic or in the Center. In permanent Wright-stained preparations of the vaginal contents obtained from cases of acute trichomoniasis, *T. vaginalis* constitutes the predominant cellular component of a smear. The very numerous trichomonads adhere to other cells or to each other and are subject to "mechanical deformation." In chronic types of infection the protozoa in smears are usually distributed singly, forming larger groups only very rarely.

TABLE 13.6. Frequency of symptoms and signs observed in 228 girls with unifocal urogenital trichomoniasis, including consideration of the correlation between the clinical manifestations and *T. vaginalis* infections.

Clinical manifestations	Frequency % ± SD*	Correlation coefficient (r)
Symptoms		
Discharge	92.5 ± 1.7	+0.609[†]
Pruritus in vagina and/or vulva	38.6 ± 3.2	+0.391[‡]
Burning sensation in vagina and/or vulva	50.9 ± 3.3	+0.497[†]
Pruritus and/or burning sensation in urethra	4.8 ± 1.4	+0.096
Pain in urethra	2.2 ± 0.9	+0.111
Polyuria and dysuria	2.9 ± 0.9	−0.072
Signs		
Edema and redness of vulva	96.5 ± 1.2	+0.537[†]
Edema and redness of external urethral orifice	3.1 ± 1.1	−0.117
Erosion of vulva	4.8 ± 1.4	+0.208
Colpitis diffusa	51.3 ± 3.3	+0.361[‡]
Colpitis maculosa	35.1 ± 3.2	+0.567[†]
Colpitis granulosa	14.0 ± 2.3	+0.193
Erosion of cervix uteri	3.9 ± 1.3	−0.106
Pain in lower abdominal region	15.8 ± 2.4	+0.138

*Sample standard deviation.
[†]Correlation is statistically significant.
[‡]Correlation at the border of statistical significance.

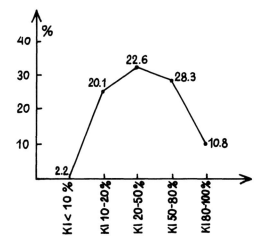

FIGURE 13.3. Distribution of the karyopyknosis index (KI) indicating the degree of estrogenization of the vaginal epithelium in 623 girls from 10 to 17 years old.

Multifocal Trichomoniasis

Trichomonas vaginalis infections in children, like those in adults, can occur in various parts of the sex organs and urinary tract. In a project on which we worked many years ago[50] and which was concerned with the diagnostic indicators of trichomoniasis in men and women, we considered various methods for obtaining material as well as various techniques useful in differentiation of the unifocal infection from multifocal infections. More recently, this study has been advanced to a limited degree in our Center. In light of the paucity of material and of clinical experience, diagnostic guidelines as dependable as those available for trichomoniasis in adults are still lacking for *T. vaginalis* infections in children.

The vagina has been recognized as the most frequent site of *T. vaginalis* infection in women. The second, much less common, infection focus is the urethra. Skene's paraurethral and Bartholin's gland constitute the third possible site.[7,50-53] In men, urogenital trichomoniasis usually starts in the glans penis and prepuce, or urethra; subsequently, the infection can spread to the higher parts of the sex organs or urinary tract.[54-61a] (For additional references see those given in the chapters of this book dealing with the clinical and pathologic aspects of *T. vaginalis* infection in women, Chaps. 11 and 14 and in men, Chaps. 12 and 15.)

In undertaking studies of trichomoniasis in children, it was anticipated that multifocal infections in these patients would have a course similar to that noted in adults. It was found, however, that in addition to the patient's sex, the course of *T. vaginalis* trichomoniasis depended on the age and also on developmental defects, especially those of the urinary tract, which contribute to retention of urine. These defects contribute effectively also to the maintenance of infection.

In newborns and infants multifocal infections are typically acute. Furthermore, a higher percentage of girls than boys void purulent urine, the ratio being 10:1 according to Schaffler.[61b] A high percentage of infections of the lower segments of the urinary tract in girls has a certain anatomic basis — these girls usually have a short urethra and a wide neck of the urinary bladder. This pertains also to bacterial infections which, however, in children of school age can be symptomless for certain periods, a condition seen more frequently in girls than in boys. Depending on the primary focus of *T. vaginalis* infection, there are dominant signs in the genitalia or the urinary organs. However, in some cases, acute infection occurs simultaneously in more than one infection site.

As in unifocal trichomoniasis, the possibility of infection during delivery exists also in multifocal infections with *T. vaginalis*. This kind of infection is illustrated by the following case histories.

Case No. 4

MC, a boy 10 days of age, carried to term, weighing 3600 g. The baby was taken home. He was restless, with a tendency to cry. During examination swelling and reddening of the external urethral orifice were noted, in which a drop of purulent material was observed. Microscopic examination of the droplet revealed numerous *T. vaginalis* and white blood cells; occasional erythrocytes were also seen. A similar picture was observed on microscopic examination of the urine sediment. In Gram-stained preparation, there were some gram-positive cocci (*Streptococcus* sp.). Strains of *T. vaginalis* were isolated in cultures of the urethral content and urinary sediment. However, no fungi could be isolated in Sabouraud's medium.

The mother suffered from chronic trichomoniasis, found to be resistant to local treatment with metronidazole which was administered several times in the second and third trimester of pregnancy. A *T. vaginalis* strain with low sensitivity to metronidazole was isolated in culture from the vaginal contents of this woman. No fungi could be grown in Sabouraud's medium.

Diagnosis: Urogenital trichomoniasis. The infection was acquired from the mother, likely during parturition (family infection). After local and oral treatment with ornidazole of the child and mother, a complete cure of both was achieved.

Case No. 5

GH, a girl 21 days of age, delivered in the 37th week of pregnancy, weighing 2300 g. The baby was taken home. She was disturbed, with a tendency to cry. Swelling and erosion of the vulva and vestibule were noted on examination. There were also distinct pathologic changes around the external orifice of the urethra. Scanty vaginal discharge was mucopurulent.

During an interview, the mother reported having chronic trichomoniasis before pregnancy and stated that she was considered cured before pregnancy. Colpitis maculosa was diagnosed in the mother, and *T. vaginalis* was found in her vaginal contents and in those of the child in fresh and permanent Giemsa-stained preparations. The parasites were found also in the child's urine, in addition to numerous leukocytes and occasional red blood cells. In Gram-stained smears, there were gram-negative rods (*Klebsiella* sp.). A *T. vaginalis* strain was isolated in Simić's medium, but no fungi could be grown in Sabouraud's agar.

Diagnosis: Urogenital trichomoniasis. The infection was acquired from the mother (family infection) probably during delivery. Cure was achieved in the child and mother following oral and local administration of tinidazole.

Case No. 6

DE, a girl 4 weeks of age, carried to term, weighing 2800 g. The child was taken home. The baby appeared disturbed with a tendency to cry. On examination, there was found swelling and reddening of the vulva, the vestibule, and of the area around the external orifice of the urethra from which exuded purulent material. The vaginal content was scanty and purulent. On routine urinary analysis albuminuria, abundant leukocytes, and bacteriuria were noted. It was possible to cultivate *E. coli* and *A. aerogenes* (the bacterial count exceeded 10^6/ml). In fresh preparations of the urinary sediment there were many *T. vaginalis*. In permanent preparations of the Giemsa-stained sediment and also in preparations of the vaginal contents occasional trichomonads and some leukocytes and erythrocytes were noted. Gram-stained preparations revealed gram-negative rods (*E. coli*). A *T. vaginalis* strain was isolated in Simić's medium from the urine sediment as well as from the vaginal contents of the infant. There was no growth of fungi in Sabouraud's medium.

No symptoms or signs of urogenital trichomoniasis were noted in the girl's mother; however, a strain of *T. vaginalis* was isolated from the mother's vaginal contents in Roiron-Ratner medium. No cultures of mycotic agents were obtained.

Diagnosis: Urogenital trichomoniasis. The infection was acquired from the mother (family infection), undoubtedly during birth. Treatment with metronidazole administered locally and systemically in the infant and orally in the mother eliminated *T. vaginalis*. The child's treatment was supplemented with antibiotics to which the bacteria were found to be sensitive.

In the foregoing three cases of multifocal *T. vaginalis* infections in newborns acquired during birth, trichomoniasis was manifested in the boy primarily by inflammation of the urethra. On the other hand, in both girls there was severe inflammation of the vagina and vaginal vestibule accompanied by pathologic changes around the external urethral orifice. The most severe disease was noted in the girl described as Case No. 5, who had the lowest weight. However, this patient also had bacteriuria, which exacerbated the trichomoniasis.

Using a group of 263 girls (irrespective of their age) with multifocal urogenital trichomoniasis, we analyzed correlation and estimated the correlation coefficient between the frequency of subjective symptoms and objective signs, on the one hand, and of *T. vaginalis* infections, on the other. The data obtained in this investigation are presented in Table 13.7.

The values of the correlation coefficients (r) in multifocal *T. vaginalis* infection indicated the existence of statistically significant positive correlation ($P < 0.05$) between the presence of this parasite and discharges, pruritus, or a burning sensation of the vulva and vagina, as well as

TABLE 13.7. Frequency of symptoms and signs observed in 263 girls (irrespective of their age) with multifocal urogenital trichomoniasis, including consideration of the correlation between the clinical manifestations and *T. vaginalis* infections.

Clinical manifestations	Frequency % ± SD*	Correlation coefficient (r)
Symptoms		
Discharge	92.9 ± 1.9	+0.576†
Pruritus in vagina and/or vulva	69.8 ± 3.4	+0.522†
Burning sensation in vagina and/or vulva	74.2 ± 3.2	+0.533†
Pruritus and/or burning sensation in urethra	48.4 ± 3.7	+0.471‡
Pain in urethra	41.8 ± 3.6	+0.115
Polyuria and dysuria	10.4 ± 2.3	+0.117
Signs		
Edema and redness of vulva	94.5 ± 1.7	+0.201
Edema and redness of external urethral orifice	29.1 ± 3.4	+0.518†
Erosion of vulva	67.0 ± 3.5	+0.317
Colpitis diffusa	50.4 ± 4.7	+0.241
Colpitis maculosa	43.4 ± 3.7	+0.698†
Colpitis granulosa	9.1 ± 2.1	+0.198
Erosion of cervix uteri	6.0 ± 1.7	+0.011
Pain in lower abdominal region	55.0 ± 3.7	+0.496‡

*Standard sample deviation.
†Correlation is statistically significant.
‡Correlation at the border of statistical significance.

with a burning sensation in the urethra. There was also a significant correlation between trichomoniasis and edema of the external urethral orifice, colpitis maculosa, and pain in the lower abdominal area.

If we accept as 100% all the diagnosed infections of *T. vaginalis*, irrespective of the clinical picture, multifocal trichomoniasis was diagnosed in about 20% of the girls (20.6 ± 2.4%) and boys (17.2 ± 4.0%). To illustrate the clinical data we shall describe two cases of multifocal trichomoniasis, one from a 12-year-old boy and the other from a 14-year-old girl.

Case No. 7

HS, a boy, 12 years of age. He came to the clinic because of severe inflammation of the glans (balanitis). Abundant purulent discharge exuded from under the prepuce. The patient complained also about pain in the urethra that increased during micturition, and reported polyuria accompanied by painful pressure. Balanitis was confirmed on objective examination, and there were also pathologic changes noted around the urethral orifice, that is, edema, reddening, and small widely dispersed spots. There was severe pain during palpation and in the course of collecting material for laboratory examination. Examination of the subprepucial material in fresh smears did not reveal *T. vaginalis*. On the other hand, in permanent Wright-stained preparations there were found occasional flagellates. In Gram-stained preparations, there were gram-negative rods, presumably *E. coli*. In fresh preparations obtained from the contents of the urethra and from urine, there were seen numerous *T. vaginalis* as well as many white blood cells and occasional erythrocytes. Two strains of *T. vaginalis*, one from the discharge obtained from under the prepuce and another from the urine sediment, were isolated in culture. *Escherichia coli* (10^6/ml of urine) was isolated also from the urine sediment. No cultures of fungi were obtained in Sabouraud's medium. Both parents were examined, but no *T. vaginalis* was found in either.

Diagnosis: Urogenital trichomoniasis. The source of infection was unknown. After treatment with ornidazole, both locally and orally, a cure was achieved. We recommended additional treatment with antibiotics to which the strain of *E. coli* isolated from urine was found sensitive.

Case No. 8

MW, a girl, 14 years of age, a virgin, was referred to us for consultation because of suspected trichomoniasis and recurrent inflammation of the urethra and urinary bladder as well as pelvic nephritis of 1 year's duration. The girl reported moderate discharges and temporary itching and a burning sensation in the vulva and vaginal vestibule. She complained also of a burning sensation and pain in the urethra appearing independently of micturition. There was also polyuria and bilateral pain of the thighs in the area of both kidneys radiating toward the pubic symphysis. Objective examination revealed no inflammation of the genital organs; only the vestibule contained a small amount of mucopurulent material. There were evident, however, edema, reddening, and small erosions in the area around the external urethral orifice. On palpation, the patient felt pain in the lower abdominal area; pain was felt also during palpation of the area around the kidneys. Fresh preparations of the urine sediment revealed occasional *T. vaginalis* and numerous white blood cells; there were also frequent individual erythrocytes and rouleaux as well as a small amount of albumin and significant bacteriuria. No *T. vaginalis* was noted on direct examination in fresh preparations of the material from the vaginal vestibule. On the other hand, in Wright-stained preparations examined at 1000 x magnification, there were seen numerous bacteria (rods and cocci), white blood cells, and occasional *T. vaginalis*. Also, Gram-stained preparations revealed occasional gram-positive rods (*Lactobacillus*) and cocci (*Streptococcus*). Two presumably different strains of *T. vaginalis*, one from the vaginal contents and the other from the urine sediment, were isolated in Roiron-Ratner medium, but no cultures of fungi could be obtained in Sabouraud's medium. Inocula of the vaginal contents into appropriate media yielded cultures of type-B *Streptococcus;* both cocci and *E. coli* could be isolated from the urine sediment. Numerous trichomonads were found in urethral urine collected during cystoscopy. Examination of both parents for *T. vaginalis* gave negative results.

Diagnosis: Urogenital trichomoniasis with pyonephritis chronica recurrens. The source of infection was unknown. Treatment with ornidazole, locally and orally, eliminated *T. vaginalis*. Additional antibacterial treatment was recommended in consultation with a urologist.

The aforementioned cases of multifocal trichomoniasis in a boy and a postmenarcheal girl were characterized by severe clinical manifestations. During periods of several to over 10 days, the children were unable to participate in their school work. This was especially true with regard to the girl, who was referred to us after 1 year of illness. Treatment with antitrichomonal drugs significantly improved the condition within only a few days.

In choosing these cases for description we were guided by the severity of the disease. It seemed also far more interesting to discuss the course of infection in a postmenarcheal girl who has not yet been involved in sexual contacts than in similar individuals who no longer were virgins. In this latter group trichomoniasis can follow a course identical to that observed in adult women, with whom we have had far more experience.

Trichomoniasis Complicated by Mycosis

Pursuing our diagnostic procedures in material collected from children with suspected trichomonal infection we searched not only for *T. vaginalis* but also for fungi. In all instances, the final diagnosis was arrived at by taking into account the results of all investigations, parasitologic as well as mycologic. In laboratory tests we examined also the morphologic and biochemical characteristics of the strains isolated from mixed infections.[62,68]

Among 2905 patients, ranging 0 to 18 years of age, who were seen in our Center because of suspected trichomoniasis or mycosis of the urogenital organs, there were 2599 girls and 306 boys. Among the girls, we found *T. vaginalis* in 182 (7.0 ± 0.50%) and fungi in 2009 (77.3 ± 0.99%). Thus, mycotic infections were about 11 times more frequent than *T. vaginalis* trichomoniasis. It should be added, however, that from 1967 to 1985, when the frequency of trichomoniasis in girls was reduced by half, that of mycosis increased about eight times. In the aforementioned material from among the 182 girls in whom *T. vaginalis* was found, in 113 (62.1 ± 3.6%) there was also mycosis of the urogenital organs.

It should be pointed out that in fixed preparations fungi were diagnosed more frequently (61.1 ± 4.6%) than urogenital trichomonads (53.1 ± 4.7%). Mixed infections were significantly rarer ($P < 0.001$), having been found in only 19 (16.8 ± 3.5%). If we failed to employ appropriate media for the cultivation of the protozoa and fungi and depended for diagnosis on microscopic examination alone, the number of diagnosed mixed infections would have been six times lower.

Among 306 boys examined, *T. vaginalis* was found in 87 (28.4 ± 2.58%), but fungi in as many as 282 (92.2 ± 1.53%). It should be noted, however, that this group of boys came from families in which earlier at least one member was found to harbor either trichomonads or fungi in genital organs or the urinary tract. This may explain the high percentages of mixed trichomonad and mycotic infections found in 72 boys (87.8 ± 4.06%).

Genera and Species of Fungi Found in Children Infected with T. vaginalis

Many strains of fungi were isolated from the vaginal contents of 113 girls and from the urogenital tracts of 72 boys harboring *T. vaginalis*. The genera and species of these strains as well as their frequencies (percentages) in the two groups of patients are presented in Table 13.8. The classification system of the yeastlike organism, Cryptococcaceae, and of yeasts, Saccharomycetaceae, was that recommended by Kreger van Rij.[62]

It is evident from the data given in Table 13.8 that in mixed infections in girls, *T. vaginalis* was associated with 16 species of fungi. In boys the trichomonad coexisted with only eight mycotic species. In both sexes, *Candida albicans* was the organism isolated most often and *Candida tropicalis* was the next most frequently found species. All the remaining fungi were noted in fewer patients.

The clinical symptoms and signs accompanying mixed trichomoniasis and mycosis were studied in 113 girls, and their frequency was recorded. Among the subjective symptoms, vaginal discharges were reported by all patients. The majority of the patients complained of itching in the vulva and vagina. Many girls complained also of pain in the urethra and of frequently painful micturition (Table 13.9). A high degree of correlation between the aforementioned subjective symptoms and the presence of mixed infection was revealed by the value of the correlation coefficient (r) estimated for the available data, which included also those derived from 263 girls examined in the course of mass screening.

Among the objective signs noted in 113 girls suffering from mixed trichomoniasis and mycosis, edema and reddening of the vulva were noted in all patients, and in the majority of

TABLE 13.8. Fungi isolated from girls and boys infected with *T. vaginalis*.

Family, genus, and species	Strains isolated from 113 girls		Strains isolated from 72 boys	
	No.	% ± SD*	No.	% ± SD
Cryptococcaceae				
Candida albicans	41	36.3 ± 4.52	32	44.4 ± 5.75
C. tropicalis	12	10.6 ± 2.90	18	25.0 ± 5.10
C. guilliermondii	8	7.1 ± 2.41	3	4.2 ± 2.35
C. pseudotropicalis	8	7.1 ± 2.41	4	5.6 ± 2.70
C. krusei	6	5.3 ± 2.11	4	5.6 ± 2.70
C. humicola	5	4.4 ± 1.93	0	
C. stelatoidea	5	4.4 ± 1.93	0	
C. utilis	3	2.7 ± 1.51	0	
C. famata	2	1.8 ± 1.24	0	
C. glabrata	2	1.8 ± 1.24	2	2.8 ± 1.84
C. castelli	0		1	1.3 ± 1.23
Rhodotorula rubra	6	5.3 ± 2.11	0	
Mucedinaceae				
Geotrichum candidum	8	7.1 ± 2.41	8	11.2 ± 3.72
Aspergillaceae				
Aspergillus flavus	1	0.8 ± 0.78	0	
A. candidus	2	1.8 ± 1.24	0	
Penicillium spinulosum	1	0.8 ± 0.78	0	
Saccharomycetaceae				
Saccharomyces cerevisiae	3	2.7 ± 1.51	0	

*Sample standard deviation.

TABLE 13.9. Frequency of symptoms and signs noted in 113 girls with trichomoniasis complicated by mycosis, including consideration of the correlation between the clinical manifestations and the mixed infection.

Clinical manifestations	Frequency % ± SD*	Correlation coefficient (r)
Symptoms		
Discharge	100.0 ± 0.0	+0.798†
Pruritus in vagina and/or vulva	94.7 ± 2.1	+0.615†
Burning sensation in vagina and/or vulva	66.4 ± 4.4	+0.419†
Burning sensation in urethra	56.6 ± 4.7	+0.672†
Pain in urethra	86.7 ± 3.2	+0.593†
Polyuria and dysuria	52.2 ± 4.7	+0.504†
Signs		
Edema and redness of vulva	100.0 ± 0.0	+0.677†
Edema of external urethral orifice	72.6 ± 4.2	+0.484†
Erosion of vulva	15.9 ± 2.7	+0.201‡
Colpitis diffusa	64.6 + 5.6	+0.511†
Colpitis maculosa	34.6 ± 4.5	+0.411
Colpitis granulosa	37.2 ± 4.7	+0.205
Erosion of cervix uteri	7.2 ± 2.4	+0.218
Pain in lower abdominal region	77.0 ± 4.0	+3.17‡

*Sample standard deviation.
†Correlation is statistically significant.
‡Correlation at the border of statistical significance.

cases there was swelling of the external orifice of the urethra. Pain in the lower abdominal area during examination was reported by a large proportion of patients. For the aforementioned signs as well as for the presence of colpitis diffusa, the value of the coefficient (r) suggests a positive correlation between these signs and finding *T. vaginalis* in the vaginal secretion and in urine.

Clinical symptoms of trichomoniasis complicated by mycosis were studied in 72 boys. The majority of the patients (94.4 ± 7.7%) complained of itching or burning sensation in the urethra (90.3 ± 2.9%) as well as pain, which became more intense during micturition (80.5 ± 4.7%); in some cases the pain radiated toward the perineum (58.3 ± 5.8%). Polyuria was frequent (91.7 ± 3.2%). Some boys (19.3 ± 4.2%) reported that in the morning before the first micturition, there appeared a purulent secretion in the urethral orifice. As far as objective signs were concerned, there was in all boys edema and reddening of the urethral orifice around which a certain amount of erosion was noted in some cases (50.0 ± 5.9%). Some of the boys presented also with phimosis (8.5 ± 3.3%). Because of the too small numbers of boys examined and since there was no control group, for example one obtained from screening of large groups, a statistical analysis of the data was not possible.

It should be added that developmental defects of the urinary tract were found in four girls and seven boys of the aforementioned group of children with trichomoniasis complicated by mycotic infections. The following defects, discovered on objective examinations as well as by x-ray and ultrasound methods were recorded: congenital hydronephrosis (two cases), primary displacement of the kidney (two cases), double kidney (ren duplex) (one case), nephromegaly (one case), and defective orifice of the urethra (five cases). The foregoing abnormalities, found in children of various ages, caused greater or lesser difficulties in micturition or even a total inability to urinate. Both these conditions favor retention of *T. vaginalis* infection not only in newborns and infants but also in preschool children, school-age children, and adolescents. The presence of fungi complicated seriously the clinical course of trichomoniasis. The children were subjected first to prolonged treatment with antibiotics against bacteria and fungi, because of the usually very early diagnosis of bacteriuria. In children with trichomoniasis complicated by mycosis, there were many interesting cases, of which only a few can be summarized here.

Case No. 9

PS, a boy, 19 days old, carried to term, weighing 4000 g. The child was taken home. The boy was restless and had distinct pain in the abdominal area. True phimosis was diagnosed on objective examination. There was a suspicion of dislocation of the right kidney, and this was confirmed by x-ray and ultrasound examinations. *Escherichia coli* and *Proteus morganii* were found in the urine, the number of bacteria exceeding 10^6/ml. No growth of *T. vaginalis* or of fungi was obtained in the appropriate culture media. The newborn was subjected to corrective surgery for phimosis. He was given cephalosporin and subsequently other antibiotics to which the bacteria isolated from the urine were found sensitive. In the sixth week of life the boy was brought to our Center with a diagnosis of pyelonephritis acuta. We saw him because of finding of *T. vaginalis* in the urine. In a fresh preparation of the urine sediment there were numerous protozoa and bacteria, as well as germinating cells and pseudohyphae of *Candida*. In Gram-stained preparations numerous gram-negative rods (*Escherichia*) were observed. An axenic culture of *T. vaginalis* was isolated in Roiron-Ratner medium; the strain was very sensitive to ornidazole. On special culture media it was possible to identify a strain of *Candida guilliermondii*, which was sensitive to 5-fluorocytosine.

Examination of both parents of the newborn failed to confirm the suspicion of symptomless trichomoniasis or mycosis.

Diagnosis: Urogenital trichomoniasis complicated by mycosis. The source of infection was unknown. After administration of ornidazole and 5-fluorocytosine, the infant showed significant improvement in 3 days. At the end of the treatment, neither *T. vaginalis* nor *Candida* could be found. The treatment lasted 3 months, but the child remained under observation for 6 months.

Case No. 10

AK, a virgin girl, 13 years of age. She was referred to us after chronic inflammation of the vagina and urethra was diagnosed. The patient complained of copious discharges, as well as of an itching and burning sensation in the area of the vulva and external urethral orifice, which was intensified after each micturition; micturition was always accompanied by intense pain in the urethra. The vulva and vestibule were red. Distinct swelling was noted in the area of the external urethral orifice. Purulent material exuded from the urethra. In fresh preparations occasional *T. vaginalis* were observed. In both fresh and permanent Wright-stained preparations obtained from the vaginal discharge, some *T. vaginalis*, numerous white blood cells, and occasional erythrocytes were seen; bacteria were also numerous. In Gram-stained smears there were gram-negative rods and gram-positive cocci. A *T. vaginalis* strain was isolated in Simić's medium and fungi were found on Sabouraud's agar.

Only negative results were obtained in attempts at isolation of *T. vaginalis* or fungi from the mother, who was asymptomatic. The father of the girl complained of polyuria and of pain and a burning sensation during micturition. He also reported pains in the lower abdominal region that radiated toward the area of the right testis and in the direction of the rectum. The man was treated for 15 years by a urologist for chronic prostatitis. In a drop taken early in the morning from the urethra and also in the urine sediment there were numerous fungi as well as many white and red cells; *T. vaginalis* was not seen. Few trichomonads were noted in fresh preparations of the prostatic secretion obtained by massaging. On the other hand, cultures of this flagellate were obtained in Roiron-Ratner medium. Furthermore, *C. albicans* was cultivated in Sabouraud's medium and it was seen also in histologic preparations of a needle biopsy.

Diagnosis: Urogenital trichomoniasis, complicated by a mycotic infection. The infection was acquired from the girl's father (family infection). After the girl was treated with ornidazole, both locally and orally, and after administrations of natamycin locally and orally, a cure was achieved. The father of the child was subjected to treatment with ornidazole and 5-fluorocytosine. He was observed by us and a urologist every 6 months for a 3-year period, during which there was no recurrence either of trichomoniasis or of mycosis of the urinary system.

Case No. 11

JH, a virgin girl 14 years of age, ailing for 2 years. She was treated in the Department of Nephrology and Urology in connection with chronic nephritis and urethritis. Polyuria, a burning sensation in the urethra, and other typical symptoms were recorded. The severity of the symptoms prevented the girl from attending school during the entire period of her illness. Reddening, edema, and small foci of erosion covered by white material in the area of the external urethral orifice were observed. No changes were noted in the vulva and in the vestibule; the vaginal secretion was scanty. Very numerous *T. vaginalis* and occasional budding fungi cells were seen in fresh preparations. Numerous white blood cells, individual erythrocytes, and quite numerous bacteria were also found. No *T. vaginalis* were observed in fresh preparations of the vaginal secretion or in permanent preparations of this material using Wright's stain. There were, however, individual budding cells and pseudohyphae of *Candida*. In Gram-stained preparations there were occasional gram-positive cocci. An axenic culture of *T. vaginalis* was established in Roiron-Ratner medium. In Sabouraud's medium we isolated a strain of *C. albicans*. *Trichomonas vaginalis* could not be cultivated from the urethral secretion; however, *C. albicans* was isolated in Sabouraud's medium. In the course of urologic examination, after insertion of the cystoscope, urine was taken from each ureter. In the urine from the right ureter we found very numerous *T. vaginalis*. The urine from the left ureter was uninfected and otherwise normal. On further examinations conducted with the aid of x-rays and ultrasound, the girl was found to have a double right kidney (ren duplex).

Only negative results for trichomoniasis and mycosis were obtained from urogenital examination of both parents of the child.

Diagnosis: Urogenital trichomoniasis, complicated by mycosis. The source of infection was unknown. Treatment with metronidazole (Flagyl) administered orally and also locally and subsequent administration of nephrecil and nitrofurazone were found to be effective.

Case No. 12

SG, a virgin girl, 14 years old. The patient was referred to us after a diagnosis of chronic vaginitis. She complained of discharges as well as an itching and burning sensation in the vulva and around the external urethral orifice. There was edema and reddening of the vulva and the vestibule. Distinct pathologic changes were noted also around the external urethral orifice. The vaginal discharge was copious and purulent. In fresh and permanent Wright-stained preparations individual trichomonads, scanty white blood cells, and numerous bacteria were found. In Gram-stained preparations numerous gram-negative rods as well as gram-positive cocci were noted. A strain of *T. vaginalis* was isolated in Simić's medium and *Candida tropicalis* and *Geotrichum candidum* were isolated on Sabouraud's agar. No cultures of trichomonads or fungi could be recovered in the appropriate media from the mother, who did not report any ailments of the genital organs.

Diagnosis: Trichomonas vaginalis infection complicated by mycosis. The source of infection was unknown. Treatment with natamycin and ornidazole resulted in cure.

Case No. 13

WR, a virgin girl 17 years of age. This patient was sent to us after being diagnosed as suffering from acute vaginitis. The girl complained of very copious vaginal discharges as well as an itching and burning sensation in the area of the vulva and the vagina that intensified at night.

She was treated, without results, with various drugs over the course of 3 years. The patient was unable to list the medications to which she was subjected. We observed swelling and erosion on the surface of the vulva and in the vestibule. The vaginal discharge was abundant and purulent. In fresh and permanent Wright-stained preparations of the vaginal discharge, there were numerous *T. vaginalis* and leukocytes as well as many bacteria. In Gram-stained preparations we noted gram-positive cocci (*Staphylococcus* sp.). A strain of *T. vaginalis* was isolated in Simić's medium and fungi were found on Sabouraud's agar.

In the mother, who was being treated for chronic vaginitis, individual *T. vaginalis* were seen in Wright-stained smears of the vaginal discharge. Budding cells and pseudohyphae of *Candida* were also noted. A strain of *T. vaginalis* was isolated by inoculation of the vaginal discharge into Simić's medium and growth of *Candida* was obtained on Sabouraud's agar.

Diagnosis: Trichomonas vaginalis infection complicated by mycosis. The infection was probably acquired from the mother (family infection). After treatment with tinidazole locally and orally as well as administration of natamycin, the girl was cured. The same medication regimen administered both orally and locally to the mother also resulted in elimination of *Trichomonas* and *Candida*. (It should be added here that the differentiation of the fungal strains is quite difficult. Therefore, we used for this purpose scanning electron microscopy, which frequently facilitated diagnosis.)

Many authors studying mixed infections of *T. vaginalis* and mycotic agents in the urogenital organs of humans considered only *C. albicans*.[63-67] The reason for this is that *C. albicans* can be most easily identified among the species of fungi isolated from these organs, its identification being based on the ability of this species to form chlamydospores, for example, on the cornstarch medium (Difco). Perhaps the emphasis placed on *C. albicans* is also the result of the assumption that only this species is of importance among the etiologic factors of sexually transmitted diseases. In our opinion, this belief is incorrect, because in the course of our long-term investigations, we found that over 10 species of various fungi play an important role in infections of women and men.

In materials isolated from the genitalia and urinary tract of girls with trichomoniasis, we found in certain cases a variety of fungi. Differentiation of these agents was based on 17 morphologic and biochemical criteria.[68] With the aid of these criteria, *C. albicans* was diagnosed in 36.5 ± 4.5% of strains belonging to 16 species isolated from vaginal or urethral secretions or from urine. In boys, *C. albicans* was found in 44.4 ± 5.8% of strains belonging to eight species.

A comparison was made of the mycotic species isolated from the vaginal secretions of girls attending the Outpatient Pediatric Gynecology Clinic in Łódź, with the species diagnosed in a corresponding consultation unit in Budapest (in collaboration with Prof. Örley). Only six of the 17 species were found in patients attending both centers. There were *C. albicans, Candida krusei, Candida pseudotropicalis, Candida tropicalis, Geotrichum candidum,* and *Saccharomyces cerevisiae*. Of interest was the observation that while many girls from Łódź harbored a high percentage of *C. pseudotropicalis* and *G. candidum,* those from Budapest were infected most often with *Candida glabrata* and the usually rare *Trichosporon cutaneum*. As far as the prevalence of mycotic infections in Poland is concerned, significant differences in the distribution of the species was observed in a mass survey of women in three geographically distant areas: Kotlina Kłodzka, Wrocławek, and Łódź.[69]

Mixed infections involving the urogenital trichomonad and fungi were diagnosed in about one-half (62.1 ± 3.6%) of all girls found positive for the protozoon; this percentage was even higher in boys (87.0 ± 4.1%). Among the girls seen in our Clinic who complained of various ailments of the genital organs and urinary tract, the frequency of mixed infections varied between 5.2% and 23.1%, depending on the degree of sexual maturity, with an average of 7.0%. The highest percentage of such infections was found in newborn girls and in female infants up to 1 year of age, and the lowest was recorded from the period of sexual quiescence, that is, until 10 years of age. Calculation of the frequency of infections of the genitourinary organs with fungi in addition to *T. vaginalis* revealed a ratio of 1.1. On the other hand, in the girls examined in our Center in the course of 15 years, the ratio varied from 1.5 to 1.9.[13,47,70] Similar analysis of the data published by some of the other workers yielded similarly low values of the ratio in question, that is, from 1.1 to 1.5.[63,65] Somewhat higher ratios, 3.5 to 5.8, were recorded by others.[21,42,71]

In children, trichomoniasis complicated by mycosis has manifestations of an acute disease; more rarely it is chronic. We have not observed asymptomatic mixed infections involving *T. vaginalis* and fungi. The clinical picture is clearly more serious and diagnosis and treatment are much more difficult than in infections by either *T. vaginalis* or mycotic species alone. The high percentages of obvious clinical symptoms and their correlation coefficients with mixed infections are shown in Table 13.9. Two actual cases are described also, namely, Cases No. 12 and 13. We could not find similar analyses in papers published by others. In our earlier reports dealing with adult women[72,73] it was shown that in mixed infections involving *T. vaginalis* there can be a larger number of mycotic species. In clinical investigations of adults, one can differentiate with more confidence unifocal and multifocal trichomoniasis complicated by mycosis than in such investigations of children.

Al-Salihi et al,[74] seeking the presence of *T. vaginalis* in secretions of the vestibule in 984 newborn females, found the trichomonad in three patients only. They found also oral thrush in two girls. This latter infection required treatment with nystatin in addition to metronidazole. It can be assumed that infection with mycotic species existed also in the sexual organs; therefore, these would be mixed infections involving fungi and *T. vaginalis*. Our clinical experience with mycotic infections during birth indicates that fungi usually occupy several sites, notably, the oral cavity, intestinal tract, genital organs, and skin.

Trichomoniasis complicated by mycosis requires further studies, laboratory as well as clinical. In in vitro experiments involving 20 to 90 axenic *T. vaginalis* strains, and reference strains of mycotic species isolated from women with mixed trichomonal-fungal infection, it was found that certain fungi stimulated growth of the urogenital trichomonad.[75,76]

Other Forms of Trichomoniasis

In addition to *T. vaginalis* infections of the genitourinary organs acquired by children during birth as reported by other workers and by us, there are in the literature reports of this species in the respiratory system of newborns.

Hiemstra and his collaborators[77] described an infection of the respiratory passages by *T. vaginalis* in a newborn brought to term. The child, weighing 3690 g, was born in a state of asphyxiation. In the white contents of the trachea obtained by intubation for microscopic observations, *T. vaginalis* and leukocytes were found in fresh and stained preparations. During pregnancy the mother was found to have vaginal infection with *C. albicans;* she had negative results for the presence of *T. vaginalis*. Following appropriate treatment, the child was released after 14 days in a good state of health. Intubation was kept up for 36 hours, and an x-ray examination indicated that the pathologic changes of the lungs had regressed after 7 days. The authors treated this case as a "during birth" infection with *T. vaginalis*. They questioned, however, the efficiency of the examination of the mother in the course of pregnancy.

Two similar cases, dealing with prematurely born infants, were described by McLaren et al.[78] One was a female, weighing 1300 g, born in the 29th week of pregnancy with serious respiratory insufficiency. In the course of intubation, greenish-whitish material was aspirated from the trachea. No *T. vaginalis* was found in this secretion on the first examination, and no other potentially pathogenic microorganisms were noted. Pneumonia developed, however, during which the child was given antibiotics against bacteria. *Trichomonas vaginalis* was noted on microscopic examination of the secretion taken on the 24th day post partum. After oral treatment with metronidazole, the pneumonia regressed. The child's state of health improved, and on the 47th day, the patient was allowed to go home.

The mother of this newborn suffered from vaginitis for 3 years; however, despite this diagnosis, she was not treated with antitrichomonal drugs.

The second case concerned a male newborn, weighing 1389 g, who was born in the 37th week of pregnancy. There were symptoms of serious gestosis observed in his mother. The respiratory difficulties of the infant required intubation. In the third week of life, he was sent home in what appeared to be good health. However, 8 weeks later the infant was returned to the hospital vomiting and suffering from severe rhinitis. In the secretions taken from the nose, throat, and larynx, no viruses or other pathogenic agents were demonstrated. In Gram-stained preparations there was no *T. vaginalis*. However, this parasite was cultivated in tissue cultures (normally used for growing of viruses) from each of several foci. In the 11th week of life, an x-ray examination revealed bilateral inflammatory infiltration in the lower pulmonary lobes. Thirteen weeks after treatment with metronidazole, the infant no longer showed any symptoms or harbored the trichomonad. This case was followed for 2 years; no evidence of an upper respiratory tract inflammation was noted during this period. The authors pointed out that trichomoniasis of the respiratory system, acquired during birth, may develop slowly—even for as long as 10 weeks.

The mother of the infant had been treated for 2 years with antitrichomonal drugs because of recurrent trichomoniasis. It is important to note that despite numerous controls, in no description of these infants was *T. vaginalis* found in the urogenital system.

It should be pointed out that many years ago Walton and Bacharach[79] described cases of trichomoniasis of the respiratory passages in adults. Although in the infections of the respiratory system acquired during birth, one can be certain that they involved *T. vaginalis*, in adults one should consider also the possibility of infection with *Trichomonas tenax*. Indeed, the prevalence of the latter species in the oral cavity of people with paradenopathies or with diseases of the oral mucosa approaches 50%.[80] This finding has been confirmed by Teras,[81] who reported the oral trichomonad in 10% of samples obtained by bronchoscopy from 370 patients. However, not all of these persons had trichomonads in the oral cavity contents and in sputum. Caution has been advised with regard to uncritical acceptance of causal relationships between *T. tenax* and oral and bronchial pathology.[82]

There are in the literature suggestions of *T. vaginalis*-caused conjunctivitis acquired during birth. However, Norn et al[83] examining fresh preparations by phase-contrast microscopy failed to find even a single case of urogenital trichomonads in the conjunctiva of newborns, even though 3.7% of the mothers were diagnosed as harboring this parasite in the vagina. Furthermore, according to Hardy et al,[65] among 39 young mothers infected with *T. vaginalis* and *Chlamydia trachomatis,* conjunctivitis was observed in three cases; possibly *Chlamydia* was responsible for this latter disease. Clearly, this problem can be resolved only by additional carefully controlled studies.

Treatment

Treatment of urogenital trichomoniasis in children is more difficult than in adults. First of all, it places more responsibility on the clinician to ascertain the reaction of the patient to the prescribed medication. It is necessary also to investigate the extent of infection by examining, with the aid of standard methods, material collected from the various likely sites of the parasite. As was mentioned previously, irrespective of the child's sex or age, urine is potentially the most useful diagnostic material. In all instances, a search for *T. vaginalis* in the urine of newborn girls, female infants less than 1 year old, and premenarcheal girls must be attempted. Since, however, there is always doubt about the exact infection site, it is often assumed that one deals with a unifocal infection of the genital organs. Indeed, this is the most common kind of infection in girls, and it should be emphasized that the knowledge of the site allows local treatment known to have fewer detrimental side effects than drugs administered systemically.

Typically, local treatment is initiated with the aid of swabs saturated with the drug of choice or with suspensions containing it. Then the drug is introduced through the opening in the hymen. In addition, it is applied in a cream or suspension to the vulva and the area around the urethra. If no improvement is noted by the end of 5 days of treatment or if *T. vaginalis* continues to be present in the urine, the drug is administered orally. In newborn and infant boys, oral, or more rarely intravenous, administration of trichomonacidal compounds is employed from the beginning. Even in these instances, however, local application should not be neglected. In adolescents, where the exact site of trichomonad infection can be ascertained with greater confidence, the site, age, and weight of the patient should be considered in prescribing treatment.

Oral or intravenous administration of an antitrichomonal drug must be preceded by evaluation of hematologic findings. Liver function as well as the serum levels of creatinine, bilirubin, and other parameters should also be ascertained to establish the baseline against which to follow changes occurring during the course of therapy.

The progress of treatment should be monitored for the presence of *T. vaginalis* (by the usual methods), symptoms and signs, and possible systemic changes as reflected in the aforementioned parameters. After a single-dose regimen, control tests are performed 3 to 5 days after termination of treatment. Similar control examinations are carried out after more prolonged treatment. In either case, the patients are checked every 2 weeks during a period of 3 months following treatment. Possible side effects are also sought during that period.

Among many antitrichomonal drugs, the most important are the 5-nitroimidazoles, that is, metronidazole, tinidazole, and ornidazole (see Table 13.10). Metronidazole, the oldest of these compounds, has been employed successfully in the treatment of trichomoniasis in humans. Also it is the drug most often administered to children. Although with metronidazole

TABLE 13.10. Dosages of metronidazole, tinidazole, and ornidazole administered to children with unifocal and multifocal trichomoniasis.

Age (yr)	Mean weight (kg)	Metronidazole daily dosages (mg) administered*		Tinidazole daily dosages (mg) administered†		Ornidazole daily dosages (mg) administered‡	
		Orally	Locally	Orally	Locally	Orally	Locally
0–1§	10.0	150	10	250	—	125	10
1–6	20.5	250	50	500	—	250	50
7–12	40.0	500	150	1000	—	375	150
>12	>40.0	500–1000	250–500	1000–1500	—	500	250–500

*Daily dosage: *orally*—one-third of the dosage administered every 8 hr, for a total of 5 to 10 days; *locally* (intravaginally)—once (the total dosage) or twice (one-half of the total dosage) administered during a 24-hr period for a total of 5 to 10 days.
†Daily dosage: *orally only*—one-half of the dosage administered every 12 hr for a total of 5 days, exceptionally for up to 10 days.
‡Daily dosage: *orally*—one-half of the dosage administered every 12 hr for a total of 5 days, exceptionally for up to 10 days; *locally* (intravaginally)—once (the total dosage) or twice (one-half of the total dosage) administered during a 24-hr period for a total of 5 days.
§Newborns and infants receive *locally* (intravaginally) 1 mg of a drug per 1 kg of body weight.

there is a tendency to give larger dosages for adults over shorter periods of time, even in very serious cases of trichomoniasis in children we had to try small dosages, and in the beginning we limited treatment to local application, for example, in unifocal infection of the vagina in girls. We administered 1 mg/kg of the drug for overnight (total period of administration, 5 to 10 days). In infections of the urinary system in girls as well as in boys, we administered also metronidazole orally about 10 to 20 mg/kg for 5 to 7 days. There have recently appeared many publications dealing with the use of metronidazole in humans; however, only a few concern the treatment of children. For example, Metze and Brandau[35] described the high efficacy of metronidazole in children treated locally and orally; the dosages were calculated in relation to body surface area. These workers also pointed out the necessity for additional treatment in the case of trichomoniasis of the urinary passages, because of the presence of accompanying pathogenic bacteria. Hüber and Hiersche[29] suggested a short (2-day) metronidazole treatment for children of 20 mg/kg/day. They also recommended a longer treatment (5 to 7 days), administering 10 mg every 12 hours, and, in severe cases, 20 mg/kg. Similar dosages of metronidazole were tried in children by Scharp and Amon[21] in short treatments (2 days) in which 20 mg/kg/day were administered. They also employed weekly treatments for children with severe trichomoniasis of 10 mg/kg every 12 hours.

Tinidazole was recommended for short-term, usually single-time, treatment. It was tried first in patients over 12 years of age. In our investigations[84] involving in vitro tests of axenic cultures of *T. vaginalis* strains with regard to their sensitivity to metronidazole and tinidazole, a certain percentage of these strains was sensitive to smaller dosages of tinidazole, which appeared to suggest a possibly higher antitrichomonal activity of tinidazole over metronidazole. Soyka and Milek[22] recommended single tinidazole treatments (30 mg/kg) for children suffering from trichomoniasis; they found no toxic side effects of this drug. Boschitsch et al[85] recommended single treatments with tinidazole for girls suffering from trichomoniasis. A dosage of 1000 mg was suggested for patients weighing less than 20 kg, 1500 mg for those weighing 20 to 40 kg, and 2000 mg for girls over 40 kg. In our Clinic, after administering 2000 mg of tinidazole, we examined various morphologic and biochemical parameters in the peripheral blood of patients who presented with trichomoniasis uncomplicated by any other diseases of the urogenital system[86]; no side effects were noted in any of these persons. The foregoing results encouraged us to employ tinidazole in the treatment of urogenital trichomoniasis in children.

Drawing from our experience obtained in

very extensive trials, extending over 10 years, of metronidazole, tinidazole, and, more recently ornidazole, we have summarized data pertaining to the treatment of trichomoniasis in children in Table 13.10.

In children of both sexes who, because of some defects of the urogenital system, tended to retain urine, elimination of *T. vaginalis* proved much more difficult—treatment had to be extended to 20 and even to 30 days, especially with ornidazole, which appears to be the drug least toxic for children. In youngsters above 12 years of age more often than in younger children, and especially in the early discovered cases of family infections, dosages corresponding to those used for adults were administered in single treatments.

The list of drugs that are useful in the treatment of trichomoniasis, especially multifocal trichomoniasis, is very short. Also the sporadic decrease in sensitivity of *T. vaginalis* strains to the presently available drugs, especially metronidazole, indicates the need for development of new effective antitrichomonal compounds. Treatment of trichomoniasis complicated by mycosis requires the application, in addition to antitrichomonal drugs, of mycostatic or mycocidal agents. The antibiotics used for this purpose are nystatin, natamycin, and polyfungin. Widely employed agents with antimycotic activity also include clotrimazole, ketoconazole, and nifuratel. Various forms of these drugs facilitate the elimination of fungi from infection sites.[87,88] In a comparative study of efficacy in eliminating mycotic agents from the vaginae of 284 girls (0 to 18 years of age) with the aid of antibiotics that were administered in the form of a suspension, we obtained the following results: for nystatin, $73.1 \pm 3.75\%$; for natamycin, $89.4 \pm 3.34\%$; and for polyfungin, $77.5 \pm 4.67\%$.[85] Because of significantly better results ($P < 0.05$) observed with natamycin, we checked its effectiveness in unifocal ($86.0 \pm 3.3\%$) and multifocal ($43.5 \pm 4.8\%$) infections; the difference was statistically significant ($P < 0.01$).[89] The results of chemotherapy indicated also a certain antitrichomonal activity especially on the part of clotrimazole and nifuratel.[71,91] Clotrimazole (96%) was also significantly more effective than nystatin (76%) in eliminating mycotic agents from girls. Ketoconazole, introduced by us most recently for the treatment of urogenital trichomoniasis complicated by mycosis, proved to be an extremely valuable drug, because the products excreted in urine do not lose their antimycotic activity. We have also used 5-fluorocytosine. It should be pointed out that even to date nifuratel remains a very important drug in certain complicated cases of urogenital trichomoniasis in children, because it has strong trichomonacidal as well as antimycotic activity; to a lesser degree it is also bactericidal.[90,91] In a recently prepared monograph about mycoses and antimycotic drugs, we included many new data that we hope will stimulate further investigations and aid clinicians in choosing effective therapeutic measures.[92]

Conclusions

Urogenital trichomoniasis of children is a problem of social significance, especially in those countries in which the disease is widely spread in human populations. *Trichomonas vaginalis* must be counted among the important etiologic agents of infections accompanying birth, as well as being a disease common in groups sharing living quarters (e.g., the family unit); it should continue to be the subject of intensive basic and applied investigations.

References

1. Kurnatowska A: Studies on *Trichomonas vaginalis* Donné. IV. Size of the population of trichomonads in the vaginal secretion of women. *Acta Med Pol* 9:175-182, 1968.
2. Kurnatowska A: On the value of the methods of identification of *Trichomonas vaginalis* Donné in the vaginal secretion. *Ginekol Pol* 29:139-150, 1958.
3. Kurnatowska A: Studies on *Trichomonas vaginalis* Donné. I. Clinical aspects and pathogenesis of trichomoniasis of the female urogenital organs. Staining of *Trichomonas vaginalis* in the diagnosis of trichomoniasis. *Acta Med Pol* 6:459-464, 1965.
4. Kurnatowska A: Studies on *Trichomonas vaginalis* Donné. III. Biometric features of *T. vaginalis* from different clinical forms of trichomoniasis. *Acta Protozool* 4:185-211, 1966.
5. Kurnatowska A: Comparison of some biometrical characteristics of *Trichomonas vaginalis* Donné in light and scanning microscope. Proc Cong Trichomoniasis with Int Particip, Bratislava, Czechoslovakia, Sept 1977. *Folia Fac Med Univ Comenianae Bratislava, Príloha.* 17 (fasc 2):343-357, 1979 (1981).

6. Kurnatowska A, Hajdukiewicz G: Further investigation on variability of *T. vaginalis* Donné in the scanning electron microscope. *Wiad Parazytol* 27:169-171, 1981.
7. Chappaz G: Recherches cliniques et biologiques recentes sur la trichomonase urogénitale. *Wiad Parazytol* 10:105-109, 1964.
8. Gorzędowska E: The treatment of vaginal trichomonadosis in girls with Terrarsol. *Wiad Parazytol* 10:225-226, 1964.
9. Peter R, Komorowska A, Kurnatowska A: The problem of trichomonadosis in different periods of a woman's life. *Wiad Parazytol* 12:244-247, 1966.
10. Sipowicz I, Kuczyńska K: Clinical investigations of trichomonadosis in girls. *Wiad Parazytol* 15:337-339, 1969.
11. Gorzędowska E, Wachnik S: Trichomonadosis in girls in the city of Wrocław and in the Wrocław voivodeship. *Wiad Parazytol* 23:575-576, 1977.
12. Bregun-Dragić N, Dordević M: Diagnosis and treatment of vulvo-vaginitis adolescentes. *Mat II Symp Polonum Paedogynaecologorum cum Particip Int,* pp 39-43. Łódź: Medical Academy, 1972.
13. Liniecka J, Komorowska A, Woźniak M: The frequency of gynecological disease in girls in the material of the Outpatient Gynecological Clinic for Girls in Łódź, in the years 1976-1979. *Mat IV Symp Polonum Paedogynaecologorum cum Particip Int,* pp 191-199. Łódź Med Acad, 1980.
14. Saygi G: An epidemiological survey of *T. vaginalis* infection in three different groups of women in Erzurum, Turkey. *Wiad Parazytol* 27:201-205, 1981.
15. Valent M, Čatar G, Klobusický M, Valentová M, Porubenová A, Gavač P: Laboratory and clinical investigations of vaginal trichomoniasis and mycosis in the material of the Department of Clinical Parasitology in Bratislava. *Wiad Parazytol* 29:131-136, 1983.
16. White S, Loda FA, Ingram DL, Pearson A: Sexually transmitted diseases in sexually abused children. *Pediatrics* 72:17-25, 1983.
17. Peter R, Vesely K: *Kindergynäkologie.* Leipzig: Georg Thieme, 1966.
18. Decker K: Diagnosis and therapy of genital infections in childhood, puberty and adolescence. *Gynaekologie* 16:56-60, 1983.
19. Dewhurst J: *Practical Pediatric and Adolescent Gynaecology.* New York: Marcel Dekker, 1980.
20. Heinz M, Hoyme S: *Gynäkologie des Kindes- und Jugendalters.* Leipzig: Georg Thieme, 1972.
21. Scharp H, Amon K: Zur Therapie der Vulvovaginitis bei Kindern und Jugendlichen mit Metronidazol. *Zentralbl Gynäkol* 106:374-378, 1984.
22. Soyka E, Milek E: Trichomoniasis vaginalis i Kindesalter: Vorkommen, Frequenz, Symptomatologie. *Gynaekol Praxis* 4:473-478, 1980.
23. Gross MT, Buckman MI: Vaginitis in children. *Am J Dis Child* 97:613-615, 1959.
24. Paradise IE, Campos JE, Friedman HM, Frishmuth G: Vulvovaginitis in premenarcheal girls: Clinical features and diagnostic evaluation. *Pediatrics* 70:193-198, 1982.
25. Hammerschlag MR, Alpert S, Rosner I, Thurston P, Semine D, McComb D, McCormack WM: Microbiology of the vagina in children: Normal and potentially pathogenic organisms. *Pediatrics* 62:57-62, 1978.
26. Heimrath T: The vaginal bacteriological examination of girls with infections of the urinary tract. *Mat II Symp Polonum Paedogynaecologorum cum Particip Int,* pp 141-144. Łódź: Medical Academy, 1975.
27. Heller RH, Joseph JM, Davis JH: Vulvovaginitis in the premenarcheal. *J Pediatr* 74:370-377, 1969.
28. Wszelaki-Lass E, Kuźmińska A: Bacteriuria and vulvovaginitis in girls. *Mat II Symp Polonum Paedogynaecologorum cum Particip Int,* pp 160-164. Łódź: Med Acad, 1972.
29. Hüber A, Hiersche HD: *Praxis der Gynäkologie im Kindes- und Jugendalter.* Stuttgart: Georg Thieme, 1977.
30. Komorowska A, Malinowski H: Enterobiosis and vaginal fluor in girls. *Wiad Parazytol* 11:587-590, 1965.
31. Komorowska A, Kurnatowska A, Liniecka J: Ocurrence of *trichomonas vaginalis* Donné in girls depending on hygienic conditions. *Wiad Parazytol* 8:247-251, 1962.
32. Gorzędowska E: Contribution to the epidemiology of trichomonadosis in girls. *Wiad Parazytol* 15:353-354, 1969.
33. Valent M, Čatár G, Klobosický M, Holková R, Solar G: Gynaecological and parasitological findings in delinquent girls. *Mat II Symp Polonum Paedogynaecologorum cum Particip Int,* pp 47-49. Łódź: Med Acad 1972.
34. Sagone I: Trichomonadosis in the virgin: Clinical and therapeutic aspects. *Wiad Parazytol* 19:329-333, 1973.
35. Metze H, Brandau H: Trichomonaden-Kolpitis im Kindesalter. *Fortschr Med* 92:978-980, 1974.
36. Komorowska A, Kurnatowska A: Investigations concerning possibilities of getting infected with *T. vaginalis* and fungi by women during a cure at Duszniki. *Wiad Parazytol* 29:343-346, 1973.
37. Neistein LS, Goldenring J, Carpenter S: Nonsexual transmission of sexually transmitted diseases: An infrequent occurrence. *Pediatrics* 74:67-76, 1984.
38. Ślaska M, Więckowska A: *T. vaginalis* and occurrence of fungi and *Neisseria gonorrhoeae* in population of Łódź. *Wiad Parazytol* 27:281-284, 1981.

39. Caterall RD: Some observations on the epidemiology and diagnosis of *T. vaginalis*. *Wiad Parazytol* 19:86-92, 1983.
40. Soszka S, Kuczyńska K: Influence of *T. vaginalis* on the physiological flora of the vagina. *Wiad Parazytol* 23:519-524, 1977.
41. Gerstner GJ, Grunberg W, Boschitsch E, Rotler M: Vaginal organisms in prepubertal children with and without vulvovaginitis: A vaginoscopic study. *Arch Gynecol* 231:247-252, 1982.
42. McCormack WN, Evrard JR, Laughlin CF, Rosner B, Alpert S, Crockett VA, McComb D, Zimmer SH: Sexually transmitted conditions among women college students. *Am J Obstet Gynecol* 139:130-133, 1981.
43. Goldstein L, Zawadzka-Krzywańska E: Acute infestation of *T. vaginalis* in two infants. *Wiad Parazytol* 27:229-232, 1981.
44. Schwarz R, Rosegger H, Trauner F: Trichomoniasis in Neugeborenenalter. *Paediatr Paedol* 12:319-322, 1977.
45. Sieroszewski J, Chrzanowski J, Komorowska A, Lisner M, Kularska I: The effect of infectious diseases on vaginal discharges in young girls, pp 51-55. *Mat 14th Meet Polish Gynaecol Soc*. Warsaw: Państwowy Zakład Wydawnictw Lekarskich (Polish Medical Publishers), 1960.
46. Komorowska A: *On the Problems of Girls' Puberty*. Warsaw: Państwowy Zakła Wydawnticw Lekarskich (Polish Medical Publishers), 1967.
47. Komorowska A: Causes and diagnosis of fluors in girls: Part I. *Przegl Ped* 4:433-439, 1974.
48. Sieroszewski J, Komorowska A, Kurnatowska A: Les causes des inflammations de la vulvae et de l'introitus du vagin chez les jeunes filles. *Obstet Ginecol* 18:271-279, 1970.
49. Komorowska A, Kurnatowska A: Estrogenic reactions of the vaginal epithelium and *T. vaginalis* infection in girls at puberty. *Wiad Parazytol* 10:136-138, 1964.
50. Kurnatowska A: The project of guiding lines in the diagnosis of trichomonadosis. *Wiad Parazytol* 10:120-121, 1964.
51. Grys E: Topography of trichomonadosis in the reproductive organ of the woman. *Wiad Parazytol* 10:122-124, 1964.
52. Jírovec O, Petrů M: *Trichomonas vaginalis* and trichomoniasis. *Adv Parasitol* 6:117-188, 1968.
53. Soszka S, Łotocki W, Kuczyńska K: On the coexistence of *T. vaginalis* in the vagina and urinary tract in women. *Wiad Parazytol* 19:463-468, 1973.
54. Hoffmann B, Kilczewski W, Małyszko E: Studies on trichomoniasis in males. *Br J Vener Dis* 37:172-175, 1961.
55. Jíra J: Trichomoniasis in men. *Wiad Parazytol* 6:519-527, 1959.
56-60. Kadłubowski R (ed): *Zarys Parazytologii Lekarskiej (Outline of Medical Parasitology)*, 1st to 5th ed. Warsaw: Państwowy Zakład Wydawnictw Lekarskich (Polish Medical Publishers), 1967, 1972, 1975, 1979, 1983.
61a. Krupicz J: Trichomonal inflammation of glans penis and preputium. *Wiad Parazytol* 10:148-151, 1964.
61b. Schaffler AG: *Diseases of the Newborn*. Philadelphia: WB Saunders, 1968.
62. Kreger-van Rij NJW (ed): *The Yeasts: A Taxonomic Study*. Amsterdam: Elsevier, 1984.
63. Calonescu M, Larose G, Birry A, Roy J, Kasatiyae SS: Genital infection in juvenile delinquent females. *Br J Vener Dis* 49:72-77, 1973.
64. Erikson G, Wanger L: Frequency of *N. gonorrhoeae*, *T. vaginalis* and *C. albicans* in female venereological patients: A one-year study. *Br J Vener Dis* 51:192-197, 1975.
65. Hardy PH, Hardy JB, Nell EE, Graham DA, Spence MR, Rosenbaum RC: Prevalence of six sexually transmitted disease agents among pregnant inner-city adolescents and pregnancy outcome. *Lancet* 2:333-337, 1984.
66. Nielsen R, Sondergaard J, Ullman S: Simultaneous occurrence of *Neisseria gonorrhoeae*, *Candida albicans*, and *Trichomonas vaginalis*. *Acta Derm Venereol* 54:413-415, 1974.
67. Schneider GT: Vaginal infections—how to identify and treat them. *Postgrad Med* 73:255-262, 1983.
68. Kurnatowska A: Basic methods for the classification and identification of pathogenic fungi. In Kowszyk-Gindifer Z, Sobiczewski W (eds): *Mycoses and Their Elimination*, 1st ed, pp 91-106. Warsaw: Państwowy Zakład Wydawnictw Lekarskich (Polish Medical Publishers), 1986.
69. Kurnatowska A: Mycoses of urogenital organs. In Kowszyk-Gindifer Z, Sobiczewski W (eds): *Mycoses and Their Elimination*, 1st ed, pp 207-220. Warsaw: Państwowy Zakład Wydawnictw Lekarskich (Polish Medical Publishers), 1986.
70. Komorowska A, Kurnatowska A: Some diagnostic problems of mycosis and trichomoniasis in girls. *Mat II Symp Polonum Paedogynaecologorum cum Particip Int*, pp 50-55. Łódź: Medical Academy, 1972.
71. Örley J: The indication field and therapeutic use of Canestan in pediatric gynecology. *Ther Hung* 28:32-35, 1980.
72. Kurnatowska A: Influence of hologenesulphonamides on vaginal microorganisms of women, including consideration of trichomonadosis complicated by mycosis. Clinical part. *Wiad Parazytol* 15:145-160, 1969.
73. Kurnatowska A: Studies on *Trichomonas vaginalis* Donné. V. Trichomonadosomycosis. *Acta Med Pol* 14:333-342, 1973.
74. Al-Salihi FL, Curran JP, Wang JS: Neonatal *Trichomonas vaginalis*: Report of 3 cases and review of literature. *Pediatrics* 53:196-200, 1974.
75. Kurnatowska A, Horwatt E: Comparison of

75. the density of population of *T. vaginalis* Donné in cultures with fungi strains isolated from cases of trichomonadosis and trichomonadosomycosis. *Wiad Parazytol* 23:529-534, 1977.
76. Kurnatowska A, Horwatt E: Quantitative study of *T. vaginalis* population in vitro with *Candida albicans*. *Wiad Parazytol* 29:13-18, 1983.
77. Hiemstra I, van Bel F, Berger HM: Can *Trichomonas vaginalis* cause pneumonia in newborn babies? *Br Med J* 289:355-356, 1984.
78. McLaren LC, Davis LE, Healy GR, James CG: Isolation of *Trichomonas vaginalis* from the respiratory tract of infants with respiratory disease. *Pediatrics* 71:888-890, 1983.
79. Walton BC, Bacharach T: Occurrence of trichomonads in the respiratory tract: Report of three cases. *J Parasitol* 49:35-38, 1963.
80. Kurnatowska A, Kurnatowska A: Analysis of cases of infection with *T. tenax* (O.F. Müller, 1773) Dobell, 1939, *Wiad Parazytol* 29:155-161, 1983.
81. Teras J: Comparison of certain biological properties of *Trichomonas* species of humans. *Wiad Parazytol* 29:53-56, 1983.
82. Honigberg BM: Trichomonads of importance in human medicine. In Kreier JP (ed): *Parasitic Protozoa*, vol 2, pp 275-454. New York: Academic Press, 1978.
83. Norn MS, Lundvall F, Paerregaard P: May *Trichomonas vaginalis* provoke conjunctivitis? *Acta Ophthalmol* 54:574-578, 1976.
84. Kurnatowska A: Treatment of multifocal trichomoniasis. *Pol Tyg Lek* 37:1469-1472, 1978.
85. Boschitsch E, Gerstner G, Grunberger W: Die Vulvovaginitis in Kindesalter. *Fortschr Med* 100:1703-1708, 1982.
86. Komorowska A, Kurnatowska A: Some parameters of the peripheral blood in trichomonadosis treated with tinidazole (Fasigyn). *Wiad Parazytol* 25:427-432, 1979.
87. Kurnatowska A, Kwaśniewska J: The efficacy of different forms of natamycin in treatment of mucous membrane and skin candidosis. *Przegl Dermatol* 48:101-104, 1981.
88. Raab WP: *Natamycin (Piramycin)*. Stuttgart: Georg Thieme, 1972.
89. Kurnatowska A, Komorowska A: Behandlungsprobleme von Mykosen der Geschlechtsorgans bei Mädchen. *Zentralbl Gynäkol* 101:723-732, 1979.
90. Kurnatowska A: Macmiror-Polichimica in the treatment of chronic multifocal trichomonadosis of the genitourinary organs. *Wiad Parazytol* 19:495-498, 1973.
91. Stramezzi P, Arbasino M: Four years of clinical investigations with nifuratel. *Wiad Parazytol* 15:367-370, 1969.
92. Kowszyk-Gindifer Z, Sobiczewski W (eds): *Mycoses and Their Elimination*, 1st ed. Warsaw: Państwowy Zakład Wydawnictw Lekarskich (Polish Medical Publishers), 1986.
93. Simić C: Biologie et la culture de *Trichomonas intestinalis* Leuckart, 1879. *Glasnik Tsentral Khig Zavoda, Beograd* 7:81-88, 1929.
94. Roiron-Ratner V: Etude comparative des principaux milieux de culture de *Trichomonas vaginalis:* Les infestations à *Trichomonas*. *I Symp Europ Reims,* May 1957, pp 244-252. Paris: Masson, 1957.
95. Feo LG: Incidence of *Trichomonas vaginalis* in various age groups. *Am J Trop Med Hyg* 5:786-790, 1956.
96. Kutcher E, Keller L, Groy LA: A microbiological study of pediatric vaginitis. *Pediatrics* 53:210-218, 1958.
97. Larry WR: Premenarcheal vaginitis. *Obstet Gynecol* 13:723-727, 1959.
98. Bartoszewski A, Klonowski H, Radomański T, Stępkowski S: Attempt to treat vaginal trichomonadosis with nystersol (Polfa). *Wiad Parazytol* 10:209-215, 1964.
99. Stopa T: Activity of the Outpatient Gynecological Clinic for Girls in Lublin in the years 1966-1969. *Mat Sci Conf Poznań*, pp 241-246. Poznań: Medical Academy, 1970.
100. Örley J: Clinico-epidemiological observations on children's trichomoniasis. *Resumes I Cong Panam Ginecol Infantilo-Juvenil*, pp 105-108. Buenos Aires: Ed. Panamericana, 1978.
101. Borowicz C, Trebicka-Kwiatkowska B: Incidence and diagnosis of vulvovaginitis in girls. *Mat II Symp Polonum Paedogynaecologorum cum Particip Int*, pp 35-37. Łódź: Med Acad, 1972.
102. Foltynowicz-Mańkowa J: Inflammation of genital tract on the basis of the material of the Outpatient Clinic for Girls in Poznań, in the years 1964-1971. *Mat II Symp Polonum Paedogynaecologorum cum Particip Int*, pp 59-63. Łódź: Med Acad, 1972.
103. Dramusić V, Zecević B, Marković M: Urinary tract during inflammatory infections of adolescents. *Mat II Symp Polonum Paedogynaecologorum cum Particip Int*, pp 64-69. Łódź: Medical Academy, 1972.
104. Nowosad K, Kwoczyński M, Penar S: Epidemiological studies of trichomoniasis in factories in Wrocław. *Wiad Parazytol* 19:283-296, 1973.
105. Örley J, Baranyai P: Experiences gained in the diagnosis of *T. vaginalis* by different culture and staining methods. *Wiad Parazytol* 19:437-439, 1973.
106. Moraes M: Clinique de gynecologie infantile et juvenile Santos (Bresil): Des rapports. *Proc II Symp Int Pediatric Adolesc Gynaecol*. Lausanne: Fédération Internationale de Gynécologie Infantile et Juvenile, 1976.
107. Berić I, Vujić M, Dordević M: Modern therapy of sexually transmitted diseases in family, chil-

dren and adolescents. *Mat IV Symp Polonum Paedogynaecologorum cum Particip Int,* pp 101-103. Łódź: Medical Academy, 1980.
108. Moreira AJ: Puberty and infections. In Bruni V, Dewhurst J, Gaspari F, Rey-Stocker I (eds): *Pediatric and Adolescent Gynaecology,* pp 269-291. Rome: Serono-Symposia, 1982.

Additional Relevant References

Altchek A: Pediatric vulvovaginitis. *Pediatr Clin North Am* 19:559-580, 1972.

Berić B, Bregun-Dragić N, Popović D, Stretenović Z: Les problèmes de la trichomonase uro-génitale chez les fillettes prépubères. *Gynecol Pract* 21:217-222, 1970.

DeJong AR: Sexually transmitted diseases in children. *Am Fam Physician* 30:185-193, 1984.

Diller C, Murphy G, Lauchlan SC: Cervicovaginal cytology in patients 16 years of age and younger. *Acta Cytol* 27:426-428, 1983.

Harper JR: *Trichomonas vaginalis* urinary infection in a boy. *Proc R Soc Med* 60:897, 1967.

Heller RH, Joseph JM, Davis HJ: Vulvovaginitis in the premenarcheal child. *J Pediatr* 74:370-377, 1969.

Huber A: Entzündliche Erkrankungen der weiblichen Genitalorgane im Kindesalter. *Wien Med Wochenschr* 119:385-391, 1969.

Kern A: Bedeutung und Diagnostik venerischer Infektionen im Kindes- und Jugendalter. *Paediatr Grenzgeb* 13:147-162, 1974.

Littlewood JM, Kohler HG: Urinary tract infection by *Trichomonas vaginalis* in a newborn baby. *Arch Dis Child* 41:693-695, 1966.

Swincow G, Pytel-Dąbrowska T, Hinz K, Gizińska K: Trichomoniasis in children with symptoms of urinary tract infection. *Pol Tyg Lek* 37:453-455, 1982.

14

Cytopathology and Histopathology of the Female Genital Tract in *Trichomonas vaginalis* Infection

Prabodh K. Gupta and John K. Frost

General Considerations

Four species of trichomonads inhabit humans: *Trichomonas vaginalis,* described initially by Donné,[1] *Trichomonads tenax,* first seen by O.F. Müller,[2] *Pentatrichomonas hominis,* reported originally by Davaine,[3] and *Trichomitus fecalis,* isolated by Cleveland.[4]

These organisms are highly site specific. *Trichomonas vaginalis* occurs almost exclusively in the lower genital tract,[5] *T. tenax* is restricted primarily to the oral cavity, and *P. hominis* lives only in the large bowel. The latter two appear to be unable to infect the genital tract (for information about the oral and intestinal trichomonads, see Chap. 19.)

Trichomonas vaginalis is considered to be the only typical trichomonad with definite pathogenicity for humans. Besides lower genital tract disease, rare cases have been reported to involve other sites and cause infection. These include neonatal pneumonia,[6,7] perinephric abscesses,[8] cutaneous lesions, acute and subacute arthritis not responding to antirheumatic therapy, gastrointestinal diseases including gastritis, as well as ascites, hepatomegaly and renal disease.[9,10] Most of these nongenitourinary diseases caused by *T. vaginalis* occur either in neonates or in postmenopausal women.

Many host factors affect clinical disease of the genitourinary tract. These include the endocervical glands and their normally alkaline mucus with its physiochemical properties; various complements and their concentrations in the local tissues of the female genital tract; immunoglobulins, especially IgA and IgE; tissue mast cells; the phagocytic activity of polymorphonuclear leukocytes and macrophages; and the concurrent polymicrobial vaginal environment.[11-14] Gonadal, adrenal, and other hormones affect the lower genital tract epithelium during the active reproductive ages. Their cyclic and periodic changes bring about physiologic alterations in the epithelia and their underlying supportive tissues. These increase susceptibility or resistance to infection by *T. vaginalis,* depending on the endocrinologic state (e.g., follicular phase, luteal phase).

Numerous factors of the parasite may also affect or be related to clinical disease, including chemotactic properties, size, flagellar form and activity, and strain pathogenicity.[15-18] There are also other correlated parameters such as adhesiveness, phagocytosis, and presumably some enzymes and toxins.

Female Genital Tract Disease

Although described elsewhere (see Chap. 11), a brief clinical description is essential to appreciate the cytologic and histologic changes of *T. vaginalis* infection. The clinical infection by *T. vaginalis* is concisely described in its acute, chronic, and latent phases.[19-21] Nearly 50% of women harboring this parasite are in the last, asymptomatic, phase.[22,23]

During the symptomatic *acute* and *chronic* infections, the organisms are most frequently found in the vagina, occasionally in the secretion from Skene's or Bartholin's glands.[24] Lower urinary tract infection may occur in approximately 10% to 12% of women with vaginal trichomoniasis.[25] In nearly 50% of the patients with *T. vaginalis* who have dysuria and urethral discharge, organisms can be recovered from the urethra[26] or from clean-catch urine specimens.[27] The urinary tract may be an important source of persistent *Trichomonas* infection, both among prepubescent girls and postmenopausal women.

The organisms have been identified from purulent tubal material in the fallopian tubes of women with lower genital tract trichomoniasis.[28,29] A similar etiologic role has been proposed for *Trichomonas* in salpingitis and pelvic inflammatory disease (PID), and these conclusions have been strengthened by extensive scanning electron microscopic observations.[30]

The precise incubation period of *T. vaginalis* infection is not known. On the basis of experimental data, however, it is believed to vary from 4 to 28 days.[20,31] A foamy vaginal discharge is reported by 10% to 25% of patients, which may be malodorous, copious, frothy, and greenish-yellow; vulvovaginitis is a common accompaniment.[32] Dyspareunia, dysuria, and symptoms of cystitis may also be present. Lower abdominal pain, although not a common feature of urogenital trichomoniasis, is reported by 2% to 15% of infected women.[20,22,23] This may represent PID and pelvic lymphadenitis. Inguinal lymphadenitis and lymphadenopathy may occur. Reddening of the vagina with small punctate hemorrhagic spots (strawberry vagina) may be noted,[31,33] but the "strawberry cervix," classically considered typical of trichomoniasis, is observed in fewer than 5% of infected patients.[34]

The disease usually becomes *latent* during the follicular, ovulatory, and early luteal phases of the menstrual cycle. The organism retreats up the more basic endocervical canal, away from the ever increasing acidity of the vagina, until it finds a pH better suited to its strict demands around neutrality. It can seek refuge in the glands, and lie, without causing any symptoms, in the painless cervix of the unknowing host.

Clinical disease may be *exacerbated* in the late luteal phase, during or immediately following the menstrual period.[19,20,24,31] *Trichomonas vaginalis* infection also appears to be more common during pregnancy when, in reality, it just becomes clinically manifest, owing to the hormonal milieu rendering the environment in the vagina more favorable for the protozoa. English workers reported about a 25% occurrence of *T. vaginalis* among pregnant women. From a study involving 2828 pregnant women at the Johns Hopkins Hospital clinics, the urogenital trichomonads were identified in 19%.[35] Deleterious effects of trichomoniasis on the outcome of human pregnancy, although reported on rare occasions, are not universally accepted.[36,37] Similar skepticism exists with regard to the occurrence of postpartum *T. vaginalis* endometritis.[38]

Cytopathologic Features

Trichomonas vaginalis can be identified in cervicovaginal specimens in a number of ways (see Chap. 16), and the accompanying tissue reactions studied in biopsies or surgically removed specimens at that time. All observations presented here were made on Papanicolaou-stained vaginopancervical (Fast) smears[19] that were promptly wet-fixed in 95% ethyl alcohol. This polychromatic staining reveals excellent morphologic details of the tissue reactions of *T. vaginalis*.

Papanicolaou stain has become the most popular and dependable clinical method of trichomonad diagnosis in a permanent preparation. When properly utilized,[19,39] it is an extremely accurate and sensitive method for identification of the organism. It must be realized that the broad term "Pap smear" includes a variety of gynecologic cervical or vaginal specimens that are obtained by various methods, and are representative of different anatomic sites and secretions. Obviously, the diagnostic value of these different "Pap smears" is not comparable and varies widely. The problem is further compounded by such variables as inclusion of inadequately representative material from the posterior fornix or cervical os, technical quality of the preparation, questionable fixation and staining, and the level of expertise and diligence of the examiner. Occasionally, heavy inflammation and excessive bleeding may obscure or destroy the organisms, making diligent search for well-preserved, clearly identifiable trichomonads extremely time-consuming.

Vaginal pool material is satisfactory for identifying the organism in symptomatic vaginitis. However, in women with asymptomatic latent infection of the endocervix, the best opportunity for detecting *T. vaginalis* is by examination of cervical material that has been obtained from a pancervical scraping[19,39] of the ectocervix, external os, and endocervical canal.

In satisfactory Fast smears stained according

to a satisfactory Papanicolaou technique, *Trichomonas* infections can often be suspected by observing certain features commonly associated with such infections. The most valuable of the presumptive signs of *Trichomonas* infection is the presence of aggregates of leukocytes covering the surface of isolated mature squamous epithelial cells.[19,39-41] These leukocytic agglomerations, usually associated with trichomoniasis, have been referred to as "buckshot," "B-B shot," or "cannonballs" (Fig. 14.1a,b). They represent the end stages of a series of events that starts with trichomonads feeding while attached by their axostyles to the surface of an individual host squamous epithelial cell (Fig. 14.1c,d) followed by phagocytosis of the protozoa by polymorphonuclear leukocytes and, occasionally, macrophages (Fig. 14.1a,b). The evolution of these cannonballs can be demonstrated in appropriately stained cellular preparations. Trichomonads attach to the margin and upper surface of the mature squamous cells and, in cultures, penetrate the cell membranes.[42] Numerous organisms accumulate, first at the outer margin (Fig. 14.1c) but gradually tending to cover the entire cell surface (Fig. 14.1d). The trichomonads, in turn, are ingested and destroyed by the polymorphonuclear leukocytes, which remain attached to each other and to the surfaces of the squamous epithelial cells (Fig. 14.1a,b), producing a cannonball.

Most of the inflammatory exudate associated with *Trichomonas* infection comes from the endocervix and its glands. Endocervicitis with varying degrees of squamous metaplasia is common in women with asymptomatic *Trichomonas* infection. In women harboring *T. vaginalis* following hysterectomy but in whom the ovaries have been preserved, the parasites may be intermingled with mature squamous epithelial cells and may occur without any accompanying inflammation (Fig. 14.2a). "Pure" trichomonad vaginitis appears to be uncommon in such women. These patients may be asymptomatic carriers of the trichomonads. Organisms may resemble degenerating small parabasal cells in women with atrophic vaginitis, and thus may be extremely difficult to diagnose accurately in such a clinical situation, which, however, requires accurate identification of the etiologic agent for effective therapy.

Another feature that may help the examiner to suspect the presence of *Trichomonas* is the presence of *Leptothrix* in the vaginal smear. These are long, nonbranching curved bacilli, usually larger than and distinct from lactobacilli (Fig. 14.2b,c). Bibbo et al[43] observed an association of *Leptothrix* and *Trichomonas* in 95% of 1003 consecutive lower genital tract smears. Most of the women were symptomatic and complained of a frothy, foul-smelling vaginal discharge.

In a vaginal smear with heavy inflammation and extensive proteolysis, intact trichomonads may be extremely sparse. In such situations, it is often difficult to find organisms with well-preserved nonhematoxylinophilic cytoplasm, and a distinct weakly hematoxylinophilic nucleus, characteristics essential for identification of *T. vaginalis* (Fig. 14.2d). If the parasites have blurred or obliterated cellular details, they can still be positively identified through the use of immunodiagnostic tagging (Fig. 14.2e,f).

The organisms multiply by binary fission (Fig. 14.2e). In active clinical disease, the most pathogenic and rapidly reproducing protozoa are usually small and pleomorphic.[19,44,45]

In immunosuppressed persons there may be minimal accompanying inflammatory response. In such cases, multiplication of the trichomonads can be most active, the parasites being packed tightly together to form "microcolonies" (Fig. 14.3a,b).

Under favorable environmental conditions, the trichomonads may enlarge to become giant forms.[28] Women harboring such forms may be clinically free of any obvious symptoms. The parasites may become inactive, round up, and have inconspicuous flagella. The inactive forms may appear to be "hibernating," and have been referred to as pseudocysts or, even, cysts. No true cysts, however, have been found to date in *T. vaginalis*.[22]

Histopathologic Features

Light Microscopy

Tissue studies of human trichomoniasis had been somewhat obscured by the generally polymicrobial infection of the lower genital tract and indistinct morphology of the protozoa in tissue. Recent developments (e.g., culture, antibody tagging, DNA probes) have increased the ability to detect accompanying infections, and thus better to identify and characterize the histopathologic effects of *Trichomonas* infections.

In routine hematoxylin and eosin (H & E)-

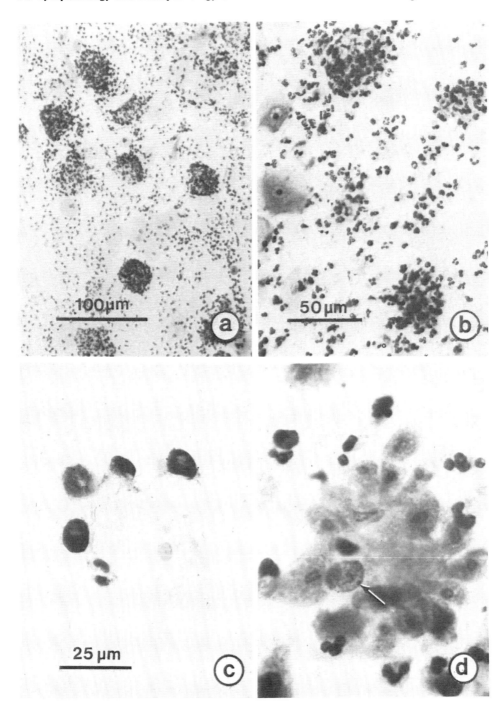

Fig. 14.1. "Cannonballs" or "B-B shots." **a.** Mostly rounded aggregates of leukocytes covering the surfaces of squamous epithelial cells. **b.** Stages of "cannonball" formation earlier than that seen in **a**, showing outlines of squamous cells and varying numbers of leukocytes. **c.** Developmental stage of a cannonball. Note a single squamous cell with four trichomonads attached to it. Scale applies also to **d**. **d.** Later developmental stage of a cannonball with numerous *T. vaginalis* attached to its surface. The epithelial cell nucleus situated at about six o'clock contains a prominent Barr body (*arrow*) in its chromatin ring. **a, b, d**: Fast smears (Fast), Papanicolaou stain (Pap); **c**: Fast, Immunoperoxidase stain (Perox).

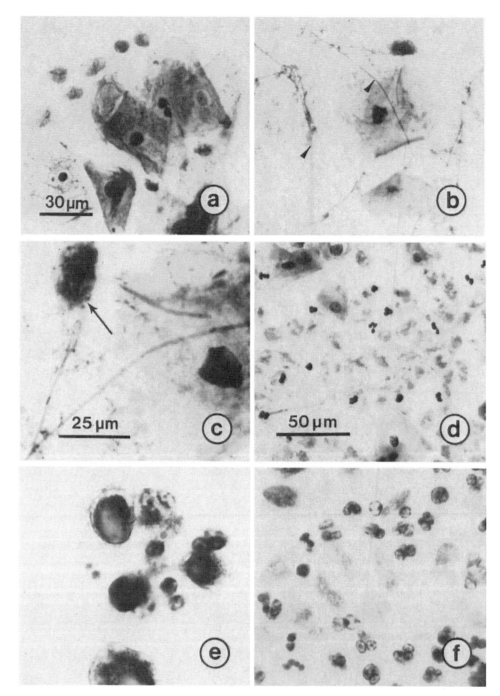

Fig. 14.2. a. Trichomonads in a posthysterectomy smear. The background is clean and there are no signs of inflammation and degeneration. (Scale applies also to **b** and **f**. **b** and **c**. Nonbranching bacterium *Leptothrix* (**b**, *arrowheads*) is evident in both figures. Epithelial cells (**b**) and a trichomonad (**c**, *arrow*) are also present. Scale in **c** applies also to **e**. **d**. Numerous poorly defined trichomonads showing extensive cytolysis and other degenerative changes cover the object field. Note also the neutrophils and squamous epithelial cells. **e**. Trichomonads in division. A late division stage is seen in the center of the field. **f**. Flagellates varying in structure and staining intensity are evident. **a–d**: Fast, Pap; **e,f**: Perox.

14. Cytopathology and Histopathology of the Female Genital Tract in *T. vaginalis* Infection 279

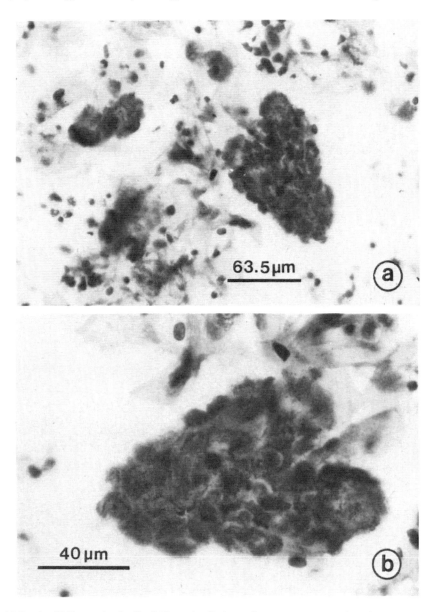

Fig. 14.3 a,b. "Microcolonies" of *T. vaginalis* in an immunosuppressed patient with leukemia. The well-preserved individual parasites are seen at higher magnification in **b**. Fast, Pap.

stained paraffin sections, the organisms are inconspicuous and may be difficult to detect. However, they can be more readily identified in Masson's trichrome-stained sections and in those treated according to the periodic acid–Schiff's (PAS) stain or appropriate immunodiagnostic methods.

Trichomonas vaginalis is generally seen lying on the surface of the squamous epithelium (Fig. 14.4a). According to Koss and Wolinska,[46] no detectable tissue changes are noted in about one-third of the female patients harboring genital trichomonads. Although principally located on the surface, these parasites have been observed within the epithelial cell layer in ultrastructural and tissue culture studies.[42,47] Under certain conditions dependent, for example, on pathogenicity of the trichomonad strains, on

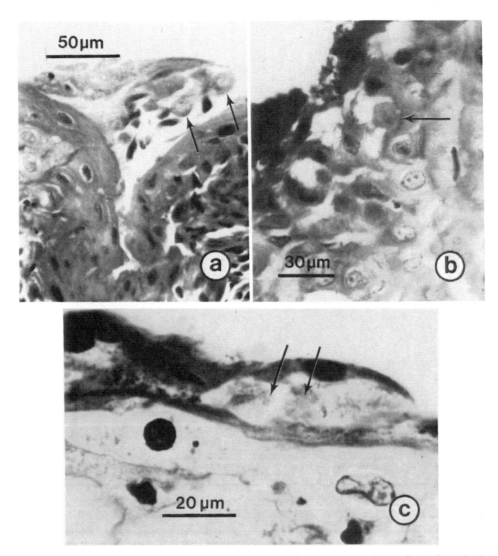

Fig. 14.4. Sections of cervical biopsies. **a.** Trichomonads (*arrows*) on the surface of cervical squamous epithelium. Hematoxylin & Eosin (H & E). **b.** Intercellular flagellates (*arrows*) in squamous epithelial cell layer. Periodic Acid-Schiff's reaction (PAS). **c.** Intracellular parasites (*arrow*) within the cytoplasm of squamous epithelial cells. Masson's trichrome stain (Mass Trich).

the hormonal and immunologic state of the host, or on concomitant infections, the parasites may penetrate the surface epithelium and be lodged beneath the luminal surface within the epithelium,[48] either intercellularly (Fig. 14.4b) or intracellularly (Fig. 14.4c). Recently, epithelial penetration with tissue invasion has been demonstrated in human prostatic tissues by Gardner and colleagues[49] (see Chap. 15).

Biopsies of the cervix and vagina from patients with active *Trichomonas* infection reveal areas of surface necrosis and abscess formation in the epithelium (Fig. 14.5a–c), with erosion of the epithelium and infiltration by leukocytes in severe cases and in the presence of concomitant infections (Fig. 14.5a–c). The infiltrates include numerous polymorphonuclear leukocytes and macrophages, with some lymphoid and plasma cells.

Local pH, redox potential, and immunologic responses may affect tissue reaction and invasion by the parasites. Inherent host tissue resist-

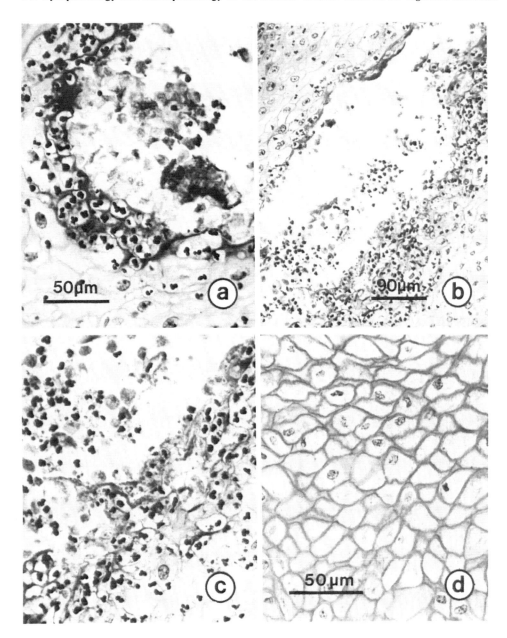

Fig. 14.5. Sections showing histologic details of trichomonas cervicitis. **a.** Necrosis with a microabscess is evident in the surface of the cervical squamous epithelium. Note the trichomonads and neutrophils in the lumen of the abscess. Scale applies also to **c**. **b.** Low-power photomicrograph showing extensive cell metaplasia, edema, and leukocyte infiltration. **c.** At a higher magnification of a part of the area seen in **b**, note inter- and intracellular edema of the squamous cervical epithelium, leukocyte infiltration, and necrosis associated with large numbers of parasites. **d.** "Chickenwire" appearance of squamous cervical epithelium, although not specific for trichomoniasis, is commonly observed in *T. vaginalis* infections. All figures: H & E.

ance to the trichomonads and the parasites' pathogenicity are additional important factors influencing the pathologic changes in *T. vaginalis* infection.

The occurrence of intercellular (i.e., pericellular) edema may lead to blunting of the sharp corners of the polygonal squamous cells. This produces an apparent condensation of the cytoplasm in the peripheral areas of the cells in the tissue, giving a so-called "chicken wire" intraepithelial appearance (Fig. 14.5d), not to be confused with koilocytic changes of the human papillomavirus (HPV) infection in condyloma acuminata.

Intracellular edema is also commonly present in the mature squamous cells of stratified squamous epithelium of the cervix and vagina (Fig. 14.6b-d). The perinuclear edema results in a thin, clear "halo" surrounding the nucleus, which resembles that of chronic irritation and inflammation[50] except that it tends to be somewhat wider and may appear to be more clearly "punched out" (Fig. 14.6a).[40] It must be differentiated from koilocytosis, observed in association with human papillomavirus infection of the genital tract. The perinuclear clearing in koilocytosis is characteristically much wider and has somewhat granular contents. It has also sharper, more membranelike lateral limits separating it from the peripherally displaced cytoplasm.

Reserve cell hyperplasia of the endocervical columnar epithelium and parabasal cell hyperplasia of the squamous epithelium as well as squamous metaplasia, with varying degrees of atypia, may accompany chronic *Trichomonas* cervicitis; however, none is specific for this infection.

Increased epithelial and subepithelial vascularity is observed within the squamous epithelium (Fig. 14.6b-d). Papillary growth of subepithelial tissue extends into the epithelium toward the luminal surface. The papillae contain small blood vessels at their core, which is surrounded by loose, edematous connective tissue (Fig. 14.6c). This arrangement results in capillary formations that are described variously as "fork-shaped," "antler-like," or "double-crested." They also produce the appearance of the granular cervix that is typically seen in active, florid trichomonad infection.[51]

In the subepithelial tissues, interstitial edema is present in an acute and chronic inflammatory response including both polymorphonuclear leukocytes and histiocytes, with a resulting papillitis (Fig. 14.6c,d). Small subepithelial hemorrhages are observed commonly; they appear clinically as a "strawberry" mucous membrane of the cervix or vagina.

When trichomonads are present on top of (Fig. 14.4a) and within (Fig. 14.4b,c) the epithelium, they are generally associated with necrosis and purulent exudate (Fig. 14.5a-c).[46] Intact trichomonads may invade an injured and reparative ("reactive") columnar cell at the squamocolumnar junction (Fig. 14.7a). The underlying subepithelial region in such cases is usually heavily inflamed and edematous. In cervical smears, similar changes may be seen, the organisms being found within the infected columnar cells (Fig. 14.7b).

More often, however, the host responses are more aggressive, the trichomonads being phagocytosed by individual (Fig. 14.7c) or jointly by several leukocytes (Fig. 14.7d), or by macrophages. Multinucleated giant histiocytes are frequently formed,[40,48,52] on rare occasions with the ingested parasites visible within their cytoplasm (Fig. 14.7e). In some instances, the appearance of trichomonads may suggest that they are attempting to ingest each other (cannibalism) (Fig. 14.7f); however, these findings probably represent bizarre forms, possibly in division.

Endocervical glands (clefts) and probably others (e.g., Skene's, Bartholin's) whose openings have been closed by inflammation and edema offer safe harbor for the trichomonads to reside, to feed off the serum and debris from the endocervicitis, and to enjoy the approximately neutral pH of their environment (Fig. 14.8a). When scraped off in the process of obtaining a Fast vaginopancervical specimen, microcolonies of these trichomonads can be accompanied by variable amounts of inflammatory and necrotic debris (Fig. 14.8b).

Cervicovaginitis emphysematosus is an unusual disease that has been ascribed etiologically to vaginal trichomoniasis.[53,54] It is believed that gas production by the flagellates causes cystic dilation in the subepithelial region, with dilation of the lymphatics and subepithelial veins. While the disease disappears following antitrichomonal therapy with metronidazole, and some experimental work supports the etiologic relationship, trichomonads have not been identified within the *cervicovaginitis emphysematosus* lesions.[55]

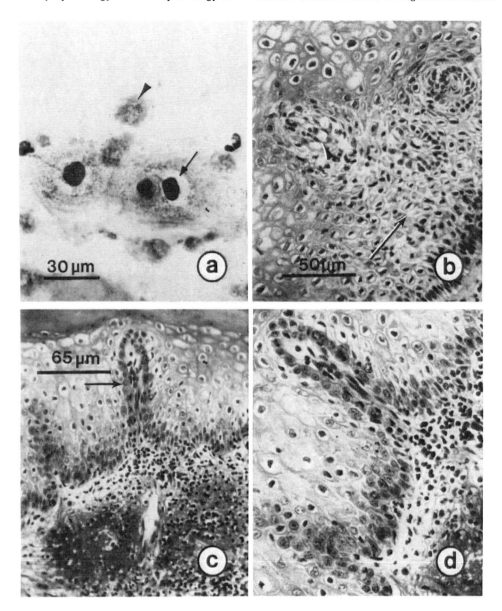

Fig. 14.6 a–d. Preparations from cases of trichomonad cervicitis. **a.** Perinuclear "halos" (*arrow*) surrounding nuclei of squamous epithelial cells are the result of perinuclear edema. Numerous trichomonads (*arrowheads*) are seen in the field. Fast, Pap. Scale applies also to **d**. **b.** Hyperplasia of the parabasal (spinal) cells (*arrowhead*) and intraepithelial vascularity (*large arrow*) are evident in a section of a cervical biopsy. **c.** Thin vascular papillae (*arrow*) extending into the epithelium explain the mechanism of formation of intraepithelial vascularity seen in **b**. Intracellular edema, papillitis, congestion, and subepithelial hemorrhage found in sections of cervical biopsies from cases of trichomonad cervicitis (such as the one presented here) cause the so-called "strawberry" appearance of the epithelium noted on clinical examination. **d.** Details of the subepithelial changes can be analyzed more satisfactorily at higher magnification of part of the area of the section in **c**. Note the subepithelial edema, chronic inflammation, congestion, and hemorrhage, with resulting papillitis. **b–d**: H & E.

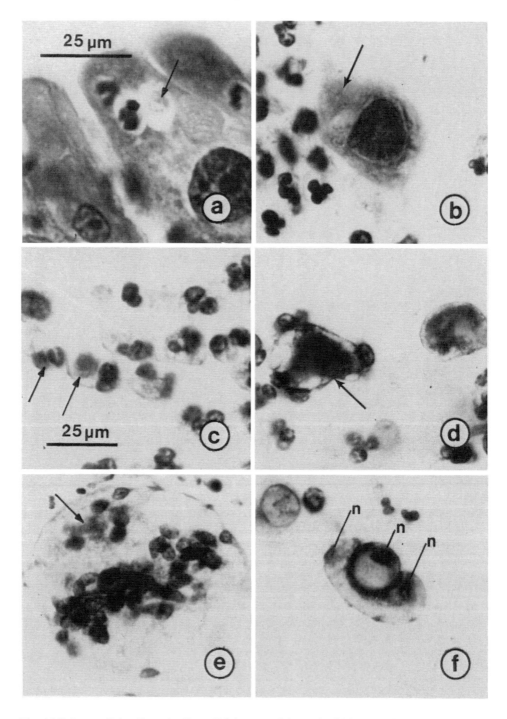

Fig. 14.7. Intracellular *T. vaginalis.* **a.** Trichomonad (*arrow*) within an endocervical cell from the squamo-columnar junction of a patient with severe chronic cervicitis. The infected cell is characterized by inflammatory nuclear atypia, with features of both degeneration and regeneration. H & E. Scale applies also to **b, e. b.** Parasite (*arrow*) within an exfoliated endocervical cell. The host cell has a hyperchromatic and atypical nucleus, with blurring and degeneration of chromatin. Note the narrow perinuclear halo. **c.** Trichomonads (*arrows*) that have

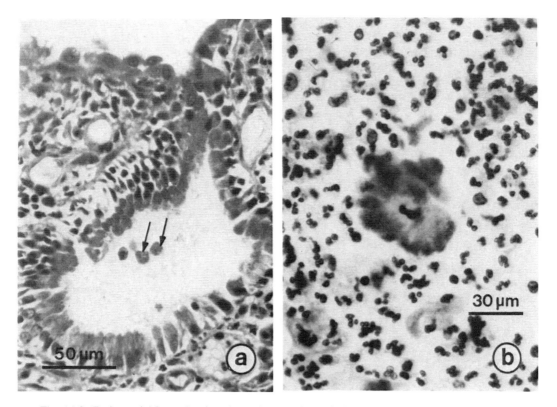

Fig. 14.8. Endocervicitis. **a.** Section through an endocervical gland in a cervical biopsy. Two trichomonads (*arrow*) and a neutrophil lie in the lumen of the gland. Note the narrowed opening and the dilated part of the gland. H & E. **b.** Microcolony of *T. vaginalis* is seen in the center of the figure. The content of the vaginopancervical smear reflects severe inflammation and shows much necrotic debris. Fast, Pap.

Electron Microscopy

Ultrastructural studies of *Trichomonas* infection of the human cervix have been reported by Nielsen and Nielsen[47] (see Fig. 9.9, this book) and more recently by Tamayo et al.[52] In the cases examined, organisms were observed on the surface of squamous epithelial cells, with areas of contact involving interdigitation of the parasite and host cell cytoplasms. The precise location of the organism and especially its ability to invade the tissues of its host in infections accompanied by cellular necrosis, edema, and inflammation appear to be related to the inherent pathogenicity of the strain and to various virulence-affecting factors.[56-60]

For many years it has been believed possible that the trichomonads may act as vectors for introducing and holding their symbiotes in the most susceptible areas of the genital tract. Plasmids and various microorganisms have been reported in cultivated trichomonads,[61] but these were thought to represent ingested food. A

been engulfed by individual polymorphonuclear leukocytes (PMNs). **d.** Flagellates (*arrow*) that have been phagocytosed jointly by several PMNs. **e.** Trichomonad (*arrow*) within a multinucleate giant histiocyte. **f.** Putative "cannibalism" seen in this micrograph probably represents three closely apposed organisms [note the three nuclei (*n*)] rather than cannibalism. **b-f:** Fast; **b,c,e,f:** Pap; **d:**Perox.

(Fig. 14.7a and e reproduced with permission of the American Society of Parasitologists, from Frost JK et al: Intracellular *Trichomonas vaginalis* and *Trichomonas gallinae* in natural and experimental infections. *J Parasitol* 47:302-303, 1961.)

valuable and most interesting ultrastructural observation was the presence of intracytoplasmic mycoplasma, bacteria, and epithelial cellular fragments.[52] We have observed *Chlamydia* associated with *Trichomonas* and lying within the protozoa. While the initial observations were made with a light microscope (Fig. 14.9a,b), they have been subsequently verified by immunologic (Fig. 14.9c) and ultrastructural studies. Naib[41] and Keith et al[30] have reported similar findings.

Trichomoniasis and Synchronous Pathologic Conditions

An association between *Trichomonas* and PID was suggested over 4 decades ago by Allen and Baum.[58] That *T. vaginalis* may be responsible for this disease was indicated also by Gallai and Sylvestre.[28] Recently, interest in the role of this protozoan parasite in the causation of PID has been renewed. On the basis of elegant scanning electron microscopic studies, Keith et al[30] have demonstrated the presence of numerous infective organisms on the surface of *Trichomonas*. They have suggested a vector role for *Trichomonas* in ascending genital tract infections. Mardh and Westrom[29] also reported such an association. These workers were able to cultivate the trichomonads from uterine tube specimens obtained by laparoscopy from a patient with clinically confirmed lower genital tract trichomoniasis.

An etiologic role of *T. vaginalis* in the causation of cervical neoplasia has been long suspected and suggested by a number of investigators.[17,34,35,46,61,62] Trichomonads may be merely opportunistic, living off serous exudate, but there is increasing evidence (clinical and experimental) that they may prepare and maintain epithelium in a state that is more receptive and

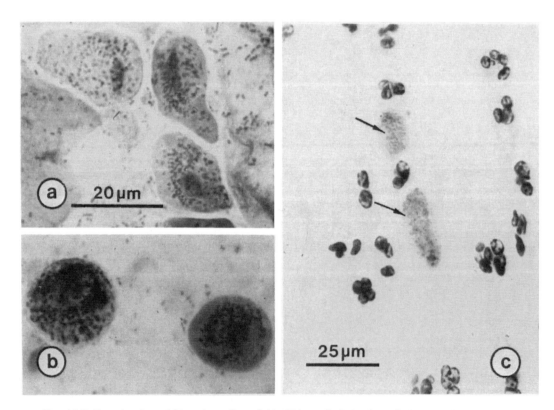

Fig. 14.9. Putative (**a** and **b**) and confirmed (**c**) *Chlamydia* infection of *T. vaginalis*. **a** and **b**. Numerous intra- and extracytoplasmic granules (probably *Chlamydia*) in association with trichomonads. The granule-filled flagellates were found in two separate cases with *Trichomonas* infection of the lower genital tract. **c**. Two elongate trichomonads in the center of the field are filled with *Chlamydia* bodies treated with anti-*Chlamydia* monoclonal antibodies. **a**, **b**: Fast, Pap; **c**: Fast, Perox.

susceptible to carcinogens. Meisels[61] and Berggren[34] reported on the prevalence of *T. vaginalis* in patients with carcinoma of the cervix. They found 39.4% and 38.6% of patients with cervical cancer to have *T. vaginalis* in Papanicolaustained vaginal smears, respectively. Meisels also analyzed cases with histologic cervical lesions and found *T. vaginalis* in 28% of patients with cervical dysplasia and in 35% of patients having cervical carcinoma with various degrees of invasion. In a study done at Johns Hopkins by Frost,[35] 2828 consecutive pregnant women seen at the clinics were analyzed for the presence of *T. vaginalis*. This parasite was reported in 19% of the cases who did not have any atypical epithelial cellular changes, in 51% of cases with epithelial atypia, and in 88% of cases with cervical cancer.

Experimental work using the subcutaneous mouse assay[15-18] supports association between *T. vaginalis* virulence as estimated by this assay and the severity of cervical epithelial changes. According to recent findings,[17] " . . . 15 (approximately 65%) of the 23 patients infected with *T. vaginalis* strains producing mouse abscesses greater than or equal to 144.5 ± 8.11 mm^3 (N = 41) had cervical epithelial dysplasia ranging from mild to . . . cervical carcinoma" (Fig. 14.10. In fact, all the cases of cervical intraepithelial neoplasia (CIN I, II, III) fell into this group. These results may suggest oncogenic potential, operating directly or indirectly, of certain strains of the urogenital flagellate.

It is obvious that there is need for additional carefully planned studies including DNA ploidy analysis of atypical epithelial cells occurring in association with *T. vaginalis*, in situ hybridization investigation of atypical epithelial lesions for *T. vaginalis* DNA, and studies to localize human papillomavirus or its genome in the protozoa. Improved immunodiagnostic procedures may be valuable in developing more cost-effective and sensitive detection and diagnostic methods for the protozoa, in elucidating the role of urogenital trichomonads in the causation of PID, and in better characterizing its role in cervical neoplasia.

Trichomoniasis and Other Sexually Transmitted Infections

Other sexually transmitted infections are frequently associated with trichomoniasis; they have been discussed in more detail in Chapter 10. In our experience, over 9% of patients with

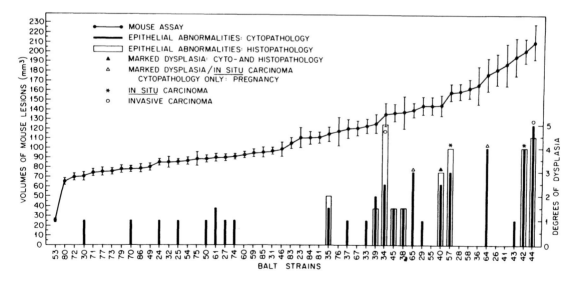

Fig. 14.10. Relationship between cervical epithelial changes found in women from whom the *T. vaginalis* strains were isolated and the putative inherent virulence of these strains as estimated by the mean volumes (in cubic millimeters) of 6-day subcutaneous lesions caused by axenic cultures in 6- to 9-week-old female C57BL/6J mice.
(Reproduced with permission of The American College of Obstetricians and Gynecologists from Honigberg BM et al: Cytopathologic and histopathologic changes of the cervical epithelium. *Obstet Gynecol* 64:179-184, 1984.)

T. vaginalis infection had "other findings." *Chlamydia* and human papillomavirus (HPV) are perhaps the most common pathogens found in the routine vaginal smear in association with *Trichomonas* (39%). Other infective agents like *Candida*, herpes, adenovirus, and *Actinomyces* less often accompany the protozoa.

Naib[59] suggested that *Chlamydia* can be ingested by *Trichomonas*. The commonly seen red intracytoplasmic granules (Fig. 14.9a,b), which are held by some to be food vacuoles, may at times represent ingested microbial structures. As mentioned before in this chapter, our preliminary studies, including monoclonal antibody staining for *Chlamydia trachomatis* of the intracytoplasmic granules within *Trichomonas* (Fig. 14.9c) and electron microscopic observations on such organisms, support such a contention. In fact, a micrograph orginally published in 1975 by Nielsen and Nielsen[47] contains *Trichomonas*-associated structures that probably represent *Chlamydia* elementary bodies. These observations, if proved correct, may have clinical and therapeutic implications.

Of special interest is the association of HPV infection, which was seen in 16% of our trichomoniasis cases. Rodgerson[60] reported *T. vaginalis* from all 31 cases he studied of vulvar and vaginal papillomas. He used wet preparations to identify *Trichomonas* and clinical features to diagnose HPV infection. Treatment of all patients for the trichomonad resulted in condyloma cure in 93% at 6 months. Subsequent recurrence of genital condyloma corresponded with *Trichomonas* infection.

Although the observations detailed in this and the preceding section are extremely interesting, they cannot be taken completely at face value. However, the recurring suggestion of a virus-parasite or carcinogen-cocarcinogen interrelationship cannot be dismissed at this time.

References

1. Donné MA: Animacules observés dans les matières purulentes et le produit des sécrétions des organes génitaux de l'homme et de la femme. *C R Hebd Seances Acad Sci (Paris)* 3:385-386, 1836.
2. Müller OF: *Vermium Terrestrium et Fluviatilium seu Animalium Infusorium Helminthicorum et Testaceorum, non Marinorum, Succinta Historia*, vol. 1, pt. 1 and 2. Havnia et Lipsiae: Heineck et Taber, 1773.
3. Davaine CJ: *Traité des Entozoaires et des Maladies Vermineuses de l'Homme et des Animaux Domestiques*, 1st ed. Paris: Baillière, 1860.
4. Cleveland LR: *Tritrichomonas fecalis* nov. sp. of man: Its ability to grow and multiply indefinitely in faeces diluted with tap water and in frogs and tadpoles. *Am J Hyg* 8:232-255, 1928.
5. Llewellyn-Jones I: *Fundamentals of Obstetrics and Gynecology*. London: Farber and Farber, 1970.
6. Hiemstra I, Van Bel F, Berger HM: Can *T. vaginalis* cause pneumonia in newborn babies? *Br Med J* 289:355-356, 1984.
7. McLaren LC, Davis LE, Healy GR, James CG: Isolation of *T. vaginalis* from the respiratory tract of infants with respiratory disease. *Pediatrics* 71:888-890, 1983.
8. Suriyanon V, Nelson KE, Ayudhya VCN: *T. vaginalis* in a perinephric abscess. *Am J Trop Med Hyg* 24:776-780, 1975.
9. Csonka GW: Trichomoniasis and the dermatologist. *Br J Dermatol* 90:713-714, 1974.
10. Kilman J: *Trichomoniasis, Diagnose und Therapie*. Vienna: Wilhelm Maurdrich, 1973.
11. Ackers JP, Lumsden WH, Catterall RD, Coyle R: Antitrichomonal antibody in the vaginal secretion of women infected with *T. vaginalis*. *Br J Vener Dis* 51:319-323, 1975.
12. Chipperfield EJ, Evans BA: The influence of local infection in immunoglobulin formation in the human endocervix. *Clin Exp Immunol* 11:219-223, 1972.
13. Green RL, Scales RW, Kraus SJ: Increase of serum immunoglobulin E concentration in veneral disease. *Br J Vener Dis* 52:247-260, 1976.
14. Jaakmees HP, Teras JK: Intradermal reaction with specific antigen in cases of genito-urinary trichomoniasis. *Wiad Parazytol* 12:385-391, 1966.
15. Frost JK, Honigberg BM: Comparative pathogenicity of *Trichomonas vaginalis* and *Trichomonas gallinae* for mice. II. Histopathology of subcutaneous lesions. *J Parasitol* 48:898-918, 1962.
16. Honigberg BM, Livingston MC, Frost JK: Pathogenicity of fresh isolates of *T. vaginalis*: "the mouse assay" versus clinical and pathologic findings. *Acta Cytol* 10:353-361, 1966.
17. Honigberg BM, Gupta PK, Spence MR, Frost JK, Kuczyńska K, Choromański L, Wartoń A: Pathogenicity of *Trichomonas vaginalis*: Cytopathologic and histopathologic changes of the cervical epithelium. *Obstet Gynecol* 64:179-184, 1984.
18. Kulda J, Honigberg BM, Frost JK, Hollander DH: Pathogenicity of *Trichomonas vaginalis*. *Am J Obstet Gynecol* 108:908-918, 1970.
19. Frost JK: Gynecologic and obstetric cytopathology. In Novak ER, Woodruff JD (eds): *Novak's Gynecologic and Obstetrical Pathology With Clin-*

ical and Endocrine Relations, 8th ed, pp 634-728. Philadelphia: WB Saunders, 1979.
20. Jírovec O, Petrů M: *T. vaginalis* and trichomoniasis. *Adv Parasitol* 6:117-186, 1968.
21. Trussell RE: Trichomonas vaginalis *and Trichomoniasis,* pp 113-124. Springfield, Ill: Charles C Thomas, 1947.
22. Honigberg BM: Trichomonads of importance in human medicine. In Kreier JP (ed): *Parasitic Protozoa,* vol 2, pp 276-454. New York: Academic Press, 1978.
23. Wisdom AR, Dunlop EMC: Trichomoniasis: Study of the disease and its treatment in women and men. *Br J Vener Dis* 41:90-96, 1965.
24. Brown MT: Trichomoniasis. *Practitioner* 209:639-644, 1972.
25. Mason PR: Trichomoniasis. New ideas on an old disease. *S Afr Med J* 58:857-858, 1980.
26. Wallin JE, Thompson SE, Saidi A, Wong KH: Urethritis in women attending an STD clinic. *Br J Vener Dis* 57:50-54, 1981.
27. Peterson WF, Stauch JE, Ryder CD: Metronidazole in pregnancy. *Am J Obstet Gynecol* 94:343-349, 1966.
28. Gallai Z, Sylvestre L: The present status of urogenital trichomoniasis: A general review of literature. *Appl Ther* 8:773-778, 1966.
29. Mardh PA, Westrom L: Tubal and cervical cultures in acute salpingitis with special reference to *Mycoplasma hominis* and *T.*-strain mycoplasmas. *Br J Vener Dis* 46:179-186, 1970.
30. Keith LG, Berger GS, Edelman DA, Newton W, Fullan N, Bailey R, Friberg J: On the causation of pelvic inflammatory disease. *Am J Obstet Gynecol* 149:215-223, 1984.
31. Catterall RD: Trichomonal infections of the genital tract. *Med Clin North Am* 56:1203-1209, 1972.
32. Rein MF, Müller M: *Trichomonas vaginalis.* In Holmes KK, Mardh PA, Weisner PH: *Sexually Transmitted Diseases,* pp 525-536. New York: McGraw-Hill, 1984.
33. Fouts AC, Kraus SJ: *Trichomonas vaginalis:* Reevaluation of its clinical presentation and laboratory diagnosis. *J Infect Dis* 141:137-143, 1980.
34. Berggren O: Association of carcinoma of the uterine cervix and *Trichomonas vaginalis* infestations: Frequency of *Trichomonas vaginalis* in preinvasive and invasive cervical carcinoma. *Am J Obstet Gynecol* 105:166-168, 1969.
35. Frost JK: *Trichomonas vaginalis* and cervical epithelial changes. *Ann NY Acad Sci* 97:792-799, 1962.
36. Grice AC: Vaginal infection causing spontaneous rupture of the membranes and premature delivery. *Aust NZ J Obstet Gynaecol* 14:156-158, 1974.
37. Mason PR, Brown IM: Trichomoniasis in pregnancy. *Lancet* 2:1024-1026, 1980.
38. Robinson SC, Halifax NS: Observations on vaginal trichomoniasis. I. In pregnancy. *Can Med Assoc J* 84:948-949, 1961.
39. Gupta PK, Frost JK: Human urogenital trichomoniasis: Epidemiology, clinical and pathological manifestations. In Kulda J, Čerkasov J (eds): Proc Int Symp Trichomonads & Trichomoniasis, Prague, July, 1985 (Post-Symp Publ, Pt 2). *Acta Univ Carolinae (Prague) Biol* 30:339-410 (1986) 1988.
40. Hypes RA, Ladewig PP: Leukocytic clusters on epithelial cells in cervicovaginal smears: A presumptive test for *Trichomonas* infection. *Am J Clin Pathol* 26:94-97, 1956.
41. Naib ZM: *Exfoliative Cytopathology,* 3rd ed, pp 95-99. Boston: Little, Brown & Co, 1985.
42. Sun T: Trichomoniasis. In Sun T (ed): *Sexually Related Infectious Diseases: Clinical and Laboratory Aspects,* pp 179-185. New York: Field, Rich and Associates, 1986.
43. Bibbo M, Harris MJ, Wied GL: Microbiology of the female genital tract. In *Compendium on Diagnostic Cytology,* 5th ed, pp 50-51. Chicago: Tutorials of Cytology, University of Chicago Press, 1983.
44. Winston RML: The relation between size and pathogenicity of *Trichomonas vaginalis. J Obstet Gynaecol Br Commonw* 81:399-404, 1974.
45. Mehta SH, Verma K: Relationship between size and pathogenicity of *Trichomonas vaginalis. Indian J Med Res* 74:231-235, 1981.
46. Koss LG, Wolinska WH: *Trichomonas vaginalis* cervicitis and its relationship to cervical cancer. *Cancer* 12:1171-1193, 1959.
47. Nielsen MH, Nielsen R: Electron microscopy of *Trichomonas vaginalis* Donné: Interaction with vaginal epithelium in human trichomoniasis. *Acta Pathol Microbiol Immunol Scand (B)* 83:305-320, 1975.
48. Frost JK, Honigberg BM, McLure MT: Intracellular *Trichomonas vaginalis* and *Trichomonas gallinae* in natural and experimental infections. *J Parasitol* 47:302-303, 1961.
49. Gardner WA, Culberson DE, Bennett BD: *Trichomonas vaginalis* in the prostate gland. *Arch Pathol Lab Med* 110:430-432, 1986.
50. Frost JK: Morphologic characteristics of retroplasia: Decreased general activity. In Frost JK: *The Cell in Health and Disease,* 2nd edn., Chapter 7, pp. 97-109. In Wied GL (ed): *Monographs in Clinical Cytology,* vol 2, Basel, New York: S. Karger, 1986.
51. Kolstad P: The colposcopic picture of *Trichomonas* vaginitis. *Acta Obstet Gynecol Scand* 43:388-398, 1964.
52. Tamayo JG, Nuñez Monteil JT, de García HP: An electron microscopic investigation on the pathogenesis of human vaginal trichomoniasis. *Acta Cytol* 22:447-455, 1978.

53. Gardner HL, Fernet P: Etiology of *vaginitis emphysematosa. Am J Obstet Gynecol* 88:680-794, 1964.
54. Wilbanks GD, Carter B: *Vaginitis emphysematosa. Obstet Gynecol* 22:301-309, 1963.
55. Newton WL, Reardon LV, DeLeva AM: A comparative study of the subcutaneous inoculation of germ-free and conventional guinea pigs with two strains of *Trichomonas vaginalis. Am J Trop Med Hyg* 9:56-61, 1960.
56. Honigberg BM: Comparative pathogenicity of *Trichomonas vaginalis* and *Trichomonas gallinae* to mice. I. Gross pathology, quantitative evaluation of virulence, and some factors affecting pathogenicity. *J Parasitol* 47:545-571, 1961.
57. Honigberg BM, King VM: Structure of *Trichomonas vaginalis* Donné. *J Parasitol* 50:345-364, 1964.
58. Allen E, Baum H: The treatment of vaginitis. *Am J Obstet Gynecol* 45:246-254, 1943.
59. Naib ZM: Stepping up the search for vaginal pathogens. In Richart RM (mod): Symposium. *Contemp Obstet/Gynecol,* Oct/Dec:139-149, 194-215, 1984.
60. Rodgerson EB: Vulvovaginal papillomas and *Trichomonas vaginalis. Obstet Gynecol* 40:327-333, 1972.
61. Meisels A: Dysplasia and carcinoma of the uterine cervix. IV. A correlated cytologic and histologic study with special emphasis on vaginal microbiology. *Acta Cytol* 13:224-231, 1969.
62. Kaarma H, Teras JK, Podar U: Histological changes in the vaginal part of the uterus in cases of genitourinary trichomonadosis. *Wiad Parazytol* 15:319-321, 1969.

15

Pathology of Urogenital Trichomoniasis in Men

William A. Gardner, Jr. and Donald E. Culberson

Introduction

History

Despite the frequency with which *Trichomonas vaginalis* has been identified in genitourinary fluids, secretions, inflammatory exudates, and epithelial scrapings, neither the gross pathology nor the microscopic appearance of the associated inflammatory reaction has been well characterized in the male genitourinary tract. Strain[1] in 1945 described his treatment of small chronic abscesses of the prostate ducts associated with urinary trichomonads. Multiple lesions of prostatic urethra and prostatic ducts were illustrated. Roth[2] suggested in 1944 that histopathologic studies would provide valuable clues to the pathogenicity of trichomonads and stated that "prostatic tissue removed by transurethral prostatic resection, by perineal prostatectomy and at autopsy are under consideration at present." Eight years later Herbert[3] in his classic text on urologic pathology noted that Roth's studies had not been published and commented that histologic changes of trichomoniasis had yet to be described in either the prostate or seminal vesicles. Van Laarhoven[4] obtained needle biopsies of the prostate gland in two patients with microscopic evidence of trichomonads in the prostatic secretion. He described a periductular lymphocytic infiltrate with few plasma cells and eosinophils extending into the adjacent perivascular and perineural stroma. Using modified periodic acid–Schiff's (PAS) and methenamine silver stains, Van Laarhoven also observed positively staining structures within the prostate that were diagnosed as *T. vaginalis* on the basis of size and staining characteristics, but without recognizable details of either internal or external morphology (Van Laarhoven, personal communication). More recently Krieger[5] summarized succinctly current knowledge of this aspect of male trichomoniasis when he stated "no relevant histopathologic studies exist."

Constraints

Tissue Biopsy

Studies of the basic pathology of infectious diseases require a technique of specific identification of organisms coupled with availability of adequate specimens. Several factors have limited studies that might have defined the histopathogenesis of trichomoniasis: (1) The "questioned status" of the role of *T. vaginalis* as a pathogen in the male[6] may have resulted in a reluctance to pursue study of this potential host-parasite relationship. (2) In many males the organisms are either asymptomatic or have self-limited clinical manifestations. (3) Tissue biopsy has not been a standard practice in the clinical evaluation of inflammatory processes of the male genitourinary system. This situation is likely to change somewhat with the advent of fine needle aspiration biopsy (see section, Male Genitourinary Cytology below). This technique is increasingly indicated for evaluation of chronic inflammatory diseases, especially in the prostate where chronic prostatitis can mimic the physical findings of carcinoma.[7] (4) Autolytic changes in autopsy tissues would likely result in unsatisfactory preservation of the relatively fragile *T. vaginalis*. (5) Identification of organisms may be difficult in fluids and exudates from the male genitourinary tract. These latter problems are compounded by the fixation, dehydration, paraffinization, and other steps in usual histologic preparation. The routinely used hematoxylin and eosin staining method is recognized as among the worst for demonstration of *T. vaginalis* in tissues.[8]

Organism Identification

Investigation of the histopathogenesis of trichomonal prostatitis and trichomonal infection requires both adequate tissue and a means for specific localization of the organisms. In view of the relative unreliability of microscopic recognition of *T. vaginalis* in both wet mounts and stained preparations from men, it is not surprising that identification of organisms within tissue sections is difficult if not impossible. The variety of stains that are useful for delineating the structure of the organism in smears lack the specificity necessary for their positive identification in tissue sections. Many microscopic components of the prostate gland, for example, have the staining characteristics and size of trichomonads when such methods are applied.[9] The development of an immunoperoxidase procedure for the specific recognition of trichomonads in tissues[10] has led to the first unequivocal documentation of *T. vaginalis* in tissues of the male genitourinary tract (Fig. 15.1a).[11] In men the only anatomic sites in which *T. vaginalis* organisms have been specifically identified in histologic preparations are the prostate gland and prostatic urethra.

Prostate and Prostatic Urethra

Organism Identification

Trichomonas vaginalis was seen in hematoxylin and eosin (H & E) stain as amorphous eosinophilic structures even in tissues with excellent histologic preservation. Such nondiagnostic structures subsequently were shown to be immunoperoxidase-positive against trichomonal antisera and were present in the urethral ductular and glandular lumina and epithelium. Occasionally trichomonads were recognizable in the subepithelial connective tissue and in the stromal inflammatory infiltrate.

Associated Histopathology

Without complete exclusion of the presence of any other organism the identification of *T. vaginalis* in histologic sections does not absolutely establish the associated histopathology as being due to the trichomonad. Nonetheless, the following changes have been found in association with the immunoperoxidase identification of *T. vaginalis*.[11] In both prostatic urethra and parenchyma, the spectrum of inflammatory changes varied from focal aggregates of polymorphonuclear leukocytes within glandular epithelium and lumina to the formation of microabscesses with focally denuded glandular epithelium. Inflammatory infiltrate often extended into the surrounding connective tissue and smooth muscle. Subepithelial hypervascularity, which is frequently seen in the infected vaginal and cervical epithelium, was not observed although subepithelial accumulation of polymorphonuclear leukocytes was a characteristic histologic appearance (Fig. 15.1b). Associated with the acute inflammatory reaction were intraepithelial vacuoles appearing to contain either organisms or remnants thereof (Fig. 15.1b,c,d). Subacute inflammation with glandular lumina containing mixtures of neutrophils and macrophages was also noted. Such macrophages were often the predominant inflammatory cell in areas of chronic inflammation either filling the glandular lumen or extending into the stroma with a mixture of chronic inflammatory cells (Fig. 15.1d). No granulomatous foci were identified in the few cases studied. In both glands and ducts focal epithelial hyperplasia was noted in association with the inflammatory patterns described above.

Interpretation

The histologic patterns described above are those of "nonspecific acute and chronic urethritis and prostatitis." Although it cannot yet be stated that these changes constitute the breadth of histopathology of male trichomoniasis, this study represents the first step in this histopathologic definition. Positively staining structures beneath the epithelium suggest several possibilities: (1) a potential active invasiveness of the organism in the prostate gland; (2) the role of *Trichomonas* as a secondary invader as suggested by Sun[12]: with necrosis of cells and change in pH produced by bacterial infection, a favorable environment could be provided for access to tissues by a normally noninvasive organism; and (3) positively staining structures could represent organisms or fragments which have been passively swept along in the inflammatory process, for example, by phagocytosis.

Penis

Several authors[13,14] have illustrated the histopathologic changes associated with trichomonal infection of the median raphe of the penis. In both reports a draining sinus was present. In one 24-year-old man a cyst lined by stratified squamous epithelium demonstrated slight keratoacanthosis and a chronic inflammatory infiltrate.[13] Sowmini et al[14] reported the case of a 35-year-old man who demonstrated sinus tracts or penile cysts also lined with squamous epithelium, which manifested an irregular acanthosis with lymphocytes, polymorphonuclear leukocytes, and chronic inflammatory cells extending into the subcutaneous tissues. A 41-year-old man described by Ayyangar[15] had a penile lesion in which histologic examination revealed an inflamed cyst of the prepuce caused by a nonspecific inflammatory reaction.

Michalowski[16] identified *T. vaginalis* in 16 patients with balanoposthitis. A long prepuce with phimosis was a frequent complication and trichomonal urethritis was present in seven cases. In most cases there was superficial erosion of either the glans or prepuce although in one case an ulcerative lesion mimicked a chancre. Edema was prominent in a number of lesions producing an elevated appearance, and in some cases the epidermis was undermined forming a collarette. Biopsy specimens of glans and foreskin lesions contained intracellular edema, acanthosis, and papillomatosis in addition to ulceration. These were associated with a cellular infiltrate in the superficial dermis composed of lymphocytes, plasma cells, histiocytes, and polymorphonuclear leukocytes along with the occurrence of capillary dilatation and proliferation.

Lopatin[17] described trichomonal ulcers of the foreskin and glans penis in a 22-year-old man. The ulcers were characterized by edema, marginal hyperemia, and a yellowish-gray film of purulent material covering the ulcer bed. In this case, both trichomonads and gram-positive cocci were isolated.

Although the above cases involve histopathologic manifestations apparently associated with *T. vaginalis* infection, organisms were never identified histologically either in lumina or adjacent tissues.

Other Male Genitourinary Sites

Although there is strong circumstantial evidence to associate trichomonads with other sites in the male genitourinary tract[5,18] (see also Chap. 12), including the bladder, seminal vesicles, epididymis, and testes, in no case has the associated histopathology been described or illustrated.

Secondary Pathology

Kuberski[19] described a 35-year-old man in whom erosion and sclerosis of the sacroiliac joint in a clinical setting of ankylosing spondylitis was strongly associated with *T. vaginalis* prostatitis. The mechanism for *T. vaginalis* to serve as precipitating agent for ankylosing spondylitis is unknown although Kuberski speculated a possible cross-reactivity between *T. vaginalis* and the HLA-B27 histocompatibility antigen, which is associated with this type of arthritis.

Inguinal lymphadenopathy was present in many of the cases of penile trichomoniasis and lymph nodes biopsied by Michalowski[16] demonstrated the histologic features of nonspecific reactive hyperplasia.

Pathologic morphology of spermatozoa has been associated with nonspecific chronic inflammation of male accessory sex glands,[20] and *T. vaginalis* is known to interfere with human spermatozoal motility.[21] Correlation of *T. vaginalis* with spermatozoal pathology awaits confirmation.

Future Studies

Enigmatic Genitourinary Diseases

There are a number of inflammatory processes in the male genitourinary tract that remain etiologic enigmas. The use of specific immunoperoxidase technique[9-11] should allow identification or exclusion of *T. vaginalis* as a pathogen in these diseases. These include nonspecific ure-

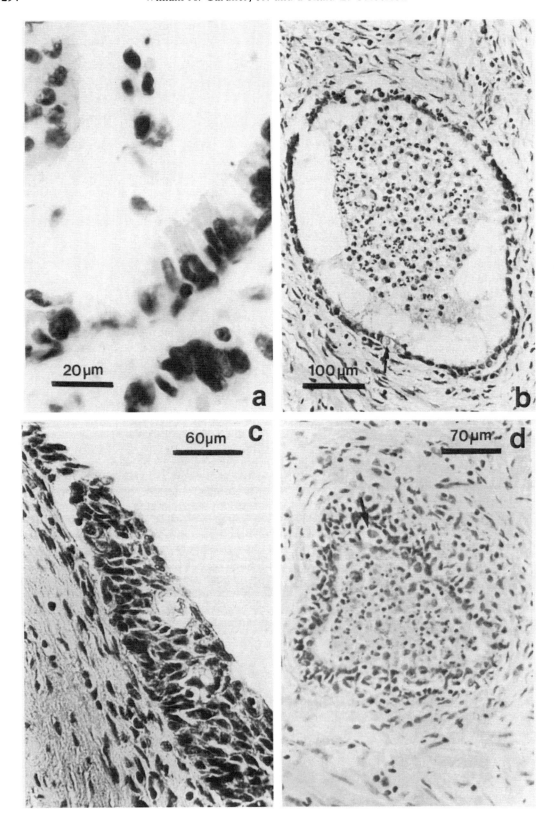

thritis and prostatitis, various cystidides, nonspecific granulomatous prostatitis, eosinophilic prostatitis, and plastic induration of the penis. This latter, clinically uncommon condition, also called Peyronie's disease, is characterized by fibrosis within the penis. Smith[22] has demonstrated the inflammatory nature of this condition, presented evidence to suggest a relationship between urethritis and Peyronie's disease,[23] and noted that subclinical urethritis is relatively common in American men. In none of these inflammatory processes have there been histopathologic studies to confirm or exclude *T. vaginalis* as an etiologic agent.

Male Genitourinary Cytology

Although the literature on the histopathology of male trichomoniasis is quite sparse, the literature on the cytopathology of the male genitourinary tract in trichomoniasis is virtually nonexistent. This is in sharp contrast to the female, in which cytopathologic studies have provided the underpinnings of much of our basic understanding of trichomoniasis. With the increasing use of fine-needle aspiration biopsy, diagnostic cytopathology of the male genitourinary tract, especially the prostate and seminal vesicles, is one of the most rapidly growing areas of diagnostic cytopathology. We may thus expect to see increasing application of this technology in identifying the cytopathology of male trichomoniasis, particularly when coupled with the immunocytochemical identification of *T. vaginalis*. The utility of this latter technique has already been demonstrated in the study of cytologic specimens.[24]

Although the occurrence of epithelial hyperplasia and basal cell proliferation was observed in the prostate in association with trichomoniasis, it may be considered a nonspecific reaction to the inflammatory process. In view of the well-known association of *T. vaginalis* with cervical epithelial dysplasia, studies should be carried out to determine the long-term effects of the organism on the prostatic epithelium. For example, in a situation reminiscent of some of the early observations of cervical cytopathology, it is now well documented that aspiration cytology of inflamed prostates often shows major and minor criteria of malignancy.[25] The possible role of trichomonads contributing to these cytopathologic changes in a manner comparable to their behavior in the uterine cervix is unknown.

Although the histopathology of male trichomoniasis has thus far been studied only very superficially, we believe that the new availability of immunocytochemical techniques for identification of the organism coupled with the increasing use of diagnostic aspiration biopsy cytology should allow a rapid growth in the knowledge of the male disease states and thus in understanding the pathogenesis of *T. vaginalis* in general.

References

1. Strain RE: Trichomoniasis in the male. *J Urol* 54:483-485, 1945.
2. Roth RB: *Trichomonas* urethritis and prostatitis: A preliminary report on incidence and an analysis of 44 cases of this common venereal infection. *Ven Dis Inf* 25:163-166, 1944.
3. Herbert PA: *Urological Pathology.* Philadelphia: Lea & Febiger, 1952.
4. Laarhoven PHA van: *Trichomonas vaginalis,* a pathogen of prostatitis. *Arch Chir Neerl* 5:263-273, 1967.
5. Krieger JN: Urologic aspects of trichomoniasis. *Invest Urol* 18:411-417, 1981.
6. Krieger JN: Prostatitis syndromes: Pathophysiology, differential diagnosis, and treatment. *Sex Transm Dis* 11:100-112, 1984.
7. Leistenschneider W, Nagel R: *Atlas of Prostatic*

Fig. 15.1.a. *Trichomonas vaginalis* adjacent to glandular epithelium of prostate. Immunoperoxidase stain. b. Early inflammatory exudate within the lumen of the prostate gland. Cells are predominantly polymorphonuclear leukocytes. Note inflammatory cells immediately beneath the epithelium and the presence of intraepithelial vacuoles (*arrow*). Hematoxylin and eosin (H & E). c. Prostatic urethral epithelium with intraepithelial vacuole containing immunoperoxidase-positive *T. vaginalis*. d. Subacute and chronic prostatitis. Note intraepithelial vacuoles (*arrow*) and mixed acute and chronic inflammatory infiltrate within the lumina and adjacent connective tissue. H & E. (**a,c:** Reprinted with permission of the Archives of Pathology and Laboratory Medicine, American Medical Association, from Gardner WA et al: *Trichomonas vaginalis* in the prostate gland. *Arch Pathol Lab Med* 110:430-432, 1986.)

Cytology: Techniques and Diagnosis. New York: Springer-Verlag, 1984.
8. Frost JK, Honigberg BM: Comparative pathogenicity of Trichomonas vaginalis and Trichomonas gallinae for mice. II. Histopathology of subcutaneous lesions. J Parasitol 48:898-918, 1962.
9. Gardner WA, Culberson DE: Histopathological correlates of male trichomoniasis. In Kulda J, Čerkasov J (eds) Proceedings of the International Symposium on Trichomonads & Trichomoniasis, Prague, July 1985 (Post-Symp Publ, Pt 2). Acta Univ Carolinae (Prague) Biol 30(5,6): 427-432, (1986) 1988.
10. Bennett BD, Bailey J, Gardner WA: Immunocytochemical identification of trichomonads. Arch Pathol Lab Med 104:247-249, 1980.
11. Gardner WA, Culberson DE, Bennett BD: Trichomonas vaginalis in the prostate gland. Arch Pathol Lab Med 110:430-432, 1986.
12. Sun J: Pathology and Clinical Features of Parasite Diseases. New York: Masson, 1982.
13. Soendjojo A, Pindha S: Trichomonas vaginalis infection of the median raphe of the penis. Sex Transm Dis 8:255-257, 1981.
14. Sowmini CN, Vijayalakshmi K, Chellamuthiah C, Sundaram SM: Infection of the median raphe of the penis. Br J Ven Dis 49:469-474, 1972.
15. Ayyangar MCR: Trichomonas vaginalis infestation of suburethral canal. J Indian Med Assoc 31:133-134, 1958.
16. Michalowski R: Balano-posthites a Trichomonas. Ann Dermatol Venereol 108:731-738, 1981.
17. Lopatin AI: Trichomonad ulcers of the penis. Vestn Dermatol Venerol 44:78-79, 1970.
18. Krieger JN: Epididymitis, orchitis, and related conditions. Sex Transm Dis 11:173-181, 1984.
19. Kuberski TT: Ankylosing spondylitis associated with Trichomonas vaginalis infection. J Clin Microbiol 13:880-881, 1981.
20. Boström K: Chronic inflammation of human male accessory sex glands and its effect on the morphology of the spermatozoa. Scand J Urol Nephrol 5:133-140, 1971.
21. Tuttle JP, Holbrook TW, Derrick FC: Interference of human spermatozoal motility by Trichomonas vaginalis. J Urol 118:1024-1025, 1977.
22. Smith BH: Peyronie's disease. Am J Clin Pathol 45:670-678, 1966.
23. Smith BH: Subclinical Peyronie's disease. Am J Clin Pathol 52:385-390, 1969.
24. O'Hara CM, Gardner WA, Bennett BD: Immunoperoxidase staining of Trichomonas vaginalis in cytologic material. Acta Cytol 24:448-451, 1980.
25. Kline TS: Guides to Clinical Aspiration Biopsy: Prostate. New York: Igaku-Shoin, 1985.

16

Laboratory Diagnostic Methods and Cryopreservation of Trichomonads

Alexander McMillan

Laboratory Diagnostic Methods

Trichomonas vaginalis

Because specific treatment for trichomoniasis is available, it is important to identify infected people so that disease in the individual can be terminated and the sexual transmission of the organism prevented. The classic symptoms of the disease in women are a thin yellow offensive vaginal discharge, pruritus vulvae or vulval soreness, dyspareunia, and dysuria. A reddened vaginal wall with frothy thin yellow exudate in the posterior fornix and punctate red spots on the ectocervix ("strawberry cervix") may be seen.[1] These features, however, are not always present and are not pathognomonic of trichomoniasis. Although about 50% of women infected with *T. vaginalis* notice an increased vaginal discharge, a similar proportion of noninfected women give a similar history.[2,3] Since about 18% and 12% of infected and noninfected women respectively have dysuria, this is not a helpful diagnostic feature.[2] An abnormal discharge in the vaginal fornices may be seen more frequently in infected women (about 50%) than in noninfected individuals (about 22%),[3] but this is described as "frothy" in only 12% of cases.[2] The strawberry cervix is rarely seen.

Although nongonococcal urethritis may be the presenting feature, the majority of men infected with *T. vaginalis* are symptomless (see Chap. 12).

It is clear that a diagnosis of trichomoniasis cannot be made with certainty on clinical grounds alone. Laboratory investigations are essential and these are now discussed.

Specimens Required for Protozoologic Examination

Women

The vagina is the most commonly infected site in women. Material is usually collected from the posterior fornix by cotton wool–tipped applicator sticks, but the amount of exudate absorbed may be small. Considerably better uptake of genital secretion is obtained by the use of polyester sponges.[4,5] Cervicovaginal material for Papanicolaou staining is best obtained by the use of an Ayre's spatula, with the slide immediately immersed in a suitable cytologic fixative.

Although the lower urinary tract can be colonized by *T. vaginalis*, this is seldom the sole site of infection and, except in women with urinary symptoms, examination of the urine or urethral smears for the protozoon is not indicated. Rare sites of infection include the paraurethral and greater vestibular glands and these secretions need not be examined unless indicated by clinical signs.

Men

Trichomonal infection in men is difficult to detect. Because organisms are more likely to be found in semen than in urine or urethral smears,[6] a fresh sample obtained by masturbation into a clean sterile container should be examined. Urethral material taken with a polyester sponge,[5] and the deposit from a centrifuged (600 g for 5 minutes) 20-ml initial urine sample taken preferably after the patient has held his urine overnight should be examined. Expressed prostatic secretions or sub-

---

Note Added in Proof: Monoclonal antibodies are now commercially available for diagnostic purposes.

preputial material collected on sponges moistened with normal saline may also be studied.

Transport of Specimens

Ideally, saline mount preparations should be examined and the appropriate culture medium inoculated immediately after collection of material from the patient. Many physicians, however, work at considerable distances from the diagnostic laboratory and, since the protozoan is susceptible to dehydration and changes in redox potential, specimens should be sent in a suitable transport medium. Stuart's transport medium[7] or Amies' modification[8] have proved useful. During the first 24 hours, trichomonads survive well in Stuart's medium but there is a significant decline in the recovery rate of the organisms beyond this time.[9] For the first 24 hours, temperature has no influence on the subsequent cultivation of the organism but beyond this time they survive better when the medium is maintained at 4 °C or 36 °C than at 20 °C. The use of transport media leads to dilution of the number of organisms in the original sample and this should be remembered if negative results are obtained by culture or microscopy.

Provided fixation has been adequate, a delay in the submission of smear preparations for staining methods is of little importance.

Laboratory Methods for Detection in Secretions

Microscopy

Although cultivation is the most sensitive method for the diagnosis of trichomonal infection, this takes time, therefore, laboratory methods that give rapid results are necessary so that appropriate treatment can be given without delay.

Wet Smears. Although direct microscopy of expressed prostatic fluid or the deposit from centrifuged urine is useful, the density of polymorphonuclear leukocytes and epithelial cells in vaginal exudate tends to obscure the trichomonads and particularly the movement of the flagella. Microscopy of exudate diluted with isotonic saline (0.154 M) is preferable. This remains the only widely accepted routine procedure for making an "on-the-spot" diagnosis of vaginal trichomoniasis. Indeed, 44% of physicians working in sexually transmitted disease (STD) clinics in England and Wales diagnose the infection by this method alone.[10]

Compared with the results of culture, the reported sensitivity of wet smear microscopy by transmitted light, phase contrast, and darkfield illumination has varied widely, from as low as 38% to as high as 82% (Table 16.1). This variation probably reflects the use of different culture media (see below) and variability in the numbers of organisms in the samples tested. There is marked variability from patient to patient in the number of trichomonads per unit volume of vaginal exudate and a prolonged search may be necessary when few organisms are present. Vaginal douching lowers significantly the sensitivity of wet smear microscopy but not that of cultivation methods.[2] Although most workers have noted that *T. vaginalis* is not commonly found by wet smear microscopy on vaginal exudate when cultivation for the protozoon has yielded negative results, McCann[18] found that 22% of 175 cases would not have been identified if culture had been the only diagnostic method used. Feinberg and Whittington[12] found that the results obtained by microscopy of urethral material from men were superior to those obtained by culture. Others, however, have identified more infected men when culture methods have been used.[22]

A method for the concentration of *T. vaginalis* in vaginal exudate was described by Robertson et al.[4] From the posterior fornix they collected material on a polyester sponge measuring 1 × 1 × 4 cm that was then mixed with 1 ml of isotonic buffered saline contained in a bijou bottle. The material was then centrifuged into a second bijou bottle connected to the first by means of a filter adapter and the deposit was examined microscopically. Among 116 women designated uninfected by direct microscopy, nine infections were identified.

In an attempt to improve the sensitivity of direct microscopy various stains have been added to the saline mount. Although trichomonads are not stained by dyes such as safranin, malachite green, methylene blue, and brilliant cresyl blue, other cellular material takes up the dye and the organisms stand out against this colored background.[23-25] This staining method, however, has not been shown convincingly to improve the detection rate of trichomonads in secretions[6] and cannot be recommended for the routine clinical laboratory.

TABLE 16.1 Sensitivity and specificity of microscopic examination of saline mount preparations of vaginal exudate in the diagnosis of trichomoniasis.

Sensitivity* of saline mount microscopy	Specificity† of saline mount microscopy	Culture medium used	Reference
76	100.0	STS‡	11
81	98.5	Feinberg-Whittington	12
46	99.7	Modified STS	13
65§		Feinberg-Whittington	14
69ǁ		Feinberg-Whittington	14
38	96	Feinberg-Whittington	15
79	98	Feinberg-Whittington and CPLM#	16
59	75	Modified Feinberg-Whittington	17
82	NS**	Oxoid No. 2	18
77	100.0	Diamond's	19
73	99.6	Modified Bushby's	20
50	100.0	Modified Diamond's	2
54	NS	Hollander's	21

*Sensitivity = percentage of culture-positive specimens identified as positive by microscopy.
†Specificity = percentage of culture-negative specimens identified as negative by microscopy.
‡Simplified trypticase medium.
§ Hanging drop method.
ǁSuspension in appropriate diluent.
#Cysteine-peptone-liver-maltose.
**Not stated.

Vital staining of *T. vaginalis* can be accomplished by the addition of fluorescein to the saline mount preparation[26]; in darkfield illumination these organisms appear emerald green. Direct staining of the protozoon in vaginal exudate mounted in a crystal violet solution was described by Barchet.[27] Although this method was not helpful when abundant trichomonads were present, the organisms were identified more easily than in saline mount preparations when they were present in small numbers. The staining of the organisms was variable, the less active protozoa staining most intensely; flagella, axostyle, and the deep-blue eccentric nucleus were clearly evident. Further studies on the efficacy of this stain have not been reported.

Although the sensitivity of wet smear microscopy is generally low, the specificity of this diagnostic method has been considered to be high (Table 16.1). Doubts, however, were expressed by Robertson et al[4] who compared the results of immediate wet smear light microscopy with phase-contrast microscopy and culture of the centrifuged deposit of vaginal exudate collected as described above. They could not confirm the diagnosis in 10 of 44 specimens that had been identified as infected on examination by immediate microscopy. As a diagnostic method, they concluded that phase-contrast microscopy of the centrifuged deposit was more sensitive and more accurate than light microscopy of a wet smear.

Fixed Smears. Because cultivation methods are relatively slow and wet smear microscopy lacks sensitivity, the staining of protozoa in fixed smears of secretions has been investigated in detail to determine the usefulness of such methods in the diagnosis of trichomoniasis. Some of these methods are now discussed.

Gram-stained smears. The attractions of Gram-smear microscopy are that it is simple, cheap, rapid, requires only a standard microscope and stains, and is used widely in STD clinics for the presumptive diagnosis of gonorrhoea, candidiasis, and anaerobic vaginosis.

In well-made smears the organisms are pear-shaped, 15 to 18 μm long and 5 to 15 μm wide. The cytoplasmic edge is stained deeply, although the edges in any given cell may not be continuous. Tiny vacuoles within the cytoplasm give it a lacy appearance; the nucleus is eccentric and often elongated. The axostyle and flagella may be seen.[28] Cree[29] detected *T. vaginalis* in Gram-stained smears of exudate from 66% of 249 patients with culture-proven infection; the false-positive rate in that series was, surprisingly, only 7%. Twice the number of infected patients were identified by Gram-smear microscopy than by wet smear examination in the small study reported by Sobrepena.[28] There is great inconsistency in the size and shape of the organisms, which may resemble polymorphonuclear leukocytes. Therefore few workers have found the method useful in the routine diagnosis of trichomonal infection.

Papanicolaou-stained smears. In carefully taken and well-fixed Papanicolaou-stained smears (Pap smears) of cervicovaginal material, *T. vaginalis* appears as an ovoid structure varying in size from 10 to 30 µm with a discrete outline. The cytoplasm is gray-green and contains minute eosinophilic granules; the eccentric nucleus stains blue.

There are conflicting data on the efficacy of Pap smear microscopy in the diagnosis of vaginal trichomoniasis (Table 16.2). Since the sensitivity and specificity of Pap smear microscopy vary according to the experience of the microscopist, and because different media were used in different studies, it is difficult to compare these data directly. Although some workers have found the sensitivities to be similar to most of the wet smear examinations,[24,32] others have found that the results of Pap smear microscopy are more reliable and, indeed, even superior to those of culture.[17,31] Serious doubts about the specificity of this diagnostic method, however, have been expressed. In a series of 1199 women whose vaginal secretions were examined by wet smear, Pap smear, and cultivation in simplified trypticase serum medium, Perl[34] could not confirm infection in 37% of 666 patients whose Pap smears were thought to contain trichomonads. Similarly, 126 infections diagnosed by Pap smears could not be confirmed by the other diagnostic methods used by Mason and his colleagues.[33]

Summers and Ford[35] found Pap staining of smears of expressed prostatic secretions to be useful in the diagnosis of trichomoniasis in men. Routine Pap staining is lengthy, and Kurpiel[36] described a rapid hematoxylin-eosin staining method. He did not, however, compare the results with other methods.

When a routine cervical cytologic smear has been taken, Pap smear microscopy may be useful in the diagnosis of unsuspected infection. Because the organisms are often difficult to see and differentiate from other cells, I feel this method has little place as a routine procedure in the diagnosis of vaginal trichomoniasis.

Romanowsky-stained smears. Fixed smears of vaginal exudate stained with Leishman's or Giemsa's stain have been used for some 50 years in the diagnosis of trichomoniasis. In carefully prepared smears the trichomonads are stained bright blue with a darker-staining small eccentric nucleus that may be elongated, fusiform, or oval. The axostyle and flagella may be seen.[37] Overall, more infections are identified by microscopic examination of Romanowsky-stained smears than by wet smear microscopy,[38-40] and in some series the sensitivity of this diagnostic method has approached that of cultivation.[33,38,41]

Acridine orange–stained smears. Acridine orange is a fluorescent nucleic acid stain that has been used to differentiate DNA from RNA in

TABLE 16.2 Comparison of results obtained by examination of Papanicolaou (Pap)-stained and nonstained wet preparations and by cultivation in diagnosis of urogenital trichomoniasis.

Pap preparations		Wet preparations		Cultivation		
%	Total no. diagnosed by all methods	%	Total no. diagnosed by all methods	%	Total no. diagnosed by all methods	Reference
72	39	76	80	100	80	11
78	107	48	107	86	107	30
98	171	33	171	75	171	31
91	119	94	69	94	34	24
82	240	54	292	69	292	17
85	198	88	198	79	198	32
52*	231	ND†		47	231	33
65	54	52	54	89	54	21

*Percentage of Pap smears examined for trichomonads.
†Not determined.

cells. Under ultraviolet light, DNA- and RNA-containing structures fluoresce yellow-green and brick-red, respectively. In stained vaginal smears *T. vaginalis* appears as a round or pear-shaped orange-red structure with a yellow-green nucleus. There is only light-green fluorescence of epithelial cells and leukocytes.[42]

The results of acridine orange staining of vaginal material for trichomonads compare favorably with other methods (Table 16.3). In reported series the sensitivity of this method has always exceeded that of wet smear microscopy. Although more infections were diagnosed by the acridine orange method than by cultivation in the series reported by Mason et al,[33] a significant difference in the detection rates by these two methods was not found in two other studies.[14,44] In stained vaginal smears, however, the organism may be difficult to distinguish from other cells and one might question the specificity of the method. Indeed, Mason et al[33] did not confirm by other methods 7% of trichomonal infections diagnosed by the acridine orange staining procedure.

Although a rapid method has been described,[47] the staining schedule is time-consuming and, because a fluorescence microscope is also required, this technique is unsuitable for the immediate diagnosis of trichomoniasis in a busy clinical setting.

Silver staining. Although a silver staining (e.g. protargol) method,[48] can yield excellent preparations suitable for morphologic studies, the procedure is regarded as unsuitable for diagnostic purposes; it is time-consuming and requires careful attention to detail. Nagesha et al,[17] however, found that trichomonads could be identified more easily in silver stained (Fontana's method) than in wet smears. A new staining schedule was reported by Franceschini[49] and, although precise data were not presented, Klugiewicz[50] found that the results of this method were superior to those obtained with Giemsa-stained smears.

Diff-Quik staining. A rapid method based on the Diff-Quik (Harleco, Herstal, Belgium) staining set in use in hematology laboratories was described by Balsdon et al.[51] Consecutive treatment with acrylmethane, xanthene, and thiazine dyes stained the cytoplasm blue, the nucleus purple, and the flagella, axostyle, and undulating membrane red. The authors found the results comparable to those of wet smear microscopy and culture. This staining method, however, although yielding results comparable to those obtained by cultivation and observation of wet smear, compared less favorably with acridine orange staining in the study reported by Levett.[41]

Periodic acid–Schiff staining. In vaginal smears fixed in 10% formalin and stained by the McManus periodic acid–Schiff method, without diastase treatment, *T. vaginalis* is easily identified by its intense magenta coloration; the flagella and the undulating membrane are not seen.[52] Although it was thought that this stain-

TABLE 16.3. Comparison of percentages of *T. vaginalis* infections diagnosed by acridine-orange staining and by other methods.

| Total no. of samples examined | Percentage of positive results | | | | Reference |
	Acridine orange staining	Wet preparation	Cultivation	Other methods	
100	64	34	—	—	43
100	44	40	51	—	14
495	33	—*	19	22†, 24‡	33
203	6	3	2	6†	39
120	97	73	93	—	44
181	26	20	—	—	45
247	9	2	5	5†, 4§	41
105	15	14	—	—	46
397	7	5	—	—	47
1102	5.8	4.5	—	5.5†	40

*Not determined.
†Giemsa or Leishman staining.
‡Papanicolaou staining.
§Diff-Quik staining.

ing method would prove useful in laboratory diagnosis of trichomoniasis, it has not been developed further.

Overall, the staining of fixed smears of exudate for *T. vaginalis* is unsatisfactory for routine diagnosis. Although the above methods may seem attractive to laboratory workers who handle material submitted from peripheral areas, cultivation of samples sent in an appropriate transport medium probably yields better results in terms of specificity and sensitivity.

Immunocytochemical. A major disadvantage of the microscopic examination of fixed smears stained by the above methods is the lack of specificity. In light of the fact that fixation methods almost always produce some degree of distortion of the organism, and because staining patterns can be variable, these trichomonads, particularly when they are present in low numbers, may be difficult to identify among other cellular material. The use of fluorescein- or enzyme-labeled antibodies against specific antigens on the surface of the protozoon might be expected to be useful in the identification of the organism in clinical material. Although McEntegart et al,[53] using polyclonal antisera, described a direct immunofluorescent antibody test that distinguished between *T. vaginalis* and *T. foetus* in a mixed culture, this method has not been helpful in the diagnosis of vaginal trichomoniasis. Because epithelial cells and other material in the exudate show nonspecific fluorescence, it is often difficult to detect the protozoon, especially if few organisms are present. Nonetheless, Hayes and Kotcher[54] diagnosed by direct immunofluorescence trichomonal infection in 40% of 225 women; organisms were detected by wet smear microscopy and culture in 23% and 40%, respectively, of these women.

The use of monoclonal antibodies (MAbs) has several advantages: the need for extensive purification of antigen for immunization is obviated and by selecting MAbs against epitopes shared by different strains of the organism, their use in diagnostic methods should result in greater sensitivity and specificity than when polyclonal antibodies are used.

Using two MAbs in an immunofluorescence (IF) method, Krieger et al[96] detected trichomonads in the vaginal exudate of each of ten patients with infection proved by cultivation. In an IF test a combination of three MAbs produced against epitopes specific for cytoplasm, flagella, and nucleus was used to detect *T. vaginalis* in vaginal material from 231 women[97]: 34 specimens were positive by both methods, 1 was positive on wet mount microscopy but negative in the IF test, 8 were positive in the IF test but negative by microscopy and culture, and 188 were negative by both methods.

Enzyme-labeled immunocytochemical methods offer certain advantages over fluorescent antibody tests in the diagnosis of any infection—there is less background staining, preparations are permanent, and a fluorescence microscope is not required. Using a rabbit polyclonal antibody conjugated with peroxidase, Bennett et al[55] described the specific staining of *T. vaginalis* in paraffin sections of subcutaneous tissue of rabbits previously infected with the organism. This study was extended to the detection of *T. vaginalis* in cervical cytologic smears.[56] In these preparations the organisms were easily identified and infection was diagnosed in 16% of 100 patients, only four of whose Pap smears were reported as positive. The immunoperoxidase method, however, is time-consuming and unlikely to be useful in on-the-spot diagnosis of trichomoniasis. It may have a place in the detection of trichomonads in fixed material submitted to the laboratory from physicians working without a hospital clinic.

Cultivation

Detailed aspects of the cultivation of *T. vaginalis* are considered in Chapter 7, and we are concerned here only with its place in the routine diagnosis of infection.

Many culture media for *T. vaginalis* have been described,[57] which have in common nutrients (e.g., liver extracts, peptone, trypticase), sera, essential salts, reducing agents, carbohydrates, and antibiotics to inhibit bacterial growth. Many diagnostic laboratories routinely use liquid media, that is, media that do not contain agar. These undefined media include those of Feinberg and Whittington (FW),[12] Lumsden et al,[58] and Oxoid No. 2 (OXO, LTD, London, England). Since *T. vaginalis* grows better when the medium is incubated under partial or complete anaerobic conditions,[16] culture vessels should be filled as full as possible with liquid media. Some laboratories routinely use semiliquid media such as CPLM (cysteine-peptone-liver-maltose)[59] or TYM (trypticase-yeast extract-maltose)[60] and Lowe's.[38] The incorporation of 0.05% to 0.1% agar into the medium reduces the diffusion of oxygen into the medium and allows better growth. Although some semisolid media (up to 0.38% agar) are thought to be useful for the selection of clones, they are seldom used in the laboratory diagnosis of trichomoniasis.

There are differences in the efficiency of the various media. For example, in a comparison of four media (Stenton's, Oxoid No. 2, FW, and Bushby's), Cox and Nicol[61] found that the latter two gave the best results. Rayner[16] found CPLM to be superior to FW medium, and he detected more than twice the number of vaginal infections by cultivation in the former. Whittington (cited by Rayner) showed that about 5 × 10^4 trichomonads in 10 ml of FW medium were necessary to initiate growth, which becomes established in 24 to 48 hours. A smaller inoculation, however, is needed if the medium is incubated under partial or complete anaerobic conditions.[16] Small numbers of *T. vaginalis* (between 1 and 40 organisms) can initiate growth in 5 ml of CPLM, but this may not be observed until 96 hours after inoculation. Rayner[16] concluded that cultures should not be described as negative until at least 72 hours after inoculation.

As is expected, cultivation is a more sensitive diagnostic method than microscopy (see Table 16.1), particularly when there are low numbers of organisms in the secretions. When laboratory facilities are adequate, cultivation should be employed routinely in the diagnosis of trichomoniasis. Since vaginal exudate or other secretions may contain nonviable trichomonads, cultivation of that material may yield negative results. Under these circumstances microscopic methods are more sensitive than culture. When few but viable organisms are present, the converse is true. Not surprisingly then, the greatest diagnostic yield is obtained when a combination of methods is used (Table 16.4).

Immunodiagnosis

Serologic Tests. Because diagnostic tests based on the detection of specific antibody in the sera of patients with other protozoal infections have proved useful, similar tests have been applied to the diagnosis of trichomoniasis. Although complement fixing (CF) antibodies can be detected in the sera of 48% to 80% of women with trichomoniasis,[62,63] similar antibodies are found in the sera of 10% to 40% of noninfected persons. Jaakmees et al[64] but not Hoffmann[65] found that positive results in the complement fixation test were more common in patients with chronic trichomoniasis than in those with acute infection. Using antigens prepared from isolates of *T. vaginalis*, Jaakmees' group[64] detected serum antibody to at least one antigen in each of 170 infected patients but in none of the control group. Beginning about 3 months after treatment, they noted a steady decline in the titer of CF antibody, which becomes undetectable generally within 1 year.

The need for the use in serologic tests of multiple antigens of *T. vaginalis* was also noted in the agglutination reaction. Because sera from noninfected persons often show low-titered antitrichomonal reactivity, it is necessary to dilute serially the sera to be tested and establish a titer above which the test is considered positive. In the test system described by Teras et al[66] a variety of antigens were used routinely. These workers did not detect antibody at a titer of greater than 160 in the sera of noninfected men and women, but sera from each of 171 women and 83 men harboring *T. vaginalis* contained antibody at a titer of greater than 200. Within 12 to 16 months of treatment, the titer of trichomonal antibody had fallen below 160 in both men and women.

Using a formamide extract of *T. vaginalis* as antigen, a passive hemagglutination method for the detection of trichomonal antibody was described by McEntegart.[67] Antibody at a titer of greater than 10 was found in 84% of 50 infected women, 6% of 50 "normal" female blood donors, and 6% of "normal" male blood donors (but see the discussion in Honigberg[80]). Hoffmann[68] found serum trichomonal antibody in over 90% of infected women, but in only 55% of infected men. Using antigens extracted in a similar fashion from four cloned strains of *T. vaginalis*, Kuberski[69] found antibody at a titer of greater than 10 in the sera of 97% of 35 women and eight of nine men with active infection. Although an antibody titer of more than 80 correlated well with active *T. vaginalis* infection in men and women, a titer below 80 did not exclude a positive diagnosis.

Using formalin-treated axenically cultivated organisms in an indirect immunofluorescent antibody test, antibody reactive with *T. vaginalis* was detected in the sera of 19 patients with

TABLE 16.4 Comparison of percentages of T. vaginalis infections diagnosed by laboratory methods employed singly or in combination with cultivation.

Total no. of samples examined	Percentage of positive results							Reference
	Stained smears			Cultivation	Wet preparations	Cultivation and		
	Wet preparations	Various stains	Pap* stain			Stained smears (various stains)	Pap-stained smears	
100	74	86	ND	83	91	99	ND	38
107	48	ND†	78	86	100	ND	99	30
175	86	ND	ND	78	100	ND	ND	18
198	88	ND	85	79	98	ND	95	32
54	52	ND	65	89	93	ND	100	21

*Papanicolaou staining.
†Not determined.

trichomoniasis, but not in the sera of 9 noninfected persons.[70] The antibody detected, however, was present at low titer. Surprisingly, follow-up studies have not been reported.

A sonicate of three isolates of *T. vaginalis* was used as antigen in an enzyme-linked immunosorbent assay (ELISA) for serum antibody.[71] Serum IgG at a titer of greater than 32 was detected in the sera of 68% of 41 infected women and 12% of 168 women who had no evidence of trichomonal infection; IgM antibody at a titer of greater than 8 was found in the sera of 22% and 3% of these women, respectively.

As diagnostic procedures, serologic tests for many infections, including trichomoniasis, have several inherent disadvantages. It is clear from the data presented above that serum antibody may not be detectable in infected patients by a particular test. This may be so because the test system is too insensitive to detect low levels of specific antibody or because a serum humoral response has not yet been elicited. The specificity of serologic tests is also variable, but since trichomonal antibody may persist for a long time after treatment or spontaneous cure, one would caution against the term "false positive" to describe a positive reaction in an apparently noninfected person. It is perhaps this persistence of antibody, however, and the time-consuming nature of the serologic tests that militate against their use in the routine diagnosis of trichomoniasis. As suggested by many workers they may, however, have a place in epidemiologic studies.

Tests Based on the Detection of Specific Antibodies in Secretions. The detection of specific antibody in secretions has been suggested as useful in the diagnosis of infection of mucosal surfaces. For example, in the study of Treharne et al,[72] the most sensitive of the serologic tests for chlamydial infection of the uterine cervix was the detection by immunofluorescence of IgG antibody in the cervical secretions. Using an ELISA with antigen prepared from three isolates of *T. vaginalis,* IgA or IgG, or both were detected in the cervicovaginal secretions of 73% of 47 infected patients and 41% of 100 apparently uninfected patients who attended an STD clinic.[71] The persistence of antibody from a previous infection may explain the high proportion of antibody-positive currently noninfected women.

The lack of sensitivity and specificity of these tests in women and the failure to detect antibody in genitourinary secretions of infected men[73] argue against their use in the diagnosis of trichomoniasis. The development of a test based on the detection of *antigen* in secretions, however, is awaited with interest.

Intradermal Tests. The supernatant fluid of a *T. vaginalis* lysate was used by Adler and Sadowsky[74] in an intradermal test for trichomonal infection. A positive reaction, which was defined as the development within 48 hours of an area of definite redness of 1 to 2 cm in diameter around the injection site, was found in 81% and 33% of 43 infected and 58 noninfected women, respectively. Using extracts of *T. vaginalis,* Kawai et al[75] found that these extracts produced specific reaction when infected intradermally into the forearm of infected women. The authors noted that the test became positive within 2 weeks of acquisition of infection and became negative within a few days of successful treatment. Respectively, only 28% and 38% of women and men infected with *T. vaginalis* gave a positive reaction to testing with a trichomonal lysate in one series reported by Jaakmees and Teras.[76] They found a higher proportion of positive results in patients tested with four rather than one antigenic type of *T. vaginalis.*

Because the intradermal test, based on the development of delayed type hypersensitivity to the trichomonad, lacks sensitivity and specificity, its use as a diagnostic test is limited.

Pentatrichomonas hominis

This intestinal flagellate is detected by microscopic examination of a saline mount preparation of feces. Organisms are more likely to be found in loose than in formed stools, but repeated examinations are often necessary to detect the protozoon. *Pentatrichomonas hominis* is a pyriform or ellipsoid organism, 5 to 15 μm long and 4 to 7 μm wide, which moves in a rapid, jerky fashion. Since the structure of the organism may not be discernible in saline mount preparations, examination of fecal smears fixed in Schaudinn's fluid and stained with iron hematoxylin[77] is helpful in its differentiation from other intestinal flagellates.

Since few protozoa may be found in fecal samples, cultivation has proved helpful in the identification of infected persons. Many media including CPLM and TYM (see *Trichomonas vaginalis* section, above) support the growth of *P. hominis,* but in my experience Robinsons's medium[78] is particularly useful, being employed routinely in the diagnosis of intestinal amebiasis.

Immunodiagnostic tests for *P. hominis* are not available.

Trichomonas tenax

Trichomonas tenax is detected by the microscopic examination of gingival scrapings emulsified in a drop of isotonic saline. The organisms, which move erratically, are ellipsoidal or ovoid with a length of 4 to 13 μm and a width of 2 to 9 μm. Although many of the media described for the cultivation of *T. vaginalis* support the growth of *T. tenax,* egg yolk infusion media such a Balamuth's[79] are used widely.

Cryopreservation of Trichomonads

The preservation of trichomonads for research purposes, including immunologic and epidemiologic studies, is essential.

Trichomonas vaginalis and Pentatrichomonas hominis

Although continuous cultivation of *T. vaginalis* has been used for research purposes, continuous changes with time in the antigenic characteristics and virulence of these organisms have been suggested.[80,81] Cryopreservation (for early methods, see reports cited in reference 87) has, however, overcome some of these difficulties. Ivey[82] found that all six isolates of *T. vaginalis* that had been maintained by continuous serial cultivation for 10 months had lost most of their virulence as assessed by intraperitoneal infection of white mice. By contrast, each isolate that had been preserved in liquid nitrogen for 10 months retained most of its original virulence. Lumsden and Hardy[83] refer to a viably preserved population of protozoa in which reproduction is arrested as a "stabilate." Protozoal populations in established cultures, that is, in cultures that can be maintained by serial passages, are referred to as "strains."

Stabilates have been prepared from axenic cultures of *T. vaginalis* in various media such as TYM, CPLM, and Lumsden's[58,84,85] but Ivey[86] found that the survival rate of organisms grown in agarless media was lower than that of protozoa grown in agar-containing media.

The capacity of the containers used varies according to the requirements of the worker. Lumsden et al[58] found that capillary tubes were satisfactory; others have used 1-ml glass ampules[85] or a number of other glass and plastic vessels. The containers are sealed or otherwise hermetically closed after having been filled with a known volume of a suspension of trichomonads at a known concentration in culture medium.

In many laboratories, the cooling is accomplished by placing hermetically closed or sealed containers with the trichomonad suspensions into an aluminum insulating jacket and storing them overnight in a mechanical freezer operating at a temperature of −70 °C. Thereafter, the containers are transferred rapidly to a liquid nitrogen refrigerator (e.g., dewar) filled with liquid nitrogen.[90] In other laboratories, cryogenic equipment, consisting of a freezing chamber, controller, and a sensitive temperature recorder, is employed. In the last few years it has become possible to program the equipment to freeze particular cells suspended in specific volumes in given culture media. In all instances, successfully cryopreserved trichomonads maintain their numbers, motility, and virulence over long periods.[58,85,88]

Trichomonads are best preserved at −196 °C (i.e., submerged in liquid nitrogen), but the rate of cooling to this temperature is important to the survival of the stabilate.[86] Different rates were found more favorable by some investigators than by others. For example, Ivey[86] found that a cooling rate of 1 °C per minute resulted in poorer survival than cooling at 2 °C or 5 °C per minute; however, using dimethyl sulfoxide (DMSO) as the cryoprotectant and Linde biological freezing equipment, Diamond[88] noted that equally satisfactory recovery of trichomonads was obtained when the rate of 8 °C was employed in cooling the protozoa down to the zone of release of the latent heat of fusion and either 1 °C or 8 °C cooling rate was used after the removal of the flagellates out of the heat of fusion zone. According to Honigberg et al[89] (and Honigberg, personal communication), when Linde cryogenic equipment is employed with trichomonads in CPLM, TYM, and similar media with 5% (vol/vol) of DMSO, the highest recovery (over 95%, as judged by motility) can be obtained by using a cooling rate of about 5 °C per minute to the latent heat of fusion release zone and about 1 °C to 2 °C after the flagellates are removed from this zone by rapid supercooling (1 to 1.5 minutes) and cooled to between −50 °C and −65 °C before being immersed in liquid nitrogen (−196 °C). Actually, good results can be obtained also by storing the stabilates in liquid nitrogen vapors (−170 °C).

The cryopreservation methods described above are applicable equally to *T. vaginalis* and *P. hominis*.

Trichomonas tenax

The cryopreservation of *T. tenax* has proved difficult. Müller[91] and Kasprzak and Rydzewski[92] described the cryopreservation of this organism by freezing agnotobiotic cultures in dry ice and ethanol, with glycerol serving as cryoprotectant. LeCorroller et al[93] found that *T. tenax* survived freezing at −70 °C in a medium containing dextran, sorbitol, and polyvinylpyrrolidone. DMSO has been used as cryoprotectant, but recovery of viable organisms is variable.

Trichomonas tenax can be grown in axenic culture in a medium used by Diamond and Bartgis[94] for *Entamoeba histolytica*. This medium is a modification of the complex medium devised by Diamond[95] for axenic cultivation of *T. tenax*. The oral trichomonads grown in the former medium can be frozen with relative ease, as evidenced by their motility on thawing. However, only a certain percentage of stabilates containing actively motile flagellates gives rise to viable cultures (Honigberg, personal communication). The reasons for the difficulties in recovering stabilates of *T. tenax* remain obscure. There is little doubt, however, that further investigations will help in eliminating these difficulties.

References

1. King A, Nicol C: *Venereal Diseases*. London: Baillière, Tindall, 1969.
2. Fouts AC, Kraus SJ: *Trichomonas vaginalis*: Reevaluation of its clinical presentation and laboratory diagnosis. *J Infect Dis* 141:133-143, 1980.
3. McLellan R, Spence MR, Brockman M, Raffel PA-C, Smith JL: The clinical diagnosis of trichomoniasis. *Obstet Gynecol* 60:30-34, 1982.
4. Robertson DHH, Lumsden WHR, Fraser KF, Hosie DD, Moore DM: Simultaneous isolation of *Trichomonas vaginalis* and collection of vaginal exudate. *Br J Vener Dis* 45:42-43, 1969.
5. Oates JK, Selwyn S, Breach MR: Polyester sponge swabs to facilitate examination for genital infection in women. *Br J Vener Dis* 47:289-292, 1971.
6. Hoffman B, Małyszko E: Studies on the way of laboratory diagnosis of trichomonadosis in men. *Wiad Parazytol* 8:179-189, 1962.
7. Stuart RD: Transport problems in public health bacteriology. *Can J Public Health* 47:114-122, 1956.
8. Amies CR: A modified formula for the preparation of Stuart's transport medium. *Can J Public Health* 58:296-300, 1967.
9. Nielsen R: *Trichomonas vaginalis* I. Survival in solid Stuart's medium. *Br J Vener Dis* 45:328-331, 1969.
10. O'Connor BH, Adler MW: Current approaches to the diagnosis, treatment and reporting of trichomoniasis and candidosis. *Br J Vener Dis* 55:52-57, 1979.
11. Kean BH, Day E: *Trichomonas vaginalis* infection: An evaluation of three diagnostic techniques with data on incidence. *Am J Obstet Gynecol* 68:1510-1518, 1954.
12. Feinberg JG, Whittington MJ: A culture medium for *Trichomonas vaginalis* Donné and species of *Candida*. *J Clin Pathol* 10:327-329, 1957.
13. Clark DH, Solomons E: An evaluation of routine culture examination for *Trichomonas vaginalis* and *Candida*. *Am J Obstet Gynecol* 78:1314-1319, 1959.
14. Martin RD, Kaufman RH, Burns M: *Trichomonas vaginalis* vaginitis: A statistical evaluation of diagnostic methods. *Am J Obstet Gynecol* 87:1024-1027, 1963.
15. Hulka BS, Hulka JF: Dyskaryosis in cervical cytology and its relationship to trichomoniasis therapy. *Am J Obstet Gynecol* 98:180-187, 1967.
16. Rayner CFA: Comparison of culture media for the growth of *Trichomonas vaginalis*. *Br J Vener Dis* 44:63-66, 1968.
17. Nagesha CN, Ananthakrishna NC, Sulochana P: Clinical and laboratory studies on vaginal trichomoniasis. *Am J Obstet Gynecol* 106:933-935, 1970.
18. McCann JS: Comparison of direct microscopy and culture in the diagnosis of trichomoniasis. *Br J Vener Dis* 50:450-452, 1974.
19. Eriksson G, Wanger L: Frequency of *N. gonorrhoeae*, *T. vaginalis* and *C. albicans* in female venereological patients. *Br J Vener Dis* 51:192-197, 1975.
20. Rothenberg RB, Simon R, Chipperfield E, Catterall RD: Efficacy of selected diagnostic tests for sexually transmitted diseases. *JAMA* 235:49-51, 1976.
21. Spence MR, Hollander DH, Smith J, McCaig L, Sewell D, Brockman M: The clinical and laboratory diagnosis of *Trichomonas vaginalis* infection. *Sex Transm Dis* 7:168-171, 1980.
22. Hoffman B, Kilczewski W, Małyszko E: Studies in trichomoniasis in males. *Br J Vener Dis* 37:172-175, 1961.
23. Miller JR: Contrast stain for the rapid identifi-

24. Eddie DAS: The laboratory diagnosis of vaginal infections caused by *Trichomonas* and *Candida* (*Monilia*) species. *J Med Microbiol* 1:153-159, 1968. cation of *Trichomonas vaginalis*. *JAMA* 106:616, 1936.
25. Perju A, Strîmbeanu I: Valeur de la spermoculture dans le diagnostic de la trichomonase urogénitale chez l'homme. *Wiad Parazytol* 12:475-480, 1966.
26. Coutts WE, Silva-Inzunza E: Vital staining of *Trichomonas vaginalis* with fluorescein. *Br J Vener Dis* 30:43-46, 1954.
27. Barchet S: A new look at vaginal discharges. *Obstet Gynecol* 40:615-617, 1972.
28. Sobrepena RL: Identification of *Trichomonas vaginalis* in Gram-stained smears. *Lab Med* 11:558-560, 1980.
29. Cree GE: *Trichomonas vaginalis* in Gram-stained smears. *Br J Vener Dis* 44:226-227, 1968.
30. Oller LZ: Routine exfoliative cytology for cancer and *Trichomonas* detection at a clinic for venereal diseases. *Br J Vener Dis* 41:304-308, 1965.
31. Hughes HE, Gordon AM, Barr GTD: A clinical and laboratory study of trichomoniasis of the female genital tract. *J Obstet Gynaecol Br Commonw* 73:821-827, 1966.
32. Thin RNT, Atia W, Parker JDJ, Nicol CS, Canti G: Value of Papnicolaou-stained smears in the diagnosis of trichomoniasis, candidiasis and cervical herpes simplex virus infection in women. *Br J Vener Dis* 51:116-118, 1975.
33. Mason PR, Super H, Fripp PJ: Comparison of four techniques for the routine diagnosis of *Trichomonas vaginalis* infection. *J Clin Pathol* 29:154-157, 1976.
34. Perl G: Errors in the diagnosis of *Trichomonas vaginalis* infection as observed among 1199 patients. *Obstet Gynecol* 39:7-9, 1972.
35. Summers JL, Ford ML: The Papanicolaou smear as a diagnostic tool in male trichomoniasis. *J Urol* 107:840-842, 1972.
36. Kurpiel M: Pictures of *T. vaginalis* obtained with modified method of staining cytological preparations. *Wiad Parazytol* 23:491-495, 1977.
37. Sorel C: Trois techniques de recherche du 'Trichomonas vaginalis' leurs valeurs comparees. *Presse Méd* 62:602-604, 1954.
38. Lowe GH: A comparison of current laboratory methods with a new semisolid culture medium for the detection of *Trichomonas vaginalis*. *J Clin Pathol* 18:432-434, 1965.
39. Rogers S, Goldsmid JM: A study of the possible value of acridine orange-O stain in the diagnosis of *Trichomonas vaginalis* infection. *Cent Afr J Med* 23:56-58, 1977.
40. Buharowski K, Wolańska M: Usefulness of samples stained with acridine orange for diagnosis of women's trichomonadosis. *Wiad Parazytol* 30:499-501, 1984.
41. Levett PN: A comparison of five methods for the detection of *Trichomonas vaginalis* in clinical specimens. *Med Lab Sci* 37:85-88, 1980.
42. Fripp PJ, Mason PR, Super H: A method for the diagnosis of *Trichomonas vaginalis* using acridine orange. *J Parasitol* 61:966-967, 1975.
43. van Niekerk WA: Acridine-orange fluorescence microscopy in the detection of pathogenic vaginal flora. *Obstet Gynecol* 20:596-601, 1962.
44. Maciejewski Z, Dziecielski H: Method of secondary fluorescence in diagnostics of trichomonadosis and in control of the effectiveness of antitrichomonal therapy. *Wiad Parazytol* 23:617-619, 1977.
45. Hipp SS, Kirkwood MW, Gaafar HA: Screening for *Trichomonas vaginalis* infection by use of acridine orange fluorescent microscopy. *Sex Transm Dis* 6:235-238, 1979.
46. Greenwood JR, Kirk-Hillaire K: Evaluation of acridine orange stain for detection of *Trichomonas vaginalis* in vaginal specimens. *J Clin Microbiol* 14:699, 1981.
47. Ridge AG: A rapid method for detection of *Trichomonas vaginalis*. *Med Lab Sci* 39:193-194, 1982.
48. Honigberg BM, Davenport HA: Staining flagellate protozoa by various silver protein compounds. *Stain Technol* 29:241-246, 1954.
49. Franceschini Ph: Nouvelle méthode de coloration du *Trichomonas*. *Presse Méd* 79:486-487, 1971.
50. Klugiewicz A: A comparison of two methods of staining of *Trichomonas vaginalis* for the needs of medical diagnosis and didactics. *Wiad Parazytol* 20:517-520, 1974.
51. Balsdon MJ, Green N, Andrew CW, Jackson DH: Rapid staining technique for *Trichomonas vaginalis*. *Br J Vener Dis* 55:289-291, 1979.
52. Rodriguez-Martínez HA, Rosales de la Luz M, de Bello LG, Ruiz-Moreno JA: Adequate staining of *Trichomonas vaginalis* by McManus periodic acid-Schiff stain. *Am J Clin Pathol* 59:741-746, 1973.
53. McEntegart MG, Chadwick CS, Nairn RC: Fluorescent antisera in the detection of serological varieties of *Trichomonas vaginalis*. *Br J Vener Dis* 34:1-3, 1958.
54. Hayes BS, Kotcher E: Evaluation of techniques for the demonstration of *Trichomonas vaginalis*. *J Parasitol* 46(suppl):45, 1960.
55. Bennett BD, Bailey J, Gardner WA: Immunocytochemical identification of trichomonads. *Arch Pathol Lab Med* 104:247-249, 1980.
56. O'Hara CM, Gardner WA, Bennett BD: Immunoperoxidase staining of *Trichomonas vaginalis* in cytologic material. *Acta Cytol* 24:448-451, 1980.

57. Taylor AER, Baker JR: *The Cultivation of Parasites in Vitro.* Oxford: Blackwell, 1968.
58. Lumsden WHR, Robertson DHH, McNeillage GJC: Isolation, cultivation, low temperature preservation, and infectivity titration of *Trichomonas vaginalis.* Br J Vener Dis 42:145-154, 1966.
59. Johnson G, Trussel RE: Experimental basis for the chemotherapy of *Trichomonas vaginalis* infestations. Proc Soc Exp Biol Med 54:245-249, 1943.
60. Diamond LS: The establishment of various trichomonads of animals and man in axenic cultures. J Parasitol 43:488-490, 1957.
61. Cox PJ, Nicol CS: Growth studies of various strains of *T. vaginalis* and possible improvements in the laboratory diagnosis of trichomoniasis. Br J Vener Dis 49:536-539, 1973.
62. Trussel RE: *Trichomonas vaginalis and Trichomoniasis.* Springfield, Ill: Charles C Thomas, 1947.
63. Hoffman B, Fiedoruk J: Studies of the carbohydrate metabolism of *T. vaginalis.* Wiad Parazytol 12:299-303, 1966.
64. Jaakmees HP, Teras JK, Rõigas EM, Nigesen UK, Tompel HJ: Complement fixing antibodies in the blood sera of men infested with *Trichomonas vaginalis.* Wiad Parazytol 12:378-384, 1966.
65. Hoffmann B, Kazanowska W, Kilczewski W, Krach J: Serologic diagnosis of *Trichomonas* infection. Med Dósw Mikrobiol 15:91-99, 1963.
66. Teras JK, Jaakmees HP, Nigesen UK, Rõigas EM, Tompel HJ: The dependence of serologic reactions on the serotypes of *Trichomonas vaginalis.* Wiad Parazytol 12:364-369, 1966.
67. McEntegart MG: The application of a haemagglutination technique in the study of *Trichomonas vaginalis* infections. J Clin Pathol 5:275-280, 1952.
68. Hoffmann B: An evaluation of the use of the indirect hemagglutination method in the serodiagnosis of trichomonadosis. Wiad Parazytol 12:392-397, 1966.
69. Kuberski T: Evaluation of the indirect hemagglutination technique for study of *Trichomonas vaginalis* infections, particularly in men. Sex Transm Dis 5:97-102, 1978.
70. Kramář J, Kučera K: Immunofluorescence demonstration of antibodies in urogenital trichomoniasis. J Hyg Epidemiol Microbiol Immunol 10:85-88, 1966.
71. Street DA, Taylor-Robinson D, Ackers JC, Hanna NF, McMillan A: Evaluation of an enzyme-linked immunosorbent assay for the detection of antibody to *Trichomonas vaginalis* in sera and vaginal secretions. Br J Vener Dis 58:330-333, 1982.
72. Treharne JD, Darougar S, Simmons PD, Thin RN: Rapid diagnosis of chlamydial infection of the cervix. Br J Vener Dis 54:403-408, 1978.
73. Ackers JC, Catterall RD, Lumsden WHR, McMillan A: Absence of detectable local antibody in genitourinary tract secretions of male contacts of women infected with *Trichomonas vaginalis.* Br J Vener Dis 54:168-171, 1977.
74. Adler S, Sadowsky A: Intradermal reaction in trichomonas infection. Lancet 1:867-868, 1947.
75. Kawai N, Ishibashu J, Watanabe A, Etsura H: Intradermal test for trichomoniasis in women. Sanfujinka No Sekai (World of Obstetrics and Gynecology) 13:1177-1181, 1961.
76. Jaakmees HP, Teras JK: Intradermal reaction with specific antigens in cases of genitourinary trichomonadosis. Wiad Parazytol 12:385-391, 1966.
77. Lumsden WHR, McMillan A: Protozoa. In Collee JG, Duguid JC (eds): *Mackie and McCartney: Medical Microbiology,* 13th ed, vol 2. Edinburgh: Churchill Livingstone, 1987.
78. Robinson GL: The laboratory diagnosis of human parasitic amoebae. Trans R Soc Trop Med Hyg 62:285-294, 1968.
79. Markell EK, Voge M: *Medical Parasitology,* Philadelphia: WB Saunders, 1976.
80. Honigberg BM: Trichomonads of importance in human medicine. In Kreier JP (ed): *Parasitic Protozoa,* vol 2, pp 275-454. New York: Academic Press, 1978.
81. Lindgren RD, Ivey MH: The effect of cultivation and freezing on the virulence of *Trichomonas vaginalis* for mice. J Parasitol 50:226-229, 1964.
82. Ivey MH: Virulence preservation of recent isolates of *Trichomonas vaginalis.* J Parasitol 61:550-552, 1975.
83. Lumsden WHR, Hardy GJC: Nomenclature of living parasite material. Nature 205:1032, 1965.
84. Diamond LS, Bartgis, IL, Reardon LV: Virulence of *Trichomonas vaginalis* after freeze-preservation for 2 years in liquid nitrogen vapor. Cryobiology 1:295-297, 1965.
85. Wasley GD, Rayner CFA: Preservation of *Trichomonas vaginalis* in liquid nitrogen. Br J Vener Dis 46:323-325, 1970.
86. Ivey MH: Use of solid medium techniques to evaluate factors affecting the ability of *Trichomonas vaginalis* to survive freezing. J Parasitol 61:1101-1103, 1975.
87. McEntegart MG: The maintenance of stock strains of trichomonads by freezing. J Hyg 52:545-550, 1954.
88. Diamond LS: Freeze-preservation of protozoa. Cryobiology 1:95-102, 1964.
89. Honigberg BM, Farris VK, Livingston MC: Preservation of *Trichomonas vaginalis* and *Trichomonas gallinae* in liquid nitrogen. Prog Protozool, Proc II Int Conf Protozool, 1965.

Excerpta Med Found, Int Congr Ser 91:199, 1965.
90. Lumsden WHR, Herbert WJ, McNeillage GJC: *Techniques with Trypanosomes.* Edinburgh: Churchill Livingstone, 1973.
91. Müller W: Zur Frage der Konservierung von Laborstämmen verschiedener Trichomonas und Trichomonasarten durch tiefe Temperaturen. *Z Tropenmed Parasitol* 17:100-102, 1966.
92. Kasprzak W, Rydzewski A: Preservation of some parasitic protozoa at low temperatures. *Z Tropenmed Parasitol* 21:198-201, 1970.
93. LeCorroller Y, Gysin J, l'Herete P: Une technique simple de conservation des protozoaires par congélation. *Ann Inst Pasteur (Algérie)* 48:109-124, 1970.
94. Diamond LS, Bartgis IL: Axenic cultures for in vitro testing of drugs against *Entamoeba histolytica. Arch Invest Med* 2(suppl):339-348, 1971.
95. Diamond LS: Axenic cultivation of *Trichomonas tenax,* the oral flagellate of man. I. Establishment of cultures. *J Protozool* 9:442-444, 1962.
96. Krieger JN, Holmes KK, Spence MR, Rein MF, McCormack WM, Tam MR: Geographic variation among isolates of *Trichomonas vaginalis:* Demonstration of antigenic heterogeneity by using monoclonal antibodies and the indirect immunofluorescence technique. *J Infect Dis* 152: 979-984, 1985.
97. Chang TH, Tsing SY, Tzeng S: Monoclonal antibodies against *Trichomonas vaginalis. Hybridoma* 5:43-51, 1986.

17

Epidemiology of Urogenital Trichomoniasis

Joseph G. Lossick

Introduction

Trichomonas vaginalis is a major cause of sexually transmitted vaginitis throughout the world. Although the organism is not an invasive pathogen associated with serious sequelae, its frequency and ability to produce a communicable vaginitis has given this agent a deserved notoriety. *Trichomonas vaginalis* was first discovered by Donné[1] in 1836. For many decades the organism was considered a commensal in the urogenital tract. In 1916, Hoehne[2] demonstrated a relationship between *T. vaginalis* and symptomatic vaginal discharge. It was not until 24 years later that Trussell and Plass[2] fulfilled Koch's postulates by transmitting the disease to volunteers by the intravaginal inoculation of axenic cultures of the organism. These data were substantiated 2 years later by Hesseltine et al.[4]

For many years, the pathogenicity of the organism was questioned, and it has been only in the last few decades that a greater understanding of *T. vaginalis'* role in vaginal disease has been achieved. Despite considerable research attention over the years, not much is known about its life cycle and epidemiology.

Epidemiology is the study of the causes, determinants, and distribution of disease. Both apparent and inapparent disease results from a triad of agent, host, and environmental factors that favor its occurrence. Control of disease requires an intimate knowledge of the interrelationship of this epidemiologic triad and a knowledge of the time, place, and personal characteristics of the populations affected.

Various aspects of the agent are discussed in some detail in other chapters of this book: structural details can be found in Chapter 3; pathogenicity and virulence in human (female and male) and experimental hosts in Chapters 8, and 11 to 15; and effects of *T. vaginalis* in tissue cultures in Chapter 9. Identification and differentiation of strains by immunologic methods are analyzed in Chapter 4. The latter chapter, which is concerned with all immunologic aspects of urogenital trichomoniasis, deals also with host-parasite interactions and possible host defense mechanisms. The bacterial flora accompanying *T. vaginalis* are considered in Chapter 10. Finally, a long chapter (Chap. 13) is devoted to unifocal and multifocal trichomoniasis in children.

There are, however, certain aspects of *T. vaginalis* trichomoniasis that may have a profound impact on the epidemiology of this disease and that have not been considered in sufficient detail in the other chapters. Some of them are discussed first before going into the primary topic of the present chapter, that is, descriptive epidemiology.

Survival Outside the Human Host

To perpetuate itself, *T. vaginalis* depends on the human host for survival. Because the organism has no cyst phase it is susceptible to desiccation and high temperatures, but can survive for surprisingly long periods outside the body under conditions of high humidity. *Trichomonas vaginalis* has been isolated from swimming pools, baths, and poorly chlorinated water.[5-7] The organism has survived in vaginal exudates at 10°C for up to 48 hours,[8] in voided urine for as long as 3 hours, in semen ejaculates for six hours,[9] and in 35°C water in wet washcloths for up to 24 hours.[10,11] About one-third of contaminated washcloths yielded viable organisms after 2 or 3 hours and 10% even after 24 hours.[11] Viable organisms have been isolated also from vaginal exudates that have been allowed to dry for 3 to 6 hours.[12,13] Whittington[8] was of the opinion that moisture is essential for the survival of the organism. Viability of the urogenital trichomonad has been preserved in Stuart's transport medium at room temperature for up to 3 days and at refrigerator temperature for 9 days.[14]

Trichomonas vaginalis has been found to contaminate toilet seats.[14,15] In a unique experiment, Whittington[14] found that 11 of 30 (37%) infected women left urine on toilet seats after use and that four (36%) of these samples yielded viable trichomonads. Also, after seeding toilet seats with *Trichomonas*-laden vaginal exudate, it was found that the organisms survived from less than 10 to as long as 45 minutes.

Trichomonas vaginalis in serous material is readily disinfected by contact with 0.42 mM hexachlorophene, 0.42 mM bithionol, and 0.93 mM dichlorophene in 24 hours.[16] Organisms in serum-free environments were destroyed within 10 minutes, with 0.0025 mM hexachlorophene. They were destroyed also by heating to 44°C.[10] Vaginal contraceptives of the nonoxynol-9 type have been shown to be trichomonacidal.[17]

Strain Variability

Strain variability expressed in variability of virulence, antigenic properties, and geographic distribution was found by many workers with the aid of tissue culture[18-24] and immunologic[25-32] techniques (for details, see Chaps. 4 and 9). Such variability has clear epidemiologic implications.

Another approach to strain differentiation was taken by Soliman et al.[33] Using isoenzyme analysis and four enzymes, lactate dehydrogenase (LDH), malate dehydrogenase (MDH), hexokinase (HK), and glucose phosphate isomerase (GPI), they identified five different *T. vaginalis* isoenzyme patterns (zymodemes) in 32 parasite isolates. Our own studies using LDH, MDH, and HK in the analysis of electrophoretic isoenzyme patterns of 160 isolates of the urogenital trichomonad, have revealed six major zymodeme groups and 17 distinguishable and reproducible enzyme interaction patterns (mi-

nor zymodeme groups) (Müller and Lossick, unpublished data). These isolates included 40 significantly metronidazole-resistant *T. vaginalis* strains from 26 U.S. states. The remaining strains were isolated from patients seen in a Columbus, Ohio, sexually transmitted disease (STD) clinic. For identification purposes, the zymodemes were labeled by the letters of the alphabet. The distribution of the zymodemes of the metronidazole-sensitive and metronidazole-resistant strains were not statistically different (Fig. 17.1); however, there were racial differences in the distribution of the zymodemes (Fig. 17.2). Metronidazole susceptibility of the isolates under aerobic testing conditions also varied among the zymodeme groups.

The development of formal serotyping methods, possibly combined with isoenzyme analysis, opens new possibilities in the study and understanding of epidemiology of *T. vaginalis* infections. Further refinement of these procedures is necessary and a better understanding of the distribution and stability of the aforementioned markers is needed. Chýle et al[34] recently raised questions about the stability of isoenzyme markers after storage and their interaction with a variety of viruses.

Evolution of the Parasite over Time

It is not known whether genetic or phenotypic changes have occurred in *T. vaginalis* over the last few decades. Surely the worldwide use of

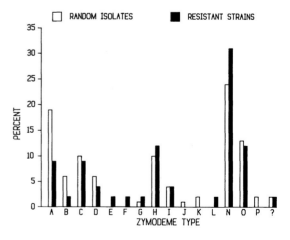

Fig. 17.1. Isoenzyme types of *T. vaginalis* random isolates ($n = 100$); metronidazole-resistant strains ($n = 41$).

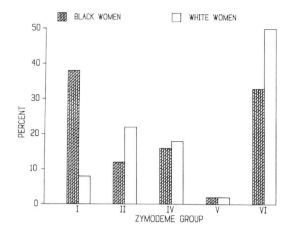

Fig. 17.2. Rates of isoenzyme groups of *T. vaginalis* in black ($n = 60$) and white ($n = 40$) women.

nitroimidazole treatment over the last 25 years has exerted considerable evolutionary pressures on the organism to ensure its survival and perpetuation of the species. Indeed, soon after the advent of metronidazole treatment, drug resistance in *T. vaginalis* emerged.[35] During the next 20 years only sporadic metronidazole-resistant cases were reported in the medical literature.[36,37] In the last 6 years alone nitroimidazole-resistant cases of vaginal trichomoniasis have been reported from Austria,[38] Sweden,[39] the United Kingdom,[40] the United States and Canada[41-44] and Czechoslovakia.[45] The genetic basis for this resistance has not been defined. The isoenzyme patterns in the 40 U.S. strains that we have studied (Lossick and Müller, unpublished data) failed to exhibit significant clustering of specific zymodemes, suggesting that they were probably not related and were independent events. Before 1982 the U.S. Centers for Disease Control (CDC) received only rare phone calls from American physicians regarding metronidazole-resistant vaginal trichomoniasis cases. Of these, 60% were tested for metronidazole susceptibility, and 62% of these (31 cases) demonstrated drug resistance. The racial distribution of these patients was 90% white; 75% were single, separated, or divorced, and the patients were nonpromiscuous.

It is not possible to say whether there has been a real increase in the incidence of metronidazole-resistant cases of trichomoniasis in the last few years. Certainly, the recent recognition by the medical community and the publication of reports of metronidazole resistance account for much of the apparent increase of resistant cases. Although the lack of clustering provides some reassurance the sharp increase in reported cases should at least make us vigilant and monitoring of reported cases should continue, to determine whether a true increased incidence in nitroimidazole resistance actually is occurring.

In our STD clinic program (Columbus [Ohio] City Health Department), in which about 1200 vaginal trichomoniasis cases are treated annually, we estimate the frequency of mildly metronidazole-resistant cases to be 1:100, moderately resistant about 1:400, and strongly resistant greater than 1:1500.

Krieger et al[31] have demonstrated the antigenic diversity of *T. vaginalis* by geographic region of the United States. One of their isolates was an organism isolated in 1963 and it displayed a similar antigenic composition to that of contemporary isolates. This suggests that there is at least some antigenic stability in the trichomonad strains. Because antigenic stability may be a significant factor in potential vaccine development and important to epidemiologic investigations, further studies are needed in this area.

Transmission of Trichomonads

The available evidence indicates that urogenital trichomoniasis is transmitted almost exclusively through coital activities.[9,46-49] Like other sexually transmissible agents that can survive for a time outside the human host (e.g., herpes and gonococcus), there has always been speculation about nonsexual transmission of urogenital infection. As yet, there has been no dependable documentation of this occurring. There remains a theoretical possibility of nonsexual disease transmission that, as of this writing, appears no more likely than that for gonorrhea or genital herpes.

Neonatal trichomoniasis in girls is acquired during passage through an infected birth canal. Theoretically, most of these cases should be self-limited. Although possible, it is unlikely that the organism would persist for years in an asymptomatic state. As a consequence, urogenital trichomoniasis infection in prepubertal children should prompt a search for possible child abuse and must not be dismissed as a congenital infection.

Definitive risk factors for infection other than sexual promiscuity have not been well defined. Socioeconomic factors such as low education and income have been associated with the prevalence rates of trichomoniasis.[50] The prevalence of trichomoniasis in women who use oral contraceptives has been found to be lower than in those who use no contraceptives.[18] Our own data substantiate these findings. This observation and the tendency for symptomatic disease to occur around the menses suggests a hormonal component to susceptibility.

Because of the varied clinical presentations of *T. vaginalis* infections in both men and women, control of this infection is dependent on identification of infection in symptomatic patients, screening of sexually active women, and the treatment of sexual contacts.

Descriptive Epidemiology

Prevalence of Urogenital Trichomoniasis

Our perception of the frequency and distribution of *T. vaginalis* is derived from scattered reports from various clinical settings and demographic populations. Few of these data were designed to provide epidemiologically relevant information. Many are biased by the lack of random sampling and have a bias toward populations likely to be diseased, such as symptomatic patients or persons with other sexually transmitted disorders. True randomly sampled or screening data are the exception rather than the rule, and those that are available are usually dated and poorly representative of the contemporary period. Studies also vary in the sensitivity and specificity of the testing procedures employed. With this in mind, I shall attempt to piece together the descriptive epidemiology of *T. vaginalis* infections and hope that the results will at least resemble reality.

A summary of prevalence data from published studies derived from nonsexually transmitted disease clinical facilities in nine countries is presented in Table 17.1. This is not meant to be all-inclusive, but rather a fair sampling of the data found in the medical literature. For further details on the pre-1975 studies the reader should consult the fine work of Krieger[46] that deals with this topic. There is a significant range in the prevalence of disease from various clinical settings and countries (0% to 65%). Similar data from 19 STD clinics in ten countries are presented in Table 17.2. Like the non-STD data (Table 17.1), the range of disease prevalence is wide. It is not known how representative these data are of the geographic regions served, but one gains the impression that aside from problems of inappropriate sampling, wide differences exist in the prevalence of *T. vaginalis* infections worldwide.

Seroepidemiology

Interpretation of *Trichomonas* culture results is certainly more secure than that of serologic testing. Yet serologic surveys can provide rapid and inexpensive screening of large numbers of

TABLE 17.1. Studies of the prevalence of *T. vaginalis* in women examined in facilities for nonsexually transmitted diseases.

Facility, country, year	No. of patients	Prevalence of *T. vaginalis* (%)	Reference
Data from 26 clinics in Czechoslovakia, England, Japan, Poland, Sweden, and U.S.A., 1942–1974	Median: 403 Range: 46–222,003	Median: 23 Range: 5–65	46
Sex offenders, health department, U.S.A., 1978	237	10.6	61
Juvenile detention facility, U.S.A., 1981	80	48.0	62
Pregnant inner-city girls, U.S.A., 1982	115	33.9	63
Antenatal hospital clinic, Finland, 1984	104	0	64
Factory worker clinic, Poland, 1984	430	4.0	65
Prison, Austria, 1985	397	20.7	66
Various clinics, Bratislava, C.S.S.R, 1975–1982	78,263	9.8	67

TABLE 17.2. Studies of the prevalence of *T. vaginalis* in women examined in facilities for sexually transmitted diseases.

Country, year	No. of patients	Prevalence of *T. vaginalis* (%)	Reference
Data from 9 clinics in the U.S.A., England, and Poland (1954–1980)	Median: 400 Range: 126–1424	Median: 40 Range: 8–88	46
U.S.A., 1977–1978	159	39.6%	68
Sweden, 1979	390	12.0	69
Spain, 1977–1979	345	12.7	70
Zimbabwe, 1979–1980	234	0.9	71
Nigeria, 1977–1981	435	11.0	72
U.S.A., 1981–1982	331	14.0	73
Canada, 1982	128	10.2	74
Kenya, 1982	122	34.4	75
Saudi Arabia, 1984	613	20.1	76
U.S.A., 1985	3984	25.4	Author's unpublished data

people. In addition, they may provide incidence data if the timing of testing is appropriate and repetitive. The sensitivity of the available serologic testing procedures for detecting current and recent (within 12–18 months) infection in women is probably at least as good as direct microscopy for motile trichomonads in active infection, and possibly better. In men, the data are less dependable and more ambiguous; nonetheless, they constitute a valuable adjunct when so few choices are available. The specificity of these procedures appears to be good if appropriate titer cutoff points are used.

Kuberski[51] used an indirect hemagglutination (IHA) procedure to study the recent occurrence of trichomoniasis in men suffering from urethritis. Although he found evidence of *T. vaginalis* by cultivation in only 1.2% of the patients, serologic evidence of recent infection was present in 11% of them. In another study,[52] the same investigator found that 36% of men with a urethritis-prostatis syndrome had trichomonal antibodies and that these antibodies were present in 14% of 146 men with nongonococcal urethritis. Regrettably, he did not test control groups of adequate sizes further to evaluate the relevancy of these findings. Mason[53] using an indirect fluorescent antibody test found that 52% of 200 prenatal patients had serologic titers sufficient to suspect recent *T. vaginalis* infection. In the same study, only 1 of 30 children under 11 years was found serologically positive. With the aid of a similar procedure Mason and Forman,[54] in a study of 451 serum samples of African patients, found seroreactivity of 31% of the blacks and only 17% of the whites, with the highest rates in the 20- to 24-year-old group. Mathews and Healy[55] used an IHA procedure in a random test of 251 female employees of a mental health facility; they found 30% to have antibody titers consistent with recent trichomoniasis. In the same study, 69% of patients attending a vaginitis clinic were found to be seropositive. These workers confirmed the specificity of their results by back adsorbance with *T. vaginalis* antigen of many of the samples.

Risk Factors Associated with *T. vaginalis* Infection

Sexual Activity

As with any sexually transmitted disease, there is little doubt that the absence of coital activities precludes the risk of acquiring trichomoniasis under ordinary circumstances. Most experts agree that there is a relationship between the risk of acquisition of disease and the numbers of different sexual partners to whom a person is exposed[10,47,48]; it is, of course, assumed at the outset that this relationship requires the presence of the infectious agent in at least some of the sexual partners. In reality the risk of exposure to infection not only involves the frequency of new exposures but predominantly relates to the underlying prevalence of disease in those to whom one is exposed. It may be helpful to compare the risk of sexual activity and spread of trichomoniasis to that of gonorrhea. Table 17.3 shows the relationship between the number of different sexual partners during the

TABLE 17.3. Relationship between sexual activity during 30 days preceding the examination and the frequency of gonorrhea and urogenital trichomoniasis.

Disease		No. of sexual partners during 30 days immediately before the examination			Relative risk (95% confidence interval)
		>1	0-1	Total	
Gonorrhea	Present	1,555 (22.9)[a] [32.0][b]	5,250 [26.7][a]	6,805	1.41 (1.30–1.51)
	Absent	3,303 (18.7)[c]	14,406	17,709	
	TOTAL:	4,858	19,656	24,514	
Trichomoniasis	Present	1,503 (21.3)[a'] [31.1][b']	5,552 [28.2][a']	7,055	1.15 (1.08–1.23)
	Absent	3,323 (19.0)[c']	14,151	17,474	
	TOTAL:	4,826	19,703	24,529	
				31,584	

[a-d] Data on gonorrhea; [a'-d'] data on trichomoniasis:
[a] Percent of gonorrhea-infected women who had more than one sexual partner during the 30 days preceding the examination; [a'] the same data for trichomoniasis.
[b] Percent of those who had more than one sexual partner during the 30 days infected with *Neisseria gonorrhoeae*; [b'] the same data for trichomoniasis.
[c] Percent of women without gonorrhea who had more than one sexual partner during the 30 days; [c'] the same for trichomoniasis.
[d] Same as [a], except that the patients had none or one sexual partner during the 30 days; [d'] the same data, but for trichomoniasis.

30 days before testing to the relative risks of acquiring either gonorrhea or vaginal trichomoniasis and is derived from the experiences of 24,500 women seen in our STD program. To obtain a high degree of statistical validity, the population was divided into two groups: those with one or no sexual partners and those with more than one partner during this 30-day period. The relative risk (RR) of disease acquisition was statistically increased in the more sexually active group for both diseases but more so for gonorrhea than for trichomoniasis.

Age

With only a few objections, there is general agreement that the peak incidence of trichomoniasis coincides with the period of maximal heterogeneous sexual activity between 20 and 30 years of age.[10,47] Also, there is evidence indicating that women with trichomoniasis tend to be older than women with gonorrhea.[47] In fact, some workers contend that peak trichomoniasis rates occur in women in their thirties and forties.[10,11] Figure 17.3 shows the age distribution of gonorrhea and *Trichomonas* infection rates in 24,500 patients seen in our STD program. In contrast to gonorrhea, which showed a progressive decrease in frequency with age and heavy disease clustering in the younger age-groups, the frequency of trichomoniasis in all the age-groups was remarkably homogeneous, and showed its peak at the ages of 25 to 29 years.

We also found no statistical differences between the frequency of infection beyond age 30, but the numbers were too small to yield meaningful statistical data. These data confirm previous observations that *T. vaginalis* occurs more often than gonorrhea in older women, but does not suggest that older women are at any higher risk of disease than younger ones. The biologic or social differences that may account for these findings are unclear. Thus, although younger age is a clear risk factor in the acquisition of gonorrhea, younger age is comparatively less of a risk factor in trichomoniasis.

Race

Black women have consistently been shown to have higher rates of trichomoniasis than white women,[10,47,48] with case rate ratios as much as eight-fold higher reported.[11] Figure 17.4 shows the race-specific trichomoniasis and gonorrhea case rates in 24,500 women seen in our STD program between 1980 and 1985. Black race is a clear risk factor for both diseases. Interestingly, race was a stronger risk factor for trichomoniasis (RR = 2.2) than gonorrhea (RR = 1.6). It is not known whether endemic gonorrhea and trichomoniasis in the black population has biologic or sociologic origins. The epidemiologic effect of sexual social segregation on the long-term trends of these endemic diseases warrants further exploration.

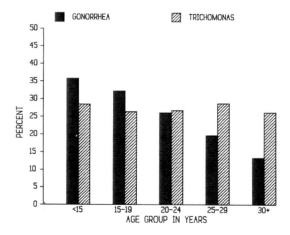

Fig. 17.3. Rates of gonorrhea and urogenital trichomoniasis in different age-groups (n = 29,239).

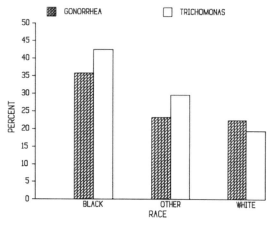

Fig. 17.4. Rates of gonorrhea and urogenital trichomoniasis in black and white patients (n = 24,500).

Contraception

It has been postulated that the use of oral contraceptives (OC) is associated with a decreased risk of trichomoniasis.[56,57] In our program OC use appears to have a substantial effect on the frequency of trichomoniasis. Table 17.4 shows the *Trichomonas* infection rates as related to contraceptive methods in 23,756 patients in our program. The infection rate was highest in persons using no contraception and significantly lower in OC users. The RR for the OC users compared with that of the women using no contraceptives was only 0.49 (95% confidence interval [CI] = 0.46 to 0.53). Thus nonuser women appear to be about two-fold more likely to acquire *Trichomonas* than those who use oral contraceptives. In our program a similar effect is not seen with gonorrhea. The biologic nature of this effect is currently unknown. It has been suggested that high levels of estrogen may not be conducive to *T. vaginalis* infection.[56] If this is true, its effect must be subtle given the homogeneity of disease distribution by age in our patient population. The effect is independent of vaginal yeast (unpublished data) and we have not found a significant variation in promiscuity reflected by contraceptive use patterns.

Previous Sexually Transmitted Infections

Women with trichomoniasis are more likely to have a prior history of sexually transmitted disease. Forty percent of 6838 patients with vaginal trichomoniasis seen in our clinic gave a prior history of STD compared with 25% of 16,912 women who were free from infection. The RR for women with a history versus no history of prior STD to acquire trichomoniasis was 1.96 (95% CI = 1.8 to 2.1). The RR for gonorrhea under the same circumstances was 1.2 (95% CI = 1.12 to 1.27).

To eliminate the effect of confusing the above risk factors, a logistic regression analysis was done, controlling for multiple factors in the model simultaneously. The factors that emerged as predictive of higher risks of disease remained the same: race, nonoral contraceptive use, and prior history of STD.

Interaction with Other Genital Microflora

As indicated by Spiegel in Chapter 10 of this book, "there are relatively few studies that have discussed or examined the bacterial flora associated with trichomoniasis and yet there is a great deal of variability in the results reported." In this connection, certain quantitative data that I have assembled appear to be of epidemiologic interest and therefore are included here (Table 17.5).

It has been reported[47] that urogenital trichomoniasis was associated with gonorrhea between 8% and 50% of the time and that the prevalence of gonorrhea was 1.4 to 1.9 times higher in women harboring *T. vaginalis* than in *Trichomonas*-free patients. The frequency with which *T. vaginalis* is associated with other important sexually transmitted diseases reflects the background frequency of the disease. Frequency of *Neisseria gonorrhoeae* and *Chlamydia trachomatis* in 3507 trichomoniasis cases is shown in Table 17.5. The cases seen in our STD clinic in Columbus, Ohio, were divided also by race: 48% of the black and 42% of the white women had concomitant infections with *Neisseria* and *Chlamydia*. Furthermore, about twice as many black women with either of these agents also harbored *T. vaginalis*. A good correlation was evident between gonorrhea and trichomoniasis, but not between *Chlamydia* and *T. vaginalis*. The frequency of trichomoniasis in patients infected with *Chlamydia* alone was 26.1%, compared with 24.6% in *Chlamydia* and *N. gonorrhoeae*-negative persons in the combined populations (χ^2 = 1.8, P =

TABLE 17.4. Effect of contraceptive methods on *T. vaginalis* rates.

Trichomonas	Contraceptive method				
	IUD	None	Other	Pill	Total
Present	249	4,047	1,053	1,535	6,884
	(30.5)*	(34.5)	(28.1)	(20.6)	(29.0)
Absent	567	7,770	2,696	5,609	16,872
TOTAL:	816	11,747	3,749	7,444	23,756

*Percentages of infected patients in the groups using specific methods and in those using no contraceptive methods.

TABLE 17.5. Urogenital trichomoniasis and distribution of accompanying STDs in black and in white women.

		Accompanying sexually-transmitted pathogens				
Race	Trichomonas	Chlamydia & Neisseria gonorrhoeae	Chlamydia	N. gonorrhoeae	Neither	Total
Black	Present	312 (13.9)[a] [45.2][b]	301 (13.4) [39.8]	455 (20.3) [42.2]	1,175 (52.4) [38.5]	2,243 [40.2]
	Absent	379 (11.4)[c] [54.8][d]	455 (13.7) [60.2]	624 (18.7) [57.8]	1,874 (56.2) [61.5]	3,332 [59.8]
	TOTAL:	691 (12.4)[e]	756 (13.6)	1,079 (19.4)	3,049 (54.7)	5,575
White	Present	156 (12.3)[a'] [26.1][b']	181 (14.3) [16.6]	194 (15.4) [23.0]	733 (58.0) [15.6]	1,264 [17.5]
	Absent	442 (7.4)[c'] [73.9][d']	908 (15.2) [83.4]	648 (10.8) [77.0]	3,978 (66.6) [84.4]	5,976 [82.5]
	TOTAL:	598 (8.3)[e']	1,089 (15.0)	842 (11.6)	4,711 (65.1)	7,240

For black population:
[a]Percent of sexually transmitted infectious agents accompanying *T. vaginalis*.
[b]Percent of sexually transmitted infectious agents in women with *T. vaginalis* and in *Trichomonas*-free patients.
[c]Same as *a* above, but in *Trichomonas*-free women.
[d]Same as *b* above, but in the absence of *T. vaginalis*.
[e]Percent of sexually transmitted infectious agents in all (both *Trichomonas*-positive and *Trichomonas*-negative) patients examined during this study.
For white population:
The footnotes marked *a'* to *e'* refer to the corresponding parameters listed for the black population.

0.18). The RR of having trichomoniasis in association with *Chlamydia* (without concomitant gonorrhea) compared with the situation in which both *Neisseria* and *Chlamydia* were absent equaled 1.09 (95% CI = 0.97 to 1.2). In contrast, women who had gonorrhea without *Chlamydia* showed a trichomoniasis frequency of 33.8%, compared with only 24.6% of patients negative for *Neisseria* or *Chlamydia*. The RR of having simultaneous trichomoniasis and gonorrhea versus trichomoniasis without gonorrhea and *Chlamydia* was 1.6 (95% CI = 1.4 to 1.7). Therefore, although *N. gonorrhea*, *C. trachomatis,* and *T. vaginalis* are transmitted sexually, it is apparent that the risk factors for contracting *Chlamydia* are somewhat different from those involved in infection with gonorrhea or urogenital trichomoniasis; however, the latter two appear similar with regard to these factors.

Seasonality of Trichomoniasis

It has been suggested that *T. vaginalis* infection may have a seasonal pattern with peak periods of infection in the fall of the year, similar to gonorrhea.[10] Figure 17.5 shows the quarterly gonorrhea and urogenital trichomoniasis rates of women seen in our program over a 4-year period between 1982 and 1985. Although not consistently, gonorrhea tended to have peak case rates in the third quarter of the year, but not always statistically different from the others. In contrast, *T. vaginalis* did not appear to be seasonal.

Geographic Distribution of Disease

As discussed earlier, worldwide trichomoniasis data suggest that there may be significant differences in the distribution of disease. Geographic variation in disease distribution is also

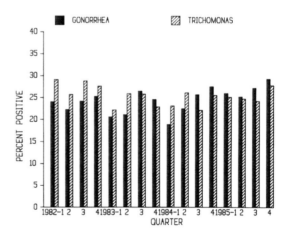

Fig. 17.5. Seasonal rates of gonorrhea and urogenital trichomoniasis in women.

evident at a city level. Figure 17.6 displays the *Trichomonas* case rates among the female patients seen in our clinic during 1985 ($n = 6018$) by home location in the city. There was significant variation in case rates by region of the city. The relative distribution of disease by city region is the same as for gonorrhea. Although the data are biased somewhat by health-care-seeking patterns of the patients, particularly inner-city patients, a similar pattern of gonorrhea exists in the gonorrhea screening programs of the private sector.

Trend of Trichomoniasis

Because urogenital trichomoniasis is not a reportable disease in most countries, accurate disease morbidity and trend data are difficult to obtain. The National Disease and Therapeutic Index (NDTI) is an annual survey of office-based physicians in the United States designed to provide data on a wide variety of health problems. Figure 17.7 shows the unadjusted trend in office visits for trichomoniasis in the United States between 1966 and 1984. It should be noted that between 1965 and 1974 there was a 9% increase in office visits for all medical conditions and that between 1974 and 1980 there was a general decrease in office visits by about 23% as a consequence of unfavorable times.[58]

The number of trichomoniasis cases peaked in 1972 and began a progressive decline in 1975, similar to the pattern seen for gonorrhea.[59] Compared with the peak in disease morbidity in 1972, there has been a 42% decrease in the

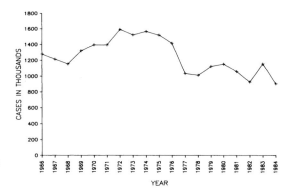

Fig. 17.7. Urogenital trichomoniasis cases seen by U.S. physicians between 1966 and 1984.

number of trichomoniasis cases seen in U.S. private physicians' offices.

On a more local level, Figure 17.8 shows the case rates of vaginal trichomoniasis in our clinic patient population since 1980. Since that year, trichomoniasis rates have declined by 22% and 18% in black and white patients, respectively.

A recent report from Białystok, Poland, has shown a significant drop in trichomoniasis in female residents.[60] There was a progressive decrease in disease prevalence from 19.7% in 1966 to 1.8% in 1984. In contrast to the decrease of disease in Poland and the United States, the annual reports of STD morbidity data in England indicate that there has been no change in the trichomoniasis case rates since 1974. Thus, the

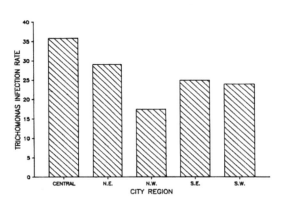

Fig. 17.6. Geographic distribution of urogenital trichomoniasis in Columbus, Ohio (U.S.A.) between 1984 and 1985 ($n = 6,018$).

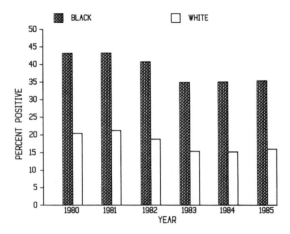

Fig. 17.8. Rates of urogenital trichomoniasis seen in STD clinic at Columbus, Ohio (U.S.A.) between 1980 and 1985.

available data suggest that although the trend in *T. vaginalis* infection rates tends to be downward over the last decade, the trend of the disease in different geographic areas may depart from that pattern.

Summary

The descriptive epidemiology of trichomoniasis is characteristic of a sexually transmissible infection. There are several antigenic varieties of *T. vaginalis* and considerable variation in strain virulence. The infective dose of the organism in women appears small and the attack rate is high, considerably higher than in men. Inapparent disease in common in both sexes and almost the rule in men; a rather large reservoir of "innocent" transmitters of the disease exists. Although a strong host response occurs in infections, it is generally ineffective in preventing recurrent disease.The sequelae historically associated with trichomoniasis probably reflect its coexistence with other important pathogens. Metronidazole-resistant strains have a worldwide distribution, but fortunately are still rare and epidemic-resistant disease has not been documented. Metronidazole resistance appears to evolve from metronidazole pressure in selected cases. *Trichomonas vaginalis* shares many of the risk factors of the other sexually transmitted agents but also differs from both gonorrhea and *Chlamydia* in a number of ways. The available data indicate that trichomoniasis is declining in frequency in some areas. Because of the wide spectrum of clinical disease, control of this infection must rely on continued screening of women and prompt and effective treatment of their sexual partners.

References

1. Donné MA: Animacules observés dans les matières purulentes et le produit des sécrétions des organes génitaux de l'homme et de la femme. *C R Acad Sci (Paris)* 3:385-386, 1836.
2. Hoehne O: Trichomonas vaginalis als häufiger Erreger einer typischen Colpitis prurulenta. *Zentralbl Gynäkol* 40:4, 1916.
3. Trussell RE, Plass ED: The pathogenicity and physiology of a pure culture of *Trichomonas vaginalis. Am J Obstet Gynecol* 40:833, 1940.
4. Hesseltine HC, Wolters JL, Campbell AJ: Experimental human vaginal trichomoniasis. *J Infect Dis* 71:127, 1942.
5. Santler R, Thurner J: Untersuchungen über die Ansteckungsmöglichkeit durch *Trichomonas vaginalis. Wien Klin Wochenschr* 86:46-49, 1974.
6. Santler R, Thurner J, Poitschek C: Trichomoniasis. *Z Hautkr* 51:757-761, 1976.
7. Kozłowska D, Wichrowska B: The effect of chlorine and its compounds used for disinfection of water on *Trichomonas vaginalis. Wiad Parazytol* 22:433-435, 1976.
8. Whittington MJ: The survival of *Trichomonas vaginalis* at temperatures below 37°C. *J Hyg* 49:400-409, 1951.
9. Gallai Z, Sylvestre L: The present status of urogenital trichomoniasis: A general review of the literature. *Appl Ther* 8:773-778, 1966.
10. Jírovec O, Petrů M: *Trichomonas vaginalis* and trichomoniasis. *Adv Parasitol* 6:117-188, 1968.
11. Burch TA, Rees CW, Reardon LV: Epidemiological studies on human trichomoniasis. *Am J Trop Med Hyg* 8:312-318, 1959.
12. Swift BH: *Trichomonas vaginalis* vaginitis as a cause of pruritus vulvae. *Med J Aust* 1:123-132, 1937.
13. Kessel JF, Thompson CF: Survival of *Trichomonas vaginalis* in vaginal discharge. *Proc Soc Exp Biol Med* 74:755-758, 1950.
14. Whittington JM: Epidemiology of infections with *Trichomonas vaginalis* in the light of improved diagnostic methods. *Br J Vener Dis* 33:80-91, 1957.
15. Burgess JA: *Trichomonas vaginalis* infection from splashing in water closets. *Br J Vener Dis* 39:248-250, 1963.
16. Takeuchi T, Kobayashi S, Tanabe M, Fujiwara T: Inhibition of *Giardia lamblia* and *Trichomonas vaginalis* growth by bithionol, dichlorophene, and hexachlorophene. *Antimicrob Agents Chemother* 27:65-70, 1985.
17. Bolch OH, Warren JC: In vitro effects of Emko on *Neisseria gonorrhoeae* and *Trichomonas vaginalis. Am J Obstet Gynecol* 115:1145-1148, 1973.
18. Krieger JN, Ravdin JI, Rein MF: Contact-dependent cytopathogenic mechanisms of *Trichomonas vaginalis. Infect Immun* 50:778-786, 1985.
19. Hogue MJ: The effect of *Trichomonas vaginalis* on tissue culture cells. *Am J Hyg* 37:142-152, 1943.
20. Kotcher E, Hoogasian AC: Cultivation of *Trichomonas vaginalis* Donné, 1837, in association with tissue cultures. *J Parasitol* 43(suppl):39, 1957.
21. Christian RT, Miller NF, Ludovici PP, Riley GM: A study of *Trichomonas vaginalis* in human cell culture. *Am J Obstet Gynecol* 85:947-954, 1963.
22. Kulda J: Effect of different species of trichomonads on monkey kidney cell cultures. *Folia Parasitol (Prague)* 14:295-310, 1967.

23. Farris VK, Honigberg BM: Behavior and pathogenicity of *Trichomonas vaginalis* Donné in chick liver cell cultures. *J Parasitol* 56:849-882, 1970.
24. Kott H, Adler S: A serological study of *Trichomonas* sp. parasitic in man. *Trans R Soc Trop Med Hyg* 55:333-344, 1961.
25. Teras JK: Differences in the antigenic properties within strains of *Trichomonas vaginalis*. *Wiad Parazytol* 12:357-363, 1966.
26. Teras JK, Jaakmees HP, Nigesen UK, Rõigas EM, Tompel HJ: The dependence of serologic reactions on the serotypes of *Trichomonas vaginalis*. *Wiad Parazytol* 12:364-369, 1966.
27. Alderete JF: Antigen analysis of several pathogenic strains of *Trichomonas vaginalis*. *Infect Immun* 39:1041-1047, 1983.
28. Su-Lin K-E, Honigberg BM: Antigenic analysis of *Trichomonas vaginalis* strains by quantitative fluorescent antibody methods. *Z Parasitenkd* 69:161-181, 1983.
29. Torian BE, Connelly RJ, Stephens RS, Stibbs HH: Specific and common antigens of *Trichomonas vaginalis* detected by monoclonal antibodies. *Infect Immun* 43:270-275, 1984.
30. Alderete JF, Suprun-Brown L, Kasmala L, Smith J, Spence M: Heterogeneity of *Trichomonas vaginalis* and discrimination among trichomonal isolates and subpopulations with sera of patients and experimentally infected mice. *Infect Immun* 49:463-468, 1985.
31. Krieger JN, Holmes KK, Spence MR, Rein MF, McCormack WM, Tam MR: Geographic variation among isolates of *Trichomonas vaginalis:* Demonstration of antigenic heterogeneity by using monoclonal antibodies and indirect immunofluorescence technique. *J Infect Dis* 152:979-984, 1985.
32. Mathews HM, Walls KW, Huong AY: Microvolume, kinetic-dependent enzyme-linked immunosorbent assay for amoeba antibodies. *J Clin Microbiol* 19:221-224, 1984.
33. Soliman MA, Ackers JP, Catterall RD: Isoenzyme characterization of *Trichomonas vaginalis*. *Br J Vener Dis* 58:250-256, 1982.
34. Chýle M, Schneiderka P, Štěpán JJ, Chýle P: Some enzyme and isoenzyme activities in *Trichomonas vaginalis* influenced by low temperature preservation and virus infection. *Abstracts Int Symp Trichomonads & Trichomoniasis,* Prague, July 1985, p 20, Prague: Dept Parasitol, Faculty Sci, Charles Univ, 1985.
35. Robinson SC: Trichomonal vaginitis resistant to metronidazole. *Can Med Assoc J* 86:665, 1962.
36. Diddle AW: *Trichomonas vaginalis:* Resistance to metronidazole. *Am J Obstet Gynecol* 98:583-584, 1967.
37. Nielsen R: *Trichomonas vaginalis*. II. Laboratory investigations in trichomoniasis. *Br J Vener Dis* 49:531-535, 1973.
38. Meingassner JG, Thurner J: Strain of *Trichomonas vaginalis* resistant to metronidazole and other 5-nitroimidazoles. *Antimicrob Agents Chemother* 15:254-257, 1979.
39. Forsgren A, Forsmann L: Metronidazole resistant *Trichomonas vaginalis*. *Br J Vener Dis* 55:351-353, 1979.
40. Heyworth R, Simpson D, McNeillage GJ, Robertson DH, Young H: Isolation of *Trichomonas vaginalis* resistant to metronidazole. *Lancet* 2:476-478, 1980.
41. Müller M, Meingassner JG, Miller WA, Ledger WJ: Three metronidazole resistant strains of *Trichomonas vaginalis* from the United States. *Am J Obstet Gynecol* 138:808-812, 1980.
42. Smith RF, DiDomenico A: Measuring the in vitro susceptibility of *Trichomonas vaginalis* to metronidazole. *Sex Transm Dis* 7:120-124, 1980.
43. Lossick JG: Metronidazole. *N Engl J Med* 304:735, 1981.
44. Lossick JG, Müller M, Gorrell TE: In vitro drug susceptibility and doses of metronidazole required for cure in cases of refractory vaginal trichomoniasis. *J Infect Dis* 153:948-955, 1986.
45. Kulda J, Vojtechovská M, Techezy J, Demeš P, Kunzová E: Metronidazole resistance of *Trichomonas vaginalis* as a cause of treatment failure in trichomoniasis. *Br J Vener Dis* 58:394-399, 1982.
46. Krieger JN: Urologic aspects of trichomoniasis. *Invest Urol* 18:411-417, 1981.
47. Rein MF, Müller M: *Trichomonas vaginalis*. In Holmes KK, Mardh PA, Sparling PF, Weisner PJ (eds): *Sexually Transmitted Diseases,* pp 525-535. New York: McGraw-Hill, 1984.
48. Catterall RD: Trichomonal infections of the genital tract. *Med Clin North Am* 56:1203-1209, 1972.
49. Catterall RD, Nicol CS: Is trichomonal infestation a venereal disease? *Br Med J* 1:1177-1179, 1960.
50. Alderete JF, Garza GE: Specific nature of *Trichomonas vaginalis* parasitism of host cell surfaces. *Infect Immun* 50:701-708, 1985.
51. Kuberski T: Evaluation of the indirect hemagglutination technique for the study of *Trichomonas vaginalis* infections, particularly in men. *Sex Transm Dis* 5:97-102, 1978.
52. Kuberski T: *Trichomonas vaginalis* associated with nongonococcal urethritis and prostatitis. *Sex Transm Dis* 7:135-136, 1980.
53. Mason PR: Serodiagnosis of *Trichomonas vaginalis* infection by the indirect fluorescent antibody test. *J Clin Pathol* 32:1211-1215, 1979.
54. Mason PR, Forman L: Serological survey of trichomoniasis in Zimbabwe Rhodesia. *Cent Afr J Med* 26:6-8, 1980.
55. Mathews HM, Healy GR: Evaluation of two serological tests for *Trichomonas vaginalis* infection. *J Clin Microbiol* 17:840-843, 1983.

56. Bramley M, Kinghorn G: Do oral contraceptives inhibit *Trichomonas vaginalis? Sex Transm Dis* 6:261-263, 1979.
57. von Birnhaum H, Kraussold E: Blastomyces and Trichomonas infections during the use of hormonal and intrauterine contraception. *Zentralbl Gynäkol* 97:1636-1640, 1975.
58. Johnson RE, Aral SO, Blount JH, Reynolds GH: A comparative analysis of STD trends in the United States. *Proc 6th Int Meet Int Soc STD Res,* Brighton, England, July 31 to August 2, 1985, p 54, Heidelberg: Int Soc Sex Trans Dis, 1985.
59. Lossick JG: Epidemiology of sexually transmitted diseases. In Spagna VA, Prior RB (eds): *Sexually Transmitted Diseases,* pp 21-58. New York: Marcel Dekker, 1985.
60. Soszka S, Kazanowska W, Kuczynska K: Detectability of trichomoniasis in women treated in Institute of Obstetrics and Gynecology, Medical Academy Białystok, Poland. *Abstracts Int Symp Trichomonads & Trichomoniasis,* Prague, July 1985, p 24, Prague: Dept Parasitol, Faculty Sci, Charles Univ, 1985.
61. Conrad GL, Kleris GS, Rush B, Darrow WW: Sexually transmitted diseases among prostitutes and other sexual offenders. *Sex Transm Dis* 8:241-244, 1981.
62. Bell TA, Farrow JA, Stamm WE, Critchlow CW, Holmes KK: Sexually transmitted diseases in females in a detention center. *Sex Transm Dis* 12:144, 1985.
63. Hardy PH, Hardy JB, Nell EE, Graham DA, Spence MR, Rosenbaum RC: Prevalence of six sexually transmitted disease agents among pregnant inner-city adolescents and pregnancy outcome. *Lancet* 2:333-337, 1984.
64. Paavonen J, Heinonen PK, Aine R, Laine S, Groonroos P: Prevalence of nonspecific vaginitis and of other cervicovaginal infections during the third trimester of pregnancy. *Sex Transm Dis* 13:5-8, 1986.
65. Zrubek H, Szymański CZ, Łopucki M: The incidence of trichomoniasis vaginae in working women—Study based on prophylactic gynaecologic examination. *Abstracts Int Symp Trichomonads & Trichomoniasis,* Prague, July 1985, p 71, Prague: Dept Parasitol, Faculty Sci, Charles Univ, 1985.
66. Dobernig H: Trichomoniasis bei weiblichen Strafgefangenen. *Wien Med Wochenschr* 135:89-91, 1985.
67. Valent M, Klobušický M, Quang LB, Valentová M: The epidemiological situation of urogenital trichomoniasis in recent years in Bratislava. *Abstracts Int Symp Trichomonads & Trichomoniasis,* Prague, July 1985, p 72, 1985.
68. Wallin JA, Thompson SE, Zaidi A, Wong KH: Urethritis in women attending an STD clinic. *Br J Vener Dis* 57:50-54, 1981.
69. Persson K, Hansson H, Bjerre B, Svanberg L, Johnsson T, Forsgren A: Prevalence of nine different microorganisms in the female genital tract: A comparison between women from a venereal disease clinic and from a health-control department. *Br J Vener Dis* 55:429-433, 1979.
70. Perea EJ, Álvarez-Dardet C, Borobio MV, Bedoya JM, Escudero J, Gallardo RM, González-Gabaldón B, deMiguel C, Moreno JC, Pérez-Bernal A, Rodríguez-Pichardo A: Three years' experience of sexually transmitted diseases in Seville, Spain. *Br J Vener Dis* 57:174-177, 1981.
71. Latif AS: Sexually transmitted disease in clinic patients in Salisbury, Zimbabwe. *Br J Vener Dis* 57:181-183, 1981.
72. Bello CSS, Elegba OY, Dada JD: Sexually transmitted diseases in northern Nigeria: Five years' experience in a university teaching hospital clinic. *Br J Vener Dis* 59:202-205, 1983.
73. Short SL, Stockman DL, Wolinsky SM, Trupei MA, Moore J, Reichman RC: Comparative rates of sexually transmitted diseases among heterosexual men, homosexual men, and heterosexual women. *Sex Transm Dis* 11:271-274, 1984.
74. Hill LH, Ruparelia H, Embil JA: Nonspecific vaginitis and other genital infections in three clinic populations. *Sex Transm Dis* 10:114-118, 1983.
75. Mirza NB, Nsanze H, D'Costa LJ, Piot P: Microbiology of vaginal discharge in Nairobi, Kenya. *Br J Vener Dis* 59:186-188, 1983.
76. Omer EF, Catterall RD, Ali MH, el-Naeem HA, Erwa HH: Vaginal trichomoniasis at a sexually transmitted disease clinic at Khartoum. *Trop Doct* 15:170-172, 1985.

18

Therapy of Urogenital Trichomoniasis

Joseph G. Lossick

Historical Background

Trichomonas vaginalis was discovered in 1836[1] and although it was known to cause vaginitis since 1916,[2] a half-century passed before effective treatment was found for this common infection. During the interim period, the infection went untreated, occasionally resolved spontaneously, or it was treated with a variety of intravaginal preparations for long periods of time. These preparations, although helpful for some patients, were cumbersome to use for both patient and clinician, did not eradicate periurethral gland infections, and were unpredictably effective. A comparative study of some of these treatments revealed cure rates of only 22% to 40%.[3] Also there was no treatment for trichomoniasis in the male. It became evident that a systemic treatment for this infection was needed.

Metronidazole

In 1954, the French pharmaceutical laboratory of Rhône-Poulence set out to find a new antitrichomonal agent by screening a variety of antibiotics, antimalarials, and amebicides. From a sample of earth taken from the island of La Réunion, a strain of *Streptomyces* was found that produced a substance with antitrichomonal activity.[4] This compound was found to be azomycin (2-nitroimidazole), a substance previously discovered by Japanese scientists.[5] Through manipulations of the chemical structure of azomycin, metronidazole, or 1-(2-hydroxyethyl)-2-methyl-5-nitroimidazole, was synthesized in 1957, and was found highly effective against *T. vaginalis* infections.[6] It was marketed in France in 1960 under the trade name of Flagyl. The introduction of metronidazole ushered in a new era in effective systemic treatment of *T. vaginalis* infections. The United States Food and Drug Administration delayed the introduction of metronidazole in the United States until 1963 because of initial concerns regarding its potential toxicity. These concerns apparently were related to the presence of a nitro group in the drug that was similar to that found in chloramphenicol and also to reports of leukopenia associated with its use.[7]

Soon after the introduction of metronidazole, another nitroheterocyclic drug of the nitrofuran family, nifuratel, was introduced in the United Kingdom for the treatment of trichomoniasis. In the United States, nifuroxime, in combination with furazolidone (Tricofuron), was marketed as suppositories and powder for the local treatment of the disease. Because of the success of metronidazole, the nitrofurans never achieved widespread acceptance. The second nitroimidazole, nimorazole (Naxogin), was discovered in Italy in 1969.[8] Soon after, tinidazole (Fasigyn)[9,10] was produced. The latter two drugs are still not available in the United States despite more than 15 years of use in the rest of the world. Other more recent nitroimidazole drugs available for the treatment of trichomoniasis worldwide include ornidazole, secnidazole, carnidazole, and misonidazole. In spite of the variety of nitroimidazoles available worldwide, metronidazole remains the most frequently used and has become the standard treatment for *T. vaginalis* infections. To date, it is the only systemic trichomonacide available in the United States.

Trichomonacidal Activity

Because trichomoniasis is a multifocal urogenital infection in women, systemic treatment is necessary to eradicate it. Metronidazole enters the trichomonad by a passive diffusion process driven by the transmembrane cellular gradient, which is determined by the intracellular and extracellular concentrations of the unchanged drug.[11] Inside the cell, its nitro group is reduced to a reactive cytotoxic intermediate that reacts with DNA interrupting nucleic acid synthesis and ultimately resulting in the death of the organism.[12] The intracellular reduction of metronidazole lessens its concentration, facilitating the entry of more drug into the cell. Oxygen inhibits metronidazole uptake by *T. vaginalis*[13] suggesting that the redox milieu of drug-organism interaction may be a contributing factor to the drug's efficacy. This deleterious effect of oxygen on metronidazole uptake is more evident in drug-resistant strains,[14] and the transient cytotoxic intermediate product is more short-lived in these strains.[15] In vitro studies indicate that susceptible trichomonads stop cell division after one hour of exposure to metronidazole, motility ceases at one to two hours, and cell death occurs in 7 to 8 hours.[16]

Pharmocokinetics

Oral absorption of metronidazole is almost complete and the bioavailability is estimated at 93% to 95%.[17] Peak serum levels are reached in 1 to 3 hours. Variation of metronidazole pharmacokinetics between human subjects can be considerable.[18] Postprandial absorption is delayed and peak blood levels attained are somewhat lower than when the drug is taken after fasting.[19,20] Metronidazole is metabolized in the liver by side chain oxidation and glucuronide conjugation into at least five products.[21] Biologically active metabolites are detected in the serum soon after ingestion, increasing from 20% in the first hour, to 66% of the total drug level at 24 hours after a single oral dose.[22] Up to six different nitro-containing metabolites have been identified in the urine of patients receiving the drug.[23] Hydroxymetronidazole and acetylmetronidazole are the two major stable metabolites, both of which (especially the former), retain some degree of biologic activity against susceptible microorganisms.[24] Hydroxymetronidazole is about 20% as active against *T. vaginalis* as the parent compound, whereas the acetylmetronidazole metabolite has marginal biologic activity.[25] There is evidence that in the first few hours after ingestion some of the oxidative metabolites are biologically more active than the parent compound.[26]

Less than 20% of metronidazole is protein bound; about 20% to 30% of the daily dose is excreted in the urine and small amounts in the feces. About 75% and 15% of the total drug dose are excreted in the urine and feces, respectively, within 5 days.[22,27] Small amounts of mutagenic compounds and acetamide (hepatocarcinogen in rats) have also been identified in metronidazole-treated patients.[28,29]

Because of its low protein binding and light molecular weight, metronidazole is well distributed throughout the body. Therapeutic tissue levels have been identified in seminal fluid, pelvic tissues, cerebrospinal fluid, bone, bile, brain, and pulmonary exudates.[30-36] Vaginal secretion levels achieved with metronidazole treatment are currently not well defined. A study of one of my patients with metronidazole-resistant vaginal trichomoniasis who was maintained on high oral doses of the drug revealed vaginal drug levels to be about half of the serum levels (unpublished data on case report).[37] In another study, involving a single patient without clinical disease, vaginal metronidazole levels were similar to those found in serum and saliva.[38] In animal experiments, metronidazole has been shown to penetrate exudate chambers at levels somewhat less than 50% of that seen in the serum.[39] In vitro bacterial inactivation of metronidazole has been demonstrated.[40,41] The clinical relevance of these findings remains to be proved, but it raises questions about the determinants of local levels of trichomonacidal drug activity.

Studies of metronidazole pharmacokinetics indicate that the drug follows the usual linear pharmacokinetics, and that it is eliminated by first-order kinetics.[24] The drug does not accumulate significantly during multiple dose therapy, and a steady state is achieved in about 2 to 3 days.[27] The serum levels achieved after metronidazole administration and the drug half-life vary according to the assay procedures used. Radiolabel and polarographic assays fail to differentiate metronidazole from its biotransformation by-products, and are poor measures of its biologic activity. Chromatography assays, although capable of differentiating metronidazole and its by-products, fail to provide a measure of the biologic potency of the drug levels. Microbiologic assays using organisms other than *T. vaginalis* may give misleadingly high values, because some of the test organisms are more susceptible to metronidazole by-products than is *T. vaginalis*; bioassay with *T. vaginalis* as the test organism would be ideal. It has generally been found that the half-life of metronidazole is about 8 hours (oral and intravenous).[22,27,42] Peak serum levels and the area under the curve are dose proportional. Except for somewhat higher peak levels seen in intravenous treatment, the pharmacokinetics of orally administered metronidazole approximate that of equal dosages given intravenously.[16] Figure 18.1 shows the approximate serum metronidazole levels achieved with various oral or intravenous dosages. These data represent an extrapolation of data from published studies,[25,26,42] and are based on a metronidazole half-life of 8 hours and a linear decay model. Rectal suppository administration of the drug results in delayed peak drug levels, reaching about 50% of those achieved by the intravenous route in about 4 hours; there is a more prolonged plateau level and less pronounced troughs than those seen in intravenous or oral treatment.[42,43] Vaginal absorption of metronidazole has been documented.[42,44,45] Peak levels of about 2 mg/l have been achieved in 12 to 24 hours after a 500-mg vaginal dose, absorption occurring with cream preparations about twofold faster than that following the use of the suppository form.[45]

The serum half-life of a single metronidazole dose is not significantly altered in patients with diminished or absent renal function.[46] In patients with renal failure on multidose metronidazole treatment dialysis removes the drug and its metabolites so that the dosages usually need not be changed.[47-50] However, because of the potential for metabolite accumulation,[49,50] monitoring of serum metronidazole levels may be advisable. Metronidazole's serum concentrations may be higher in patients with enteric disease.[51] The drug's half-life and subsequent serum levels may be decreased by more than 50% by the concomitant use of phenobarbital.[52,53] Cimetidine at dosages of 400 mg twice daily have been shown to decrease metronidazole clearance by about 30%, resulting in elevated drug levels.[54] It is possible that other drugs that affect hepatic enzymes may change the metabolism of metronidazole. Further studies are needed in this area.

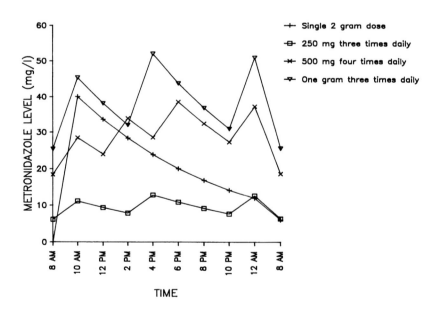

Fig. 18.1. Approximate serum levels of metronidazole after various treatment regimens. Multiple dose regimen values are steady-state levels. (Data taken from Bergman,[25] Wood and Monro,[26] and Mattila et al.[42])

Adverse Drug Reactions

Metronidazole has a wide therapeutic index. Side-effects to metronidazole tend to be dose related and self-limited. Nausea, emesis, or anorexia are the most common side-effects. The etiology of the gastric symptoms have not been well defined, but probably are mediated by both the peripheral and central nervous systems. At normal to moderate metronidazole doses nausea and vomiting are less frequently seen during intravenous therapy, although anorexia is common.[55] It is my experience that at high doses of metronidazole, nausea and vomiting may occur regardless of the dosage route if treatment is continued for more than 3 to 4 days. The drug freely penetrates the central nervous system. Placidi et al[56] found that radiolabeled metronidazole accumulated in the cerebellum and hippocampus in mice. Minor side-effects may be more common in single high-dose treatment than in multidose, multiday treatment. In a placebo double-blind study comparing the treatment efficacy of a single 2-g dose of metronidazole with that of 250 mg administered three times a day for 7 days, Hager et al[57] recorded nausea or emesis in 16% and 8% of the patients, respectively. A metallic taste in the mouth, headache, dizziness, and darkening of the urine are also common dose-related side-effects associated with metronidazole treatment. In another double-blind study that compared placebo with metronidazole at doses of 200 mg three times a day for 10 days, 8% of the metronidazole-treated group complained of one or more of the above symptoms compared with none in the placebo group.[58] In the same study, two of 22 patients (9%) receiving metronidazole developed significant leukopenia that disappeared within a week after termination of therapy. In another study,[59] the authors reported reversible leukocyte depression in 3 of 41 patients (7.3%) treated according to a similar regimen. No leukopenia has been associated with single-dose treatment. There appears to be a significant individual variability in susceptibility to side-effects to metronidazole. Some patients tolerate large doses without problems, whereas others develop symptoms after receiving relatively low doses.

Urticaria is occasionally associated with metronidazole treatment. No research has been done on this phenomenon. In my experience it is benign, not dose related, and unpredictable. In a given patient it may occur during one treatment and not another. One metronidazole-resistant patient in my practice manifested urticaria twice in seven different metronidazole treatment regimens. There have been no reports of anaphylactic reactions associated with metronidazole use. I have had one report of reproducible severe bronchospasm associated with standard-dose metronidazole treatment, which suggests that true allergy to the drug may occur. My policy with regard to patients with a history of metronidazole-associated urticaria is to pretreat them with 50 mg of hydroxyzine every 6 hours for four to eight doses before retreatment with a single 2-g dose of metronidazole. This policy is simply based on anecdotal experience, and further studies are needed.

Reversible peripheral neuropathy, seizures, and ataxia have been reported in patients given large oral and intravenous dosages of metronidazole.[55,59-63] Severe idiosyncratic neurologic dysfunction has occurred with a metronidazole dose of only 250 mg.[64] The available data suggest that central nervous system toxicity occurs more frequently when chronic daily metronidazole dosages exceed 3 to 4 g.

When combined with moderate alcohol intake, metronidazole may produce disulfiram-like effects.[65] Potentiation of anticonvulsants[66] and warfarin activity[67] can occur.

George et al[63] found significant decreases in serum glutamic-oxaloacetic transaminase (SGOT) levels during intravenous metronidazole therapy—in 40% of the patients, the SGOT level dropped to zero. The clinical significance of these findings is not yet clear. Theoretically, metronidazole treatment of patients with hepatic disease may be associated with decreased drug clearance and drug dosages may need to be decreased. Long-term metronidazole treatment has been associated with a high frequency of side-effects. Brandt et al[68] found that 50% of patients with Crohn's disease treated with 1.0 to 1.5 g of metronidazole daily developed distal limb paresthesias after 1 to 14 months of treatment (average, 6.5 months). In most patients the paresthesias disappeared after decreasing the drug dosage; in one patient, they persisted for 23 months after stopping treatment. Jensen and Gugler[69] have shown that the serum half-life of the metronidazole metabolite, hydroxymetronidazole, may be about 1.5 times longer than that of metronidazole and that during long-term metronidazole treatment the clearance of this metabolite decreases. They have suggested that a progressive accumulation of this metabolite may be responsible for the high frequency of paresthesia seen in long-term metronidazole treatment; however, this assumption has not been confirmed.

Mutagenic and Carcinogenic Properties

After over 20 years of widespread use, there is no indication that the standard metronidazole treatment regimens for trichomoniasis are unsafe for use in humans. Although like chloramphenicol, metronidazole contains a nitro group, no serious blood dyscrasias have been reported. Despite this, there is still some equivocation with regard to the drug's long-term safety.

The nitroimidazole compounds and their metabolites have been shown to produce chromosomal alterations in bacteria.[70-72] Rust's studies[73] of susceptible animals exposed to large doses of metronidazole suggested that the drug increased the incidence of mammary tumors. Rustia and Shubik[74] found an increased incidence of mammary, testicular, pituitary, and liver tumors in rats given metronidazole orally. Because the metronidazole-treated animals lived longer than the controls, this study has been criticized for its lack of age-adjustment of the data. Increased rates of hepatocarcinoma, lung tumors, and other tumors have also been noted in some other studies,[74,75] and identifiable animal carcinogens have been recovered in the urine of metronidazole-treated patients.[28] There is little doubt that metronidazole and its byproducts are capable of inducing DNA damage in bacteria, and may increase the incidence of neoplasia in animals. Of major concern, of course, is whether it causes problems in humans. Chromosomal abnormalities have been detected in some patients exposed to high doses of this drug,[76] but these findings could not be replicated by other investigators.[77,78]

Although many of the aforementioned animal studies suffered from some methodologic problems and are not directly extrapolated to humans, without convincing proof of the drug's safety it must be assumed that metronidazole may be a direct or indirect human carcinogen and is capable of causing DNA alterations.

Most animal studies of metronidazole's teratogenicity have been done by the manufacturer and its associates, and have failed to delineate fetal abnormalities in rabbit and rat models.[79] Studies by other investigators have produced conflicting evidence. One worker using rat, mouse, and guinea pig models demonstrated teratogenicity with metronidazole doses corresponding to those used in humans.[80] However, another investigator, who employed even higher doses of the drug in mice, failed to confirm these findings.[81] Thus, only one of nine studies using various experimental animals has shown teratogenic effects. This indicates that if metronidazole is a teratogen, it is probably a very weak one.

The available data indicate that exposure of the developing human fetus to metronidazole during pregnancy entails little to no short-term risks. Studies of infants born to mothers treated with metronidazole during pregnancy have generally failed to show an increase in teratogenesis, low birth rates, prematurity, or excess fetal deaths.[82-85] Another study[86] identified minor congenital abnormalities in 7.3% of the infants born to 55 women treated with metronidazole during their first trimester, and only 2.7% of the infants exposed to the drug in the last two trimesters showed similar abnormalities. There have been no studies to determine the long-term cancer rates in infants exposed in utero to metronidazole.

Recent retrospective studies have addressed the issue of the causal relationship between human carcinoma rates and the use of metronidazole.[87-89] In a study of pharmacy users, Friedman[88] evaluated the carcinoma incidence (over a 3- to 7-year period) in patients exposed to metronidazole compared with those who were not exposed. Except for an excess incidence of cervical cancer, no other increase of cancer incidence was noted in the exposed group. Because the latency period of clinical carcinogenesis may extend up to 30 years, no serious conclusion can be drawn from such a short-term study. Also, type II error associated with accepting the null hypothesis was not identified.

Beard et al[87] retrospectively studied the cancer incidence in 771 women exposed to metronidazole with a follow-up period that averaged 11 years. They concluded that excess cancer rates could not be shown. The control subjects were 237 *Trichomonas*-infected patients not receiving the drug, and the data were also compared with age- and site-specific cancer rates nationally and locally. Of the ten cancer types evaluated, the risk ratio for metronidazole users exceeded 1.0 in nine of them. The individual risk ratios were dismissed as unimportant because of lack of statistical significance. Although type II errors were not presented in the original paper, this issue was addressed later.[90] The power to detect an aggregate two-fold risk in total cancer rates was 97%; in breast cancer 65%, and in all others, 15%. In the metronidazole-treated subgroup of patients followed for at least 10 years, the observed lung cancer rate was 10.5 times that expected (95% confidence interval, 2.2 to 3.1). This obvious increase in lung cancer rates in women with prior exposure to metronidazole, although confounded by a smoking history, is disturbing, particularly in light of the pulmonary neoplasms associated with prior studies of metronidazole in animals. Follow-up data and a later reanalysis of the expected bronchogenic carcinoma risks based on proxy estimates in smokers versus nonsmokers eliminated the increased lung cancer rate in the metronidazole users.[91] This study continues and further analysis of the data may provide more information on this important topic.

A report by Danielson et al[89] that did not detect an excess cancer rate in humans deserves little attention because of its short (2½-year) follow-up period.

The human retrospective studies to date that have been widely quoted as proof for the safety of metronidazole provide only little solace for the serious epidemiologist. The unanswered questions regarding metronidazole's potential for causing human oncogenesis are similar to those confronting the study of environmentally induced carcinogenesis. Long study periods and large patient numbers are needed to define small, but significant, increases in carcinogenicity-related risks. Considering the high prevalence of trichomoniasis, its clinical morbidity, and the lack of alternative treatment for this disease, the benefits of metronidazole treatment appear to outweigh the currently undefined theoretical risks associated with its use. Nonetheless, indiscriminate use of the drug should be discouraged.

It should be noted that two of the aforementioned studies found an unequivocal excess of cervical carcinoma cases in the patients exposed to metronidazole.[87,88] *Trichomonas vaginalis* had been linked to cervical cancer before the widespread use of metronidazole.[92,93] Thus, *T. vaginalis* infection, or some factor associated with infection other than metronidazole exposure, could have accounted for the excess cervical cancer rates.

In summary, there are no data to indicate that metronidazole is a significant human carcinogen. Although studies on humans performed to date indicate that the drug does not have significant carcinogenic properties, the available data do not preclude the possibility that metronidazole may be a weak carcinogen. It is probable that metronidazole's mutagenic activity is related to its antitrichomonal effects[11,94]; thus it is unlikely that other nitroimidazoles will be free of mutagenic potential.

Treatment Regimens

In 1982, I[95] reviewed 41 studies involving 15 different metronidazole treatment regimens for *T. vaginalis* that involved total drug dosages from 1 to 10 g. With the exception of the single 1-g dose, which appeared unacceptable, the remainder of the regimens were uniformly successful in eradicating *T. vaginalis* infections. The most extensively studied metronidazole treatment regimens have been 200 to 250 mg three times daily for 7 days, and the 2-g single dose. These treatments had median cure rates of 92% and 96%, respectively. The 2-g single dose treatment minimizes noncompliance, uses 62% less drug, and is more practical for treatment of the sexual partner. It does not provide protection against prompt reinfection by untreated male sexual partners and is associated with slightly more side-effects. The single 2-g dose appears to be the treatment of choice for the average patient.

Untreated asymptomatic vaginal trichomoniasis may result in acute clinical exacerbations, produce cellular atypia, and serve as a reservoir for continuing disease transmission. Consequently, asymptomatic women with documented *T. vaginalis* should be treated.

Because of the apparently transient nature of infection in men, the need for concomitant treatment of male sexual partners is less clear. Studies have shown relapse rates from 6.2% to 23.7%[82,86,96] in women whose sexual partners were not treated simultaneously. In a double-blind study, Lyng and Christensen[96] reported a reinfection rate of 23.7% in such women, but it was only 5.2% in those women whose partners received treatment. In addition, untreated *T. vaginalis* infections in men may result in chronic infection and continuing disease transmission. The practice of concomitant treatment of male sexual partners of women with trichomoniasis should be continued. In this manner, the risk of reinfection is minimized, symptomatic disease in the male is curtailed, and the pool of asymptomatic male transmitters of disease is diminished.

The treatment efficacy of the other 5-nitroimidazoles in both single and multiple dose treatment regimens has been comparable to that achieved with metronidazole.[96-106]

Based on bacterial activity, there appears to be no antagonism between metronidazole and other antibiotics.[107,108] Naladixic acid, clindamycin, and rifampin actually may potentiate the bactericidal activity of metronidazole. Similar studies have not been done with *T. vaginalis* infections.

Prenatal and Postnatal Treatment

Small quantities of metronidazole cross the placental barrier during pregnancy.[109] Although it is an unlikely teratogenic agent in humans, because of the mutagenic potential of this drug it appears prudent to withhold metronidazole treatment during pregnancy. Regrettably, there are currently no effective systemic alternatives to metronidazole for the treatment of trichomoniasis infections during pregnancy.

Vaginal trichomoniasis cure rates of 41% to 81% have been reported for intravaginal clotrimazole administered at doses of 100 mg for 6 days[110-112] Since I failed to produce culture-documented cures in any of the first few patients I treated with clotrimazole, I have abandoned its use except for palliative treatment. Further controlled studies of this treatment regimen using larger local doses of the drug need to be done, but my preliminary experience with it indicates that it has little value in the treatment of trichomoniasis during pregnancy.

There are few data available with regard to the risk of infection for infants born to women infected with *T. vaginalis*. Trussell et al[113] found two (5%) of 41 female infants born to such mothers to be infected. Because vaginal colonization may be self-limited owing to the transient nature of the maternally derived hormonal effect (3 to 6 weeks), treatment of neonatal infection should be limited to significantly symptomatic infants, or those with persistent infection beyond 6 weeks. Farouk et al[114] found oral metronidazole dosages of 10 to 30 mg/kg daily for 5 to 8 days to be an effective treatment for neonatal trichomoniasis.

After a single 2-g maternal dose of metronidazole, the breast-fed infant would consume about 25 mg of the drug.[115] Breast-feeding women with trichomoniasis should be treated with a 2-g single dose of the drug, and the infant should be removed from breast-feedings for at least 24 hours after treatment.

For additional comments on this subject, see Chapter 13, section V.

Organism Susceptibility and Treatment Outcome

The minimal metronidazole levels required for cure of vaginal trichomoniasis and the determinants of in situ trichomonacidal drug activity have not been well defined. The lack of standardized practical procedures for the in vitro evaluation of *T. vaginalis* susceptibility to metronidazole and its biologically active metabolites has been a major impediment to a better understanding of the trichomonacidal activity of this drug. Simple comparisons of metronidazole blood levels with drug susceptibility in vitro done under standardized laboratory conditions (which may be unlike the biologic conditions) have equivocal clinical predictive value. Despite this, in vitro studies of the organism's susceptibility to metronidazole provide a measure of relative drug susceptibility under standardized conditions. As such, they can be helpful in defining significant departures from the average and serve as a basis for a better understanding of the dynamics of treatment efficacy.

In vitro susceptibility testing of *T. vaginalis* is similar to that used in studies of antibacterial drugs. Comparative drug efficacy is determined by the lowest drug concentration at which the organism's growth inhibition or death occurs. Testing is usually done in cultures of *T. vaginalis* rendered axenic by the use of antibacterial and antifungal agents. Serial dilutions are usually done in test tubes or microwell trays. The minimal inhibitory concentration (MIC) is the lowest drug concentration at which organism multiplication stops, as determined by serial cell counts or culture transfer to antibiotic-free culture media. The minimal lethal concentration (MLC), or trichomonacidal level, is that which causes death as indicated by cell lysis or inability to grow on subculture. Although cell death has been customarily correlated with the loss of the organism's motility,[116] a recent study[117] has shown that this is not always true and that about 13% of immotile organisms may still remain viable. Because *T. vaginalis* is an aerotolerant anaerobic protozoan, most investigators have studied its drug susceptibility under anaerobic conditions. Meingassner et al[118] questioned this practice by showing that aerobic conditions were needed to identify significant in vitro resistance to metronidazole in a strain of *Tritrichomonas foetus* proved resistant to the drug in vivo. This phenomenon was later correlated in *T. vaginalis* by Meingassner and Thurner[119] and other investigators[37,120,121] Until recently, there have been no readily available reference standards for drug susceptibility testing of *T. vaginalis*. Because of a lack of standardization of testing procedures, it is difficult to interpret and compare the susceptibility data published by various investigators. Most published data obtained from tests performed under anaerobic or microaerophilic conditions revealed that the average trichomonad was susceptible to ≤1.0 mg/l of metronidazole, with MIC values up to 16 mg/l.[116,122,123] For the most part, *T. vaginalis* susceptibility studies have been done on random isolates without regard to the clinical picture observed in the individual patients. It was simply assumed that susceptibility values within the serum levels usually achieved with standard dosage regimens were normal. Recently, attempts have been made to correlate prospectively *T. vaginalis* drug susceptibility and treatment outcome.[124,125] Ralph et al[125] using a broth (anaerobic-microaerophilic) testing procedure studied single-dose 1-g and 2-g metronidazole treatments and found a good correlation between decreasing metronidazole susceptibility and the frequency of treatment failures, which was particularly evident with the 1-g single-dose treatment. The level at which increased treatment failures occurred was ≥2.0 mg/l. Müller et al[124] used a microwell anaerobic and aerobic testing procedure to study the correlation between metronidazole susceptibility of *T. vaginalis* and the result of a single 2-g treatment. They also found an increased treatment failure rate in *T. vaginalis* isolates with anaerobic MLC values >1.8 mg/l. Thus it appears that normal anaerobic metronidazole susceptibility for *T. vaginalis* is <3.0 mg/l. This is much lower than was previously assumed using achievable serum levels (about 10 mg/l). Using this testing procedure, these investigators also defined the normal values for aerobic metronidazole susceptibility to be ≤25 mg/l. This is similar to although slightly lower than that postulated by Meingassner and Stockinger.[126] Using a similar testing procedure, these workers found aerobic MLC levels of ≥25 mg/l to be associated with metronidazole-resistant trichomoniasis. These differences may be due to minor variations in the procedures used by the two groups of investigators. In contrast to the conclusions reached by Ralph et al,[125] Müller et al[124] did not show a clear correlation between organism MLC values and treatment outcome until the values exceeded the normal ones of 1.8 mg/l and 25 mg/l, under anaerobic and aerobic conditions respectively. Figure 18.2 shows the distribution of aerobic and anaerobic metronidazole MLC values derived from the latter study[124] and differentiated by treatment outcome. The results were obtained from *T. vaginalis* strains isolated from 124 cured patients, 46 simple treatment failures, and 53 metronidazole-resistant cases. The data reveal considerable overlap in the MLC values of the treatment-susceptible and treatment-insusceptible cases. They indicate also that at anaerobic levels of ≥6.3 mg/l and aerobic levels of ≥200

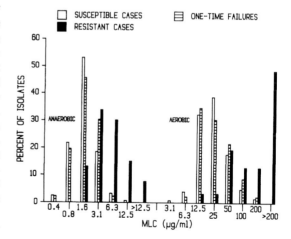

Fig. 18.2. Relationship of minimal lethal concentration (MLC) values of metronidazole to the success of treatment. (From Müller M et al: Metronidazole susceptibility of *Trichomonas vaginalis* and results of treatment. *Proc 83rd Ann Meet Am Soc Microbiol*, New Orleans, March 6-11, 1983.)

mg/l significant clinical resistance is virtually the rule. Some isolates associated with mildly clinically intractable disease had normal MLC values. It is possible that these cases represented treatment failure caused by drug malabsorption or competitive metronidazole inactivation by vaginal bacteria flora. It is evident that the Meingassner susceptibility procedure produces values of comparative susceptibility that are biologically impractical and cannot be readily extrapolated to the required tissue or secretion levels needed for cure. Furthermore, no susceptibility procedure to date has produced values that have been proved to be reliable estimates of the tissue levels required in the human for treatment efficacy. Thus, these values must be viewed as guides to relative drug susceptibilities and not serum or tissue levels needed to be achieved.

Treatment failures should be treated again with similar doses of metronidazole. Follow-up of the treatment failures in Müller's study[124] indicated that 85% were cured with a repeat single 2-g dose of this drug.

Several animal models have been used to test the efficacy of antitrichomonal drugs (see also Chap. 8 for detailed documentation of this topic). Intravaginal T. vaginalis infections have been nonuniformly established in mice,[127] rats,[128,129] and monkeys.[130,131] Intraperitoneal and subcutaneous mouse inoculation assays[127,129,132-135] have proved to be the most successful, and are used now as reference testing procedures. Using a murine model, Meingassner et al[134] demonstrated that subcutaneous infections of T. vaginalis susceptible to metronidazole can be cured with oral doses of the drug of $\leq$ 35 mg/kg given three times daily.[134] Of the in vitro testing procedures, Meingassner's is the only one that has been verified by correlation with in vivo animal testing results.[118,130,134]

Treatment of Drug-Resistant Trichomoniasis

Metronidazole-refractory vaginal trichomoniasis was first reported in 1962 by Robinson.[136] The organism associated with this infection showed decreased in vitro susceptibility to this drug. The patient was ultimately cured by doubling the usual doses. In a prison population, Pereyra and Lansing[137] reported cases of vaginal trichomoniasis that resisted repeated treatment with metronidazole. These infections responded to a repeat treatment of three daily 250-mg oral doses of the drug, supplemented with a 250-mg tablet administered intravaginally for 5 days. The same investigators concluded that an abbreviated 3-day treatment with 250 mg three times daily produced a failure rate of about 30%, and that prolonging treatment beyond 5 to 7 days did not improve cure rates. In 1967, Diddle[138] reported another metronidazole-resistant case of vaginal trichomoniasis. This infection required a 6-week course of vaginal acetarsone treatment plus estrogens for cure. Despite these reports and the in vitro and in vivo induction of metronidazole resistance in laboratory strains of T. vaginalis,[139] a relationship between altered metronidazole susceptibility and intractable clinical disease was not readily accepted. McFadzean et al,[122] studying 25 T. vaginalis strains associated with treatment failures, concluded that the major reasons for the failures were drug malabsorption and its local inactivation by bacteria. These hypotheses were accepted until Meingassner and Thurner[119] demonstrated both in vitro and in vivo nitroimidazole resistance in a T. vaginalis strain isolated from a woman with therapeutically intractable urogenital trichomoniasis. This group of workers identified a disproportionate degree of in vitro resistance to metronidazole if the organism was tested under aerobic rather than the traditional anaerobic conditions used by other investigators. Since then, several other workers have identified metronidazole-resistant T. vaginalis strains associated with treatment-refractory vaginal trichomoniasis.[37,120,121,140-144] Many of these infections were ultimately cured with larger doses of metronidazole. Lossick et al[145] evaluated T. vaginalis strains associated with treatment-intractable cases and found most of these organisms to be from 2-fold to 40-fold more resistant to metronidazole under aerobic conditions, and equal to or 12-fold more resistant under anaerobic conditions than susceptible strains. Of the cases that were followed, 90% were ultimately cured with increased doses of the drug ranging from 6 to 52 g over a 3- to 14-day period. Many simultaneously received metronidazole by the intravaginal route. It appears that metronidazole resistance in T. vaginalis strains is relative, and not an all-or-none phenomenon. Resistance varies from slight to strong, and most cases can be cured with increased dosages of the drug. Treatment dosages in resistant cases should be based on in vitro susceptibility studies of the organism to metronidazole and the patient's clinical response to prior treatments. Adequate drug concentration and duration of treatment are the critical determinants of cure. Concomitant intravaginal treatment, using metronidazole tablets or suppositories, may be helpful in achieving cure in some cases. The average doses of 5-nitroimidazole needed to cure resistant cases based on the aerobic MLC values are shown in Table 18.1. The data are derived from 31 cases of metronidazole-resistant vaginitis that I have followed.[145] I have encountered a few cases (about 8% to 10% of the total) that have proved resistant to the maximum recommended dosage shown in Table 18.1. Some have been cured by extending treatment a few days longer (from 14 to 18 days) and some remain uncured. In an occasional case, I have managed to cure the periurethral gland infection (but not the vaginal infection) with systemic treatment and have cured the vaginal infection with intravaginal treatment alone. My preference has been high-dose oral treatment, but intravenous metronidazole treatment may be required in patients who cannot tolerate the higher oral doses needed for cure. Significant nausea and other benign side-effects are commonly associated with multiple-day treatment that exceeds 2 g daily (even on intravenous treatment), and it often requires a strong commitment on the part of the patient to complete such high-dose treatments. Peak side-effects generally occur at 72 to 96 hours and level off thereafter. My experience with antiemetics in these patients has been limited and disappointing. Perhaps significantly higher doses (4 to 8 g daily) for shorter periods, such as 3 to 4 days, may be tolerated and effective in some of these resistant infections.

Other Antitrichomonal Agents and Their Properties

Other Antitrichomonal Agents

Although metronidazole is the only systemic trichomonacide available in the United States, other highly effective nitroimidazoles are available throughout the world.

Nimorazole, known by some as nitrimidazine, has the chemical formula 1-(*N*-beta-ethylmorpholino)-5-nitroimidazole and is marketed under the trade name of Nagoxin. This drug differs from metronidazole by the presence of morpholine as the basic constituent. It appears to be more rapidly absorbed than metronidazole and its two major metabolites retain significant antiprotozoal activity.

Ornidazole differs from metronidazole by replacement of a hydrogen by a chlorine radical and has the chemical formula A-(chloromethyl)-2-methyl-5-nitroimidazole-1-ethanol. It is marketed under the trade name of Tiberal. After an oral dose, peak serum values are seen in 2 to 4 hours and the half-life is about 14 hours.[22] It may have a higher frequency of side-effects than metronidazole when given in single doses of 1.5 to 2.0 g.[97,146]

Carnidazole, which possesses a thiocarbamate system, has a chemical formula 0-methyl [2-(2-methyl-5-nitro-1-H-imidazole-2-yl) ethyl] carbamothioate. It has been shown to be effective in single-dose therapy[147] and has side-effects similar to metronidazole.

Tinidazole, marketed under the trade name of Fasigyn[9,10] has a chemical structure of i-[2-(ethylsulfonyl)-ethyl]-2-methyl-5-nitroimidazole. It has been shown to inhibit *T. vaginalis* strains at levels similar to those of metronidazole, but may have more cidal activity.[9,148] After

TABLE 18.1. Approximate dosages of metronidazole required for cures of urogenital trichomoniasis.

Aerobic MLC (mg/l)	Resistance level	Total daily dose (g) required	Duration of treatment (days)
<50	Practically none	2.0	1
50	Very low	2.0	3
100	Low	2.0	3–5
200	Moderate	2.0–2.5	7–10
>200	High	3.0–3.5	14*

Selected data from Lossick JG et al: In vitro drug susceptibility and dosages required for cure in metronidazole resistant vaginal trichomoniasis. *J. Infect Dis* 153: 948-955, 1986
*Concomitant intravaginal treatment with 500 to 1000 mg of metronidazole may be necessary.
MLC Minimum lethal concentration.

oral administration, peak levels are attained in 2 to 3 hours, and the drug is widely distributed in body tissues achieving vaginal levels comparable to those attained in the serum.[149] Its half-life is 11 to 14 hours[42,150] and little of the drug undergoes metabolism, excretion occurring in the urine.

Secnidazole is a metronidazole homologue with a half-life of about 17 hours; it has the chemical formula 1-(2-hydroxypropyl)-2-methyl-5-nitroimidazole.[98]

Go 10213, a new nitroimidazole developed by CIBA-Geigy, differs from the currently marketed drugs in having a nitrogen instead of a carbon atom attached to position 2 of the nitroimidazole molecule.[151] Its chemical formula is 1-methanesulfonyl-3-(1-methyl-5-nitroimidazole-2-yl)-2-imidazolidinone and if it proves to be safe for human use it may be a promising drug for treatment of *T. vaginalis* infections.

Nifuratel, a non-nitroimidazole nitrofuran derivative, is used in some areas of the world for the combined systemic and local treatment of urogenital trichomoniasis.[152-154] Although moderately effective, the efficacy and popularity of the nitroimidazoles have precluded its widespread use.

Recently, a novel approach to the treatment of trichomoniasis in the form of *vaccine therapy* (one dose every 2 weeks for a total of three treatments) has become available in some parts of the world. A lyophilizate of selected strains of *Lactobacillus acidophilus* is being marketed as a vaccine for the treatment and prevention of *T. vaginalis* infections by Solco Basle Ltd. in Switzerland under the trade names of Solco-Trichovac and Gynatren. Cure rates after three to four doses of the vaccine approximating those of the nitroimidazoles have been reported[155,156] in open (nonblind) studies, and significantly decreased reinfection rates have been shown in a double-blind study.[157] Some investigators have achieved cure rates as low as 34% with this therapy.[158] Certain workers have found that this vaccine stimulates the production of *T. vaginalis* antibodies.[159,160] Studies involving human antibody responses have been difficult to interpret because of the presence of *T. vaginalis* infection and the lack of unvaccinated controls in studies published to date. The scientific logic behind this approach to treatment is currently poorly defined, especially since natural infection does not appear to confer immunity against reinfection, and well-controlled studies using cultures and blind procedures need to be done. The future value of this approach to treatment is questionable. (Doubts with regard to SolcoTrichovac have been expressed also in Chaps. 4 and 10 of this book.)

Comparative Susceptibility and Cross-Resistance of *T. vaginalis* to 5-Nitroimidazole Drugs

Are some nitroimidazoles better than others in treating trichomoniasis? Comparative clinical studies have failed to show significant differences between the efficacy of the various nitroimidazoles. In one laboratory study, tinidazole was shown to be 4 to 16 times more cidal against *T. vaginalis* than metronidazole, although the static activity of both drugs was similar.[10,148] Some,[161] but not all investigators[98,116,162] have found tinidazole to be more active than metronidazole under in vitro testing conditions. However, most of the above and other investigations[97,163] failed to show significant differences in the in vitro activities of a variety of nitroimidazoles against susceptible *T. vaginalis* isolates. In vivo studies have been more consistent in their findings. In one such study,[163] nitroimidazole doses able to cure 50% of the subcutaneously infected mice were 11.0, 7.5, 34.0, and 11.0 mg/kg, for metronidazole, tinidazole, nimorazole, and ornidazole, respectively. Another study, using a similar animal model,[151] although showing some variability among isolates, indicated the following trend in nitroimidazole activity in descending order: Go 10213, tinidazole, ornidazole, secnidazole, metronidazole, and nimorazole.

Although opinions vary,[164] there appears to be a high correlation between the trichomonacidal activity of the nitroimidazoles and their mutagenic potential.[165] While metronidazole is not the most active of the available nitroimidazoles, its relatively high activity and intermediate mutagenicity may render it the drug of choice for routine treatment of vaginal trichomoniasis.

Susceptibility studies with nifuratel yielded conflicting results. In one study comparable activities for both metronidazole and nifuratel (MLC values about 3 mg/l)[116] were demonstrated. In contrast, other investigators reported nifuratel to be only about one-tenth as active against *T. vaginalis* as metronidazole in both in vitro and in vivo studies.[162]

Meingassner and Thurner[119] showed cross-resistance with other 5-nitroimidazole compounds in the metronidazole-resistant strain they encountered. Although cross-resistance with other nitroimidazoles was noted in their in vitro studies, both tinidazole and nimorazole appeared to be significantly more active in subcutaneous mouse infections than metronidazole. In another study of a metronidazole-resistant strain of *T. foetus*,[166] a similar pattern was observed with tinidazole, ornidazole, and some of the other nitroimidazoles. In still another study a *T. vaginalis* strain was found resistant in vivo and in vitro to secnidazole, ornidazole, tinidazole, and nimorazole, but not to Go 10213.[167] My own studies show that there may be significant variability in the in vitro nitroimidazole susceptibility of metronidazole-resistant *T. vaginalis* isolates, particularly under aerobic conditions. Figures 18.3 and 18.4 show the tinidazole and ornidazole susceptibility values of 16 metronidazole-resistant isolates of *T. vaginalis* that we have evaluated by the Meingassner aerobic and anaerobic susceptibility procedure. It is evident from these data that the most aerobically active drugs against the strains studied in descending order were tinidazole, ornidazole, and metronidazole, with the geometric mean aerobic MLC values of 81.7, 134.7, and 273.1 mg/l, for each of the drugs respectively. Over 50% of the isolates had tinidazole susceptibility values equal to or greater than two dilutions of metronidazole. Under anaerobic conditions, no differences between the drugs were noted.

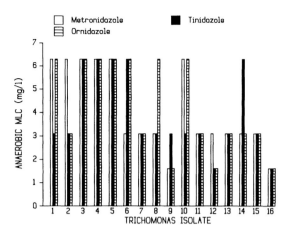

Fig. 18.4. Susceptibility of metronidazole-resistant *T. vaginalis* strains to tinidazole and ornidazole under anaerobic conditions.

The foregoing data confirm Meingassner's initial observation of nitroimidazole cross-resistance for metronidazole-resistant *T. vaginalis*. They suggest also that the cross-resistance may be incomplete and that for some strains, other nitromidazoles such as tinidazole and ornidazole may be more active than metronidazole. Clinical studies using treatment with alternate nitroimidazoles guided by susceptibility testing are needed to confirm these observations and to test further the value of aerobic susceptibility testing.

The choice of treatment for male sexual partners of women with metronidazole-resistant trichomoniasis is not clear. It is our policy to provide the same treatment for the male sexual partner as is used in the infected woman, unless the male is adequately tested and shown to be free of infection. It is possible that the drug may be more effective in men than in women[168,169] and that lower doses may be used.

Summary and Conclusions

The discovery of azomycin and the subsequent synthesis of metronidazole heralded a new era in the treatment of trichomoniasis. Chemotherapy of trichomoniasis has evolved from multiple-day to single-dose treatment and new nitroimidazoles have been introduced throughout the world. Regrettably, the continuing lingering doubts about the oncogenic properties of the nitroimidazole drugs have not served as an impetus to the drug industry to develop new prod-

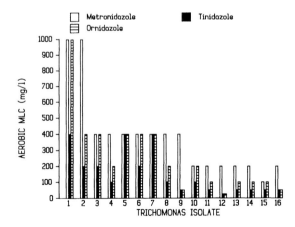

Fig. 18.3. Susceptibility of metronidazole-resistant *T. vaginalis* strains to tinidazole and ornidazole under aerobic conditions.

ucts to treat this very common disorder. Nitroimidazole-resistant *T. vaginalis* infections pose a new therapeutic dilemma for both patient and clinician because of the lack of effective alternative treatment. Fortunately, nitroimidazole resistance is relative, not absolute, ranging from slight to strong. In most cases, treatment resistance can be overcome by using larger doses of the available drugs. The ultimate clinical outcome of these cases is dependent on the susceptibility of the organism and the safety and patient tolerance of the drug doses needed for cure. The limiting factor in these cases is drug toxicity. The emergence of nitroimidazole-resistant *T. vaginalis* strains introduces a new and challenging era in the treatment of trichomoniasis. Efforts to understand better the pharmacodynamic determinants of cure in this disease must be sharply increased. In the short term, therapeutic regimens using available drugs must be innovatively applied in treatment-refractory cases to effect cure. In the long term, there is a need to increase our efforts to find alternatives to the nitroimidazoles for the treatment of trichomoniasis.

References

1. Donné MA: Animacules observés dans les matières prurrulentes et le produit des sécrétions des organes génitaux de l'homme et de la femme. *C R Acad Sci (Paris)* 3:385-386, 1836.
2. Hoehne O: Trichomonas vaginalis als häufiger Erreger einer typischen Colpitis purulenta. *Zentralbl Gynäkol* 40:40, 1916.
3. Clark DH, Solomons E, Siegal SA: Drugs for vaginal trichomoniasis: Results of a comparative study. *Obstet Gynecol* 20:615-620, 1962.
4. Despois R, Pinnert-Sindico S, Ninet L, Preud'homme J: Three antibiotics of different groups produced by the same strain of streptomyces. *J Microbiol* 21:76-90, 1956.
5. Maeda K, Osato T, Umezawa H: A new antibiotic: Azomycin. *J Antibiot* 6A:182, 1953.
6. Cosar C, Julou L: Activité de 1(2-hydroxyethyl)-2-methyl-5-nitroimidazole (8823 R.P.) vis-à-vis des infections expérimentales à *Trichomonas vaginalis*. *Ann Inst Pasteur (Paris)* 96:238-241, 1959.
7. Flagyl, a systemic trichomonacide. *Med Lett* 5:33-34, 1963.
8. DeCarneri I, Cantone A, Emanueli A, Giraldi PN, Logemann W, Meinardi C, Monti G, Nannini G, Tosolini G, Vita G: Nitrimidazine: A new systemic trichomonacide. In *Proc Sixth Int Cong Chemother,* Tokyo, Japan, 1969. In Umezawa H (ed): *Progresss in Antimicrobial and Anticancer Chemotherapy* 1:149-154, Baltimore: University Park Press, 1970.
9. Miller MS, Howes HL, English AR: Tinidazole, a potent new antiprotozoal agent. *Antimicrob Agents Chemother* 1969:257-260, 1970.
10. Howes HL, Lynch JE, Kirlin JL: Tinidazole, a new antiprotozoal agent: Effect on *Trichomonas* and other protozoa. *Antimicrob Agents Chemother* 9:261-266, 1969.
11. Müller M: Action of clinically utilized 5-nitroimidazoles on microorganisms. *Scand J Infect Dis (Suppl)* 26:31-41, 1981.
12. Ings RMH, McFadzean JA, Ormerod WE: The mode of action of metronidazole in *Trichomonas vaginalis* and other micro-organisms. *Biochem Pharmacol* 23:1421-1429, 1974.
13. Müller M, Lindmark DG: Uptake of metronidazole and its effects on viability in trichomonads and *Entamoeba invadens* under anaerobic and aerobic conditions. *Antimicrob Agents Chemother* 9:696-700, 1976.
14. Müller M, Gorrell TE: Metabolism and metronidazole uptake in *Trichomonas vaginalis* isolates with different metronidazole susceptibilities. *Antimicrob Agents Chemother* 24:667-673, 1983.
15. Lloyd D, Pedersen JZ: Metronidazole radical anion generation in vivo in *Trichomonas vaginalis:* Oxygen quenching is enhanced in a drug-resistant strain. *J Gen Microbiol* 131:87-92, 1985.
16. Nielsen MH: *In vitro* effect of metronidazole on the ultrastructure of *Trichomonas vaginalis* Donné. *Acta Pathol Microbiol Scand (B)* 84:93-100, 1976.
17. Houghton GW, Smith J, Thorne PS, Templeton R: The pharmacokinetics of oral and intravenous metronidazole in man. *J Antimicrob Chemother* 5:621-623, 1979.
18. McGilveray IJ, Midha KK, Loo JC, Cooper JK: The bioavailability of commercial metronidazole formulations. *Int J Clin Pharmacol Biopharm* 16:110-115, 1978.
19. Urtasun RC, Sturmwind J, Rabin H, Band PR, Chapman JD: "High dose" metronidazole: A preliminary pharmacological study prior to its investigational use in clinical radiotherapy trials. *Br J Radiol* 47:297-299, 1974.
20. Levison ME: Microbiological agar diffusion assay for metronidazole concentrations in serum. *Antimicrob Agents Chemother* 5:466-468, 1974.
21. Stambaugh JE, Feo LG, Manthei RW: The isolation and identification of the urinary oxidative metabolites of metronidazole in man. *J Pharmacol Exp Ther* 161:373-381, 1968.
22. Schwartz DE, Jeunet F: Comparative pharmacokinetic studies of ornidazole and metronidazole in man. *Chemotherapy* 22:19-29, 1976.
23. Stambaugh JE, Feo LG, Manthei RW: Isolation and identification of the major urinary me-

tabolites of metronidazole. *Life Sci* 6:1811-1819, 1967.
24. Ralph ED, Kirby WM: Bioassay of metronidazole with either anaerobic or aerobic incubation. *J Infect Dis* 132:587-591, 1975.
25. Bergman T: The pharmacokinetics of metronidazole. In Finegold SM (ed): *Proc First US Metronidazole Conf, Tarpon Springs, Florida*, pp 83-111. New York: Biomedical Information Corporation, 1982.
26. Wood BA, Monro AM: Pharmacokinetics of tinidazole and metronidazole in women after single large oral doses. *Br J Vener Dis* 51:51-53, 1975.
27. Ralph ED, Clarke JT, Libke RD, Luthy RP, Kirby WM: Pharmacokinetics of metronidazole as determined by bioassay. *Antimicrob Agents Chemother* 6:691-696, 1974.
28. Speck WT, Stein AB, Rosenkranz HS: Mutagenicity of metronidazole: Presence of several active metabolites in human urine. *J. Natl Cancer Inst* 56:283-284, 1976.
29. Koch RL, Beauliew BB, Chrystal EJT, Goldman, P: A metronidazole metabolite in human urine and its risk. *Science* 211:398-400, 1981.
30. Gray MS, Kane PO, Squires S: Further observations on metronidazole (Flagyl). *Br J Vener Dis* 37:278-279, 1961.
31. Elder MG, Kane JL: The pelvic tissue levels achieved by metronidazole after single or multiple dosing—oral and rectal metronidazole. In Phillips I, Collier J (eds): *Metronidazole: Proc Second Int Symp Anaerobic Infect*, pp 55-58. London: The Royal Society of Medicine and Academic Press, 1979.
32. Davies AH: Metronidazole in human infections with syphilis. *Br J Vener Dis* 43:197-200, 1967.
33. Rood JP, Collier J: Metronidazole levels in alveolar bone. In Phillips I, Collier J (eds): *Metronidazole: Proc Second Int Symp Anaerobic Infect*, pp 45-47. London: The Royal Society of Medicine and Academic Press, 1979.
34. Nielsen ML, Justeson T: Excretion of metronidazole in human bile. Investigations of hepatic bile, common duct bile, and gallbladder bile. *Scand J Gastroenterol* 12:1003-1008, 1977.
35. George RH, Bint AJ: Treatment of a brain abscess due to *Bacteroides fragilis* with metronidazole. *J Antimicrob Chemother* 2:101-102, 1976.
36. Douglas-Smith BJ, Wellingham J: Metronidazole in treatment of empyema. *Br Med J* 1:1074-1075, 1976.
37. Lossick JG: Metronidazole. *N Engl J Med* 304:735, 1981.
38. Davis B, Glover DD, Larsen B: Analysis of metronidazole penetration into vaginal fluid by reversed-phase high-performance liquid chromatography. *Am J Obstet Gynecol* 149:802-803, 1984.
39. Hofstad T, Sveen K: Penetration of metronidazole into preformed cavities in rabbits. *J Antimicrob Chemother* 6:275-278, 1980.
40. Edwards DI, Thompson EJ, Tomusange J, Shanson D: Inactivation of metronidazole by aerobic organisms. *J Antimicrob Chemother* 5:315-316, 1979.
41. Ingham HR, Hall CJ, Sisson PR, Tharagonnet D, Selkon JB: Inactivation of metronidazole by aerobic organisms. *J Antimicrob Chemother* 5:734-735, 1979.
42. Mattila J, Männistö PT, Mäntylä R, Nykänen S, Lamminsivu U: Comparative pharmacokinetics of metronidazole and tinidazole as influenced by administration route. *Antimicrob Agents Chemother* 23:721-725, 1983.
43. Ioannides L, Somyogi A, Spicer J, Heinzow B, Tong N, Franklin C, McLean A: Rectal administration of metronidazole provides therapeutic plasma levels in postoperative patients. *N Engl J Med* 305:1569-1570, 1981.
44. Opperman JA: Comparative disposition of 14-C-metronidazole after its oral and intravaginal administration to female subjects. Program & Abstracts *18th Intersci Conf Antimicrob Chemother*, Atlanta, Georgia, 1978, Abstract 800. Wash DC: Am Soc Microbiol, 1978.
45. Alper MM, Barwin BN, McLean WM, McGilveray IJ, Sved S: Systemic absorption of metronidazole by the vaginal route. *Obstet Gynecol* 65:781-784, 1985.
46. Cerat GA, Cerat LL, McHenry MC, Wagner JG, Hall PM, Gaven TL: Metronidazole in renal failure. In Finegold SM (ed): *Metronidazole: Proc Int Metronidazole Conf*, pp 404-414. Amsterdam: Excerpta Medica, 1977.
47. Gabriel R, Page CM, Collier J, Houghton GW, Templeton R, Thorne PS: Removal of metronidazole by haemodialysis. *Br J Surg* 67:553, 1980.
48. Cassey JG, Clark DA, Merrick P, Jones B: Pharmacokinetics of metronidazole in patients undergoing peritoneal dialysis. *Antimicrob Agents Chemother* 24:950-951, 1983.
49. Guay DR, Weatherall RC, Baxter H, Jacyk WR, Penner B: Pharmacokinetics of metronidazole in patients undergoing continuous ambulatory peritoneal dialysis. *Antimicrob Agents Chemother* 25:306-310, 1984.
50. Kreeft JH, Ogilvie RI, Dufresne LR: Metronidazole kinetics in dialysis patients. *Surgery* 93:149-153, 1983.
51. Bergan T, Bjerke PE, Fausa O: Pharmacokinetics of metronidazole in patients with enteric disease compared to normal volunteers. *Chemotherapy* 27:233-238, 1981.
52. Mead PB, Gibson M, Schentag JJ, Ziemniak JA: Possible alteration of metronidazole metabolism by phenobarbital. *N Engl J Med* 306:1490, 1982.
53. Gupte S: Phenobarbital and metabolism of metronidazole. *N Engl J Med* 308:529, 1983.

54. Gugler G, Jensen JC: Interaction between cimetidine and metronidazole. *N Engl J Med* 309:1518-1519, 1983.
55. Pereyra M, Chipping PM, Noone P: Intravenous metronidazole in the treatment and prophylaxis of anaerobic infection. *J Antimicrob Chemother* 6:105-112, 1980.
56. Placidi GF, Masuoka D, Alcaraz A, Taylor JAT, Earle R: Distribution and metabolism of 14C-metronidazole in mice. *Arch Int Pharmacodyn Ther* 188:168-179, 1970.
57. Hager WD, Brown ST, Kraus SJ, Kleris GS, Perkins GJ, Henderson M: Metronidazole for vaginal trichomoniasis: Seven-day vs single-dose regimens. *JAMA* 244:1219-1220, 1980.
58. Csonka GW: Longterm aspect of treatment with metronidazole (Flagyl) in trichomonal vaginitis. *Br J Vener Dis* 39:258-260, 1963.
59. Rodin P, King AJ, Nicol CS, Barrow J: Flagyl in the treatment of trichomoniasis. *Br J Vener Dis* 36:147-151, 1960.
60. Roe JFC: A critical appraisal of the toxicology of metronidazole. In Phillips I, Collier J (eds): *Metronidazole: Proc Second Int Symp Anaerobic Infect,* pp 215-222. London: The Royal Society of Medicine and Academic Press, 1979.
61. Frytak S, Moertel CE, Childs DS, Albers JW: Neurologic toxicity associated with high-dose metronidazole therapy. *Ann Intern Med* 88:361-362, 1978.
62. Kusumi RK, Plouffe JF, Wyatt RH, Fass RJ: Central nervous system toxicity associated with metronidazole therapy. *Ann Intern Med* 93:59-60, 1980.
63. George WL, Kirby BD, Sutter VL, Wheeler LA, Mulligan ME, Finegold SM: Intravenous metronidazole for treatment of infections involving anaerobic bacteria. *Antimicrob Agents Chemother* 21:441-449, 1982.
64. Dainer MH: Untoward reaction to Flagyl (metronidazole). *Am J Obstet Gynecol* 133:939-940, 1979.
65. Winter D, Stănescu C, Sauverd E: The effect of metronidazole on the toxicity of ethanol. *Biochem Pharmacol* 18:1246-1248, 1969.
66. Iyer KS, Kutty AG: Influence of metronidazole and chloral hydrate on the activity of other drugs. *Indian J Physiol Pharmacol* 18:49-51, 1974.
67. O'Reilly RA: Sterospecific interaction of warfarin and metronidazole. *Fed Proc* 34:259, 1975.
68. Brandt LJ, Bernstein LH, Boley SJ, Frank MS: Metronidazole therapy for perineal Crohn's disease: A follow-up study. *Gastroenterology* 83:383-387, 1982.
69. Jensen JC, Gugler R: Single and multiple-dose metronidazole kinetics. *Clin Pharmacol Ther* 34:481-487, 1983.
70. Voogd CE, Van Der Stel JJ, Jacobs JAA: The mutagenic action of nitroimidazoles. I. Metronidazole, nimorazole, dimetridazole and ronidazole. *Mutat Res* 26:483-490, 1974.
71. Legator MS, Conner TH, Stoeckel M: Detection of mutagenic activity of metronidazole and niridazole in body fluids of humans and mice. *Science* 188:1118-1119, 1975.
72. Conner TH, Stoeckel M, Evrard J, Legator MS: The contribution of metronidazole and two metabolites to the mutagenic activity detected in urine in treated humans and mice. *Cancer Res* 37:629-633, 1977.
73. Rust JH: An assessment of metronidazole tumorigenicity: Studies in mouse and rat. In Finegold SM (ed): *Metronidazole: Proc Int Metronidazole Conf,* pp 138-144. Amsterdam: Excerpta Medica, 1977.
74. Rustia M, Shubik P: Experimental induction of hepatomas, mammary tumors, and other tumors with metronidazole in noninbred Sas: MRC(WI)BR rats. *J Natl Cancer Inst* 63:863-868, 1979.
75. Rustia M, Shubik P: Induction of lung tumors and malignant lymphomas in mice by metronidazole. *J Natl Cancer Inst* 48:721-729, 1972.
76. Mitelmann F, Hartley ASPB, Ursing B: Chromosome aberrations and metronidazole. *Lancet* 2:802, 1976.
77. Mitelmann F, Strömbeck B, Ursing B: No cytogenic effect of metronidazole. *Lancet* 1:1249-1250, 1980.
78. Lambert B, Lindblad A, Ringborg U: Absence of genotoxic effects of metronidazole and two of its urinary metabolites on human lymphocytes in vitro. *Mutat Res* 26:483-490, 1974.
79. Bost RG: Metronidazole: Toxicology and teratology. In Finegold SM (ed): *Metronidazole: Proc Int Metronidazole Conf,* pp 112-117. Amsterdam: Excerpta Medica, 1977.
80. Ivanov I: The effect of "Trichomonicid" on pregnancy in experimental animals. *Akush Ginekol* 8:241-244, 1969.
81. Cella PL: Experimental studies on the teratology of metronidazole. *Riv Patol Clin* 24:529-537, 1969.
82. Sands RX: Pregnancy, trichomoniasis, and metronidazole: A novel dose schedule. *Am J Obstet Gynecol* 94:350-353, 1966.
83. Perl G: Metronidazole treatment of trichomoniasis in pregnancy. *Obstet Gynecol* 25:273-276, 1965.
84. Rodin P, Hass G: Metronidazole and pregnancy. *Br J Vener Dis* 42:210-212, 1966.
85. Morgan I: Metronidazole in pregnancy. *Int J Gynaecol Obstet* 15:501-502, 1978.
86. Peterson WF, Stauch JE, Ryder CO: Metronidazole in pregnancy. *Am J Obstet Gynecol* 94:343-349, 1966.
87. Beard MC, Noller KL, O'Fallon WM, Kurkland LT, Dockerty MB: Lack of evidence for

cancer due to use of metronidazole. *N Engl J Med* 301:519-522, 1979.
88. Friedman GD: Cancer after metronidazole. *N Engl J Med* 302:519, 1980.
89. Danielson DA, Hannan MT, Jick H: Metronidazole and cancer. *JAMA* 247:2498-2499, 1982.
90. Beard MC: Cancer after metronidazole. *N Engl J Med* 302:520, 1980.
91. Beard MC, Noller KL, O'Fallon WM: Metronidazole and subsequent malignant neoplasms. In *Proc 18th Meet Soc Epidemiol Res*, Chapel Hill, NC, June 1985, *Am J Epidemiol* 122:529, 1988.
92. Koss IG, Wolinska WH: *Trichomonas vaginalis* cervicitis and its relationship to cervical cancer. *Cancer* 12:1171-1193, 1959.
93. Simeckova M, Lonser E, Nicholas EE, Rubenstein I: Chronic trichomoniasis and cervical cancer. *Obstet Gynecol* 20:410-412, 1962.
94. Lindmark DG, Müller M: Antitrichomonal action, mutagenicity and reduction of metronidazole and other nitroimidazoles. *Antimicrob Agents Chemother* 10:476-482, 1976.
95. Lossick JG: Treatment of *Trichomonas vaginalis* infections. *Rev Infect Dis* 4(suppl):5801-5818, 1982.
96. Lyng J, Christensen J: A double-blind study of the value of treatment with a single dose tinidazole of partners to females with trichomoniasis. *Acta Obstet Gynecol Scand* 60:199-201, 1981.
97. Sköld M, Gnarpe H, Hillström L: Ornidazole: A new antiprotozoal compound for the treatment of *Trichomonas vaginalis* infections. *Br J Vener Dis* 53:44-48, 1977.
98. Videau D, Niel G, Siboulet A, Catalan F: Secnidazole. A 5-nitroimidazole with a long half-life. *Br J Vener Dis* 54:77-80, 1978.
99. Sawyer PA, Brodgen RN, Pinder RM: Tinidazole: A review of its antiprotozoal activity and therapeutic efficacy. *Drugs* 2:423-440, 1976.
100. Hillström L, Pettersson L, Pálsson E, Sandström SO: Comparison of ornidazole and tinidazole in single-dose treatment of trichomoniasis in women. *Br J Vener Dis* 53:193-194, 1977.
101. Tinkler AE: Nimorazole compared to metronidazole in the treatment of vaginal trichomoniasis. *Practitioner* 212:115-119, 1974.
102. Wigfield AS: Trichomonal vaginitis: A 24 hour regimen of nimorazole compared to a 7-day regimen of metronidazole. *Br J Vener Dis* 51:54-56, 1975.
103. Hayward MJ, Roy RB: Two-day treatment of trichomoniasis in the female: Comparison of metronidazole and nimorazole. *Br J Vener Dis* 52:63-64, 1976.
104. Weidenbach A, Leix H: Treatment of trichomonal vaginitis with a single dose of tinidazole. *Curr Med Res Opin* 2:147-152, 1974.
105. Mahony JDH, Harris JRW, Farrer CJ: Nimorazole and metronidazole in the treatment of trichomonal vaginitis. *Br J Clin Pract* 29:71-72, 1975.
106. Chaudhuri P, Drogendijk AC: A double-blind controlled clinical trial of carnidazole and tinidazole in the treatment of vaginal trichomoniasis. *Eur J Obstet Gynecol Reprod Biol* 10:325-328, 1980.
107. Salem AR, Jackson DD, McFadzean JA: An investigation of interactions between metronidazole ("Flagyl") and other antibacterial agents. *J Antimicrob Chemother* 1:387-391, 1975.
108. Ralph ED, Amatnieks YE: Potentially synergistic antimicrobial combinations with metronidazole against *Bacteroides fragilis*. *Antimicrob Agents Chemother* 17:379-382, 1980.
109. Amon K, Amon I, Muller H: Maternal-fetal passage of metronidazole. In Hejzlar M, Semonsky M, Masak S (eds): *Advances in Antimicrobial Antineoplastic Chemotherapy*, vol 1/1, pp 113-115. Baltimore: University Park Press, 1972.
110. Schnell JD: The incidence of vaginal *Candida* and *Trichomonas* infections and treatment of *Trichomonas* vaginitis with clotrimazole. *Postgrad Med J* 50(suppl 1):79-81, 1974.
111. Legal HP: The treatment of trichomoniasis and *Candida* vaginitis with clotrimazole vaginal tablets. *Postgrad Med J* 50:(suppl 1):81-83, 1974.
112. Lohmeyer H: Treatment of candidiasis and trichomoniasis of the female genital tract. *Postgrad Med J* 50(suppl 1):78-79, 1974.
113. Trussell RE, Wilson ME, Longwell FH, Laughlin KA: Vaginal trichomoniasis: Complement fixation, puerperal morbidity, and early infection of newborn infants. *Am J Obstet Gynecol* 44:292, 1942.
114. Farouk L, Salihi A, Curran JP: Neonatal *Trichomonas vaginalis:* Report of three cases and a review of the literature. *Pediatrics* 53:196-200, 1974.
115. Erickson SH, Oppenheim GL, Smith GH: Metronidazole in breast milk. *Obstet Gynecol* 57:48-50, 1981.
116. Korner B, Jensen HK: Sensitivity of *Trichomonas vaginalis* to metronidazole, tinidazole, and nifuratel in vitro. *Br J Vener Dis* 52:404-408, 1976.
117. Krieger JN, Dickins CS, Rein MF: Use of a time-kill technique for susceptibility testing of *Trichomonas vaginalis*. *Antimicrob Agents Chemother* 27:332-336, 1985.
118. Meingassner JG, Mieth H, Czok R, Lindmark DG, Müller M: Assay conditions and the demonstration of nitroimidazole resistance in *Tritrichomonas foetus*. *Antimicrob Agents Chemother* 13:1-3, 1978.

119. Meingassner JG, Thurner J: Strain of *Trichomonas vaginalis* resistant to metronidazole and other 5-nitroimidazoles. *Antimicrob Agents Chemother* 15:254-257, 1979.
120. Müller M, Meingassner JG, Miller WA, Ledger WJ: Three metronidazole-resistant strains of *Trichomonas vaginalis* from the United States. *Am J Obstet Gynecol* 138:808-812, 1980.
121. Smith RF, DiDomenico A: Measuring the in vitro susceptibility of *Trichomonas vaginalis* to metronidazole. *Sex Transm Dis* 7:120-124, 1980.
122. McFadzean JA, Pugh IM, Squires SL, Whelan JPF: Further observations on strain sensitivity of *Trichomonas vaginalis* to metronidazole. *Br J Vener Dis* 45:161-162, 1969.
123. Nielsen R: *Trichomonas vaginalis*. II. Laboratory investigations in trichomoniasis. *Br J Vener Dis* 49:531-535, 1973.
124. Müller M, Gorrell TE, Lossick JG: Metronidazole susceptibility of *Trichomonas vaginalis* and results of treatment. *Proc 83rd Ann Meet Am Soc Microbiol,* New Orleans, March 6-11, 1983, p 9. Washington DC: American Society for Microbiology.
125. Ralph ED, Darwish R, Austin TW, Smith EA, Pattison FLM: Susceptibility of *Trichomonas vaginalis* strains to metronidazole: Response to treatment. *Sex Transm Dis* 10:119-122, 1983.
126. Meingassner JG, Stockinger K: Untersuchungen zur Identifikation metronidazolresistenter *Trichomonas vaginalis*–Stämme in vitro. *Z Hautkr* 56:7-15, 1981.
127. Meingassner JG: Comparative studies on the trichomonacidal activity of 5-nitroimidazole derivatives in mice infected s.c. or intravaginally with *T. vaginalis*. *Experientia* 33:1160-1161, 1977.
128. Asami K: Effects of metronidazole on *Trichomonas vaginalis* in culture and in an experimental host. *Am J Trop Med Hyg* 12:535-538, 1963.
129. Michaels RM, Peterson LJ, Stahl GL: Substituted dithiocarbamates and related compounds as trichomonacides. *J Pharm Pharmacol* 15:107-113, 1963.
130. Cuckler AC, Kupferberg AB, Millman N: Chemotherapeutic and tolerance studies of amino-nitro-thiazoles. *Antibiot Chemother* 5:540-550, 1955.
131. Street DA, Taylor-Robinson D, Hetherington CM: Infection of squirrel monkeys (*Saimiri sciureus*) with *Trichomonas vaginalis* as a model of trichomoniasis in women. *Br J Vener Dis* 59:249-254, 1983.
132. Benazet F, Guillaume L: Activités antiamibienne et trichomonacide du secnidazole au laboratoire. *Bull Soc Pathol Exot Filiales* 69:309-319, 1976.
133. Richle R, Scholer HJ, Angehrn P, Fernex M, Hummler H, Jeunet F, Schärer K, Schüpach K, Schwartz DE: Grundlagen der chemotherapie von trichomoniasis und amoebiasis mit ornidazole. *Arzneim-Forsch* 28:612-625, 1978.
134. Meingassner JG, Havelec L, Mieth H: Studies on strain sensitivity of *Trichomonas vaginalis* to metronidazole. *Br J Vener Dis* 54:72-76, 1978.
135. Tsai YH, Price KE: Experimental models for the evaluation of systemic trichomonacides. *Chemotherapy* 18:348-357, 1973.
136. Robinson SC: Trichomonal vaginitis resistant to metronidazole. *Can Med Assoc J* 86:665, 1962.
137. Pereyra AJ, Lansing JD: Urogenital trichomoniasis: Treatment with metronidazole in 2,002 incarcerated women. *Obstet Gynecol* 24:499-508, 1964.
138. Diddle AW: *Trichomonas vaginalis:* Resistance to metronidazole. *Am J Obstet Gynecol* 98:583-584, 1967.
139. de Carneri I, Achilli G, Mont G, Trane F: Induction of in-vivo resistance of *Trichomonas vaginalis* to metronidazole. *Lancet* 2:1308-1309, 1969.
140. Forsgren A, Forsmann L: Metronidazole resistant *Trichomonas vaginalis*. *Br J Vener Dis* 55:351-353, 1979.
141. Heyworth R, Simpson D, McNeillage GJ, Robertson DH, Young H: Isolation of *Trichomonas vaginalis* resistant to metronidazole. *Lancet* 2:476-478, 1980.
142. Mattern CFT, Spence MR, Navovitz PC: Properties of metronidazole resistant *Trichomonas vaginalis:* Possible therapeutic approaches. *Program & Abstracts 21st Intersci Conf Antimicrob Agents Chemother,* Chicago, Illinois, 1981. Abstract No. 766, Washington DC: American Society for Microbiology, 1981.
143. Waitkins SA, Thomas DJ: Isolation of *Trichomonas vaginalis* resistant to metronidazole. *Lancet* 2:590, 1981.
144. Kulda J, Vojtechovská M, Tachezy J, Demeš P, Kunzová E: Metronidazole resistance of *Trichomonas vaginalis* as a cause of treatment failure in trichomoniasis—A case report. *Br J Vener Dis* 58:394-399, 1982.
145. Lossick JG, Müller M, Gorrell TE: In vitro drug susceptibility and dosages required for cure in metronidazole resistant vaginal trichomoniasis. *J Infect Dis* 153:948-955, 1986.
146. Matheson I, Hernborg Johannessen K, Bjorkvoll B: Plasma levels after a single dose of 1.5 g ornidazole. *Br J Vener Dis* 53:236-239, 1977.
147. Notowicz A, Stolz E, DeKoning GAF: First experiences with single-dose treatment of vaginal trichomoniasis with carnidazole (R 25831). *Br J Vener Dis* 53:129-131, 1977.
148. Taylor JA, Migliardi JR, Schach von Wittenau M: Tinidazole and metronidazole pharmacoki-

netics in man and mouse. *Antimicrob Agents Chemother* 10:267-270, 1970.
149. Ripa T, Weström L, Mårdh PA, Anderson KE: Concentrations of tinidazole in body fluids and tissues in gynaecological patients. *Chemotherapy* 23:227-235, 1977.
150. Nilsson-Ehle I, Ursing B, Nilsson-Ehle P: Liquid chromatographic assay for metronidazole and tinidazole: Pharmacokinetic and metabolic studies in human subjects. *Antimicrob Agents Chemother* 19:754-760, 1981.
151. Ray DK, Chatterjee DK, Tendulkar JS: Comparative efficacy of Go 10213 and some nitroimidazoles against *Trichomonas vaginalis* and *T. foetus* in mice infected subcutaneously. *Ann Trop Med Parasitol* 76:175-178, 1982.
152. Fowler W, Hussain M: Nifuratel (Magmilor) in trichomonal vaginitis. *Br J Vener Dis* 44:331-333, 1968.
153. Evans BA, Catterall RD: Nifuratel compared to metronidazole in the treatment of trichomonal vaginitis. *Br Med J* 1:335-336, 1970.
154. Pathak UN, Sur SK, Farrand RS: Comparison of metronidazole/nystatin and nitrofuratel in the treatment of vaginitis. *Br J Clin Pract* 29:270-272, 1975.
155. Ripmann ET: SolcoTrichovac in medical practice: An open multicentre study to investigate the antitrichomonal vaccine SolcoTrichovac. *Gynäkol Rundsch* 23(suppl 2):77-84, 1983.
156. Elokda HH, Andrial M: The therapeutic and prophylactic efficacy of SolcoTrichovac in women with trichomoniasis: Investigations in Cairo. *Gynäkol Rundsch* 23(suppl 2):85-88, 1983.
157. Litschgi M: SolcoTrichovac in the prophylaxis of trichomonad reinfection: A randomized double-blind study. *Gynäkol Rundsch* 23(suppl 2):72-76, 1983.
158. Hatala M, Kejda J, Kveton J, Liska M: Solco-Trichovac in the treatment of vaginitis episodes. *Abstracts Int Symp Trichomonads & Trichomoniasis,* Prague, July 1985, p 68. Prague: Local Organizing Comm, Dept Parasitol, Faculty Sci, Charles Univ, 1985.
159. Stojković L, Bossart W: A *Lactobacillus acidophilus* inactivated vaccine active against trichomoniasis in women: Some experimental data. *Abstracts Int Symp Trichomonads & Trichomoniasis,* Prague, July 1985, p 63. Prague: Local Organizing Comm, Dept Parasitol, Faculty Sci, Charles Univ, 1985.
160. Gombošová A, Demeš P, Valent M: Immunotherapeutic effect of the Lactobacillus vaccine SolcoTrichovac in trichomoniasis is not mediated by cross-reacting antibodies against *Trichomonas vaginalis. Abstracts Int Symp Trichomonads & Trichomoniasis,* Prague, July 1985, p 65. Prague: Local Organizing Comm, Dept Parasitol, Faculty Sci, Charles Univ, 1985.
161. Lorgren T, Salmela I: In vitro sensitivity of *Trichomonas vaginalis* and *Candia albicans* to chemotherapeutic agents. *Acta Pathol Microbiol Scand* 86:155-158, 1978.
162. Meingassner JG, Heyworth PG: Intestinal and urogenital flagellates. *Antibiot Chemother* 30:163-202, 1981.
163. Paredes FR, Hawkins DF: Sensitivity of *Trichomonas vaginalis* to chemotherapeutic agents. *J Obstet Gynaecol Br Commonw* 80:86-88, 1973.
164. Bruckner DA, Bueding E, Voge M: Lack of obligatory association between mutagenic and antitrichomonal effects of metronidazole. *J Parasitol* 65:473-474, 1979.
165. Cantelli-Forti G, Aicardi G, Guerra MC, Barbaro AM, Biagi GL: Mutagenicity of a series of 25 nitroimidazoles and two nitrothiazoles in *Salmonella typhimurium. Teratog Carcinog Mutagen* 3:51-63, 1983.
166. Meingassner JG, Mieth H: Cross-resistance of trichomonads to 5-nitroimidazole derivatives. *Experientia* 32:183-184, 1976.
167. Ray DK, Tendulkar JS, Shrivastava VB, Datta AK, Nagarajan K: A metronidazole-resistant strain of *Trichomonas vaginalis* and its sensitivity to Go 10213. *J Antimicrob Chemother* 14:423-425, 1984.
168. Kawamura N: Studies on *Trichomonas vaginalis* in the urological field. Pt. I-V. *Jpn J Urol* 60:15-49, 1969.
169. Kawamura N: Studies on *Trichomonas vaginalis* in the urological field. VI. Clinical efficacy of various drugs. *Jpn J Urol* 64:281-286, 1973.

19

Trichomonads Found Outside the Urogenital Tract of Humans

B.M. Honigberg

Introduction

The trichomonad species found outside the urogenital tract of humans are *Trichomonas vaginalis,* which, under certain circumstances, can establish itself in a variety of sites; *Trichomonas tenax,* found typically in the oral cavity, but presumably also in other sites; *Pentatrichomonas hominis,* inhabiting the large intestine of humans and various other mammals; and *Trichomitus fecalis,* which has been reported from only one person and which may not be primarily a parasite of humans.[1] The last species will not be discussed in this chapter.

Trichomonas vaginalis Donné, 1836

All aspects of this species are discussed throughout the present book. Therefore, only the reasonably well documented cases of *T. vaginalis* infections in sites other than the urogenital tract (especially the lower parts of this tract) are dealt with here.

An early record of *T. vaginalis* as the causal agent of chronic recurrent bronchitis in a man was published by Rebhun.[2] The identification of the flagellate was based primarily on the case history and not on morphologic characteristics or cultivation. The disease was traced to the practice of cunnilingus by the patient with his presumably infected sexual partners before coitus. Also Grein et al[3] described lesions discharging foul-smelling purulent material from the mouth of a man who engaged in cunnilingus. The lesions, located on the tongue, palate, and mucosa of the lips, were caused by a flagellate diagnosed in the Department of Odontology, Federal University of Paraná, Brazil. Oral treatment with metronidazole was fully effective. According to a more recent report,[4] *T. vaginalis* has been recovered in a number of cases in which persons practiced cunnilingus or fellatio with sexual partners who had trichomonal vaginitis or urethritis; the flagellates were found in throat swab material. In such cases, the mucosa showed strawberry inflammation from vasodilation and the lesion surfaces were covered with exudate rich in polymorphonuclear leukocytes and mononuclear cells. Examination of fresh preparations in warm saline with methylene blue revealed motile trichomonads. Diagnosis could be made also by cultivation. An interesting example of such an oral infection was described by Brenner and Simon,[5] who found sheets of motile flagellates in wet fresh preparations of scrapings of the tongue of a patient with a sore throat, dysgeusia, and severe glossitis. The patient admitted having multiple orogenital contacts with a new sexual partner. Complete cure was achieved after treatment with metronidazole. On the basis of epidemiologic and certain biologic evidence, for example, relative survival times of *T. tenax* and *T. vaginalis* outside the host, the authors concluded, undoubtedly correctly, that the flagellates in question were the urogenital trichomonads. They allowed, however, that the patient could have had "an anaerobic infection of the tongue" and that the *Trichomonas* was simply a commensal associated with this infection. They added that if this were so, the presence of the trichomonad could be used as a signal of such anaerobic infections; the treatment regimen would remain unaffected. Therefore they suggested that fresh preparations of tongue scrapings ought to be made in cases of sexually active patients suffering from glossitis.

There are also some reasonably well documented cases of bronchitis, pneumonia, and rhinitis in newborn babies. A case of trichomonal pneumonia was reported by Hiemstra et al,[6] who also found *T. vaginalis* in the gastric contents of two babies with respiratory distress. The protozoa were identified by microscopic examination of fresh preparations. The babies responded well to treatment with metronidazole. Culture tests for bacteria were negative, but the patient with pneumonia responded to treatment with antibiotics. Although the authors believed that the respiratory tract disease was caused by *T. vaginalis*, they thought that "bacterial pneumonia was also a possibility." In another study,[7] two newborns out of 1366 examined during 5 years were found to suffer from *Trichomonas*-caused pneumonia and rhinitis. The mothers of these babies had intermittent trichomonad vaginitis. In both cases cultures for bacteria and viruses were negative, but trichomonads were found in various kinds of cell cultures used for growing viruses, in which the trichomonads caused typical cytopathic changes. In one instance, that of pneumonia, the flagellates were identified as *T. vaginalis* by experts in the Centers for Disease Control (CDC), Atlanta, Georgia. When inoculated intraperitoneally into mice, these organisms caused large hepatic lesions characteristic of virulent strains of the urogenital trichomonad. Although the organisms from the case of rhinitis were not identified by CDC or inoculated into mice, they were structurally identical with those from the case of pneumonia. The authors[7] report that *T. vaginalis* pulmonary infection appears to be uncommon. However, they conclude that this parasite "is capable of infecting the mouth and respiratory tract of infants and the possibility that . . . on occasion, it may cause respiratory disease needs further evaluation."

According to Kurnatowska and Komorowska (see Chap. 13), the respiratory tract infections caused by *T. vaginalis* in newborns are acquired from infected mothers during birth.

A trichomonad, presumably *T. vaginalis*, as diagnosed by microscopy and in some instances by cultivation, was found in the renal pelvis.[8-11] For example, Suriyanon et al[11] described *T. vaginalis* isolated from a 58-year-old woman with chronic trichomonad vulvovaginitis and with obstructive uropathy related to renal calculi. The patient developed a perihepatic abscess after trauma caused by a vehicular accident. The trichomonad was noted in smears and in cultures obtained from the purulent drainage of the abscess. The authors suggested that the presence of *T. vaginalis* reflected ascending infection by the parasite. That *T. vaginalis* is found on occasion in purulent material in the fallopian tubes of women suffering from urogenital trichomoniasis and is thus capable of invading the upper parts of the urogenital tract appears to be documented in a number of published reports.[12-14]

The occurrence of *T. vaginalis* in unusual sites, especially in the respiratory tract, is of significance in discussing the invasion of respiratory passages by allegedly virulent strains of *T. tenax*, a topic considered in the following section.

Trichomonas tenax (O.F. Müller, 1773)

Oral Cavity

Geographic and Host Distribution

Although studied mostly in Europe and in the United States, as discussed later in the chapter, *T. tenax* is a cosmopolitan parasite of the human oral cavity. In addition to humans, certain nonhuman primates have been reported as natural hosts of this species (for relevant references see Ref. 15,16). Aside from epidemiologic surveys of oral cavity trichomonads of various nonprimate animals, studies of the protozoa in hosts other than humans have been scarce[17](see also reference 18 for relevant citations). Indeed, despite the fact that in some instances the oral trichomonads harbored by various hosts may be closely related, to date the possibility that other animals may serve as reservoirs of the human *T. tenax* has not been seriously entertained.

Sites in the Human Host

Trichomonas tenax has been reported from a variety of sites in or closely connected with the oral cavity.[17,19-24] No attempt will be made to indicate all the specific sites that have been mentioned in the literature, but I shall try to list as many as practicable, especially in connection with the pathologic changes found in these sites that appear to favor development of the oral trichomonad. As might be anticipated all the sites are in the periodontium and adjacent areas. Among them are alveolar gingivae, periodontal or gingival pockets, interdental papillae, maxillary sinuses, and paranasal sinuses; under certain circumstances, the ear or throat can also serve as a site of *T. tenax*.

Immunologic Aspects

Information on the antigenic properties and other immunologic aspects of the oral flagellates is limited. The available data were detailed by Honigberg[1]; they are summarized only very briefly here. Furthermore, certain facts pertaining to the antigenic properties of the O. F. Müller species are considered in connection with the possible existence of its strains that have been at various times associated with invasion of the respiratory organs of humans.

Tokura,[25] using flagellates grown in xenic cultures and sera developed in rabbits by intravenous inoculation of formalin-fixed trichomonads, observed stronger cross-agglutination between *T. tenax*, on the one hand, and either *P. hominis* or *T. vaginalis*, on the other, than between the latter two species. These results suggested to the author that, antigenically, the oral flagellates occupied a position intermediate between the intestinal and the urogenital trichomonads of humans. Unfortunately, the very low titers obtained by Tokura render the evaluation of his results uncertain.

Subsequently, Kott and Adler[26] used xenic cultures of three strains of *T. tenax* in agglutination experiments involving sera developed in rabbits by intravenous inoculation of living flagellates grown in xenic cultures and adsorbed with the bacteria present in the cultures. Results of agglutination experiments suggested that two of the strains were more closely related to each other than either was to the third. None of the strains tested seemed to share antigens with *P. hominis* or *T. vaginalis*. However, Oleinik,[27] using agglutination, reported *P. hominis* and *T. tenax* to be antigenically identical, but to differ from *T. vaginalis*. The difficulties with the results reported by the latter worker might have been due to the fact that of the species she studied, only *T. vaginalis* was grown in axenic culture. The same criticism can be leveled at the results of Kott and Adler,[26] who employed xenic cultures of *T. tenax*.

Stępkowski and Honigberg,[28] using gel diffusion and immunoelectrophoresis, observed common and unique antigens in *T. tenax, T. vaginalis, T. gallinae, Tetratrichomonas gallinarum,* and *P. hominis*. They separated the last two species from the members of the genus *Trichomonas* on the bases of the relative proportions of thermostable (resistant to a 30-minute exposure to 70°C) and thermolabile (inactivated by a 30-minute exposure to 70°C) antigens. In the genus *Trichomonas,* the thermostable antigens were either predominant, as in *T. tenax,* or about as abundant as the thermolabile ones, as in *T. vaginalis* and *T. gallinae*. In contrast, thermolabile antigens were more abundant in *Tetratrichomonas* and *Pentatrichomonas*.

In an abstract Rõigas et al[29] reported that results of agglutination and complement fixation tests revealed the presence of common and unique antigens in *T. tenax, T. vaginalis,* and *P. hominis*. Furthermore, each of these species contained various antigenic types, *T. tenax* having four (A, B, C, and D).[30] However, using sera from naturally infected persons and from rabbits, these workers observed that the trichomonads from the oral cavity appeared to differ in their antigenic composition from the flagellates isolated from the respiratory passages of persons suffering from nonspecific bronchitis. Furthermore, antigenic differences evidently could be demonstrated in single individuals between trichomonads infecting their oral cavity and bronchi. The aforementioned findings suggested a need for further immunologic studies to ascertain if the trichomonads from the two sites represented merely different antigenic types of *T. tenax* or if the parasites from the respiratory passages actually belonged to a separate species. In light of the findings of *T. vaginalis* described in the preceding section, the

Biochemical Attributes

Only limited published information is available on the biochemical attributes of the oral trichomonad.[31,32] There is also an unpublished dissertation by Probst on the physiology and nutrition of *T. tenax*.[33] However, the results included in this dissertation deal with axenic cultivation of the parasite and with some factors influencing its growth, rather than with its general physiologic and biochemical attributes. Wantland et al[31] used *T. tenax* grown in xenic cultures in the cytochemical Feulgen reaction for DNA; the periodic acid–Schiff and Best's carmine–Harris' hematoxylin procedures for glycogen; the Spirit (aniline blue) technique of Lillie and Lasky for lipids; and Romien's modification of the Liebermann-Burchardt method for cholesterol (for references to the methods, see reference 31). The reaction for DNA was demonstrated in the "nuclear wall" chromatin, in the chromatin network distributed throughout the nucleus, and in the "tripartite endosome." Glycogen was found in the form of large ovoid or irregularly shaped clear areas, "glycogen vacuoles," ranging from 2 to 3 μm in their greatest diameter, and containing purple-red to bright red granules. Strains differed in the levels of stored glycogen, which decreased during division. Treatment with Best's carmine and Harris' hematoxylin stains imparted a dark blue color to chromatic granules (hydrogenosomes), costa, pelta, and nuclear structures. Lipids appeared as small grayish blue rounded areas, located most frequently toward the posterior part of the cell; the reaction for lipids was localized also in the parabasal body (Golgi complex) and in the axostyle. Chromatic granules, nuclei, the kinetosome complex, as well as cytoskeletal structures such as the costa and pelta, stained purple-red. Cholesterol accumulations were localized in pale violet or red spherical regions, with diameters ranging from 1 to 1.5 μm; these areas tended to turn blue-green in a few minutes.

Acid phosphatase was demonstrated by Ohashi[32] with the aid of Barka and Anderson's modification[34] of Gomori's procedure. By this method a strong reaction for the enzyme could be localized in the perinuclear area; a moderate reaction was present in vacuolar bodies near the nucleus and in some vesicles, presumably phagosomes, noted near the cytoplasmic membrane. Furthermore, a weak reaction was observed in the cell membrane, but none in the nucleus. The parabasal body appeared to be positive for the enzyme. Acid phosphatase activity is expected in phagosomes, but its significance in many of the other sites remains enigmatic.

Admittedly, the dependability of cytochemical results obtained with trichomonads by employing techniques devised for metazoan cells has been questioned. For comparison with the biochemical data available on other trichomonads, the biochemical attributes of *T. tenax* will have to be studied with the aid of these methods.

Pathogenicity

Human Hosts

The inherent pathogenicity of *T. tenax* has not been generally accepted by parasitologists and clinicians. There appear to be valid arguments on both sides of this issue. Those who believe in pathogenicity of the oral trichomonad point to the fact that although prevalence of this parasite as high as 7.5% has been demonstrated in persons with evidently healthy mouths,[35] this prevalence is invariably much higher in persons who have diseases of the oral cavity. Jaskoski[36] found the prevalence of *T. tenax* to be 66% in inhabitants of Illinois (U.S.A.), who exhibited "poor oral conditions," and Musaev[37] estimated the frequency of this trichomonad at 74% among the citizens of Azerbaidzhan, U.S.S.R., suffering from inflammatory dystrophic periodontosis. Furthermore, Abramova and Voskresenskaia[35] reported a 43% prevalence of *T. tenax* in patients with periodontosis suffering from skin and/or venereal diseases, and Čatár et al[38] found the highest prevalence in persons with periodontosis, but a lower one in those suffering from gingivitis. The correlation between oral diseases is also evident from the data included in the reports published by investigators from several countries.[24,37,39-42] Some of these data are shown in Tables 19.1 to 19.4. It must be remembered that the actual

TABLE 19.1. Dependence of the prevalence of *T. tenax* and *Entamoeba gingivalis* on pathologic changes in the oral cavity (data from Prague and Zlin, Czechoslovakia).

Groups representing sets of conditions in the oral cavity	Prague			Zlin	
	Number examined	*T. tenax* (%)	*E. gingivalis* (%)	Number examined	*T. tenax* (%)
I*	140	3.6	8.0	24	12.5
II[†]	169	22.0	46.0	168	20.2
III[‡]	60	21.0	50.0	53	34.0
IV[§]	131	32.0	60.0	29	38.0

Modified from Jírovec O: Contributions à la microbiologie de la cavité buccale humaine. *Paradentologie* 2:117-123, 1948, with permission of l'Arpa Internationale, Association pour les Recherches sur les Parodontopaties, Geneva.
*Healthy and well-cared-for dentition and oral cavity.
[†]Minor dental caries, or some dentures, or abnormal occlusions, or small amount of dental calculus and some gingivitis without gingival pockets.
[‡]Advanced dental caries; numerous exposed dental roots; abnormal occlusions, dentures, false teeth, but no severe gingivitis.
[§]Severe periodontosis; acute gingivitis with suppurating gingival pockets; much dental calculus; poor oral hygiene and grossly neglected dentition.

numbers included in these tables depend, to a considerable degree, on the diagnostic methods employed by the various investigators, that is, fresh preparations, stained preparations, cultures, and combinations of these techniques. Of interest regarding correlation with oral disease and prevalence (see section on epidemiology below) is the diagnostic method used by Čatár et al.[38] These investigators employed cultivation in three different media of saliva from 200 persons attending the Stomatologic Clinic of the Medical Faculty, University of Bratislava. Examinations of the cultures after 3 days of incubation revealed the parasites in over 90% of the 52 positive cases in which the oral trichomonad was diagnosed by cultivation in the same media of materials obtained from "teeth and periodontium, gingival sulci, and from . . . periodontal pockets." Because I find it difficult to accept uncritically the idea of inherent

TABLE 19.2. Dependence of the prevalence of *T. tenax* on pathologic changes in the oral cavity (data from Clermont-Ferrand, France).

Conditions in the oral cavity	Number examined	Individuals positive for trichomonads	
		No.	%
Intact	25	7	28
Three or more occluded	169	51	30
Three or more caries	71	19	27
Three or more roots	28	14	50
Ten or more missing teeth	189	57	30
Tartar (calculus)			
Absent	123	28	23
Present	386	127	33
Gingival disease			
Absent	243	61	25
Present	266	93	35
Periodontal disease			
Absent	347	96	27
Present	152	58	38

Modified from Cambon M, Pétavy A-F, Guillot J, Glandier I, Deguillaume J, Coulet M: Etude de la fréquence des protozoaires et des levures isolés du parodonte chez 509 sujets. *Pathol Biol* 27:603-606, 1979, with permission of Société d'Édition de l'Association d'Enseignement Médical des Hôpitaux de Paris.

TABLE 19.3. Dependence of the prevalence of *T. tenax* on conditions in the oral cavity (data from Bloomington, Indiana, U.S.A.).

Conditions in the oral cavity	Individuals with trichomonads (%)
Amount of dental calculus	
Slight	9
Medium	30
Heavy	35
Condition of tongue surface	
Clean	18
Coated	45
Conditions of periodontal tissue	
Clean	11
Dirty	33
Pyorrhetic	50

Modified from Wantland WW, Lauer D: Correlation of some oral hygiene variables with age, sex, and incidence of oral protozoa. *J Dent Res* 49:293–297, 1970, with permission of the American Association for Dental Research, Washington, D.C.

pathogenicity of *T. tenax*, the following discussion attempts to present arguments for and against this idea.

A relevant discussion of some ecologic aspects of the oral microorganisms, including protozoal and bacterial species, for example, *T. tenax* and the organisms involved in causing necrotic ulcerative gingivitis (Vincent's angina), has been presented by Blake.[43] His contribution contains interesting data dealing with interactions of the oral protozoal, bacterial, and mycotic species with one another and with the environment. It is pointed out also that the environment can be altered by the metabolic products of the microorganisms. Blake considers the ways in which these interactions can cause the appearance of pathologic states. The question of the relationships between *T. tenax* and bacteria and fungi, for example, gram-positive and gram-negative rods, diphtheroids, and some fungi, was discussed also by Abramova and Voskresenskaia.[35]

As stated earlier in this chapter, the oral trichomonad, although found in healthy mouths, is far more prevalent in patients showing abnormal changes, especially in those with characteristics of periodontal diseases. Furthermore, Teras et al (1972, as cited in reference no. 35) asserted that *T. tenax* cultivated in his axenic medium[99] caused trichomonal peritonitis with its typical lesions in white mice inoculated by the intraperitoneal route. However, these results could not be verified by other workers who at that time did not have media suitable for the axenic cultivation of the oral trichomonad (e.g., De Carneri, Oleinik, and Tumka, as cited in reference 35). A survey of the literature reveals that opinions about the pathogenicity potential of *T. tenax* range widely. Some investigators considered this flagellate to be a true pathogen (see the preceding discussion; also references 21 and 44), others thought it to be a commensal,[45] while still others believed the oral flagellate was an opportunistic saprophyte[46] or an irritant favored by and favoring congenital and acquired diseases of the oral cavity.[47] However, Cambon et al[20] thought that congenital "dento-maxillo-facial" abnormalities played no role in the initiation of oral trichomoniasis.

Musaev,[37,48-50] unable to obtain axenic cultures of *T. tenax*, was unwilling to express a definite opinion about the inherent pathogenicity of these flagellates. He believed, rather, that the

TABLE 19.4. Dependence of the prevalence of *T. tenax* on pathologic changes in the oral cavity (data from Kirobad, Azerbaidzhan, U.S.S.R.).

		Individuals positive for trichomonads	
Conditions in the oral cavity	Number examined	Number	%
Catarrhal gingivitis and stomatitis	45	13	29
Ulcerative gingivitis and stomatitis	45	22	49
Aphthous (mycotic) stomatitis	45	6	13
Periodontitis			
Dystrophic form	53	20	38
Inflammatory-dystrophic form	100	74	74
Osteomyelitis of the jaw	50	8	16
Swelling of facial maxillary area	50	12	24
Tartar (dental calculus)	8	2	25

Modified from Musaev FA: Parasitic protozoa in the human oral cavity and their clinical importance. *Stomatologiia* 50:60–62, 1971, with permission of Izdatel'stvo Meditsina, Moscow.

constant finding of *T. tenax* in periodontosis and successful results of treatment by Trichopol of patients having the inflammatory-dystrophic form of this disease, with abundant purulent exudate from periodontal pockets, suggested an active role of the trichomonad in providing a favorable environment for the inflammatory processes in the oral cavity. Rousset and Lauvergeat[51] surmised that if the factors irritating the gums provided a suitable substrate for the parasites, it was possible to consider the buccal protozoa an aggravating factor in periodontal diseases.

There seems also to be a certain degree of correlation between broken facial bones or facial bruises and wounds[39,40] and the occurrence of *T. tenax*. The conditions prevalent in the oral cavity and the adjacent areas (i.e., maxillary and frontal tissues) following surgery, and the installation of orthodontic devices, appear to favor the establishment and growth of the protozoa. Of interest in this connection is the finding that the incidence of *Trichomonas* depends on the presence of teeth, a complete lack of teeth precluding growth of the parasites[38,41]; according to some authors at least one tooth[52] or three reduced to the root state[20] (Cambon as cited in reference 20) are needed for the survival of the parasites.

The entire problem was discussed in more detail by Feki et al[53] who referred to many of the aforementioned workers. The "pathogenic possibilities" considered by Feki and his colleagues fell into three groups: (1) epidemiologic, (2) therapeutic, and (3) "biologic."

1. The correlation between the prevalence of the oral trichomonad and the lesions observed in the periodontium and adjacent areas, discussed at the beginning of this section, has been emphasized by numerous investigators. In light of these correlations, it could be concluded that *T. tenax* contributes to or causes the pathologic changes of various oral cavity diseases, especially those affecting the periodontium. As pointed out earlier, other workers believed that the trichomonads were inoffensive commensals that constituted components of the buccal ecology and were not directly related to any pathologic state. The frequent association of these protozoa with certain diseases could then be explained by the fact that they found the environmental conditions caused by these diseases especially favorable. It could also be surmised that the metabolic products of the protozoa create conditions favorable for the development of other intrinsically pathogenic microorganisms. Such a milieu could, for example, aid in the maintenance of chronic irritating conditions. Yet, according to some workers, including Feki et al,[53] all the epidemiologic arguments adduced to date do not appear strong enough to support unequivocally the idea of lack of inherent pathogenicity of the oral trichomonads.

2. The arguments dealing with pathogenicity of *T. tenax* concern therapeutic responses. Disappearance or perceptible improvement of certain oral cavity diseases caused by treatment with metronidazole could suggest a relation of cause and effect between trichomonads and certain diseases of the oral cavity, for example, ulcerative necrotic gingivitis,[54-56] pericoronitis,[57] and experimental gingivitis in a dog.[58] It should be emphasized, however, that, in addition to its well known antiprotozoal activity, metronidazole has a wide action spectrum against anaerobic bacteria incriminated in most of the aforementioned diseases.[59,60] Furthermore, the role of these bacteria in the etiology of these infections, judged important by some authors,[43,61] is not yet well defined. All the arguments notwithstanding, some authors[47] asserted that "the efficacy of certain antiparasitic medications in periodontopathies [periodontal diseases] proves pathogenicity of these [oral] protozoa."

The foregoing arguments based on therapeutic results, although plausible, are insufficient to incriminate the protozoa as direct causes of orodental infections, because the position that they occupy in the environment characteristic of these infections has not been adequately defined.

3. As far as arguments of the "biologic" kind are concerned, studies of many aspects of pathogenicity of the oral protozoa performed to date remain fragmentary. The phagocytic capacity of the protozoa vis-à-vis the red blood cells has often been adduced as a strong argument in favor of pathogenicity. Yet in light of the presently available information, this capacity cannot be considered to prove pathogenicity potential. Tissue invasion has not been demonstrated to date and, for the most part, pathogenicity assays using laboratory animals have not been convincing. Finally, immunologic studies have been limited to demonstration of circulating antibodies to *T. tenax*.[1] To gain credibility, the foregoing approaches will have to be pursued in greater depth.

Despite all the arguments they have adduced, Feki et al[53] felt that since *T. tenax* is often associated with severe lesions of the periodontal tissues and, according to their tests, shows correlation with pathogenicity, it may well play an active role in oral cavity diseases. To confirm this role, an in-depth biologic and immunologic study of *T. tenax* by Feki et al is said to be now in progress.

Except for isolated instances, experimentation in humans is impossible; only prospective studies remain to the investigator, in which for several years parasitized and nonparasitized subjects would have to be examined at regular intervals when all modifications of their oral conditions, diseases, habits, and so on, would be noted. A comparison of the conditions of the periodontium in these subjects after a definite time period would help to determine the role played by the oral trichomonad in periodontal diseases.

Laboratory Assays

Shinohara and Iwai[62] were able to produce large intramuscular abscesses in mice by inoculating *T. tenax* in the presence of bacteria and with antibiotics. Smaller lesions were caused by inoculation of bacteria with antibiotics, but without trichomonads. As was pointed out before,[1] "even if the report of the Japanese investigators included a statistical analysis of a very large number of determinations, which it does not, the data would be difficult to evaluate."

Kulda[63] noted only very small subcutaneous lesions in CBA mice inoculated with an axenic culture of *T. tenax*. The only other common trichomonad species that caused similarly small lesions was *P. hominis*. The abscesses produced by *Trichomonas vaginalis, Trichomonas gallinae, Tetratrichomonas gallinarum, Tritrichomonas foetus,* and *Tritrichomonas suis,* known to contain virulent strains, were significantly larger than those caused by the oral trichomonad. (For details of the subcutaneous mouse assay, see Chap. 8 of this book.)

According to Teras et al (1972, as cited in reference 35), *T. tenax* cultivated in his 1970 axenic medium[99] produced peritonitis in intraperitoneally inoculated mice; the resulting lesions involved liver, pancreas, and spleen. As far as can be ascertained, these results were not reproduced by other workers. Delaunay et al[64] observed that only mice subjected to treatment with antibiotics survived for a sufficiently long time to allow the finding of living oral trichomonads 4 days after intraperitoneal inoculation. Although evidently adapted to survival in the peritoneal cavity of the primarily infected mice, the trichomonads could not be transferred into the peritoneal cavity of other, parasite-free mice. The results could not be improved through the use of serum-containing medium to produce artificial ascitic fluid. Neither did injection of irritating particles of mesothelial cells render conditions favorable for survival of the trichomonads, despite the fact that abundant ascitic fluid was produced by this procedure.[64]

Subcutaneous abscesses obtained by injection of xenic cultures provided good environment for *T. tenax*—living organisms could be found up to 18 days after infection. However, the abscess healed spontaneously on the 25th day following inoculation.[64]

Alderete and Garza,[65] who assayed virulence of *T. tenax* in cell cultures, indicated that this parasite "did not adhere to cell monolayers and did not cause host cell damage," implying its lack of inherent virulence. (For the cell culture assay, see Chap. 9 of this book.)

Admittedly, the employment of the standard virulence tests for *T. tenax* causes certain difficulties, the main one being that axenization of strains of this organism takes a long time and involves extensive use of antibiotics, both procedures typically resulting in reduction of the inherent virulence levels of trichomonads and other protozoan and nonprotozoan parasites.

Epidemiology

This section is concerned almost exclusively with *T. tenax,* although many reports on which it is based also dealt with *Entamoeba gingivalis,* and even fungi, especially of the genus *Candida.*

Prevalence

In light of the numerous host-dependent factors and the different diagnostic procedures employed by various authors, I shall not deal with the prevalence of *T. tenax* in more general terms, as was done previously by numerous investigators, for example, Wenrich[16] or Honigberg.[1] Rather, I shall try to adduce sufficient information about the factors controlling the prevalence of this species to provide the reader with some insight into the subject.

We shall consider the factors irritating the

oral cavity tissue as either "natural" (e.g., poor dentition, periodontal diseases) or "acquired" (e.g., nutrition, oral hygiene, tobacco smoking, and chemical substances). It must be realized also that without a doubt in each instance the infection itself may be capable of maintaining or enhancing the irritating effects of factors that favor growth of the oral protozoa. Furthermore, host age, together with other factors, strongly affects the prevalence of these parasites. Because of its relative importance, the role of host age is discussed first.

Effect of Host Age on the Prevalence of the Oral Protozoa. Information concerning this relationship involving trichomonads (in some instances also oral amebae) can be found in many reports.* Many of the relevant data from several countries are summarized in Tables 19.5 to 19.10.

In general, there is sufficient agreement among the data provided by the workers from any one country to allow pooling of the results and thus to obtain higher and therefore statistically more meaningful quantitative data. In many instances diagnosis is based on the employment of xenic cultures.

TABLE 19.5. Relationship between host age (in years) of individuals examined and prevalence of the oral protozoa (*Entamoeba gingivalis* and *Trichomonas tenax*) in the Peoples' Republic of China.

Age (yr)	Number examined	Positive for protozoa*	
		Number	%
0–10	19	3	15.8
11–20	22	10	45.5
21–30	39	20	51.3
31–40	42	16	38.1
41–50	42	18	42.9
51–60	29	5	17.2
60 and older	17	11	64.7

Reproduced from Su KX: Oral protozoiasis. *Chung Hua Kou Chiang Ko Tsa Chih* 20:112–114, 1985, with permission of Chung Hua Kou Chiang Ko Tsa Chih, Beijing, China.
*The flagellates and amebae were counted together. The relatively high numbers are due to the high numbers of amebae in relation to the flagellates (86%:20% according to Cambon et al[42] and 48%:19% according to Lapierre and Rousset[47]).

*References 21, 37, 38, 40–42, 46, 47, 50, 53, 66–69.

Of importance in considering the relationship between the prevalence of *T. tenax* in the oral cavity and the age of the host is the analysis of the factors responsible for this relationship. Perusal of the literature reveals the prevalence to be dependent on the conditions obtaining in the oral cavity and adjacent areas. As might be anticipated, younger subjects have healthier mouths. It was pointed out by Aradi[70] that infection with oral protozoa increased with age and with the accompanying deterioration of the conditions prevailing in the oral cavity, that is, with the increased frequency of dental caries, gingivitis, stomatitis ulcerosa, and so on (Tables 19.1 to 19.4). He thought that the last two pathologic conditions provided an especially favorable environment for infection with these protozoa. Such an increase in patients up to 50 years of age was noted also by Čatár et al[38]; however, these investigators found no correlation between oral protozoal infection and "the condition of teeth," the prevalence being equal in persons with and without dental caries. It was demonstrated also by Cambon et al,[42] among others, that toothless infants and young children without periodontal and gingival diseases did not harbor *T. tenax*; even in the early teens, infection with the oral trichomonad was either very low or nonexistent (Tables 19.6 to 19.10). Low numbers or lack of teeth seen in old people also provide an unfavorable milieu for the oral protozoa.[20,38,41,52] It was indicated by Feki et al[53] and other investigators that the frequency of oral disease and of oral trichomonads is highest between over 40 to over 50 years of age, but beyond very young and old subjects the infection rate is subject to considerable variation depending on some known and many unknown factors. Among the important factors appear to be those related to hygienic habits and cultural customs; in some instances the former may be related to the latter.

Oral Hygiene. Oral hygiene undoubtedly affects prevalence. Some workers speak of "clean," "dirty," and "pyorrhetic" conditions of periodontal tissue, as well as of "clean" and "coated" conditions of the tongue surface.[24] The accumulation of dental calculus is also often related to oral hygiene (see Tables 19.2 and 19.3). Green and Vermillion[71] and Feki et al[53] evaluated oral hygiene in terms of the simplified oral hygiene index (OHIS), the sum of mean indices of debris and dental calculus es-

TABLE 19.6. Relationship between host age and prevalence of the oral protozoa in Prague and Zlin, Czechoslovakia.

Age(yr)	Prague			Zlin	
	Number examined	T. tenax (%)	E. gingivalis (%)	Number examined	T. tenax (%)
0–10	10	0.0	0.0	198	1.0
11–20	73	4.0	20.0	76	2.6
21–30	167	14.0	42.0	106	17.9
31–40	109	25.0	56.0	84	33.0
41–50	81	28.0	38.0	24	29.0
51–60	38	39.0	45.5	11	45.0
61+	22	22.8*	33.3	7	71.4*

Modified from Jírovec O: Contribution à la microbiologie de la cavité buccale humaine. *Paradentologie* 2:117-123, 1948, with permission of l'Arpa Internationale, Association pour les Recherches sur les Paradontopathies, Geneva.
*These high percentages suggest that none or only a few of the subjects in the advanced age-group were toothless.

tablished after numerical evaluation of their deposits on certain predetermined teeth. Evidently there is a direct correlation between the OHIS value and infection level, especially by *T. tenax*.[53]

The question of oral hygiene entails also the effects of toothbrushing (assuming, of course, that the employment of the same brush by several members of a family is not practiced). Actually, most protozoa are outside the areas accessible to brushing, for example, tonsillar crypts or areas below the tooth collars. Nevertheless, medicated toothpastes can have a certain value in the improvement of oral hygiene. In general, treatment of teeth to suppress irritating factors seems to be unfavorable for the parasites, but the importance of this effect remains open to question.[47]

Although for various reasons there is no unanimity on the importance of oral hygiene in infection with oral protozoa, especially in older persons,[47] many investigators, for example, Palmieri et al[72] working in Borneo, and Trofimova et al,[22] in the U.S.S.R., considered oral hygiene an important factor in the epidemiology of the oral protozoa.

Ethnic Origin. According to Lapierre and Rousset,[47] black East Africans show a lower frequency of infection with oral protozoa in comparison with Europeans of the same age. Actually more than 75% of the Africans were free of *T. tenax*. Healthy dentition and lower infection risk could explain this relatively low prevalence of the oral flagellate.

Comparative investigation of other ethnic groups, such as other Africans and Asians, and above all, of American blacks with regard to their customs and westernized diet would be

TABLE 19.7. Relationship between host age and prevalence of *Trichomonas tenax* in Paris and Bobigny (I) and Clermont-Ferrand, France (II).

Age			Positive for trichomonads (pooled)	
I*	II†	Pooled numbers	Number	%
1–10	0–9	294	6	2.0
11–20	10–19	572	68	11.9
21–30	20–29	730	135	18.5
31–40	30–39	476	176	36.9
41–50	40–49	354	128	36.2
51–60	50–59	217	84	38.7
60 and older	60–69	86(I) 52(II)	34(I) 19(II)	39.5(I) 36.5(II)
	70–79	39(II)	11(II)	28.2(II)
	80 and older	6(II)	3(II)	50.0(II)

*From Lapierre J, Rousset J-J: L'infestation à protozoaires buccaux. *Ann Parasitol Hum Comp* 48:205-216, 1973, with permission of Masson S.A., Paris.
†From Cambon M, Pétavy A-F, Guillot J, Glandier I, Deguillaume J, Coulet M: Etude de la fréquence des protozoaires et des levures isolés du parodonte chez 509 sujets. *Pathol Biol* 27:603-606, 1979, with permission of Société d'Édition de l'Association d'Enseignement Médical des Hôpitaux de Paris.

TABLE 19.8. Relationship between host age and prevalence of *T. tenax* in Como,* Milan,*,† Pavia,‡ and Messina,§ Italy.

Age (yr)	Number examined	Positive for *T. tenax*	
		Number	%
0–0.5	29	0	0.0
0.5–5.0	71	0	0.0
8–12	60	2	3.3
14–20	21	5	23.8
20–29	179	68	37.9
30–39	169	73	43.2
40–49	121	51	42.1
50–59	48	22	45.8
60–69	37	24	64.9
70–79	41	17	41.5
80–89	33	12	36.4
90 and older	5	1	20.0

*Modified from Grisi AM, De Carneri I: Frequenza dell' infezione da *Trichomonas tenax* ed *Entamoeba gingivalis* in soggeti di età avanzata viventi in comunità. *Parassitologia* 3:151-154, 1961, with permission of Lombardo Editore, Rome.
†Modified from Bonvini E, De Carneri I: Ricerca di "Entamoeba gingivalis" e di "Trichomonas tenax" nei bambini de età inferiore ai cinque anni. *G Mal Infett Parassitol* 14:361-362, 1962, with permission of the Società Italiana per lo Studio delle Malattie Infettive e Parassitarie, Milan.
‡Modified from De Carneri I, Giannone R: Frequency of *Trichomonas vaginalis, Trichomonas tenax,* and *Entamoeba gingivalis* infections and absence of correlation between oral and vaginal protozooses in Italian women. *Am J Trop Med Hyg* 13:261-264, 1963, with permission of the American Society of Tropical Medicine and Hygiene, San Antonio, Texas.
§Modified from Miligi G, Magaudda-Borzì L, Mento G: Le frequenza dei portatori di *"Trichomonas tenax,"* e di *Entamoeba gingivalis* in alcune zone della provincia de Messina. *Arch Ital Sci Med Trop Parassitol* 45:95-99, 1964, with permission of Istituto di Clinica Delle Malattie Tropicali e Infettive dell'Universita di Roma/Policlinico Umberto I, Rome.

TABLE 19.9. Relationship between age of individuals examined and prevalence of *T. tenax* in Bloomington, Indiana.

Age (yr)	Number examined	Positive for *T. tenax*	
		Number	%
6–12	23	0	0
13–19	145	9	6.2
20–30	175	34	19.4
31–40	122	40	32.8
41–50	145	54	37.2
51–60	68	20	29.4
61–80	22	5	22.7

Data taken from Wantland WW, Wantland EM: Incidence, ecology, and reproduction of oral protozoa. *J Dent Res* 39:863, 1960, with permission of the American Association for Dental Research, Washington, D.C.

TABLE 19.10. Relationship between age (months or years) of individuals examined and prevalence of *Trichomonas tenax* in Kirobad (Azerbaidzhan), Leningrad, and Moscow (R.S.S.R), U.S.S.R.

Age*	Number of individuals examined	Positive for *T. tenax*	
		Number	%
1–9 mo	50	0	0
15–20	31	4	12.9
21–30	89	24	26.9
31–40	138	72	52.8
41–50	69	33	47.8
51–60	37	15	40.5
61 and older	32	9	28.1

Reproduced from Musaev FA: Trichomonads of the oral cavity of man. *Parazitologiia*, 6:185-188, 1972, with permission of Parazitologiia, Leningrad, U.S.S.R.
*In years, except as indicated.

useful in elucidating the importance of ethnic origin in the epidemiology of the oral protozoa. However, according to some investigators, experiments dealing with food habits are nearly impossible to manage. Also, in light of some of the findings, the ethnic factors and debatable; for example, whereas immigrants from East Africa harbored few oral parasites, in Dakar (West Africa) the infection rate was statistically higher than the infection mean found in the French. The level of dental cleanliness was thought to be responsible for this condition.[52]

Sex. Although many investigators believed that the sex of the subject did not have any effect on the frequency or course of infection with the oral protozoa,[20,39,50,66] Wantland and Lauer[24] found the disease distribution somewhat different between members of the two sexes. Actually the prevalence pattern remained sex-independent until the peak infection in patients, from 31 to 40 years of age. In persons older than 50 years, the frequency distribution became sex-dependent, especially with regard to *T. tenax*. In women older than 30 years there was a steady and irreversible drop in the infection level (31 to 40 years, 31.5%; 41 to 50 years, 23%; 51 to 60 years, 12.6%; over 60 years, 1.9%). Although a drop in the infection level was seen also in men, this was followed by a complete recovery of the infection level (up to 33% in persons over 60 years). According to Lapierre and Rousset,[47] prevalence of *T. tenax* showed statistically significant differences between men and women, the infection rate being higher in men. Also, Čatár et al[38] found the prevalence of this parasite to be higher in men. No explanation was offered by any of the investigators for the aforementioned differences between the sexes.

Rural versus Urban Environment. In Romania, the prevalence of the oral protozoa was 24.8% in the rural environment, while in the city it equaled 34.1%.[73] Higher prevalence of *T. tenax* (approximately 11.0%) was reported also from urban dwellers than from inhabitants of rural regions (approximately 3.6%) in France.[51] The reasons for these differences were not explained, but they probably depended on social habits and customs, including oral hygiene, of the persons examined.

Socioeconomic Factors. There appears to be a significant statistical difference in the prevalence of *T. tenax* among populations belonging to different socioeconomic groups: agricultural workers, students, laborers, professionals, and unemployed. According to Cambon et al[42] the highest prevalence of the oral parasites was found in agricultural workers (41%). They were followed by laborers (37%) and unemployed persons (34%). Self-employed professionals had a much lower prevalence of *T. tenax* (28%). By far the lowest prevalence was noted among students (15%).

The authors attributed the aforementioned differences to the fact that the populations having the highest infection rates did not pay as much attention to their oral hygiene and state of dentition as did the professionals and students. As far as the last group was concerned, it must also be remembered that its members were the youngest individuals examined; thus, as pointed out before, their oral cavities would be relatively healthy.

Tobacco Smoking. It is difficult to differentiate the groups of smokers and nonsmokers among people under 25 years of age, although Rousset and Lauvergeat[51] called attention to the fact that tobacco smoking favored the establishment of the oral protozoa. The local gingival congestion caused by nicotine appears to be similar to the lesions associated with various dental changes or to those dependent on biting of fingernails.[47] Rousset and Lauvergeat[51] indicated that prevalence of *T. tenax* in young, apparently healthy persons who smoked more than one pack of cigarettes per day was 20%, while only 5.2% of nonsmokers harbored this flagellate in their oral cavity. On the other hand, no such differences were reported by Pechtorová et al,[41] but the very small numbers of subjects examined by these latter investigators preclude drawing any valid epidemiologic conclusions. However, a similar conclusion was reached by Čatár et al[38] on the basis of examination of 200 persons.

Alcohol Consumption. According to some investigators, regular intake of significant volumes of alcohol corresponds to higher prevalence of *T. tenax* (Omnes, as cited in reference 42). However, these conclusions, based on small numbers of samples, need further verification.

Correlation with Fungal Infections. There does not appear to be any ecologic relationship between the presence of *Candida* spp., of which *Candida albicans* is predominant, and *T. tenax*. The yeastlike fungi do not seem to be associated with periodontal disease, but rather with poor oral hygiene, poor state of dentition, and, above all, with diabetes.[42] In one of their reports Cambon et al[20] stated that the mycotic agents they studied were very often associated with oral protozoa. All the fungi behaved like saprophytes; none were found to cause clinical oral candidiasis in their hosts. Evidently, only a lack of oral hygiene seemed to favor the development of *Candida*. Curiously enough, dental calculus had an unfavorable effect. It does not appear from the presently available information that *Candida* is an agent favoring *T. tenax* infections.

Transmission

In light of the fact that *T. tenax* and *E. gingivalis* are not known to form cysts in the course of their life cycles, the statements pertaining to the means of transmission hold true for both species—the absence of resistant cysts necessitates direct transmission of the trophozoites. However, on occasion, rapid transmission on objects soaked in infected saliva or in droplets of such saliva is also possible.[52,74]

Direct Transmission. In his study, Lauvergeat (as quoted in reference 17) compared the prevalence of oral protozoan infections in young men brought up in Western culture, but who did not practice the "erogenic" custom of deep kissing, with the prevalence of these protozoa in men of the same age who practiced this custom. By analyzing his data, Lauvergeat demonstrated that osculation was the primary means of transmission of oral protozoa: 40% of the men who practiced deep kissing harbored the parasites, which, however, could be found in only 11% of persons who abstained from this habit.

Indirect Transmission. The finding of oral protozoa in children who were never involved in deep kissing strongly suggests the possibility of indirect transmission. This assumption finds support in the observation that cultures of oral protozoa can be initiated from saliva containing the protozoa and from objects soaked in such saliva.[52] Futhermore, survival of *T. tenax* in tap water for up to 64 hours was reported by Beatman[74] and for shorter periods by Hinshaw[75] and Hogue.[76] According to Lapierre and Rousset[47] the following customs can be the source of infecting a child with oral protozoa: use of the same toothbrush by several members of a family; tasting food before giving it to a young child; prechewing food by the mother before offering it to the child, a habit customary in certain ethnic groups. These parasites can be transmitted also by passing around a bottle or by biting on the same piece of fruit.

Respiratory Tract

Pathogenicity

Although the question of the potential pathogenicity of *T. tenax* has already been discussed in this chapter, it should be noted that in several abstracts the members of the Estonian group[44,77] state that "pathogenicity of trichomonads isolated from bronchi was demonstrated experimentally [in] white mice inoculated intraperitoneally with axenic cultures of this flagellate."[77] According to Sardis et al,[78] there are only minor differences in structure and pathogenicity for mice between *T. tenax* and trichomonads from bronchi. Some of the older workers cited in Table 19.11 as well as Skipina,[79] Turgel and Balode,[80] Shevchenko et al,[81] and the Estonian school[82] considered the oral trichomonads pathogenic. Khanin and Staroverova,[83] on the basis of a single case, related the presence of the oral trichomonad in the bronchi with the exceptional case of a (recurrent) bronchopulmonary inflammatory process accompanied by a strong eosinophilic allergic host response. Evidently, the protozoa played some role in pathogenesis. However, agreement is lacking with regard to the potential pathogenicity of the putative *T. tenax* strains inhabiting the respiratory passages of humans. Some workers considered the flagellates commensals without any pathologic significance,[84] while others expressed the view that the trichomonads may support or increase rather than initiate the inflammatory process,[85] or they may be pathogenic in an already diseased host.[86] A survey of the relevant literature suggests, however, that for the most part investigators tend not to reject unequivocally the pathogenic potential of the pulmonary trichomonads pending further studies.[87-92]

Prevalence

In this subsection I shall consider only prevalence of organisms identified as *Trichomonas* and leave all speculation regarding the specific identity of these flagellates to the next subsection. Although it is impossible to be certain that one has seen references to all the cases, a literature search performed with the aid of computers as well as comparison of these references with those in the lists published previously by various authors, especially Walton and Bacharach[89] and Hersh,[92] should provide a reasonably complete record of the cases of respiratory tract trichomoniasis reported since 1867 (i.e., a period of about 120 years) (Table 19.11). Although, to date, the largest number of persons with respiratory tract trichomoniasis has been reported from the U.S.S.R., that is, Tumka,[93] 19 (23%) of 82 cases of pulmonary disorders examined, or Teras et al,[91] 37 (10%) of 370 cases examined, the geographic distribution of respiratory tract infection by these flagellates appears to be cosmopolitan. During the past 120 years, 107* cases have been reported from the U.S.S.R., 21 from Germany, 11 from the United States, four from France, two each from Canada and Japan, and one each from Argentina, Brazil, Holland, India, and Switzerland. Undoubtedly, a more systematic and intensive search by clinicians would have revealed more cases of respiratory tract trichomoniasis in many countries. However, certain indigenous customs can affect the prevalence of this kind

---

*If the strains examined by Rõigas et al[94] are counted separately from those studied by Kazakova et al[95] and Teras et al.[91]

TABLE 19.11. Trichomonads, other than *T. vaginalis*,* found in the respiratory organs.†

Name applied to the flagellates	Material in which demonstrated	Number of cases	Associated with disease or condition	Reference‡
Infusorien	Sputum	2	Lung gangrene and purulent bronchitis	Leyden and Jaffee (1867)§,‖ (Germany)
Monas lens	Sputum	5	Lung gangrene	Kannenberg (1879)§,‖ (Germany)
Cercomonas	Lung abscess	6	Lung gangrene	Kannenberg (1880)§,‖ (Germany)
Paramecium	Sputum	1	Hemoptysis-purulent sputum	Stockvis (1884)§,‖ (Holland)
Cercomonas	Pleural exudate	1	Tuberculous hydropneumothorax	Litten (1886)§,‖ (Germany)
Monaden	Sputum; histologic section	3	Exudative pleuritis with abscess; lobar pneumonia	Streng (1892)§,‖ (Germany)
Cercomonaden	Sputum; pleural exudate	1	Purulent pleuritis; lung abscess	Roos (1893)§,‖ (Germany)
Flagellaten	Sputum; liver abscess	1	Lung and liver abscess	Grimm (1894)§,‖ (Japan)
Trichomonas pulmonalis Schmidt, 1895	Sputum coughed up from lungs; Dittrich's plugs; bronchus	3	Carcinoma of the larynx; pneumonia; secondary infection bronchiectasis; chronic pleuritis	Schmidt (1895)§,‖ (Germany)
Trichomonas pulmonalis Schmidt, 1895	Sputum	2	Lung gangrene	Artault (1898)§,‖ (France)
Trichomonas intestinalis Leuckart, 1879¶	Sputum; bronchi; non-necrotic parenchyma	1	Pneumonia; lung gangrene	Dolley (1910)§,‖ (U.S.A.)
Trichomonas hominis (Davaine, 1854)¶	Sputum; Dittrich's plugs	1	Chronic bronchitis; bronchiectasis; emphysema	Honigman (1921)§,‖ (Germany)
Trichomonas intestinalis Leuckart, 1879¶	Sputum; Dittrich's plugs; abscess; non-necrotic parenchyma	1	Lung gangrene	Parisot and Simonin (1921)§,‖ (France)
Trichomonas pulmonalis Schmidt, 1895	Sputum; bronchial lumen	1	Putrid bronchitis; bronchiectasis; chronic pneumonia	Marx (1927)§,‖ (Switzerland)
Trichomonas hominis (Davaine, 1854)¶	Sputum; pus from chronic abscess	1	Hepatic and thoracic abscess	Navarro and de Alzaga (1933)§,‖ (Argentina)
"Similar to *Trichomonas buccalis*" Goodey & Wellings, 1917	Sputum	1	Pneumonia; pyorrhea	Glaubach and Guller (1942)§,‖,# (U.S.A.)97
"*Trichomonas tenax* (O.F. Müller, 1773)"(?)	Sputum	1	Chronic bronchiectasis (cylindric variety)	Lehman & Prendiville (1946)§,# (India)84

(continued)

TABLE 19.11. Continued

Name applied to the flagellates	Material in which demonstrated	Number of cases	Associated with disease or condition	Reference[‡]
Trichomonas sp.	Sputum; gingiva; bronchial washings	1	Lung abscess; chronic bronchiectasis; gingivitis	Barbosa and Amaral (1950)[#] (Brazil)[87]
Trichomonas sp.	Sputum; bronchial washings	1	Lung abscess; chronic bronchiectasis; meningitis; trichomonanemia	Khrushcheva and Kriazheva (1951) (U.S.S.R.)[85]
Trichomonas elongata (Steinberg, 1862)	Bronchial washings; resected lung tissue; sputum	19	Lung cancer; lung abscess; bronchiectasis; pneumonia; chronic bronchitis	Tumka (1956)[#] (U.S.S.R.)[93]
Trichomonas tenax (O.F. Müller, 1773)?	Bronchial washings; sputum (in fresh material or Papanicolaou-stained preparations)	3	Pulmonary fibrosis; bronchogenic carcinoma	Walton and Bacharach (1963)[§,#] (U.S.A.)[89]
Trichomonas sp.	Pleural empyema fluid	1	Empyema; bronchopleural fistula	Abed et al (1966)[§,#] (France)[104]
Trichomonas tenax (O.F. Müller, 1773)?	Pleural fluid obtained by thoracocentesis (empyema fluid)	1	Hydropneumothorax; lung abscess	Memik (1968)[§,#] (U.S.A.)[88]
Trichomonas sp.	Sputum	1	Emphysema; chronic productive cough	Skipina (1968)[§,#] (U.S.S.R.)[79]
Trichomonas sp.	Resected lung tissue	2	Cavitating tuberculosis	Fardy and March (1969)[§,#] (Canada)[90]
Trichomonas hominis (Davaine, 1860)**	Subphrenic abscess; pleural empyema fluid	1	Stomach cancer; gastrectomy; subphrenic abscess; thoracic empyema; bronchopleural fistula	Houin et al (1973)[§,#] (France)[156]
Trichomonas elongata (Steinberg, 1862)	Bronchial washings	1	Endobronchitis	Turgel and Balode (1973)[§,#] (U.S.S.R.)[80]
Trichomonas tenax (O.F. Müller, 1773) or *Trichomonas bronchopulmonalis* Kazakova, Rõigas, & Teras, 1973	Bronchial washings	? (several)	Nonspecific chronic bronchitis (pneumonia)	Kazakova et al (1977) (U.S.S.R.)[96]
Trichomonas tenax (O.F. Müller, 1773)?	Pleural empyema fluid	1	Empyema secondary to aspiration pneumonia	Walzer et al (1978)[§,#] (U.S.A.)[177]
Trichomonas tenax (O.F. Müller, 1773)	Bronchial washings	37[††]	Chronic pneumonia; chronic bronchitis; lung tuberculosis	Kazakova et al (1980) (U.S.S.R.)[95]

Species	Specimen	Number	Disease/Finding	Reference
Trichomonas tenax (O.F. Müller, 1773)	Bronchial washings		Bronchitis	Rõigas et al (1980)[94] (U.S.S.R.)
Trichomonas sp.	Bronchial washings	27††(?)	Chronic pneumonia; chronic bronchitis; lung tuberculosis; other bronchial and pulmonary diseases	Teras et al. (1980) (U.S.S.R.)[91]
		37††		
Trichomonas elongata (Steinberg, 1862)	Bronchial washings		Catarrhal diffuse endobronchitis	Shevchenko et al (1980) (U.S.S.R.)[81]
"Oral trichomonad"	Bronchial washings and sputum		Recurrent bronchitis involving an allergic reaction (eosinophilia)	Khanin and Staroverova (1982) (U.S.S.R.)[83]
Trichomonas tenax (O.F. Müller, 1773)?	Pleural empyema fluid	1	Gastric carcinoma; gastrectmy; esophagealpleural fistula	Miller et al. (1982)[98] §,# (U.S.A.)
Trichomonas sp.	Pleural fluid (turbid yellow exudate)	1	Purulent pleuritis; left pleural effusion (with the trachea deviated to the right)	Osborne et al (1984)[86] § (U.S.A.)
Trichomonas tenax (O.F. Müller, 1773)?	Purulent pleural fluid	1	Purulent pleuritis	Hersh (1985)[92] § (U.S.A.)
Trichomonas tenax (O.F. Müller, 1773)?	Pleural effusion	1	Pleuritis	Okhura et al (1985) (Japan)[178]
Trichomonas hominis (Davaine, 1854)**	Sputum	1	Chronic diffuse bronchitis (possibly bronchiectasis and protracted pneumonia in the lower segment of the right lung)	Polivoda et al (1987) (U.S.S.R.)[157]

*As identified on epidemiologic basis, cultivation, and pathogenicity potential in mice.

†The list is not necessarily complete, but it undoubtedly includes the preponderant majority of reported cases.

‡I saw all the references in the original; however, to be able to shorten the already very long reference list, I ask the reader to follow the procedure outlined in footnotes ‖ and # to this table.

§No cultivation.

‖These references are listed in the report of Walton and Bacharach, 1963,[89] which should be consulted for the complete citations.

¶As stated by Walton and Bacharach[89]: "No reliable identification [of the pulmonary trichomonads] is available from the earlier investigators. Half of these reports assign the organism to the genus *Trichomonas*, and half of these presume it to be the same species which inhabits the intestine of man." As indicated in the text, most of the identifications of *P. hominis*, and also many of *T. tenax* (= *T. pulmonalis* (?), *T. buccalis*, *T. elongata*) were not based on morphologic and/or antigenic characteristics sufficient to provide for their assignment to one of the species found in humans. This is not surprising, since Wenyon,[154a] one of the important authorities on parasitic protozoa, was not convinced (even in 1926) that there existed in humans three different trichomonad species; he had the strongest reservations with regard to the differences between *T. tenax* and *P. hominis*.

#These references are listed in the report of Hersh, 1985,[92] which should be consulted by the reader for the complete citations.

**These appear to be actual intestinal flagellates of humans; therefore, they would properly belong in the species *P. hominis*.

††It is likely that these organisms were obtained from the same group of patients.

of infection. Furthermore, the employment of more exact criteria in defining infection of the respiratory organs and in identification of the trichomonad species involved undoubtedly would influence the results of the epidemiologic investigations. To date the majority of the trichomonads from the respiratory tract could not be assigned to species with any degree of confidence (Table 19.11).

As far as the definition of trichomonad infection of the respiratory tract is concerned, the report of Teras et al[91] is of interest. Although the meaning of some of the data is not entirely clear, in light of the fact that the invasion of the respiratory tract by trichomonads need not be specific, it is not surprising that the authors found different percentages of a variety of diseases in patients harboring flagellates in their respiratory tracts. For example, trichomonads were noted in $14.5 \pm 6.1\%$* of cases of chronic pneumonia, in $8.5 \pm 6.1\%$ of chronic bronchitis, in $6.6 \pm 4.2\%$ of pulmonary tuberculosis, and in $11.5 \pm 9.6\%$ of all other cases of respiratory tract diseases. On the other hand, the cases in which the trichomonads of a given antigenic type were found in bronchi but not in the sputum and vice versa are not easy to understand, especially since sputum was the material from which the parasites were isolated in many instances of respiratory tract infections (Table 19.11). In some instances, the sputum could have been contaminated with trichomonads from a widespread oral infection that belonged to different serotypes. Such an explanation need not be valid in other instances, and the entire situation regarding differences in antigenic types among these trichomonads from the oral cavity, sputum, and bronchi are difficult to interpret in light of our present knowledge; consequently no such explanation is attempted. To facilitate the following presentation and discussion, these trichomonads are arbitrarily divided into two groups, those found in or isolated from the respiratory passages (e.g., bronchi) and those encountered in "other sites" (i.e., sputum and oral cavity). It is also not readily apparent in light of the available records why the parasites reported by Teras et al,[91] Kazakova et al,[95] and Rõigas et al[94] from the material aspirated from the respiratory passages could be demonstrated primarily or exclusively by cultivation. It should be pointed out, however, that Tumka[93] was able to find trichomonads in only 3 out of 16 pulmonary infections; in the other cases he had to use cultures.

Even when the "pulmonary" trichomonads could be identified with relative confidence, information about their antigenic properties has been rather incomplete; it has been accumulated primarily by the Estonian group. The results published by the members of this group constitute most of the remaining part of this section.

Antigenic Types and Taxonomic Considerations

Kazakova et al[96] briefly reported a comparative study of the antigenic composition of *T. tenax* from the oral cavity and respiratory tract of humans. As determined previously,[30] *T. tenax* from the oral cavity had four antigenic types; A, B, C, and D. Using sera from *Trichomonas*-infected patients and hyperimmune sera prepared in rabbits for agglutination tests, Kazakova et al[96] found significant antigenic differences between the parasites from the oral cavity and the respiratory organs. Such differences existed also between the oral-cavity and respiratory-passage populations in a single patient. The Estonian investigators considered the possibilities that the antigenic differences observed by them might represent intraspecific strain differences or could even provide sufficient grounds for placing the trichomonads from the respiratory organs in a new species, "e.g. *Trichomonas broncho-pulmonalis*."

In a more extensive study, Kazakova et al[95] analyzed by agglutination some of the antigenic properties of trichomonads she found in fresh preparations and by cultivation of the materials obtained from the bronchi (by bronchoscopy), sputum, and oral cavity of patients suffering from various respiratory diseases, including chronic pneumonia and chronic bronchitis (Table 19.12). These parasites were found in the respiratory passages in 37 (10%) of 370 patients examined (Table 19.12). Because the presence of the trichomonads in the respiratory organs need not coincide with their occurrence in the sputum and oral cavity, henceforth referred to as the "other sites," finding the flagellates in all the sites required examination of the material obtained by bronchoscopy. The opinion that the trichomonads need not occur simulta-

---

*Mean ± sample standard deviation (SD).

TABLE 19.12. Distribution of trichomonads in patients harboring these protozoa in the respiratory tract.

Disease	Total number of patients examined	Trichomonads detected (number of cases)	
		In bronchi	In sites other than the respiratory system*
Chronic pneumonia	109	16	9
Chronic bronchitis	82	7	3
Pulmonary tuberculosis	135	9	8
Various other diseases of the respiratory tract	44	5	1
TOTAL	370	37	21

Modified from Kazakova I et al: Trichomonad invasion in respiratory tract and specific agglutinins in the blood sera of patients suffering from lung diseases. *Izv Akad Nauk Est SSR* (Biologiia) 29:87-94, 1980, with permission of the Academy of Sciences the Estonian SSR.
*This category includes sputum and oral cavity.

neously in the respiratory system and in the other sites was not limited to the Estonian workers. Indeed, it has been suggested also by other investigators. Thus Glaubach and Guller[97] found a trichomonad, presumably resembling "*Trichomonas buccalis,*" in the sputum, but not in the periodontium, of a patient presenting with pneumonia (evidently no material was aspirated from the bronchi); Memik[88] noted flagellates he thought to be *T. tenax* in the pleural effusion, but not in mouth washings, of a patient with a chronic pulmonary disease; and Skipina[79] observed trichomonads in the sputum, but not in mouth washings, from a patient with emphysema and abnormal changes in the lower lobes of the lungs. Also, Miller et al[98] reported a trichomonad (more likely *T. tenax* but possibly *P. hominis*) in pleural fluid; their search for the flagellates in oropharyngeal washings and sputum gave only negative results, and the authors suggested that "oropharyngeal trichomonads were too few in number to be found in oral washings and proliferated only in the environment of the pleural fluid."

For the purpose of their immunologic studies, Kazakova et al[95] and Rõigas et al[94] employed organisms cultivated axenically in a little-known (in the West) complex diphasic medium "TT," whose liquid overlay included tryptone, yeast extract, beef liver infusion, and ascitic fluid; the solid phase consisted of egg or serum slants (see Teras et al[99,100] as cited by Kazakova et al[95]) or a modification of this medium.

Before discussing the results of the Estonian workers, it ought to be mentioned that in many instances their experimental evidence points to a trend rather than providing solid support to some statements expressed with considerable caution in the Russian text, although not necessarily in the English abstracts.

As indicated before, of a total of 370 patients suffering from respiratory tract diseases (the authors evidently did not examine healthy patients) 37 (10%) harbored trichomonads in the respiratory tract. Furthermore, among these patients, 21 had the flagellates in sites other than the respiratory system, that is, the sputum and/or oral cavity (Table 19.12). When only one of the known serotypes (A, B, C, and D) of *T. tenax*[30] was used with the sera obtained from the patients harboring the trichomonads in their respiratory tract (18 cases) or other sites (30 cases), fewer agglutination reactions with specific titers, that is, titers greater than or equal to 1:320, were recorded than when two or more of the serotypes were reacted with the same sera (Table 19.13). A possible explanation of the specific titers observed in reactions between sera from the persons in whom no trichomonads were actually detected and *T. tenax* antigenic types is given in the final footnote to Table 19.13. It is evident from the foregoing results that differences among the trichomonad strains found in the various sites depend on the antigenic constitution of the parasites. Similar results were obtained and similar conclusions could be drawn from the experiments in which one or more than one trichomonad strains from various sites were reacted with their homologous and heterologous sera (Table 19.14).

Among the comparative studies of the antigenic relationships of trichomonad strains found in the respiratory passages, sputum, and oral cavity of humans, that of Kazakova et al[101] appears to be one of the most extensive. The

TABLE 19.13. Agglutination reactions involving sera from patients with respiratory diseases in whom the trichomonads were or were not detected.

Trichomonads	Number of sera employed	Number of sera with titers ≥ 1:320* in reactions with each of the four known antigenic types of T. tenax[†]				Number of sera showing titers >1:320 in reactions with two or more T. tenax antigenic types[‡]
		A	B	C	D	
In bronchi	18	4	12	10	2	13[§]
In other sites[ll]	30	3	21	18	7	26[§]
Not detected	34	4	21	22	0	28[¶]

Modified from Kazakova I et al: Trichomonad invasion in respiratory tract and specific agglutinins in the blood sera of patients suffering from lung diseases. *Izv Akad Nauk Est SSR (Biologiia)* 29:87–94, 1980, with permission of the Academy of Sciences of the Estonian SSR.
*Titers <1:320 are considered nonspecific.
[†]According to Kumm et al[30]; see also Kazakova et al.[96]
[‡]In most instances, simultaneous employment of all the antigens gave the largest number of positive reactions.
[§]The authors[95] were uncertain of the reasons for these discrepancies (13 vs. 18 and 30 vs. 26) specific agglutination reactions. Their explanation suggested the presence of major antigenic differences between some of the strains isolated from the patients and the four antigenic types (A to D) of *T. tenax*.
[ll]This category includes sputum and oral cavity.
[¶]Kazakova et al[95] hypothesized that, although actually present, the trichomonads might have been undetected on single examinations necessitated by the conditions of their experiments. If this were the case, specific agglutinins would be found in the sera from the hosts of these parasites.

study, based on agglutination reactions, involved nine strains isolated in axenic cultures (in modified TT-1 medium) from four patients, Nos. 33, 53, 206, and 220, suffering from respiratory tract diseases. Three strains (53 III, 206 III, 220 III) were isolated from bronchi; four strains (33 II, 53 II, 206 II, 220 II) from sputum; and two strains (33 I, 53 I) from the oral cavity. Difficulties encountered in establishing axenic cultures limited the investigation to relatively few strains. Antisera against the axenically cultivated strains were prepared in rabbits by intravenous inoculations of living organisms. The antisera, nonadsorbed and cross-adsorbed, were used in agglutination reactions with their homologous and heterologous strains. Antisera against the four established *T. tenax* serotypes (A, B, C, D) were also employed in the aforementioned reactions.

It was established on the basis of homologous agglutination reactions that strains 53 I and 53 III were characterized by strong aggluti-

TABLE 19.14. Dependence of the agglutination titers of sera from individuals harboring trichomonads on the sites of origin of the strains employed as antigens.

Sites of trichomonads	Number of sera employed	Number of sera with titers ≥ 1:320* in reactions with trichomonad strains isolated from each			Number of sera with titers ≥1:320 in reactions with two or more trichomonad strains employed[†]
		Oral cavity	Sputum	Bronchi	
Bronchi	12	8	4	10	10
Bronchi and sputum	4	3	3	3	4
Bronchi, sputum, and oral cavity	4	3	3	2	4
Bronchi and oral cavity	6	4	1	6	6
TOTAL	26	18	11	21	24

Modified from Kazakova I et al: Trichomonad invasion in respiratory tract and specific agglutinins in the blood sera of patients suffering from lung diseases. *Izv Akad Nauk Est SSR (Biologiia)* 29:87–94, 1980, with permission of the Academy of Sciences of the Estonian SSR.
*Titers <1:320 are considered nonspecific.
[†]In most instances, simultaneous employment of all the antigens gave the largest number of positive reactions.

nation, strains 53 II, 206 II, and 206 III by medium agglutination, and strains 33 I and 33 II by weak agglutination (Table 19.15). (The significance of this finding, which need not depend on the properties of the strains, is difficult to assess.) There also appeared to be differences in titers in the reactions involving strains from the three sites in the same and different hosts and from the same sites in different hosts. When antisera against the four serotypes of *T. tenax* were reacted with the nine antigens isolated from the bronchi, sputum, and/or oral cavity of the four patients (not shown in any table in this chapter), it became apparent that the strongest reactions (although at different titers) usually occurred between the anti–*T. tenax* sera and the strains obtained from the oral cavity. These results suggested that, as might have been anticipated, the trichomonads inhabiting the oral cavity were antigenically closely related to authentic *T. tenax*.

For a more meaningful antigenic comparison of trichomonads from the three different sites of a single host and different hosts, all the experimental strains as well as strains representing the four specific antigenic types of *T. tenax* were reacted with cross-adsorbed antisera against the nine trichomonad strains employed in the investigation (Table 19.16). To facilitate analysis of the data, the antigens found in the strains were arbitrarily designated by lowercase letters. For the species-specific antigens the letters **a** and **b** were used; the strain (type)-specific antigens were designated by the letters **c** to **j**.

It is evident from Table 19.16 that strain 33 I had a unique strain-specific antigen **c**; strain 33 II, antigen **d**; strain 53 I, antigen **e**; strain 53 II, antigens **e**, **i**, and **j**; strain 53 III, antigens **e** and **i**; strain 206 II, antigen **f**; strain 206 III, antigen **g**; and strains 220 II and 220 III, antigen **h**. Among all these strains, only 220 II and 220 III showed no differences in their unique strain-specific antigens. Strains 53 I, 53 II, and 53 III appeared to have a similar strain-specific constitution, antigens **c** through **e** being present in all three strains irrespective of the site of their origin in the host. Strains 206 II and 206 III were distinguished from each other by a single unique strain-specific antigen present in each; these two strains shared only the species-specific antigens **a** and **b**. On the other hand, strains 33 I and 33 II had also many common strain-specific antigens in addition to some unique ones. As far as the absence of differences between strains 220 II and 220 III was concerned, the authors suggested two possible explanations of which one appears to be more plausible, that is, the 220 II trichomonads isolated from the sputum could have actually originated in the bronchi.

Among the four established serotypes of *T. tenax*, only three (A, B, D) had strain-specific antigens (**c**, **f**, and **g**) in common with some of the strains isolated by the authors. Type C evidently contained no such antigens.

Antigenic differences existing among trichomonads from various sites in one host or different hosts or in the same sites of different hosts

TABLE 19.15. Agglutination reactions between trichomonads isolated from respiratory tract (by bronchoscopy) (III)* and other sites (I and II).*

Antisera against strain	Antigen titer,[†] with strains								
	33[‡]-I	53-I	33-II	53-II	206-II	220-II	53-III	206-III	220-III
33-I	800	640	100	100	100	200	100	960	200
53-I	100	6400	100	160	100	240	10240	100	100
33-II	100	200	480	960	400	480	100	100	480
53-II	1600	480	240	1600	240	160	1920	960	100
206-II	100	160	320	800	1920	480	100	960	400
220-II	100	100	240	160	240	240	400	100	200
53-III	2560	10240	100	240	160	320	7680	160	200
206-III	240	100	320	2560	800	480	100	1600	1600
220-III	200	100	160	100	100	100	100	100	320

Reproduced from Kazakova I et al: Intraspecific antigenic differences between *Trichomonas* strains isolated from the respiratory tract and oral cavity of man. *Med Parazitol* 4:34–38, 1985, with permission of Izdatel'stvo Meditsina, Moscow.
*I, oral cavity; II, sputum; III, bronchi.
[†]Expressed as reciprocal.
[‡]Arabic numbers designate individual patients.

TABLE 19.16. Agglutination reactions between adsorbed rabbit antisera and the trichomonad strains (antigens) used in this study as well as the four basic serotypes of *T. tenax*.

Antisera	Strains used for adsorption	Agglutination obtained with												
		Experimental strains									*T. tenax* serotypes			
		33 I abcgij	33 II abdfh	53 I abcde	53 II abcdegij	53 III abcdei	206 II abef	206 III abg	220 II abh	220 III abh	A a...cf	B a...g	C —	D a...g
33* I†	33 II abdfh‡	+	−	−	−	−	−	−	−	−	+	+	−	+
ABCGIJ‡	33 I abcgij	−	+	+	+	+	+	+	−	+	+	−	−	−
33 II†	53 II abcdegij	−	−	−	−	−	−	−	−	−	−	−	−	−
ABDFH	53 III abcdei	−	+	−	−	−	−	−	−	−	−	−	−	−
53 I	53 II abcde	−	−	−	+	+	−	+	−	−	−	+	−	+
ABCDE	53 III abcdei	−	−	−	+	+	−	−	−	−	−	+	−	+
53 II	53 I abcde	+	−	+	+	+	+	−	−	−	−	+	−	+
ABCDEGIJ	53 III abcdei	+	−	−	+	−	−	+	−	−	−	−	−	−
53 III†	53 I abcde	+	−	−	+	−	−	−	−	−	−	−	−	−
ABCDEI	53 II abcdegij	−	−	−	−	−	−	−	−	−	−	−	−	−
206 II	206 III abg	−	−	+	+	+	+	−	−	−	+	+	−	+
ABEF	206 II abef	−	−	−	−	−	−	−	−	−	−	−	−	−
206 III	206 II abef	−	−	−	−	−	−	+	−	+	−	+	−	+
ABG	220 III abh	−	−	−	−	−	−	−	−	−	−	−	−	−
220 II		−	−	−	−	−	−	−	+	−	−	−	−	−
ABH	220 III abh	−	−	−	−	−	−	−	−	−	−	−	−	−
220 III		−	−	−	−	−	−	−	−	−	−	−	−	−
ABH	220 II abh	−	−	−	−	−	−	−	−	−	−	−	−	−

Reproduced from Kazakova I et al: Intraspecific antigenic differences between *Trichomonas* strains isolated from the respiratory tract and oral cavity of man. *Med Parazitol* 4:34-38, 1985, with permission of Izdatel'stvo Meditsina, Moscow.

*Arabic numbers designate individual patients.

†I, oral cavity; II, sputum; III, bronchi.

‡To facilitate the analysis of the data, the *antigenic structure* of the trichomonad strains is represented arbitrarily by *lowercase letters*; that of the corresponding *antibodies* is represented by *capital letters*. In addition, the following designations are used: *species-specific* antigens—a, b; *type-specific antigens* for strain 33 I—c; strain 33 II—d; strain 53 I—e; strain 53 II—e,i,j; strain 53 III—e, i; strain 206 II—f; strain 206 III—g; strains 220 II and 220 III—h.

are evident also from the results of quantitative complement fixation (CF) reactions reported by Rõigas et al.[94] Unfortunately some of the antigens prepared from trichomonad strains occurring in the bronchi, sputum, and oral cavity were found to have anticomplementary activity that could not be eliminated by any of the known methods. Therefore, the authors had to carry out the CF reactions only with antigens that lacked this activity, that is, from the four known serotypes of *T. tenax* and from one strain, 53 III, isolated from the bronchi.

To establish the minimum specific CF titer, Rõigas et al[94] tested sera from 32 children, 5 to 12 years in age, who had no diseases of the oral cavity or the respiratory tract. This titer was found to be lower than 1:40.

The unexpected findings of specific CF reactions (with one or more strains) in sera of 10 out of 27 patients in whom trichomonads could not be demonstrated in any site (Table 19.17) were assumed by the authors to be caused by the fact that the donors of the sera actually harbored these parasites, which, however, could not be demonstrated on only the single examinations possible under the conditions of their investigation.

Using sera of 21 patients harboring the trichomonads in sites other than the bronchi, Rõigas et al[94] were able to obtain positive CF reactions (titers greater than or equal to 1:40) in 18 patients, and these only by simultaneously employing all four *T. tenax* antigenic types (Table 19.17). Evidently with a few sera a positive reaction could not be obtained even when all the antigens were used. In any event, the titer of a given serum depended on the antigenic composition of the trichomonad strain involved and thus on the properties of the antibodies whose formation this antigen elicited in the serum. Further to prove this point, the authors employed the *T. tenax* serotype C and strain 53 III from the bronchi in the analysis by CF reaction of sera from patients with trichomonads present in their bronchi (Table 19.18). Of the 27 sera, positive reactions were obtained with 9 sera through the use of the *T. tenax* antigenic type C, while the use of strain 53 III yielded 15 positive CF reactions. When both strains were employed simultaneously such reactions were recorded for 18 sera (Table 19.18).

In addition to the CF reactions, Rõigas et al[94] used the indirect fluorescent antibody test (IFAT) in the antigenic analysis of trichomonads from the respiratory tract and other sites. In 18 serum donors with pulmonary diseases no trichomonads were found in any site, while in 17 donors these flagellates were detected in bronchi and in other sites (sputum and the oral cavity) (Table 19.19).

TABLE 19.17. Quantitative complement fixation reactions (CF) involving the four antigenic types of *T. tenax* and sera from patients with respiratory disease and either with trichomonads found in sites other than the respiratory tract or from those presumably negative for trichomonads.

| Trichomonads | Number of sera employed | Number of sera with CF titers $\geq 1:40$* with each of the four known antigenic types of *T. tenax*[†] | | | | Number of sera with CF titers of no less than 1:40 noted with two or more antigenic types of *T. tenax* |
		A	B	C	D	
In sites other than bronchi[‡]	21	7	14	12	3	18[§]
Not detected	27	5	6	3	0	10[ǁ]

Modified from Rõigas E et al: Results of quantitative complement fixation and indirect immunofluorescence method with blood sera of persons in cases of trichomonad invasion of lungs. *Izv Akad Nauk Est SSR (Biologiia)* 29:97-102, 1980, with permission of the Academy of Sciences of the Estonian S.S.R.
[‡]This category includes sputum and oral cavity.
*Titers <1:40 are considered nonspecific.
[†]According to Kumm et al[30]; see also Kazakova et al.[96]
[§]The authors[94] were uncertain of the reasons for this discrepancy (18 vs. 21) in CF reactions. Their explanation suggested the presence of major antigenic differences between some of the strains isolated from the patients and the four antigenic types (A to D) of *T. tenax*.
[ǁ]The authors[94] hypothesized that, although actually present, the trichomonads might have been undetected on single examinations necessitated by the conditions of their experiments. If this were the case, finding of positive CF reactions would be anticipated.

TABLE 19.18. Dependence of the quantitative complement fixation (CF) titers of sera from individuals harboring trichomonads on the sites of origin of the strains employed as antigens.

Sites of trichomonads	Number of sera employed	Number of sera with titers ≥1:40* in reactions with trichomonad strains		Number of sera with titers ≥1:40 in reactions with both strains
		T. tenax serotype C[†]	53 III from bronchi	
Bronchi	14	4	7	9
Bronchi and sputum	4	0	3	3
Bronchi, sputum, and oral cavity	3	1	1	1
Bronchi and oral cavity	6	4	4	5
TOTAL	27	9	15	18

Modified from Rõigas E et al: Results of quantitative complement fixation and indirect immunofluorescence method with blood sera of persons in cases of trichomonad invasion of lungs. *Izv Akad Nauk Est SSR (Biologiia)* 29:97-102, 1980, with permission of the Academy of Sciences of the Estonian S.S.R.
*Titers <1:40 are considered nonspecific.
[†]According to Kumm et al[30]; see also Kazakova et al.[96]

With two exceptions, discussed briefly in the final footnote to Table 19.19, sera from patients without trichomonads gave none (−) or only weak (±) fluorescence reactions with all the antigens employed. On the other hand, strong fluorescence (+ or ++) was observed in the IFAT involving sera from the 17 patients who harbored the parasites in the respiratory passages and in the other sites. The authors claimed to have observed in these reactions a dependence of the results on the antigenic properties of the trichomonad strains employed. It was demonstrated, for example, that among these strains, the one isolated from the sputum (53 II) was unique in that it emitted only strong fluorescence on treatment with all 17 sera from *Trichomonas*-positive patients (Table 19.19). It was also of interest that with strains 53 I and 53 III isolated, respectively, from the oral cavity and bronchi of a single patient, high fluorescence was observed on treatment of the former strain with 14 sera and of the latter with 15 sera. Furthermore, evidently among the 17 sera only three imparted fluorescence of the same intensity to trichomonads belonging to all strains and only eight gave a positive IFAT with all strains (not shown in Table 19.19). The aforementioned findings, although they strongly suggest a correlation between strong fluorescence and the presence of trichomonads in the host, do not seem to constitute strong evidence for dependence of the differences in fluorescence intensity on the antigenic constitution of the test strains.

Of some interest was the finding that no correlation existed among the reactions obtained by the several immunologic methods. According to Kazakova et al,[102] "the reactions [agglutination, CF, and IFAT] depended to a [high] degree on the type-specificity of the antigens used, but there was no correlation between the results obtained by the several reactions, i.e. some sera with 'negative' CF contained specific agglutinins or gave 'positive' [IFAT]. At the same time, in some sera with 'positive' CF, the titer of agglutinins was not higher than . . . [nor-

TABLE 19.19. Indirect fluorescent antibody test (IFAT) with sera from patients suffering from respiratory diseases.

Trichomonads	Number of sera employed	Fluorescence intensity* with trichomonad strains from								
		Bronchi			Sputum			Oral cavity		
		(±)	+	++	(±)	+	++	(±)	+	++
From bronchi	17	2	12	3	0	15	2	3	11	3
Not detected	18	17	1[†]	0	17	1[†]	0	18	0	0

Modified from Rõigas E et al: Results of quantitative complement fixation and indirect immunofluorescence method with blood sera of persons in cases of trichomonad invasion of lungs. *Izv Akad Nauk Est SSR (Biologiia)* 29:97-102, 1980, with permission of the Academy of Sciences of the Estonian S.S.R.
*±, Weak fluorescence; +, strong fluorescence; ++, very strong fluorescence.
[†]Possibly the "+" fluorescence was due to the fact that the trichomonads, although actually present, could not be detected on the single examination feasible under the conditions of the experiments.

mal]." Furthermore, it was stated by Rõigas et al[94] that there was no correlation among the results obtained by IFAT, agglutination, and CF with the sera from individual patients and diseases of the respiratory tract found in the donors of these sera. Also, no relationship was noted between the results of the aforementioned immunologic tests and the trichomonad strains (antigens) isolated from the respiratory passages or simultaneously from those passages and the oral cavities of individual patients.

More recently, Sardis et al[103] performed sensitization experiments using *T. tenax* antigenic type C for subcutaneous and intracutaneous sensitization of guinea pigs; antigenic types A, B, C, and D of this species were employed for intraperitoneal sensitization. In the animals sensitized with type C antigen, one strain from a case of pulmonary disease, one from the oral cavity, one from sputum, and two from bronchi were used as the allergens, as were the strains employed originally for sensitization and also *P. hominis* (strain II) and *T. vaginalis* (strain TN). After obtaining baseline data with known allergens (especially from pulmonary infections) by the intracutaneous test, the investigators tested the reaction in 30 patients with pulmonary diseases. Among these patients, 16 harbored trichomonads in the bronchi, and in six patients the parasites were found in the bronchi as well as in the oral cavity. On the basis of the results obtained in their experiments, Sardis et al[103] concluded that "trichomonads isolated from the human respiratory tract have allergenic properties." They were able to induce specific allergic reactions in laboratory animals sensitized experimentally, as well as in human hosts harboring the parasites in their respiratory organs. According to the authors[103] the allergic response appeared to be dependent on the specific antigenic types of the protozoa. Therefore the success of the sensitivity test was predicated on the simultaneous employment of more than one antigenic type of the respiratory tract trichomonads in preparing the allergens.

It was concluded from all the experiments that

1. Most strains, irrespective of their site in the host, shared, in addition to species-specific antigens (present in all strains), also some strain-specific antigens. However, some of the latter, found in the parasites from one site, could be found also in those from other sites of the same host or various sites in different hosts.[101]

2. Most strains shared some strain-specific antigens with at least three (A, B, D) of the known serotypes of *T. tenax*. Some results suggested that the organisms isolated from the oral cavity were the ones most closely related to the previously identified serotypes of "authentic" *T. tenax*.

3. It appeared that some antigenic differences typically existed between the trichomonads inhabiting the respiratory tract and those from the "other sites." Therefore, it seemed highly probable to the Estonian workers that the parasites adapted to parasitizing the respiratory tract represented either a subspecies or at least a different serotype of *T. tenax*. However, as indicated by Sardis,[103] on the basis of the available evidence, the second alternative appears to be much more likely.

Treatment

The question still remains whether pulmonary trichomonads should be treated with 5-nitroimidazoles. Some workers felt that the large numbers of trichomonads in the respiratory organs interfered with recovery even in patients in whom the primary cause of the disease might have been eliminated by other treatments. In light of this, they administered to these patients the 5-nitroimidazole compounds and obtained good results.[81,83,86,88,104] In light of the possible carcinogenic effects of these drugs other workers were less willing to administer nitroimidazoles indiscriminately, at least pending the establishment of the actual pathogenic potential of the parasites.

Conclusions

At this time conclusive evidence is lacking on whether strains of all the species of *Trichomonas* found in the bronchi include strains with a true pathogenic potential. There can be little doubt that, as indicated earlier in the section on *T. vaginalis* Donné in this chapter and also in Chapter 13, *T. vaginalis* can cause pathologic changes in the lungs and in other organs of humans. As far as *T. tenax* strains are concerned, the evidence for their pathogenicity in the oral cavity has not been proved, and many workers who believed in the virulence of the strains found in the respiratory tract failed to provide indisputable evidence of the ability of these parasites to be *per se* harmful to their hosts. Yet under favorable conditions *T. tenax* evidently can migrate into the respiratory passages, and it appears that the trichomonads found in the respiratory organs belong to strains antigenically different from those that inhabit the oral cavity.[78,94,95,99] If, indeed, many or even some of the flagellates found in the human respiratory tract actually belonged to the urogenital species, their alleged pathogenicity or their antigenic differences (as revealed by relatively simple immunologic methods) would not be surprising.

In any event, it seems likely that large numbers of any flagellates in the respiratory passages aggravate diseases caused by other agents[85] and this idea is suggested also by the investigators who prefer to reserve judgment until more is learned about pulmonary trichomoniasis.[87-90,92] It is therefore not surprising that treatment with 5-nitroimidazoles often has been found helpful.

It should be mentioned that in many instances the route of invasion of the respiratory tract remains in doubt. It is likely that *T. tenax* enters by the bronchi from the mouth and, at least in adults, that *T. vaginalis* is most probably introduced by the practice of oral sex. There appear to be few creditable cases of pulmonary infection by *P. hominis*.

In addition to very careful morphologic comparisons in suitable preparations (see Chap. 3), antigenic relationships among the strains of the member species of the genus *Trichomonas* inhabiting the respiratory organs ought to be examined by modern immunologic techniques (see Chap. 4).

Pentatrichomonas hominis (Davaine, 1860)

Nomenclature

The rather confusing nomenclature of the human intestinal trichomonad was discussed in detail by several investigators,[16,105-107] and a list of its synonyms was provided by Honigberg.[107]

Evidently only one trichomonad species characterized by a "4 + 1" arrangement of the anterior flagella can be found in the large intestine of humans. There is still some question if the "independent" flagellum (see Figs. 4 to 7 in reference 107a) is a true anterior flagellum. Discussion of the structure and nomenclature of the human intestinal trichomonad usually includes an explanation, based mainly on the studies of Flick,[108] of the variability of the number of "anterior flagella" in the species in question. In any event, there can be now little doubt that the intestinal trichomonad of humans should be referred to as *Pentatrichomonas hominis* (Davaine, 1860) Wenrich, 1931, and that all the other names that have been used for this species ought to be regarded as synonyms of *P. hominis*.

Geographic and Host Distribution

Pentatrichomonas hominis appears to have a very wide host distribution. Flagellates structurally indistinguishable from the species found in the human colon have been reported to occur naturally in the large intestine and cecum of many nonhuman primates, cats, dogs, and a variety of rodents.[108-112] Experimental infections of various mammalian hosts with *P. hominis* were recorded by Levine.[15] Furthermore, an organism difficult to distinguish from the intestinal trichomonad of humans has been reported from kangaroos, namely, *Pentatrichomonas macropi*,[113] and a morphologically similar flagellate was described in detail from the rumen of cattle.[114]

Pentatrichomonas hominis has a cosmopolitan distribution. According to a number of investigators,[16,115,116] this trichomonad appears to be more prevalent in tropical and subtropical regions of the world. Since transmission of *P. hominis* is contaminative, it must depend primarily on the hygienic customs and dietary habits of the human host. Indeed, the customs and habits prevalent among the populations of the developing countries in the subtropical and tropical parts of the globe, combined with the warm and humid climatic conditions, are often especially favorable for transmission of this intestinal flagellate.

Antigenic Identity and Other Immunologic Aspects

As is the case with other trichomonad species parasitic in humans, the earliest immunologic experiments were published by Tokura,[25] who used xenic cultures and immune sera from rabbits injected intravenously with formalinized *P. hominis.* Tokura considered titers of 1:6 to be specific. Cross-reactions among *P. hominis, T. tenax,* and *T. vaginalis* were recorded with indiluted or 1:2 diluted anti-*Pentatrichomonas* serum; they were stronger between *P. hominis* and *T. tenax* than between *P. hominis* and *T. vaginalis.* In light of the very low titers, which were lower than those typically attributed to "natural" antibodies, a meaningful interpretation of Tokura's results is not possible.

MacDonald and Tatum,[117] using sera from rabbits immunized intravenously with formalinized trichomonads as antigens in agglutination reactions, concluded that *P. hominis* and *T. vaginalis* were antigenically identical. However, these results differed from experimental data obtained by all subsequent investigators.

Kott and Adler[26] reported antigenic similarities and differences between *P. hominis* and *T. vaginalis.* With the aid of agglutination and two hemagglutination methods, these workers demonstrated that one of the antigenic types (Type I) of *P. hominis,* which included three strains, had certain common antigens with some strains of the urogenital parasite; however, it could be distinguished from them by means of cross-adsorbed sera. Subsequently, Kott and Adler[26] showed by hemagglutination involving red blood cells sensitized with Fuller's "polysaccharide" preparation that Type II antigens of *P. hominis* failed to react either with Type I of this species or with strains of *T. vaginalis* available to the investigators. However, cross-reactions between Type I and Type II antigens of *P. hominis* as well as with *T. vaginalis* strains were observed in hemagglutination tests involving red blood cells treated with tannic acid before being sensitized with the trichomonad extract. On the basis of their results, Kott and Adler[26] concluded that the antigens demonstrated in reactions using the tanned red blood cells were neither agglutinogenic nor "polysaccharide" in nature.

No cross-reacting antigens between *P. hominis* and *T. tenax* were observed by Kott and Adler[26] in agglutination reactions. It should be noted, however, that the oral trichomonad was grown in xenic cultures and that the anti-*T. tenax* sera raised in rabbits by intravenous inoculation of living cells were adsorbed with the bacteria cocultivated with the flagellates. Somewhat different results were reported by Oleinik[27] who, using agglutination and gel diffusion methods, could differentiate antigenically *T. vaginalis* from *P. hominis* and *T. tenax,* but was unable to distinguish between the latter two species by these methods. In light of the fact that of the three species, only *T. vaginalis* was grown in axenic culture, Oleinik's results could have been affected by the difficulties encountered in immunologic tests employing parasites from xenic cultures.

Mannweiler and Oelerich[118] using antisera produced in rabbits against *T. vaginalis* grown axenically in the serum-free C7 medium of Samuels[119] observed cross-reactions between the urogenital trichomonad and *"Trichomonas hominis"* (both from axenic cultures) by means of complement fixation and gel diffusion tests. Two "partial antigens" revealed by gel diffusion were apparently responsible for these reactions. No immunologic reactions were noted by Mannweiler and Oelerich[118] between the anti-*T. vaginalis* serum and either *"Pentatrichomonas ardindelteili"* or *T. foetus.* Subsequently these results were adduced by Michel and Westphal[120] as additional evidence in support of the existence of separate four and five anterior-flagella-bearing, but otherwise morphologically indistinguishable, genera and species of trichomonads inhabiting the human large intestine. However, the presently available information, briefly mentioned in the section dealing with nomenclature and supported by much evidence (the important references are cited in that section), renders most unlikely the presence of more than one species of *P. hominis*-like flagellates in the human colon. It seems more probable that the differences in cross-reactivity of the two allegedly different *P. hominis*-like intestinal trichomonads with *T. vaginalis* depended on those in the serotypes reported from *P. hominis.*[21,121-123] This assumption finds support in the results reported for the skin test, in which the delayed-type hypersensitivity reaction was studied in mice injected intracutaneously or intramuscularly with *T. vaginalis* and then challenged by intradermal inoculations of *T. vaginalis, "T. hominis,"* or *"P. ardindelteili."*[120] All mice challenged with either *T. vaginalis* or *"T. hominis"* gave positive delayed-type hypersensitivity reactions, but such reactions were recorded in fewer mice challenged with two strains of *"P. ardindelteili." Trichomonas vaginalis*-sensitized mice challenged with *Tritrichomonas foetus* gave only negative reactions.

The results obtained by Michel and Westphal[120] in agglutination reactions involving anti-"*Trichomonas hominis*" and anti-*T. vaginalis* sera raised in mice with the homologous and heterologous antigens, including "*P. ardindelteili*" and *T. foetus,* are of interest. Indeed, they lead to conclusions essentially similar to those noted with delayed-type hypersensitivity reactions regarding the antigenic relationships between *T. vaginalis* and *P. hominis*-like trichomonads and the absence of such relationships between the aforementioned serotypes and species of the human parasites and *T. foetus.*

Apparently, the only species-specific,[120] although not strain (serotype) specific,[124] test for the trichomonad species discussed in this section is the cell-mediated "peritoneal cell reaction."

With the aid of agglutination[29,121-123] and complement fixation tests[29] in which immune sera raised in rabbits by intravenous inoculations of living *P. hominis* or *T. vaginalis* were reacted with homologous and heterologous antigens, Estonian workers noted common and unique antigens in these two species. The urogenital and intestinal species also shared some antigens with *T. tenax.*[29] Furthermore, through the use of agglutination with nonadsorbed and cross-adsorbed sera, these workers[121,122] demonstrated the presence of four antigenic types among 23 strains of *P. hominis.*

The most recent report on the antigenic composition of *P. hominis* strains is that of Kazakova.[123] She used agglutination for the antigenic analysis of axenically grown flagellates, and provided more details on the preparation of the antigens and antisera used in the reactions than have been available in the previous papers published on this subject by the Estonian workers. It would have been welcome if Kazakova[123] cited a reference to, or provided an account of, "minimum heterogeneity, EBM 'Razdan-Z' method devised in the Tallinn Polytechnic Institute," which she appeared to have employed for grouping antisera on the basis of their titers and thus for dividing the *P. hominis* strains she isolated into four antigenic types.

After a long discussion leading to the conclusion that the differences among the trichomonad strains were based on their antigenic composition, Kazakova performed a crucial agglutination experiment using two strains that showed major differences in the titers with their homologous antisera. These strains were reacted with their homologous and heterologous antisera, nonadsorbed and cross-adsorbed. Agglutination was obtained with homologous sera adsorbed with heterologous antigens; sera adsorbed with their homologous antigens failed to react with either strain. It should be assumed that these strains were the same as those designated previously as the first two serotypes of *P. hominis.*[121,122] It ought to be assumed further that the four antigenic types (among which were included the two mentioned in the previous sentence) determined in the agglutination reactions involving the remaining 23 strains isolated by Kazakova[123] were also the same as those already reported by Teras et al.[122] As far as can be ascertained, Kazakova's paper[123] is the first to include data showing cross-reactions (in this instance, cross-agglutinations) between the four known basic antigenic types (serotypes) I, II, III, and IV, of *P. hominis* and their homologous antisera, each adsorbed with one of the three heterologous strains (Table 19.20). To facilitate the analysis of the four strains, their antigenic components were arbitrarily designated by lowercase letters and their antibody sets by capital letters. The species-specific antigens, found in all strains, were represented by the letters **a** and **b**, and the type-specific antigens by the letters **c** to **i**. Of these latter, antigen **c** was found to be unique to type I, **d** to type II, **e** to type III, and **f** to type IV. Appropriate species-specific (**A** and **B**) and type-specific (**C** to **I**) antibodies were found in the corresponding antisera. Finally, Kazakova[123] reacted 20 strains with antisera to the basic antigenic types of *P. hominis,* each of which was adsorbed with each of its three heterologous antigenic types separately. This time, the results of all the reactions indicated the presence of six antigenic groups among the test strains. As before, antigen **c** was unique to the type I serotype, antigen **d** to type II, antigen **e** to type III, and antigen **f** to type IV. The new antigenic type V reacted with antibodies to two serotypes and the new type VI failed to react with any of the antibodies present in the four tested antisera. All strains of type VI were found in patients hospitalized in Tallinn. Among these patients were noted also strains belonging to type III. On the other hand, with one exception, strains of antigenic type II were isolated from persons examined in Tbilisi. It was concluded that in addition to strains belonging to different serotypes (see above) there were found distributed in the Estonian and Georgian Republics strains with similar antigenic composition.

TABLE 19.20. Reactions between each of the four known antigenic types of *P. hominis* and immune sera against them adsorbed with each of the heterologous antigens.

Immune serum against the four antigenic types	Specific antigenic type used for adsorption	Agglutination reaction with specific antigenic types			
		I (abcgh)	II (abdhi)	III (abegi)	IV (abf)
I* (ABCGH)[§]	II (abdhi)[†] III (abegi) IV (abf)	+[‡] + +	(−) + +	+ (−) +	(−) (−) (−)
II (ABDHI)	I (abcgh) III (abegi) IV (abf)	(−) + +	+ + +	+ (−) +	(−) (−) (−)
III (ABEGI)	I (abcgh) II (abdhi) IV (abf)	(−) + +	+ (−) +	+ + +	(−) (−) (−)
IV (ABF)	I (abcgh) II (abdhi) III (abegi)	(−) (−) (−)	(−) (−) (−)	(−) (−) (−)	+ + +

From Kazakova I: Agglutinogenicity and agglutinability of strains of *Trichomonas hominis* (Davaine). *Izv Akad Nauk Est SSR (Biologiia)* 24:130-140, 1975, with permission of the Academy of Sciences of the Estonian S.S.R.
*I to IV: The four specific antigenic types of *P. hominis*.
[§] Antibodies are designated by capital letters: A,B—species-specific (common) antibodies; C-H—type-specific antibodies.
[†] Antigens are designated by lowercase letters: a,b—species-specific (common) antigens; c-h—type-specific antigens.
[‡] +, Positive reaction; (−), negative reaction.

The findings of Rõigas and his collaborators[125,126] confirmed the presence of several antigenic types (serotypes) of *P. hominis*. By using agglutination and complement fixation techniques, they demonstrated the occurrence of antigenic changes in the course of 40 in vivo (in mice) and in vitro passages of clone-derived populations of strains belonging to the four basic serotypes. They noted also that ". . . even when, after in vitro and in vivo passages, changes occurred in the biological properties including antigenic properties of the serotypes of *T.* [= *P.*] *hominis*, the intraspecific antigenic variations of the protozoan were still detectable."[125,126] It appears thus that the changes in the antigenic characteristics occurring in the course in vitro and in vivo passages do not affect the basic serotypes of *P. hominis* and support the view that the serotypes characterize groups of strains rather than individual strains. Rõigas and his group failed to demonstrate positive correlations between the antigenic changes, on the one hand, and pathogenicity for mice and hexokinase activity, on the other; however, there appeared to be a certain degree of positive correlation between the latter two parameters. The details of the experiments, which entailed only agglutination and complement fixation methods, are not given in either paper. Therefore, the results of Rõigas et al[125,126] are difficult to compare with those obtained by other investigators, especially on *T. vaginalis* about which most information is available (see the references to the reports of Teras and his collaborators).[1,127] It is of interest, however, that the reports of Rõigas et al[125,126] contain the first indication by the members of the Estonian group that, as recognized previously by other investigators,[128-130] antigenic changes occur during in vitro cultivation of trichomonads.

Stępkowski and Honigberg,[129] using gel diffusion and immunoelectrophoresis, found *P. hominis* and *T. gallinarum*, from the large intestine of primarily galliform birds, to have more of the thermolabile antigens (inactivated by exposure to 70°C for 30 minutes) than of the thermostable antigens (resistant to 30-minute exposure to 70°C). In contrast, *T. vaginalis, T. tenax*, and *T. gallinae* have as many or more thermostable antigens as thermolabile ones. Evidently, major antigenic differences exist between *P. hominis* and *T. gallinarum* on the one hand, and the *Trichomonas* spp. on the other. There are also common and unique antigens in *P. hominis* and *T. gallinarum*, but fewer antigens are shared by either of the latter species and those of the genus *Trichomonas*. Among *Trichomonas* spp., *T. vaginalis* seems to be antigenically closest to *P. hominis*, as could be anticipated from previously published results.[26,117]

Biochemical Attributes

There are only a few published accounts dealing with the biochemistry of *P. hominis*, a species known for over 100 years. As with immunology, most information has been compiled on *T. vaginalis* and *Tritrichomonas foetus*, both of which contain strains pathogenic, respectively, for humans and cattle. In light of this, only a limited amount of information can be summarized here on *P. hominis*.

Solomon,[131] using "chemical analysis," found *P. hominis* suspended in Krebs-Ringer phosphate solution, pH 6.5, to utilize the following carbohydrate energy substrates: glucose, galactose, maltose, lactose, and sucrose. In washed organisms obtained from 12- to 48-hour cultures, the rate of glucose utilization increased linearly with culture age. By the triphenyl tetrazolium assay, *P. hominis* was shown to oxidize citrate, α-ketoglutarate, malate, and β-hydroxybutyrate as well as glutamate. Thus this species has some of the tricarboxylic cycle enzymes. Indeed, as indicated by Shorb,[132] the finding of these enzymes suggests that the intestinal trichomonad of humans utilizes a different metabolic pathway than the typical Krebs cycle. Furthermore, hexokinase activity was shown in cell-free homogenates of *P. hominis*,[125,126] this activity increasing parallel to virulence on serial intraperitoneal passages in mice.

Mehra et al,[133] using column chromatography, reported the following amino acids in hydrolysates of the intestinal trichomonads: alanine, arginine, aspartic acid, glutamic acid, glycine, histidine, isoleucine, leucine, phenylalanine, proline, serine, threonine, tyrosine, and valine; some unidentified compounds were also found. Among the identifiable amino acids aspartic and glutamic acids had the highest concentrations, while histidine had the lowest. The same amino acids were demonstrated in seven other species of Trichomonadida, there being, however, some differences in the levels of these compounds among the several species.

Pathogenicity

Human Hosts

In my earlier work on *P. hominis*[1] I might have overemphasized the evidence against even a limited level of potential virulence of the strains of this typically extracellular intestinal parasite. In light of the present-day tendency to underemphasize the virulence of protozoa such as *T. vaginalis*, which unquestionably contains inherently virulent strains, caution must be exercised not to negate all possibility (even if remote) of potential pathogenicity among the less thoroughly investigated species. In light of this, I shall approach this subject more cautiously.

Admittedly, the problem of the inherent virulence potential of *P. hominis* strains, found in many instances in asymptomatic patients and in some in whom the symptoms of intestinal disease must be ascribed to another coexisting etiologic agent, has not been satisfactorily resolved until now, that is, nearly 130 years after its first description by Davaine.[134] Many of the earlier investigators[135-137] held the view that, at least under favorable conditions (conditions of unbalance) prevailing in the host digestive tract, *P. hominis* strains were able to express their inherent virulence, causing a series of intestinal symptoms and signs differing in severity. Also, as indicated by Teras,[82] the Estonian workers considered this trichomonad to be pathogenic. Even those investigators who thought the intestinal flagellate of humans to be inherently pathogenic admitted that in many instances the infections were asymptomatic,[138,139] whereas in other instances the symptoms ranged from mild to severe[116,138-150] (see especially reference nos. 138, 139, and 147 for additional reports). (Consult also Kazakova,[123] especially for some Russian papers that appeared in publications, such as proceedings of local meetings, unavailable to me.)

At this point it seems appropriate to list some of the symptoms and signs as well as pathologic changes described in association with human intestinal trichomoniasis. It must be emphasized, however, that identical manifestations have been observed in patients suffering from amebiasis, giardiasis, and bacterial infections. It is essential, therefore, to eliminate the possible involvement of protozoan parasites other than *P. hominis* and to exclude the possibility that the observed changes are caused in part or in whole by pathogenic bacteria. One of the lines of evidence supporting the role played by the intestinal trichomonad in causing various gastrointestinal disturbances is the disappearance of the symptoms following the elimination of the parasite by specific drugs, for example, one of the 5-nitroimidazoles. However, even in the absence of *Giardia* or the dysentery ameba, such a treatment would eliminate not only the trichomonads, but also anaerobic bacteria[1,18]; therefore, even this line of evidence must be viewed with caution.

One of the most common manifestations of symptomatic intestinal trichomoniasis has been diarrhea of differing severity, the most severe resembling that characteristic of dysentery. In severe cases the stool may contain mucus, blood, or both. Chronic enteritis and chronic or acute colitis have also been reported. Migrating abdominal pain establishing itself in the right inguinal region is not uncommon, and appendicitis, accompanied at times by periappendicular ulcers, has been reported on a number of occasions. Acute hepatitis, with trichomonads present in hepatic ulcers, has been described by some clinicians, and inflammation and edema of the intestinal mucosa have been observed by rectosigmoidoscopy and sigmoidoscopy. The manifestations of intestinal trichomoniasis are said to include nausea and vomiting, anorexia, and dysuria; body temperature may be elevated, ranging between 38°C and 39°C.

Even those authors who subscribed to the idea of inherent virulence of the intestinal trichomonad of humans equivocated somewhat. A few examples should suffice to illustrate the point. Caruso[138] believed that, especially under favorable conditions prevalent in the host's digestive tract, *P. hominis* may be virulent per se. Yet he suggested in his long review that the question of pathogenicity of this flagellate has not yet been resolved on the basis of the capacity of the parasite to "invade" tissues and phagocytize red blood cells (with regard to this alleged pathogenicity indicator, see Dobell[151]) in the course of acute, subacute, or chronic colitis. According to Caruso[138] (see also Manson-Bahr[152] below), it is often difficult to decide if the presence of *P. hominis* is the cause or the consequence of the disease. The presence of high numbers of the flagellates in the intestine might be due in large part of their swarming in an abnormal (liquid) environment, as for example in the case of bacillary dysentery. Brisou,[116] who believed that in some instances serious prolonged intestinal disturbances are caused by *P. hominis,* conceded that this parasite would cause little damage in a generally healthy host. Symptomatic diseases may occur in those instances in which changes in the intestinal chemistry equilibrium stimulate swarming of the trichomonad-bacterial complex. In light of this, both the protozoan fauna and bacterial flora ought to be considered in planning treatment. It was thought by Cziráki and Dobías[146] that since *P. hominis* is rarely found in humans and is even less frequently associated with symptomatic infection, the question may be raised whether strains of this protozoon are virulent per se or if they depend on the presence of other environmental factors for expression of their potential virulence.

Although Manson-Bahr[152] is often cited as categorically opposed to the idea of pathogenicity of *P. hominis,* that impression is not quite correct. Actually, this author agreed with Lynch[153] that inflammatory states associated with a variety of intestinal disturbances (in the case of Manson-Bahr, primarily with severe chronic bacillary dysentery) that result in the presence of liquid stools rich in exudates provide excellent culture medium for the trichomonads. As stated by Manson-Bahr,[152] bacillary dysentery causes pathologic changes in the intestinal mucosa resulting in secretion of mucus by the goblet cells. Moreover, in this infection the bowel, damaged by the dysentery toxins, takes a long time to recover and "this is where the flagellates come in." They must therefore be looked on as secondary invaders and steps should be taken to eliminate them. Admittedly, Manson-Bahr,[152] like many other workers (see above), felt that since the majority of persons harboring the intestinal trichomonads present no symptoms, the flagellates are mostly commensal. Yet he conceded that they could constitute an aggravating factor. On the other hand, Lynch[153] considered *P. hominis* essentially nonpathogenic.

As mentioned before, some of the older workers and even a few recent ones[149,150] have been convinced of the inherent virulence of *P. hominis.* These latter investigators judged the flagellate to be pathogenic primarily because they were able to eliminate the symptoms and manifestations of the intestinal disease by metronidazole treatment of patients whose fecal samples included also *Ascaris, Trichuris,* and hookworm eggs. The difficulties in accepting as proof of virulence elimination of the symptoms by eliminating the intestinal trichomonad with the aid of drugs have been pointed out earlier in this chapter.

As is usual with non-tissue-dwelling protozoan parasites, a number of investigators denied any virulence potential of *P. hominis.*

Among them were Whittingham,[154] Wenyon,[154a] Dobell,[151] Lynch,[153] and authors such as G. A. Alexiev (1937), G. V. Epstein (1941) (quoted in reference 148), and Faust et al.[155]

Although all claims regarding the lack of inherent pathogenicity of *P. hominis* remain to be proved, it appears likely that this flagellate is not an important pathogen of humans.

There are some reports of *P. hominis* from the respiratory tract of humans (Table 19.11). As was indicated by Walton and Bacharach[89] and as will be pointed out here, the identifications of most of these pulmonary trichomonads were based on morphologic, antigenic, and physiologic characteristics insufficient for their assignment to one of the species found in humans. There are, however, two instances in which the identification of the pulmonary trichomonad as *P. hominis* seems justified. One of these, documented by Houin et al,[156] dealt with trichomonads abundant in a subphrenic abscess and thoracic empyema fluid following gastrectomy. The flagellates, which could be isolated in culture, had the structural characteristics of *P. hominis* and were associated with bacteria and fungi typically found in the intestine. The authors considered the trichomonads nonpathogenic inhabitants of the respiratory system. In the second record, Polivoda et al[157] stated that "during a laboratory examination of sputum, there were found trichomonads which, on the basis of their structure, were diagnosed as *Trichomonas hominis*." Although no details of the techniques employed were given, it can perhaps be assumed that the authors saw the intestinal trichomonads.

Experimentally Infected Rodents

Subcutaneous, intramuscular, and intraperitoneal routes were used by Shinohara and Iwai[62] for inoculation of mice with 9×10^4 *P. hominis* grown in axenic cultures. Inocula of this size caused subcutaneous and intramuscular abscesses, but no intraperitoneal lesions. Since no living organisms were noted in the purulent contents of the subcutaneous and intramuscular abscesses and since no cultures could be initiated from these contents after the seventh day following inoculation, it was concluded that the lesions healed rapidly. Their experimental results suggested to the Japanese authors that *P. hominis* had much lower pathogenicity for mice than was exhibited by *T. vaginalis*.

Relatively low pathogenicity of *P. hominis* for mice inoculated by oral and rectal routes were reported by Foresi[158]; the intracecal route was also used, but it presented certain technical difficulties. The most dependable results were obtained by intrarectal infection in which typical trophozoites were employed. The oral route also resulted in good infection, but the "cyst-like" forms (presumably pseudocysts) previously described by Foresi[159] were the primary infectious stages. Of 43 mice inoculated by the several routes, 67.4% became infected, with the parasitemia lasting from 2 days to 11 weeks. In general, the mice infected with *P. hominis* exhibited no pathologic manifestations, except for occasional diarrhea in seven animals (24.1%). In no case was diarrhea observed before the third postinfection day. In mice infected for a longer time, the diarrhea was intermittent. The health of the hosts was generally good. Of some interest was the finding that one control animal had an episode of diarrhea during the first 3 days of the experiment. The immune response was weak, agglutination tests with sera of mice being negative on the 24th, 50th, and 90th postinoculation days. On the other hand, agglutination was noted with antitrichomonad rabbit sera at a 1:16 titer. However, an immobilization reaction was obtained after 30 minutes with sera, diluted 1:24, from the infected mice.

The experimental results reported by Kulda[63] and Teras et al[160,161] who employed axenic cultures of *P. hominis* for subcutaneous and intraperitoneal inoculations of mice, respectively, suggested that this species has a very low pathogenicity level. The subcutaneous mouse lesions caused by the intestinal trichomonad had significantly smaller mean volumes than those produced by strains of *T. vaginalis, T. gallinae, Tetratrichomonas gallinarum, Tritrichomonas foetus,* and *Tritrichomonas suis*. On the other hand, the abscesses caused by *P. hominis* were of a size similar to that recorded for the lesions in mice inoculated with *Trichomonas tenax*.[63] Teras et al[160,161] had to employ inocula containing 2.5×10^8 organisms in the intraperitoneal mouse assay to demonstrate differences among the virulence levels of *P. hominis* strains; inocula of 2×10^7 intestinal trichomonads failed to result in the formation of any lesions. An intraperitoneal inoculum of 2.5×10^8 organisms is quite large in comparison with that of 4×10^6 flagellates used in the assay for virulence of *T. vaginalis* strains. Histologic examination of the hepatic lesions from the mice injected intraperitoneally with strains of *P. hominis* showed that these lesions were superficial and had a tendency to heal, as evidenced by encapsulation of the necrotic foci.[160] The relatively low virulence of the *P. hominis* strains notwithstanding, Teras and his collaborators[161] divided the 31 strains they employed into three groups on the grounds of the extent and character of the pathologic changes the parasites caused in the intraperitoneal cavity of mice.

Al-Dabagh and Shafiq[162] used a *P. hominis* strain isolated in xenic culture from a preschool child. Splenectomized rats, 2 to 3 months old, presumably free of rodent trichomonads, were inoculated orally, rectally, and intracecally with about 3×10^5 trichomonads in 1 ml of culture medium. The rats were divided into four groups as follows: group 1—four infected with *Trichomonas muris* (controls) and four infected orally with *P. hominis;* group 2—four infected with *T. muris* (controls) and four infected rectally with *P. hominis;* group 3—two infected with *T. muris* (controls) and six inoculated intracecally with *P. hominis;* and group 4—eight uninfected. No abnormal manifestations were noted in the *T. muris*-infected controls. Among the animals inoculated experimentally with *P. hominis* no infection was noted in group 1 in the course of 3 months of observation. In group 2, the feces were semiformed and contained intestinal trichomonads, red blood cells, and pus cells. The parasites were eliminated 5 to 11 days after infection and all morbid signs disappeared in 1 to 2 months. In group 3, diarrhea with the stools containing *P. hominis* as well as red and white blood cells was observed 3 to 5 days after infection. Two of these rats died; on necropsy, severe inflammatory lesions were noted in the colon and motile trichomonads were seen in intestinal secretions.

Al-Dabagh and Shafiq[162] believed that *P. hominis* can acquire pathogenicity for splenectomized rats or after stress, which interferes with the immune response of the host. They thought that in the human host the intestinal trichomonad is generally nonpathogenic, but its strains may become virulent when the host-parasite balance is upset. However, their results can be viewed also in a somewhat different light. It is not surprising that *T. muris,* a common commensal of rats, caused no intestinal disease in the course of relatively long-lasting infections. On the other hand, the results with *P. hominis,* a species often found in healthy rats and other rodent hosts,[110] are somewhat unexpected. Admittedly, it is not known if the flagellates from rats, although structurally identical with those of humans, are also antigenically and physiologically identical. The fact that no infection with *P. hominis* was established in splenectomized rats by oral inoculation might be construed as supporting the view of the existence of antigenic and/or physiologic differences among various strains. The same could be said about the short-lived infections reported from rectally infected animals—the appearance of red blood cells and pus cells in semiformed feces that contained the trichomonads might have been caused by injuries resulting from experimental manipulations. Such procedures could have also been responsible for the appearance of abnormal stools and of pathologic changes in the colon of the experimental animals subjected to laparotomy before receiving direct injections of the trichomonad suspension into the cecum. Perhaps factors other than virulence of *P. hominis,* that is, the experimental procedures, were responsible for the morbid manifestations and pathologic changes reported from the experimental animals of groups 2 and 3.

Tissue Cultures

Kulda[63] tested the cytopathic effect of *P. hominis* in monkey kidney cell cultures. Despite the fact that the trichomonads survived for up to 48 hours in the presence of the cell cultures, they failed to multiply and appeared to cause no pathologic changes in the monkey cells.

Epidemiology

Prevalence

The prevalence of *P. hominis* is summarized in Table 19.21. The data presented in this table are not strictly comparable because of differences in the nature of the groups examined (e.g., random populations or selected groups, such as prisoners, soldiers, hospital patients, patients in asylums); age of the subjects (children, adults,

TABLE 19.21. Prevalence of *P. hominis*.

Geographic area	Total number of individuals examined	Groups examined (various relevant data)	Methods of sampling	Individuals positive for *P. hominis* (%)	Reference
Argentina (La Plata)	716	Random group of subjects	(Stool examination)[a]	1.5	Castex and Greenway (1928) (as cited in ref. no. 179)
Argentina (La Plata)	171	Patients with intest. disorders, Med. Faculty, University of La Plata	(Stool examinations)	8.2	Greenway (1930) (as cited in ref. no. 179)
Brazil (Rio de Janeiro)	82	Infants, almost all with intestinal disturbances	Examination[b] of fresh[c] stool preparations; fixed and stained preparations[d]; cultivation[e]	23.2	Da Cunha and Pacheco (1923)[180]
Brazil (Rio de Janeiro)	86	Adults or children (4 to 12 yr) with some protozoan or helminth parasites	Examination of fresh and concentrated[f] stool preparations	1.2	De Padua Vilela et al (1962)[181]
England	1680	Veterans with history of various intestinal disorders: dysentery (914), others (766)	Probably examination of fresh stool preparations	0.8	McKinnon (1918)[182]
England (Lancashire County)	504	Asylum patients; 285 ♂ 219 ♀; average age 42 yr; majority without gastrointestinal manifestations	Examination of fresh stool preparations; flagellates found in semiformed or liquid stools; 1 exam./patient	0.8	Smith (1919)[183]
France (Paris)	174	Hospitalized patients	(Stool examinations)	0.5	Labbé et al (1912) (as cited in ref. no. 179)
France (Vichy)	562	Random group of private patients	(Stool examinations)	1.6	Nepveux (no date) (as cited in ref. no. 179)
France (Paris)	2000	Hospitalized patients	(Stool examinations)	0.5	Le Noir and Deschiens (1923) (as cited in ref. no. 179)
France (Toulon)	500	Random group of individuals (?)	(Stool examinations)	1.2	Pirot (1932) (as cited in ref. no. 184)
Germany	444	Random group of 13-to-25-yr-old individuals examined in Tropeninstitut, Hamburg, entering Germany from South and Central America, Africa, China, Spain, Greece	Examination of fresh stool preparations, unstained and iodine-stained[g]; fixed and stained preparations[c]; cultivation[g]	2.5	Gönnert and Westphal (1936)[185]
Hungary	5545	Group of patients with cutaneous diseases	(Stool examinations)	0.1	Hargita (1962)[186]

Location	N	Subjects	Method	Prevalence (%)	Reference
Hungary	—[h]	Random groups of apparently healthy individuals	(Stool examinations)	0.5 to 3.0	Quoted (without details) from several sources in ref. no. 146
Indonesia [South Kalimantan (Borneo), 7 villages]	2169	Random group of 1- to 50+-yr-old persons; 1316♂, 853♀	Examination of formalinized stool preparations; examination of concentrated fecal matter	~0.1	Cross et al (1975)[187]
Indonesia (West Kalimantan [Borneo], 8 villages)	2101	Random groups of 1- to 89+-yr-old persons; 1180♂, 921♀	Examination of formalinized stool preparations, nonstained and iodine stained; examination of concentrated fecal matter	~0.1	Cross et al (1976)[188]
Indonesia (Central Sulawesi, North Lore District, 11 villages in Napu and Besoa valleys)	1003	Random groups of 0- to 50+-yr-old persons; ♂ and ♀ about equal	Examination of formalinized stool preparations; examination of concentrated fecal matter	<2.0	Carney et al (1977)[189]
Iran (Fars province)	24,287	Group of hospitalized patients from entire province	Examination of fresh stool preparations; fixed and stained preparations; 1 stool/patient	2.0	Nazarian (1973)[190]
Italy (Naples)	994	Presumably healthy 20-yr-old army inductees quartered in Naples; 338 from northern, 219 from central, 387 from southern and insular Italy	Examination of fresh stool preparations, unstained and iodine-stained; examination of concentrated fecal matter	0.2 (0.9, Central Italy)	Chieffi and Basso (1951)[191]
Italy (Naples)	200	Children 4 to 8 years from Institute of Preventive Medicine	Examination of fresh stool preparations, then of concentrated fecal matter	1.0	Lojodice and Mazzitelli (1956)[192]
Italy (Naples?)	140	Group of children	(Stool examinations)	5.0	Caruso (1958) (as cited in ref. no. 138)
Kampuchea (Cambodia) (Takeo Province)	1407	Outpatients and in-patients, 0 to 18+ yr, examined in hospital	Examination of fresh stool preparations, unstained and iodine-stained; examination of concentrated fecal matter	12.5 [(3 to 5 yr) 20.1 (>18 yr) 10.3]	Giboda (1985)[64]
Kampuchea (Cambodia) (Takeo Province)	322	Random group of presumably healthy preschool and school children (3 to 18 yr)	As above	10.8 [(3 to 5 yr) 19.3 (15 to 18 yr) 9.8]	Giboda (1985)[64]
Laos (Vientiane)	2493	Random group of 0- to 30+-yr-old subjects from Vientiane; about equal ♂ and ♀	Examination of fresh stool preparations, unstained and iodine-stained	6.1	Sornmani et al (1972)[193]

(continued)

TABLE 19.21. *Continued*

Geographic area	Total number of individuals examined	Groups examined (various relevant data)	Methods of sampling	Individuals positive for *P. hominis* (%)	Reference
Mexico	866	Groups of Indians (largest), of mixed parentage (smaller), some whites; from 8 communities in various parts of country differing in climate; from various socioeconomic strata; 0 to 14 yr (mostly 7 to 13 yr); in most communities about equal ♂ and ♀	Examination of fresh stool preparations, unstained and iodine-stained; mostly after purgation	31.9	Hegner et al (1940)[194]
Mexico (Jalisco, Guadalajara)	380	Group of children, 5 to 17 yr, in an orphanage; 214♂, 166♀	Examination of fresh stool preparations after purgation; unstained or iodine-stained	21.3 [(5 yr) 50 (10 to 13 yr) 13.6]	Ruiz Sanchez (1943)[195]
New Guinea	341	War refugees (1 to 29 yr); 196 Papuans, 132 Indonesians, 13 Chinese; 179♂, 162♀	Examination of fresh stool preparations; concentrated fecal matter	4.7	Burrows (1945)[196]
Nigeria (Bendel State, Benin City)	6213	Hospitalized patients	Examination of fresh stool preparations, unstained and iodine-stained	3.4	Obiamiwe (1977)[197]
Nigeria (Ibadan)	360,000	Sick children and adults (from all parts of country) examined in University College Hospital	Examination of fresh stool preparations, unstained or iodine-stained; also concentrated fecal matter	6.0	Ogunba (1977)[198]
Nigeria (Ibadan)	4,021	Random group of apparently healthy school-age (5 to 16 yr) children (3400); adults (teachers, food handlers, parents of some children) (621)	As above	11.0	Ogunba (1977)[198]
Nigeria (Lagos)	515	Government (blue-collar and lower grades of white-collar) workers, 16 to 57 yr; 513♂, 2♀; all apparently healthy	Examination of fresh stool preparations, unstained and iodine stained; concentrated fecal matter	11.0	Okpala (1961)[199]

Location	N	Population	Method	%	Reference
Philippines (Malaybalay, Bukindon, Mindanao)	831	Random group of persons 0 to 50+ yr; larger no. under 20 yr; about equal ♂ and ♀	Examination of formalinized stool matter directly and after concentrated	<1.0	Carney et al (1981)[200]
Philippines (Manila and other provinces)	100	Hospitalized children, 7 mo to 13 yr; from 17 districts of Manila (77%) and from 9 provinces (23%); presenting with various diseases, esp. intestinal and pulmonary; 47♂, 53♀	Examination of fresh stool preparations, unstained and iodine-stained; also concentrated fecal matter; fixed and stained preparations	10	Haughwout and Horrilleno (1920)[201]
Philippines (Victoria, Oriental Mindanao)	1058	Random group of persons to 0 to 50+ yr; 586♂, 472♀	Examination of formalinized stool samples directly and after concentrated	<1.0	Carney et al (1981)[202]
Puerto Rico (Vega Baja and nearby country)	125	Children of all classes, urban and rural population; ages 6 mo to 6 yr	2 to 3 fresh stool preparations examined/child; 2 in saline with eosin, 1 iodine stained; cultivation	20.9	Hill and Hill (1927)[203]
Romania (Moldavia)	124	Apparently healthy persons (in 59 families) from rural (river meadow, Danube delta) community	Cultivation during 2 consecutive winters and 1 summer	8.0–10.4	Boldescu and Stratulat (1968)[204]
Spain (Barcelona)	650	Group of persons submitting stool samples to hospital	Examination of stool samples by MIFD[k] and MIFC[l]	0.2	Portús and Prats (1981)[205]
Spain (Caceres, Campo Lugar)	130	Children <15 yr	(Stool examinations)	26.2	Hill and Niño Astudillo (1932) (as cited in ref. no. 179)
Spain (Granada)	96	Persons with diarrhea examined in hospital	(Stool examinations)	8.3	Fernández (1933)[206]
Spain (Granada)	100	Diarrheic patients, Hospital San Lazaro and Provincial Institute of Hygiene; climate hot and humid; poor hygienic habits; poor sanitary conditions (no water closets, mixing of drinking and waste water)	Examination of fresh stool preparations (2 to 3/patient), unstained and iodine-stained; except in cases of severe diarrhea, saline purgative before taking samples; concentrated fecal matter; fixed and stained preparations	8.0	Fernández Martinez and Suarez Peregrín (1933)[179]

(continued)

TABLE 19.21. Continued

Geographic area	Total number of individuals examined	Groups examined (various relevant data)	Methods of sampling	Individuals positive for P. hominis (%)	Reference
Spain (city of Granada)	100	Group of children, 0 to 13 yr, primarily grade school students, all with diarrhea; attending various clinics; 49♂, 51♀	As above	5.0	Fernández and Duarte Salcedo (1933)[163]
Taiwan (Taihoku)	616	Random groups of Chinese laborers and Japanese officials of Taihoku Municipal Sanitation Department	Examination of centrifuged fecal matter, unstained and iodine stained	2.8	Kawai et al (1934)[207]
Taiwan (Wulai district)	439	Random group of aborigines	Examination of fresh stool preparations	1.6	Huang et al (1966)[208]
U.S.A. (Los Angeles, California)	10,655	Group of patients (incl. many children) with various gastrointestinal disorders, esp. dysentery and diarrhea, examined in hospital	Fecal samples or swabs from sigmoidoscopic or proctoscopic examinations; examination of fresh stool preparations	3.4	Kessel et al (1936)[209]
U.S.A. (San Diego, California)	513 children 1339 adults	Random group of private patients, some with gastrointestinal complaints; many stool examinations part of routine geneal exams[m]	Usually 3 fecal specimens examined on successive days; 1st sample liquid after saline purgative; examination of fresh stool preparations stained with iodine-eosin; fixed and stained permanent preparations in special cases	0.4 3.5	Summerlin (1934)[167]
U.S.A. (San Quentin Prison, California)	1000	Apparently healthy ♂ inmates 28.5 (20 to 30 yr); persons sampled: 750 freshly admitted, 250 on mess force	Fixed and stained permanent preparations sent for examination to Pacific Institute of Tropical Medicine	0.2	Johnstone et al (1933)[210]
U.S.A. (Athens, Georgia)	729	Freshmen from 2 units of the state university; 360♂, 369♀; students from 20 states, 658 from Georgia	Examination of formalinized stools concentrated by centrifugation, unstained and iodine stained; fresh stool preparations in some cases	0.3	Byrd (1937)[211]

Location	N	Population	Method	%	Reference
U.S.A. (Indiana)	771	Patients in Evansville Hospital for the Insane; all Indiana residents at admission; most from 19 counties in SW part of state; 457♂, 314♀	Examination of fresh stool preparations; concentrated fecal matter	0.1	Hopp (1944)[212]
U.S.A. (New Orleans, Louisiana)	4270	Random group of patients from the white outpatient clinic of the city; group represented fair sample of city charity cases	Examination of fresh stool preparations, unstained and iodine stained; some samples collected after purgation or enemas"; up to 4 exams/patient	1.2^o	Faust and Headlee (1936)[166]
U.S.A. (90% from Louisiana; 76.5% from New Orleans)	202	Random group of cases (accidents and suicides) autopsied in New Orleans; 156♂, 46♀; 1 to 50+ yr (most >21 yr)	Exam. of fresh preparation of fecal matter from each level of digestive tract between posterior ileum and anal sphincter; matter nonconcentrated or concentrated, unstained or iodine stained; also fixed and stained preparations	1.0	Faust (1941)[213]
U.S.A. (Oklahoma City, Oklahoma)	953	University of Oklahoma Medical School students (219♂, 12♀); food handlers (26♂, 55♀); hospital patients (370♂, 253♀); nurses (18♀); 1 to 72 yr, many 16 to 25 yr	Examination of fresh stool preparations, nonstained or iodine stained; also concentrated fecal matter; fixed and stained preparations	2.1	Canavan and Hefley (1937)[214]
U.S.A. (Philadelphia, Pennsylvania)	368	Unselected group of patients, with stool specimens taken regardless of complaints, on admission to U. Penn. Hospital; additional specimens from persons with suspected intestinal diseases	Fixed and stained fecal smears typically from unpurged patients	1.1	Hinshaw and Showers (1934)[215]
U.S.A. (Philadelphia, Pennsylvania)	1060	U. Penn. freshmen; 709♂, 351♀; average age for 965 students 18.4 (16 to 33) yr; of 1005 students, 73.6% from Pa. remainder from 15 states, many from N.J.	Examination of fresh stool prparations, unstained and iodine stained; 1 stool/subject; fixed and stained preparations	0.0^o	Wenrich et al (1935)[216]

(continued)

TABLE 19.21. Continued

Geographic area	Total number of individuals examined	Groups examined (various relevant data)	Methods of sampling	Individuals positive for P. hominis (%)	Reference
U.S.A. (Philadelphia, Pennsylvania)	53	Small group (35 ♀) of residents of Old Ladies' Home; average age 72.2 (63 to 93) yr; also 9 ♂, 9 ♀ food handlers in the home, average age 36.9 (19 to 60) yr; no symptoms of gastrointestinal disorders in any subject	As above	1.9	Wenrich and Arnett (1937)[217]
U.S.A. (Philadelphia Pennsylvania)	200	Random groups of 100 white and 100 black women attending antepartum and postpartum clinics, Dept. Obstet., Jefferson Medical School	Fecal samples obtained by digital manipulation; subsequent samples, if any, obtained by defecation); examination of fresh stool preparations; fixed and stained preparations; cultivation	1.5 (1.0, white; 2.0, black)	Bland and Rakoff (1937)[218]
U.S.A. (South Carolina)	142	White women from South Carolina State (Mental) Hospital; patients of varying age but comparable mental level; untidy in personal habits	Two stool specimens from each patient; examination of fresh stool preparations, unstained and iodine stained	20.0	Young and Ham (1941)[219]
U.S.A. (Nashville, Tennessee)	2112	Random group of persons; feces examined at Vanderbilt School of Medicine; 1231 adult medical outpatients; 881 outpatients and in-patients with intestinal disorders (incl. children), few similar private patients; subjects primarily from Nashville and surrounding rural districts	Examination of fresh stool preparations; concentrated fecal matter; in doubtful cases fixed and stained preparations; 1 exam/person	1.7	Meleney (1933)[220]
U.S.A. (Virginia)	460	Native white population of Wise County; 2 areas strictly rural, 2 areas urban or semiurban; all corners of county represented; 0 to 36+ yr; 202 ♂, 258 ♀	Examination of fresh stool preparations, unstained and iodine stained; most stools <24 hr old; in doubtful cases fixed and stained preparations	2.4 (6-15 yr) 1.3 (26-35 yr) 5.3	Faust (1930)[221]

Location	N	Population	Method	%	Reference
U.S.S.R. (Armenia, Transcaucasus, Idjewan region)	692	Peasants in 12 villages; in nearly every case family of 4 to 15 lived in 1 room; poor hygienic conditions, no water closet, latrines close to houses; highest infection in oldest group	Examinations of fresh stool preparations, unstained and iodine stained; fixed and stained preparations in doubtful cases; most persons examined once	15.0	Mathewossian (1928)[165]
U.S.S.R. (Azerbaidzhan)	330	Patients suffering from intestinal disorders of different, primarily protozoan, etiologies	(Stool examinations)	4.5	Nazirov (1965)[145]
U.S.S.R. (Moscow)	125	Random group of persons with stool samples examined in "practical diagnostic laboratories"	Liquid stools obtained by purgation; examination of fresh stool preparations, unstained and iodine stained; also examination of stools preserved before and after concentration	3.2	Turdeyev (1967)[222]
Yugoslavia (Skolpje)	110	Schoolchildren, 8 to 12 yr; apparently healthy	Examination of fresh stool preparations after purgation; cultivation	40.0[a]	Simić (1935)[223]
Zimbabwe [Rhodesia] (12 urban and rural areas, incl. Harare)	1373	Random group (no history of any disease, age distribution, etc. given)	Examination of fresh stool preparations; also after concentration	~0.9[b]	Goldsmid et al (1976)[224]

General Statement About the Stool Examination Techniques and Their Applicability to Diagnosis of Intestinal Protozoa (Especially P. hominis)

There are numerous original reports, many rather old (e.g., Ref nos. 225-229) in which various methods of fecal examinations for intestinal parasites are described and evaluated. Often undigested food components render diagnosis of protozoa difficult, especially for less experienced laboratory workers. Such persons can refer to an illustrated description of various undigested food elements published by Denison.[230]

As far as protozoa are concerned, formed stools are more useful for finding cysts than trophozoites. In all instances, the success of using formed stools depends on several important factors: (1) the number of samples examined; (2) evenness of cyst distribution; (3) the time elapsed between collecting the samples and examination—this time must not be long enough to allow degeneration of any stages of the parasites. In general, as will be pointed out elsewhere in this section, liquid stools, the result of natural or induced (purgatives or enemas) diarrhea are most suitable for finding trophozoites, although they also have certain advantages for cysts.

In all instances, sample size and the numbers of methods employed in the course of an examination affect the results. Since no particular time appears to be more suitable than any other for finding the protozoa, six examinations performed at various times would probably be the safest means to demonstrate the highest percentage of positive cases. Furthermore, because of the uneven distribution of cysts and other forms, various parts of a fecal sample should be examined.

[a]Details of the examination methods are not available to this writer. There can be little doubt, however, that they must have included some of the basic techniques of stool examination; in most instances examination of fresh stool preparations was most likely one of these methods.

[b]In all instances "examination" refers to a microscopic examination. The highest magnification employed depended on the technique used in the preparation of the material and to some degree on the preference of the investigator.

(continued)

[a] Stools formed or water; whenever necessary, especially in the case of formed stools, the material was suspended in physiologic (~0.9% NaCl) salt solution or in a solution such as Ringer's.

[b] Permanent preparations, fixed in] Schaudinn's fluid and stained with iron hematoxylin, appear to have been used most often; also fecal smears processed according to Giemsa's technique, unmodified, or modified, have been employed successfully by many investigators. The use of permanent preparations represents the method of choice if the material is to be examined some time after having been collected, and especially if the examination is to be carried out in a distant center.

[c] Growth of xenic or axenic cultures in a variety of media, monophasic or diphasic, has been used successfully by many investigators for diagnosis of intestinal protozoa. The culture methods are more valuable for finding some protozoan species; however, their employment is useful in diagnosing most intestinal protozoa. Many suitable media have been described or merely listed in Taylor and Baker,[233] Honigberg,[1] and also in Chapter 7 of this book.

[f] Various concentration methods were used by different investigators. Note that these methods include flotation techniques, e.g., zinc sulfate flotation[231] in addition to water and formol-ether centrifugation (the latter sometimes modified, as by Ridley and Hawgood[232]). The concentration methods, devised primarily for protozoan cysts and helminth eggs, are of very limited use in searching for trophozoites of parasitic protozoa, including the motile forms of P. hominis. Since most reports dealt primarily with E. histolytica and helminths, the authors found the concentration procedures helpful in their surveys.

[g] Iodine staining. "Iodine solution" or "saturated iodine" in potassium iodide, with or without eosin, was employed by Faust[221] and Summerlin[167]; Lugol's iodine, with or without eosin, was used by many workers listed in Table 19.21; Donaldson's iodine-eosin was employed by Faust[221] and Summerlin[167]; Lugol's iodine, with or without eosin, was used by many workers listed in Table 19.21; e.g., Grisolia y Juristo[184] Turdyev,[222] Giboda[164]; Gram's iodine, used by Haughwout and Horrilleno,[201] was adopted less often. It should be noted that iodine staining is not useful for finding trophic stages of protozoa.

[h] Data not available for various reasons.

[i] Purgatives, e.g., MgSO$_4$, to render the stools watery and to increase the chances of finding trophic forms of various intestinal protozoan species, such as P. hominis.

[j] Curiously, children from environments characterized by the best sanitary conditions were more or less parasitized. On the other hand, some juvenile patients from the most unsanitary conditions were found to be parasite free.

[k] MIFD, a technique involving fixation and direct observation.[234]

[l] MIFC, a biphasic concentration method.[235]

[m] The results obtained in this study represent a fair estimate of the prevalence of intestinal protozoa in a general practice sample of population belonging to a higher socioeconomic stratum than those for whom prevalence data are usually reported.

[n] See footnote i.

[o] Prevalence increases with the age of the subjects, up to 16 to 20 years old (5.8% to 7.7%), then falls again (~1.5%).

[p] Evidently, prevalence was 0 among the freshmen. Although no cultivation was attempted, Wenrich's examinations have always been conducted very carefully, and this fact lends credence to his results.

[q] Forty percent prevalence seems inordinately high, especially among reputedly healthy children.

[r] The data provided by these authors do not constitute a dependable basis for the estimation of prevalence of P. hominis. The figure given in Table 19.21 represents a possible very approximate prevalence deduced from the available information. Actually Goldsmid et al[224] state: "The commensal flagellates Chilomastix mesnili and Trichomonas hominis are not uncommonly encountered in uniformed stools at Harare Central Hospital. . . T. hominis is sometimes found in great numbers in children with severe diarrhoea."

[s] In Table 19.21, the total percentage of Chilomastix in several geographic locations was 0.9%. In light of this, it could be assumed that the prevalence of Pentatrichomonas is about equal to that of Chilomastix.

old people); socioeconomic status of the groups examined; the number of stool samples examined from individual subjects; and methods of stool examinations. However, they provide a general idea of the worldwide prevalence of *P. hominis.*

It is evident from Table 19.21 that general hygienic standards (e.g., method of disposal of excrement, purity of drinking water, and crowding) as well as personal habits affect the frequency of infection with *P. hominis.* According to many workers, for example, Fernández and Duarte Salcedo,[163] the prevalence is higher in 2- to 6-year-old children who play in the soil, place everything in their mouths, and take no hygienic precautions (see also Giboda[164]). Furthermore, asylum populations are apt to have a higher prevalence of *P. hominis* than other institutionalized groups. According to some workers, prevalence increases with the age of the subjects.[165-167] However, except for special cases, even in tropical countries where the prevalence of infection is higher, the frequency of *P. hominis* infection tends to be rather low in human populations (Table 19.21).

Transmission

Hypotheses regarding the possible origin of infection by the intestinal trichomonad from other sites, that is, the oral cavity or the urogenital tract, need no longer be considered. As indicated by Honigberg,[1] there is overwhelming evidence based on morphologic (see Chap. 3), immunologic, and physiologic evidence that distinct species are limited to the oral cavity, large intestine, and urogenital tract of humans. Also the results of cross-infection experiments involving these species (see for example those of Westphal[168]), many of which have been summarized by Wenrich,[16] provide evidence that "the trichomonads of the mouth, intestine, and genital tract of human hosts are unable to establish lasting infections in any part of the body other than the one in which they are customarily found."[16] It is also Wenrich's review[16] that includes summaries of experiments involving attempts at transfers of the trichomonads of humans into different sites of various nonhuman hosts. Another line of evidence for the site specificity of human trichomonad species is found in epidemiologic studies[115,169] (see also reference 16 for additional citations).

The mode of transmission of *P. hominis* is contaminative. The parasite is acquired by eating contaminated food, placing in the mouth either the fingers or other objects soiled with infected feces (the last two ways of transmission are most frequent among young children), and/ or, under certain circumstances, by drinking water that contains the trichomonads. Since *P. hominis,* like most trichomonads, does not form true cysts, typical trophic forms or, according to a few workers,[135,159] pseudocysts must survive for a time in the feces.

A number of workers studied survival of *P. hominis* trophozoites under various environmental conditions. Only some of the relevant experiments are cited here. Thus Hegner,[170] who tested the viability of the flagellates in the feces, obtained the following results: (1) At 5°C, 21°C, 25°C, and 31°C many organisms survived for a few hours in undiluted feces; some survived up to 8 days. No living parasites were noted, however, after 4 hours at 40°C and 44°C. (2) In serum-saline-citrate medium the intestinal trichomonads remained viable for 2 days at 4°C and 21°C, but for only 2 hours at 40°C and 44°C. (3) Fecal samples rich in *P. hominis* that were kept on moist garden soil during cloudy periods at 12°C to 19°C for 7 days could initiate cultures. Shorter survival times were noted for trichomonads kept in feces on sandy soil. (4) In samples in which 1 g of infected feces was diluted in 99 ml of tap water, the parasites remained viable for 5 to 6 hours; viability was retained for up to several days in samples in which 25 to 75 g of infected feces were suspended in 75 and 25 ml of tap water, respectively. Similar viability levels were noted when distilled or spring water was used in place of tap water. It is evident from the foregoing results that *P. hominis* trophozoites are sufficiently resistant to thermal and osmotic changes in the environment to be transmitted by contamination. Hegner[170] also obtained cultures of the intestinal trichomonads using vomit and feces of flies, for example, *Musca domestica, Lucilia sericata,* and *Cynomyia cadaverina* for up to 4 hours after the insects were fed human feces containing these parasites. Also, Simić and Kostić[109] demonstrated experimentally that *M. domestica* can be involved in transmission of *P. hominis.*

Transmission by flies may be direct or indirect. In the first instance, it is accomplished by the flagellates carried on the fly legs or proboscises contaminated with feces contained in liquid or moist materials; this kind of transmission must be accomplished within about 10 minutes, a period during which the trichomonads remain viable. Indirect transmission involves the excrement of flies fed on contaminated food. Although they do not multiply in the insect's alimentary tract, the flagellates survive there for at least 8 hours.

Thermal resistance of *P. hominis* was investigated also by other workers. Giboda and Čatár[171] noted survival of *P. hominis* for 7 days at 4°C and for 2 days at 25°C in moist stools. In aqueous suspension, the survival period was extended to 3 days. Furthermore, Tumka[172] demonstrated survival of the intestinal trichomonads in the feces for 50 to 79 hours at temperatures ranging from −2°C to 4°C; this would provide for retention of viability by the parasites at temperatures below freezing.

Dobell[151] obtained resistant forms of *P. hominis*—"rounded-up and motionless, devoid of flagella and undulating membrane"; their ingested starch had typically been assimilated. These organisms, many of which remained viable, were produced by being grown for about 2 days in the "Ehs + S" medium containing rice starch grains. Rich 48-hour cultures of organisms were transferred from 37°C to room temperature for a few hours, then placed at about 10°C and left there for 1 month. These resistant forms appeared closely to resemble pseudocysts. Dobell's data[151] with regard to the length of survival of the rounded forms at room temperature (7 to 10 days or longer) and at constant 10°C temperature suggest that they may play a role in transmission of the intestinal flagellates of humans.

Foresi[159] thought that the stages involved in transmission of *P. hominis* were the "cyst-like" forms. These forms were described as rounded with rigid contours, nonmotile, and very rich in "glycogen." They were said to develop at room temperatures and still better at 4°C. When inoculated into fresh culture medium they gave rise to normal vegetative forms (motile trophozoites); evidently they were not found in human feces. These cystlike forms were more resistant than trophozoites to antimicrobial, physical, and chemical factors found under natural conditions, for example, tap water, elevated temperature, freezing, and hydrochloric acid (HCl). In tap water, the typical trophozoites remained viable at 4°C for 24 hours, at 20°C for 3 days, and at 37°C for 2 days. However, the cystlike forms yielded positive cultures after 14 days at 4°C and after 12 days at 20°C; these forms were absent at 37°C. There was a progressive reduction in the number of the organisms from the original 10^4/ml, the decrease being especially marked during the first 2 to 3 days of cultivation. Thermal resistance of the cystlike forms was reflected in the finding that they remained viable for 50 minutes at 50°C, while typical trophozoites survived in Locke's solution for only 5 minutes at this temperature. Motile flagellates suspended in small volumes of Locke's solution survived for only 3 minutes in a frozen state (−4°C), but the cystlike forms remained viable for 12 hours in these thermal conditions. Finally, the trophozoites were able to withstand exposure to 1.1% HCl for only 10 minutes, while the cystlike stages withstood exposure to this concentration of the acid for 120 minutes.

Judging from their description the cystlike forms of *P. hominis* described by Dobell[151] and Foresi[159] were actually equivalents of pseudocysts, as reported from other trichomonads by many authors on the basis of light microscopy and by Mattern et al[173] and Mattern and Daniel[174] on the basis of electron microscope observations. It is surprising, however, that the resistant forms of the human intestinal flagellate were not found in the intestine or in the feces of their hosts. If they were, one could compare them with the pseudocysts reported for *Tritrichomonas muris* from hamsters[174] said to have been involved in natural transmission of *T. muris* to baby hamsters. Additional studies of *P. hominis* pseudocysts are needed to confirm and further to elucidate their role in transmission of this species.

Resistance of *P. hominis* to the gastric juice and various concentrations of HCl has been investigated by a number of workers. Thus Shinohara[175] examined the effect of saliva, artificial gastric juice, and bile (culture medium for *Salmonella*) on survival of this trichomonad. Viability was ascertained by cultivation at room temperature. Saliva had no effect on the flagellates, the organisms surviving for 48 hours or longer in its presence. *Pentatrichomonas hominis*, like *T. tenax*, exposed to artificial gastric juice remained viable for a maximum of 30 minutes. In diluted gastric juice, the former species survived a little longer. In the presence of bile, *T. tenax* lost motility almost instantly and died within 10 minutes. On the other hand, the intestinal trichomonad exhibited occasional movements for about 15 minutes of exposure to bile and remained viable for up to 30 minutes. Furthermore, in a 10% solution of bile *P. hominis* remained motile for twice as long (60 minutes) as the oral trichomonad. The need of *P. hominis* to pass through the stomach may explain its greater resistance to HCl. Also, according to Caruso[138] transmission of the intestinal trichomonad is mediated by trophozoites that are very resistant to the action of the gastric juice, being thus better adapted to traversing the stomach without sustaining any damage. Dobell[151] succeeded in infecting himself with one of the strains of *P. hominis* from macaques with a rich suspension of the protozoa from a xenic culture mixed with sterile milk. The infective suspension, swallowed on an empty stomach, persisted for 4½ years without causing any symptoms. The parasites seemed to be viable after only a short exposure (usually no longer than 5 minutes) to N/20 HCl at 37°C (Ann Bishop as quoted in reference 151).

A more extensive experimental study of the in vitro resistance of *P. hominis* to HCl and gastric juice was reported by Głębski.[176] No differences in results were noted among the various trichomonad strains. It became evident from the results that at about 0.055% HCl concentration the trichomonads survived and even multiplied in the course of 4 days of observation; at an HCl concentration of about 0.11% the flagellates lived for 72 hours but did not multiply; at about 0.22% concentration of HCl, representing the normal acidity of gastric juice, the trichomonads were dead after 5 minutes and were falling apart after 24 hours. At higher acidities, the destruction of the parasites was even faster; after 3 hours only a few damaged protozoan cells could be found. The destructive effect of the gastric juice on the intestinal trichomonads is more rapid than that noted with HCl at a comparable acidity, for example, in an acidity equivalent of about 0.1% HCl the trichomonads were killed in 10 minutes; they were completely destroyed after 48 hours. Evidently the activity of the gastric juice depends not only on the acidity level but also on the presence of other active biologic components, perhaps enzymes.

Diagnosis and Treatment

Definitive diagnosis is accomplished primarily by finding the flagellates in naturally liquid stools or in stools rendered liquid by purgatives or enemas. Furthermore, repeated examination of fresh fecal preparations and of cultures are the most dependable means for diagnosing *P. hominis*.

If treatment is deemed necessary, and many investigators question this need, one of the 5-nitroimidazoles, metronidazole (Flagyl) or tinidazole (Fasigyn), administered orally, is effective. Only a few authors provided information concerning the effective dosages of these drugs. Among those who did, Zhang[150] recommended 600 to 900 mg of metronidazole daily for 3 to 4 days, while Tanev and Tzvetkova[148] used a single 2-g dose of tinidazole.

References

1. Honigberg BM: Trichomonads of importance in human medicine. In Kreier JP (ed): *Parasitic Protozoa*, vol 2, pp 276-454. New York: Academic Press, 1978.
2. Rebhun J: Pulmonary trichomoniasis associated with a fever of unknown origin. *Calif Med* 100:443-444, 1964.
3. Grein NJ, Tetu E, Neves JF, Piazzetta CM: Trichomoníase bucal. *Ars Curandi Odontol* 6:22-23, 1979.
4. Terezhalmy GT: Oral manifestations of sexually related diseases. *Ear Nose Throat J* 62:287-296, 1983.
5. Brenner BE, Simon RR: Glossitis and dysgeusia. *Am J Emerg Med* 2:147, 1984.
6. Hiemstra J, Van Bel F, Berger HM: Can *Trichomonas vaginalis* cause pneumonia in newborn babies? *Br Med J* 289:355-356, 1984.
7. McLaren LC, Davis LE, Healy GR, James CG: Isolation of *Trichomonas vaginalis* from the respiratory tract of infants with respiratory disease. *Pediatrics* 71:888-890, 1983.
8. Lewis B, Carroll G: A case of *Trichomonas vaginalis* infection of the kidney pelvis. *J Urol* 19:337-339, 1928.
9. Madsen AC: A case of *Trichomonas* infection of the pelvis. *W Va Med J* 29:356-357, 1933.
10. Littlewood JM, Kohler HG: Urinary tract infection by *Trichomonas vaginalis* in a newborn baby. *Arch Dis Child* 41:693-695, 1966.
11. Suriyanon V, Nelson KE, Ayudhya VCN: *Trichomonas vaginalis* in a perinephric abscess. *Am J Trop Med Hyg* 24:776-780, 1975.
12. Gallai Z, Sylvestre L: The present status of urogenital trichomoniasis: A general review of literature. *Appl Ther* 8:773-778, 1966.
13. Mardh PA, Westrom L: Tubal and cervical cultures in acute salpingitis with special reference to *Mycoplasma hominis* and T.-strain mycoplasmas. *Br J Vener Dis* 46:179-186, 1970.
14. Keith LG, Berger GS, Edelman DA, Newton W, Fullan N, Bailey R, Friberg J: On the cau-

sation of pelvic inflammatory disease. *Am J Obstet Gynecol* 149:213-223, 1984.
15. Levine ND: *Protozoan Parasites of Domestic Animals and Man,* 2nd ed. Minneapolis: Burgess, 1973.
16. Wenrich DH: The species of *Trichomonas* in man. *J Parasitol* 33:177-178, 1947.
17. Delaunay P, Rousset JJ: Amoeba trichomonas e patologia orale. *Dental Cadmos* 6:37-40, 1976.
18. Honigberg BM: Trichomonads of veterinary importance. In Kreier JP (ed): *Parasitic Protozoa,* vol 2, pp 163-275. New York: Academic Press, 1978.
19. Burkova T, Nikova M: Comparative histological and enzyme-histochemical studies on gingival papillae in periodontosis patients invaded or not by *Tr. tenax* and *E. gingivalis. Stomatologiia* 65:25-31, 1983 (in Russian).
20. Cambon M, Pétavy A-F, Bourges M, Deguillaume J, Glandier Y: Etude de la fréquence des protozoaires et levures de la cavité buccale chez l'homme. *Actual Odonto-Stomatol* 34:279-286, 1980.
21. Su KX: Oral protozoiasis. *Chung Hua Kou Chiang Ko Tsa Chih* 20:112-114, 1985 (in Chinese).
22. Trofimova LV, Siirde EK, Lentzer HP: Presence of a trichomonad in the oral cavity, nose and ears. *Zh Ushn Nos Gerl Bolezn* 38:97-98, 1978 (in Russian).
23. Tumka AF: Protozoa in cases of chronic tonsillitis. *Vestn Otorinolaringol* 19:78-83, 1957 (in Russian, English summary).
24. Wantland WW, Lauer D: Correlation of some oral hygiene variables with age, sex, and incidence of oral protozoa. *J Dent Res* 49:293-297, 1970.
25. Tokura N: Biologische und immunologische Untersuchungen über die menschenparasitären Trichomonaden. *Igaku Kenkyu* 9:1-13, 1935.
26. Kott H, Adler S: A serological study of *Trichomonas* sp. parasitic in man. *Trans R Soc Trop Med Hyg* 55:333-344, 1961.
27. Oleinik GI: On the study of antigenic properties of human trichomonads. *Mikrobiol Zh* 26:50-56, 1964 (in Ukrainian, English summary).
28. Stępkowski S, Honigberg BM: Antigenic studies of some members of the subfamily Trichomonadidae. *Wiad Parazytol* 29:209-211, 1983.
29. Rõigas EM, Teras JK, Kazakova I, Kumm R, Ellamaa M: The antigenic properties of trichomonads. *Proc III Int Cong Parasitol,* 1974, vol 2, p 1098, Vienna: Facta Publication, H Egermann, 1974.
30. Kumm RA, Teras JK, Kallas EV: Serotypes of *Trichomonas tenax. Proc VI Baltic Scientific Conf on Parasitological Problems,* Vilnius, June 21-22, 1973, pp 91-93, 1973.
31. Wantland WW, Wantland EM, Weidman TA: Cytochemical studies on *Trichomonas tenax. J Parasitol* 48:305, 1962.
32. Ohashi O: Studies on acid phosphatase in trichomonads. I. Cytochemical demonstration of the enzyme activity in *Trichomonas tenax* and *T. vaginalis. Jpn J Parasitol* 20:399-405, 1971 (in Japanese, English summary).
33. Probst RT: *Studies on the Physiology and Nutrition of the Oral Protozoan* Trichomonas tenax, Dissertation. University of Maryland, College Park, Dissertation Abstracts, vol 27B (2207-3354), p 3203-B, 1967.
34. Barka T, Anderson PJ: Histochemical methods for acid phosphatase using hexazonium pararosanilin as coupler. *J Histochem Cytochem* 10:751-753, 1962.
35. Abramova EI, Voskresenskaia GA: The role played by the oral *Trichomonas* in pathogenesis of periodontal disease. *Stomatologiia* 59:28-30, 1980 (in Russian, English summary).
36. Jaskoski BJ: Incidence of oral protozoa. *Trans Am Microsc Soc* 82:418-420, 1963.
37. Musaev FA: Parasitic protozoa in the human oral cavity and their clinical importance. *Stomatologiia* 50:60-62, 1971 (in Russian, English summary).
38. Čatár G, Vráblic J, Vodrážka J, Masárová E: Concerning occurrence and diagnosis of oral flagellate *Trichomonas tenax. Abstracts Int Symp Trichomonads & Trichomoniasis,* Prague, July 1985, p 85. Prague: Local Organizing Committee, Dept Parasitol, Faculty Sci, Charles Univ, 1985.
39. Reczyk J, Głębski J, Nierychlewska A: Incidence of *Ameba gingivalis* and *Trichomonas buccalis* in pathological conditions of the oral cavity of man. *Czas Stomatol* 29:697-701, 1976 (in Polish, English summary).
40. Reczyk J, Głębski J, Nierychlewska A: Presence of protozoa in the mouth in relation to maxillofacial operations and diseases. *Czas Stomatol* 33:237-244, 1980 (in Polish, English summary).
41. Pechtorová J, Zajicová S, Palická P: Occurrence of *Trichomonas tenax* in specific population groups. *Česk Epidemiol Mikrobiol Immunol* 34:123-127, 1985 (in Czech).
42. Cambon M, Pétavy A-F, Guillot J, Glandier I, Deguillaume J, Coulet M: Etude de la fréquence des protozoaires et des levures isolés du parodonte chez 509 sujets. *Pathol Biol* 27:603-606, 1979.
43. Blake GC: The microbiology of acute ulcerative gingivitis with reference to the culture of oral trichomonads and spirochaetes. *Proc R Soc Med* 61:131-136, 1968.

44. Teras JK: Comparison of certain biological properties of trichomonad species of humans. *Wiad Parazytol* 29:53-56, 1983.
45. Franjola R, Puga S, Matamala F: Investigación preliminar sobre *Trichomonas tenax* en la ciudad de Valdivia, Chile. *Bol Chil Parasitol* 33:37-38, 1978.
46. Miligi G, Magaudda-Borzì L, Mento G: Le frequenza dei portatori di "*Trichomonas tenax*," e di *Entamoeba gingivalis* in alcune zone della provincia di Messina. *Arch Ital Sci Med Trop Parassitol* 45:95-99, 1964.
47. Lapierre J, Rousset J-J: L'infestation à protozoaires buccaux. *Ann Parasitol Hum Comp* 48:205-216, 1973.
48. Musaev FA: Distribution of oral protozoa in Azerbaijan. In Strelkov AA, Sukhanova KM, Raikov IV (eds): *Prog Protozool, Abstracts Proc III Int Cong Protozool,* Leningrad, 1969, 217-218, Leningrad: Publ House "Nauka", 1969.
49. Musaev FA: On the laboratory diagnosis of the parasitic protozoa of the oral cavity of man. *Izv Akad Nauk Azerbaidzhan SSR (Ser Biol)* 1:74-78, 1970 (in Russian).
50. Musaev FA: Trichomonads of the oral cavity of man. *Parazitologiia* 6:185-188, 1972 (in Russian, English summary).
51. Rousset J-J, Lauvergeat J-A: Protozoaires buccaux: Enquête épidémiologique. *Presse Méd* 79:1495-1497, 1971.
52. Delaunay P, Omnes N, Rousset J-J: Protozooses buccales: Possibilités expérimentales. I. Chez l'homme. *Actual Odonto-Stomatol* 116:725-734, 1976.
53. Feki A, Molet B, Haag R, Kremer M: Les protozoaires de la cavité buccale humaine: Corrélations épidémiologiques et possibilités pathogéniques. *J Biol Buccale* 9:155-161, 1981.
54. Shinn LDS, Squires S, McFadzean JA: The treatment of Vincent's disease with metronidazole. *Dent Pract* 15:275-280, 1965.
55. Lacour M, Grappin G: Traitement de la gingivostomatite ulcéreuse par la métronidazole. *Bull Soc Med Afr Noire* 13:408-410, 1968.
56. Ingham HR, Hood FJC, Brandumpp Tharagonnet D, Selkon JB: Metronidazole compared with penicillin in the treatment of acute dental infection. *Br J Oral Surg* 14:264-269, 1977.
57. McGowan DA, Murphy KJ, Sheiham A: Metronidazole in the treatment of severe acute pericoronitis (a clinical trial). *Br Dent J* 142:221-223, 1977.
58. Heijl L, Lindhe J: The effect of metronidazole on development of plaque and gingivitis in the beagle dog. *J Clin Periodontol* 6:197-209, 1979.
59. Guillermet FN, Bertoye A, Boureay M: Sensibilité des bactéries anaérobies au métronidazole. *Rev Inst Pasteur (Lyon)* 4:43-48, 1971.
60. Videau D: Association métronidazole-spiramycine sur les germes anaérobies. *Pathol Biol* 19:661-666, 1971.
61. Gottlieb DS, Diamond LS, Fedi PF: Prevalence of oral protozoa in acute necrotizing ulcerative gingivitis. *Program & Abstracts of Papers, 46th Gen Meet Int Assoc Dental Res,* San Francisco, 1968, Abstract 54, Am Dental Assoc Council on Dental Res: Washington DC, 1968.
62. Shinohara T, Iwai S: Experimental inoculation of *Trichomonas hominis* and *T. tenax* into mice. *Keio Igaku* 35:383-387, 1958 (in Japanese, English summary).
63. Kulda J: Effect of different species of trichomonads on monkey kidney cell cultures. *Folia Parasitol (Prague)* 14:295-310, 1967.
64. Delaunay P, Teboul P, Rousset JJ: Protozooses buccales: Possibilités expérimentales. III. Possibilités d'étude en laboratoire. *Actual Odonto-Stomatol* 116:737-740, 1976.
65. Alderere JF, Garza GE: Specific nature of *Trichomonas vaginalis* parasitism of host cell surfaces. *Infect Immun* 50:701-708, 1985.
66. Jírovec O: Contribution à la microbiologie de la cavité buccale humaine. *Paradentologie* 2:117-123, 1948.
67. Wantland WW, Wantland EM: Incidence, ecology, and reproduction of oral protozoa. *J Dent Res* 39:863, 1960.
68. Grisi AM, De Carneri I: Frequenza dell'infezione da *Trichomonas tenax* ed *Entamoeba gingivalis* in soggetti di età avanzata viventi in comunità. *Parassitologia* 3:151-154, 1961.
69. Bonvini E, De Carneri I: Ricerca di "Entamoeba gingivalis" e di "Trichomonas tenax" nei bambini di età inferiore ai cinque anni. *G Malat Infet Parassitol* 14:361-362, 1962.
70. Aradi MP: The frequency of infections by oral protozoa in Budapest. *Parasitologia Hungarica* 1:69-76, 1968.
71. Green JC, Vermillion JR: The simplified oral hygiene index. *J Am Dent Assoc* 68:7-13, 1964.
72. Palmieri JR, Halverson BA, Sudjadi S-T, Purnomo, Masbar S: Parasites found in the mouths of inhabitants of three villages of South Kalimantan (Borneo), Indonesia. *Trop Geogr Med* 36:57-59, 1984.
73. Gherman I, Plecias M, Sendroiu L: Frecvenţa protozarelor bucale *Trichomonas buccalis* si *Entamoeba gingivalis* în diferite colectivităţi din mediul urban. *Microbiol Parasitol Epidemiol (Bucureşti)* 5:543-545, 1960 (in Romanian).
74. Beatman LH: Studies on *Trichomonas buccalis*. *J Dent Res* 13:339-347, 1933.
75. Hinshaw HC: Correlation of protozoan infections of human mouth with extent of certain lesions in pyorrhea alveolaris. *Proc Soc Exp Biol Med* 24:71-73, 1926.

76. Hogue MJ: Studies of *Trichomonas buccalis*. *Am J Trop Med* 8:75-87, 1926.
77. Rõigas E, Teras JK, Kazakova I, Sardis H: On the role of *Trichomonas tenax* in genesis of pulmonary inflammation. In Dryl S, Kazubski SL, Płoszaj J (eds): *Prog Protozool, Abstracts VI Int Cong Protozool*, Warsaw, 1981, p 315, Warsaw: Nencki Inst Exp Biol, 1981.
78. Sardis H, Rõigas E, Teras JK, Kazakova I, Kumm R: On the taxonomy of trichomonads inhabiting the human respiratory tract. In Dryl S, Kazubski SL, Płoszaj J (eds): *Prog Protozool, Abstracts VI Int Cong Protozool*, Warsaw, 1981, p 324, Warsaw: Nencki Inst Exp Biol, 1981.
79. Skipina LV: On trichomonads in the lungs. *Vrach Delo* 4:134-135, 1968 (in Russian).
80. Turgel ES, Balode VK: A case of endobronchitis caused by *Trichomonas elongata*. *Klin Med* 51:127-128, 1973 (in Russian).
81. Shevchenko BI, Pishak MN, Iasinskaia LV: Chronic trichomonal tracheobronchitis. *Klin Med* 58:83-84, 1980 (in Russian).
82. Teras JK: Extraurogenital infections of man by Trichomonadidae. *Abstracts Int Symp Trichomonads & Trichomoniasis*, Prague, July 1985, p 83. Prague: Local Organizing Committee, Dept Parasitol, Faculty Sci, Charles Univ, 1985.
83. Khanin AL, Staroverova LL: Recurrent eosinophilic pneumonia and eosinophilia (of the blood) in trichomonad bronchitis. *Klin Med* 60:95-97, 1982 (in Russian).
84. Lehmann GD, Prendiville JT: Occurrence of a flagellate in the sputum of a case of bronchiectasis. *Br Med J* 1:158-160, 1946.
85. Khrushcheva EA, Kriazheva VI: A case of trichomonads in the sputum with bronchietactic abscesses of the lung. *Med Parazitol* 20:502-503, 1951.
86. Osborne FT, Giltman LI, Uthman EO: Trichomonads in the respiratory tract: A case report and literature review. *Acta Cytol* 28:136-138, 1984.
87. Barbosa AG, Amaral ADF: Sôbre a presença de flagelados do gênero *Trichomonas* no pulmão. *Folia Clinica et Biologica* (*São Paolo*) 16:169-179, 1950.
88. Memik F: Trichomonads in pleural effusion. *JAMA* 204:1145-1146, 1968.
89. Walton BC, Bacharach T: Occurrence of trichomonads in the respiratory tract: Report of three cases. *J Parasitol* 49:35-38, 1963.
90. Fardy PW, March S: Trichomonads in resected lung tissue. *Am Rev Resp Dis* 100:893-894, 1969.
91. Teras JK, Rõigas EM, Kazakova I, Ranne HR, Trapido LE, Sardis H, Kaal VA: Detection of *Trichomonas* in the bronchi, sputum and oral cavity in patients with diverse pulmonary abnormalities. *Ter Arkh* 52:123-125, 1980.
92. Hersh SM: Pulmonary trichomonas and *Trichomonas tenax*. *J Med Microbiol* 20:1-10, 1985.
93. Tumka AF: Trichomonad invasion of the lung. *Klin Med* 34:35-40, 1956 (in Russian).
94. Rõigas E, Teras JK, Kaal V, Kazakova I, Sardis H: Results of quantitative complement fixation and indirect immunofluorescence method with blood sera of persons in cases of trichomonad invasion of lungs. *Izv Akad Nauk Est SSR (Biologiia)* 29:97-102, 1980 (in Russian, English summary).
95. Kazakova I, Teras JK, Rõigas E, Sardis H, Kumm R, Kaal V: Trichomonad invasion in respiratory tract and specific agglutinins in the blood sera of patients suffering from lung diseases. *Izv Akad Nauk Est SSR (Biologiia)* 24:87-94, 1980 (in Russian, English summary).
96. Kazakova I, Rõigas E, Teras JK: Comparative investigation of the antigenic properties of trichomonads isolated from the human oral cavity and bronchial tubes. In Hutner SH (ed): *Abstracts V Int Cong Protozool*, New York, 1977, p 177, Society of Protozoologists, 1977.
97. Glaubach N, Guller EJ: Pneumonia apparently due to *Trichomonas buccalis*. *JAMA* 120:280-281, 1942.
98. Miller MJ, Leith DE, Brooks JR, Fencl V: *Trichomonas* empyema. *Thorax* 37:384-385, 1982.
99. Teras JK, Tompel HJ, Mirme EA, Kallas EV: Development of media for axenic cultivation and obtaining of axenic cultures of *Trichomonas hominis* and *Trichomonas tenax*. In *Problems of Parasitology in Baltic Countries*, pp 251-252. Riga: 1970 (in Russian).
100. Teras JK, Tompel HJ, Kallas EV, Kumm RA: Cultivation and obtaining of axenic cultures of trichomonads inhabiting the human organism. *Proc I Meet Soc Protozool USSR*, Baku, October 1971, pp 156-166, 1971 (in Russian).
101. Kazakova I, Teras J, Rõigas E, Sardis H: Intraspecific antigenic differences between *Trichomonas* strains isolated from the respiratory tract and oral cavity of man. *Med Parazitol* 4:34-38, 1985 (in Russian, English summary).
102. Kazakova I, Rõigas E, Teras JK, Sardis H: On the specific antibodies in blood sera of patients with trichomonads in respiratory tract. In Dryl S, Kazubski SL, Płoszaj J (eds): *Prog Protozool, Abstracts VI Int Cong Protozool*, Warsaw, 1981, p 129, Warsaw: Nencki Inst Exp Biol, 1981.
103. Sardis H, Teras JK, Rõigas E, Kazakova I,

Trapido L: Allergenic properties of trichomonads inhabiting human respiratory tract. *Izv Akad Nauk Est SSR (Biologiia)* 32:135-141, 1983.
104. Abed L, Delemotte J, Marill R, Ripert C, Tordjman G: Localisation pleuro-pulmonaire du *Trichomonas*. *Bull Soc Pathol Exot Filiales* 59:962-964, 1966.
105. Wenrich DH: Morphological studies on the trichomonad flagellates of man. *Arch Soc Biol Montevideo Suppl* 5:1189-1199, 1931.
106. Kirby H: The structure of the common intestinal trichomonad of man. *J Parasitol* 31:163-175, 1945.
107. Honigberg BM: Evolutionary and systematic relationships in the flagellate order Trichomonadida Kirby. *J Protozool* 10:20-63, 1963.
107a. Honigberg BM, Mattern CFT, Daniel WA: Structure of *Pentatrichomonas hominis* as revealed by electron microscopy. *J Protozool* 15:419-430, 1968.
108. Flick EW: Experimental analysis of some factors influencing variation in the flagellar number of *Trichomonas hominis* from man and other primates and their relationship to nomenclature. *Exp Parasitol* 3:105-121, 1954.
109. Simić T, Kostić D: Rôle de la mouche domestique dans la propagation du *Trichomonas intestinalis* chez l'homme. *Ann Parasitol Hum Comp* 15:323-325, 1937.
110. Wenrich DH: Comparative morphology of the trichomonad flagellates in man. *Am J Trop Med* 24:39-51, 1944.
111. Wenrich DH, Saxe LH: *Trichomonas microti* n. sp. (Protozoa, Mastigophora). *J Parasitol* 36:251-269, 1950.
112. Reardon LV, Rininger BF: A survey of parasites in laboratory primates. *Lab Anim Care* 18:577-580, 1968.
113. Tanabe M: Morphological studies on *Trichomonas*. *J Parasitol* 12:120-130, 1926.
114. Jensen EA, Hammond DM: A morphological study of trichomonads and related flagellates from the bovine digestive tract. *J Parasitol* 11:386-394, 1964.
115. Červa L, Červová H: Occurrence of *Trichomonas intestinalis* in women and connection of this infection with vaginal trichomoniasis. *Česk Epidemiol Mikrobiol Immunol* 10:128-133, 1961 (in Czech, English summary).
116. Brisou B: *Trichomonas intestinalis* parent pauvre de la parasitologie intestinale (A propos de quelques observations personelles). *Bull Soc Pathol Exot Filiales* 57:1058-1064, 1965.
117. MacDonald EM, Tatum AL: The differentiation of species of trichomonads by immunological methods. *J Immunol* 59:309-316, 1948.
118. Mannweiler E, Oelerich S: Serologische Untersuchungen mit Trichomonaden. *Z Tropenmed Parasitol* 19:308-316, 1968.
119. Samuels R: Growth of axenic trichomonads in a serum-free medium. In Neal RA (Vuysje P): *Prog Protozool, Abstracts II Int Cong Protozool*, 1965, p 200, Amsterdam, New York, London: Int Cong Ser No. 91, Excerpta Medica Foundation, 1965.
120. Michel R, Westphal A: Die Spezifität der Dermal- und Peritonealzellreaktion sowie der Eosinotaxis bei der durch *Trichomonas vaginalis* sensibilisierten Maus. *Z Tropenmed Parasitol* 20:151-161, 1969.
121. Kazakova I, Teras JK: A comparative study of the antigenic properties of the strains of *Trichomonas hominis* Davaine. In Strelkov AA, Sukhanova KM, Raikov IV (eds): *Prog Protozool, Abstracts III Int Cong Protozool*, Leningrad, 1969, pp 299-300, Leningrad: Publ House "Nauka", 1969.
122. Terax JK, Kazakova I, Ellamaa M: Comparison of the antigenic properties of *Trichomonas vaginalis* Donné and *Trichomonas hominis* Davaine. *Wiad Parazytol* 15:241-243, 1969.
123. Kazakova I: Agglutinogenicity and agglutinability of strains of *Trichomonas hominis* Davaine. *Izv Akad Nauk Est SSR (Biologiia)* 24:130-140, 1975.
124. Michel R: Nachweis des allergischen Spätreaktionstyps bei Mäusen nach Sensibilisierung mit *Trichomonas vaginalis* und ein weiterer Beitrag zur Spezifität der Peritonealzellreaktion. *Z Tropenmed Parasitol* 22:91-97, 1971.
125. Rõigas EM, Tompel H, Kazakova I, Mirme E, Ellamaa M: Effects of in vitro and in vivo passages on the intraspecific variations in *Trichomonas hominis*, preprint. *Acad Sci Estonian SSR, Inst Zool Bot*, Tallinn, 1973.
126. Rõigas EM, Tompel H, Kazakova I, Mirme E, Ellamaa M: Effects of in vitro and in vivo passages on the intraspecific variation in *Trichomonas hominis*. In De Puytorac P, Grain J (eds): *Prog Protozool, Abstracts IV Int Cong Protozool*, Clermont-Ferrand, 1973, p 351, Clermont-Ferrand: Université de Clermont, 1973.
127. Honigberg BM: Trichomonads. In Jackson GJ, Herman R, Singer L (eds): *Immunity to Parasitic Animals*, vol 2, pp 469-550. New York: Appleton, 1970.
128. Honigberg BM, Goldman M: Immunologic analysis by quantitative fluorescent antibody methods of the effects of prolonged cultivation on *Trichomonas gallinae*. *J Protozool* 15:176-184, 1968.
129. Stępkowski S, Honigberg BM: Antigenic analysis of virulent and avirulent strains of *Trichomonas gallinae* by gel diffusion methods. *J Protozool* 19:306-315, 1972.

130. Honigberg BM, Lindmark DG: Trichomonads and *Giardia*. In Soulsby EJL: *Immune Responses in Parasitic Infections: Immunology, Immunopathology, and Immunoprophylaxis,* vol 4, pp 99-139. Boca Raton, Fla: CRC Press, 1987.
131. Solomon JM: Studies on the physiology of *Trichomonas hominis*. *J Parasitol* 43(suppl):39-40, 1957.
132. Shorb MS: The physiology of Trichomonads. In Hutner SH, Lwoff A (eds): *Biochemistry and Physiology of Protozoa,* vol 3, pp 383-457. New York: Academic Press, 1964.
133. Mehra KN, Levine ND, Reber EF: The amino acid composition of trichomonad protozoa. *J Protozool* 7(suppl):12, 1960.
134. Davaine CJ: *Traité des entozoaires et des maladies vermineuses de l'homme et des animaux domestiques*. Paris: Baillière, 1860.
135. Billet A: De la dysentérie à Trichomonas. *J Chir Méd d'Armée* 7:215-217, 1907.
136. Escomel E: Quelques remarques à propos de trichomoniases intestinale et vaginale. *Bull Soc Pathol Exot Filiales* 10:553-557, 1917.
137. Paranhos U: Nota sôbre o tratamento da trichomonoses intestinal. *Brasil Med* 32:314, 1918.
138. Caruso G: Le tricomoniasi umane. *Acta Med Ital Malatt Infett Parassitol* 15:231-239, 1960.
139. Croce J, Campos R: Ação da Cabimicina (Tricomicina) sôbre *Trichomonas hominis* e *Chilomastix mesnili*. *Hospital (Rio de Janeiro)* 63:169-173, 1963.
140. Cohen SO, Stapiński SM: Enteritis and urticaria associated with *Trichomonas* infection. *Am Pract Diag Treat* 3:238-239, 1954.
141. Vaingrib LG: Concerning the clinical picture of infestation by *Trichomonas intestinalis*. *Med Parazitol* 26:311-316, 1957.
142. Gamet A, Brottes H, Essomba E: Etiologies parasitaires et microbiennes des syndromes dysentériformes observés au centre Cameroun, région de Yaoundé. *Bull Soc Pathol Exot Filiales* 57:233-239, 1964.
143. Maurus M, Staib F: Zum Vorkommen von *Trichomonas* bei Enterokolitis. *Muench Med Wochenschr* 100:1647-1648, 1958.
144. De Oliveira SB, De Sá GS: Tricomoníase intestinal associada a apendicite aguda: Apresentação de 1 caso. *Hospital (Rio de Janeiro)* 67:215-218, 1965.
145. Nazirov MR: Some questions with regard to the protozoal and bacterial colitis and hepatocholecystitis considering the organism as a habitat. *Azerbaidzhan Med Zh* 42:77-82, 1965 (in Azerbaidzhani, Russian summary).
145a. Nazirov MR, Malikova TA: Protozoal and bacterial diseases of the intestines considering the organism as a habitat. *Azerbaidzhani Med Zh* 41:54-59, 1964 (in Azerbaidzhani, Russian summary).
146. Cziráki L, Dobías G: Clinical picture of *Trichomonas hominis* colitis and its differential diagnosis. *Orv Hetil* 108:1850-1852, 1967.
147. Vasallo Matilla F, Sicilia Enriquez JJ, Diaz Fernández MC: Hallazgo de un caso de parasitación humana por *Pentatrichomonas hominis*. *Rev Clin Española* 135:323-324, 1979.
148. Tanev IH, Tzvetkova AD: The problem of trichomonad colitis and its treatment. *Folia Med (Plovdiv)* 22:25-28, 1980.
149. Zhunhua Y: Treatment of trichomonal enteritis by Chinese medicinal herbs: Report of 16 cases. *J Tradit Chin Med* 1:63-64, 1981.
150. Zhang C: Clinical analysis of 253 cases of intestinal trichomoniasis. *Chi Sheng Chung Hsueh Yu Chi Sheng Chung Ping Tsa Chih* 1:73, 1983 (in Chinese).
151. Dobell C: Research on the intestinal protozoa of monkeys and man. VI. Experiments with the trichomonads of man and the macaques. *Parasitology* 24:531-577, 1934.
152. Manson-Bahr P: *The Dysenteric Disorders: The Diagnosis and Treatment of Dysentery, Sprue, Colitis, and Other Diarrhoeas in General Practice,* with an Appendix by J Muggleton. 15th ed. Baltimore: Williams & Wilkins, 1943.
153. Lynch KM: *Protozoan Parasitism of the Alimentary Tract: Pathology, Diagnosis, and Treatment*. New York: MacMillan, 1930.
154. Whittingham WE: Observations on the pathogenicity and treatment of flagellate dysentery. *Br Med J* 1:799-801, 1923.
154a. Wenyon CM: *Protozoology: A Manual for Medical Men, Veterinarians and Zoologists,* 2 vols. New York: William Wood, 1926.
155. Faust EC, Russell PF, Jung CR: *Craig and Faust's Clinical Parasitology,* 8th ed. Philadelphia: Lee & Febiger, 1970.
156. Houin R, Deniau M, Romano P, Poirot J-L: Trichomoniase pleurale. *Bull Soc Pathol Exot Filiales* 66:627-631, 1973.
157. Polivoda NG, Demchuk ND, Krivonos ZP: A case of lung disease caused by the intestinal trichomonad. *Vrach Delo* 2:33, 1987 (in Russian).
158. Foresi C: Studi sulla biologia di *Trichomonas hominis*. Nota III: Prove di invezione sperimentale nel topo. *Riv Ital Ig* 25:505-513, 1965.
159. Foresi C: Studi sulla biologia di *Trichomonas hominis*. Nota II: Resistenza delle varie forme ad agenti fisici e chimici. *Riv Ital Ig* 25:179-190, 1965.
160. Teras JK, Tompel H, Rõigas EM, Podar U, Laan I: Comparison of the patho-morphological lesions of the abdominal organs of white mice infected with axenic cultures of *Trichomonas vaginalis* Donné and *Trichomonas hominis* Davaine. *Wiad Parazytol* 15:311-313, 1969.
161. Teras JK, Tompel H, Podar U: Experimental

investigation of the pathogenicity of *Trichomonas hominis* and its changeability in vitro and in vivo. In Strelkov AA, Sukhanova KM, Raikov IV (eds): *Prog Protozool, Abstracts III Int Cong Protozool*, Leningrad, 1969, pp 293-295, Leningrad: Publ House "Nauka", 1969.
162. Al-Dabagh MA, Shafiq MA: Pathogenicity of *Trichomonas hominis* to splenectomized rats. *Trans R Soc Trop Med Hyg* 64:826-828, 1970.
163. Fernández F, Duarte Salcedo R: Estudio parasitológico de cien casos de diarrea infantil. *Med Países Cálidos* 8:326-337, 1933.
164. Giboda M: Intestinal parasites in Kampuchea, Takeo Province. *J Hyg Epidemiol Microbiol Immunol* 29:377-386, 1985.
165. Mathewossian ST: Untersuchung der Bevölkerung des Rayons von Idjewan (Armenien, Transkaukasien) auf Darmprotozoen. *Arch Schiffs Tropenhyg* 32:320-324, 1928.
166. Faust EC, Headless WH: Intestinal parasite infections of the ambulatory white clinic population of New Orleans. *Am J Trop Med* 16:25-38, 1936.
167. Summerlin HS: Amoebiasis: Incidence in private practice. *JAMA* 102:363-364, 1934.
168. Westphal A: Zur Morphologie, Biologie und Infektionsfähigkeit der viergeisseligen Trichomonasarten des Menschen. *Zentralbl Bakteriol Mikrobiol Hyg (A)* 137:363-376, 1936.
169. Panaitescu D, Gavrilescu M: Contributions to the study of parallel infection with *Trichomonas vaginalis* and *Trichomonas hominis*. *Microbiol Parazitol Immunol* 8:155-158, 1963 (in Romanian, English summary).
170. Hegner R: Experimental studies on the viability and transmission of *Trichomonas hominis*. *Am J Hyg* 8:16-34, 1928.
171. Giboda M, Čatár G: Contribution to the laboratory diagnostics of the rarer species of intestinal protozoa. *Bratisl Lek Listy* 57:199-201, 1972 (in Slovak, English summary).
172. Tumka AF: *Parasitology, Epidemiology and Laboratory Diagnosis of Intestinal Protozoa*. Leningrad: Meditzina, 1967 (in Russian).
173. Mattern CFT, Honigberg BM, Daniel WA: Fine-structural changes associated with pseudocyst formation in *Trichomitus batrachorum*. *J Protozool* 20:222-229, 1973.
174. Mattern CFT, Daniel WA: *Tritrichomonas muris* in the hamster: Pseudocysts and the infection of newborn. *J Protozool* 27:435-439, 1980.
175. Shinohara T: Effect of several digestive fluids upon the survival of *Trichomonas tenax* and *Trichomonas hominis*. *Keio Igaku* 35, 383-387, 1958 (in Japanese, English summary).
176. Głebski J: Effect of hydrochloric acid and gastric juice in the in vitro survival rate of *Trichomonas intestinalis* L. *Wiad Parazytol* 8:31-35, 1967 (in Polish, English summary).
176a. De Carneri I, Giannone R: Frequency of *Trichomonas vaginalis, Trichomonas tenax,* and *Entamoeba gingivalis* infections and absence of correlation between oral and vaginal protozooses in Italian women. *Am J Trop Med Hyg* 13:261-264, 1963.
177. Walzer PD, Rutherford I, East R: Empyema with *Trichomonas* species. *Am Rev Respir Dis* 118:415-418, 1978.
178. Ohkura S, Suzuki H, Hashiguchi Y: Letter to the editor. *Am J Trop Med Hyg* 34:823, 1985.
179. Fernández Martínez F, Suarez Peregrin E: Estudio parasitológico de cien casos de diarrea en Granada. *Med Países Cálidos* 6:177-195, 1933.
180. Da Cunha A, Pacheco G: Recherches sur les flagellés intestinaux de l'homme. *C R Soc Biol (Paris)* 89:765-767, 1923.
181. De Padua Vilela M, Rodrigues Dias L, Iglesias Capell J, Alves Brandão J, Martirani I, Zucato M: O emprêgo do Tiabendazol no tratamento da estrongiloidíase e de outras parasitoses humanas. *Hospital (Rio de Janeiro)* 62:691-709, 1962.
182. MacKinnon DL: Notes in the intestinal protozoal infections of 1680 men examined at the University War Hospital, Southampton. *Lancet* 195(vol.2):386-389, 1918.
183. Smith AM: Cases of acute amoebic dysentery in asylum patients never out of England. *Ann Trop Med Parasitol* 13:177-185, 1919.
184. Grisolia y Juristo M: Parasitismo intestinal en 200 habitantes de un distrito de Granada. *Med Países Cálidos* 6:455-487, 1933.
185. Gönnert R, Westphal A: Zur Technik der Stuhluntersuchung auf Protozoen. *Arch Schiffs Tropenhyg* 40:5-16, 1936.
186. Hargita G: Observations en connection avec l'infection par des vers intestinaux et des protozoaires, rencontrées chez les malades cutanés. *Borgyogy Venerol Sz* 43:63-68, 1962 (in Hungarian, French summary).
187. Cross JH, Clarke MD, Durfee PT, Irving GS, Taylor J, Partono F, Joesoef A, Hudojo, Oemijati S: Parasitology survey and seroepidemiology of amoebiasis in South Kalimantan (Borneo), Indonesia. *Southeast Asian J Trop Med Public Health* 6:52-60, 1975.
188. Cross JH, Clarke MD, Cole WC, Lien JC, Partono F, Djakaria, Joesoef A, Oemijati S: Parasitic infections in humans in West Kalimantan (Borneo), Indonesia. *Trop Geogr Med* 28:121-130, 1976.
189. Carney WP, Masri S, Stafford EE, Putrali J: Intestinal and blood parasites in the North Lore District, Central Sulawesi, Indonesia. *Southeast Asian J Trop Med Public Health* 89:165-172, 1977.
190. Nazarian I: Intestinal parasitic infection in Fars Province, Iran. *Z Tropenmed Parasitol* 24:45-50, 1973.
191. Chieffi G, Basso M: Indagine sulla diffusione

delle parassitosi intestinali: Osservazioni in una collettività militare. *Acta Med Ital Malatt Infett Parassitol* 6:344-348, 1951.
192. Lojodice C, Mazzitelli L: Indagine sulla diffusione delle parassitosi intestinali in una collettività infantile preventoriale. *Arch Ital Sci Med Terap Parassitol* 37:629-636, 1956.
193. Sornmani S, Pathammavong O, Bunnag T, Impland P, Intarakhao C, Thirachantra S: An epidemiological study of human intestinal parasites in Vientiane, Laos. *Southest Asian J Trop Med Public Health* 5:541-546, 1972.
194. Hegner R, Beltrán E, Hewitt R: Human intestinal protozoa in Mexico. *Am J Hyg* 32:27-44, 1940.
195. Ruiz Sanchez F: Flagelados intestinales en Guadalajara. *Med Rev Mexicana* 23:312-323, 1943.
196. Burrows RB: A survey of intestinal parasites in natives in Dutch New Guinea. *Am J Hyg* 42:262-265, 1945.
197. Obiamiwe BA: The pattern of parasitic infection in human gut at the Specialist Hospital, Benin City, Nigeria. *Ann Trop Med Parasitol* 71:35-43, 1977.
198. Ogunba EO: The prevalence of human intestinal protozoa in Ibadan, Nigeria. *J Trop Med Hyg* 80:187-191, 1977.
199. Okpala I: A survey of the incidence of intestinal parasites amongst government workers in Lagos, Nigeria. *West Afr Med J* 10:148-156, 1961.
200. Carney WP, de Veyra VU, Cala EM, Cross JH: Intestinal parasites of man in Bukidnon, Philippines, with emphasis on schistosomiasis. *Southeast Asian J Trop Med Public Health* 12:24-29, 1981.
201. Haughwout FG, Horrilleno FS: The intestinal animal parasites found in one hundred sick Filipino children. *Philippine J Sci* 16:1-73, 1920.
202. Carney WP, Banzon T, de Veyra V, Papasin MC, Cross JH: Intestinal parasites of man in Oriental Mindoro, Philippines, with emphasis on schistosomiasis. *Southeast Asian J Trop Med Public Health* 12:13-18, 1981.
203. Hill CM, Hill RB: Infection with protozoa and the incidence of diarrhoea and dysentery in Porto Rican children of the preschool age. *Am J Hyg* 7:134-146, 1927.
204. Boldescu I, Stratulat C: Investigations on infestation with *Trichomonas intestinalis* in a rural community. *Rev Med Chir Soc Med Nat Iasi* 72:391-394, 1968 (in Romanian, English summary).
205. Portús M, Prats G: Contribución al conocimiento de las protozoosis intestinales en la población hospitalaria barcelonesa. *Med Clin (Barcelona)* 76:203-205, 1981.
206. Fernández F: Clínica de la chilomastosis intestinal. *Med Países Cálidos* 6:302-306, 1933.

207. Kawai T, Nagayoshi Y, Koo C: A survey of the human intestinal protozoa in North Formosa. *J Med Assoc Formosa* 33:115-116, 1934.
208. Huang S-W, Lin C-Y, Khaw O-K: Studies on *Taenia* species prevalent among the aborigines in Wulai district, Taiwan. Part I. On the parasitological fauna of the aborigines in Wulai district. *Bull Inst Zool Acad Sinica (Taipei)* 5:87-91, 1966.
209. Kessel JF, Blakely L, Cavell K: Amebiasis and bacillary dysentery in the Los Angeles County Hospital 1929-1935. *Am J Trop Med* 16:417-430, 1936.
210. Johnstone HS, David NA, Reed AC: A protozoal survey of one thousand prisoners with clinical data on ninety-two cases of amoebiasis. *JAMA* 100:728-731, 1933.
211. Byrd EE: The intestinal parasites observed in fecal samples from 729 college freshmen. *J Parasitol* 23:213-215, 1937.
212. Hopp WB: On the epidemiology of human parasite infections in a state hospital in Indiana. *Am J Hyg* 39:138-144, 1944.
213. Faust EC: Amoebiasis in the New Orleans population as revealed by autopsy examination of accidental cases. *Am J Hyg* 11:371-384, 1941.
214. Canavan WPN, Hefley HM: Investigation of intestinal parasitic infections of a selected population of Oklahoma City. *Am J Trop Med* 17:363-383, 1937.
215. Hinshaw HC, Showers EM: A survey of human intestinal protozoan parasites in Philadelphia. *Am J Med Sci* 188:108-116, 1934.
216. Wenrich DH, Stabler RM, Arnett JH: *Endamoeba histolytica* and other intestinal protozoa in 1060 college freshmen. *Am J Trop Med* 15:331-345, 1935.
217. Wenrich DH, Arnett JH: A protozoological survey of 35 guests and 18 employees at a home for the aged. *J Parasitol* 23:318-319, 1937.
218. Bland PB, Rakoff AE: The incidence of trichomonads in the vagina, mouth and rectum: Evidence that vaginal trichomonads do not originate in the mouth or intestine. *JAMA* 108:2013-2016, 1937.
219. Young MD, Ham C: The incidence of intestinal parasites in a selected group at a mental hospital. *J Parasitol* 27:71-74, 1941.
220. Meleney HE: The relative incidence of intestinal parasites in hospital patients in Nashville and in rural Tennessee. *J Lab Clin Med* 19:113-119, 1933.
221. Faust EC: A study of the intestinal protozoa of a representative sampling of the population of Wise County, Va. *Am J Hyg* 11:317-384, 1930.
222. Turdeyev AA: A contribution to methodology of fecal examination for intestinal protozoa.

223. Simić T: L'infestation á protozoaires intestinaux des écoliers de Skoplje. *Ann Parasitol Hum Comp* 13:231-233, 1935.
224. Goldsmid JM, Rogers S, Mahomed K: Observations on the intestinal protozoa infecting man in Rhodesia. *S Afr Med J* 50:1547-1550, 1976.
225. Meyer KF, Johnstone HG: Laboratory diagnosis of amebiasis. *Am J Public Health* 25:405-414, 1935.
226. Svensson R: Studies on human intestinal protozoa, especially with regard to their demonstrability and the connexion between their distribution and hygienic conditions. *Acta Med Scand Suppl* 70:1-115, 1935.
227. Faust EC, D'Antoni JS, Odom V: A critical study of clinical laboratory methods for the diagnosis of protozoan cysts and helminth eggs in faeces. I. Preliminary communication. *Am J Trop Med* 18:169-183, 1938.
228. Faust EC: Clinical significance of newer methods for the diagnosis of intestinal parasites. *New Orleans Med Surg J* 91:447-451, 1939.
229. Swartzwelder JC: A comparison of five laboratory techniques for the demonstration of intestinal parasites. *J Trop Med Hyg* 42:185-187, 1939.
230. Denison N: Food remnants as a cause of confusion in the diagnosis of intestinal parasites. *J Lab Clin Med* 27:1036-1042, 1942.
231. Sawitz WG, Faust EC: The probability of detecting intestinal protozoa by successive stool examinations. *Am J Trop Med* 22:131-136, 1942.
232. Ridley DS, Hawgood BC: The formol-ether method of concentrating fecal cysts. *Trans R Soc Trop Med Hyg* 50:305, 1956.
233. Taylor AER, Baker JR: *The Cultivation of Parasites* In Vitro. Oxford: Blackwell, 1968.
234. Sapero JJ, Lawless DF: The "MIF" stain preservation technic for the identification of intestinal protozoa. *Am J Trop Med Hyg* 2:613-619, 1953.
235. Blagg W, Schloegel EL, Mansour NS, Khalaf GI: A new concentration technic for the demonstration of protozoa and helminth eggs in feces. *Am J Trop Med Hyg* 4:23-28, 1955.

Relevant Literature Not Cited in the Text

Ferrara A, Conca R, Grassi L, de Carneri I: Rilievi su un possible ruolo patogenico di *Trichomonas tenax* nella parodontite cronica. *Ann Ist Super Sanità* 22:253-256, 1986.

Jakobsen EB, Friis-Møller A, Friis J: *Trichomonas* species in a subhepatic abscess. *Eur J Clin Microbiol* 6:296-297, 1987.

Sato M, Hayashi A, Kato M, Hiroshi N, Namikawa I, Shiraki M, Katsutani Y, Iwayama S, Hirata K, Kimura K: Incidence of the oral trichomonads from subgingival plaque materials. *Nippon Shishubyo Gakkai Kaichi* 27:407-415, 1985 (in Japanese, English summary).

Vráblic J, Čatár G, Staník R, Hromada J, Žitňan D: Concerning the occurrence of the oral flagellate *Trichomonas tenax* in children with chronic tonsillitis. *Bratisl Lek Listy* 88:64-70, 1987.

20

Symptomatology, Pathology, Epidemiology, and Diagnosis of *Dientamoeba fragilis*

Günter Ockert

Symptomatology

In their 1918 publication of the first account of detailed investigations of *Dientamoeba fragilis*, Margaret W. Jepps and Clifford Dobell[1] expressed the opinion that there was no reason to believe that this parasite was potentially pathogenic. Yet in three of the seven cases they investigated, there were certain clinical histories long before the demonstration of the protozoon that could not be explained on other grounds. Later, several authors[2-6] described various symptoms in the carriers of *Dientamoeba* and pointed out the pathogenic effects of this species. They also described examples in which, after the introduction of specific antiparasitic therapy, the clinical manifestations disappeared together with the infection.

In his detailed account of the life history of *D. fragilis*, Dobell[7] pointed out the possible relationship of this protozoon to the pathogenic flagellate *Histomonas meleagridis*. He therefore did not rule out the possibility of a pathogenic potential of *Dientamoeba*, although he doubted it. Later workers[8-19] established the causal relationship with many well-documented examples, so that today there is no longer any doubt about the potential pathogenicity of *D. fragilis*. Gastrointestinal disturbances were seen in children as well as in adult patients infected with this parasite alone. The following list of complaints registered by 100 carriers of *D. fragilis* was given by Kean and Malloch[12]: abdominal pain in 20 patients; diarrhea in 40; flatulence and eructation in 5; fatigue and faintness in 5; nausea and vomiting in 3; anal pruritus in 3; headache in 2; indigestion in 2; and appearance of "worms" in stool in 2 patients. Other complaints unrelated to intestinal tract disease were noted in 44 cases. In agreement with these findings, we observed most frequently abdominal pain in 21 of 37 adult patients.[14,15] This was followed by changing stool consistency in 20 patients, diarrhea in 16, meteorism and mucus in stools in 11, loss of weight in 9, and bloody stools in 5 patients. In a large proportion of patients the complaints were long-lasting. Steinitz,[16] on the basis of his clinical observations, also expressed the opinion that *D. fragilis* by itself can be the cause of intestinal disturbances. He pointed to the frequently reported observations that *Dientamoeba* often accompanies *Entamoeba histolytica* and, because of its faster growth, can actually mask the presence of the latter. Steinitz[16] observed in 19% of patients infected with *D. fragilis* alone the CRIA (chronic recurrent intestinal amebiasis) syndrome, manifested by at least three of the following symptoms: abdominal pain, nausea, meteorism, and diarrhea; in contrast, he found these manifestations in 50% of infections involving *E. histolytica*. Very acute disease was not found in *D. fragilis* infections. The disease had a primarily chronic, recurrent character, with symptoms for the most part milder than those seen in *E. histolytica* amebiasis; only rarely was there chronic enlargement of the liver, as in amebic hepatopathy. Interesting in this connection was the fact that in many cases *D. fragilis* occurred in the gallbladder and in the duodenal fluid (Talis et al, cited in Steinitz[16]). Such findings require, however, further

---

Translated from German by B.M. Honigberg.

detailed investigation, because of corresponding observations, for example, the presence of *Entamoeba coli,* which is related to a specific clinical syndrome referred to as primary *E. coli* cholecystopathy.[20,21] Additional symptoms characteristic of *D. fragilis* infection were observed by Yang and Scholten.[17] More than half the persons infected only with *Dientamoeba* had diarrhea or abdominal pain, or both. The frequency of all recorded gastrointestinal and other symptoms is shown in Table 20.1, which includes also data published by others.

According to Millet et al,[19] there were significant differences in the frequencies of symptoms between persons with and without *D. fragilis* infection. Eighty-five percent of 26 infected persons exhibited more than one of the following symptoms: flatus, irritability, and weakness in 54% of the patients; abdominal pains lasting longer than 1 month in 46%; anorexia in 27%; nausea and vomiting in 19%; constipation in 15%; and diarrhea in 8%. Furthermore, many observations on children infected with *D. fragilis* reveal the pathogenic potential of this parasite. Our investigations[22] of 387 kindergarten children, with or without diarrhea, yielded a significantly higher number ($P < 0.001\%$) of patients infected with *Dientamoeba* in children with diarrhea than in the asymptomatic group (Table 20.2).

In a review of the clinical data collected from 35 patients published by Spencer et al[18] the most frequent symptoms were abdominal pain in 60%, diarrhea in 51%, and anorexia in 31% of the patients. Furthermore, the following less frequent symptoms were noted: fever in 26%, irritability in 20%, fatigue in 17%, constipation in 14%, and malaise. Only 6% of the persons in this group had no complaints. Of interest also was the finding that in more than half the children, the symptomatic disease lasted for a month or longer. In comparison to the *Dientamoeba*-free group, those infected with this parasite had a significantly higher percentage of persons with eosinophilia. In numerous cases, fecal smears from infected persons contained white cells, red cells, eosinophils, macrophages, and Charcot-Leyden crystals. In most cases the symptoms could be eliminated by therapy. The following drugs were used in treatment: diiodohydroxyl, 30 to 40 mg/kg for 21 days; metronidazole, 250 mg three times daily for 1 week; tetracycline hydrochloride 250 mg twice daily for 5 days.

According to the investigations of Ockert et al,[23] in a group consisting of 520 newborns and

TABLE 20.1. Symptoms in *D. fragilis* infections.

	Percentage of Patients	
Symptoms	Ontario Public Health Laboratory (Number examined = 225)	Data from the literature* (Number examined = 186)
Diarrhea	58.4	42.5
Abdominal pain	53.7	46.2
Anal pruritus	11.0	2.7
Abnormal stools (bloody, containing mucus, and watery)	9.8	22.6
Urticaria	6.7	0.0
Flatulence	5.9	19.9
Fatigue	5.9	13.4
Eosinophilia	5.1	4.3
Alternating diarrhea and constipation	3.9	13.4
Nausea and vomiting	3.5	20.4
Loss of weight	3.1	10.2
Constipation	2.4	6.5
Belching	2.0	5.4
Tenesmus	1.2	5.9
Anorexia	1.2	5.4
Other symptoms	2.0	18.3

Reproduced from Yang J, Scholten TH: *Dientamoeba fragilis:* A review with notes on its epidemiology, pathogenicity, mode of transmission and diagnosis. *Am J Trop Med Hyg* 26:16-22, 1977, with permission of The American Society of Tropical Medicine and Hygiene.
*For sources of individual data, see the report of Yang[17] and Scholten, above.

TABLE 20.2. Diarrhea in *D. fragilis*-positive and -negative preschool children.

Diarrhea	Number	D. fragilis	
		Positive	Negative
Present	168	69	99
Absent	219	54	165

Data from author's Dr Sc Dissertation.[22]
$x^2 = 11.85; P < 0.001$.

children ranging from 6 months to 15 years of age, those suffering from recurrent enteritis harbored *D. fragilis,* the infection rate being about 60%, on the average. Among the clinical symptoms the dominant one was the repeated occurrence of diarrheic stools. In contrast to the numerous symptoms observed in adults (e.g., abdominal pain, feeling of fullness, and nausea) none were noted in children. Only in exceptional cases was growth affected. In general, the prolonged duration of diarrhea was shown as a characteristic symptom of the infection; among the patients, about one-third had diarrheic stools for more than 2 years. The different durations of attacks of diarrhea were as follows: up to 1 month, 6 patients; 3 months, 4 patients; 12 months, 6 patients; 2 years, 11 patients; longer than 2 years, 15 patients.

Additional investigations carried out in the patients for pathogenic intestinal bacteria yielded negative results in all cases. Following antiparasitic therapy in which metronidazole was employed, the infection could no longer be demonstrated and the clinical manifestations disappeared completely in most instances. This result provides further proof of the pathogenic properties of the parasite.

Pathology

To date only a limited amount of research has been done on the pathogenic effects of *D. fragilis* and on the pathologic changes it causes. Burrows et al[9] and Swerdlow and Burrows[10] published histologic findings obtained by examinations of appendices removed from 15 patients infected with *Dientamoeba*. In 10 of these patients the appendix contained also *Enterobius vermicularis*. The histopathologic changes in the wall of the appendix ranged from suppurative appendicitis to lymphoid hyperplasia and marked fibrosis. The most common change was the increased expansion of the fibrous area of the connective tissue in the submucosa, which in most cases displaced lymphoid tissue of the mucosa or surrounded this tissue. To some degree, such changes occurred also in the internal layer of the muscular coat, or bandlike or finger-shaped eversions extended to the mucosa. It was shown in more detailed examinations that, although significantly less extensive, fibroses occur also in other parasitic infections. It was suggested that fibrosis of the appendix may occur with increased age and perhaps also as the result of repeated inflammation. Still, the patients examined by Burrows and Swerdlow[10] were young (4 to 15 years of age); thus in this instance *D. fragilis* might have acted as the causal agent of the inflammatory process. The authors assumed that a low-grade irritation, evoked by the protozoa over a longer period, leads to an only insignificant impairment. It was postulated that a result of this is a minimal inflammatory change in the appendix that ultimately leads to increased fibrosis of the connective tissue. Heretofore, the occurrence of active penetration of the mucosa by the parasites has not been established. Notable, however, was the incorporation of red cells by the ameboid organisms.

Although the diameter of the protozoa was usually less than 10 μm, three to six intracellular erythrocytes were found at one time. According to Shein and Gelb,[24] *D. fragilis* causes colitis by an invasive ulcerative process. With the aid of a flexible sigmoidoscope (35 cm), numerous punctiform ulcers, about 0.2 cm in diameter, were found in the intestine of a female patient infected with *D. fragilis*. These ulcers had yellowish centers and erythematous halos. The nonfriable areas of the mucosa located between the ulcers also showed erythematous changes. Flat ulcerations and acute as well as chronic inflammation could be demonstrated in a biopsy. The presence of *D. fragilis* could be confirmed only in direct suspensions of polyvinyl alcohol–fixed organisms in saline. Aspirates from the areas of ulceration were negative for the protozoa. Additional data on the pathology of *D. fragilis* infections are neither numerous nor informative. In the patients we examined,[14,15] there appeared in many instances inflammatory changes of the mucosa of the rectum and sigmoid colon. The areas of inflammation had the form of superficial rounded infiltrations of cells in loosened stroma yielding a picture of low-grade chronic proctitis. Hemorrhagic proctitis (a localized form of colitis ulcerosa) was seen only rarely. Steinitz,[16] using rectoscopy, found isolated cases of mild hypermia in persons infected with *D. fragilis.*

Until now it could not be demonstrated through animal investigations if the flagellate causes ulcers in the colon. It can be assumed that through histologic investigations of the higher secretions of the large intestine in the predilection sites of the parasite one will be better able to visualize the typical pathologic changes. On the basis of the aforementioned findings, the inclusion of *D. fragilis* with harmless saprozoic intestinal dwellers such as *Endolimax nana, Iodamoeba buetschlii,* and *Enteromonas hominis* appears unjustified. More appropriately, an infection with *D. fragilis* should be considered a latent infection (Doerr, cited by Piekarski[25]) in which the potential pathogenicity of the flagellate ought to be assumed. Similar relationships were said to exist under certain conditions in infection with the species* *E. coli,*[20,21] *Pentatrichomonas (Trichomonas) hominis,* and *Trichomonas tenax.*[26] Lamy[11] asserted that multiplication of *Dientamoeba* by itself cannot cause disturbance of intestinal functions, but that the parasite may play a role in such disturbances in concert with suitable bacterial intestinal flora. Such a combination of bacteria and the parasite, which possibly develops because of an amplified secretion, may lead to the development of an optimal environment for the colonization by and multiplication of *Dientamoeba.* An example of such a situation is the irritability of the colon. Under these conditions *Dientamoeba* could assume the role of a pathogen.

Important factors one should discuss here are, on the one hand, the high multiplication rate of the protozoon and, on the other hand, the possible increase in extracellular activity of the parasite enzymes in the optimal environment. When the number of protozoa exceeds the threshold, their pathogenic potential can be achieved. In this way, under the influence of the appropriate intestinal bacterial flora, a gradual physiologic change of the protozoan cell could cause a qualitative alteration of the parasite's enzyme activity. Also, in cases of repeated host passages, one would have to consider the virulence increase of the parasite. Finally, there must be taken into account the host-specific standard responses to pathologic stimuli as well as the individual susceptibility of the host as essential factors determining the progression of the disease. It is thus determined if such an infection can cause a shift of balance in the host-parasite relationship and thereby a disease in the host.[27] For example, morphologic damage to the intestine, disturbances in the microbiology of the large intestine, and/or serious primary diseases as well as colonization and multiplication of *D. fragilis* give rise to the appearance of clinical symptoms. Further detailed investigations are needed to obtain decisive evidence for and full understanding of the multifaceted nature of this relationship.

On the basis of the findings presented here it can be generally recommended that in cases of ill-defined intestinal disturbances, especially in the presence of long-lasting diarrhea, stool examinations be carried out. Such tests should be used especially when no explanation of the clinical symptoms by bacteriologic methods is possible, and, if necessary, specific antiparasitic therapy should be employed.

Epidemiology

Distribution

Reports of *D. fragilis* in humans come from all parts of the world. In the data that document the distribution of the species, one can note great differences in the prevalence of the infection. Most authors report comparatively small numbers.[28-36] It is noteworthy that Červa[37] could not demonstrate this species in his extensive studies in the district of Prague. This is even more surprising because other, in some cases older, reports (which many authors[12-35] do not take into account), cited unusually high incidence rates. According to Svensson[38] in 1928, in the population of Scandinavia, as in other parts

---

*As indicated in Chapter 19 of this book, the potential pathogenicity of *T. tenax* and *P. hominis* remains to be established. Usually, *E. coli* is considered a harmless commensal.

of the globe, there was high prevalence of the disease. In selected small population groups, *Dientamoeba* was reported to range from 26.0% to 32.6%. In some hospitalized mental patients over 50% were infected. Large numbers were reported also by other investigators.[4,11,39,40-43] Noble and Noble[44] assumed at least 20% of *D. fragilis* carriers among the world population and estimated 51% as the highest percentage. In some of the more recent reports, the parasite is assumed to be frequent; for example, a 41% prevalence indicated by Millet et al.[19] These somewhat contradictory data can be explained by the fact that in many cases the numbers or percentages would be significantly higher if a laxative were administered before examination.[38,40] Among other factors, one should take into consideration the different sensitivities of individual research methods. *Dientamoeba fragilis* can survive for a longer time in stools than was generally believed.[3,7] It can be isolated relatively easily from older material. However, abnormal, difficult-to-diagnose forms are seen in wet mounts and in stained fecal smears, especially when the material is no longer fresh. Additional data were obtained by examination of fecal smears fixed and stained according to the method of Lawless.[45] In samples from 6749 representatives of various population groups in the German Democratic Republic (Halle District), the infection rate was only 0.6%, which corresponds with the previously cited lower values.[20,46,47] Similarly, the average figures calculated from the data obtained from various laboratories in the period 1975 to 1987 lie between 0.93% and 3.61% (Ockert, unpublished data). Here, too, the limited diagnostic effectiveness of the most commonly employed method, that is, stained smears, must be considered responsible for the apparent lower frequency. In carrying out direct fecal examinations, supplemented by cultivation, in groups of children and also in adults in the district of Halle an unusually high infection rate was noted when stringent controls were employed. Of 376 children, 1 to 3 years old, 30.1% were found infected with *D. fragilis*. In 200 children from 3 to 6 years old, the frequency was 46.5%. Other intestinal protozoa occurred in significantly smaller numbers of children examined. For example, in all children, *Giardia lamblia* was found in only 11.1%; for *E. coli* this percentage was 7.8%, and for *E. histolytica*, 1.7%.[47] This picture of the disease was confirmed by the data gathered during continuation of this study, which ultimately included 1440 persons, 1066 children and 374 adults. The distribution of the frequencies among individual age-groups is seen in Table 20.3. The infection peak in the group of 6- to 10-year-olds is striking. In contrast, the other intestinal protozoa show the highest prevalence in the group of 3- to 6-year-olds. The difference between *D. fragilis* and other species of intestinal protozoa is also clear from the epidemiology of *Dientamoeba*. This is reflected in the finding that children in residential (boarding) and Monday-through-Friday child care centers are not always more frequently infected with this parasite than are children in day-care centers. In contrast, a correlation of the kinds of institutions can be established, for example, with *G. lamblia* and *E. histolytica* infections. The sum total of children of all ages shows an infection rate of 39%; the infection rate for adults is 25%. The portion of cases with multiple protozoan infections amounted to 8.4% of infected persons in the investigated groups. *Dientamoeba fragilis* was most frequently associated with *G. lamblia* (4.2%); less frequently with *E. coli* (2.3%) and with *E. histolytica* (0.4%). In individual communities of children, the frequencies of *D. fragilis* infection varied between 10% and 82%; most of these groups showed infection rates exceeding 20%. References to the existence of similarly large infection frequencies cite the numbers calculated on the basis of routine findings in diagnostic laboratories in the period extending from 1975 to 1987. In these laboratories, in addition to the direct stool examination, in some of the cases the cultivation technique was also used; furthermore, targeted groups of children were investigated. The maximum numbers were found to range between 10.4% and 27.0% (Ockert, unpublished observation).

In summary, the prevalence of *D. fragilis* differs in several respects from that of other intestinal protozoa. It indicates that the epidemiologic factors responsible for transmission of *Dientamoeba* cannot be equally favorable for transmission of the other protozoan species.

Transmission

The high frequencies of *D. fragilis* described by various authors led to the earlier assumption expressed occasionally that the cause of this prevalence is a transmision method unique to

TABLE 20.3. Percentage of infection with intestinal protozoa in various age-groups.

Protozoan species	n*	> 0 – <3 Positive		3 – <6 Positive			6 – <10 Positive			10 – <14 Positive			≥ 18 Positive	
		Number	%	n	Number	%	n	Number	%	n	Number	%	n	Number	%
Giardia lamblia	3177	416	13.1	1882	322	17.1	498	56	11.2	114	4	3.5	2100	95	4.8
Chilomastix mesnili	"	6	0.2	"	9	0.5	"	0	0.0	"	0	0.0	"	0	0.0
Entamoeba histolytica	"	16	0.5	"	51	2.7	"	4	0.8	"	0	0.0	"	23	1.1
Entamoeba hartmanni	"	6	0.2	"	38	2.0	"	0	0.0	"	0	0.0	"	17	0.8
Entamoeba coli	"	130	4.1	"	200	10.6	"	38	7.6	"	1	0.9	"	176	8.4
Iodamoeba buetschlii	"	10	0.3	"	32	1.7	"	4	0.8	"	0	0.0	"	25	1.2
Endolimax nana	"	25	0.8	"	68	3.6	"	5	1.0	"	1	0.9	"	40	1.9
Dientamoeba fragilis	382	143	37.4	200	81	40.5	338	159	47.0	114	46	40.4	374	94	25.1

*Total number of persons examined.

this intestinal parasite. In view of the fact that the possibility of cyst formation by *Dientamoeba,* entertained by Greenway,[48] Wenrich,[49] and Piekarski,[50] was not confirmed, contaminative "disseminating-georeceptive" transmission,[27] characteristic of most other intestinal protozoa, could be excluded. This raises a question about the reason for the unusually high infection rates of an intestinal protozoon unable to form a cyst.

With respect to its prevalence, *D. fragilis* assumes a unique position because other intestinal trichomonads* in which cysts do not form generally occur only infrequently. In the extensive studies carried out by the author, *P. hominis* was diagnosed only five times under special conditions in institutes for child psychiatry.[20] Also, according to the findings of various diagnostic laboratories, in work carried out between 1975 and 1987, the intestinal trichomonads were found only rarely. The first indications of the possibilities of transmission of *D. fragilis* by vectors, that is, intestinal helminths, are found in the work of Dobell.[7] This author pointed to the observations of Tyzzer[51] and also to those of Graybill and Smith,[52] who recognized the vector role of the eggs of the bird nematode *Heterakis gallinarum* in transmission of *Histomonas meleagridis,* a flagellate parasitic in turkeys. Because this species is related to *Dientamoeba,* Dobell assumed a similar relationship of this protozoon with worms, such as *Trichuris,* the whipworm; however, *Dientamoeba* has high prevalence in areas with low prevalence of the whipworm, for example, in Scandinavia.[38] Burrows and Swerdlow[53] have postulated for the first time a connection between *Dientamoeba* and the pinworm, *E. vermicularis,* and tried to prove this hypothesis. In appendices excised from infected persons they could ascertain double infections with both the pinworm and *Dientamoeba* 20 times more frequently than was theoretically expected. In such cases there were in the *Enterobius* eggs many small organisms similar to amebae whose nuclei closely resembled those of mononucleate or binucleate *Dientamoeba* from the lumen of formalin-fixed appendices. Investigations aimed at a conclusive confirmation of these relationships failed to yield positive results. The authors tested three possibilities for the demonstration of *Dientamoeba* in the eggs, or in the larvae enclosed in pinworm eggs.

1. *Enterobius* eggs from patients positive for *Dientamoeba* were crushed, then inoculated into culture media suitable for the protozoa.
2. Before being inoculated into the culture medium, the eggs were stimulated to develop in artificial digestive juices.
3. The egg material was inoculated into white mice.

None of these methods yielded cultures of the protozoa. The primary reason for the failure of the experiments was considered the inability to obtain sufficiently large numbers of worm eggs for the initiation of individual experiments. Burrows (personal communication) tested further possibilities for cultivation of *D. fragilis* from *Enterobius* eggs. However, these tests also failed to yield positive results. Our studies included a total of 37 isolation attempts, from 11 patients (children 3 to 6 years of age) who harbored simultaneously *Enterobius* and *Dientamoeba.* Eggs obtained from worms were allowed to develop in artificial digestive juices according to the method of Penfold et al[54]; they were then incubated in culture medium at 37°C after the method of Dobell and Laidlaw.[55] In no case was it possible to demonstrate growth of *Dientamoeba* with certainty, despite repeated subpassages (up to seven). Only twice, after the third or fourth subinoculation, were doubtful positive findings recorded, and those could not be confirmed on subsequent subpassages.

The foregoing data yield a contradiction to the presumed relationship between the two intestinal parasites that is difficult to explain and also raise doubts about its accuracy. However, one finds statements confirming this relationship in the literature. For example, *D. fragilis* infections, as studied for a prolonged period by Svennson,[38] are readily understandable if one accepts transmission of the protozoon in *Enterobius* eggs. An infection connected with fecal contamination can be excluded, because the author[38] was very mindful of personal hygiene and also could not diagnose other intestinal protozoa in her stool.[7,53] Burrows and Swerdlow[53] reported a similar indirectly confirmed case. Additional examples can be adduced here in which protozoal infection could be demonstrated shortly after diagnosing enterobiasis and could thus be interpreted as confirming the presumed interrelationship of the two parasites. Also, in

---

*Beyond doubt *Dientamoeba* is a modified trichomonad.[50a]

these examples one can safely exclude direct or indirect transmission of trophozoite-containing materials.

Furthermore, the results of statistical analysis of quantitative data related to the prevalence of both species speak in favor of such a relationship. Analysis of the incidence of infection shows that both pinworms and *Dientamoeba* are more frequent in 6- to 10-year-old than in 3- to 6-year-old children. In contrast, in giardiasis, for example, the frequencies of the protozoal and helminth infections have an inverse relationship. Furthermore, in studies of various groups of children it is possible to show with certainty a close positive correlation ($r = 0.71$) between the frequency of the pinworm and *Dientamoeba* infections (Table 20.4). The relationship of all *Enterobius*- and *Dientamoeba*-positive and -negative patients in a random sample is seen in Table 20.5.

Among 217 children with a positive cellophane-tape test for pinworm, 99 (45.6%) were infected also with *Dientamoeba*. In the second group the *Dientamoeba*-carrying fraction of persons amounted to only 81 of 296 (27.4%). Statistical comparison by the chi-square test indicates that this difference ($\chi^2 = 18.3$; $P < 0.001$%) is highly significant. Demonstration of *Dientamoeba* infection in *Enterobius*-negative children may be explained by the fact that the finding of eggs need not represent the actual incidence of the helminth infection. Even if the findings for *Enterobius* are negative, helminthiasis can exist. Often the hygienic condition of the patient can preclude positive findings. When regular and careful cleansing of the anal region is practiced by the patient, it is often difficult to collect pinworm eggs with cotton swabs or cellophane tape. Furthermore, the results of the infection experiments described below indicate that the infection with pinworms that caused the *Dientamoeba* infection could have, under certain circumstances, occurred a year earlier. Thus, by the time of the diagnosis of the protozoon, pinworm eggs are not necessarily found. Therefore, it is questionable if the 19 patients infected with *Dientamoeba*, who were investigated by Kean and Malloch,[12] could actually be considered as *Enterobius*-negative. With regard to the data of these authors[12] who, on the basis of their results, disagreed with Burrows and Swerdlow,[53] it should be pointed out that stool examinations alone do not provide a dependable diagnosis of *Enterobius* infections. Similar data that attest a close positive correlation between infections with *E. vermicularis* and *D. fragilis* were published by Krelle.[43] The correlation suggested by the quantitative prevalence relationship of the two parasites becomes clearer when similar comparisons of the infection rates of the protozoa closest in their frequencies to *Dientamoeba*, that is, *G. lamblia* and *E. coli*, are made (Table 20.6). The two latter species are transmitted by a contaminative method involving moist feces, foodstuffs, water, and the like, their prevalence having no relationship to infection with *Enterobius*. Therefore, in contrast with the previous comparison, the differences are not statistically significant ($\chi^2 = 2.0$ or 5.0; $P > 0.05^{47}$). The relationship between *D. fragilis* and *E. vermicularis* is established on statistical grounds that support also the assumption of Burrows and Swerdlow[53] with regard to the epidemiology of this intestinal protozoon. The apparent confirmation of the relationship between the protozoon and helminth also stimulated an attempt at its further clarification with the aid of three self-infection experiments which, together with the necessary preparation periods, extended over about 5 years and which, to a large extent, were con-

TABLE 20.4. Correlation* between *Enterobius* and *Dientamoeba* infection (expressed in percent) in various groups of children.

Group Number	% infection	
	Enterobius	Dientamoeba
1	29.4	8.8
2	47.1	35.3
3	40.9	42.0
4	13.3	16.3
5	25.0	55.0
6	64.8	37.4
7	75.0	86.7
8	37.5	50.0
9	21.9	12.5
10	13.4	37.1
11	28.9	50.0
12	17.1	12.5
13	48.9	41.4
14	41.6	42.3
15	9.5	11.1
16	44.4	26.3
17	28.0	30.0
18	40.0	43.5
19	70.0	63.3
20	63.3	45.0

Data from author's Dr Sc Dissertation[22]
*In all instances n (sample size) = 20; r (correlation coefficient) = 0.71.

TABLE 20.5. Correlation between frequencies of *Enterobius* and *Dientamoeba* infections.

Enterobius	Total Number	Dientamoeba	
		Positive	Negative
Positive	217	99	118
Negative	296	81	215

Reproduced from Ockert G: Zur Epidemiologie von *Dientamoeba fragilis* Jepps et Dobell 1918. I. Mitteilung: Die Verbreitung der Art in Kinderkollektiven. *J Hyg Epidemiol Microbiol Immunol* 16:213-221, 1972, with permission of Avicenum, Prague, Czechoslovakia.

ducted according to identical methods. The pinworm eggs used in the experiments were obtained with the aid of cotton swabs from the perianal skin of children harboring simultaneously *Enterobius* and *Dientamoeba*. In each experiment the prepatent periods were about 50 days for *Enterobius* and about 28 days for *Dientamoeba*. When considered together, the results of the three experiments carried out over long periods under carefully controlled conditions lend support to the idea of the presumed vector-mediated transmission of *D. fragilis*.

To clarify the still open questions of dependable identification of the protozoon in the eggs of the helminth vector, special investigations were carried out with the aid of various methods. Previously Burrows and Swerdlow[53] endeavored to demonstrate the presence of *Dientamoeba* in *Enterobius*. In the course of histologic examination the authors found in female pinworms from appendices removed from patients with double infections (*Dientamoeba-Enterobius*) from one to three small, predominantly ameboid forms, from 2 to 3 μm in diameter, in about 30% of nonsegmented eggs. Some of these forms were situated close to one another, which could be considered an indication of division within eggs. Since the material was fixed in formalin, whose effect frequently leads to agglomeration of the nuclear granules of *Dientamoeba*, the structures typical of the species could not be clearly visualized. Most of the structures were mononucleate forms. These forms could not be demonstrated in eggs of pinworms from stool samples of patients negative for *Dientamoeba*.

The foregoing findings led the authors to assume that the small questionable forms were mononucleate *D. fragilis*. In our studies we attempted to demonstrate intrahelminthic *Dientamoeba* by histologic methods (unpublished data). *Enterobius* eggs were collected from the perianal regions of patients found to be infected with both parasites. They were fixed in formalin. Further preparation of the eggs followed routine histologic methods, hematoxylin-eosin being used for staining. None of the experiments was successful in demonstrating *Dientamoeba* with certainty, which might be explained by the previously mentioned drawbacks of formalin fixation. Subsequently, in a single instance, the presence of cellular elements whose structure

TABLE 20.6. Correlations between infections with *Enterobius* and *Giardia lamblia* or *Entamoeba coli*.

	Patients					
	Enterobius-positive			Enterobius-negative		
Protozoan species	n	Protozoan infections[†]		n	Protozoan infections	
		Number	%		Number	%
Giardia lamblia	217	24	11.1	256	22	7.4
Entamoeba coli	11	20	9.2	11	22	7.4

Modified from Ockert G: Zur Epidemiologie von *Dientamoeba fragilis* Jepps et Dobell 1918. 1. Mitteilung: Die Verbreitung der Art in Kinderkollektiven. *J Hyg Epidemiol Microbiol Immunol* 16:213-221, 1972, with permission of Avicenum, Prague, Czechoslovakia.
*Total number of persons examined.
[†]Other than *D. fragilis*.

closely resembled that of *D. fragilis* could be ascertained in *Enterobius* larvae enclosed in eggs.

In a continuation of the study, identification of the aforementioned cellular inclusions was attempted with the aid of isoelectric point (IEP) determinations[55a]. It is known that there is a close relationship between electrical charges of cells or tissues and the adsorption of certain stains by these cells. Thus, two hypotheses were proposed for the solution of the problems:

1. Differential IEP ranges corresponding to the species-specific electrostatic charge levels of the nucleus and cytoplasm of *D. fragilis* and *E. vermicularis* would be demonstrable.
2. Equal or only insignificantly different electrostatic charge levels of the nucleus and cytoplasm of *D. fragilis* in cultures and *Enterobius* eggs would correspond also to equal IEP ranges of these cell components.

The IEP studies test and compare behavior of *D. fragilis* culture forms and of the cells of uncertain identity found in *Enterobius* eggs with regard to dye adsorption at various pH values. *Dientamoeba* grown in Dobell-Laidlaw medium[55] were used for the preparation of smears to be fixed in Carnoy's fluid and stained with methylene blue. The worm eggs were collected with the aid of cotton swabs from the perianal skin area of children harboring simultaneously *Dientamoeba* and *Enterobius*. The eggs were fixed in Carnoy's fluid and embedded in paraffin, benzine serving as the clearing medium. The paraffin sections were cut at a thickness from 6 to 8 μm; mounting was in Eukitt.

An old preparation method for methylene blue (1943), from Grübler, Leipzig) at a 4-mM concentration in Sörensen's phosphate buffer $\frac{1}{15}$ M, pH = 3.0, 3.5, 4.0, and 4.5) was employed for IEP determination in culture forms of *D. fragilis* and in the sectioned egg material; the length of staining was 10 hours. Rinsing was done in a buffer with corresponding pH, dehydration in tertiary butanol, clearing in xylene, and mounting in Eukitt. Because one depends on an optimal contrast between the protozoan nucleus and cytoplasm, IEP determination was not carried out with a pair of dyes used customarily for this purpose,[56,57] instead we used the so-called basophilia test employing only one dye (methylene blue). The pH level of 3.5 proved to be especially favorable, as seen from the data on color determinations in *D.*

TABLE 20.7. Isoelectric point determination in culture forms of *D. fragilis* using methylene blue at various pH values.

pH	Staining intensity	
	Nucleus	Cytoplasm
3.0	Pale blue (+)	Very pale blue (±)
3.5	Stronger blue + +	Blue +
4.0	Dark blue + + +	Blue +
4.5	Very dark blue + + + +	Stronger blue + +

Reproduced from Ockert G and Schmidt TH: Zur Epidemiologie von *Dientamoeba fragilis* Jepps et Dobell 1918. 4. Mitteilung: Nachweis von *Dientamoeba fragilis* in Enterobius-Eiern mit Hilfe von IEP Bestimmungen. *J Hyg Epidemiol Microbiol Immunol* 20:76-81, with permission of Avicenum, Prague, Czechoslovakia.

fragilis culture forms observed at various pH levels (Table 20.7). In the IEP test, the cells found in pinworm eggs and culture forms of *Dientamoeba* had a similar color intensity of the nucleus and cytoplasm at pH 3.5. They were rounded to oval, and occasionally stretched lengthwise (Fig. 20.1). The cell diameter ranged from 2.5 to 5.0 μm. In many instances one or two darkly stained condensations were noted that could be nuclei. The cytoplasm was weakly granular. At × 1300 magnification, the outlines of the cell usually appeared irregularly wavy or lobed. During focusing, the individual optical sections of the cells had variable outlines, an indication of the ameboid nature of these stages. In a few sections, cells whose cytoplasm showed stronger adsorption of methylene blue were found. This could have been interpreted as an expression of altered functional conditions of the cell.

In the egg stages there were an average of 11 to 13 of the previously described cells in the area between the larvae and the egg membrane. That these might have been fungi was contradicted by the fact that the latter fluoresce light-green on staining with 0.01% acridine orange, while *D. fragilis* in culture forms and the *Dientamoeba*-like stages found in worm eggs fluoresce gold-yellow on blue excitation (BG ¾-mm and OG 1-mm filters). Moreover, the ameboid form and the lack of a distinct membrane recognizable at a higher magnification are not typical of mycotic cells. With the aid of the results of the aforementioned physicochemical investi-

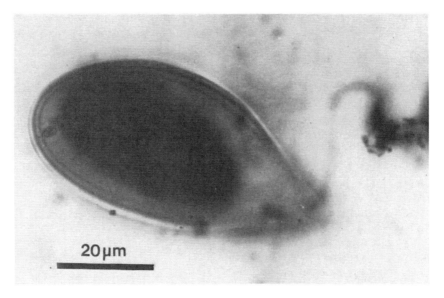

Fig. 20.1. Longitudinal section through an *Enterobius vermicularis* egg subjected to the methylene blue test for IEP determination. Note the cells lodged in the egg between the larva and the egg membrane.

gations, because of the agreement in the IEP results, it can be asserted that the forms demonstrated in the pinworm eggs are *D. fragilis*. It should be emphasized that these cells are smaller than *Dientamoeba* culture forms, this difference being probably adaptation to the vector.

In continuation of the aforementioned study, investigations of histopathologic material from patients infected with *Enterobius* were initiated to identify in stained preparations the typical cell structure of the *Dientamoeba* stages localized in the worms. To do this, sections of appendices excised from the patients infected with pinworms were examined. It was shown that *D. fragilis* remained intralarval in the young developmental stages of *Enterobius* eggs, while in the eggs containing fully developed infectious larvae, the protozoa were found in the area between the larvae and the egg membranes. Figure 20a shows a transversely sectioned *Enterobius* female lying in the appendix. In the area seen here, a section of the uterus with numerous eggs as well as part of the ovarian tube are evident. In young larvae enclosed in the eggs one is impressed by the ameboid structures with distinct pseudopodial extensions of the lobopodial type. These forms contain finely granular cytoplasm and a nucleus whose content consists of several chromatin fragments very similar to those seen in the nuclei of *D. fragilis*. Three, four, or five of these fragments can be identified with some assurance. Multinucleate cells cannot be demonstrated. The sizes of the ameboid stages vary between 5.2 and 7.0 μm; the diameter of their nuclei ranges between 1.2 and 1.8 μm. Of interest is the observation that several such structures can be found also in the growth zone of immature eggs. According to the investigations reported to date, multiple infections of larvae could not be recognized with certainty, and the same is true of the occurrence of ameboid forms in other organs or tissues of the helminths. Whether the complete developmental cycle of *Dientamoeba* can be shown in the appropriate *Enterobius* cells will depend on the results of further histologic studies (Fig. 20.2b,c). Current knowledge can be summarized as follows:

1. In comparison with other intestinal protozoa, *D. fragilis* shows unusually high prevalence and wide distribution, which suggest a unique mode of transmission, especially since this species, like the rarer *P. hominis*, forms no cyst.
2. The infection frequencies of *D. fragilis* and *E. vermicularis* show a high degree of positive correlation. In similar comparisons of *Dientamoeba* frequency with other intestinal protozoa, this kind of relationship cannot be demonstrated. Similar findings were reported by Burrows and Swerdlow,[53] according to whom the two species could be dem-

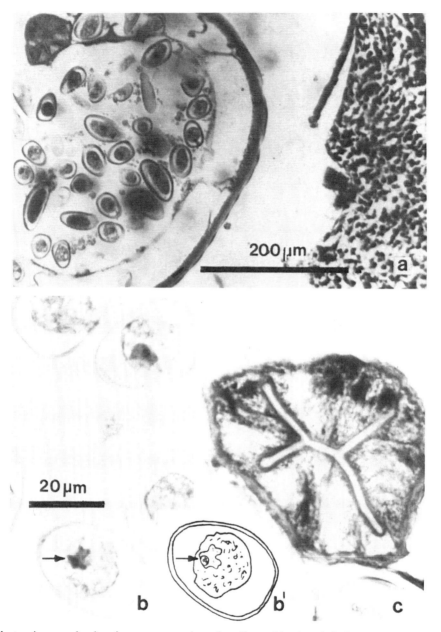

Fig. 20.2. Photomicrographs showing a cross-section of an *Enterobius* female lodged in an excised human appendix (**a**) and parts of this section at higher magnifications (**b,c**). **a.** General view taken at a low magnification. **b.** Transverse section of an egg. Ameboid cells are seen within the larva. The cell marked with the arrow is represented diagramatically in **b'**. **c.** Transverse section through the growth zone of an ovary. Note the indistinct ameboid cells with nuclei resembling those of the forms found with larvae. **b,c**: Same magnification.

onstrated together in excised appendices 20 times more frequently than would be expected on theoretical grounds.
3. The observed onset of *D. fragilis* invasion after three self-infection experiments provided further support for the presumed vector role of *E. vermicularis*.
4. The establishment of an agreement in IEP range between culture forms of *Dientamoeba* and the ameboid structures found in pinworm eggs proves the localization of *D. fragilis* within these eggs.
5. Supplementation of the still uncertain findings of Burrows and Swerdlow[53] is found in

the histologic demonstration of ameboid cells with nuclei typical of *D. fragilis* in immature *Enterobius* eggs and young larvae. These studies speak for the intrahelminthic development of the protozoon. The segment of the cycle of the species demonstrated by these findings makes it clear that *Dientamoeba* leaves the larvae in the course of ripening of eggs and presumably is disseminated in the intestine of the host by the gliding infected larval stages. Further investigations are necessary to describe the complete course of *D. fragilis* development in the helminth.

In addition, one should refer to the interesting report of Sukanahaketu,[58] according to whom *Dientamoeba* could be demonstrated also in *Ascaris* eggs. Similarities could be established in the nuclear structure of the parasite cells, most of which were mononucleate. The author asserted that the environment within the eggs is not optimal for the complete development of the protozoon and thus does not provide adequate conditions for nuclear division. It was assumed that *Ascaris lumbricoides* represented a vector of *D. fragilis;* however, the route which the protozoa follow in their wanderings in this helminth's eggs is unclear. Entrance through the digestive tract or through the cloacal region was assumed. Further investigations are necessary to provide proof that this nematode can serve as a facultative intermediate host for the development of *D. fragilis*.

Diagnosis

The great differences in the frequency of *D. fragilis* found in various laboratories emphasize the methodology problems encountered in diagnosing this species. Direct examination of fresh wet preparations which, because of its simplicity, is used in most diagnostic laboratories yields meaningful results only when fresh stool is examined microscopically immediately after its voiding. In such preparations, *Dientamoeba* often appears rounded. Its size is variable, the cells usually having a diameter from 5 to 15 μm. In some organisms, the pseudopodial activity is striking, so that confusion with *Entamoeba hartmanni* and *Endolimax nana,* for example, is quite possible. In contrast to the enteric amebas, the pseudopodia of *Dientamoeba* are typically attenuated or serrated, and these shapes can be used as dependable differential characteristics. However, one also can recognize broad or leaf-shaped lobopodia. The employment of phase-contrast or darkfield microscopy can be helpful in recognizing pseudopodial activity. Even after only about 1 hour in wet stool preparations, *Dientamoeba* cells become hard to recognize and can no longer be identified with certainty. However, their viability is retained for several hours,[4,7] as is shown by the results of cultivation. Addition of Lugol's iodine to fresh preparations does not reveal, for the most part, nuclear structure and is therefore of almost no value in diagnosis. Burrows,[59] on the basis of his own experiences, recommended the crystal violet–hematoxylin method of Velat et al[60] as especially useful for staining *D. fragilis* and trophozoites of other flagellates in fresh wet preparations. An advantage of this method is the specificity of the dyes employed, and stabilized pH values (4.6 or 5.2). *Dientamoeba* can be identified reliably in stained permanent fecal smears if fresh material is employed in their preparation. In a few hours in old stools, the protozoon is changed to such a degree that staining no longer suffices for demonstration of the typical cell structures. The hematoxylin method is especially well suited for staining of *Dientamoeba* nuclei. For fixation one should use mixtures with a large amount of glacial acetic acid, which guarantees a high quality of nuclear staining.[61] Fast methods of preparation of stained permanent fecal smears, for example, according to the procedure of Lawless,[45] often lead to unsatisfactory results of the damage-prone *D. fragilis*. In many cases the nuclear contents coalesce, and the individual chromatin bodies can no longer be identified. However, the extranuclear spindles are recognizable in some preparations as streaks or band-type strutures whereby an unequivocal diagnosis can be made. To avoid the often unsatisfactory results of staining when the simpler fast methods are employed, certain important steps in the preparation of stained fecal smears ought to be taken. These are as follows:

1. Fixation of thin (20 μm) fecal smears
2. Sufficiently large amounts of a fixative or a mixture of stain and fixative
3. Rapid placement of the fecal smears into the stain-fixative mixture or rapid overlay of the smears with this mixture

4. Complete rinsing out of the fixative
5. Employment of very pure solutions of reagents

The surest way to demonstrate *D. fragilis* is by cultivation. The dependence of the *D. fragilis* demonstration upon methodology was indicated earlier in the section on epidemiology. It was stated also that the frequently reported low incidence should be considered the result of examinations carried out exclusively by the direct method (fresh wet stool preparations or stained fecal smears). For example, in stool examinations of 3670 children in the area of Halle in the German Democratic Republic an infection rate of 0.6% could be demonstrated through the use of the fast staining method of Lawless[45] alone. In an additional sample of 576 children, the same method revealed 3% of *Dientamoeba* infections. However, an investigation of this random sample employing stained fecal smears as well as cultivation demonstrated the fraction of infected persons to be about 35%. A corresponding efficiency comparison of diagnostic methods carried out in a large random sample of 1066 persons led to a similar conclusion. Here, the incidence of *D. fragilis* obtained by examination of stained fecal smears alone equaled 1.97%. In contrast, combined methods (stained fecal smears and cultivation) showed *Dientamoeba* infection in 39.31% of persons. The corresponding relationships for all the intestinal protozoa demonstrated in this group of children are shown in Table 20.8.

The diphasic medium of Dobell and Laidlaw[55] is especially well suited for cultivation of *D. fragilis*. We employed a somewhat modified substrate whose solid phase consisted mainly of coagulated human serum. Furthermore, good growth could be obtained in a completely liquid medium. This latter consists of Ringer's solution supplemented with 5% to 10% bovine serum. In both media, rice starch granules constitute the chief particulate nutrient of the protozoon.

Because cell multiplication in primary cultures can be limited to such an extent as to render demonstration of the protozoa impossible even with the most careful sampling, to obtain dependable results it is always necessary to make at least one subinoculation into fresh medium. Among others, Magyar[62] and Burrows[59] have pointed out the unfavorable conditions of the primary cultures and recommended the employment of blind passages. According to our results, the primary cultures should always be followed by one or two passages in fresh culture medium, to ensure utilization of the full potential of a cultivation method. These relationships are shown in Figure 20.3, which includes 1571 cases of *D. fragilis*, 48 of *E. histolytica*, 63 of *E. coli*, and 18 of *P. hominis*. It is evident that with the aid of the primary culture demonstration of *D. fragilis* was successful in only about 40%; it was comparatively lower for *E. histolytica* (23%). The primary cultures and first subcultures together provided for diagnosis of over 80% of infections. A more complete demonstration is, however, possible when a second subculture is made. The employment of two subpassages must always be considered when a

TABLE 20.8. Comparison of detection efficiency (expressed as percents of infection frequencies) of stained fecal smears (A) with a combination of such smears and cultivation (B).

Protozoan species	Percentage of infection frequency*		
	A	B	B/A
Giardia lamblia	9.29	9.29	1
Chilomastix mesnili	0.00	0.28	—
Pentatrichomonas hominis	0.00	0.66	—
Entamoeba histolytica	1.31	2.06	1.57
Entamoeba hartmanni	0.19	0.28	1.47
Entamoeba coli	4.32	5.72	1.32
Iodamoeba buetschlii	0.56	0.56	1.00
Endolimax nana	0.75	0.75	1.00
Dientamoeba fragilis	1.97	39.31	19.95

Reproduced from Ockert G: Zu einigen Fragen des mikroskopischen Nachweises intestinaler Protozoeninfektionen. *Z Gesamte Hyg* 26:726-729, 1980, with permission of VEB-Verlag Volk und Gesundheit, Berlin, D.D.R.
*n (sample size) = 1066.

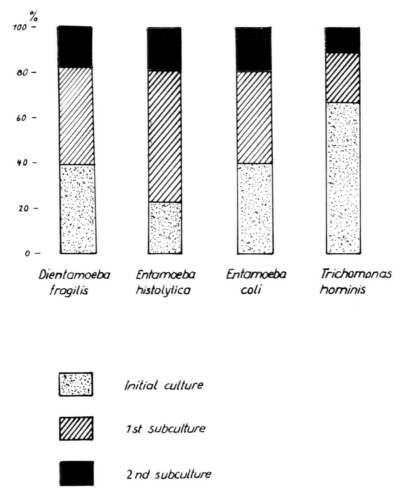

Fig. 20.3. Sensitivity of primary cultures and the first two sequential subcultures in detecting intestinal protozoa [*D. fragilis, E. histolytica, E. coli,* and *Pentatrichomonas* (= *Trichomonas*) *hominis*].

dependable demonstration of *Dientamoeba* is especially important in clinically suspected cases, such as in etiologically doubtful gastrointestinal disturbances.[63]

References

1. Jepps MW, Dobell C: *Dientamoeba fragilis* n. g., n. sp., a new intestinal amoeba from man. *Parasitology* 10:352-367, 1918.
2. Gittings JC, Waltz AD: *Dientamoeba fragilis. Am J Dis Child* 34:542-546, 1927.
3. Hakansson EG: *Dientamoeba fragilis,* a cause of illness: Report of a case. *Am J Trop Med* 16:175-185, 1937.
4. Hakansson EG: *Dientamoeba fragilis:* Some further observations. *Am J Trop Med* 17:349-362, 1937.
5. Wenrich DH: Studies on *Dientamoeba fragilis* (Protozoa). 2. Report of unusual morphology in one case with suggestions as to pathogenicity. *J Parasitol* 3:183-196, 1937.
6. Mollari M, Anzulovic JV: Cultivation and pathogenicity of *Dientamoeba fragilis,* with a case report. *J Trop Med Hyg* 41:246-247, 1938.
7. Dobell C: Researches on the intestinal protozoa of monkeys and man. X. The life-history of *Dientamoeba fragilis:* Observations, experiments and speculations. *Parasitology* 32:417-461, 1940.
8. Knoll EW, Howell KM: Studies on *Dientamoeba fragilis:* Its incidence and possible pathogenicity. *Am J Clin Pathol* 15:178-183, 1945.

9. Burrows RB, Swerdlow MA, Frost JK, Leeper CK: Pathology of *Dientamoeba fragilis* infections of the appendix. *Am J Trop Med Hyg* 3:1033-1039, 1954.
10. Swerdlow MA, Burrows RB: *Dientamoeba fragilis:* An intestinal pathogen. *JAMA* 158:176-178, 1955.
11. Lamy L: *Dientamoeba fragilis:* Recherche, culture, frequence, intérêt et caractères pathogènes. *Bull Soc Pathol Exot Filiales* 53:505-509, 1960.
12. Kean BH, Malloch CL: The neglected ameba; *Dientamoeba fragilis:* A report of 100 "pure" infections. *Am J Dig Dis* 11:735-746, 1966.
13. Machado JO, DePinho AL, Silva S, Gomes FJR: Aspects parasitaires de *Dientamoeba fragilis:* Action pathogène et thérapeutique. *Bull Soc Pathol Exot Filiales* 61:30-35, 1968.
14. Schulz U, Ockert G: Darmprotozoen als Ursache intestinaler Störungen. *Dtsch Med Wochenschr* 96:1963-1967, 1971.
15. Ockert G, Schulz U: Zur pathogenetischen Rolle von *Dientamoeba fragilis. Dtsch Gesundheitswes* 27:1156-1158, 1972.
16. Steinitz H: Klinische Beobachtungen bei Infektionen mit *Dientamoeba fragilis. Z Gastroenterol* 11:170-182, 1973.
17. Yang J, Scholten TH: *Dientamoeba fragilis:* A review with notes on its epidemiology, pathogenicity, mode of transmission and diagnosis. *Am J Trop Med Hyg* 26:16-22, 1977.
18. Spencer MJ, Garcia LS, Chapin MR: *Dientamoeba fragilis*, an intestinal pathogen in children? *J Dis Child* 133:390-393, 1979.
19. Millet VM, Spencer MJ, Chapin U, Steward U, Yatabe JA, Brewer TH, Garcia LS: *Dientamoeba fragilis:* A protozoan parasite in adult members of a semicommunal group. *Dig Dis Sci* 28:335-339, 1983.
20. Kalk H, Wildhirt E: Über das Vorkommen von Amöben im Duodenalsaft und in der Galle. *Med Klin* 49:1466-1468, 1954.
21. Geyer E: Gibt es eine *Entamoeba coli*-Cholezystopathie. *Z Gesamte Inn Med* 14:968-972, 1959.
22. Ockert G: *Beispiele zum protozoologischen Status autochthoner Populationen mit Bemerkungen zur Epidemiologie und Ultrastruktur von* Dientamoeba fragilis, Dr Sc Dissertation B. Martin-Luther-Universität, Halle-Wittenberg, 1976.
23. Ockert G, Preiss U, Brömme S, Grävinghoff J: Zur Bedeutung enteraler Protozoeninfektionen im Säuglings—und Kleinkindalter, *Z Klin Med* 41:1885-1888, 1986.
24. Shein R, Gelb A: Colitis due to *Dientamoeba fragilis. Int J Gastroenterol* 78:634-636, 1983.
25. Piekarski G: Latente Protozoen-Infektion und chronische Protozoen-Erkrankungen. *Dtsch Med J* 18:428-433, 1967.
26. Teras J: Extraurogenital infections of man by trichomonads. *Abstracts Int Symp Trichomonads & Trichomoniasis,* Prague, July 1985, p 83, Prague: Dept Parasitol, Faculty Sci, Charles Univ, 1985.
27. Odening K: *Parasitismus—Grundfragen und Grundbegriffe.* Wiss Taschenbücher No. 112. Berlin: Akademie Verlag, 1974.
28. Burrows RB: Studies on the intestinal parasites of mental patients. *Am J Hyg* 38:293-305, 1943.
29. Burrows RB: Intestinal parasitic infections in military food handlers. *US Armed Forces Med J* 5:77-82, 1954.
30. Levine ND: *Protozoan Parasites of Domestic Animals and of Man.* Minneapolis: Burgess, 1960.
31. Csürös C, Mraz T: Belprotozoonok gyakorisaga Baranya megyeben. *Gesundh Wiss* 8:243, 1964.
32. Dufek M, Kalivoda R, Blaha R: Intestinal parasitic infections in persons arriving in Czechoslovakia from tropics and subtropics. *Bull Inst Mar Med Gdańsk* 18:31-33, 1967.
33. Müller WA: Untersuchungen über die Befallshaufigkeit mit Intestinalparasiten bei der Bevölkerung Dresdens und den Wert einiger gebräuchlicher Nachweisverfahren für Darmparasiten beziehungsweise deren Geschlechtsprodukte im Stuhl des Menschen. 2. Mitteilung. *Dtsch Gesundheitswes* 24:465-469, 1969.
34. Müller WA: *Untersuchungen über den Befall durch Zooparasiten bei Einwohnern der Stadt Dresden,* thesis. Technische Universität Dresden, 1974.
35. Müller WA: Die Infektionen mit Intestinalprotozoen bei Einwohnern Dresdens. *Z Gesamte Inn Med* 30:209-214, 1975.
36. Flentje B: Protozoeninfektionen in Kinderkollektiven. *Dtsch Gesundheitswes* 26:1216-1219, 1971.
37. Červa L: Die Darmparasiten in den Kinderheimen des Prager Bezirkes. *Angew Parasitol* 2:119-124, 1961; 3:7-16, 1962.
38. Svensson R: A survey of human intestinal protozoa in Sweden and Finland—A preliminary report. *Parasitology* 20:237-249, 1928.
39. Lörincz F: Über das Vorkommen von menschlichen Darmprotozoen in Ungarn. *Arch Schiffs Tropenhyg* 37:276-283, 1933.
40. Sapero JJ, Johnson CM: *Entamoeba histolytica* and other intestinal parasites: Incidence in variously exposed groups of the Navy. *US Nav Med Bull* 37:297-301, 1939.
41. Brug SL: *Dientamoeba fragilis. Geneesk Tijdschr Ned Indie* 76:1288-1302, 1936.
42. Karlewiczowa R, Kasprzak W: The fauna of intestinal protozoa in children and youth from Poznan. *Wiad Parazytol* 10:423-424, 1964.
43. Krelle E: Zur Rolle von *Enterobius vermicu-*

laris bei der Übertragung von *Dientamoeba fragilis*. *Dtsch Gesundheitswes* 29:167-168, 1974.
44. Noble ER, Noble GA: *Parasitology: The Biology of Animal Parasites,* 2nd ed. Philadelphia: Lea & Febiger, 1964.
45. Lawless DK: A rapid permanent-mount stain technique for the diagnosis of intestinal protozoa. *Am J Trop Med Hyg* 20:1137-1138, 1953.
46. Ockert G: Über die Verbreitung einiger Darmparasiten unter der Bevölkerung des Bezirkes Halle. *Z Arztl Fortbild* 59:850-855, 1965.
47. Ockert G: Zur Epidemiologie von *Dientamoeba fragilis* Jepps et Dobell 1918. 1. Mitteilung: Die Verbreitung der Art in Kinderkollektiven. *J Hyg Epidemiol Microbiol Immunol* 16:213-221, 1972.
48. Greenway D: *Dientamoeba fragilis* en la Argentina. *Arch Argentinos Enfermendades Aparato Digestivo* (*Buenos Aires*) 3:897, 1928.
49. Wenrich DH: Studies on *Dientamoeba fragilis* (Protozoa) 1. Observations with special reference to nuclear structure. *J Parasitol* 22:76-83, 1936.
50. Piekarski G: Zur Frage der Cystenbildung bei *Dientamoeba fragilis*. *Z Gesamte Hyg* 127:496-500, 1948.
50a. Honigberg BM: II. Taxonomic position and revision of the genus *Dientamoeba* Jepps & Dobell, pp 79-80. In Camp RR, Mattern CFT, Honigberg BM: Study of *Dientamoeba fragilis* Jepps & Dobell. I. and II. *J Protozool* 21:69-82, 1974.
51. Tyzzer EE: Studies on histomoniasis or "blackhead" infection in the chicken and the turkey. *Proc Am Acad Arts Sci* 69:189-264, 1934.
52. Graybill HW, Smith T: Production of fatal blackhead in turkeys by feeding embryonated eggs of *Heterakis papillosa*. *J Exp Med* 31:647-655, 1920.
53. Burrows RB, Swerdlow MA: *Enterobius vermicularis* as a probable vector of *Dientamoeba fragilis*. *Am J Trop Med Hyg* 5:258-265, 1956.
54. Penfold WJ, Penfold HD, Phillips M: Artificial hatching of *Taenia saginata* ova. *Med J Aust* 2:1039-1042, 1937.
55. Dobell C, Laidlaw PP: On the cultivation of *Entamoeba histolytica* and some other entozoic amoebae. *Parasitology* 18:283-318, 1926.
55a. Ockert G, Schmidt T: Zur Epidemiologie von *Dientamoeba fragilis* Jepps et Dobell, 1918. 4. Mitteilung: Nachweis von *Dientamoeba fragilis* in *Enterobius*. Eiern mit Hilfe von JEP-Bestimmungen. *J Hyg Epidemiol Microbiol Immunol* 20:76-81, 1976.
56. Drawert H: Untersuchungen über die-pH-Abhangigkeit der Plastidenfärbung mit Säurefuchsin und Toluidinblau in fixierten pflanzlichen Zellen. *Flora* 131:341-354, 1936.
57. Drawert H: Das Verhalten der einzelnen Zellbestandteile fixierter pflanzlicher Gewebe gegen saure und basische Farbstoffe bei verschiedenen Wasserstoffionenkonzentration. *Flora* 132:91-124, 1937/38.
58. Sukanahaketu S: The presence of *Dientamoeba fragilis* in the *Ascaris lumbricoides* ova: The first report from Thailand. *J Med Assoc Thailand* 60:256-258, 1977.
59. Burrows RB: *Microscopic Diagnosis of the Parasites of Man*. New Haven: Yale University Press, 1965.
60. Velat CA, Weinstein PP, Otto GF: A stain for the rapid differentiation of the trophozoites of the intestinal amoebae in fresh wet preparations. *Am J Trop Med* 30:43-51, 1950.
61. Markell EK, Voge M: *Medical Parasitology,* 5th ed. Philadelphia: WB Saunders, 1981.
62. Magyar E: Biologische Besonderheiten einiger *Entamoeba histolytica*—Stämme und ihre Laboratoriumsdiagnose unter besonderer Berücksichtigung serologischer Methoden. *Zentralbl Bakteriol Mikrobiol Hyg (A)* 186:126-131, 1962.
63. Ockert G: Zu einigen Fragen des mikroskopischen Nachweises intestinaler Protozoeninfektionen. *Z Gesamte Hgy* 26:726-729, 1980.

Index

Abscess
  animal model and, 133, 134
  female genital tract and, 280, 282
  intramuscular infection of mouse and, 144, 146
  lung and, 355
  subcutaneous, 141–142
    rupture of, 141, 144
Accessory filament
  Pentatrichomonas hominis, 26, 27
  Trichomitus fecalis and, 30
  Trichomonas tenax and, 22
  Trichomonas vaginalis and, 9, 11
Acetamide, 73
Acetate, 57, 61, 62, 69
Acetic acid, 94
Acetyl-CoA, 59, 62
  thioester bond and, 70
Acid phosphatase
  Trichomonas tenax and, 345
  Trichomonas vaginalis and, 66
Acidic end product, 58
Acridine orange-stained smear, 300–301
Actin, 17, 71
Adenine, 85
Adenosine, 85
Adenosine kinase, 85
Adenosine phosphorylase, 85
Adherence, 67, 184–192
  microflora and, 213
  Trichomonas vaginalis and, 67, 182–190
Adverse drug reaction, 327–328
Aerobic carbohydrate metabolism, 63–64
Aerobic condition
  hydrogenosomes and, 69
  polymorphonuclear leukocytes and, 204–205
Aerobic protozoa, 70

Age, host
  Trichomonas tenax and, 350
  Trichomonas vaginalis and, 249, 317
Agglutinating factor, 44–46
Agglutination
  Pentatrichomonas hominis and, 368–369
  respiratory infection and, 360, 361, 362
  Trichomonas tenax and, 344
  Trichomonas vaginalis and, 37
Alanine, 57
Alanine aminotransferase, 64
Alcohol
  metronidazole and, 327
  Trichomonas tenax and, 353
Allopurinol, 84
Amino acid, 56, 59, 64–65
  vagina and, 76
Aminotransferase, 64
Amphotericin B, 92, 93
Anaerobic carbohydrate metabolism, 59–63
Anaerobic metabolism, 74
Anaerobic organism, 70, 74, 217
Anaphylactic reaction, 327
Angina, Vincent's, 347
Animal model
  flora and, 218
  laboratory rodent and, 120–149
    ectopic infections and, 129–146. *See also* Ectopic infection
    hamster and, 120, 122
    male urogenital system and, 128–129
    mouse and, 125–128
    rat and, 122–125
    virulence and, 146–148
  Pentatrichomonas hominis and, 372–373
  Trichomonas tenax and, 349
  Trichomonas vaginalis and, 41–42, 112–154

Ankylosing spondylitis, 293
Anorexia, 327
Antibiotic in axenic cultivation, 91–92
Antibody
  complement fixing, 303
  fluorescent monoclonal, 231
  immunocytochemical examination and, 302
  monoclonal
    Trichomonas vaginalis and, 40
    tubulins and, 71
  in secretion, 305
  in sera, 303, 305
  Trichomonas vaginalis and, 40–41, 44–46
    monkey and, 117
    vaccination and, 219–221
Anticytoskeleton compound, 185–187
Antifungal antibiotic, 92
Antigen
  absorbed and shed, 39–40
  pathogenicity and, 189
Antigenic property
  Pentatrichomonas hominis and, 367–369
  Trichomonas tenax and, 358–365, 367
  Trichomonas vaginalis and, 365, 367
Antigenic structure, 36–40
Antitrichomonal agent
  child and, 267–268
  metronidazole, 324–333
Antitrichomonal antibody, 44–46
Arginine, 64
Ascitic fluid, 131
Aspartate: α-ketoglutarate aminotransferase, 64
Assay
  intraperitoneal mouse, 134–135, 137
  subcutaneous mouse, 184, 218
  virulence, 155–208
    Trichomonas tenax and *Pentatrichomonas hominis* and, 205, 207–208
    Trichomonas vaginalis and, 155–207. *See also* Host cell-trichomonad interaction, *Trichomonas vaginalis* and
Asymptomatic infection, 226–227
Ataxia, 327
dATP, 87
ATPase, 72
Atractophore
  Dientamoeba fragilis and, 31, 33
  Trichomonas vaginalis and, 20, 21
Attachment, cell culture monolayers and, 191
Autotonomy, 26
Axenic cultivation, 91–92
Axenization, 107–108, 109
Axostylar microtubule, 15, 17, 20
Axostyle
  Pentatrichomonas hominis, 26–27

Trichomonas fecalis and, 31
Trichomonas tenax and, 22
Trichomonas vaginalis and, 9, 14, 15, 17, 19
Azomycin, 324

Baby hamster kidney cell, 164–165, 167, 188
Bacteria
  adherence inhibition and, 185
  Trichomonas tenax and, 349
Bacterial flora, 253
Bacterial vaginosis, 216, 230
  clinical features of, 229
Bacteriocin, 214
Bacterioides, 216–217
Balanitis, 240
  child and, 259
Band, protein, 38
Bell-clapper rod, *See* Atractophore
Benzyl penicillin, 93
BHK. *See* Baby hamster kidney cell
Bile, 384–385
Bimanual examination, 228–229
Binding, protein, 325–326
Biochemistry of *Trichomonas vaginalis*
  drug metabolism and, 72–75
  electrophoresis and, 54–55, 65
  enzymes and, 59–66
  host environments and, 75–76
  metabolic pathways of, 59–65
  metabolism and, 59–66
  nutritional requirements and, 56–57
  organelles and, 66–72
  release of end products of metabolism and, 57–59
Biopsy, 291–293, 295
Bladder infection, 226
Bronchitis, 343
Buffered salts solution, 94

C3 compound, carboxylation of, 62–63
CACH medium, *See* Media
Cancer, 328–329
  cervical neoplasia and, 286–287
Candida, 122, 124, 216, 217, 220
  child and, 261, 263–265
  mouse and, 126, 128
  Trichomonas tenax and, 353
Candidiasis, 229–230
Capitulum
  Pentatrichomonas hominis, 26–27
  Trichomitus fecalis and, 30–31
  Trichomonas tenax and, 22
  Trichomonas vaginalis and, 7, 9, 13, 15

Carbohydrate
  Trichomonas vaginalis and, 56, 188–189
  vagina and, 76
Carbohydrate metabolism
  aerobic, 63–64
  anaerobic, 59–63
Carbon dioxide, 57, 69
Carcinogenicity, 328–329
Carnidazole, 333
Catabolism, 60
Catheterization of bladder, 226
CDF. *See* Cell detaching factor
Cecum, 120
Cell
  division of, 21–22
  epithelial
    squamous, 282–283
    wet mount examination and, 114–115
  mast, 119–120
Cell coat, 9, 14, 19
Cell detaching factor, 177, 191
Cell-free filtrate, 167, 169
Cell-mediated cytotoxicity, 194–201
Cell-mediated immunity, 46–47
Cell-mediated killing, 41
Cercomonaden, 355
Cercomonas, 355
Cervical neoplasia, 286–287, 329
Cervicitis, 281
Cervicovaginitis emphysematosus, 282
Cesium chloride buoyant density gradient, 87–88
Chemokinesis, 203
Chemotaxis, 203–205
Child, urogenital trichomoniasis and, 246–273
Chinese hamster ovary cell culture, 167, 177–178, 182–183, 186, 191–192
Chlamydia, 267, 288
  Trichomonas vaginalis and, 218, 318–319
Chloramphenicol, 93
CHO. *See* Chinese hamster ovary cell culture; culture
Cholesterol, 345
  media and, 99
  Trichomonas vaginalis and, 55, 56
Chromatin, 345
Chromosome
  metronidazole and, 328
  Trichomonas vaginalis and, 21
CLID mouse fibroblast, 178, 180, 186
Cloning, 98
Clotrimazole
  intravaginal, 330
  mycotic infection in child and, 269
CMC. *See* Cell-mediated cytotoxicity
CMRL 1066 medium. *See* Media

Coenzyne A, 56
Colchicine, 185–186
Colitis, 397
Colon, flagellate and, 120
Complement, 40–41, 43–44
Complement fixation
  antibody and, 303
  Pentatrichomonas hominis and, 344, 368
  Trichomonas tenax and, 344, 363–365
  Trichomonas vaginalis and, 344, 368
Connective tissue, 129
Contaminant, culture, 92
Contraception, 318
Cooling, *Trichomonas vaginalis* and, 107
Corynebacteria, 214
Costa
  Pentatrichomonas hominis and, 26, 28
  Trichomitus fecalis and, 30
  Trichomonas tenax and, 22, 23
  Trichomonas vaginalis and, 7, 9, 11–15, 16, 17, 18, 71
Crohn's disease, 327
Cross-resistance of *Trichomonas vaginalis*, 334–335
Cryopreservation, 106–109, 306–307
Cryptococcaceae, 261
dCTP, 87
Cultivation
  Dientamoeba fragilis and, 407–408
  host cell-trichomonad interaction and, 155–208
    Pentatrichomonas hominis, 207, 373
    Trichomitus fecalis, 30
    Trichomonas tenax, 207
    Trichomonas vaginalis, 156–205
  Pentatrichomonas hominis and, 109
  Trichomonas tenax and, 107
  Trichomonas vaginalis and, 99–106, 197–198, 302–303
  solid medium and cloning, 98
  prolonged in vitro, 183–184
Culture
  African green monkey kidney fibroblast (Vero) cells, 181–182
  Axenic, 91–92
    Pentatrichomonas hominis and, 109
    Trichomonas tenax and, 107–108
    Trichomonas vaginalis and, 94–95
  Baby hamster kidney (BHK) cells, 164–165, 167, 188
  Baby hamster lung fibroblast cells, 181–182, 191
  BS-C-1 cells, 181
  Chick embryo cells, 182
  Chick fibroblast cells, 171, 174
  Chick liver cells, 161–164, 169

Chinese hamster ovary (CHO) cells, 167, 177–178, 182–183, 191–192
CLID mouse fibroblast cells, 178, 180, 186
Darby canine cells, 169
HeLa cells, 156, 169–170, 181, 185, 187–191, 208
HEp-2 cells, 181–183, 185, 191, 207
human amnion (WISH) cells, 182, 185, 189–190, 192
human foreskin fibroblast cells, 181–182
human skin fibroblast cells, 181
L_{929} cells, 181–182, 191
McCoy cells, 169, 174, 177, 181–182, 191
MDCK cells, 181–182, 185–186, 188–189
monkey kidney cells, 181, 207, 373
monoxenic
   Trichomonas tenax and *Trypanosoma cruzi*, 107–108
   Trichomonas vaginalis and *Candida albicans*, 91, 128
normal baboon testicular cells, 181
RD cells, 181–182
Rhesus monkey kidney cells, 182
RK_{13} cells, 181–182
xenic, 91, 344, 373
Cunnilingus, 342
Cycle
   estrous, lab animals, 113–116
   menstrual, human, 215–216, 217
Cyst, 8, 293, 384, 400, 404
Cysteine, 56
Cytidine, 86
Cytochalasin B, 71, 197
Cytochemical examination, 345
Cytochrome, 72
Cytology, human male, 295
Cytopathogenicity, 182–191
Cytoskeleton, 15, 17, 71
Cytosol, 63–64
Cytotoxicity
   cell-mediated, 194–201
   T cell and, 41

DAB. *See* 3, 3′-diaminobenzidine
dATP, 87
dCTP, 87
Dental infection, 347–349
Deoxyadenosine, 85
Deoxyribonucleic acid
   Trichomonas tenax and, 345
   Trichomonas vaginalis and, 55, 86–90
   metronidazole and, 73–74
Deoxyribonucleotide, 86–89
Dexamethasone, 128
DFMO. *See* Difluoromethylornithine

dGTP, 87
3, 3′-Diaminobenzidine, 21
Dientamoeba fragilis, 394–408
   cultivation and, 407–408
   diagnosis and, 406–408
   Enterobius eggs and, 400–404
   epidemiology and, 397–398
   nomenclature of, 4
   pathology and, 396–397
   structure of, 31–33
   symptomatology and, 394–396
   transmission and, 398, 400–406
Diestrus, 113, 114–115
Diff-Quik staining, 301
Difluoromethylornithine, 190
Dimethyl sulfoxide, 107–109, 306
Discharge, vaginal
   assessment of, 229
   biochemistry of, 76
   characteristics of, 227–228
   guinea pig and, 119
   monkey and, 117
   polymorphonuclear leukocyte and, 201
DL7 medium. *See* Media
DL8 medium, *See* Media
DMSO. *See* Dimethyl sulfoxide
DNA. *See* Deoxyribonucleic acid
Double-stranded RNA, 88–89
Drug
   antitrichomonal, 267–269
   metronidazole. *See* Metronidazole
   5-nitroimidazoles, metabolism of, 72–75
Drug-resistant trichomoniasis, 332–333
Dysplastic epithelial change, 128
Dystrophic periodontosis, 345–346

Earle's balanced salt solution, 193, 197
Ecology, microbial, 213–214
Ectopic infection, 129–146
   intramuscular infection of mouse and, 144, 146
   intraperitoneal infection and, 129–140
      inoculation and, 129–130
      pathology and, 130–134, 141
      virulence and, 134–140
   subcutaneous infection and, 129, 140–144
Edema
   intracellular, 282
   subcutaneous infection and, 142
Egg, pinworm, 400–404
Electron microscopy, 174, 180, 186, 285–286
   Dientamoeba fragilis and, 31–33
   Pentatrichomonas hominis, 27–30
   Trichomonas tenax and, 22–26
   Trichomonas vaginalis and, 8–22
Electron paramagnetic resonance, 55

Electron transport, 73, 74
Electrophoresis, 38, 55–56
ELISA. *See* Enzyme-linked immunosorbent assay
Emesis, 327
Endocervical examination, 229
Endocervical gland, 282
Endometrium, 113
Endoplasmic reticulum, 9, 11, 14, 71
　rough, 19
Energy metabolism, trichomonad, 53
Entamoeba histolytica, 191
　Dientamoeba fragilis and, 394–395
　nucleic acid metabolism and, 84–85
Enterobius, 400–404
Enterococcus, 215
Environmental condition, *Trichomonas vaginalis* and, 75–76
Enzyme, 55, 64–65
　hydrogenosomes and, 68
　iron-sulfur, 57
Enzyme-labeled immunocytochemical methods, 302
Enzyme-linked immunosorbent assay, 189
　Trichomonas vaginalis and, 305
Epididymitis, 240
Epithelial cell
　wet mount examination and, 230
Epithelial vaginal membrane, 115
Epithelium
　dysplastic change in, 128
　guinea pig and, 119
　squamous, 282–283, 285
　Trichomonas vaginalis and, 280, 282
EPR. *See* Electron paramagnetic resonance
Escherichia coli, 124, 253
Estradiol, 122
Estrogen, 115–116
　Macaca arctoides and, 116
　mouse model and, 127
　rat model and, 122, 123, 124, 125
Estrous cycle, 113–114
　mouse and, 115
　rat and, 115, 122, 124
Eubacterium, 215
Evolution of *Trichomonas vaginalis*, 312–313
Extragenital infection, 226, 342–393
　Pentatrichomonas hominis, 366–385. *See also* *Pentatrichomonas hominis*
　Trichomonas tenax, 343–366
　　oral infection and, 343–354
　　respiratory infection and, 354–366
　Trichomonas vaginalis, 342–343
Exudate
　fallopian tubal, 226
　inflammatory, 276
　pleural, 355

Fallopian tube, 343
　exudate and, 226
Fatty acid
　media and, 99
　Trichomonas vaginalis and, 55, 56
Feces and *Pentatrichomonas* transmission, 383–384
FEM. *See* Freeze-fracture electron microscopy
Female sex hormone, 189
Fermentation, 57, 58
Ferredoxin, 61–62, 70, 74
　5-nitroimidazole and, 73, 74
Ferrodoxins, 70
Fibrosis, 142
Filament
　accessory
　　Pentatrichomonas hominis, 26, 27
　　Trichomitus fecalis and, 30
　　Trichomonas tenax and, 22
　　Trichomonas vaginalis and, 7, 8, 9, 11
　parabasal
　　Dientamoeba fragilis and, 31, 33
　　Pentatrichomonas hominis, 27, 29
　　Trichomonas tenax and, 22, 23
　　Trichomonas vaginalis and, 13, 15, 18
Filtrate, parasite-free, 161, 164, 169
Fixed smear, 299–302
Flagellaten, 355
Flagellum
　Pentatrichomonas hominis, 26, 27, 28, 366
　recurrent. *See* Recurrent flagellum
　Trichomitus fecalis and, 30
　Trichomonas tenax and, 22, 23
　Trichomonas vaginalis and, 5, 7, 9, 10, 11, 14
Flavin, 56
Flora, bacterial, 253
Floxacillin, 93
Fluid, ascitic, 131
Fluorescent monoclonal antibody, 231
Fly and *Pentatrichomonas* transmission, 383–384
Food vacuole, 9, 14
Foreskin, 293
Freeze-fracture electron microscopy, 8–9
Fungizone. *See* Amphotericin B
Fungus
　Candida albicans and, 217
　Trichomonas vaginalis culture and, 92
　trichomoniasis and, 260–267
Furazolidone, 324

Gardnerella vaginalis
　child and, 254
　menstrual cycle and, 215
Gastric juice, 384–385
Gastrointestinal tract
　animal model and, 120–121

Gastrointestinal tract (*cont.*)
   Dientamoeba fragilis and, 394–410
   Pentatrichomonas hominis and, 366–385
Gentamycin, 93
Giardia lamblia, 84
Giardiasis, *Trichomonas vaginalis* versus, 43
Giemsa's stain, 230–231
Gingivitis, 347
Gland
   endocervical, 282
   prostate, 239–240, 243
Glans penis, 240, 293
Glucose
   catabolism and, 60, 94
   vagina and, 76
Glucose phosphate isomerase, 312
Glyceraldehyde 3-phosphate, 62
Glycerol, 57
Glycogen
   Pentatrichomonas hominis and, 384
   Trichomonas tenax and, 345
   Trichomonas vaginalis and, 55, 59
      granule and, 19, 20
Glycolytic pathway, 59–61
Glycosidase, 66
Glycosome, 72
Go 10213, 334
Golden hamster, 120, 122
Golgi apparatus, 71
Golgi complex. *See also* Parabasal body
   Dientamoeba fragilis and, 32
   Pentatrichomonas hominis, 27, 29
   Trichomonas tenax and, 23
   Trichomonas vaginalis and, 11, 16, 17
Gonadotropin, 113
Gram stain, 230
   smear and, 299–300
Granule, 7, 19
   glycogen, 19, 20
   paracostal, 6, 7, 22, 26
   paracostylar, 7, 27
Griseofulvin, 186, 187
dGTP, 87
Guanine, 85
Guanosine, 85
Guanosine kinase, 85
Guinea pig, 115, 120
   epithelial vaginal membrane of, 115
   interperitoneal infection and, 119
Gynatren, 334

Hamster, 120, 122. *See also* Culture
   intravaginal infection and, 115
Hank's balanced salt solution, 201, 203

HeLa cell, 156, 169–170, 181, 185, 187–191, 208
Hemagglutination procedure, 315
HEp-2 cell, 181–183, 185, 191, 207
Hepatocyte, 133, 134
Heterogeneity, 37–39
Heterosexual transmission, 236–239
Hexokinase, 312
Histomonas meleagridis, 394
Hormone
   female sex, 189
      ovarian, 115–116
   microflora and, 214
Host cell-trichomonad interaction, 155–208
   pellet experiments, 167
   Trichomonas tenax and *Pentatrichomonas hominis* and, 207–208
   Trichomonas vaginalis and, 155–207
      adherence and, 182–192
      cytotoxicity and, 156–170
      effector and target cells, in vivo/vitro systems and, 180–182
      environmental conditions and, 75–76
      lectins and, 190
      lymphocytes and, 192–194
      monocytes and macrophages and, 194–201
      pathologic process and, 170–180
      polymorphonuclear leukocytes and, 203–205
Host factor, microflora and, 214
Hydrogen, 57, 74
Hydrogenase, 70, 74
Hydrogenosome, 7, 9, 16, 17, 19, 20, 22, 33, 64, 67–71
   metabolic pathway of, 61–62
Hydrolase, 65–66
2-Hydroxyethyl oxamic acid, 73
Hygiene, oral, 350–351
Hyperimmune serum, 47–48
Hysterectomy, 276

IgA, 40–41
Immune system
   pathogenicity and, 189–190
   Trichomonas tenax and, 344–345
   Trichomonas vaginalis and, 42
      monkey and, 117
Immunocytochemical examination, 302
Immunodiagnosis, 303–305
Immunofluorescence, 303, 305
Immunoglobulin, 45, 46, 305
   A, 40–41
Immunology, 36–52
   animal models and, 41–42
   antigenic structure and, 36–40

human mechanisms of, 42–47
immunoprophylaxis and, 47–48
immunosuppression and, 47
mediators and, 40–41
Immunoprophylaxis, 47–48
Immunosuppression
  mouse and, 128
  Trichomonas vaginalis and, 47
IMP dehydrogenase, 85
Indirect fluorescent antibody test, 303, 305, 363–365
  respiratory infection and, 364
  Trichomonas vaginalis and, 315
Indirect hemagglutination, 315
Infant
  metronidazole and, 330
  multifocal infections and, 257–258
  Trichomonas vaginalis and, 249–250, 254–256
    respiratory infection and, 343
    transmission of, 313
    vaginal ecology of, 215
Infestation index, 137
Inflammatory exudate, 276
Infusorien, 355
Inguinal lymphadenopathy, 293
Inoculation, 129–130
  guinea pig and, 119
  hamster and, 120
  monkey and, 116
  mouse and, 126–128
  rat and, 122–124
  subcutaneous infection and, 140–141
Inositol, 106
Intestinal infection
  animal model and, 120–122
  cell interactions and, 207–208
  Dientamoeba fragilis and, 394–408. See also *Dientamoeba fragilis*
  Pentatrichomonas hominis and, 366–385. See also *Pentatrichomonas hominis*
Intracellular edema, 282
Intradermal test, 305
Intramembranous particle, 11
Intramuscular infection of mouse, 144, 146
Intraperitoneal assay, 146, 147
Intraperitoneal infection, 129–140
  inoculation and, 129–130
  pathology and, 130–134
  virulence and, 134–140
Intratesticular infection, 129
Intraurethral inoculation, 237
Intravaginal clotrimazole, 330
Intravaginal infection. See also *Trichomonas vaginalis*; Urogenital trichomoniasis

animal model and, 122–128
  guinea pig and, 118–120
  hamster and, 120, 122
  monkey and, 116–118
  mouse and, 125–128
  rat and, 122–125
  human, 226
Iodoacetate, 188
Iron, 43, 106
Iron-sulfur enzyme, 57
Iron-transport protein, 67
Isoenzyme, 55
Isolation of DNA, 86

JH30A strain, 162–164
  pathologic process and, 170–171
JH32A strain, 164
  pathologic process and, 171
JHHR strain, 184

Kanamycin, 92, 93
Ketoconazole, 269
Kidney cell, baby hamster. See Culture
Kidney disorder, 260
Kinetosome
  Pentatrichomonas hominis, 29
  Trichomitus fecalis, 30
  Trichomonas tenax, 23
  Trichomonas vaginalis and, 9, 11, 12, 16, 17
Koilocytosis, 282

Laboratory diagnostic method, 297–306
Laboratory rodent
  Pentatrichomonas hominis and, 372–373
  sexual cycles and, 113–116
  Trichomonas vaginalis and, 120–149. See also Animal model
Lactate, 58, 59, 61, 62
  Trichomonas vaginalis and, 57, 58
Lactate dehydrogenase, 312
Lactic acid, 94
Lactobacillus, 215
  bacterial vaginosis and, 216
  child and, 253
  menstrual cycle and, 215–216
Lactobacillus acidophilus, 219, 334
Lactobacillus fermentum, 185
Lactoferrin, 67, 106
Lamella, 8, 9, 11, 13, 15
Latent infection, 117
Lectin, 191–192
Leptothrix, 276

Leukocyte
  aggregates of, 276
  polymorphonuclear, 201–206, 292
  subcutaneous infection and, 142
Light microscopy, 175
  Dientamoeba fragilis and, 31, 32
  Pentatrichomonas hominis and, 26–27
  Trichomitus fecalis and, 30–31
  Trichomonas tenax and, 22
  Trichomonas vaginalis and, 5–7, 276–284
  vaginal discharge and, 230
Lipid
  media and, 99
  Trichomonas tenax and, 345
  Trichomonas vaginalis and, 55, 65, 106
Lipoprotein, 67
Liquid nitrogen storage, 106–107
Liver, 131, 133, 134
Liver digest, 94
Local antibody, 46
Lung abscess, 355
Lung cancer, 329
Lymphadenopathy, inguinal, 293
Lymphocyte, 192–194
Lysosome, 71
Lytic factor, 44–46

Macaca species, 116–118
Macrophage
  male patient and, 292
  Trichomonas vaginalis and, 44, 194–201
Malate, 57, 58, 69
Malate dehydrogenase, 312
Male patient
  child, 246, 252–253, 259
    mycotic infection and, 261–263
  infant, 257–258
  laboratory diagnosis and, 297–298
  respiratory infection and, 266
  urogenital trichomoniasis and
    clinical manifestations of, 235–245
    pathology of, 291–296
Male urogenital system, 128–129
Malignancy, 328–329
  cervical neoplasia and, 286–287
Mast cell, 119–120
Mean infestation index, 137
Mebendazole, 186–187
Media
  antibiotics and, 93
  buffered salts solution, 94
  BSS (Earle's balanced salt solution), 193, 197
  CACH, 95–96
  CMRL 1066, modified, 97, 99–100

CPLM (cysteine-peptone-liver-maltose), 95–96, 302–303, 305–306
DL7, 100–103, 105–106
DL8, 100–101, 103–106
GMP (empirical growth medium), 177
Loeffler's saline-citrate, 30
Loeffler's serum-citrate, 30
protein hydrolysate and, 94
reducing agents and, 94–95
Ringer's solution, 407
RPMI 1640, 197
serum and, 95
solid for cloning, 98
STS (simplified trypticase serum), 95, 97
subcutaneous infection and, 140
TTY (tryptose-trypticase-yeast), 107–108, 207
TYM (trypticase-yeast-maltose), Diamond's, 5, 94–95, 97–99, 169, 187, 193, 197, 203–204, 207
  modified by Hollander and Kulda et al., 95
Membrane
  cell, 19, 66
  epithelial vaginal, 115
  Pentatrichomonas hominis, 26
  Trichomitus fecalis and, 30
  Trichomonas vaginalis and, 8–9, 11–17
Menopause and vaginal ecology, 216
Menstrual cycle
  estrous cycle and, 114
  laboratory rodent and, 113
  trichomoniasis and, 217
  vaginal flora and, 216–219
Mesocricetus auratus, 120, 122
Metabolism
  drug, 72–75
  energy, 53
  inhibitor and, 188
  Trichomonas vaginalis and, 59–66
    nucleic acid and, 84–90
Metestrus, 113, 114
Methyl viologen, 74
Metronidazole
  adverse reactions and, 327–328
  breast feeding and, 330
  cancer and, 328–329
  child and, 266–269
  Dientamoeba fragilis and, 395
  mutagenic properties of, 328
  pregnancy and, 330
  protein binding, 325–326
  Trichomonas tenax and, 348
  Trichomonas vaginalis and, 72–75, 324–333
    metabolic inhibition and, 188
MIC. *See* Minimal inhibitory concentration
Miconazole, 93

Mycoplasma hominis, 216–218
Microbial ecology, 213–215
Microfilament, 71, 180
Microflora, 213–224, 318–319
　microbial ecology and, 213–214
　vaccination against, 219–221
　vaginal environment and, 214–219
Microscopy
　diagnosing male child and, 246–247
　electron, 172, 174–176, 179, 285–286
　　Dientamoeba fragilis and, 31–33
　　Pentatrichomonas hominis, 27–30
　　Trichomonas tenax and, 22–26
　　Trichomonas vaginalis and, 8–22
　light
　　Dientamoeba fragilis and, 31, 32
　　Pentatrichomonas hominis and, 26–27
　　Trichomitus fecalis and, 30
　　Trichomonas tenax and, 22
　　Trichomonas vaginalis and, 5–7, 276–284
　vaginal secretions and, 298–302
Microtubule
　Dientamoeba fragilis and, 32
　Trichomonas vaginalis and, 71
　　peltar-axostylar complex and, 15, 17
Minimal inhibitory concentration, 330–331
Minimal lethal concentration, 330–331
Mitochondria, 72
MLC. *See* Minimal lethal concentration
Mobiluncus, 216, 217
Model, animal. *See* Animal model
Moisture, microflora and, 214
Monaden, 355
Monas lens, 355
Monitoring of estrous cycle, 113–114
Monkey, *See also Macaca* and culture
　intravaginal infection and, 116
　vaginal flora and, 218
Monocercomonas, 70
Monoclonal antibody
　antigenic structures of, 38
　fluorescent, 231
　Trichomonas vaginalis and, 40, 302
　tubulins and, 71
Monocyte, 194–201
Mononuclear cell layer, 193
Mononucleate, 8
Mouse
　inoculation and, 126–130
　intramuscular infection of, 144, 146
　intraperitoneal infection and, 129–140
　　pathology and, 130–134
　　virulence in, 134–140
　intravaginal infection and, 114

　mortality and, 137, 146
　Pentatrichomonas hominis and, 372–373
　subcutaneous assay and, 218
　subcutaneous infection and, 140–144
　Trichomonas tenax and, 349
　virulence studies and, 161
Mouth infection, 342–344
Mutagenicity, 328
Mycophenolic acid, 85
Mycoplasma hominis, 92
Mycostatin. *See* Nystatin
Mycotic infection
　child and, 260–266
　Trichomonas tenax and, 353

NAD. *See* Nicotinamide-adenine dinucleotide
NAD: ferredoxin oxidoreductase, 62
NADH, 63–64
NADH oxidase, 68
NADP. *See* Nicotinamide-adenine dinucleotide phosphate
Natamycin, 269
Nausea, metronidazole and, 327
NBT. *See* Nitroblue tetrazolium
Necrosis, 280, 282
Necrotic ulcerative gingivitis, 347
Neisseria gonorrhoea, 318–319
Neocallimastix patriciarum, 70
Neomycin, 93
Neonate. *See* Newborn; Infant
Neoplasia, cervical, 286–287, 329
Neoplasm, pulmonary, 329
Neuropathy, 327
Neutrophil
　subcutaneous infection and, 142
　Trichomonas vaginalis and, 41
Newborn. *See also* Infant
　multifocal infections and, 257–258
　respiratory infection of, 266–267
　Trichomonas vaginalis and, 249–250, 254–256
　　transmission of, 313
　vaginal ecology of, 215
Nicotinamide-adenine dinucleotide, 56
Nicotinamide-adenine dinucleotide phosphate, 56, 63–64
Nicotinic acid, 56
Nifuratel, 324, 334
Nifuroxime, 324
Nimorazole, 334
Nitroblue tetrazolium, 205
Nitrogen
　storage of liquid, 106–107
　Trichomonas vaginalis and, 59

5-Nitroimidazole, 72–75, 333–334
  cross-resistance to, 334–335
  metronidazole synthesis and, 324
Nomenclature, 3
Nongonococcal urethritis, 238–239
Nonimmunologic host defense mechanism, 43
Nucleic acid metabolism, 84–90
Nucleic acid precursor, 99
Nucleotide, 56
Nucleus
  Dientamoeba fragilis and, 32
  epithelial cell, 119
  Pentatrichomonas hominis and, 27
  Trichomitus fecalis and 30, 31
  Trichomonas tenax and, 22
  Trichomonas vaginalis and, 6, 7, 9, 21–22
Nutritional requirements
  Trichomonas vaginalis and, 56–57, 105–106
Nutrients
  microflora and, 214
Nystatin
  culture and, 92, 93
  mycotic infection in child and, 269

Obstructive uropathy, 343
Oncogenesis, 329
Oral contraception, 318
Oral hygiene, 350–352
Oral infection
  Trichomonas tenax and, 343–354
    biochemistry of, 345
    epidemiology and, 349–354
    human host and, 344
    immunology and, 344–345
    pathogenicity and, 345–349
  Trichomonas vaginalis and, 342–343
Organelle, 71–72
Ornidazole, 333
  child and, 267–268
Ornithine carbamyltransferase, 64
Ornithine decarboxylase, 65
Ovarian hormone, 115–116
Oxidation of glyceraldehyde 3-phosphate, 62
Oxygen, 63–64, 69
  vagina and, 76

Pancreas, 131, 133, 134
Pantothenic acid, 56
Papanicolaou stain, 275–276
  smear and, 300
Papilla, 282
Parabasal apparatus
  Pentatrichomonas hominis and, 27

Trichomitus fecalis and, 31
Trichomonas tenax and, 22
Trichomonas vaginalis and, 7, 11, 15
Parabasal body
  Dientamoeba fragilis and, 31, 33
  Pentatrichomonas hominis and, 27
  Trichomitus fecalis and, 30
  Trichomonas tenax and, 22, 23
  Trichomonas vaginalis and, 7, 9
Parabasasl filament
  Dientamoeba fragilis and 31, 33
  Pentatrichomonas hominis and, 27, 29
  Trichomonas tenax and, 22, 23
  Trichomonas vaginalis and, 7, 9, 13, 15, 18
Paracostal granule. See also Hydrogenosome
  Trichomonas tenax and, 24
  Trichomonas vaginalis and 6, 7, 19
Paraxostylar granule. See also Hydrogenosome
  Trichomonas vaginalis and 6, 7, 19
PBL. See Peripheral blood lymphocyte
Pellet experiments, 167
Pelta
  Pentatrichomonas hominis and, 27, 28, 29
  Trichomitus fecalis and, 30
  Trichomonas tenax and, 23, 25
  Trichomonas vaginalis and, 9, 10, 12, 14, 16, 17
Peltar-axostylar complex
  Pentatrichomonas hominis and, 29
  Trichomonas tenax and, 23, 25
  Trichomonas vaginalis and, 11, 12, 15, 17, 22
Peltar-axostylar junction
  Trichomonas vaginalis and, 9, 11, 12, 15
Pelvic inflammatory disease, 275
Pelvis, renal, 343
Penicillin, 92–93
Penis, 240, 293
Pentatrichomonas
  monkey and, 117
  Trichomonas tenax and, 344
Pentatrichomonas hominis, 366–385
  antigenic immunologic aspects of, 366–369
  biochemical attributes of, 370
  cell interactions and, 205–208
  as contaminant in *Trichomonas vaginalis* cultures, 120–122
  cryopreservation and, 306–307
  cultivation of, 109
  epidemiology and, 373–385
    prevalence and, 373–383
    transmission and, 383–385
  geographic and host distribution of, 366
  immunity and, 48

laboratory diagnosis and, 305–306
nomenclature of, 3, 366
pathogenicity and, 370–373
structure of, 26–30
Pentose phosphate shunt, 61
Peptostreptococcus, 215–216
Periodic acid-Schiff staining, 301–302
Periodontium, 348
Periodontosis, 345–346
Peripheral blood lymphocyte, 192–193
Peripheral neuropathy, 327
Peritoneal infection, 129–140
inoculation and, 129–130
pathology and, 130–134
virulence and, 134–140
Peroxisome, 72
pH
microflora and, 214
vaginal, 230
Phagocytosis, 19
Phagosome, 71
Pharmacokinetics, 325–326
Phenyl methyl sulfonyl fluoride, 167
Phimosis, 263
Phosphatase, 66
Phosphatidylethanolamine, 55–56
Phospholipid, 55–56
PID. See Pelvic inflammatory disease
Pinocytotic canal, 9, 14
Pinocytotic vesicle, 9, 14
Pinworm, 400–406
Placental barrier, 330
Pleural exudate, 355
PMN. See Polymorphonuclear leukocyte
PMSF. See Phenyl methyl sulfonyl fluoride
Pneumonia, 343
Polyamine, 64–65
Polyester sponge, 297–298
Polyfungin, 269
Polymonad organism, 8
Polymorphonuclear leukocyte
Trichomonas vaginalis and, 201–205
wet mount examination and, 230
Polymorphonuclear neutrophil
subcutaneous infection and, 142
Trichomonas vaginalis and, 41, 44
Pore, 11, 13
Posthitis, 240
Postnatal treatment, 330
Pregnancy, 217
Premenarcheal girl, 215
Prenatal treatment, 330
Proestrus, 113–114
Pronase, 187
Prostate, 239–240, 243, 292–293, 294

Protease, 17, 187–188
Protein, 54–55, 56, 59
cell membrane and, 66
iron-transport, 67
Protein binding, 325–326
Protein hydrolysate, 94
Proteinase, 65–66
Protozoa, 70
Pseudocyst, 384, 400, 404
Pentatrichomonas hominis, 26
Trichomonas vaginalis and, 8
Tritrichomonas muris and, 8
Pseudopod, 20
Puberty, 215–216
Pulmonary neoplasm, 329
Purine
metabolism of, 85
Trichomonas vaginalis and, 56, 105–106
Purine nucleoside phosphorylase, 85
Purulent discharge, 119
Putrescine, 59, 64
Pyonephritis chronica recurrens, 260
Pyridine nucleotide, 56
Pyridoxamine, 56
Pyrimidine, 56, 105–106
metabolism and, 85–86
Pyruvate, 59, 61
Pyruvate:ferredoxin oxidoreductase
anaerobic protozoa and, 70
metronidazole resistance and, 75
5-nitroimidazole and, 74
Trichomonas vaginalis and, 61, 106

Race
Trichomonas tenax and, 351–352
Trichomonas vaginalis and, 317, 319
Radioimmunoprecipitation, 38
Rat, 122–125
Rectal suppository, 326
Recurrent flagellum
Pentatrichomonas hominis, 29
Trichomitus fecalis and, 30
Trichomonas tenax and, 25
Trichomonas vaginalis and, 12, 13
kinetosome and, 9, 11
Reducing agent, 94–95
Renal disorder, 260
Renal pelvis, 343
Resistance
metronidazole, 75, 332
Pentatrichomonas hominis and, 384–385
Respiration
amino acid and, 64
Trichomonas vaginalis and, 63–64

Respiratory tract
  infection of newborn and, 266–267
  organelles and, 69
  Trichomonas tenax and, 354–366
    antigenic properties and, 358–365
    pathogenicity and, 354
    prevalence and, 354, 358
    treatment and, 365
  Trichomonas vaginalis and, 343
Reticulum, 71
Rhesus monkey, 116–117 *See also* Culture
Rhinitis, 343
Ribonucleic acid, 55
  double-stranded, 88–89
Ribonucleotide, 86–87
Ribosome, 71
RIP. *See* Radioimmunoprecipitation
Risk factors associated with *Trichomonas vaginalis* infection, 315–320
RNA. *See* Ribonucleic acid
Rodent
  Pentatrichomonas hominis and, 372–373
  Trichomonas vaginalis and, 120–149. *See also* Animal model, laboratory rodent and
Romanowsky-stained smear, 300
Rough endoplasmic reticulum, 19
Rupture of abscess, 144

Sacroiliac joint, 293
Saimiri sciureus, 116
Salpingitis, 226
Salts solution, buffered, 94
Salvage, nucleic acid metabolism and, 85, 86, 87
Sangivamycin, 85
Scanning electron microscopy, 174–176
  Pentatrichomonas hominis, 27
  Trichomonas vaginalis and, 8
Seasonality of *Trichomonas vaginalis* infection, 319
Secnidazole, 334
Secretion, antibody in, 305
Secretory IgA, 40–41
SEM. *See* Scanning electron microscopy
Seminiferous tubule, 129
Seroepidemiology, 314–315
Serologic test, 303, 305
Serotype, 37
Serum
  hyperimmune, 47–48
  Trichomonas vaginalis culture and, 95
Serum antibody, 45
Serum complement, 40–41
Serum glutamic-oxaloacetic transaminase, 327

Sex
  female hormone, 190
  Trichomonas tenax and, 352–353
Sexual activity as risk factor, 315–316
  men and, 238
  women and, 225
Sexual cycle, 113–116
Sexually transmitted disease
  men and, 238
  previous, 318
  women and, 225
Silver staining, 301
Smear
  Dientamoeba fragilis and, 406–407
  fixed, 299–302
  stained, 230–231
    diagnosing male child and, 247
  wet, 298–299, 300, 301, 304
Smoking, tobacco, 353
Sn-glycerol 3-phosphate, 66
SolcoTrichovac
  Trichomonas vaginalis and, 48, 334
  vaginal trichomoniasis and, 219–221
Solid media, 98
Specimen collection, 298
Spermidine, 64
Spermine, 64
Spleen, 131
Squamous epithelium
  guinea pig and, 119
  penis and, 293
  Trichomonas vaginalis and, 283
Squirrel monkey, 116–117
Stained smear, 230–231
  male patient and, 292
  child, 247
Staining
  Dientamoeba fragilis and, 406–407
  Trichomitus fecalis and, 30
  Trichomonas tenax and, 22
  Trichomonas vaginalis and, 276, 279, 292, 298–301
Staphylococcus, 215–216
Storage of liquid nitrogen, 106–107
Strain variability, 312
Streptococcus agalactia, 185
Streptomycin sulfate, 93
Subcutaneous abscess, 349
Subcutaneous infection, 140–144
Subcutaneous mouse assay, 146
  flora and, 218
Succinate CoA transferase, 61
Succinate thiokinase, 61
Succinyl-CoA, 59, 61
Superoxide dismutase, 64

T cell, 46–47
  cytotoxic, 41
Taxol, 186
Taxonomic status, 3
TEM. *See* Transmission electron microscopy
Temperature
  microflora and, 214
  Pentatrichomonas hominis and, 384
Teratogenicity, 328
Testicular infection, 129
Tetracycline, 395
Tetratrichomonas, 344
Tetratrichomonas macacovagina, 117
Thiamine, 56
Thioester bond of acetyl-CoA, 70
Thiol group, 65
Thymidine, 86
Tinidazole, 333–334
  child and, 267–268
Tobacco smoking, 353
Toyocamycin, 85
Transmission electron microscopy
  Pentatrichomonas hominis, 27, 29–30
  Trichomonas tenax and, 23
  Trichomonas vaginalis and, 9, 14, 16, 20
Transport of electron, 73, 74
Tricarboxylic acid cycle, 72
Trichomitus batrachorum, 31
Trichomitus fecalis, 30–31
  nomenclature of, 3–4
Trichomonas buccalis, 359
Trichomonas elongata, 356, 357
Trichomonas foetus
  hydrogenosomes and, 68–69, 70
  media and, 99
  metronidazole-resistant, 75, 335
  nucleic acid metabolism and, 84–85
  radioimmunoprecipitation and, 38
  thiamine-deficient cultures of, 56
Trichomonas gallinae, 99
Trichomonas gallinarum, 369
Trichomonas hominis, 355
Trichomonas intestinalis, 355
Trichomonas pulmonalis, 355
Trichomonas suis, 99
Trichomonas tenax, 343–366
  cryopreservation of, 307
  cultivation of, 107–109
  host cell interactions and, 207–208
  immunity and, 48
  laboratory diagnosis of, 306
  nomenclature of, 3
  oral infection and, 343–354. *See also* Oral infection
  respiratory tract and, 354–366. *See also*
    Respiratory tract, *Trichomonas tenax* and
  structure of, 22–26
Trichomonas vaginalis
  animal models and, 112–154
    ectopic infections and, 129–146. *See also*
      Ectopic infection
    intravaginal infections and, 112–128. *See also*
      Intravaginal infection, animal model and
    male urogenital system and, 128–129
  biochemistry of, 53–83
  clinical manifestations of
    in men, 235–245
    in women, 225–234
  cryopreservation and, 306–307
  cultivation of, 91–107. *See also* Cultivation
  cytopathology of, 275–276
  extragenital infection and, 226, 342–393. *See also*
    Extragenital infection
  histopathology in infection, 131–134, 276–288
    electron microscopy and, 286–287
    light microscopy and, 276, 278–284
  host cell-trichomonad interaction and, 156–205
  immune system and, 36–48
  laboratory diagnostic methods and, 297–305
  microflora and, 213–224
    microbial ecology and, 213–214
    vaccination against, 219–221
    vaginal environment and, 214–219
  nomenclature of, 3
  nucleic acid metabolism in, 84–90
  nutrition of, 105
  structure of, 5–22
Trichomoniasis, urogenital. *See Trichomonas vaginalis; Urogenital trichomoniasis*
Tricofuron, 324
Tritrichomonas mobiliensis, 117–118
Trypsin, 188
Trypticase-yeast-maltose. *See* Media
Tube, fallopian, 226–343
Tubular axostylar trunk, 9, 14
Tubulin, 15, 17, 71
Tylocine, *See* Tylosin
Tylosin, 92, 93
TYM. *See* Media

Ulceration
  Dientamoeba fragilis and, 396–397
  penis and, 293
Ulcerative gingivitis, 347
UMP. *See* Uridine monophosphate
Undulating membrane
  Pentatrichomonas hominis and, 26
  Trichomitus fecalis and, 30
  Trichomonas vaginalis and, 11–15, 16, 17

Uracil, 86
Ureaplasma urealyticum, 92
Urethra
  male patient and, 292–293, 294
  Trichomonas vaginalis and, 226
Urethritis, 238, 239, 240
Uridine, 86
Uridine monophosphate, 86
Urinary tract, 263–265
Urine
  diagnosis in child and, 246
  retention of, 269
  Trichomonas vaginalis and, 226
Urogenital system, male, 128–129
Urogenital trichomoniasis. *See also Trichomonas vaginalis*
  in children, 246–273
    diagnosis and, 246–247
    epidemiology and, 247–252
    multifocal trichomoniasis and, 257–260
    mycotic infections and, 260–266
    treatment of, 267–269
    unifocal trichomoniasis and, 254–257
  epidemiology of, 311–323
    evolution of parasite and, 312–313
    prevalence and, 314
    risk factors and, 315–320
    seroepidemiology and, 314–315
    strain variability and, 312
    survival outside human host and, 311–312
    transmission of trichomonads and, 313–314
  in men, 235–245
    pathology of, 291–296
  therapy of, 324–341
    cross-resistance to nitroimidazoles and, 334–335
    metronidazole and, 324–333
    other antitrichomonal agents and, 333–335
  in women, 225–234
    microbial ecology and, 213–214
    vaccination against, 219–221
    vaginal environment and, 214–219
Uropathy, 343
Urticaria, 327

Vaccination, 219–221, 334
Vagina
  host environment and, 75–76
  microflora and, 214–219
  pH of, 124
Vaginal discharge
  assessment of, 229
  biochemistry of, 76
  characteristics of, 227–228
  guinea pig and, 119
  monkey and, 117
  polymorphonuclear leukocyte and, 201
Vaginal trichomoniasis. *See Trichomonas vaginalis;* Urogenital trichomoniasis, in women
Vaginitis
  flora and, 216–219
  Trichomonas vaginalis and, 216–218
  trichomoniasis and, 225
Vaginosis, 216, 229
Variability of strains, 312
Vincent's angina, 347
Viridans streptococci, 215
Virulence
  antigenic structure and, 39
  indices, 137
  intraperitoneal mouse assays for, 134–140
  mouse and human infection and, 146–148
  subcutaneous mouse assay for, 143–144
Virulence assay, 156–208
  pellet experiments, 167
  Trichomonas tenax and *Pentatrichomonas hominis* and, 207–208
  Trichomonas vaginalis and, 156–205 *See also* Host cell-trichomonad interaction. *Trichomonas vaginalis* and
Virus, 88
Vitamin, 56–57, 99, 106
Vulva, 227
Vulvovaginal candidiasis, 229
Vulvovaginitis, 229–232

Wet mount, 230, 231
Wet smear, 298–299, 301, 304
Whiff test, 230

Xenic culture
  Pentatrichomonas hominis and, 373
  Trichomonas tenax and, 344
  Trichomonas vaginalis and, 91

Yeast vaginitis, 216

Zinc, 43
Zymodeme, 55

Printed by Books on Demand, Germany